Novel Food Processing

Effects on Rheological and Functional Properties

Electro-Technologies for Food Processing Series

Series Editor

Hosahalli S. Ramaswamy
McGill University
Department of Food Science
Ste-Anne-de-Bellevue, Quebec, Canada

Novel Food Processing : Effects on Rheological and Functional Properties, *edited by Jasim Ahmed, Hosahalli S. Ramaswamy, Stefan Kasapis, and Joyce I. Boye*

Electro-Technologies for Food Processing Series

Novel Food Processing

Effects on Rheological and Functional Properties

Edited by

Jasim Ahmed

Hosahalli S. Ramaswamy

Stefan Kasapis

Joyce I. Boye

CRC Press

Taylor & Francis Group

Boca Raton London New York

CRC Press is an imprint of the Taylor & Francis Group, an **informa** business

CRC Press
Taylor & Francis Group
6000 Broken Sound Parkway NW, Suite 300
Boca Raton, FL 33487-2742

First issued in paperback 2017

© 2010 by Taylor and Francis Group, LLC
CRC Press is an imprint of Taylor & Francis Group, an Informa business

No claim to original U.S. Government works

ISBN 13: 978-1-138-11511-8 (pbk)
ISBN 13: 978-1-4200-7119-1 (hbk)

This book contains information obtained from authentic and highly regarded sources. Reasonable efforts have been made to publish reliable data and information, but the author and publisher cannot assume responsibility for the validity of all materials or the consequences of their use. The authors and publishers have attempted to trace the copyright holders of all material reproduced in this publication and apologize to copyright holders if permission to publish in this form has not been obtained. If any copyright material has not been acknowledged please write and let us know so we may rectify in any future reprint.

Except as permitted under U.S. Copyright Law, no part of this book may be reprinted, reproduced, transmitted, or utilized in any form by any electronic, mechanical, or other means, now known or hereafter invented, including photocopying, microfilming, and recording, or in any information storage or retrieval system, without written permission from the publishers.

For permission to photocopy or use material electronically from this work, please access www.copyright.com (http://www.copyright.com/) or contact the Copyright Clearance Center, Inc. (CCC), 222 Rosewood Drive, Danvers, MA 01923, 978-750-8400. CCC is a not-for-profit organization that provides licenses and registration for a variety of users. For organizations that have been granted a photocopy license by the CCC, a separate system of payment has been arranged.

Trademark Notice: Product or corporate names may be trademarks or registered trademarks, and are used only for identification and explanation without intent to infringe.

Library of Congress Cataloging-in-Publication Data

Novel food processing : effects on rheological and functional properties / edited by Jasim Ahmed ... [et al.].
p. cm. -- (Electro-technologies for food processing series)
Includes bibliographical references and index.
ISBN 978-1-4200-7119-1 (alk. paper)
1. Food industry and trade--Quality control. 2. Food industry and trade--Technological innovations. 3. Food--Analysis. I. Ahmed, Jasim.

TP372.5.N685 2010
664'.07--dc22 2009024325

Visit the Taylor & Francis Web site at
http://www.taylorandfrancis.com

and the CRC Press Web site at
http://www.crcpress.com

Contents

Series Preface

Novel Food Processing: Effects on Rheological and Functional Properties, edited by Jasim Ahmed, Hosahalli S. Ramaswamy, Stefan Kasapis, and Joyce I. Boye, is the first issue under the general umbrella of edited books in the Electro-Technologies for Food Processing Book Series involving the application of electro-technologies for various aspects of food processing, from pasteurization to sterilization, food preparation to food formulations, shelf-life extension to promoting food safety, food spoilage control to enhancing safety of foods, and from alternate to novel sources of the use of energy. Electromagnetic technologies offer unlimited potential to processing applications in foods. Industrially, electro-technologies provide unique opportunities and advantages not necessarily found in other techniques. The book series will look at each of them in detail, especially from the point of view of various industrial food processing applications. Each book in this series is expected to be devoted to a specific area of electro-technology, covering all aspects of its science and engineering, chemistry and physics, biochemistry and nutrition, quality and safety, and development and technology, both basic and applied.

Notable among the novel approaches in heating and food processing techniques are microwave and radio-frequency heating, electrical resistance or ohmic heating, induction, and infrared heating applications. Use of pulsed electric fields, high-frequency magnetic fields, electric shockwaves, pulsed light, UV radiation, and ionizing irradiation offer potential nonthermal alternatives to food processing. On a different note, these also include such separation techniques as ultrasonics, electroacoustic dewatering techniques, electrodialysis, and ion-exchange systems. Stretching it further, one can look at other electromagnetic applications in spectroscopy, near infrared (NIR), Fourier transform infrared (FTIR), and nuclear magnetic resonance (NMR) techniques finding their way into analytical and imaging concepts.

This book is special in that, rather than focusing entirely on one technology, it is focused on the rheology and functionality associated with all novel methods. The leading editor, Jasim Ahmed, has done a wonderful job of bringing together contributions from global experts focusing uniquely on structure, microstructure, rheology, and functionality issues related to the various advanced technologies. This volume has been designed to be a valuable tool to graduate students and researchers as a source of scientific information and is a useful addition to any library devoted to life sciences.

Hosahalli S. Ramaswamy
Book Series Editor

Series Preface

Preface

Food processing is gradually moving in a new direction, incorporating technologies from other fields and at the same time taking into account consumers' concerns about food safety, quality, and sensory attributes. The main reason for writing this book at the present time is to make available the wealth of knowledge becoming available on novel processing and its effects on food products. The concept was initiated by the leading editor (Jasim Ahmed), who intended to publish a book on the rheological characteristics of food proteins as affected by novel processing. After discussions with the second editor (Hosahalli S. Ramaswamy) and considering feedback from reviewers of the original book, it was proposed to the publisher to broaden the concept to rheology and functionality of foods. Meanwhile, the complementary expertise of the other two editors (Stefan Kasapis and Joyce I. Boye) and their meticulous input enabled the successful handling of the project.

The chapters in this book are authored by international peers who have both academic and professional credentials. The book illustrates in a very clear and concise fashion the structure–property relationship. It is intended for scientists; technologists/engineers working in the area of food processing, process equipment design, and product development; and students of food science, technology, nutrition, health science, and engineering.

The editors thank all of the contributors and, in many cases, their students and spouses for their dedicated efforts in the production of this book. We express our appreciation to our spouses and children for their understanding and encouragement during the preparation, editing of manuscripts, and constructive discussions. The leading editor (Jasim Ahmed) gratefully acknowledges the support of Dr. Sunil K. Varshney, Director of Polymer Source Inc., in the many tasks involved in editing a book in a highly demanding industrial environment. Special thanks are due to Steve Zollo of Taylor & Francis for his constant encouragement and excellent communication.

Jasim Ahmed
Hosahalli S. Ramaswamy
Stefan Kasapis
Joyce I. Boye

Editors

Dr. Jasim Ahmed has been associated with food processing teaching, research, and industry for the last 16 years in India, the Middle East, and Canada. He has both bachelor's and master's degrees in food and biochemical engineering and a PhD in food technology. He has published more than 80 peer-reviewed research papers, 15 industry-oriented technical papers, and 11 book chapters. He has served as special editor for three peer-reviewed international journals and is currently serving as one of the editors of the *International Journal of Food Properties* (Taylor & Francis). He has worked on food product development, food rheology, structure, nanocomposites, and the thermal behavior of food and biomaterials. He has extensively studied high-pressure-assisted textural modifications of protein foods and starches.

Dr. Hosahalli Ramaswamy is a professor in the Department of Food Science at McGill University, Canada. He has developed a strong research program at McGill in the area of food process engineering. His research activities include overpressure thermal processing of foods in thin profile flexible and semi-rigid containers in still and rotary autoclaves, continuous aseptic processing of particulate low acid foods, food extrusion, food quality optimization, application of microwaves for food processing and drying, evaluation of thermophysical properties of foods, food system rheology, ultra-high pressure processing, ohmic heating and the use of artificial intelligence for characterizing and modeling food processing and quality. He has to his credit more than 250 publications and several book chapters, and has edited 5 books. He has also authored a textbook on food processing. He is currently serving as editor for the *Journal of Food Engineering.*

Dr. Stefan Kasapis is a professor at RMIT University, Australia. He has an established reputation in research on conformation, interactions, and functional properties of food hydrocolloids and co-solutes. His expertise extends from the gelation properties of high-water systems to the vitrification phenomena of high-solid mixtures. He is the author of more than 100 peer-reviewed publications in reputable food science journals and seven book chapters, and the inventor of eight international patents. He serves as editor of *Food Hydrocolloids* and is a member of the editorial board of *Carbohydrate Polymers.* He organized the 9th International Hydrocolloids Conference in June 2008.

Dr. Joyce I. Boye is a research scientist with Agriculture and Agri-Food Canada. Her research in the past 15 years has been directed at developing techniques for the isolation, extraction, and characterization of proteins from both animal and plant sources. In the past ten years, Dr. Boye and her research team have done extensive work on oilseeds and pulses, working in collaboration with industries from the private sector to develop new processing techniques and products. She is the senior author of several confidential industrial reports and has a total of over 128 scientific publications/abstracts/technical presentations, including five book chapters.

Editors

Contributors

Jasim Ahmed
Polymer Source Inc.
Dorval
Montréal, Québec, Canada
jahmed2k@yahoo.com or
jasim@polymersource.com

Sajid Alavi
Department of Grain Science and Industry
Kansas State University
Manhattan, Kansas
salavi@k-state.edu

Sally Alkhafaji
Department of Chemical and Materials Engineering
The University of Auckland
Auckland, New Zealand
salk002@xtra.co.nz

Pedro A. Alvarez
Department of Food Science
Macdonald Campus of McGill University
Ste Anne de Bellevue, Québec, Canada
Pedro.alvarez@mail.mcgill.ca

T. R. Bajgai
Department of Bioresource Engineering
Macdonald Campus of McGill University
Ste Anne de Bellevue, Québec, Canada
trbajgai@gmail.com

Paramita Bhattacharjee
Department of Food Engineering and Technology
Institute of Chemical Technology
Matunga, Mumbai, Maharashtra, India
yellowdaffofils07@gmail.com

Joyce I. Boye
Food Research and Development Centre
Agriculture and Agri-Food Canada
St Hyacinthe, Québec, Canada
joyce.boye@agr.gc.ca

Armand V. Cardello
US Army Natick Soldier R, D&E Center
Natick, Massachusetts
armand.cardello@us.army.mil

Juan A. Chávez
Sensor Systems Group
Department of Electronic Engineering
Universitat Politècnica de Catalunya
Barcelona, Spain
chavez@eel.upc.edu

P. J. Cullen
School of Food Science and Environmental Health
Dublin Institute of Technology
Dublin, Ireland
pjcullen@dit.ie

Mohammed Farid
Department of Chemical and Materials Engineering
The University of Auckland
Auckland, New Zealand
m.farid@auckland.ac.nz

Miguel J. García-Hernández
Sensor Systems Group
Department of Electronic Engineering
Universitat Politècnica de Catalunya
Barcelona, Spain
mgarcia@eel.upc.edu

Sundaram Gunasekaran
Department of Biological Systems Engineering
University of Wisconsin-Madison
Madison, Wisconsin
guna@wisc.edu

Ashraf A. Ismail
Department of Food Science
Macdonald Campus of McGill University
Ste Anne de Bellevue, Québec, Canada
ashraf.ismail@mcgill.ca

Yousef Karimi-Zindashty
Department of Food Science
Macdonald Campus of McGill University
Ste Anne de Bellevue, Québec, Canada
yousef.karimizindashty@mail.mcgill.ca

Stefan Kasapis
School of Applied Sciences
RMIT University
Melbourne, Victoria, Australia
stefan.kasapis@rmit.edu.au

Yeting Liu
Food Science and Technology Programme
Department of Chemistry
National University of Singapore
Singapore
yeting.liu@gmail.com

Andreas Lopata
School of Applied Sciences
RMIT University
Melbourne, Victoria, Australia
andreas.lopata@rmit.edu.au

Michele Marcotte
Food Research and Development Center
Agriculture and Agri-Food Canada
St Hyacinthe, Québec, Canada
michele.marcotte@agr.gc.ca

Ruben Mercade-Prieto
Department of Biological Systems Engineering
University of Wisconsin-Madison
Madison, Wisconsin
r.mercadeprieto@bham.ac.uk

A. Muthukumaran
Department of Bioresource Engineering
Macdonald Campus of McGill University
Ste Anne de Bellevue, Québec, Canada
arun.muthukumaran@mcgill.ca

Kasiviswanathan Muthukumarappan
Department of Agricultural and Biosystems Engineering
South Dakota State University
Brookings, South Dakota
muthukum@sdstate.edu

Michael Ngadi
Department of Bioresource Engineering
McGill University
Montréal, Québec, Canada
michael.ngadi@mcgill.ca

Colm P. O'Donnell
Biosystems Engineering, School of Agriculture
Food Science and Veterinary Medicine
University College Dublin
Belfield, Dublin, Ireland
colm.odonnell@ucd.ie

Valérie Orsat
Bioresource Engineering Department
Macdonald Campus of McGill University
Ste Anne de Bellevue, Québec, Canada
valerie.orsat@mcgill.ca

G. S. V. Raghavan
Department of Bioresource Engineering
Macdonald Campus of McGill University
Ste Anne de Bellevue, Québec, Canada
vijaya.raghavan@mcgill.ca

S. H. Rajamohamed
Food Research and Development Centre
Agriculture and Agri-Food Canada
St Hyacinthe, Québec, Canada
SahulHameed.RajaMohamed@agr.gc.ca

Hosahalli S. Ramaswamy
Department of Food Science
Macdonald Campus of McGill University
Ste Anne de Bellevue, Québec, Canada
hosahalli.ramaswamy@mcgill.ca

M. A. Rao
Department of Food Science and Technology
Cornell University
Geneva, New York
mar2@cornell.edu

Navin K. Rastogi
Department of Food Engineering
Central Food Technological Research Institute
Council of Scientific and Industrial Research
Mysore, India
nkrastogi@cftri.res.in or nkrastogi@yahoo.com

Syed S. H. Rizvi
Department of Food Science
Cornell University
Ithaca, New York
SRizvi@cornell.edu

Jordi Salazar
Sensor Systems Group
Department of Electronic Engineering
Universitat Politècnica de Catalunya
Barcelona, Spain
jsalazar@eel.upc.edu

Rakesh K. Singh
Food Science and Technology Department
The University of Georgia
Athens, Georgia
rsingh@uga.edu

Rekha S. Singhal
Department of Food Engineering and Technology
Institute of Chemical Technology
Matunga, Mumbai, Maharashtra, India
rsinghal7@rediffmail.com

Litha Sivanandan
Reseach and Development Food Technologist
Oceana Foods
Shelby, Machigan
LSivanandan@oceanafoods.com

Jirarat Tattiyakul
Department of Food Technology
Faculty of Science
Chulalongkorn University
Bangkok, Thailand
jirarat.t@chula.ac.th

B. K. Tiwari
Biosystems Engineering, School of Agriculture
Food Science and Veterinary Medicine
University College Dublin
Dublin, Ireland
brijesh.tiwari@ucd.ie

Antoni Turó
Sensor Systems Group
Department of Electronic Engineering
Universitat Politècnica de Catalunya
Barcelona, Spain
turo@eel.upc.edu

Alan O. Wright
US Army Natick Soldier R, D & E Center
Natick, Massachusetts
alan.o.wright@us.army.mil

Jun Xue
Food Research Center
Agriculture and Agri-Food Canada
Guelph, Ontario, Canada
Jun.Xue@agr.gc.ca

David Young
School of Biomolecular and Physical Sciences
Griffith University
Brisbane, Australia
d.young@griffith.edu.au

Mohammad R. Zareifard
Food Research and Development Center
Agriculture and Agri-Food Canada
St Hyacinthe, Québec, Canada
mohammadreza.zareifard@agr.gc.ca

Weibiao Zhou
Food Science and Technology Programme
Department of Chemistry
National University of Singapore
Singapore
chmzwb@nus.edu.sg

Songming Zhu
College of Biosystems Engineering and Food Science
Zhejiang University
Hangzhou, Zhejiang, People's Republic of China
zhusm@zju.edu.cn

1 Introduction and Plan of the Book

Jasim Ahmed
Polymer Source Inc.

Hosahalli S. Ramaswamy
Macdonald Campus of McGill University

Stefan Kasapis
RMIT University

Joyce I. Boye
Agriculture and Agri-Food Canada

CONTENTS

The growth of novel processing technologies has now become extremely commercially important, involving many new products, industrial collaborations, and academic research groups. This book covers the present status and future trends of novel technologies alongside major lines of development in relation to rheological and functional properties of food. It is important to recognize that developments in novel food processing have mostly taken place in developed countries, however developing countries could benefit from the knowledge already generated in the field.

The advantages of novel processing technologies over conventional protocols are the retention of sensory attributes, desired texture, and improved functional properties by focusing on the promising areas of next-generation food processing. Novel technologies can be classified into two categories: electrotechnologies (pulse electric field, radio-frequency (RF) heating, microwave heating, infrared heating, ohmic heating, etc.) which make use of novel methods of heating, and nonthermal technologies (pulse-light, high hydrostatic pressure, oscillating magnetic field, irradiation, ozonization, plasma, osmotic treatment, etc.). Most of the electrotechnologies focus on novel approaches to generate heat and rely on conventional thermal mechanisms for achieving preservation and processing. Nonthermal processing technologies, on the other hand, inactivate enzymes and microorganisms, and modify the functional properties of food by alternate means without substantially increasing the product temperature.

The majority of the literature on this subject is devoted to equipment design and process optimization, in conjunction with microbial aspects and safety of various foods. There is, indeed, an urgent need to study microbial survival and destruction under the novel and alternate processing technologies to achieve food sterility or food safety. On the other hand, structural modifications and functionality of ingredients and formulations under novel processes cannot be neglected since they have a profound effect on both sensory and nutritional aspects of food. The current emphasis on health-related functional foods leading an integration of food production, processing/preservation, and nutritional aspects has led to renewed interest in the texture and functional properties of materials under novel processing technologies.

Structural and functional properties are important determinants of food quality, as perceived by consumers and required by the food industry. Rheology deals with establishing predictions of mechanical behavior based on the micro- or nanostructure of the material, e.g., the molecular size and architecture of food polymers in solution or particle size distribution in a solid suspension. Rheological measurements are essential tools in the analytical laboratory for characterizing component materials and finished products, and monitoring process conditions, as well as predicting product performance and consumer acceptance. Thus, rheological properties provide fundamental insights into the structural organization of food. For example, starch/protein dispersions converted to gel under novel processing exhibit a true viscoelastic nature (Molina, Defaye, and Ledward 2002; Ahmed et al. 2007).

In addition, the functionality of food is improved significantly with a novel processing technology like high pressure (HP), which can partially reduce allergenic proteins in rice (Kato et al. 2000), and pulse electric field, which inactivates undesirable enzymes (pectin methyl esterase [Yeom et al. 2002], polyphenol oxidase [Giner et al. 2002], and alkaline phosphatase [Castro 1994]). Frequently, the academic study of food structure is conducted in isolation from the wider commercial and consumer context. A major objective of this book, therefore, is to analyze changes in food structure at the micro- and macromolecular levels when subjected to novel processing technologies and to discuss food properties within this wider context.

The contributions culminating in the present form of the book are briefly surveyed in this opening chapter so that readers can readily obtain an overview of the book and its highlights. There are 22 chapters in total, covering most of the novel technologies. It is worth mentioning that a few of the novel processing technologies have been explored mainly for their role in preservation and processing and there is not much information on their effect on rheology/texture and/or functional properties, e.g., pulsed light or pulsed electric field (PEF), and, in these cases, the focus lies on technology rather than core content.

RF dielectric heating has been used for decades in thermal processing of nonconductive materials with industrial applications for drying of wood and tempering of frozen foods. RF is promising for many heating and drying processes but there is still a need for more readily available data related to the design of the applicators, targeted product formulations with specific dielectric properties, and control systems for the development of RF heating process applications. Selection of dielectric energy input and its combination with conventional thermal technologies may lead to products that meet and often improve the quality requirements of existing food products, hence opening the field for product development. RF heating applications are focused on in Chapter 2.

Ohmic heating is an innovative heating technique employed for thermal processing. The food is placed between two electrodes serving as an electrical resistor and an alternating electric current is passed through the circuit. The electrical resistance causes heat to be generated throughout the food matrix in a uniform and volumetric fashion. The electrical energy is directly converted into heat, causing a temperature rise, and depends largely on the electrical conductivity of the product. It is theoretically possible to properly match the electrical conductivity of solid and liquid phase of particulate fluids such that under ohmic heating the particles could heat faster than the liquid. This is a unique feature that has attracted the attention of the industry, academia, and regulatory agencies to permit ohmic heating as a potential source of treatment in aseptic processing of particulate fluids. Chapter 3 highlights the principles and applications of the technique and their influence on food rheology and functional properties.

Chapter 4 deals with the application of high electric field (HEF) to food processing, especially drying. HEF technology offers many distinctive advantages over conventional processing methods. One of the major advantages is that there is insignificant change in temperature during processing. As a result, the method can be successfully applied to any temperature-sensitive fresh material to retain flavonoids. The processing efficiency and capacity mainly depends on the voltage strength used for processing. Though interest in HEF processing is high, the technique is still in its infancy in terms of industrial exploitation.

PEF treatment of liquid foods is one of the emerging nonthermal food preservation processes. It can accomplish food preservation with short treatment times and small increases in food temperature, thus, providing an alternative to thermal pasteurization. The exposure of microbial cells to electric field induces transmembrane potential on the cell membrane, which results in electroporation and electrofusion. The generation of PEF requires a system for generating high-voltage pulses and a treatment chamber that converts the pulsed voltage into PEFs. There has been a significant development in the design of PEF systems during the last 20 years. Chapter 5 reviews the various high-voltage pulse generators and the continuous treatment chambers that have been designed and constructed by different research groups.

Ultrasound refers to energy generated by sound waves of 20,000 or more vibrations per second. High frequencies in the range of 0.1–20 MHz, pulsed operation, and low power levels (100 mW) are used for nondestructive testing (Gunasekaran and Chiyung 1994). Ultrasonic excitation is being examined for nondestructive evaluation of the internal quality and latent defects of whole fruits and vegetables in a manner similar to the use of ultrasound for viewing the developing fetus in a mother's womb (Mizrach et al. 1994). Ultrasound could be used in various applications including rheology, texture, and concentration measurements of solid or fluid foods; composition determination of eggs, meats, fruits, vegetables, and dairy products; thickness, flow level, and temperature measurements for controlling several processes; and nondestructive inspection of egg shells and food packages. Two chapters deal with the technology (Chapter 6) and applications (Chapter 7) of the subject.

Ionizing radiation (IR) was the first novel technology applied to food products to enhance shelf life, although it was not easy to convince regulatory agencies and consumers of its safety. IR offers various technological benefits by reducing food losses and improving food safety. Irradiation (1.5–7.0 kGy) applied to solid and semisolid food like meat, poultry, seafood, potatoes, and onions confers the same benefits as thermal pasteurization confers on liquids. After decades of extensive research and testing have demonstrated that IR is safe within prescribed doses, it has been permitted by all specialized agencies of the United Nations (WHO, FAO, and IAEA). Chapter 8 of this book is devoted to this area.

Ozone has a wide antimicrobial spectrum that, combined with high oxidation potential, makes it an attractive processing option for the food industry. Relatively small quantities of ozone and short contact times are sufficient for the desired antimicrobial effect as it rapidly decomposes into oxygen, leaving no toxic residues in food (Muthukumarappan, Halaweish, and Naidu 2000). The interest in ozone as an antimicrobial agent is based on its high biocidal efficacy and wide antimicrobial spectrum that is active against bacteria, fungi, viruses, protozoa, and bacterial and fungal spores (Khadre, Yousef, and Kim 2001). Ozone is 50% more effective over a wider spectrum of microorganisms than chlorine. It reacts up to 3000 times faster than chlorine with organic materials and produces no harmful decomposition products (Graham 1997). Such advantages make ozone attractive to the food industry and consequently it was declared as Generally Recognized as Safe (GRAS) for use in food processing by the U.S. Food and Drug Administration (FDA) in 1997 (Graham 1997). Chapter 9 provides details on the issue.

Most of the proteins form aggregates under many destabilizing conditions, particularly at high temperatures. Further aggregation leads to a solid-like gel of three-dimensional networks. Chapter 10 discusses protein gelation at room temperatures using a two-step process, usually referred to as cold gelation. The first step involves the formation of soluble aggregates, usually with a heat treatment. The final gelation step is commonly achieved following addition of salts or by reducing pH. Studies of cold gelation procedures in emulsions, bead and microcapsule formation, and with polysaccharide mixtures are reviewed.

There are several contributions in this book dealing with HP processing and its effect on rheology, structure, and functional properties. The first of them emphasizes the significant interest among academicians and processors in the technique. In 1899, Japanese researchers attempted HP treatment of food for the first time. The related publication represents the first reported case of reduction of spoilage by means of high hydrostatic pressure. Pioneering work in 1914 reported that pressure could denature egg white (without heat) resulting in an outlook similar but not identical to that of a cooked egg, is also mentioned. Research in the area of HP food processing was further advanced following the work carried out in Japan in the 1980s. Today, HP appears to be a major processing technology and alternative to heat treatment. It has reached the consumer with a variety of products including fruit jams, jellies, sauces, juices, avocado pulp, guacamole, and cooked ham. This year the FDA approved the pressure-assisted thermal sterilization (PATS) process.

The effect of high hydrostatic pressure on the rubber-to-glass transformation of biomaterials with potential industrial interest in confectionery formulations is discussed in Chapter 11. Theoretical aspects of the application of HP to the mechanical properties of synthetic and natural polymers are elaborated to offer a sound basis for utilization of the approach in model food systems. The free volume theory in conjunction with the correct pressure dependence of the compressibility coefficient leads to the Fillers–Moonan–Tschoegl equation, which follows the combined thermorheologically and piezorheologically simple behavior of synthetic polymers. This school of thought was considered in the vitrification of high solid biomaterials but it was demonstrated that the time/temperature/pressure equivalence of synthetic materials is not operational in the glass-like behavior of high sugar systems in the presence of gelatin or gelling polysaccharides. This deviation from the "normal" course is discussed in terms of the irreversible destabilization of intermolecular aggregates that occurs mainly in polysaccharide networks following pressure-induced vitrification.

HP treatments influence the functional properties of proteins through the disruption and reformation of hydrogen bonds, as well as hydrophobic interactions and the separation of ion pairs (Hayakawa, Linko, and Linko 1996). After HP treatment rheological properties of protein foods are significantly altered with texture modification. Protein-water dispersion converts to gel that is softer than thermally treated gel, as indicated by small amplitude oscillatory shear (SAOS) measurement. After pressure treatment, the primary structure of proteins remains intact while secondary, tertiary, and quaternary structures are affected to various degrees. Various instrumental techniques (differential scanning calorimetry, Fourier transform infrared spectroscopy, circular dichromism, size exclusion chromatography, and electrophoresis) are commonly employed to detect structural changes after pressurization. Details are available in Chapters 12, 13, and 14. Chapter 13 deals with pressure effect on plant proteins (soybean, rice, wheat, and lentil) whereas Chapter 14 emphasizes the effect on animal proteins.

Chapter 15 discusses the distinct conditions of hydrostatic HP processing of starch-water suspensions, dynamic HP homogenization of starch-water suspensions, and hydrostatic HP compression of starch granules at low moisture contents. These processing treatments lead to various effects ranging from pressure-induced gelatinization to structural and functional modifications. It is found that HP compression has a direct impact on the gelatinization behavior of starches with a favorable temperature shift and enthalpy reduction.

Application of HP to fruits and vegetables (Chapter 16) has an influence on their texture due to liquid infiltration and gas displacement. Changes such as collapse of air pockets and shape distortion result in tissue shrinkage. In many cases, pressure-treated vegetables did not soften during subsequent cooking, which was attributed to the inaction of pectin methyl esterase (PME) that was only partially activated by pressure. HP has a major effect on the solid–liquid phase transition of water. Pressure shift freezing (PSF) and HP thawing are two major applications with potential benefits in food processes based on this phenomenon. PSF involves cooling the food samples under pressure (usually up to 200 MPa) to just above the freezing point (–20°C) and quickly releasing the pressure to form massive nuclei of ice crystals; the ice crystals are then allowed to grow at atmospheric pressure in a conventional freezer. PSF can thus, have a significant impact on frozen food quality, especially in terms of preserving the texture of frozen thawed foods, which is greatly affected by the conventional freezing techniques. These issues are highlighted in Chapter 17. Overall it can be concluded that HP is a potential nonthermal alternative in the processing and preservation of fruit and vegetables that allows better retention of organoleptic qualities.

A rapid expansion of research and development on novel food processes has taken place in the last 25 years and has resulted in a variety of novel processes; ultimately, the market success of these novel technologies will depend on the acceptability of these processes to consumers. Chapter 18 focuses on sensory and other consumer issues that must be addressed during the development of foods processed by novel technologies and discusses new and evolving methodologies that should be employed to maximize the sensory quality and acceptance of these foods.

The soybean processing industry has matured in the last few decades. Wide varieties of nutritional, functional, and therapeutic ingredients are extracted from soybeans and are currently commercially available. Soybeans also contain antinutritional factors such as trypsin inhibitors, phytic acid, and lectins, which require inactivation or removal prior to consumption. In recent years, novel technologies for soybean processing have emerged, which effectively inactivate antinutritional components and can be used to extract soybean components of interest or to modify functionality. Chapter 19 addresses some of the novel techniques in use or under study for the processing of soy foods and soy ingredients and their effects on product quality and functionality.

Supercritical fluid extrusion (SCFX) is an elegant new process developed at the cutting-edge interface of supercritical fluid and conventional extrusion technologies. Low-shear and low-temperature processing, formation of highly expanded products with a unique microcellular structure comprising non-porous skin and homogenous cell size distribution, and use of SC-CO_2 as a carrier for soluble flavors and an in-line process modifier are some of the defining characteristics of the SCFX process. These make it superior to the conventional high-shear and high-temperature extrusion cooking and puffing processes. Moreover, SCFX allows a high degree of control over the nucleation and cell growth phenomena, thus, imparting great flexibility in the design of microcellular biopolymer foams of desired morphology and properties such as sensory texture, thermal insulation, etc. Details are available in Chapter 20.

The continuous-flow high-pressure throttling (CFHPT) process was developed as a means of continuous microbial inactivation of fluid foods (Toledo and Moorman 2000). Throttling valve is a fine restriction such as a partially closed valve or porous plug to control the flow of the fluid through a micrometering valve. CFHPT uses HP (about 310 MPa) to pressurize liquid foods and to continuously throttle the foods from that HP to atmospheric pressure to inactivate microbes and modifying proteins. As the fluid material exits the throttling valve, temperature increases and the product is rapidly cooled and retains sensory quality. The CFHPT system is a continuous system that uses relatively low pressures. Changes in rheological characteristics of liquid foods processed through CFHPT are discussed in Chapter 21.

Frying has remained a major cooking technique and fried foods are popular despite health concerns. However, there is a major effort underway to deal with the health issues of frying with the development of novel processing technologies to reduce fat content without sacrificing taste and flavor. One particular issue relates to batters that are important in determining quality and shelf life of coated fried food products. Their formulation and ingredients impact various functionalities in coated products. Good understanding of the roles of the various ingredients and how they interact with each other are vital in designing optimal batters for specific products. Rheological and thermal properties of batters vary with the different types of flours and their combination ratios, and with the different types of hydrocolloids and other ingredients. Careful study and elucidation of these interactions can be used to control batter functionalities, as described in Chapter 22.

About 4% of adults and up to 8% of children suffer from some type of food allergy, with foods such as peanuts, tree nuts, wheat, soy, cow's milk, egg, fish, and shellfish being responsible for over 90% of reactions. While food allergy is reaching epidemic proportions, currently no curative or prophylactic treatments are available. A molecular description of the allergenic components and protein families present in the major allergenic foods is given in Chapter 23. Research into the area of food allergy is complex because allergenic proteins are exposed to digestive processes and in addition undergo extensive modifications during processing of food. Thermal processing can reduce the allergenicity of certain allergens (e.g., fruits), while roasting in the presence of sugars can dramatically enhance the allergenicity. The effects of modern processing technologies (e.g., HP processing) on allergenicity are discussed as well as opportunities for the food industry to develop successful processing strategies to effectively protect sensitive consumers to prevent unwanted allergen exposure.

A number of outstanding scientists were invited to contribute to this book by providing novelties related to their area of research. Their contributions form the backbone of this innovative work. Contributions were specifically requested to address the effects of novel technologies on rheology and functional properties. A few of our contributors faced difficulties in obtaining detailed mechanistic and rheological information for certain novel technologies, e.g., PEF and irradiation. In such cases, general overviews of related food applications are highlighted in order to include these technologies in the work. Overall, the contributions in the book illustrate in a very clear and concise fashion the structure–functionality relationships as affected by novel processing technologies.

REFERENCES

Ahmed J., H. S. Ramaswamy, A. Ayad, I. Alli, and P. Alvarez. 2007. Effect of high pressure treatment on rheological thermal and structural changes in Basmati rice flour slurry. *Journal of Cereal Science* 46 (2), 148–56.

Castro, A. J. 1994. Pulsed electric field modification of activity and denaturation of alkaline phosphatase. PhD thesis, Washington State University.

Giner, J., M. Ortega, M. Mesegué, V. Gimeno, G. V. Barbosa-Cánovas, and O. Martín. 2002. Inactivation of peach polyphenoloxidase by exposure to pulsed electric fields. *Journal of Food Science* 67, 1467–72.

Graham, D. M. 1997. Use of ozone for food processing. *Food Technology* 51, 121–37.

Gunasekaran S., and A. Chiyung. 1994. Evaluating milk coagulation with ultrasonics. *Food Technology* 48 (12), 74–78.

Hayakawa, I., Y. Y. Linko, and P. Linko. 1996. Mechanism of high pressure denaturation of proteins. *Food Science & Technology* 29, 756–62.

Kato, T., E. Katayama, S. Matsubara, Y. Omi, and T. Matsuda. 2000. Release of allergic proteins from lentil grains induced by high hydrostatic pressure. *Journal of Agricultural and Food Chemistry* 48, 3124–26.

Khadre, M. A., A. E. Yousef, and J. Kim. 2001. Microbiological aspects of ozone applications in food: A review. *Journal of Food Science* 6, 1242–52.

Mizrach, A., N. Galili, D. C. Teitel, and G. Rosenhouse. 1994. Ultrasonic evaluation of some ripening parameters of autumn and winter-grown 'Galia' melons. *Scientia Horticulturae* 56 (4), 291–97.

Molina, E., A. B. Defaye, and D. A. Ledward. 2002. Soy protein pressure-induced gels. *Food Hydrocolloids* 16, 625–32.

Muthukumarappan, K., F. Halaweish, and A. S. Naidu. 2000. Ozone. In: *Natural Food Anti-Microbial Systems*. A.S. Naidu, Eds. CRC Press, Boca Raton, FL, pp. 783–800.

Toledo, R. T., and J. E. Moorman Jr, J. E., inventors; University of Georgia, assignee. 2000. Microbial inactivation by high-pressure throttling. US patent 6,120,732.

Yeom, H. W., Q. H. Zhang, and G. W. Chism. 2002. Inactivation of pectin methyl esterase in orange juice by pulsed electric fields. *Journal of Food Science* 67, 2154–59.

2 Effect of Radio-Frequency Heating on Food

Valérie Orsat
Macdonald Campus of McGill University

CONTENTS

2.1 INTRODUCTION

The food industry is in continuous flux to meet market demands for healthy, preservative-free, high-quality, and safe food products. Processing technologies being used to ensure food quality and safety are freezing, thermal processing, sterilization, drying, refrigeration, and the controlled distribution of fresh produce. In some applications, dielectric heating/sterilization can deliver high-quality products since electromagnetic waves are able to heat the product 3–5 times faster than conventional sterilization systems (Rosenberg and Bögl 1987). The dielectrically heat-treated product is not temperature-abused due to the reduction in process time, so the food has better overall quality attributes than products processed by other available thermal technologies.

Microbial and pest reduction by dielectric heating has been well studied for several products including meat and meat products, poultry, eggs, fish, fruit and vegetable products, ready-cooked meals, milk and milk products, cereals, and spices (Harlfinger 1992; Rosenberg and Bögl 1987; Wang et al. 2003b; Lagunas-Solar et al. 2005). Wang et al. (2003b) studied the radio-frequency (RF) sterilization of a model food and macaroni and cheese. Their findings showed that the RF process (27.12 MHz) produced better-quality products, at a faster rate, while using less energy than the conventional retort process used to produce shelf-stable foods. However, conventional cooking and heating techniques generally yield lower microbial counts than microwave/RF treatments when the process is controlled by temperature only (Rosenberg and Bögl 1987). Speed and evenness of heating are influenced by the composition and mass of the food as well as by features of the heating unit. Since the heating during dielectric cooking could be uneven, the presence of relatively cool regions might account for the survival of bacteria even when very high temperatures are recorded in other parts of a food.

Many biological materials are color sensitive to temperature, and rigorous temperature control must be maintained to preserve the appearance of the material. Temperature excesses or extended periods of time at tolerable temperatures may change the appearance, taste, or quality of the food product. Research on mutual interactions between food products and dielectric processing equipment is still needed to provide a practical basis for process control and minimizing process energy costs, while ensuring microbial safety and overall product quality.

2.2 MECHANISMS OF RADIO FREQUENCY (RF) HEATING

2.2.1 Dielectric Heating

Dielectric heating lies in the electromagnetic spectrum range of frequencies from 300 kHz to 300 GHz (Table 2.1). Radio frequencies range from 300 kHz to 300 MHz and microwaves range from 300 MHz to 300 GHz (Assenheim et al. 1979).

There are differences in the electrical and thermal responses of foods to microwave and RF fields, which depend on aspects of chemical composition, physical structure, and geometry of the product (Mudgett 1985). The choice of processing frequency in RF or microwave, for a particular unit operation, may be critical since the dielectric behavior and heating characteristics of foods vary with frequency, temperature, moisture, and salt contents. Ionic losses for a particular product may be much higher and dipole losses much lower at 27.12 MHz than at 2450 MHz and vice versa.

In a RF heating system, the RF generator creates an alternating electric field between two electrodes. The material to be heated is placed between the electrodes where the alternating electric energy causes polarization, where the molecules in the material continuously reorient themselves to face opposite poles (Figure 2.1).

TABLE 2.1
Dielectric Heating Frequency Ranges

300–3000 kHz	Medium frequency (MF)
3–30 MHz	High frequency (HF)
30–300 MHz	Very high frequency (VHF)
300–3000 MHz	Ultra high frequency (UHF)
3–30 GHz	Super high frequency (SHF)
30–300 GHz	Extremely high frequency (EHF)

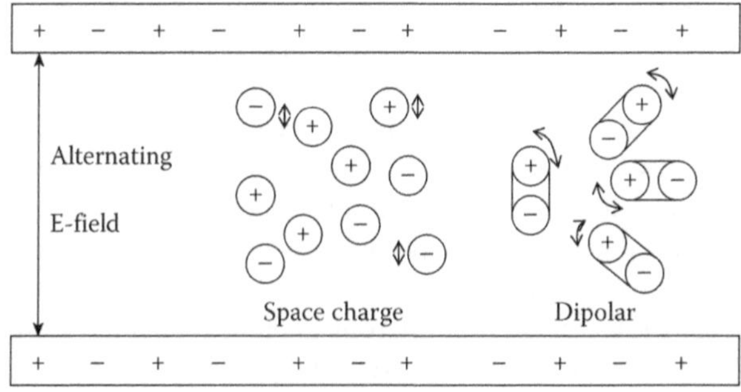

FIGURE 2.1 Space charge and dipolar polarization in between RF electrodes applying an alternating electric field.

With an electric field alternating at the 27.12 MHz radio frequency, the electric field alternates 27,120,000 cycles per second. The energy release resulting from the movement of the molecules and the space charge displacement causes the material to rapidly heat throughout. The amount of heat generated in the product is determined by the frequency, the square of the applied voltage, dimensions of the product, and the dielectric loss factor of the material, which is essentially a measure of the ease with which the material can be heated by RF energy. It is the composition of the material that governs its dielectric properties. The principle of dielectric heating is described by the general expressions for time rate of temperature rise and power dissipation in a dielectric material:

$$\frac{dT}{dt} = 0.239 \times 10^{-6} \frac{P}{c\rho} \tag{2.1}$$

and

$$P = 2\pi f E^2 \varepsilon_0 \varepsilon'' \tag{2.2}$$

which yields the following equation, since $\varepsilon_0 = 8.85 \times 10^{-12}$ F/m:

$$P = 55.61 f E^2 \varepsilon'' \times 10^{-12}, \tag{2.3}$$

where dT/dt is expressed in degrees Celsius per second, P is the power (W/m^3), c represents the specific heat of the dielectric material (J/kg·K), ρ is the density of the material (kg/m^3), f is the frequency (Hz), E is the electric field strength (V/m), and ε'' is the dielectric loss factor of the material.

When a mixture of materials is subjected to dielectric heating by high-frequency or microwave electric fields, the relative power absorption will depend upon the relative values of E and ε'' for each of the materials in the mixture (Nelson 1985).

Too high a value of ε'' results in a skin effect, which annuls the desirable volumetric heating effect. On the other hand, too low an ε'' renders the material practically transparent to the incoming energy. Experience has shown that materials with an effective loss factor in the range $10^2<\varepsilon''<2$ will be suitable for dielectric processing (Metaxas 1988). Hence, knowledge of the dielectric properties is very important when assessing the feasibility of heating any material by dielectric heating. Investigations by Zhong, Sandeep, and Swartzel (2003) indicate that uniform heating is not always possible with products having a high dielectric loss wherein the temperature of the product may indeed show a significant gradient.

The major difference between RF and microwave is in the hardware and in the method of transfer of the energy to the product. In RF, an electric field is developed between electrodes while in microwave it is a wave being propagated and reflected under the laws of optics. RF works well with larger quantities (at 27.12 MHz, the wavelength is of the order of 10 m) having higher ionic conductivity, while microwave works better with smaller quantities of a dipolar nature (Mudgett 1985; Metaxas 1996).

The success of an RF heating set-up lies in its design and in the impedance matching between the power generator and the applicator. The quality of the applicator's design is very important to ensure its efficiency. Furthermore, knowledge and control of the dielectric properties of the material to be processed are essential for optimal thermal processing and efforts need to be concentrated on product formulations. The development of new applications for RF and the design of applicators require sophisticated tools (network analyzer) and technical expertise for fine tuning and the measurement of dielectric properties for optimal product formulation (Bialod and Marchand 1986).

2.2.2 Temperature Uniformity and Arcing Problems

The electric field (in V/m) is theoretically uniform between the electrode of the RF applicator except for some edge effects. Experimental evidence has demonstrated that the electric field has a tendency to concentrate on the corner of the material placed in a parallel plate configuration. This is explained by the fact that at the corners there are conflicting boundary conditions satisfied by field concentration (Roussy and Pearce 1995). Using parallel flat plates, completely filled with a homogeneous material, the resulting electric field is assumed to be relatively homogeneous. However, the temperature distribution is often nonuniform, as reported by Marra et al. (2007), who demonstrated in their study that at low power the temperature profile is relatively uniform, whereas with increasing RF power there is an increasing temperature gradient occurring in the material placed between the electrodes.

The voltage applied on the electrodes is equal to the electric field's value between the electrodes multiplied by the distance from one electrode to the other. When the field strength E reaches a value of about 3 kV/mm (3×10^6 V/m) in dry air, the air molecules become ionized and corona discharges (arcing) occur, which can significantly damage processed products and equipment components (Elgerd 1977). Hence, the air gap between the heated product and the electrodes should be minimized. In the presence of an air gap, there are two homogeneous electric field distributions in each medium, but the corresponding values are not independent: the electric field in the air is equal to the electric field in the product multiplied by its permittivity. The voltage applied is then the sum of two voltages: one creates the electric field through the product and the other through the air. In most RF heating applications, the presence of an air gap between the product and the top electrode is nearly unavoidable, from a design point of view, however air gaps should be minimized to limit energy wastage (Awuah, Ramaswamy, and Piyasena 2002) and arcing (Clee and Metaxas 1994).

Runaway heating takes place when the warmest parts take more and more of the available power at the expense of the coldest parts. In most dielectric heating applications to food, runaway heating is unavoidable if contact with both electrodes is maintained. In practice, this is why an air gap is introduced between the top electrode and the upper surface of the material being heated (Sanders 1966). Therefore, the air gap becomes part of the equipment design considerations. In some cases it has been suggested to surround the product with water to isolate it from the electrodes and limit the occurrence of surface burning (Brunton et al. 2005; Lyng et al. 2007).

The majority of commercially available RF heating systems follow the standardized 50Ω RF technology, which offers superior frequency stability and compliance with stringent emissions regulations in force due to the overwhelming use of RF for telecommunication purposes (Marchand and Meunier 1990; Bernard 1997).

At the frequencies used in RF heating (13.56 or 27.12 MHz) high electric field amplitudes are developed in the load and applicator compared to microwave systems of the same power densities. Consequently, localized overheating, microarcs, or corona discharges are likely to occur, especially with nonhomogeneous food products (Zhao et al. 2000). These effects lower the breakdown potential, further increasing the possibility of an electrical discharge occurring, which can result in considerable damage to both equipment and material product. Furthermore, if the material has regions with high moisture content, the possibility exists of extremely sharp corners at the boundary of a void, inducing the formation of microarcs within the material due to the concentration of the electric field at a corner (Clee and Metaxas 1994). In some instances, direct RF heating of a material is impractical due to the unavoidable occurrence of arcing and burning caused by field concentration. To help alleviate this problem, Brunton et al. (2005) demonstrated the potential of RF pasteurization of cased meat products without arcing or burning by simply immersing the material in water during RF processing.

Foods of a lower density heat faster at a given level of power absorption than do denser foods of similar composition. Air is practically transparent to electromagnetic waves because of its low dielectric constant ($\varepsilon'=1$) and therefore, its presence in a material reduces the amount of power absorbed. Although the relative dielectric constants ($\varepsilon'<3$) and loss factor ($\varepsilon''<0.1$) of oils and fats are much smaller than that of water at 20°C ($\varepsilon'=80$), they heat considerably faster since their thermal capacity of 2 kJ/kg°C is less than half that of water (4.2 kJ/kg°C). The thermal conductivity influences the homogeneity of the heating process. RF heating has a direct relationship between the amount of energy for heating and the size of the material subjected to heating. If the size of an individual piece is very large in comparison to the wavelength, superficial heating is favored, whereas for sizes closer to the wavelength, temperature may be higher in the center. The more regular the shape, the more uniform the heating. Thinner parts may be overheated compared to larger parts. This effect may be controlled by reducing the power input and extending the heating time. Although the speed of heating can be easily increased by boosting the electromagnetic power, in practice this option is treated with caution as an excessive rate will lead to nonuniform temperature profiles (Marra et al. 2007). While in conventional heating the limiting factor of the heating rate is the thermal conductivity of the material, with dielectric heating the heterogeneity of dielectric properties is the limiting determinant. Each processing operation, such as cooking, baking, drying, pasteurizing, etc., requires specific optimized heating gradients to enable the desired physicochemical changes to occur adequately.

2.3 RADIO FREQUENCY (RF) HEATING APPLICATIONS

The rapid heating of foods generally delivers products of a superior quality (Harlfinger 1992; Rosenberg and Bögl 1987). For this reason, electromagnetic energy, with its rapid volumetric heating, offers a competitive edge in agricultural and food thermal process applications.

2.3.1 Dielectric Properties of Foods

Dielectric materials exhibit the property of polarization because their molecular structure has strongly bound electrons, unlike that of conductive materials, which have free or loosely bound electrons (Metaxas 1996; Kasevich 1998). In the case of heating at RF there exist two principal mechanisms of polarization, namely dipolar polarization, where polarized molecules are realigned with the alternating field, and space charge polarization, where some charge carriers migrate under the influence of the alternating field, as represented in Figure 2.1. Polarization can take place at both the atomic and molecular levels. Owing to the interaction of the dipole moment with the electric field, a polar substance has a dielectric constant, which is larger than that of a nonpolar material. The dielectric constant of a polar material is strongly dependent upon various physical parameters such as temperature, pressure, and frequency of applied field (Grant, Sheppard, and South 1978). For a polar substance, the relative permittivity ε' (dielectric constant) decreases with increasing frequency as the motion of molecular dipoles is unable to keep up with the changes in direction of the electric field. Biological molecules are polar therefore they are likely to respond well to electromagnetic stimulation. The process of heating by microwaves and RF is defined by the dissipation of electrical energy in "lossy media." The polarization effect is a function of the radiation frequency, the electromagnetic properties of the material, the viscosity of the medium, and the size of the polar molecules. The interaction between RF energy and the material being heated is principally a function of the dielectric properties, namely the dielectric constant ε' and dielectric loss factor ε''.

The permittivity is a measure of a material's ability to store electrical energy and the loss factor is a measure of its ability to dissipate electrical energy. The loss tangent $\tan\delta=\varepsilon''/\varepsilon'$ is related to the material's ability to be penetrated by an electrical field and to dissipate electrical energy as heat (Mudgett 1986).

The power penetration depth is defined as the depth below the surface of the substance where the power density of a perpendicular electromagnetic wave has reduced by $1/e$ from the surface value ($1/e=1/2.718=37\%$).

$$d_p = \frac{\lambda_o}{2\pi}\sqrt{\frac{2}{\varepsilon'[(1+\tan^2\delta)^{1/2}-1)]}} \tag{2.4}$$

Water is the major absorber of electromagnetic waves in foods and, consequently, the higher the moisture content, the better the heating. The organic constituents of foods are dielectrically inert ($\varepsilon'<3$ and $\varepsilon''<0.1$) and, compared to aqueous ionic fluids or water, may be considered transparent to electromagnetic waves. Only at very low moisture levels, when the remaining traces of water are bound and unaffected by the rapidly alternating field, do the components of low specific heat become the major factors in heating. In high carbohydrate foods, such as bakery products and alcoholic beverages, the dissolved salts/sugars and alcohol are the main susceptors (Rosenthal 1992). Foods with phases of diametrically opposed dielectric properties are likely to be heated with drastically different temperature gradients such as highly absorbing components suspended in a continuous phase of low absorbance or low absorbance components suspended in a continuous phase of high absorbance; or even layered products with alternating phases of low and high absorbance.

Selective polar or ionic additions to a low-loss host material can enhance its effective loss and render it suitable for dielectric processing. It is sometimes possible to modify a low loss factor material, without significantly altering its other properties, with a small amount of high-loss factor additive, such as carbon black added to natural rubber, sodium chloride to urea-formaldehyde glues (UIE 1992), and immersion in a saline solution for cherries (Ikediala et al. 2002).

2.3.1.1 Enhancing the Dielectric Properties

It is advisable to determine the dielectric properties of each component that is to undergo dielectric heating (Wang et al. 2003c). A problem with heating uniformity during RF heating, reported by Laycock, Piyasena, and Mittal (2003), was caused by the nonuniform distribution of ingredients in ground meat. The problem was alleviated by improved mixing of the ingredients during the formulation process. Meat dough is a mixture of meat and fat particles in a water phase that contains various salts. The actual meat dough formulation can significantly affect its performance when dielectrically heated, impacting on the texture and particle size distribution of the protein/fat matrix (van Roon et al. 1994).

Lyng, Zhang, and Brunton (2004) studied the dielectric properties of meats and their ingredients. In general, ε'' values of meat blends were dependent on the ε'' values of the nonmeat ingredients added such as salt, phosphate, nitrite, starch, fat, and gluten. However, the changes in ε'' did not correlate well with the temperature behavior of the RF-treated meat blends, which suggested the importance of the specific heat capacity (c, J/kg·K) in the actual temperature rise in combination with the dielectric properties, further highlighting the importance of developing target product formulation for RF applications.

Zhang, Lyng, and Brunton (2007) conducted a study to better understand the influence of added ingredients on the dielectric and thermal properties of meat blends for RF heating applications. Increasing the salt levels significantly increased ε'' with the increased presence of conductive charge carriers from the salts (Awuah et al. 2002; Zhang et al. 2007). The impact from added moisture and added fats was not as well defined. The dielectric loss increase following the temperature increase experienced with the presence of added ions led, in turn, to the decrease of the penetration depth of the RF electromagnetic wave (Zhong, Sandeep, and Swartzel 2003, 2004; Zhang et al. 2007).

Piyasena et al. (2003b) evaluated the dielectric properties of starch solutions (1–4% w/w) at temperatures ranging from 20 to 80°C at 10, 20, and 30 MHz. They studied the effect of added salt (0.2 and 0.5% w/w) in relation to changes in trends exhibited by the relative permittivity, loss factor, and penetration depth. They reported that the relative permittivity ranged from 46 to 308 and 65 to 92 for solutions with and without salt, respectively. The corresponding loss factor ranged from 266 to 4133 and 9 to 266, respectively. Temperature, frequency, concentration, and their interactions had different levels of significance for the dielectric properties of starch solutions. Salt enhanced the relative permittivity and its effect conformed to the anomalous dispersion phenomenon. The loss factor increased with increasing temperature and salt content, and penetration depths associated with salt-enriched samples were low compared to samples without salt. They attributed the effects of temperature, frequency, concentration, and salt on the dielectric properties of starch solutions to the complex interaction between conductivity, density, moisture content, loss angle, and the rheological properties of starch. They developed correlations that could be used for estimating the dielectric properties of starch solutions with and without salt.

Piyasena et al. (2003a) provided a review on the status of RF applications in food processing, and the influence of the properties of liquid foods and operational condition on dielectric properties during RF heating. They concluded that it is evident that frequency, temperature, and the properties of food such as viscosity, water content, and chemical composition of foods affect the dielectric properties of liquid foods. Therefore, they recommended that these parameters be taken into account when designing an RF heating system for foods.

The dielectric properties are no doubt affected by changes in physical state during processing whether it is moisture loss or phase changes (Lu, Fujii, and Kanai 1998). The physical state of a food influences the mobility of water, ions, and polar molecules, which directly influences the dielectric behavior. The onset of gelatinization and denaturation is characterized by a significant change in the dielectric properties which, when monitored, could be used for feedback process control (Bircan and Barringer 2002).

2.3.1.2 Determining the Dielectric Properties

The dielectric properties of food materials may be determined in frequency intervals from direct current to optical frequencies by various measuring techniques. Lumped circuit methods may be used to determine complex permittivity over the frequency range from zero to approximately 200 MHz. At frequencies above 10 MHz, the capacitance and dissipation factor of samples are generally measured by a capacitance bridge. At frequencies from 10^4 to 10^8 Hz, resonant circuits with fixed inductors and variable capacitors may be used, with resonance indicated by voltmeter deflection. Resonance methods also measure the material's capacitance. However, the loss component is obtained from the width of the resonance curve, determined by either susceptance or frequency variation of the material. Dielectric constants are obtained by either method from their relationship with the material's capacitance (Mudgett 1985; Venkatesh and Raghavan 2004).

The dielectric properties of semisolid food products can be determined by their chemical composition in terms of moisture, salts, and solid contents and, to a much lesser extent, by effects of physical structure. Moisture and dissolved salts are the major determinants of dielectric activity in the liquid phase of such products as modified by volumetric exclusion effects of an inert solids phase containing colloidal or undissolved lipid, proteins, carbohydrate, ash, or bound water. The dielectric properties of food materials have been studied under a variety of conditions and reported widely (Nelson and Stetson 1975; Mudgett 1985; Nelson 1991; Ryynanen 1995; Funebo and Ohlsson 1999; Venkatesh and Raghavan 2004; Wang et al. 2003a, 2003c; Ahmed, Ramaswamy, and Raghavan 2007a, 2007b).

2.3.2 Thermal Treatment of Food Products

2.3.2.1 Radio Frequency (RF) Inactivation of Microorganisms

Early research in RF technology has stated the possibility of sterilizing or pasteurizing a food product at time/temperature values much lower than those now required using conventional heating techniques. Beckwith and Olsen (1931) reported significant reductions in the numbers of *Saccharomyces ellipsoideus* and other yeast suspensions irradiated with RF heating for 15 min. at temperatures not exceeding 39°C. Fabian and Graham (1933) treated broth suspensions of *Escherichia coli* with 7.5, 10, and 15 MHz RF energy in a combination condenser-cooler apparatus which maintained the medium at about 19°C by cold water circulation. Destruction of the bacteria was found to occur at the three frequencies, with greatest lethal effect at 10 MHz. Fleming (1944) irradiated *E. coli* with RF energy of various frequencies from 11 to 350 MHz. All frequencies tested had a lethal effect on the bacteria, with the greatest effect, about 98% destruction, occurring at approximately 60 MHz. Brown and Morrison (1954) studied the effect of RF energy at 50Hz, 190 kHz, and 25 MHz on *E. coli*. The bacteria were irradiated in nutrient broth by means of a capsule electrode assembly and bacteria mortality was reported in treatments where the final temperature exceeded about 50°C.

The general opinion is that the effect of RF energy on the inactivation of microorganisms is due only to heat (Decareau 1985; Brunkhorst, Ciotti, and Fredd 2000; Geveke et al. 2002). It is however often interpreted that the lack of nonthermal effects is due to the low electric field strength tested (Geveke et al. 2002).

Lagunas-Solar et al. (2005) tested RF heating (14 MHz) of fishmeal for disinfection (*Salmonella* and *Escherichia*) at 60, 65, 70, 80, and 90°C (170 W for 50 g samples and 1.2 KW for 1 kg samples). The measured temperature profiles indicated homogeneous heating with rapid processing of 3–6 min. The results indicated that for RF heating between 70 and 90°C, a greater than 5 $\log_{10}$ reduction (>99.99%) can be obtained for *Salmonella* and *Escherichia* spp. The RF heating achieved elimination of the bacterial pathogens with low thermal loads, permitting the full retention of the fishmeal quality attributes.

Research was conducted on the electric heating of fat/muscle layers of ham for pasteurizing with capacitive dielectric heating up to 100 MHz (Bengtsson, Green, and Del Valle 1970). Power efficiency was higher at 60 MHz than at 35 MHz. Temperature distribution in the ham was improved by using lean hams of uniform salt content, by increasing the frequency from 35 to 60 MHz, and by good thermal insulation of the moulds. Horizontal layers of fat were always overheated, while vertical layers (perpendicular to the electrode) and embedded balls or cylinders of fat were not (Bengtsson et al. 1970). Significantly lower juice losses were obtained with RF processing than in hot water processing and treatment time was less than half. Sensory evaluation showed a general tendency toward better quality of RF-processed hams, particularly for juiciness. Microbiological examination after prolonged storage showed considerably higher total counts for RF-processed hams, indicating a need for higher final temperatures or supplemental heat treatment. In a comparative study by Orsat et al. (2004), RF-heating of ham reduced the total number of bacteria and helped maintain quality during refrigerated vacuum-packaged storage, with reduced drip losses and improved overall product appearance (lower surface dryness).

Using a 2 kW, 27.12 MHz RF heater, Awuah et al. (2006) studied the effectiveness of RF heating for inactivating surrogates of both *Listeria* and *Escherichia coli* cells in milk under continuous flow conditions. Depending on product residence time and RF power level, they found the RF heating to be capable of inactivating both *Listeria* and *E. coli* in milk, with *E. coli* being the more heat sensitive of the two. For a total residence time of 55.5 s (i.e., 29.5 and 26 s in the applicator and holding tube, respectively), they found up to 5 and 7 log reductions for heating *Listeria* and *E. coli*, respectively, at 1200 W, and an applicator tube exit temperature of approximately 65°C. They concluded that RF heating could be used to effectively pasteurize milk by manipulating incident power levels and flow rate.

2.3.2.2 Radio Frequency (RF) Heating Behavior and Rheology

Tang et al. (2006) studied the RF cooking of beef rolls prepared with various brine additives. The brine additives raised the dielectric loss factor while they reduced the penetration depth of the RF cooking process. Hence, the additive-free meat registered shorter cooking time when compared to the brined samples. In comparing with steam-cooked samples, RF heating yielded significant reductions in cooking time with lower moisture losses and more tender cooked meats. Tang, Cronin, and Brunton (2005) reported uniform temperatures during the RF heating (27 MHz) of ground turkey meat thoroughly mixed with 1.2% salt.

Temperature effects of whey protein gels on their dielectric properties were studied by Wang et al. (2003c) at radio frequencies. It was found that there was a sudden increase of the dielectric constant at temperatures ranging from 70 to 90°C corresponding to the denaturation of β-lactoglobulin, the main component of the whey protein gel. The phase transition from liquid mixture to gel occurred around 80°C. Furthermore, the loss factor increased sharply with increasing temperature, which is indicative of the potential for runaway heating with the combination of the increased ionic conductivity and increased temperature.

Zhang, Lyng, and Brunton (2006) studied the RF cooking of leg and shoulder hams with significant reduction in cooking duration when compared with steam-cooked hams. However, the RF-cooked hams had a lower level of protein denaturation, particularly with respect to collagen.

Awuah et al. (2002) studied the effect of system and product parameters on the temperature change (ΔT) across a 1.5 kW RF heater operating at 27.12 MHz. Starch solutions (1–4% w/w) were used at three different flow rates (0.35, 0.5, and 1 L/min) and four power levels (672, 912, 1152, and 1392 W). They reported that the average heating rate of starch solutions varied from 6 to 19°C/min, depending on flow rate, concentration, and power level. The corresponding residence times varied from 1.5 to 4.3 min. They used central composite designs involving power and starch concentration (1–4% w/w) at 0.5 L/min to study the effects of salt, pH, and sugar. As expected, fluid flow rate, power level and salt concentration had significant impact on temperature change (ΔT) across the applicator tube. Although the interaction effect of salt and concentration influenced ΔT, they observed that the trends were not clear-cut. Sugar and pH had no influence on ΔT, probably due to their relatively low conductivities. However, the interaction effect of sugar and starch concentration affected ΔT. They developed correlations for estimating ΔT across the tube as a function of power level, concentration, pH, added salt, and sugar. They also proposed dimensionless correlations involving the generalized Reynolds, Prandtl, and Grashof numbers, dimensionless power, and loss-factor ratios for estimating the temperature ratio (U) across the RF applicator.

Ahmed et al. (2007c) studied egg white denaturation and gel formation during RF heating with 27.12 MHz. A minimum of 7.5% concentration of the dried egg white dispersion was required for gel formation during RF heating. A minimum time/temperature combination was also required for gel formation and denaturation of the egg white dispersion. In the measurement of the dielectric properties, there was a sharp increase in ε'' between 85 and 90°C, which corresponds to the temperature range of denaturation. In comparison with hot-plate heating, RF heating produced significantly stronger gels.

2.3.2.3 Radio Frequency (RF) Cooking and Enzymatic Inactivation

Heating tests with sausage emulsions at varying formulations stuffed in tubes were performed at 27 MHz by Houben et al. (1991) in a heating unit consisting of a cylindrical glass tube through which a sausage emulsion was passed. The rapid heating rates resulted in considerably reduced cook values as compared to conventional heating methods (Houben et al. 1991). Sausage products heated well and had a good appearance without release or loss of moisture and fat. Laycock et al. (2003) studied the RF cooking of meat products. RF cooking resulted in reduced cooking time, lower juice losses, acceptable color and texture, and competitive energy efficiency. Similar results were reported by McKenna et al. (2006) for the RF cooking of ham products. However, they also found that the organoleptic properties can be adversely affected, although the overall quality of the product can be ensured with target product formulations to control quality during RF processing.

A study by Orsat et al. (2001) was conducted to develop a processing method for the RF treatment of fresh-like carrot sticks to reduce their microbial load and their enzymatic activity while ensuring their quality. Results showed that when compared to chlorinated water dipping and hot water dipping, the RF-treated carrot sticks had better quality in terms of color and taste.

A study by Schuster-Gajzágó et al. (2006) investigated the potential of a mild RF treatment for the heat inactivation of myrosinase enzyme in mustard flour. The RF thermal treatment produced a bland product with the reduction of the myrosinase activity while it preserved the nutritional and functional qualities of the mustard flour.

2.3.2.4 Radio Frequency (RF) Drying

RF drying has had applications mostly in textile and wood drying. With regards to food, RF drying has mainly been used for finish drying applications for cookies, crackers, and pasta (UIE 1992; Mermelstein 1998). Cookies and crackers, fresh out of the oven, have a non-uniform moisture distribution, which may yield to cracking during handling. RF heating can help even out the moisture distribution after baking, by targeting the remaining moisture. Drying application developments are investigating hybrid drying systems, involving RF with heat pumps or fluidized beds, to cater for the special needs of heat-sensitive food stuffs (Zhao et al. 2000; Chou and Chua 2001; Vega-Mercado, Gongora-Nieto, and Barbosa-Canovas 2001).

2.4 CONCLUSIONS

The successful development of the use of RF heating in food applications lies in its product-specific design and in the optimal impedance matching between the power generator and the applicator holding the carefully formulated food product. The quality of the applicator's design is very important for its efficiency while the dielectric behavior of the processed material requires optimization in careful food formulation. The development of new applications of RF heating requires well-tuned targeted equipment design and automated control systems taking advantage of monitoring the changes in dielectric properties to predict product quality. The industrial potential of RF processing is interesting, with its greater penetration depth than microwave, which has well-designed applicators in heating/drying process applications. The potential of RF is even better with hybrid systems, which take the volumetric heating advantages of dielectric heating and couple them with conventional processing for efficient, rapid, and high-quality results.

NOMENCLATURE

Symbol	Units	Term
c	J/kg·K	Specific heat
d_p	m	Field penetration depth
dT/dt	°C/s	Time rate of temperature rise
E	V/m	Electric field strength
$f = 1/T$	cycles/s, Hz	Frequency
P	W/m³	Power
T	s	Period
$\varepsilon_0 = 8.854188 \times 10^{-12}$	F/m	Absolute permittivity of vacuum
ε	F/m	Relative complex permittivity
ε'	F/m	Relative real permittivity (dielectric constant)
ε''	F/m	Relative dielectric loss factor (loss factor)
$\tan \delta = \varepsilon''/\varepsilon'$		Dielectric loss tangent
ρ	kg/m³	Density
λ	m	Wavelength
λ_0=11.1111 m at 27 MHz		Wavelength in free space

REFERENCES

Ahmed J, Ramaswamy HS, and Raghavan GSV. 2007a. Dielectric properties of Indian Basmati rice flour slurry. *Journal of Food Engineering* 80:1125–33.

———. 2007b. Dielectric properties of butter in the MW frequency range as affected by salt and temperature. *Journal of Food Engineering* 82:351–58.

Ahmed J, Ramaswamy HS, Alli I, and Raghavan GSV. 2007c. Protein denaturation, rheology and gelation characteristics of radio-frequency heated egg white dispersions. *International Journal of Food Properties* 10 (1): 145–61.

Assenheim HM, Hill DA, Preston E, and Cairnie AB. 1979. The biological effects of radio-frequency and microwave radiation. Report of the NRC Associate Committee on Scientific Criteria for Environmental Quality. National Research Council of Canada Report no. 16448.

Awuah GB, Ramaswamy HS, Economides A, and Mallikarjunan K. 2006. Inactivation of *Escherichia coli* K-12 and *Listeria innocua* in milk using radio frequency (RF) heating. *Innovative Food Science and Emerging Technologies* 6 (4): 396–402.

Awuah GB, Ramaswamy HS, and Piyasena P. 2002. Radio frequency (RF) heating of starch solutions under continuous flow conditions: Effect of system and product parameters on temperature change across the applicator tube. *Journal of Food Process Engineering* 25:201–23.

Beckwith TD, and Olsen AR. 1931. Ultrasonic radiation and yeast cells. *Proceedings of the Society for Experimental Biology and Medicine* 29 (4): 362–64.

Bengtsson NE, Green W, and Del Valle FR. 1970. Radio-frequency pasteurization of cured hams. *Journal of Food Science* 35:681–87.

Bernard JP. 1997. The 50 Ω radio-frequency technology in the fish transformation industry tempering. In *Proceedings of the Conference "Microwave and High Frequency Heating"*, S. Martino Conference Hall, Fermo, Italy, eds. Breccia A, DeLea R, and Metaxas AC. Bologna: Lo Scarabeo Printers. 279–280.

Bialod D, and Marchand T. 1986. La conception des installations industrielles de chauffage diélectrique par haute fréquence et micro-ondes: méthode, technologie, exemples. Electricité de France, rapport HE142T530, pp. 68.

Bircan C, and Barringer SA. 2002. Determination of protein denaturation of muscle foods using dielectric properties. *Journal of Food Science* 67 (1): 202–205.

Brown GH, and Morrison WC. 1954. An exploration of the effects of strong radio-frequency fields on microorganisms in aqueous solutions. *Food Technology* 8 (8): 361–66.

Brunkhorst C, Ciotti D, and Fredd E. 2000. Development of process equipment to separate nonthermal and thermal effects of RF energy of microorganisms. *Journal of Microwave Power and Electromagnetic Energy* 35 (1): 45–50.

Brunton NP, Lyng JG, Li W, Cronin DA, Morgan D, and McKenna B. 2005. Effect of radio-frequency (RF) heating on the texture, colour and sensory properties of comminuted pork meat product. *Food Research International* 38:337–44.

Chou SK, and Chua KJ. 2001. New hybrid drying technologies for heat sensitive foodstuffs. *Trends in Food Science and Technology* 12 (10): 359–69.

Clee MJ, and Metaxas AC. 1994. An investigation into frequency arcing in latex foams. *Journal of Microwave Power and Electromagnetic Energy* 29 (2): 94–100.

Decareau RV. 1985. *Microwaves in the Food Processing Industry*. London: Academic Press.

Elgerd DO. 1977. *Basic Electric Power Engineering*. Reading, MA: Addison-Wesley.

Fabian FW, and Graham HT. 1933. Influence of high frequency displacement currents on bacteria. *Journal of Infectious Diseases*, 53 (1): 76–88.

Fleming H. 1944. Effect of high frequency fields on microorganisms. *Electrical Engineering* 63:18–21.

Funebo T, and Ohlsson T. 1999. Dielectric properties of fruits and vegetables as a function of temperature and moisture content. *Journal of Microwave Power and Electromagnetic Energy* 34 (1): 42–54.

Geveke DJ, Kozempel M, Scullen OJ, and Brunkhorst C (2002) Radio-frequency energy effects on microorganisms in foods. *Innovative Food Science and Emerging Technologies* 3 (2): 133–38.

Grant EH, Sheppard RJ, and South GP. 1978. *Dielectric Behaviour of Biological Molecules in Solution*. Oxford: Clarendon Press.

Harlfinger L. 1992. Microwave sterilization. *Food Technology* 46 (12): 57–61.

Houben J, Schoenmakers L, van Putten E, van Roon P, and Krol B. 1991. Radio-frequency pasteurization of sausage emulsions as a continuous process. *Journal of Microwave Power and Electromagnetic Energy* 26 (4): 202–205.

Ikediala JN, Hansen JD, Tang J, Drake SR, and Wang S. 2002. Development of a saline water immersion technique with RF energy as a postharvest treatment against codling moth in cherries. *Postharvest Biology and Technology* 24 (1): 25–37.

Kasevich RS. 1998. Understand the potential of radio frequency energy. *Chemical Engineering Progress* 94 (1): 75–81.

Lagunas-Solar MC, Zeng NX, Essert TK, Truong TD, Pina C, Cullor JS, Smith WL, and Larrain R. 2005. Disinfection of fishmeal with radio-frequency heating for improved quality and energy efficiency. *Journal of the Science of Food and Agriculture* 85:2273–80.

Laycock L, Piyasena P, and Mittal GS. 2003. Radio-frequency cooking of ground, comminuted and muscle meat products. *Meat Science* 65:959–65.

Lu Y, Fujii M, and Kanai H. 1998. Dielectric analysis of hen egg white denaturation and cool storage. *International Journal of Food Science and Technology* 33:393–99.

Lyng JG, Cronin DA, Brunton NP, Li W, and Gu X. 2007. An examination of factors affecting radio-frequency heating of an encased meat emulsion. *Meat Science* 75:470–79.

Lyng JG, Zhang L, and Brunton NP. 2004. A survey of the dielectric properties of meats and ingredients in meat product manufacture. *Meat Science* 69:589–602.

Marchand C, and Meunier T. 1990. Recent development in industrial radio-frequency technology. *Journal of Microwave Power and Electromagnetic Energy* 25 (1): 39–46.

Marra F, Lyng J, Romano V, and McKenna B. 2007. Radio-frequency heating of foodstuff: solution and validation of a mathematical model. *Journal of Food Engineering* 79:998–1006.

McKenna BM, Lyng J, Brunton N, and Shirsat N. 2006. Advances in radio frequency and ohmic heating of meats. *Journal of Food Engineering* 77 (2): 215–29.

Mermelstein NH. 1998. Microwave and radio frequency drying. *Food Technology* 52 (11): 84–85.

Metaxas AC. 1988. RF and microwave energy hots up. *IEE Review* 34 (5): 185–87.

———. 1996. *Foundations of Electroheat: A Unified Approach.* Chichester, UK: John Wiley.

Mudgett RE. 1985. Electrical properties of foods. In *Microwaves in the Food Processing Industry*, ed. Decareau RV. London: Academic Press. 15–37.

———. 1986. Microwave properties and heating characteristics of foods. *Food Technology* 40 (6): 84–93.

Nelson SO. 1985. RF and microwave energy for potential agricultural applications. *Journal of Microwave Power* 20:65–70.

———. 1991. Dielectric properties of agricultural products: measurements and applications. *IEEE Transactions Electrical Insulation* 26:845–69.

Nelson SO, and Stetson LE. 1975. 250 Hz to 12 GHz dielectric properties of grain and seed. *Transactions of the ASAE* 18 (4): 714–18.

Orsat V, Bai L, Raghavan GSV, and Smith JP. 2004. Radio-frequency heating of ham to enhance shelf-life in vacuum packaging. *Journal of Food Process Engineering* 27:267–83.

Orsat V, Gariépy Y, Raghavan GSV, and Lyew D. 2001. Radio-frequency treatment for ready-to-eat fresh carrots. *Food Research International* 34 (6): 527–36.

Piyasena P, Dussault C, Koutchma T, and Ramaswamy HS. 2003a. Radio frequency heating of foods: Principles, applications and related properties – A review. *Critical Reviews in Food Science* 43 (6): 587–606.

Piyasena P, Ramaswamy HS, Awuah GB, and Defelice C. 2003b. Dielectric properties of starch solutions as influenced by temperature, concentration, frequency and salt. *Journal of Food Process Engineering*, 26 (1): 93–119.

Rosenberg U, and Bögl W. 1987. Microwave pasteurization, sterilization, blanching and pest control in the food industry. *Food Technology* 41 (6): 92–99.

Rosenthal I. 1992. *Electromagnetic radiations in food science*. Vol. 19, *Advanced Series in Agricultural Sciences*. Berlin: Springer Verlag.

Roussy G, and Pearce JA. 1995. *Foundations and Industrial Applications of Microwaves and Radio-Frequency Fields: Physical & Chemical Processes*. Chichester, UK: John Wiley.

Ryynanen S. 1995. The electromagnetic properties of food materials: A review of the basic principles. *Journal of Food Engineering* 26 (4): 409–29.

Sanders HR. 1966. Dielectric thawing of meat and meat products. *Journal of Food Technology* (1): 183–92.

Schuster-Gajzágó I, Kiszter AK, Tóth-Márkus M, Baráth A, Márkus-Bednarik Z, and Czukor B. 2006. The effect of radio frequency heat treatment on nutritional and colloid-chemical properties of different white mustard (*Sinapis alba* L.) varieties. *Innovative Food Science & Emerging Technologies* 7:74–79.

Tang X, Cronin DA, and Brunton NP. 2005. The effect of radio-frequency heating on chemical, physical and sensory aspects of quality in turkey breast rolls. *Food Chemistry* 93:1–7.

Tang X, Lyng JG, Cronin DA, and Durand C. 2006. Radio-frequency heating of beef rolls from biceps femoris muscle. *Meat Science* 72:467–74.

UIE, Union Internationale d'Électrothermie (The International Union for Electroheat). 1992. Dielectric heating for industrial processes. La Defense, Paris: UIE working group.

van Roon PS, Houben JH, Koolnees PA, and van Vliet T. 1994. Mechanical and microstructural characteristics of meat doughs either heated by a continuous process in a radio-frequency field or conventionally in a water bath. *Meat Science* 38:103–16.

Vega-Mercado H, Gongora-Nieto MM, and Barbosa-Canovas GV. 2001. Advances in dehydration of foods. *Journal of Food Engineering* 49 (3): 271–89.

Venkatesh MS, and Raghavan GSV. 2004. An overview of microwave processing and dielectric properties of agri-food materials. *Biosystems Engineering* 88 (1): 1–18.

Wang S, Tang J, Johnson JA, Mitcham E, Hansen JD, Hallman G, Drake SR, and Wang Y. 2003a. Dielectric properties of fruits and insect pests as related to radio-frequency and microwave treatments. *Biosystems Engineering* 85 (2): 201–12.

Wang Y, Wig TD, Tang J, and Hallberg LM. 2003b. Sterilization of foodstuff using radio-frequency heating. *Journal of Food Science* 68 (2): 539–44.

———. 2003c. Dielectric properties of foods relevant to RF and microwave pasteurization and sterilization. *Journal of Food Engineering* 57 (3): 257–68.

Zhang L, Lyng JG, and Brunton NP. 2006. Quality of radio-frequency heated pork leg and shoulder ham. *Journal of Food Engineering* 75:275–87.

———. 2007. The effect of fat, water and salt on the thermal and dielectric properties of meat batter and its temperature following microwave or radio-frequency heating. *Journal of Food Engineering* 80:142–51.

Zhao Y, Flugstad B, Kolbe E, Park JW, and Wells JH. 2000. Using capacitive (radio frequency) dielectric heating in food processing and preservation – A review. *Journal of Food Process Engineering* 23 (1): 25–55.

Zhong Q, Sandeep KP, and Swartzel KR. 2003. Continuous flow radio-frequency heating of water and carboxymethylcellulose solutions. *Journal of Food Science* 68 (1): 217–23.

———. 2004. Continuous flow radio-frequency heating of particulate foods. *Innovative Food Science & Emerging Technologies* 5:475–83.

[illegible]

3 Ohmic Heating Effects on Rheological and Functional Properties of Foods

Michele Marcotte
Food Research and Development Center

Hosahalli S. Ramaswamy and Yousef Karimi-Zindashty
Macdonald Campus of McGill University

Mohammad R. Zareifard
Food Research and Development Center

CONTENTS

3.1 INTRODUCTION

Ohmic heating is an innovative heating technique employed for thermal processing. In this method, the food is placed between two electrodes serving as an electrical resistor and an alternating electric current is passed through the circuit. Due to the electrical resistance, heat is generated throughout the food. The heat generation takes place volumetrically. The electrical energy is directly converted into heat, causing a temperature rise. This system is comparable to an electrical circuit, which is comprised of a resistance and a source of voltage and current. The food product acts as the resistance when placed between two electrodes and the current passes through it. In other words, the food is made part of an electrical circuit. It offers an alternate way to rapidly heat food, bypassing conventional heating systems. Ohmic heating is comparable to microwave heating without an intermediary step of converting electricity into microwaves through the magnetron before heating the product.

The concept of ohmic heating technology is not new and dates back to 1897 (Jones 1897). Ohmic heating, also known as Joule heating and resistive heating, is the process by which the passage of an electric current through a conductor releases heat. This was first studied by James Prescott Joule in 1841. Joule immersed a length of wire in a fixed mass of water and measured the temperature rise due to a known current flowing through the wire for a 30-min period. By varying the current and the length of the wire he deduced that the heat produced was proportional to the electrical resistance of the wire multiplied by the square of the current. This relationship is Joule's First Law (http://en.wikipedia.org/wiki/Joule_heating):

$$Q = RI^2. \tag{3.1}$$

George Simon Ohm (1789–1854) determined that there is a direct proportionality between the potential difference (voltage, V in volts) applied across a conductor and the resultant electric current (I in amperes) (Ohm's law):

$$V = RI, \tag{3.2}$$

where R is the electrical resistance measured in ohm. Since the heat results from the electrical resistance, it is referred to as "ohmic" heating. Other synonyms are used in the literature to describe this principle of heating: direct resistance heating, electroconductive heating, and electroresistive heating. Ohmic heating has attracted considerable attention in the last two decades for the thermal processing of foods due to its rapid and uniform treatment coupled with its high-energy efficiency and technical simplicity. Ohmic heating has been used for many food processes such as pasteurization, dehydration, extraction, microbial inactivation, blanching, and thawing (Sastry 1991; Palaniappan and Sastry 1990; de Alwis and Fryer 1990).

3.2 HISTORICAL PERSPECTIVE

Over the last few decades, a number of attempts have been made to use this technique in several food processing applications (Palaniappan and Sastry 1990; de Alwis and Fryer 1990). Between the 1950s and 1970s, experiments were performed to use ohmic heating for thawing purposes. Faster thawing rates were observed, but satisfactory results were obtained only if good contact between the food and electrodes was maintained. Problems occurred with complex geometry. In some instances, microwave thawing has replaced the technology. Ohmic heating experiments have been performed on cut and peeled potato slices and corn on the cob to deactivate enzymes. An industrial process, "OSCO," was developed in the 1970s to treat peeled and cut potatoes in solution before frying. Ohmic heating has been used as a rapid heating method for frankfurters, with electrodes spiked at both ends of the sausages. From 1930 to 1970, rapid heating methods for vending applications were developed for heating sausages, pizzas, and hamburgers. Early concepts in pasteurization and sterilization preceded the introduction of a reliable aseptic packaging technology. Modifications of can designs to include electrodes fixed either temporarily or permanently on containers were used to ohmically sterilize foods. Continuous ohmic heating was introduced in the United States for the pasteurization of milk in late 1920s as a successful commercial technique, referred to as the "Electro-Pure" process (Anderson and Finkelstein 1919). In the 1930s, 50 industrial electrical milk sterilizers were in operation but disappeared in the 1950s (Getchell 1935; Moses 1938). A review of problems in the early development of the ohmic heating technique is outlined by de Alwis and Fryer (1990). The problems resulted mainly from improper contact between electrodes and the food product. Electrolysis and product contamination were observed following the use of unsuitable electrode materials. As well, adhesion of the product to the electrodes often occurred. The difficulty in ensuring good contact between electrodes for complex solid geometry was the reason for poor experimental results. Before a reliable aseptic packaging technology existed, rapid and continuous sterilization techniques using ohmic heating were not practical. As with packaging, recent developments in pumping technologies have ensured that particulate foods move without any mechanical damage. This rapid and sophisticated technology requires a level of control only possible through the use of recent computer technology.

A milestone in the industrial achievement of ohmic heating includes the "ELECSTER" process for the pasteurization of milk, which is based on the "Electro-Pure" process and the "APV Baker Ohmic Heating Technology" for the sterilization of particulate foods (Skudder 1991). The latter was recognized as a commercial breakthrough at the 1996 annual meeting of the institute of food technologists (IFT). APV Baker Ltd. received the Industry Achievement Award (Giese 1996) for the development of the ohmic heating technology for the sterilization of fluid containing particles. In France, the Centre Technique de la Conservation des Produits Agricoles (CTCPA) and Université Technologique de Compiègne (UTC) joined together to install an APV pilot plant unit to help European food companies in developing sterilization processes for liquids containing particles in 1995 (Zuber 1997). In France, the Association pour le développement de l'industrie de la viande (ADIV) and Électricité de France (EDF) formed an alliance to design, build, and evaluate a static ohmic heating unit for processed meats that are pumpable (Peyron 1996). Ohmic heating units have been in commercial operation in the United Kingdom and Japan since 1990 (Parrott 1992). In 1995, an ohmic heating unit was manufactured in Japan by Yanagiya Machinery (http://www.ube-yanagiya.co.jp/Yanagiya/companyguide.htm) for tofu production. An ohmic heating unit was patented by Thomas R. Parker of the United Kingdom to produce Japanese-style breadcrumbs (Panko) (Anderson 2003). Furthermore, Anderson (2003) summarized commercialized ohmic heating systems that include Sous Chef Ltd. (in England), which processes low-acid meats and vegetables in bags; Wildfruit Products (a division of Nissei Co. Ltd. of Japan), which processes whole fruits; Papetti's Hygrade Egg Products of Elizabeth, NJ (manufactured by Raztek) to process liquid eggs; and Emmepiemme SRL (in Piacenza, Italy), which processes different kinds of foods such as baby food, artichokes, carrots, mushrooms, ketchup, fruit nectars, fruit juice, peppers, cauliflower, tomato paste, sausages, pâté, and fruit puree.

Since the 1980s, ohmic heating has been seen as a promising development to solve problems encountered in aseptic processing of low-acid liquids containing particulates. Several authors (de Alwis, Halden, and Fryer 1989; Sastry and Palaniappan 1992) have demonstrated that in ohmic heating it is possible to heat the center of the particle faster than the liquid. Therefore, the cold spot of the particle is located at the surface. From a legislative point of view, ohmic heating of the liquids containing particulates becomes a special condition (Larkin and Spinak 1996). If it can be proven that the temperature at the center of all particles during ohmic heating is always greater than that of the liquid at the inlet of the holding tube then the calculated process time is simplified as for the establishment of an aseptic process of homogeneous liquid using only the temperature of the liquid at the inlet of the holding tube. In the ohmic heating process, biological validation may not be required to establish the process as in conventional thermal processing but only to verify it. The ohmic heater assembly can be seen in the context of a complete product sterilization or cooking process where there is already a holding tube, pumping, and a cooling system (Marcotte 1999). In this case, an ohmic heating column consisting of several electrodes would replace the conventional Swept Surface Heat Exchanger (SSHE). The food product is fed vertically from the bottom to the top of the column. Heating occurs in three sections between two electrodes. The food is then held in the holding tube for temperature equilibration between the liquid and solid. It is cooled in tubular heat exchangers before being packed in an aseptic environment.

Research and development in continuous thermal processing of foods has been mainly driven toward assessing the technical feasibility of new and rapid methods, based on the use of electric energy. Many ohmic heating systems have been developed at the research level (Palaniappan and Sastry 1990; Huang, Chen, and Morissey 1997; Marcotte 1999; Farid 2001; Shirsat et al. 2004a,b;

Jun and Sastry 2005; Lei et al. 2007; Sarang, Sastry, and Knipe 2008) as well as industrial levels (Allen, Eidman, and Kinsey 1996; Zuber 1997; Anderson 2003).

3.3 ADVANTAGES OF OHMIC HEATING

The benefits of ohmic heating are numerous (Biss, Coombes, and Skudder 1989). The most important is that heating is very rapid. A large temperature gradient is not experienced within the food, i.e., heating is uniform. The process is ideal for shear-sensitive products. Ohmic heating has the ability to heat the food continuously without the need for a hot heat transfer surface of a SSHE or a tubular heat exchanger that may foul. Using an ohmic heater, particles are handled more gently and maintain their integrity when compared with a SSHE. The process is quiet in operation due to the absence of rotating parts in the system. In addition, a high percentage of solids in liquid (50–80%) can be processed. Unlike microwave heating, the depth of heat penetration in the food is virtually unlimited. A high level of control and automation ensures safety during the operation. Finally, it is easier to tailor a heating time/temperature profile to ensure sterility because heat is generated within the solids without the reliance on thermal conductivity through the liquid.

3.4 FACTORS AFFECTING OHMIC HEATING

Parameters that can affect the ohmic heating process mainly can be divided in two groups: system and product parameters. System parameters include processing temperature, current, voltage, and frequency of the power unit, as well as the specification of the piping system, electrode type, and their positions in the processing line. Product parameters, on the other hand, are directly related to the kind of commodities undergoing the process such as physical properties including size, shape, electrical conductivity (EC), and the composition of food matrix such as the ratio of solid to liquid, percentage of salt or other minerals, and the presence of fat and other ingredients. Considering that very limited industrial-scale experience exists in ohmic heating, parameters that may influence the process must be taken into account when designing a commercial system.

3.4.1 Effect of Electrical Conductivity (EC) and Temperature

Temperature has remarkable effects on food properties in general and is considered the key factor in any thermal processing operation. EC, as one of the most important parameters in ohmic heating, is highly temperature dependent. It has been reported that EC values of raw vegetable pieces follow a sigmoid curve as a function of temperature with a slope change around 50°C due to critical structural changes in biological material. For liquids, formulated products, and thermally treated solid vegetable pieces, it has been demonstrated, by several authors (Palaniappan and Sastry 1991a, 1991b; Fryer et al. 1993; Qihua, Jindal, and van Winden 1993; Yongsawatdigul, Park, and Kolbe 1995b; Legrand et al. 2007), that the EC-temperature relationship is linear within a temperature range of 20–80°C. EC increases with temperature. Fryer et al. (1993) used the following equation for viscous liquids:

$$\sigma = \sigma_0 + K_T(T), \tag{3.3}$$

which anchors the reference value of the EC at 0°C. Palaniappan and Sastry (1991b) expressed the relationship between EC and temperature for liquids in a slightly different manner:

$$\sigma = \sigma_{25}(1 + K_T(T - 25)). \tag{3.4}$$

3.4.2 Electrode Limitation

Although viewed as a promising food processing technology, ohmic heating has a few constraints. One of these, seen at AC electrical frequencies below 500 Hz and at current density up to 3000 A/m^2, is an electrolytic reaction at the electrode surface (Shiba 1992; Reznick 1996; Yongsawatdigul et al. 1995a; Wu et al. 1998). Using stainless steel 304 electrodes, Wu et al. (1998) observed the corrosion as a light brown porous film, and its formation and degree to be a function of AC frequency, current density, and product salinity. Previous investigators had reported that this corrosion could be reduced by increasing frequency (Uemura et al. 1994; Reznick 1996). Wu et al. (1998) observed that when heating Pacific whiting stabilized mince at current density of 2300 A/m^2 and at AC electrical frequencies of 5000 Hz and higher, the corrosion on the stainless steel electrode surface essentially disappeared. As this technology is further developed, some measure of electrode corrosion is needed. Zhao et al. (1999) reported that electrode corrosion may be reduced by increasing the frequency up to 5000 Hz. In order to overcome the corrosion problem in addition to the use of high-frequency power supply equipment, platinum metal or platinum coated material have been proposed and used as the electrodes in the ohmic heating systems employed for food processing applications.

3.4.3 Effect of Particle Shape and Size

It is well known that the overall resistance of food material controls the heating rate under ohmic heating conditions, as stated earlier. Therefore, heating rates should be known in order to ensure the proper design of the process from a product safety and quality point of view. In addition to the electrical conductivities of fluid and particles, factors such as particle size, shape, and concentration as well as particle orientation in the electric field can also affect the heating rate in the ohmic heating process (Kim et al. 1996). Therefore, care must be taken with the preparation of formulations for ohmic heating processes.

Several mathematical models have been developed to simulate the ohmic heating behavior of solid–liquid systems and the models generally demonstrate the importance and uniformity of mixed-phase EC in product temperature predictions (Sastry and Palaniappan 1992; Benabderrahmane and Pain 2000; Salengke and Sastry 2007). Concerning the role of particle concentration in orange and tomato juices, Palaniappan and Sastry (1991b) revealed that overall EC decreased as the percentage of solid constituents dispersed in the liquid phase increased. With respect to particle size, the EC increased as size decreased. Zareifard et al. (2003), studied the effect of particle size, concentration, and orientation of the particles in the field using a food matrix including carrot solids and observed that particle size and concentration influence the heating time and the overall EC of the two-phase food systems. Electrical conductivities were found to decrease as particle size and concentration increased. It was observed that the heating behavior of the solid and the liquid phase could be reversed depending on whether the mass of particles were placed in parallel, in series, or in a well-mixed system. Although the heating behavior was different, very slight changes in values of the overall electrical conductivities of the two-phase food systems were observed.

3.5 APPLICATIONS OF OHMIC HEATING

Qihua et al. (1993) developed an ohmic heating system for continuous thermal processing of liquid foods (sodium chloride solution of 0.1 M concentration and fresh orange juice). The nature of ions (chemical composition), ionic movement, and the viscosity of the liquid affect the electrical properties of liquids and hence the ohmic heating rate. All of these are temperature dependent (Palaniappan and Sastry 1991b). Uemura, Isobe, and Noguchi (1996) evaluated a novel continuous ohmic heating process for liquids. A 2 cm diameter tube was used. Ring electrodes were inserted into the wall of the tube and spaced. A constant voltage was applied (20 kHz and 68 V) to the two electrodes. The space between electrodes was found to be the most significant factor influencing the temperature profile. This system is another option that could be used instead of the APV system for liquids. Marcotte (1999) established a pilot-scale batch and continues ohmic heating research at a facility at the Agriculture and Agri-Food Canada Food Research Center located in Ste Hyacinthe, Quebec, Canada (Figure 3.1), where an in-depth study on ohmic heating of simulated liquid foods

FIGURE 3.1 A pilot-scale continuous ohmic heating system at the Agriculture and Agri-Food Canada Research Center, St Hyacinthe, Quebec.

was carried out (Marcotte, Trigui, and Ramaswamy 2000; Marcotte et al. 2001). The ohmic heating technique has been successfully applied for liquid food products such as orange and tomato juices (Palaniappan and Sastry 1991b), soymilk (Liu and Chang 2007), and milk (Sun et al. 2008). The speed and simplicity of the method make it suitable for aseptic processing.

Application of ohmic heating of fluid and semifluid foods has been investigated and addressed in the literature. On the other hand, considering the problem in providing proper contact between the electrodes and the food surface, applying ohmic heating directly to solid food has its own limitations (de Alwis and Fryer 1992) and remains a challenging research area, which has not been commercialized yet, as highlighted by Piette et al. (2004). However, some attempts have been made in this area. Shirsat et al. (2004a) carried out an ohmic heating experiment to study and measure the conductivities of entire lean and fat components of meat individually. They found that the conductivity was different between entire pieces of lean leg and lean shoulder. They also reported not only that the increase in fat content reduced the overall conductivity of lean muscle, but also that conductivity can be influenced by the structural differences in the meat cuts. In another study, a combined ohmic and plate cooking method was applied to hamburger patties (Özkan et al. 2004). The method successfully reduced cooking time and produced safer products.

One of the considerable applications of ohmic heating is the processing of solid–liquid mixtures with a high solid concentration, which has been investigated by several researchers (Sastry and Palaniappan 1992; Khalaf and Sastry 1996; Zareifard et al. 2003; Salengke and Sastry 2007). The advantages of an ohmic heating system over a conventional system are that the solid particles can be heated faster than the liquid (Sastry and Palaniappan 1992; Khalaf and Sastry 1996). The viscosity of the fluid phase has an important role in ohmic heating of solid–liquid mixtures. The fluid's viscosity must be adequate to produce uniform heating and prevent from cold shadow within the liquid. If the solids are surrounded by motionless viscous fluids, nonuniformities in heating within the fluid can occur (Fryer et al. 1992; Khalaf and Sastry 1996). Alternatively, more uniform heating can be achieved if fluid motion occurs, which indeed can cause mixing of hot and cold fluid regions (Sastry 1993; Khalaf and Sastry 1996). Ohmic heating was applied to the solid–liquid mixture containing potato cubes (0.7 cm surface) by Khalaf and Sastry (1996) to investigate the effect of fluid viscosity on the ohmic heating rates of highly concentrated solid–liquid mixtures. In comparison to vibrating and bath methods, they concluded that the fluid viscosity had no considerable effect on the heating rates of fluid and particles under a static ohmic heating situation. However the effect of viscosity was more evident in the case of fluid mixing.

In the case of applying ohmic heating to a solid-liquid mixture, the overall resistance of food material controls the heating rate. Therefore, heating rates plays an important role in the proper design of the process from a product safety and quality point of view. Many factors affect the heating rate of foods undergoing ohmic heating: electrical conductivities of fluid and particles, specific heat, particle size, shape, and concentration as well as particle orientation in the electric field (Kim et al. 1996; Zareifard et al. 2003). The effect of particle size, concentration, location, and temperature on the EC was investigated by Zareifard et al. (2003). The overall EC and heating time were influenced by the particle size and concentration. A low EC in food systems resulted in longer heating times.

Naveh, Kopelman, and Mizrahi (1983) applied ohmic heating for thawing of foods. Traditionally, frozen blocks of fish are thawed by immersion in warm water with the following major disadvantage: requirement of large amounts of fresh water that become wastewater. It also compromises the surface microbial quality of foods as the water temperature is 30°C. Also, valuable soluble proteins are leached. Early investigations of ohmic heating (Jason and Sanders 1962a, 1962b; Sanders 1963; Burgess et al. 1967) revealed the appearance of hot spots as enough heat is generated to cook. Henderson (1993) showed that hot spots occur simultaneously on localized regions and could be controlled by the supply of electrical current. Roberts et al. (1998) successfully designed and tested a fully automated prototype for ohmic thawing of fish blocks based on the control of the current flow within the product to eliminate the formation of hot spots.

Pereira et al. (2007) studied the effect of ohmic heating on different chemical and physical parameters of milk in comparison with conventional heating. They reported that the pH value and total fatty acids content in milk fat of goat milk samples were comparable with the ohmic and conventional methods, whereas the lactic acid of the ohmically treated samples were lower. They also reported that for cloudberry jam samples treated with both methods their results showed no significant difference for the parameters tested (total sugar content, ascorbic acid content, and titratable acidity).

3.5.1 Effect of Ohmic Heating on Biopolymers

Park et al. (1995) performed a study on the effect of ohmic heating on fish protein gel. The structure of the gel was examined using scanning electron microscopy. As well, the breaking strength and color were measured using conventional methods. Electrical properties were recorded. The electrode corrosion was examined by immersion in NaCl. It was found that over 100 Hz, gels had similar electrical properties, which would tend to prove that heating rates would be similar. However, specific heating rates increased remarkably as the frequency increased from 1 to 10 kHz. This indicates that there was a significant dielectric loss in the gel. Corrosion of aluminum electrodes increased at 50 Hz with increasing NaCl concentration. There was a negligible contamination of aluminum ions in the product as the frequency was increased to 50 kHz.

Wang and Sastry (1997) studied the effect of starch gelatinization on EC. Suspensions of corn and potato starch were prepared in water (1:5 w/w) and were ohmically treated to 90°C using a voltage gradient of 20 V/cm. Ungelatinized, partially gelatinized, and fully gelatinized suspensions were tested. Endothermic gelatinization peaks were found for both differential scanning calorimetry (DSC) thermograms and EC curves as a function of temperature. The percentage of starch gelatinization as determined by DSC thermograms and σ-T curves were correlated well in the low and mid-gelatinization ranges. Different results were found under a high percentage of gelatinization, probably due to the high ohmic heating rate. Therefore, the potential for quantifying starch gelatinization by EC changes vs temperature has been demonstrated. An online ohmic heating sensor could possibly be installed to monitor starch gelatinization in a food processing unit operation. Furthermore, it was found that EC increases with temperature but decreases with the degree of gelatinization, apparently caused by structural changes and an increase in bound water.

3.5.2 Effect on Rheological Properties

In design and operation of ohmic heating systems, rheological properties of hydrocolloids are important. Addition of salt to hydrocolloids may be desirable to enhance the efficiency of ohmic heating. Rheological properties will influence the velocity profiles in the system (Steffe 1992). A continuous ohmic heating column can be described as a flow in an ordinary pipe or holding tube. Palmer and Jones (1976) demonstrated that the estimation of the holding tube length in aseptic processing was influenced by the rheological properties of the liquid medium (Newtonian vs non-Newtonian). Based on laminar or turbulent flow conditions, a microbiologically safe design criterion would be ensured for Newtonian fluids and shear-thinning liquids but not for dilatant fluids (Rao 1992).

The rheological properties of the fluid are required for fluid mechanics studies in order to characterize the flow nature (Holdsworth 1971) as the fluid travels through the continuous ohmic heating system. Flow regime specification (laminar, transition, or turbulent) requires the calculation of the Reynolds number, which is a measure of the ratio of the inertial force to the force of internal friction (viscosity). The rheological behavior of matter that flows is characterized by the measurement of viscosity, which is defined as the internal friction or resistance experienced by the fluid as it moves over another layer of the fluid. Singh, Tarsikka, and Singh (2008) conducted a study to measure the viscosity and EC of apple, pineapple, orange, and tomato juices under different conditions with an ohmic heating system. Their results showed that among the juices studied, tomato and pineapple juices had minimum and maximum viscosity values, respectively. The viscosity of all juices was decreased with an increase in temperature. An exponential relationship between viscosity and temperature was developed for the juices studied.

The effect of ohmic heating on the rheological properties of carrageenan, pectin, starch, and xanthan solutions was investigated (Marcotte et al. 2001) at various temperatures, concentrations, and shear rates in the presence of 1% salt. They found that concentration, temperature, and shear rate effects on rheological properties were different depending on the type of hydrocolloids. Flow curves were well described by the power law model for starch and pectin. The concentration effect on apparent viscosity was more important for starch and carrageenan and less important for the pectin and xanthan.

Furthermore, viscosity of the fluid phase is an important consideration in ohmic heating. For the food system containing large particle (solid) fraction, the fluid viscosity must be sufficient to effectively entrain and convey it (Khalaf and Sastry 1996). Although some nonhomogeneities in heating within the fluid phase surrounding the solids can occur if a motionless viscous carrier is being used (Fryer et al. 1992), if fluid motion occurs, heating becomes significantly more uniform, because of the mixture of hot and cold fluid in a continuous flow heater.

The effects of fluid viscosity on the ohmic heating rates of identical solid–liquid mixtures containing potato cubes were investigated by Khalaf and Sastry (1996) and compared with batch (static and vibrating) as well as continuous flow conditions. Considering the commercial processing condition, the focus of their study was mixtures of high solids concentrations (> 30%). They reported that there were no significant effects of fluid viscosity on the heating rate of fluid and particles for static heating conditions, where the fluid was at stationery conditions with outermost limited motion. However, in vibrating batch and continuous flow ohmic heater, where adequate fluid mixing occurs, the rate of heating increased with fluid viscosity. Furthermore, for a continuous flow heater, considering both the temperature and power dissipation data, they suggested that the use of high viscosity carriers could be useful in improving the overall heater rates.

Effects of different heat transfer methods, namely ohmic heating, water bath (WB), and steam injection, on soymilk viscosity were studied by Liu and Chang (2007) using one- and two-step heating procedures to different temperatures. Their results showed that the soymilk viscosity was increased by the two-step heating procedure for all three methods by 161, 250, and 729%, for steam injection, ohmic heating, and WB, respectively. They indicated that the differences in viscosity may arise from the different heating rates of the three applied procedures.

3.5.3 Effect of Ohmic Heating on Texture

Huang et al. (1997) successfully investigated the feasibility of using a batch-type ohmic heater to coagulate fish proteins from the frozen fish mince wash water. Ohmic heating was employed as a rapid method of heating and was found to maximize gel functionality in Pacific whiting surimi as compared to conventional heat treatment, which allows the product to develop undesirable textural properties as a result of optimal enzymatic activity occurring at 55°C (Yongsawatdigul et al. 1995a).

The effect of ohmic heating on texture of food material has been studied and reported in the literature (Laycock, Piyasena, and Mittal 2003; Piette et al. 2004; Wills et al. 2006; Chai and Park 2007). In a study conducted on meat products, namely, ground, comminuted, and muscle, Laycock et al. (2003) investigated the effect of a radio-frequency (RF) (ohmic heating) cooking method on food quality. Their results indicated that the springiness of the RF-cooked ground beef and comminuted samples was much higher than that of the WB-cooked samples, which means these samples were more elastic. They reported a comparable hardness of the ground beef for both methods, however the RF-cooked comminuted samples were softer than WB-cooked samples. In a similar study conducted on meat products, Piette et al. (2004) found that textural attributes of smoke-chamber-heated products were statistically different ($P > .05$) from those of ohmically heated products. Ohmic products were softer than smokehouse products. However, considering the fact that the measured textural values fell within the range for commercial sausages, they concluded that the softer texture of ohmic products was satisfactory. The effects of ohmic heating on texture and microstructure of surimi seafood gels were investigated by Chai and Park (2007). Comparing the results with conventional WB-cooked gels, they reported that cooking method had a significant effect on texture. They reported that the type and concentration of additives had a direct effect on the surimi's properties.

The effect of ohmic heating on beef patties in comparison with the impingement oven was investigated by Wills et al. (2006). They reported significant textural differences ($P < .05$) between the two cooking methods. Their results indicated that ohmically cooked samples were harder, chewier, and more cohesive than samples cooked with the impingement oven method. Comparing RF (ohmic heating) meat products with the steam-cooked method, Zhang, Lyng, and Brunton (2004) also discovered that the texture of meat batters cooked with RF was harder, gummier, and chewier.

Wills et al. (2006) concluded that textural differences observed between the two cooking methods can be caused by the temperature distribution in impingement-cooked samples. In ohmic-heated patties heat distribution is more uniform, which results in closer exterior and interior temperatures.

3.5.4 Effect on Color

Color is one of the visual sensory attributes considered as a quality aspect of food stuff. Food color is evaluated by measuring Hunter L^* (lightness), a^* (redness), and b^* (yellowness), and H^o (hue angle) values. During thermal process, color changes are very significant, which can occur due to migration of moisture and certain chemical changes. Few studies have been conducted on selected food material to measure the effect of ohmic heating compared to conventional methods (Zhang et al. 2004; Leizerson and Shimoni 2005; Wills et al. 2006). Comparing RF (ohmic heating) with steam-cooked meat products, Zhang et al. (2004) found that the L^*, b^*, and hue angle values in samples cooked with the RF method were lower, which resulted in less color development in the RF-cooked samples. In a similar study conducted on ground, comminuted, and entire beef, Laycock et al. (2003) reported that there were no significant differences in L^* and a^* values for meat samples cooked with the RF and WB methods.

The effect of ohmic heating on the color of beef patties was explored by Wills et al. (2006) by comparing these products with the same product cooked with an impingement oven. Comparing the results of L^* values, they reported that ohmically heated patties had a lighter ($P < .05$) color than impingement patties, however there was no significant differences between a^* values (redness) for both cooking methods. However, their results indicated that the ohmic-heated patties had larger ($P < .05$) b^* values (yellow) than those cooked by the impingement method. Chai and Park (2007) studied the effect of ohmic heating on the color of surimi seafood gels. Comparing the results with those of conventional WB-cooked gels, they reported that the cooking method had very little effect on color. Comparing the color of smokehouse- and ohmic-cooked sausages or between ohmic sausages cooked with different heating rates or to different end-point temperatures, Piette et al. (2004) found that they statistically differed ($P > .05$). The differences were occasionally between the color L^* and a^* components. They reported that the differences were small in size, which not only consumers but also trained panellists would be likely be unable to detect.

3.5.5 Effect of Ohmic Heating on Microorganisms and Enzymes

A number of studies have been performed to measure the effect of electricity on microorganisms and enzymes in cell suspensions or in food directly under a variety of conditions (Palaniappan and Sastry 1990; Palaniappan, Sastry, and Richter 1992; Imai et al. 1995). So far, microbial inactivation during continuous low-voltage alternating electrical treatments has been attributed primarily to heating (Palaniappan and Sastry 1990). As well, with low voltage alternating current, enzymes are inhibited due to heat generation (Mizrahi, Kopelman, and Perlman 1975). Palaniappan et al. (1992) compared ohmic to conventional heating on suspensions of yeast (*Saccharomyces bailii*) and bacteria (*Escherichia coli*). They concluded that microbial death was primarily due to thermal effects with the electrical current having no significant effect.

Cho, Yousef, and Sastry (1996) compared a conventional and an ohmic fermentation of *Lactobacillus acidophilus* at three temperatures (30, 35, and 40°C). Using ohmic heating, a constant temperature was maintained by intermittently circulating cooling water through the water circulation coil. The electric current was found to enhance early stages but to slightly inhibit the late stages of growth. The production of bacteriocin was reduced during ohmic heating. It was also mentioned that it was possible to monitor the progress of the fermentation by monitoring the current at a constant voltage.

Studies on ohmic heating of potato (Schade 1951) and corn (Mizrahi et al. 1975) have demonstrated its applicability as a blanching treatment. Enzymes responsible for discoloration were inhibited by heating. Castro et al. (2004) conducted a study to determine the effect of ohmic heating on inactivation kinetics of polyphenoloxidase, lipoxygenase, pectinase, alkaline phosphatase, and β-galactosidase enzymes. In order to be able to make a comparison, the work was also performed under conventional conditions. They concluded that the inactivation of alkaline phosphatase, pectinase, and β-galactosidase did not increase due to the existence of an electric field. However, the electric field had a significant effect on lipoxygenase and polyphenoloxidase, and the time needed for inactivation was reduced. Icier, Yildiz, and Baysal (2006) carried out a study to investigate and determine the critical peroxidase inactivation time for ohmic blanching of pea puree and also to study the effect of method on color values in comparison to water blanching. They reported that ohmic heating inactivates the peroxidase enzyme at a shorter processing time than conventional water blanching. The critical inactivation time decreased when the voltage gradient increased during ohmic blanching. Furthermore, they reported that better color values were obtained with the ohmic blanching above 20 V/cm rather than the water-blanched puree at critical peroxidase inactivation times. The voltage gradient during ohmic blanching had a significant ($P < .01$) effect on the color of puree, where the color changes follow first-order reaction kinetics.

The inactivation effect of ohmic heating on microorganisms (aerobes and *Streptococcus thermophilus*) in milk was investigated and compared with the inactivation effect of conventional heating under identical temperature history conditions (Sun et al. 2008). It was reported that the microbial counts and calculated decimal reduction time (*D* value) were significantly lower with the ohmic heating method. Furthermore, they evaluated the quality of the milk by assessing the degree of protein denaturation in raw as well as sterilized milk by the two methods and concluded the resulting protein denaturation was in the same range for both methods.

3.6 CLOSING REMARKS

Research and technological efforts in the past two to three decades have made ohmic heating one of the leading thermal processing techniques for food preservation. Creating data banks on the EC of foods and developing process equipment are still a priority in ohmic heating research. Engineers, in their turn, need to develop the appropriate economic high-frequency power suppliers with suitable noncorrosive electrodes to enhance the application and use of ohmic heating techniques in food processing plants. Due to the relatively high energy consumption of conventional in-can thermal food processing techniques, the high cost of natural gas and petroleum, and growing environmental global warming concerns, industries have been showing more interest in adapting other technologies such as ohmic heating. Further, due to rapid and uniform heating, ohmic heating has great potential to be used in the processing of high-quality products and can be commercially adapted for the aseptic processing of particulate fluid foods.

REFERENCES

Allen, K., Eidman, V., and Kinsey, J. 1996. An economic-engineering study of Ohmic food processing. *Food Technology* 50 (5): 269–73.

Anderson, A. K., and Finkelstein, R. 1919. A study of the electropure process of treating milk. *Journal of Dairy Science* 2:374–406.

Anderson, D. R. 2003. Ohmic heating as an alternative food processing technology. MSc thesis, Johnson & Wales University.

Benabderrahmane, Y., and Pain, J. P. 2000. Thermal behavior of solid–liquid mixture in an ohmic heating sterilise-slip phase model. *Chemical Engineering Science* 55:1371–84.

Biss, C. H., Coombes, S. A., and Skudder, P. J. 1989. The development and application of ohmic heating for the continuous heating of particulate foodstuffs. In *Process Engineering in the Food Industry*, ed. R. W. Field and J. A. Howell, Chap. 2, 17–26. London: Elsevier Applied Science.

Burgess, G. H. O., Cutting, C. L., Lovern, J. A., and Waterman, J. J. 1967. Chap. 8, Thawing. In *Fish Handling and Processing*, 182–94. New York: Chemical Publishing Company, Inc.

Castro, I., Macedo, B., Teixeira, J. A., and Vicente, A. A. 2004. The effect of electric field on important food-processing enzymes: Comparison of inactivation kinetics under conventional and ohmic heating. *Journal of Food Science* 69 (9): 696–701.

Chai, P. P., and Park, J. W. 2007. Physical properties of fish proteins cooked with starches or protein additives under ohmic heating. *Journal of Food Quality* 30:783–96.

Cho, H.-Y., Yousef, A. E., and Sastry, S. K. 1996. Growth kinetics of Lactobacillus acidophilus under ohmic heating. *Biotechnology and Bioengineering* 49:334–40.

de Alwis, A. A. P., and Fryer, P. J. 1990. The use of direct resistance heating in the food industry. *Journal of Food Engineering* 11:3–27.

______. 1992. Operability of the ohmic heating process: Electrical conductivity effects. *Journal of Food Engineering* 15:21–48.

de Alwis, A. A. P., Halden, K., and Fryer, P. J. 1989. Shape and conductivity effects in the ohmic heating of foods. *Chemical Engineering Research and Design* 67:159–68.

Farid, M. 2001. New methods and apparatus for cooking. International Patent Application, #PCT/NZ02/00108.

Fryer, P. J., de Alwis, A. A. P., Koury, E., Stapley, A. G. F., and Zhang, L. 1992. Ohmic processing of solid-liquid mixtures: Heat generation and convection effects. *Journal of Food Engineering* 18:101–25.

______. 1993. Ohmic processing of solid-liquid mixtures: Heat generation and convection effects. *Journal of Food Engineering* 18:102–25.

Getchell, B. E. 1935. Electric pasteurization of milk. *Agricultural Engineering* 16:408–10.

Giese, J. 1996. Commercial development of ohmic heating garners 1996 industrial achievement award. *Food Technology* 50 (9): 114–15.

Henderson, J. T. 1993. Ohmic thawing of frozen shrimp: Preliminary technical and economic feasibility. Master's thesis, University of Florida.

Holdsworth, S. D. 1971. Applicability of rheological models to interpretation of flow and processing behavior of fluid food product. *Journal Texture Studies* 2 (4): 393–418.

Huang, L., Chen, Y., and Morissey, M. T. 1997. Coagulation of fish proteins from frozen fish mince wash water by ohmic heating. *Journal of Food Process Engineering* 20: 285–300.

Icier, F., Yildiz, H., and Baysal, T. 2006. Peroxidase inactivation and colour changes during ohmic blanching of pea puree. *Journal of Food Engineering* 74:424–29.

Imai, T. K., Uemura, K., Ishida, K., Yoshizaki, S., and Noguchi, A. 1995. Ohmic heating of Japanese white radish *Rhaphanus sativus* L. *International Journal of Food Science and Technology* 30:461–72.

Jason, A. C., and Sanders, H. R. 1962a. Dielectric thawing of fish. I. Experiments with frozen herrings. *Food Technology* 16 (6): 101–106.

______. 1962b. Dielectric thawing of fish. II. Experiments with frozen white fish. *Food Technology* 16 (6): 107–12.

Jones, F. 1897. Apparatus for electrically treating liquids. US Patent 1 592 735.

Jun, S., and Sastry, S. 2005. Modeling and optimization of ohmic heating of foods inside a flexible package. *Journal of Food Process Engineering* 28:417–36.

Khalaf, W. G., and Sastry, S. K. 1996. Effect of fluid viscosity on the ohmic heating rate of solid-liquid mixtures. *Journal of Food Engineering* 27:145–58.

Kim, H. J., Choi, Y. M., Yang, T. C. S., Taub, I. A., Tempest, P., Skudder, P., Tucker, G., and Parrott, D. L. 1996. Validation of ohmic heating for quality enhancement of food products. *Food Technology* 50 (5): 253–61.

Larkin, J. W., and Spinak, S. H. 1996. Safety considerations for ohmically heated, aseptically processed, multiphase low-acid products. *Food Technology* 50 (5): 242–45.

Laycock, L., Piyasena, P., and Mittal, G. S. 2003. Radio frequency cooking of ground, comminuted and muscle meat products. *Meat Science* 65:959–65.

Legrand, A., Leuliet, J.-C., Duquesne, S., Kesteloot, R., Winterton, P., and Fillaudeau, L. 2007. Physical, mechanical, thermal and electrical properties of cooked red bean (Phaseolus vulgaris L.) for continuous ohmic heating process. *Journal of Food Engineering* 81:447–58.

Lei, L., Zhi, H., Xiujin, Z., Takasuke, I., and Zaigui, L. 2007. Effects of different heating methods on the production of protein–lipid film. *Journal of Food Engineering* 82:292–97.

Leizerson, S., and Shimoni, E. 2005. Stability and sensory shelf life of orange juice pasteurized by continuous ohmic heating. *Journal of Agricultural and Food Chemistry* 53:4012–18.

Liu, Z. S., and Chang, S. K. C. 2007. Soymilk viscosity as influenced by heating methods and soybean varieties. *Journal of Food Processing and Preservation* 31:320–33.

Marcotte, M. 1999. Ohmic heating of viscous liquid foods. PhD thesis, McGill University.

Marcotte, M., Taherian, A. R., Trigui, M., and Ramaswamy, H. S. 2001. Evaluation of rheological properties of selected salt enriched food hydrocolloids. *Journal of Food Engineering* 48:157–67.

Marcotte, M., Trigui, M., and Ramaswamy, H. S. 2000. Effect of salt and citric acid on electrical conductivities and ohmic heating of viscous liquids. *Journal of Food Processing and Preservation* 24:389–406.

Mizrahi, S., Kopelman, I. J., and Perlman, J. 1975. Blanching by electroconductive heating. *Journal of Food Technology (Brit.)* 10:281–88.

Moses, D. B. 1938. Electric pasteurization of milk. *Agricultural Engineering* 19:525–26.

Naveh, D., Kopelman, I. J., and Mizrahi, S. 1983. Electroconductive thawing by liquid contact. *Journal of Food Technology* 10:282–88.

Özkan, N., Ho, I., and Farid, M. 2004. Combined ohmic and plate heating of hamburger patties: Quality of cooked patties. *Journal of Food Engineering* 63:141–45.

Palaniappan, S., and Sastry, S. K. 1990. Effects of electricity on microorganisms: A review. *Journal of Food Processing Preservation* 14:393–414.

———. 1991a. Electrical conductivities of selected solid foods during ohmic heating. *Journal of Food Process Engineering* 14:221–36.

———. 1991b. Electrical conductivity of selected juices: Influences of temperature, solids content, applied voltage, and particle size. *Journal of Food Process Engineering* 14:247–60.

Palaniappan, S., Sastry, S. K., and Richter, E. R. 1992. Effects of electroconductive heat treatment and electrical pretreatment on thermal death kinetics of selected microorganisms. *Biotechnology and Bioengineering* 39:225–32.

Palmer, J. A., and Jones, V. A. 1976. Prediction of holding times for continuous thermal processing of power law fluids. *Journal of Food Science* 41:1233–44.

Park, S. J., Kim, D., Uemura, K., and Noguchi, A. 1995. Influence of frequency on ohmic heating of fish protein gel. *Nippon Shokuhin Kagaku Kogaku Kaishi* 42 (8): 569–74.

Parrott, D. L. 1992. Use of ohmic heating for aseptic processing of food particulates. *Food Technology* 46 (2): 68–72.

Pereira, R., Pereira, M., Teixeira, J. A., and Vicente, A. A. 2007. Comparison of chemical properties of food products processed by conventional and ohmic heating. *Chemical Papers* 61 (1): 30–35.

Peyron, A. 1996. Perpectives et dévelopement de nouvelles techniques de cuisson des viandes. *Viandes et produits carnés* 17 (6): 255–62.

Piette, G., Buteau, M. L., De Halleux, D., Chiu, L., Raymond, Y., Ramaswamy, H. S., and Dostie, M. 2004. Ohmic cooking of processed meats and its effects on product quality. *Food Engineering and Physical Properties* 69 (2): FEP71–78.

Qihua, T., Jindal, V. K., and van Winden, J. 1993. Design and performance evaluation of an ohmic heating unit for liquid foods. *Computers and Electronics in Agriculture* 9:243–53.

Rao, M. A. 1992. Rheology of fluids in food processing. *Food Technology* 46 (2): 116–26.

Reznick, D. 1996. Ohmic heating of fluid foods. *Food Technology* 50 (5): 250–51.

Roberts, J. S., Balaban, M. O., Zimmerman, R., and Luzuriaga, D. 1998. Design and testing of a prototype ohmic thawing unit. *Computers and Electronics in Agriculture* 19: 211–22.

Salengke, S., and Sastry, S. K. 2007. Effects of ohmic pretreatment on oil uptake of potato slices during frying and subsequent cooling. *Journal of Food Process Engineering* 30:1–12.

Sanders, H.R. 1963. Electrical resistance thawing of fish. In: *Annual Report on the Handling and Preservation of Fish and Fish Products*. Torry Memoir No. 143, 16–17. Aberdeen, Scotland: Torry Research Station.

Sarang, S., Sastry, S. K., and Knipe, L. 2008. Electrical conductivity of fruits and meats during ohmic heating. *Journal of Food Engineering* 87:351–56.

Sastry, S. K. 1991. Ohmic sterilization and related safety issues. In *International Winter Meeting of ASAE*. Paper 916616.

———. 1993. Continuous sterilization of particulate foods by ohmic heating: Critical process design considerations. Presented at the *6th International Congress on Engineering and Food*, Chiba, Japan, 23–27 May.

Sastry, S. K., and Palaniappan, S. 1992. Ohmic heating of liquid-particle mixtures. *Food Technology* 46 (12): 64–67.

Schade, A. L. 1951. Prevention of enzymatic discoloration of potatoes. US Patent 2 569 075.

Shiba, M. 1992. Properties of kamaboko gels prepared by using a new heating apparatus. *Nippon Suisan Gakkaishi* 58:895–901.

Shirsat, N., Lyng, J. G., Brunton N. P., and McKenna, B. M. 2004a. Conductivities and ohmic heating of meat emulsion batters. *Journal of Muscle Foods* 15:121–37.

———. 2004b. Ohmic processing: Electrical conductivities of pork cuts. *Meat Science* 67:507–14.

Singh, S. P., Tarsikka, P. S., and Singh, H. 2008. Study on viscosity and electrical conductivity of fruit juices. *Journal of Food Science and Technology* 45 (4): 371–72.

Skudder, P. J. 1991. Industrial application of the ohmic heater for the production of high-added value food products. In *Proceedings of the VTT Symposium No.19 Technical Research*, 47–52. Espoo: Technical Research Centre of Finland.

Steffe, J. F. 1992. *Rheological Methods in Food Process Engineering*. East Lansing MI: Freeman Press.

Sun, H., Kawamura, S., Himoto, J., Itoh, K., Wada, T., and Kimura, T. 2008. Effects of ohmic heating on microbial counts and denaturation of proteins in milk. *Journal of Food Science Technology Research* 14 (2): 117–23.

Uemura, K., Isobe, S., and Noguchi, A. 1996. Estimation of temperature profile at sequence ohmic heating by finite element method. *Nippon Shokuhin Kagaku Kogaku Kaishi* 43 (11): 1190–96.

Uemura, K., Noguchi, A., Park, S. J., and Kim, D. U. 1994. Ohmic heating of food materials: Effects of frequency on the heating rate of fish protein. In *Developments in Food Engineering. Proceedings of the Sixth International Congress on Engineering and Food* (Chiba, Japan), ed. J. Yano, R. Matsuno, and K. Nakamura, 310–312. London: Blackie Academic and Professional Press.

Wang, W.-C., and Sastry, S. K. 1997. Starch gelatinization in ohmic heating. *Journal of Food Engineering* 34:225–42.

Wills, T. M., Mireles Dewitt, C. A., Sigfusson, H., and Bellmer, D. 2006. Effect of cooking method and ethanolic tocopherol on oxidative stability and quality of beef patties during refrigerated storage (oxidative stability of cooked patties). *Journal of Food Science* 71 (3): C109–14.

Wu, H., Kolbe, E., Flugstad, B., Park, J. W., and Yongsawatdigul, J. 1998. Electrical properties of fish mince during multi-frequency ohmic heating. *Journal of Food Science* 63 (6): 1028–32.

Yongsawatdigul, J., Park, J. W., Kolbe, E., Abu Dagga, Y., and Morissey, M. T. 1995a. Ohmic heating maximizes gel functionality of Pacific whiting surimi. *Journal of Food Science* 60 (1): 10–14.

Yongsawatdigul, J., Park, J. W., and Kolbe, E. 1995b. Electrical conductivity of Pacific whiting surimi paste during ohmic heating. *Journal of Food Science* 60 (5): 922–25, 935.

Zareifard, M. R., Ramaswamy, H. S., Trigui, M., and Marcotte, M. 2003. Ohmic heating behaviour and electrical conductivity of two-phase food systems. *Innovative Food Science and Emerging Technologies* 4:45–55.

Zhang, L., Lyng J. G., and Brunton, N. P. 2004. Effect of radio frequency cooking on the texture, color and sensory properties of large diameter comminuted meat product. *Meat Science* 68 (2): 257–68.

Zhao, Y., Kolbe, K., and Flugstad, B. 1999. A method to Characterize electrode corrosion during ohmic heating. *Journal of Food Process Engineering*, 22: 81–89.

Zuber, F. 1997. Le chauffage ohmique: Une nouvelle technologie pour la stabilisation des plats cuisinés. *Viandes et produits carnés* 18 (2): 91–95.

4 Effect of High Electric Field on Food Processing

A. Muthukumaran, Valérie Orsat,
T. R. Bajgai and G. S. V. Raghavan
Macdonald Campus of McGill University

CONTENTS

4.1 INTRODUCTION

Drying has been one of the single most important unit operations in food processing; it typically determines the final quality and production cost. Usually 10% of the total energy required in the food processing industry goes into drying (Li et al. 2006); hence any improvement in energy-efficient drying could drastically reduce the total cost associated with processing. Traditionally conductive, convective, and radiation modes of heat transfer are used in different drying situations. These drying techniques provide simple yet more powerful drying methods. They reign supreme when the reduction of food spoilage is of primary importance over overall nutritional quality of the product. Application of high drying temperature can adversely affect the product quality by affecting the properties of the food (physical, chemical, and biological). Products dried at higher drying temperature lose their original texture, color, and volatile compounds. Changing consumer dynamics has forced producers to develop alternative methods to produce nutritionally better and safer products. This has led to the application of superheated steam, freeze-drying, microwave, etc. These drying methods can be used either alone or in different combinations with conventional drying methods. Freeze-drying can produce best-quality dried products with excellent sensory and rehydration properties. The very high cost associated with freeze-drying significantly hinders the widespread adaptation of this method and it is limited to high valued products like coffee and pharmaceutical products.

Rising energy costs is another of the driving factors that has forced manufacturers to look for more energy efficient drying methods. Microwave and radio-frequency (RF) heating can help to overcome some of the energy issues associated with drying. But nonuniform heating and governmental regulations associated with the usage of different frequencies pose a serious problem in adapting this technology on an industrial scale. Penetration depth, dielectric loss factor, and dielectric constant of the material being dried also create challenges in applying these methods to successful commercial drying (Mujumdar 2004).

4.2 HIGH-VOLTAGE ELECTRIC FIELD (HVEF) PROCESSING

Nonthermal food processing techniques have been receiving considerable attention lately. They have the potential to improve the product quality while maintaining food safety. Application of high-intensity electric field is one of the important nonthermal processing methods and a considerable amount of research is being carried out in the development of new applications (Knorr et al. 2001). In 1976 Asakwa (Aibara and Esaki 1998) reported the application of high-voltage electric field (HVEF) in food processing (Asakawa, 1976). He conducted the pioneering experiments on the effects of electric fields on heat transfer. Since then interest in high electric field (HEF) has resulted in the development of new treatment applications in food processing for improving the postharvest quality of fruits and vegetables. Several researchers conducted experiments to study the effect of electric field on the water evaporation rate (Jones 1978; Barthakur and Arnold 1995; Cao et al. 2004). Freshness is retained, retarding microbial and enzymatic activity (Hsieh and Ko 2008). Though there is much interest in HEF processing, it is still in the research stage. Most of the review work on drying does not mention HEF drying (Vega-Mercado, Gongora-Nleto, and Barbosa-Canovas 2001; Li et al. 2006).

4.2.1 Advantages of High Electric Field (HEF) Processing

HEF offers many distinct advantages over conventional processing methods. One of the important advantages is that there is no significant change in temperature during processing. Hence, this method can be successfully applied to any temperature-sensitive fruits and vegetables. It can help in reducing the loss of volatiles during processing. The processing efficiency and capacity mainly depend on the voltage strength used for processing.

4.2.2 Factors Affecting High Electric Field (HEF) Processing

4.2.2.1 Electric Field

Michael Faraday first introduced the concept of electric field. When there is a varying electric field or an electric charge the surrounding space has a property which is called the electric field. The electric field can exert a force on electrically charged particles.

4.2.2.2 Electric Field Strength

Electric field strength can be defined as electric potential difference (V) between two given points in space divided by the distance, d, between them:

$$E = \frac{V}{d}. \tag{4.1}$$

4.2.2.3 Energy in the Electric Field

The energy stored or the energy density (u) in the electric field can be calculated using the following equation:

$$u = \frac{1}{2}\varepsilon|E_v|^2. \tag{4.2}$$

The total energy stored in a particular volume can be measured by:

$$\int_v \frac{1}{2}\varepsilon|E|^2 dV. \tag{4.3}$$

Energy stored in a unit volume (u_e) is

$$u_e = \frac{1}{2}\varepsilon_0 E^2. \tag{4.4}$$

4.3 HIGH ELECTRIC FIELD (HEF) DRYING

HEF drying is one of the many wider possible applications of the HEF processing technique. It can be a better alternative to freeze-drying where the production cost as well as the product quality is of greater importance. HEF drying of a food material mainly depends on the electric force (F) being applied (Panofsky and Phillips 1962):

$$F = E\rho. \tag{4.5}$$

4.3.1 HIGH ELECTRIC FIELD (HEF) DRYING MECHANISM

Foods are primarily composed of water, carbohydrates, proteins, minerals, and vitamins. When the food is subjected to electric field the dipole molecules go through polarization. There is also bulk movement of carriers such as ions, which create a capacitive and resistive current. The polarization of dipole molecules is also responsible for improved mass transfer rates during drying (Zhang, Barbosa-Cánovas, and Swanson 1995; Bajgai and Hashinaga 2001b):

$$C = \frac{\varepsilon_0 \varepsilon_r A}{d}. \tag{4.6}$$

The effective resistance of a food material (R) can be calculated by using the following equation,

$$R = \frac{d}{\sigma A}. \tag{4.7}$$

4.4 HIGH ELECTRIC FIELD (HEF) SYSTEM

A typical HEF system, shown in Figure 4.1, is similar to the setup used by Palanimuthu et al. (2009) for their cranberry shelf-life enhancement studies. The HEF system consists of a high-voltage power source, voltage regulator, treatment chamber, current, voltage and temperature probe, and a control panel. The most important components of the system are the transformer and the voltage regulator. The voltage regulator gives flexibility to the HEF system so that different types of food materials can be processed. Data loggers connected to the system automate the data acquisition process and the acquired data can be used for improving the design for specific applications.

4.4.1 The Treatment Chamber

The treatment chamber consists of two electrodes separated by a distance. The electrodes also serve as an enclosure for the material being processed. One of the important functions of this test chamber is to conduct HEF to the food. The distance between the two electrodes depends on the food material being used and the level of processing required. The shape and size of the electrode depend on the sample and sample volume. Bajgai and Hashinaga (2001a) used pointed needles as anodes for the drying of spinach. The test chamber design for HEF treatment is still

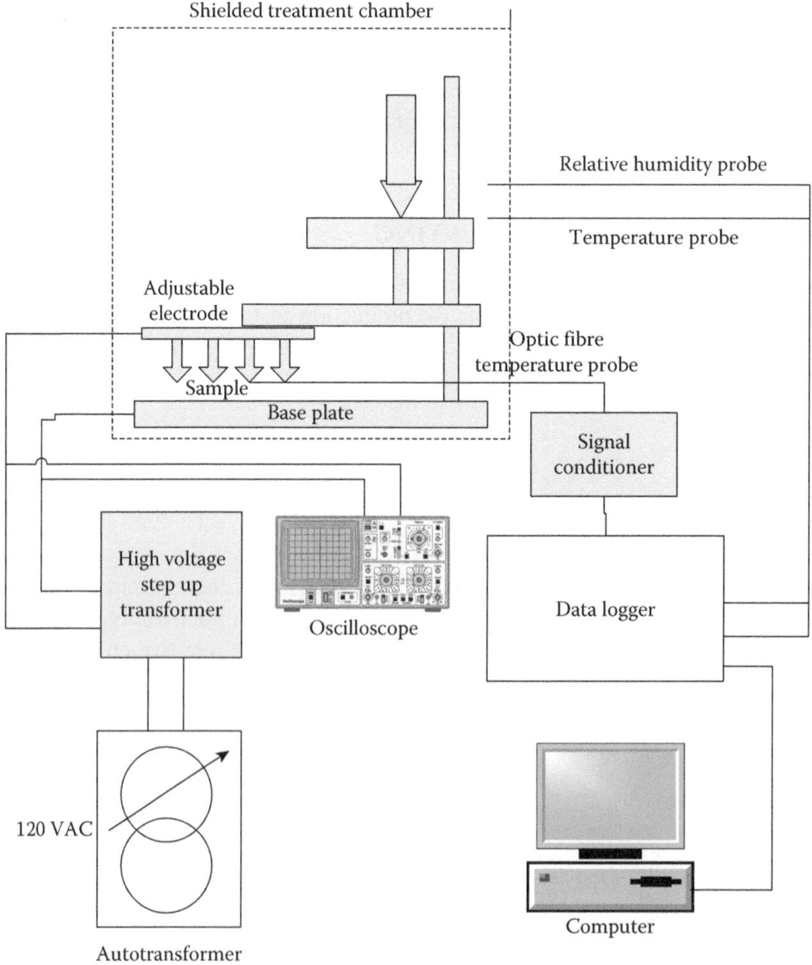

FIGURE 4.1 High-voltage electric field treatment—schematic.

in the research stage. Most of the test chambers used in present research have been designed on a trial and error basis.

4.5 ELECTROHYDRODYNAMIC DRYING (EHD)

Electrohydrodynamic drying (EHD) is similar to HEF drying. In an EHD setup the top electrode is replaced by copper needles (Bajgai et al. 2006b; Li et al. 2006). The electrohydrodynamic heating is highly localized below the pointed electrode. Hence, the use of multiple points to plate configuration can improve the drying rate as well as the uniformity of drying (Bajgai and Hashinaga 2001a). The pointed electrodes disrupt the surface boundary layer of the surrounding inert air by causing electric wind. This effect was observed by Chen, Barthakur, and Arnold (1994) during the drying of potato slabs using EHD (Li et al. 2006). Hashinaga et al. (1999) concluded by drying apple slices in an EHD system that the number of needles is not directly proportional to the rate of drying. Overall enhancement in drying depends on the size of the electrode needle, electric field intensity, and gap between the electrodes.

4.6 MAJOR RESEARCH WORK IN HIGH ELECTRIC FIELD (HEF) PROCESSING

Though the HEF processing concept has been around for a long time, there has been very limited adaptation from industry and academia. Most of the pioneering working in HEF processing has been done at the Kagoshima University of Japan. In North America, the potential of HEF has not been fully explored. A list of patents obtained in the United States is presented in Table 4.1. Some of the major ongoing research work in HEF processing is detailed in the following sections.

4.6.1 Electric Field and Wetted Surface

Carlon and Latham (1992) performed laboratory experiments to determine the effect of wetted surface on the drying rate of a material in an electric field. In their experiments they used paper discs moistened by water and measured the drying rate as a function of electric field strength (E). They came to a conclusion that drying rate increased linearly with electric field strength (E). The total drying time was reduced by a factor of 6 when E increased from 0 to 7 kV/cm.

TABLE 4.1
List of U.S. Patents on High Electric Field Technology

Patent	Reference
LC/MS method of analyzing high molecular weight proteins	Dillon et. al (2008)
Optoelectronic probe	Lin (2006)
Method for treating waste-activated sludge using elecroporation	Held and Chauhan (2006)
Process for treating vegetables and fruit before cooking	Cousin et al. (2004)
Method for electrostatically separating particles and apparatus development	Yoshiyama et al. (2006)
Process and system for electrical extraction of intercellular matter from biological matter	Moldavsky et al. (2002)
Continuous liquid pasteurizing apparatus and method thereof	Uemura (2000)
High electric pasteurization	Gupta (2000)
Uniform product flow in a high electric field treatment cell	Bushnell and Lloyd (2000)
High-strength electric field pumpable food product treatment in a serial electrode treatment cell	Bushnell et al. (2000)
Prevention of electrochemical and electrophoretic effects in high-strength electric field pumpable food product treatment systems	Bushnell et al. (1995)

4.6.2 Drying of Spinach

Spinach is rich in vitamin C, iron, and calcium but has a relatively short shelf life. It is widely used in different cuisines. The color of spinach is often considered as a quality indicator from the consumer point of view, hence it is important to preserve the color during drying of spinach. Bajgai and Hashinaga (2001a) used the EHD technique for the drying of spinach. In their experiment they used a multiple points to plate electrodes system which could generate an electric field of 430 kV/m. Moisture content, Hunter color value, chlorophyll content, levels of ascorbic acid content and organic acid, and sugar content were used to assess the effectiveness of the HEF drying.

The effectiveness of three-, five-, and seven-point electrode systems was compared with conventional oven method. There was a significant increase in moisture removal rate compared to the conventional method. There was a direct positive correlation between the number of point electrodes and the mass transfer rate. The average moisture removal rate for the HEF system was 140 mg/m^2/s where as the oven drying rate was 97 mg/m^2/s.

HEF drying produced a final dried product that had similar color attributes to fresh spinach. The Hunter *a* value for the dried product was closer to the fresh product after 7 h of drying. The absence of a significant temperature change plays a major role in retaining the color value of spinach during EHD drying. After six weeks of storage EHD-dried spinach contained 54% higher chlorophyll content compared to oven-dried spinach. The vitamin C content was also higher (28.8%) after the storage period. There was also less browning, which agrees with the results published by Yamauchi and Watada (1998). HPLC analysis revealed that there was not any new component in the dried spinach compared to the fresh spinach.

4.6.3 Drying of Japanese Radish

Radish is a very popular vegetable and is a better representative for high-moisture vegetables. Bajgai and Hashinaga (2001b) carried out a study to find out the effectiveness of EHD on radish dehydration. Water absorption, rehydration rate, shrinkage level, and Hunter color measurements were used to identify the effectiveness of HEF for drying radish. They used a modified drying chamber design compared to their previous study (Bajgai and Hashinaga 2001b). Instead of using a standard copper plate to hold the sample they used wires to enhance the moisture removal process. They used seven electrode needles and an electric field strength of 430 kV/m for the drying. From their results they concluded that it is possible to remove up to 87.5% of the total moisture within 7 h of drying. The distance between the two electrodes was maintained at 1 cm to obtain the maximum moisture removal rate.

Oven drying produced higher levels of shrinkage, almost twice that of the EHD-dried radish. Similar observations were made for apple slices by Hashinaga et al. (1999). EHD also resulted in better rehydration properties. The dried radish absorbed 365 g/100 g, more than 25 g/100 g compared to the oven-dried sample. It was also observed that there was significantly lower solid loss for the EHD-dried sample than the oven-dried sample. The color of the radish dried by HEF was very close to that of the fresh sample. The results can be attributed to lower drying temperatures encountered during HEF drying.

4.6.4 Effect of High-Voltage Electric Field (HVEF) on Storage

Hsieh and Ko (2008) reported that the application of HVEF at 100 kV/m can improve the physicochemical properties and prolong the shelf life of carrot juice during refrigerated storage at 4°C. They found out from their experiment that the application of HVEF increases the tannins and phenolics, and decreases the plate counts of carrot juice. They also observed a decrease in turbidity and total carotenoid content.

4.6.5 Improving the Shelf Life of Cranberries

Cranberries are highly nutritious and possess significant medicinal properties. They are highly perishable in nature and require processing to enhance their shelf life. Modified atmospheric (MA) storage systems can be applied (Ratti, Raghavan, and Gariépy 1996) to increase their storage life. However, the maturity at the harvesting stage can play a major role in determining the MA storage requirements and it is often difficult to maintain. The nonthermal nature of HEF could help in preserving cranberries for longer periods of time while maintaining their nutritional value. Palanimuthu et al. (2009) used an HEF system similar to that in Figure 4.1 to develop a new processing method for cranberry storage. The treatment chamber used by Palanimuthu et al. (2009) is shown in Figure 4.2. They assessed the effectiveness of two different electric field strengths (2.5 kV/cm and 8 kV/cm). Three different time durations (30, 60, and 90 min) were used for both field strengths. Respiration rate, physiological loss of mass, Hunter color measurements, skin puncture strength, and total soluble solid (TSS) content were used to determine the effectiveness of the HEF treatment. There was not any significant difference between different

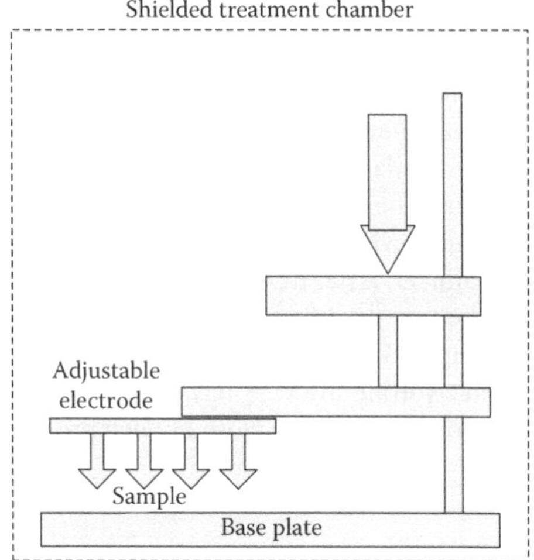

FIGURE 4.2 Test chamber for electrohydrodynamic drying.

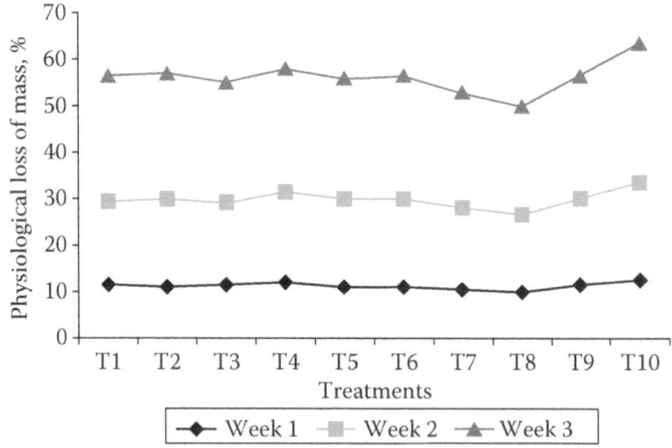

FIGURE 4.3 Cumulative physiological loss of mass of HEF-treated cranberries at ambient storage (23°C; 65% RH). T1: 2 kV/cm, 30 min; T2: 2 kV/cm, 60 min; T3: 2 kV/cm, 90 min; T4: 5 kV/cm, 30 min; T5: 5 kV/ cm, 60 min; T6: 5 kV/cm, 90 min; T7: 8 kV/cm, 30 min; T8: 8 kV/cm, 60 min; T9: 8 kV/cm, 90 min; T10: no treatment (control).

treatment methods as well as the control after three weeks of storage (Figure 4.3), although the physiological loss after three weeks was less for T8 (8 kV/cm, 60 min) compared to the control (no treatment).

The respiration rate of the HEF-treated sample was not significantly different to the control during the first week of storage. However, there was a significant change in respiration rate from the second week of storage onward. The electric field strength applied did not affect the respiration rate as compared to the HEF treatment itself. The respiration quotient increased to 1.88 and 3.48 by the end of third week. The higher respiration quotient can be attributed to the presence of high levels of ascorbic acid after three weeks of storage. Further research is necessary to identify the mechanism in which the respiration rate is affected by the HEF.

The color change after a week of storage was not significant. The color measurement values after one week of storage are shown in Table 4.2.

The skin puncture strength after a week of storage varied from 11.7 to 14.3 N, as shown in Figure 4.4. It was higher compared to the control (11.2 N). In terms of skin puncture strength there was no significant difference. However, the effect of different field strength requires further study.

4.6.6 Shelf-Life Improvement of Emblic Fruits Using High Electric Field (HEF)

Emblic (*Phyllanthus emblica* L.) fruit is widely available in India, Nepal, and other South Asian countries. It has high medicinal value and is also a rich source of ascorbic acid (Barthakur and Arnold 1991). It also improves the availability of iron from staple cereals and pulses. Bajgai et al. (2006a) conducted studies to evaluate the effectiveness of HEF on shelf-life enhancement of emblic fruit storage at different temperature levels. Electric field strength of 430 kV/m (both AC and DC fields) for 2 h was used for the treatment. After treatment, the fruits were stored at 4 and 20°C for 25 days and 35°C storage temperature was used for 15 days of storage. AC HEF was found to reduce the weight loss during storage compared to other treatments.

The percentage of emblic fruits rotting after 25 days of storage is shown in Table 4.3. It was found that HEF-treated emblics stored at 4°C in a closed pouch produced a better product after the

TABLE 4.2
Color Values (L^*, a^*, b^*) and Color Difference (ΔE^*ab) of HVEF-Treated Cranberries after a Week of Storage at Ambient Conditions (23°C; 65% RH)

Treatment	L^*	A^*	b^*	ΔE^*ab
T1	26.57	13.70	3.85	7.84
T2	26.14	13.26	3.60	4.64
T3	27.26	17.54	5.35	1.96
T4	25.88	12.53	3.22	5.50
T5	27.81	18.74	5.81	4.65
T6	26.75	16.43	5.03	4.76
T7	26.71	16.13	4.66	2.46
T8	26.75	16.44	4.73	2.68
T9	26.52	15.66	4.54	2.28
T10	26.59	15.87	4.61	2.06
F-test ($\alpha = 0.05$)	NS	NS	NS	*
CD @ 5%	–	–	–	3.26
CV%	3.86	24.62	33.97	49.32
SE_d	0.84	3.14	1.26	1.56

NS, not significant
* Significant

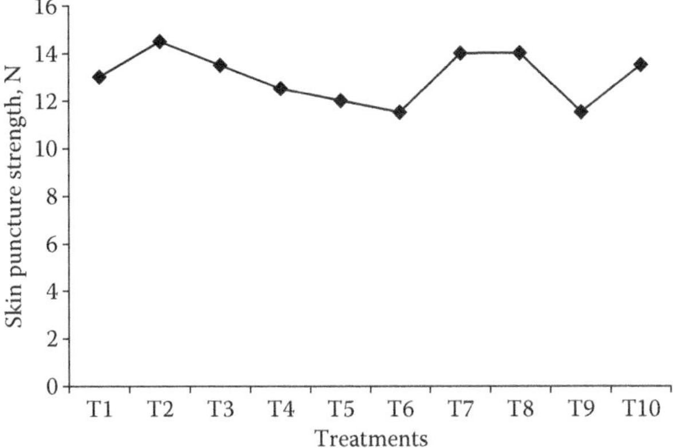

FIGURE 4.4 Skin puncture strength of HVEF-treated cranberries after one week of storage at ambient conditions (23°C; 65% RH). T1: 2 kV/cm, 30 min; T2: 2 kV/cm, 60 min; T3: 2 kV/cm, 90 min; T4: 5 kV/cm, 30 min; T5: 5 kV/cm, 60 min; T6: 5 kV/cm, 90 min; T7: 8 kV/cm, 30 min; T8: 8 kV/cm, 60 min; T9: 8 kV/cm, 90 min; T10: no treatment (control).

TABLE 4.3
Percentage of Rotting Emblic Fruits after Storage Period of 25 days

	Temperature, °C				
	HEF (Control)				
Samples	**4**	**20**	**35**	**4**	**20**
		Closed Pouch			
Small Green	25 (35)	75 (85)	95 (95)	35	85
Small Yellow	25 (35)	85 (85)	85 (90)	35	85
Medium Green	25 (35)	60 (75)	65 (80)	35	75
Medium Yellow	20 (30)	85 (95)	85 (95)	30	95
Large Green	20 (35)	85 (90)	90 (95)	35	90
		Open Pouch			
Small Green	65 (70)	55 (70)	65 (80)	70	70
Small Yellow	60 (70)	70 (80)	75 (85)	70	80
Medium Green	55 (60)	75 (80)	75 (85)	60	80
Medium Yellow	60 (75)	70 (80)	75 (80)	75	80
Large Green	65 (75)	75 (80)	75 (80)	75	80
Large Yellow	65 (75)	65 (80)	70 (85)	75	80

Note: The control values are given in parentheses

storage period. In terms of color change there was no significant difference between open-pouch and closed-pouch samples after the storage period. It was also observed that there was no significant change in the ascorbic acid due to HEF treatment.

4.6.7 Electrohydrodynamic Drying (EHD) of Okara Cake

Li, Li, and Tatsumi (2000) used HEF in an oven to dry okara cake at 105°C. They found that the drying rate was almost twice as high as that of the control and the drying directly entered into falling rate. They concluded that there was an effective area of HEF on the surface of the okara cake. The total drying time was reduced by 15% to 40% due to the application of HEF. Li et al. (2006) conducted further studies in 2006 to further understand the EHD drying characteristics of okara cake. They observed two different falling rate drying stages during drying and the critical moisture content was 29.57%. There was no constant rate of drying observed during HEF treatment. The efficiency of HEF treatment was directly proportional to the total moisture content.

4.6.8 Membrane Permeabilization

Application of high-intensity electric field pulses (HELP) in durations of nanoseconds and microseconds leads to membrane permeabilization in biological cells. This effect can be used in reversible and irreversible permeabilization applications on many biological systems (Chang et al. 1992; Lynch and Davey 1996; Ho, Mittal, and Cross 1997; Knorr and Angersbach 1998; Knorr et al. 2001). However, knowledge on the critical external electric field strength, membrane breakdown voltage, dynamics of membrane permeabilization, and the time sequence is very limited. The changes in the food systems are either reversible or irreversible during and after HELP treatment. It is important to understand these properties in order to develop a new process or design equipment for the application of HELP in food processing.

4.7 CONCLUSIONS

HEF is a new processing method with great potential. It could be used for high-value food products to produce superior quality products. The nonthermal nature of this method ensures that fresh-like qualities are well preserved during processing. HEF drying and EHD can help in retaining the product color, nutrients, and textural properties. HEF is a highly energy efficient and environmentally sound processing method compared to conventional food processing methods, especially drying. However, further research is necessary in this area to scale up the process, and establish postprocess handling methods and other problems associated with HEF.

ACKNOWLEDGMENTS

The authors would like to express their sincere gratitude to the Natural Sciences and Engineering Research Council (NSERC) Canada and Fonds de recherche sur la nature et les technologies (FQRNT) for their financial support for this work.

NOMENCLATURE

A Electrode area, cm^2
C Effective capacity, μF
d Distance between two electrodes, cm
dV Differential volume element
E Electric field strength, V/m
E_v Electric field vector, V/m
F Coulomb force, N
R Effective resistance of food, Ω
u Energy density, J/kg
u_e Energy per unit volume, J/m^3

V Electric potential difference, V
ε Permittivity of the medium, F/m
ε_0 Permittivity of vacuum, F/m
ε_r Relative permittivity
ρ Charge density, C/m^3
σ Conductivity, S/m

REFERENCES

Aibara, S., and K. Esaki (1998) Effects of high-voltage electric field treatment on bread starch. *Bioscience, Biotechnology, and Biochemistry* 62 (11): 2194–98.

Asakawa, Y. (1976) Promotion and retardation of heat transfer by electric fields. *Nature* 261 (5557): 220–21.

Bajgai, T. R., and F. Hashinaga (2001a) Drying of spinach with a high electric field. *Drying Technology* 19 (9): 2331–41.

———. (2001b) High electric field drying of Japanese radish. *Drying Technology* 19 (9): 2291–302.

Bajgai, T. R., F. Hashinaga, S. Isobe, G. S. Vijaya Raghavan, and M. O. Ngadi (2006a) Application of high electric field (HEF) on the shelf-life extension of emblic fruit (Phyllanthus emblica L.). *Journal of Food Engineering* 74 (3): 308–13.

Bajgai, T. R., G. S. V. Raghavan, F. Hashinaga, and M. O. Ngadi (2006b) Electrohydrodynamic drying—a concise overview. *Drying Technology* 24 (7): 905–10.

Barthakur, N. N., and N. P. Arnold (1991) Chemical analysis of the emblic (*Phyllanthus embilica* L.) and its potential as a food source. *Scientia Horticulturae* 47:99–105.

———. (1995) Evaporation rate enhancement of water with air ions from a corona discharge. *International Journal of Biometeorology* 39 (1): 29–33.

Bushnell, A. H., R. W. Clark, J. E. Dunn, and S. W. Lloyd (1995) Prevention of electrochemical and electrophoretic effects in high strength electric field pumpable food product treatment systems. US Patent no. 5447733.

Bushnell, A. H., J. E. Dunn, R. W. Clark, and S. W. Lloyd (2000) High strength electric field pumpable food product treatment in a serial electrode treatment cell. US Patent no. 6110423.

Bushnell, A. H., and S. W. Lloyd (2000) Uniform product flow in a high-electric-field treatment cell. US Patent no. 6027754.

Cao, W., Y. Nishiyama, S. Koide, and Z. H. Lu (2004) Drying enhancement of rough rice by an electric field. *Biosystems Engineering* 87 (4): 445–51.

Carlon, H. R., and J. Latham (1992) Enhanced drying rates of wetted materials in electric fields. *Journal of Atmospheric and Terrestrial Physics* 54 (2): 117–18.

Chang, D. C., B. M. Chassy, J. A. Saunders, and A. E. Sowers (1992) *Guide to Electroporation and Electrofusion*. San Diego: Academic Press.

Chen, Y., N. N. Barthakur, and N. P. Arnold (1994) Electrohydrodynamic (EHD) drying of potato slabs. *Journal of Food Engineering* 23 (1): 107–19.

Cousin, J.-F., F. Desailly, A. Goullieux, and J.-P. Pain (2004) Process for treating vegetables and fruit before cooking. US Patent no. 6821540.

Dillon, T., P. Bondarenko, G. Pipes, M. Ricci, D. Rehder, and G. Kleemann (2008) LC/MS method of analyzing high molecular weight proteins. US Patent no. 7329353.

Gupta, R. P. (2000) High electric pasteurization. US Patent no. 6086932.

Hashinaga, F., T. R. Bajgai, S. Isobe, and N. N. Barthakur (1999) Electrohydrodynamic (EHD) drying of apple slices. *Drying Technology* 17 (3): 479–95.

Held, J., and S. P. Chauhan (2006) Method for treating waste-activated sludge using electroporation. US Patent no. 7001520.

Ho, S. Y., G. S. Mittal, and J. D. Cross (1997) Effects of high field electric pulses on the activity of selected enzymes. *Journal of Food Engineering* 31 (1): 69–84.

Hsieh, C.-W., and W.-C. Ko (2008) Effect of high-voltage electrostatic field on quality of carrot juice during refrigeration. *Lebensmittel-Wissenschaft und Technologie* 41 (10): 1752–57.

Jones, T. B. (1978) Electrohydrodynamically enhanced heat in liquids (review). *Advances in Heat Transfer* 14:107–48.

Knorr, D., and A. Angersbach (1998) Impact of high-intensity electric field pulses on plant membrane permeabilization. *Trends in Food Science & Technology* 9 (5): 185–91.

Knorr, D., A. Angersbach, M. N. Eshtiaghi, V. Heinz, and D.-U. Lee (2001) Processing concepts based on high intensity electric field pulses. *Trends in Food Science & Technology* 12 (3–4): 129–35.

Li, F. D., L. T. Li, J. F. Sun, and E. Tatsumi (2006) Effect of electrohydrodynamic (EHD) technique on drying process and appearance of okara cake. *Journal of Food Engineering* 77 (2): 275–80.

Li, L. T., F. D. Li, and E. Tatsumi (2000) Effects of high voltage electrostatic field on evaporation of distilled water and okara drying. *Biosystem Studies* 3 (1): 43–52.

Lin, H. (2006) Optoelectronic probe. US Patent no. 7088116.

Lynch, P. T., and M. R. Davey (1996) *Electrical Manipulation of Cells*. New York: Chapman and Hall.

Moldavsky, L., M. Fichman, K. Shuster, and M. Govberg (2002) Process and system for electrical extraction of intercellular matter from biological matter. US Patent no. 6344349.

Mujumdar, A. S. (2004) Research and development in drying: Recent trends and future prospects. *Drying Technology* 22 (1): 1–26.

Palanimuthu, V., P. Rajkumar, V. Orsat, Y. Gariépy, and G. S. V. Raghavan (2009) Improving cranberry shelf-life using high voltage electric field treatment. *Journal of Food Engineering* 90 (3): 365–71.

Panofsky, W. K. H., and M. Phillips (1962) *Classical Electricity and Magnetism*, 95–117. London: Addison-Wesley.

Ratti, C., G. S. V. Raghavan, and Y. Gariépy (1996) Respiration rate model and modified atmosphere packaging of fresh cauliflower. *Journal of Food Engineering* 28 (3–4): 297–306.

Uemura, K. (2000) Continuous liquid pasteurizing apparatus and method thereof. US Patent no. 6050178.

Vega-Mercado, H., M. M. Gongora-Nleto, and G. V. Barbosa-Canovas (2001) Advances in dehydration of foods. *Journal of Food Engineering* 49 (4): 271–89.

Yamauchi, N., and A. E. Watada (1998) Ascorbic acid and β-carotene affect the chlorophyll degradation in stored spinach (*Spinacia oleracea* L.) leaves. *Food Preservation Science* 24 (1): 17–21.

Yoshiyama, E., Y. Shibata, and T. Kinoshita (2006) Method for electrostatically separating particles and apparatus development. US Patent no. 7119298.

Zhang, Q., G. V. Barbosa-Cánovas, and B. G. Swanson (1995) Engineering aspects of pulsed electric field pasteurization. *Journal of Food Engineering* 25 (2): 261–81.

5 Pulsed Electric Fields: A Review on Design

Sally Alkhafaji and Mohammed Farid
The University of Auckland

CONTENTS

5.1 INTRODUCTION

The roots of pulsed electric field (PEF) processing can be traced to the work conducted in Germany. Doevenspeck (1960) described different PEF equipment and methods ranging from PEF processing of sausage to specific electronic embodiments. Doevenspeck remained active for many years and collaborated on PEF development with later German investigators (Barbosa-Cánovas and Zhang 2001).

Sale and Hamilton (1967) and Hamilton and Sale (1967) have studied microbial cell repair after PEF treatment (Barbosa-Cánovas and Zhang 2001). They believed that when an external electric field is applied to a cell, transmembrane potential is induced across its membrane. The induced potential difference is proportional to the external field intensity. The transmembrane potential leads to membrane damage, which is the direct cause of cell inactivation (Barbosa-Cánovas et al. 1999). When this potential reaches a critical value (which is approximately 1 V for a bimolecular lipid membrane), the membrane breaks down. If the electric field is switched off, the membrane will return to the initial and normal state. But if the electrical field strength exceeds the critical value

(E_C) for a given length of time (t_c), permanent holes will form in the cell membrane. The resulting inactivation of microorganisms is thus, considered to be related to both the electric field strength and treatment time (Sale and Hamilton 1967; Hulsheger, Potel, and Niemann 1983).

PurePulse Technologies, California, owns several U.S. patents on PEF treatment units for the treatment of liquid foods, such as dairy products, fruit juices, and liquid eggs (Dunn and Pearlman 1987; Bushnell et al. 1993, 1996). The patents describe batch and continuous systems, including chamber characteristics, pulse forming network (PFN) components, and specific switching arrangements to avoid electrode fouling.

Washington State University (WSU) also has a comprehensive program to preserve foods by high-intensity PEF. The system designed and constructed by the WSU group includes a power supply capable of delivering a peak voltage of 40 kV and parallel plate and coaxial continuous treatment chambers. Numerical simulations and experimental studies also supported the chamber's design and system effectiveness (Barbosa-Cánovas et al. 1999).

Another leading group in PEF technology is that of Ohio State University. They have implemented an integrated pilot plant system with aseptic packaging (Zhang, Qui, and Sharma 1997). The system includes treatment chambers with co-field treatment zones and a PFN capable of delivering energy at rates of the order of kHz (Barbosa-Cánovas et al. 1999).

Many European groups have worked in the application of PEF technology to food processing. The microbial inactivation mechanisms and kinetics in model and real foods have been contributed by those researchers (Hamilton and Sale 1967; Hulsheger et al. 1983; Grahl and Märkl 1996).

More recently, a research group at the University of Auckland had designed a PEF system using the modern insulated gate bipolar transistors (IGBT). The main focus of the work was to design a new PEF treatment chamber that operates at high electric field intensities with minimum increase in liquid temperature and limited fouling of electrodes (Alkhafaji and Farid 2007).

The objective of this chapter is to review research activities on PEF treatment around the world with major focus on the design of the treatment chamber and pulse generator.

5.2 BASIC ASPECTS OF PULSED ELECTRIC FIELD (PEF)

PEF treatment of liquid foods is based on the application of high-intensity electric field to the food product as it flows between two electrodes that confine the treatment gap of the PEF chamber. The key variables involved in PEF are electric field strength (E), pulse duration or pulse width (τ), treatment time (t), pulse repetition rate (f), waveform of the pulse, and treatment temperature.

Average electric field strength E (kV/cm) is defined as the electric potential difference V for two given points in space divided by the distance d between them, $E = V/d$, where V (kV) is the voltage applied and d (m) is the gap between the electrodes. The treatment time in a PEF chamber is derived from the product of the number of pulses the fluid receives and pulse duration. In general, an increase in any of these variables results in an increase in microbial inactivation (Barbosa-Cánovas et al. 1999).

The treatment time is defined as $t = n\tau$, where n is the number of pulses and τ is the pulse width. The critical treatment time, t_c, is the time beyond which the microbial population decreases linearly with respect to treatment time in a log-log scale, ranging from 2 to 80 μs (Hulsheger et al. 1983). However, this depends largely on the electric field intensity, conductivity of the treated liquid, and treatment temperature.

5.3 DESIGN OF PULSED ELECTRIC FIELD (PEF) PROCESSING EQUIPMENTS

A PEF processing system for food application consists of a number of pieces of equipment, including a high-voltage pulse generator, treatment chamber/s, process monitoring and control systems, a cooling system, and a degassing unit.

5.3.1 High-Voltage Pulse Generators

Most high-voltage pulse generators have been designed to provide high-voltage pulses, which induce stress and mortality to biological cells. These pulse generators convert normal utility voltage into high-intensity voltage. A power source is used to charge a capacitor bank, where a large amount of energy is stored. A switch is used to discharge the electrical energy from the capacitor bank across the food held in a treatment chamber (Barbosa-Cánovas et al. 1999).

A typical electrical circuit for generating exponential pulses (Figure 5.1) consists of a DC power supply that charges a capacitor bank, which is connected in series with a charging resistor R_C. The charges stored in the capacitor flow through the food in the PEF chamber when a trigger signal is applied. The resistance R_1 limits the current in case of an arc discharge in the food and R_2 controls decay time in the event of food conductivity being smaller than expected. An electrical circuit used to generate the preferred square pulses as illustrated in Figure 5.2. A high-voltage transmission line connected to a matched load generates a square pulse. However, it is difficult to match the resistance of the food with the characteristic impedance of the transmission line to provide the highest energy transfer to the food. Using a PFN circuit consisting of an array of capacitors and inductors can solve this problem (Barbosa-Cánovas et al. 1999).

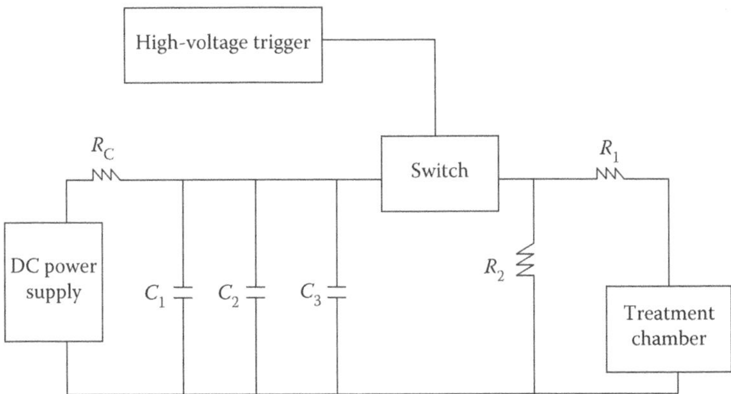

FIGURE 5.1 Exponential decay pulse-generating circuit. (Reproduced from Qin, B., Zhang, Q., Barbosa-Cánovas, G. V., Swanson, B. G., and Pedrow, P. D., *IEEE Transactions on Dielectrics and Electrical Insulation*, 1, 1047–57, 1994.)

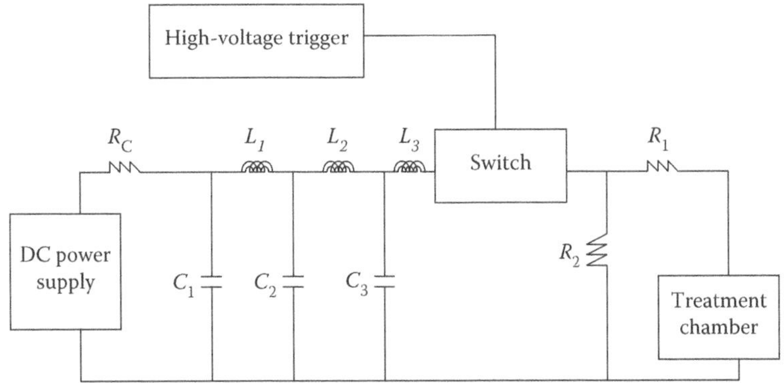

FIGURE 5.2 Square wave pulse-generating circuit. (Reproduced from Qin, B., Zhang, Q., Barbosa-Cánovas, G. V., Swanson, B. G., and Pedrow, P. D., *IEEE Transactions on Dielectrics and Electrical Insulation*, 1, 1047–57, 1994.)

Conventional high-voltage pulse generators have limitations for a number of reasons. Few of the high-voltage pulse generators can maintain the extremely high currents resulting from the low electrical resistance of the fluid in the treatment chamber. Some of the conventional high-voltage pulse generators can provide only unipolar pulses; these pulses cause electrolysis, and deposition of protein and other charge-carrying particles on the surface of the electrodes. Therefore, methods and apparatuses for providing high-voltage pulses that do not suffer from these disadvantages are desirable (Qiu and Zhang 2001).

Ho et al. (1995) demonstrated the effectiveness of a low-cost pulse generator with a maximum voltage of 30 kV and pulse width of 2 μs. Zhang et al. (1997) successfully used a versatile pulse-forming network capable of delivering square, bipolar, and exponential wave-shape pulses; the pulse generator provides voltages up to 50 kV.

Qiu and Zhang (2001) presented a high-voltage pulse generator system for effective PEF treatment, as shown in Figure 5.3. The system generates bipolar and/or unipolar high-voltage pulses. The system includes a power source and an energy storage component (either a capacitor or a PFN) for storing electrical energy from the power source. The particular composition of the energy storage component influences the shape of the high-voltage pulse that can be applied (either substantially square or exponential decay pulses). The values of the capacitors and inductors used are dependent on specific design characteristics (i.e., pulse duration, amplitude, etc.). The system also includes a plurality of switches for opening and closing periodically to discharge the energy stored, a load

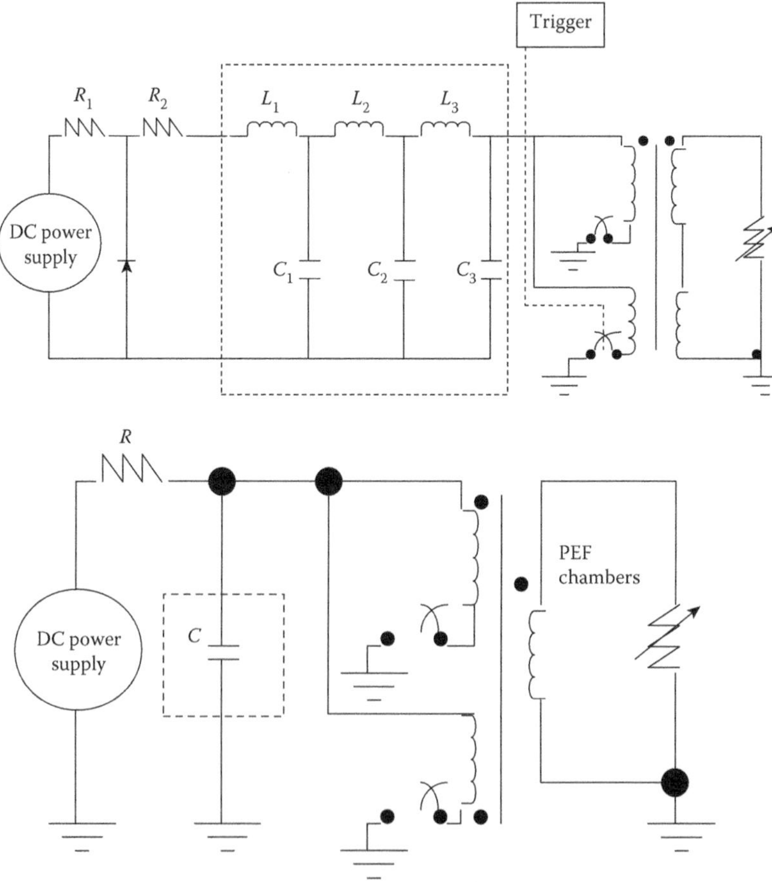

FIGURE 5.3 Circuit diagrams of a high-voltage bipolar pulse generator. (Reproduced from Qin, B., Zhang, Q., Barbosa-Cánovas, G. V., Swanson, B. G., and Pedrow, P. D., *IEEE Transactions on Dielectrics and Electrical Insulation,* 1, 1047–57, 1994.)

comprising the treatment chamber, and a pulse transformer for allowing a plurality of voltage and current levels to be generated at the load. The power source charges the energy storage component and an appropriate trigger device triggers the opening and closing of the switches. Switches with different configurations were presented including an H-bridge switch configuration.

The switches hold the high voltage stored in the PFN and control the current flow reliably when fired. The switches are controlled by a command signal generator and operate at high repetition rates (0.1–5000 Hz). Commonly used types of discharge switches include mercury-ignition, gas-spark-gap, vacuum spark-gap, thyratron or magnetic, and mechanical rotary switches (Calderon-Miranda et al. 1999).

A single thyratron can switch very large powers, but because of their high losses their frequencies are limited to a few hundred Hertz. Their main disadvantage, however, is their limited lifetime. Solid-state semiconductor switches are considered by experts to be the future of high-power switching. New generations of the solid-state semiconductor switches are the IGBTs, which combined the best features of a metal-oxide semiconductor field-effect transistor (MOFET) input and a bipolar transistor output into a newer power-switching device. The very rapid switching and small power consumption are the main features of the IGBTs (Bartos 2000).

Alkhafaji and Farid (2007) reported another high-voltage pulse generator (Figure 5.4). The generator was designed to maintain high currents through a small resistive load and to provide square bipolar pulses of 1.7 μs pulse duration, pause duration 2.5 μs, and a frequency up to 200 Hz. The system provides short pulse rising and falling time of less than 0.5 μs, which increases the efficiency of the PEF treatment. The efficiency of PEF treatment depends on the increase in the pulse rising time and can approach 100% for a very short rising time increase (Álvarez et al. 2003). The alternating changes in the movement of charged molecules using bipolar pulses increase the stress on the cell membrane and enhance its electric breakdown. Bipolar pulses also offer minimum energy utilization, reduced deposition of solids on the electrode surface, and decreased food electrolysis.

The system further consists of a DC power supply that charges electrical energy storage components, a set of four switches operating in pairs, and a transformer designed to set up the voltage from 1 to 30 kV. The switching devices used are the IGBTs. Periodic closing and opening of the pairs of switches result in the application of bipolar pulses across the treatment chamber.

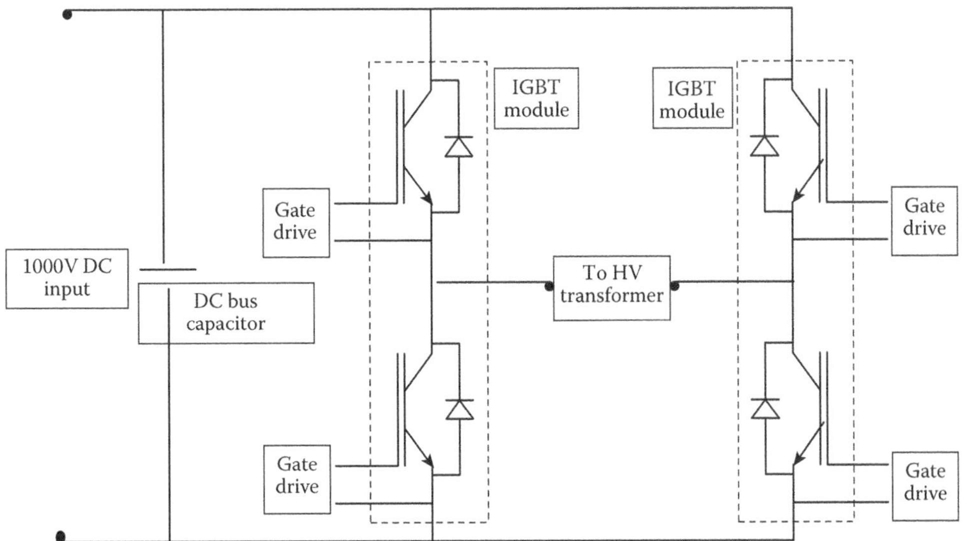

FIGURE 5.4 High-voltage pulse generator. (Reprinted from Alkhafaji, S., and Farid, M., *Innovative Food Science and Emerging Technologies,* 8, 205–12, 2007.)

There is a galvanic isolation between the cell and the switching power supply. The cell can be configured to operate symmetrically around ground potential, i.e., one terminal of the cell does not have to be grounded. This solution reduces unwanted ground currents through the processed food. Safety of the system is also enhanced, as the maximum voltage to ground is now equal to only half the maximum voltage between the cell electrodes.

A DC voltmeter connected to the DC bus bar was used to measure the voltage and an oscilloscope with current probes was used to measure the output current. The frequency of the output pulses can be varied by means of a signal generator. The output voltage can be varied by means of a three-phase input variac. The current is measured using a Rogowski coil to control any changes in the current during application.

5.3.2 Treatment Chambers

One of the key components IN the design of a PEF pasteurization process is the treatment chamber. It must be designed to operate at high electric field intensities. Dielectric breakdown inside the treatment chamber must be prevented.

Static and continuous treatment chambers are common. Sale and Hamilton (1967) were among the earliest researchers to study inactivation of microorganisms in a static PEF chamber. Carbon electrodes supported on brass blocks were used and placed in a U-shaped polythene spacer. The chamber withstands electrical field of maximum 30 kV/cm due to electrical breakdown caused by the presence of air above the food. Another static chamber was designed at WSU; it had two round-edged, disk-shaped stainless steel electrodes polished to mirror surfaces. Electric field strengths of up to 70 kV/cm have been applied successfully. Treatment chambers with parallel plate electrodes offer a uniform electric field distribution along the gap axes and electrode surfaces, but create a field enhancement problem at the edges of the electrodes. Static chambers are mainly suitable for laboratory use.

For larger-scale operations, continuous chambers are more efficient. Dunn and Pearlman (1987) designed a continuous chamber consisting of two parallel plate electrodes and a dielectric space insulator. The electrodes are separated from the food by ion conductive membranes to avoid product contact with the electrodes.

The attempts to improve the efficiency of the PEF treatment chambers are classified as below.

5.3.2.1 Designs to Concentrate the Electric Field Intensity in the Treatment Region

The intensity of the applied electric field is one of the most important factors influencing microbial inactivation by PEF. Once the applied electric field exceeds a critical value for a given length of time, transmembrane potential is induced, which results in cell death. Matsumoto et al. (1991) applied the concept of concentrating the electric field in the treatment region. They devised a converged electric field treatment chamber (Figure 5.5). The liquid food was continuously fed into a vessel through the hole of a disc electrode. An insulating plate with small holes was placed between the parallel disc electrodes to concentrate the electric field. Only the fluid inside the holes of the plate is subjected to the PEF treatment. The current density at the electrode-liquid interface is kept low to minimize electrolysis and reduce bubble formation. Three to six log reductions of selected microorganisms (*Saccharomyces cerevisiae*, *Escherichia coli*, and *Bacillus subtilis*) was achieved by 30–40 kV/cm of electric field treatments. However, this design has a limitation: the creation of stagnant zones in the 90° corners of the cell where microbes can build up and the liquid food may overheat. In addition, the stagnant zone may also be prone to trap air bubbles, which could cause sparking. If the treatment outside the disc electrode is important, then the liquid in the center of the cell will receive less treatment, since it is passing at higher velocities and receiving a lower electric field.

Another design to concentrate the electric field in the treatment region was presented by Sensoy, Zhang, and Sastry (1996). A continuous co-field flow PEF chamber (Figure 5.6) with a conical insulator shape was designed so that the actual treatment zone is the small orifice. The special

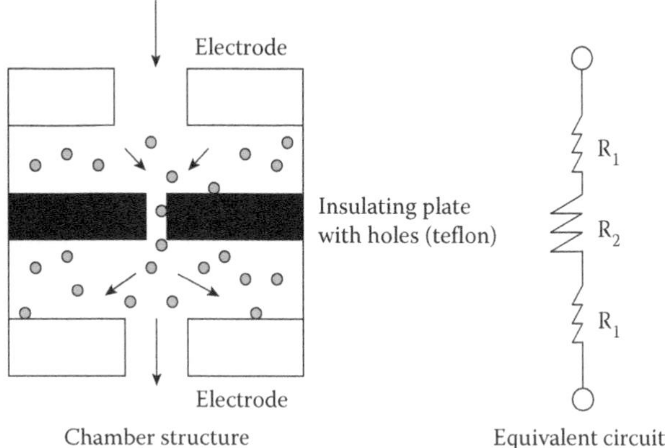

FIGURE 5.5 Cross-section view of treatment chamber designed by Matsumoto et al. (Reproduced from Matsumoto, Y., Satake, T., Shioji, N., and Sakuma, A., *Conference Record of IEEE Industrial Applications Society Annual Meeting*, 652–59, 1991.)

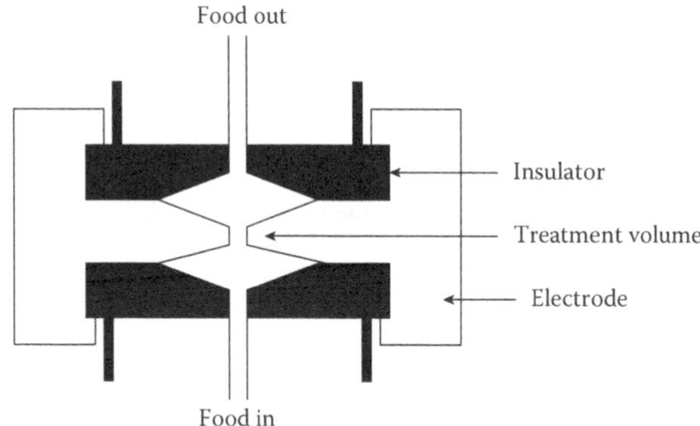

FIGURE 5.6 A co-field continuous treatment chamber. (Reproduced from Sensoy, I., Zhang, Q. H., and Sastry, S. K., *Journal of Food Process Engineering*, 20, 367–81, 1996.)

conical shaped electrodes and insulators were designed to eliminate gas bubble formation within the treatment volume. The volume of the orifice and the conical regions were designed so that the voltage across the orifice is close to the supplied voltage. Electric field strength of 15–40 kV/cm was used. The problem of a stagnant zone present in the previous design was minimized in this design. However, again if the treatment outside the disc electrode is important, then the liquid in the center of the cell will receive less treatment, reducing the efficiency of the process.

5.3.2.2 Designs to Adjust the Treatment Time

Microbial inactivation increases with the number of pulses the product receives. In general, increasing the number of pulses and/or their duration results in an increase in microbial inactivation (Barbosa-Cánovas et al. 1999).

To increase the residence time of the liquid food inside the treatment chamber, the WSU research group has constructed a parallel plate continuous flow treatment chamber of small scale (Figure 5.7) by modifying the previously designed static parallel plate electrode chamber (Barbosa-Cánovas et al. 1999). Baffled flow channels were added inside the chamber to provide a tortuous path of the fluid in the treatment zone. Cooling of the chamber was provided by circulating cooling water through jackets built within the two stainless steel electrodes. Operating conditions were as follows: chamber volume 20–8 cm^3, electrode gap 0.95 or 0.51 cm, PEF strength 35 or 70 kV/cm, pulse width

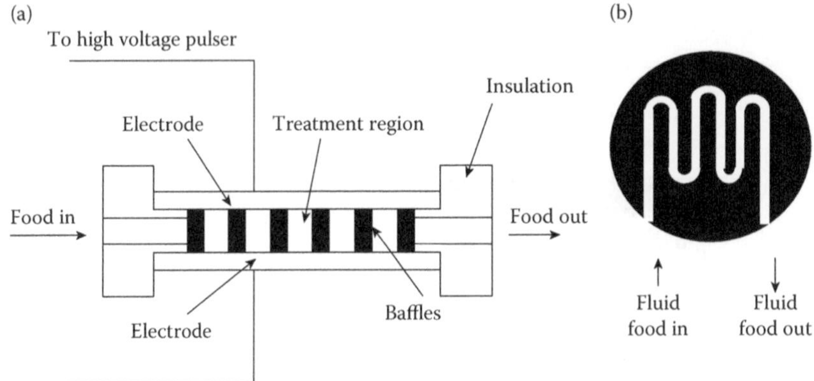

FIGURE 5.7 Continuous treatment chamber with baffles, WSU group. (a) Cross-section view. (b) Top view.

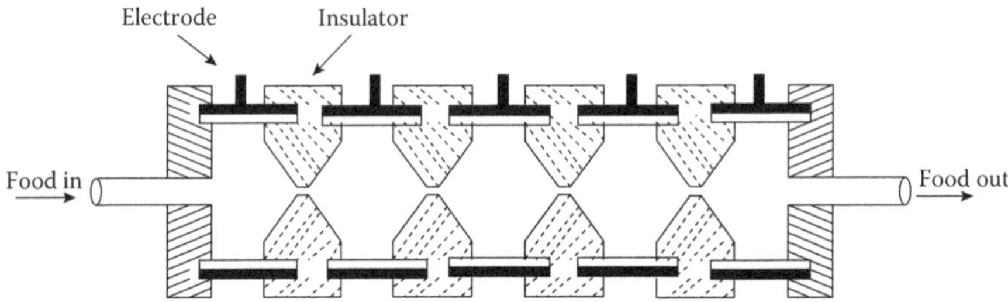

FIGURE 5.8 A cross-sectional side view of an embodiment of a continuous-current, high electric field treatment cell assembly. (Reproduced from Dunn, J. E., and Pearlman, J. S., US Patent 4695472, 1987.)

2–15 μs, pulse repetition rate 1 Hz, and liquid food flow rate 600–1200 cm^3/min. In this design the field intensity that could be applied to the food was limited because of the interface between the liquid food and the insulating spacer. This increases the chance of electrical breakdown produced by electrical tracking along the insulator surface separating the two electrodes.

Dunn and Pearlman (1987) provided a continuous treatment unit where the fluid food flows through a series of treatment chambers to achieve the desired treatment time. The chamber (Figure 5.8) comprises a plurality of electrode reservoir zones, which are electrically isolated from each other by intervening dielectric separating elements so that only the electrical current will pass through food product itself. The operating conditions were as follows: electric field of 5–25 kV/cm with a square and exponentially decaying pulses having pulse width of 1 and 100 μs and frequency of 0.1 and 100 Hz. The disadvantage of this cell is that the uniformity of the electric field distribution degrades with time, because the temperature gradient in the product will have the tendency to produce arcs or current filaments (Morshuis et al. 2001).

5.3.2.3 Designs Based on Electric Field Distribution

Numerous researchers reported treatment chamber designs based on coaxial cylinder-electrode configurations. The coaxial device is characterized by a well-controlled uniform electrical field and simple chamber structure. Coaxial chambers are basically composed of an inner cylindrical electrode surrounded by an outer annular cylindrical electrode that allows food to flow between them (Barbosa-Cánovas et al. 1999).

The coaxial device configurations are not suitable for high-intensity electric fields because of possible electrical breakdown from the insulator surface tracking or local field enhancement (Boulart 1983; Hofmann and Evens 1986; Sato and Kawata 1991). The design of these chambers does not provide a uniform treatment across the treated volume since the radial geometry causes the field strength to decrease toward the outer electrode (Qin 1995). One challenge associated with coaxial chambers is that they generally present low load resistance when used to treat most foods and the pulser system must be able to deliver high current at the voltage employed. Bushnell et al. (1993) designed another coaxial chamber, which provides a long tracking path on the insulator surface to assure complete treatment of the food product as it flows through the electrode region. The major disadvantage of the chamber is the limited width of the annulus through which the product can flow and the relatively large electrode surfaces.

Qin (1995) modified the coaxial treatment chamber design to provide uniform electric field distribution (Figure 5.9). The electrode configuration in the chamber was obtained through an optimized shape design using computational fluid dynamics (CFD) analysis. The outer electrode has a protruded contour surface. A prescribed field distribution along the fluid path without electric field enhancement points was observed. The electric field region between the high-voltage electrode and the grounded electrode in the coaxial treatment chamber is illustrated in Figure 5.10. The chamber includes cooling systems for controlling the electrode temperature. The outer electrode has a protruded contour surface obtained from the CFD analysis. In the treatment region between the two electrodes, the potential drop is nearly uniform and a strong electric field is generated. On the other hand, in the nontreatment zone most of the potential drop is found inside the dielectric spacer, thus the electric field is quite weak in the liquid food path outside of the treatment region. This modified coaxial treatment chamber may operate at an electric field exceeding 70 kV/cm without dielectric breakdown (Qin 1995).

Cornelis and Vincent (2000) described a novel modular design of a PEF treatment chamber (Figure 5.11) comprising of an open electrode structure with a large aperture. The objective of

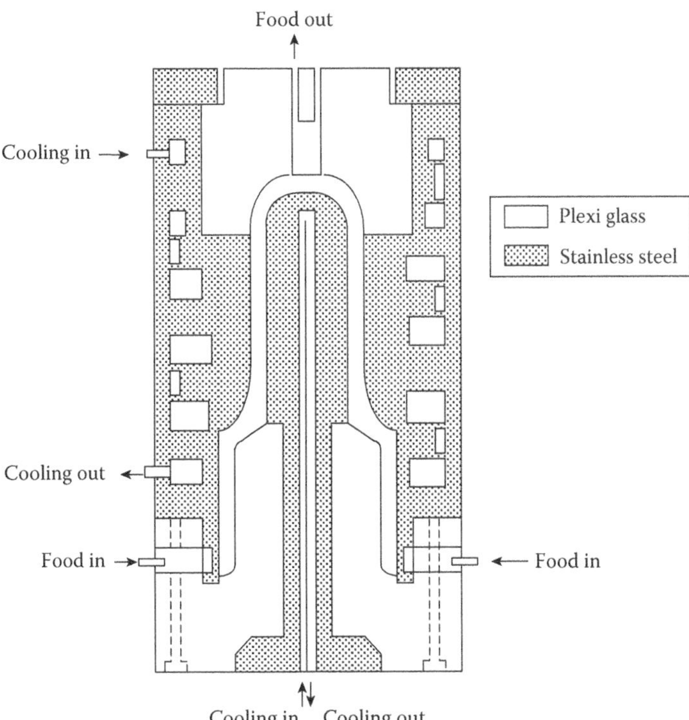

FIGURE 5.9 A cross-sectional view of a modified coaxial treatment chamber. (Reproduced from Qin, B.L., *Food Technology*, 49, 55, 1995.)

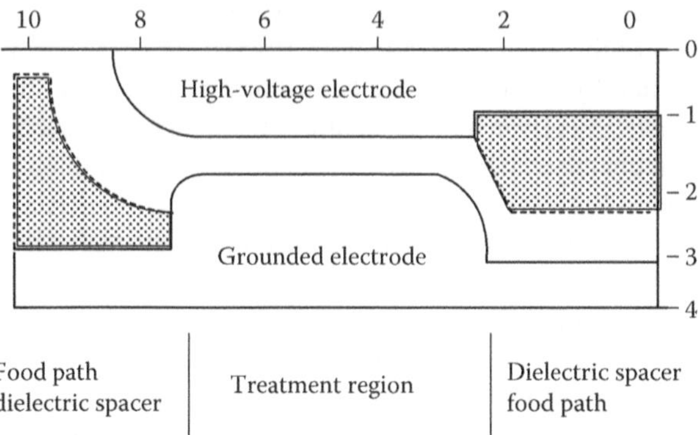

FIGURE 5.10 Electric field region between high-voltage electrode and grounded electrode in the coaxial treatment chamber. (Reproduced from Qin, B.L., *Food Technology*, 49, 55, 1995.)

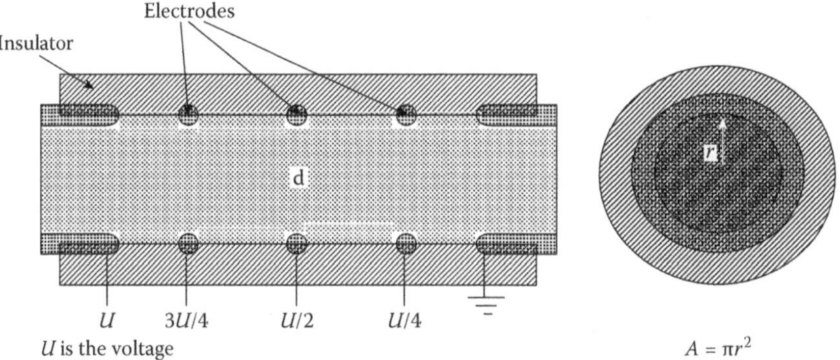

FIGURE 5.11 Modular design of PEF treatment chamber. (Reproduced from Cornelis, M. H., and Vincent, B. P., EP Patent 1000554, 2000.)

this design is to overcome the problem of nonuniformity in the electric field distribution due to a gradient in the product conductivity at steady-state flow conditions.

The treatment chamber consists of several identical modules; each module has a large aperture and a small internal volume. The aperture of each module has a large cross-sectional area with respect to the contained volume. By proper positioning of several of these modules the nonuniformity in the electrical field distribution of individual modules was minimized. As a result the overall electrical field in the treatment chamber was uniform across the volume. Due to its design the peak power that is needed can be distributed over several modules which can be fed by different auxiliary pulse power supplies. The disadvantage of this design is that the electrical field distribution of the first and the last cell are slightly nonuniform. This is due to the fact that translation symmetry in the electrode array is absent.

5.3.2.4 Designs Where the Electric Field Lines are Parallel to the Fluid Flow

Yin, Sastry, and Zhang (1997) described a co-linear treatment chamber as illustrated in Figure 5.12. The treatment chamber comprises of a minimum of two cylindrical electrodes and a cylindrical insulator to electrically insulate the electrodes from each other. PEF vector direction is parallel to the liquid product flow. The applied electric field is 30 kV/cm, frequency 500–20,000 Hz, and pulse duration 1–6 μs. The nonuniform distribution of the electrical field in the treatment zone is one of the disadvantages of this design. Using small gap distances (with respect to the diameter), the electric field distribution at the entrance and exit of the treatment zone is nonuniform. If the gap distance is enlarged, a large temperature gradient will appear across the treatment zone at steady-state conditions. Due to the heat generation in the chamber when receiving pulse treatment, the temperature of the product at the exit of the treatment zone becomes higher than at the entrance. As a result the voltage drop over the product column will change, which results in a smaller electrical field in the region near the exit of the treatment zone. This treatment offers a much smaller region of the chamber volume and therefore lasts for a shorter period of time. This leads to an increase in the spread in the effective field strength and spread in the treatment time in the product, which leads to a decrease in the microbiological inactivation. In addition to this temperature-induced effect, the release of minerals and other components by cellular membrane structures will cause a similar effect. This can cause a change in the conductivity of the product across the treatment zone (for example, due to electroporation of biological membranes, intracellular contents can be released) (Cornelis and Vincent 2000).

5.3.2.5 Designs to Reduce the Energy Input

De Jong and Van Heesch (1999) provided a treatment chamber where a 30% energy saving was achieved. The electrodes are included in the fluid flow channel, which results in a homogeneous electric field being generated. The residence time distribution of the fluid products in the fluid treatment unit is uniform. This in turn restricted the increase in temperature of the product, and minimized protein denaturation and vitamin destruction. The treatment unit (Figure 5.13)

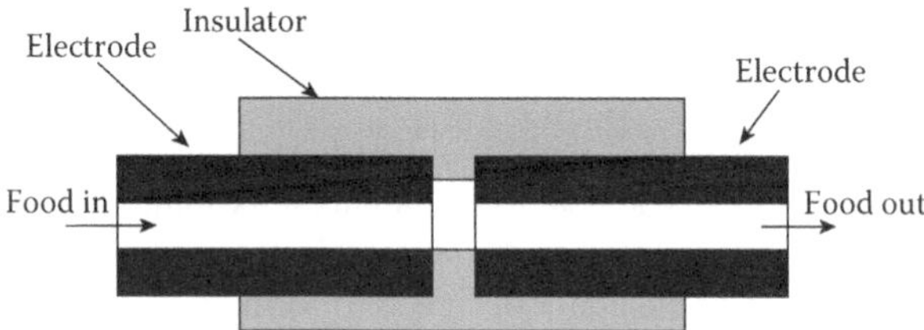

FIGURE 5.12 A cross-sectional side view of an embodiment of pulse dielectric field treatment device. (Reproduced from Yin, Y. G, Sastry, S. K., and Zhang, Q. H., US Patent 5690978, 1997.)

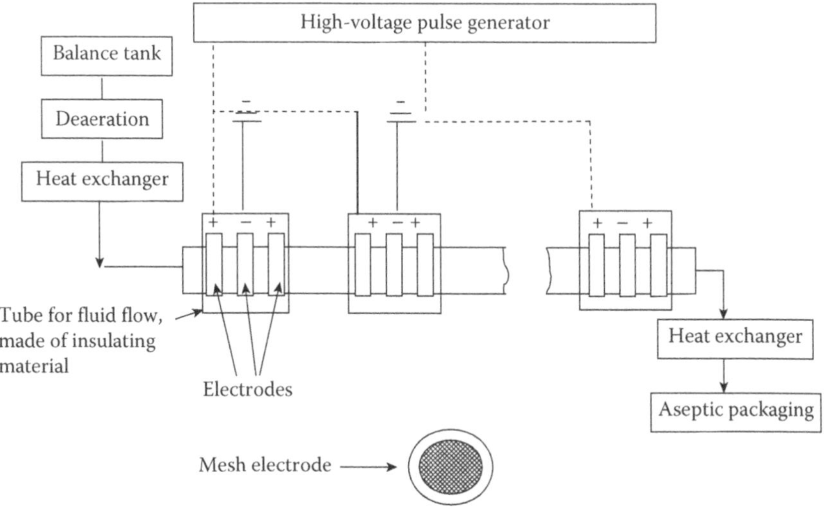

FIGURE 5.13 A cross-section view of a treatment unit. (Reproduced from De Jong, P., and Van Heesch, E. J. M., US Patent 2001013467, 1999.)

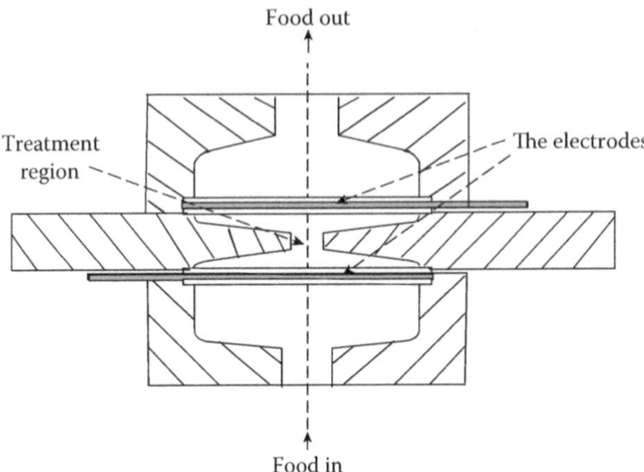

FIGURE 5.14 A cross-sectional view of the treatment chamber. (Reprinted from Alkhafaji, S., and Farid, M., *Innovative Food Science and Emerging Technologies*, 8 (2), 205–12, 2007.)

comprises of 10 treatment chambers connected in series and many electrodes, which are composed of metal wires (5 mm in diameter) to form a net. The fluid flows through the openings of the net, which have an equivalent diameter of 25 mm. It was possible to achieve electric field strength of 70 kV/cm, pulse width of 2 μs, frequencies of 40 Hz, and residence time of 0.2 sec. in each treatment chamber.

Alkhafaji and Farid (2007) designed multipass treatment chambers (Figure 5.14) to attain a relatively uniform high-intensity electric field for achieving maximum and uniform microbial inactivation. Each chamber includes two stainless steel mesh electrodes. The two electrodes are electrically isolated from each other by an insulator designed to form an orifice where the voltage across the orifice could be close to the supplied voltage. The liquid food flows through the openings of the first electrode and through the orifice where the electric field is concentrated, then through the opening of the second electrode. The surface area provided by the wires in each electrode and the distance between the electrodes and the treatment region were calculated to provide maximum electrical field strength inside the treatment region. The chamber was placed in a vertical position to ensure accurate control of the fluid residence time distribution, temperature distribution, and product conductivity (which changes with temperature).

The electrodes were removed from the high electric field region to prevent accumulations of materials such as protein molecules upon the electrodes. Such accumulations can cause fouling, which will affect the properties of the electrical field emanating from the electrodes and their interaction with the fluid being processed. This can lead to a nonuniform pulse distribution into the flowing product, which in turn can result in inadequate microbial inactivation.

5.3.2.6 Designs Based on Reducing the Risk of Dielectric Breakdowns

Barbosa-Cánovas et al. (2000) improved the PEF treatment systems by reducing the risk of dielectric variations and breakdowns that can occur between the electrodes within the processed food that contains mixed particles. A treatment chamber (Figure 5.15) comprises a cylindrical rod first electrode, a second electrode, and a treatment zone that includes a curved transition face to provide an inwardly converging second electrode face. Both electrodes are provided with a temperature stabilizer (heat-stabilizing fluids can be brought into contact with the electrode to stabilize and control the electrode temperature). The electric field magnitude is relatively smaller near the product inlets and increases up to a maximum in the principal treatment zone between the coaxially parallel

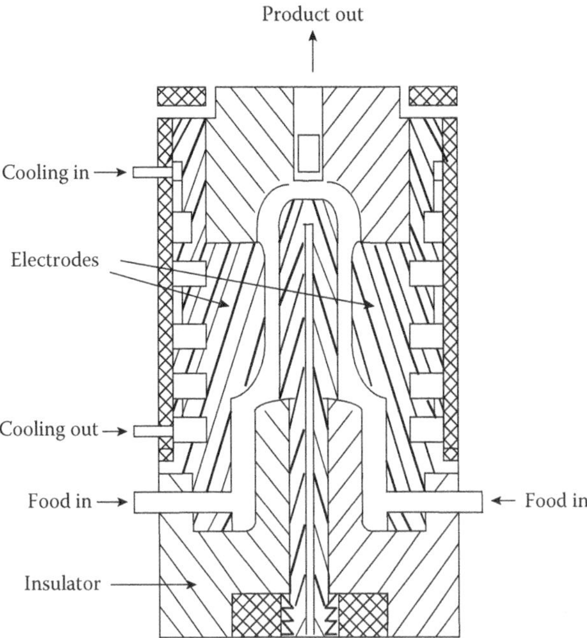

FIGURE 5.15 A longitudinal sectional view of a treatment cell. (Reproduced from Barbosa-Cánovas, G. V., Qin, B.-L, Zhang, H., Olsen, R. G., Swanson, B. G., and Pedrow, P. D., US Patent 6019031, 2000.)

faces of the first and second electrodes. The electric field strength smoothly decreases toward the product outflow port (Barbosa-Cánovas et al. 2000).

Another treatment chamber was reported by Barbosa-Cánovas et al. (2000) for the treatment of food products with high electrical conductivity. These products require more electrical power and result in higher increase in temperature during the PEF treatment. The chamber (Figure 5.16) comprises specially contoured faces of electrodes to provide a treatment zone in which the electrical field strength varies between relatively high and low values. Higher values of the electric field are associated with the points of the primary treatment zone where the complementary electrode faces are closely spaced. The lower values of the electric field are associated with the points of the primary treatment zone where the complementary electrode faces are relatively further spaced. The electrode faces in the treatment zone were provided with a longitudinally scalloped face shape. The convoluted face shapes of each side are in complementary registration with the further extension of each face in axial position. As a result, the effective electrical resistance across the treatment chamber can be increased without reducing the processed fluid path length through the treatment zone, and hence reduces the power load on the circuitry used to drive the electrodes. Additionally, these specially contoured electrode faces aid agitation of the processed fluid (Barbosa-Cánovas et al. 2000).

5.3.2.7 Designs Based on the Shape of the Insulation Spacer Separating the Electrodes

Lindgren et al. (2002) modeled and optimized the electric field strength distribution in a co-flow treatment chamber using a finite element method and genetic algorithms. The optimization was performed with respect to the shape of the insulating space that separates the electrodes from each other. The model was based on the PEF treatment chamber design described by Yin et al. (1997). Local enhancements of the electric field strength were found in some designs of the insulation spacer. This may arise as the transition from electrode to insulator approaches 180°. The local enhancement arises because equipotential lines and electric field lines are mutually perpendicular

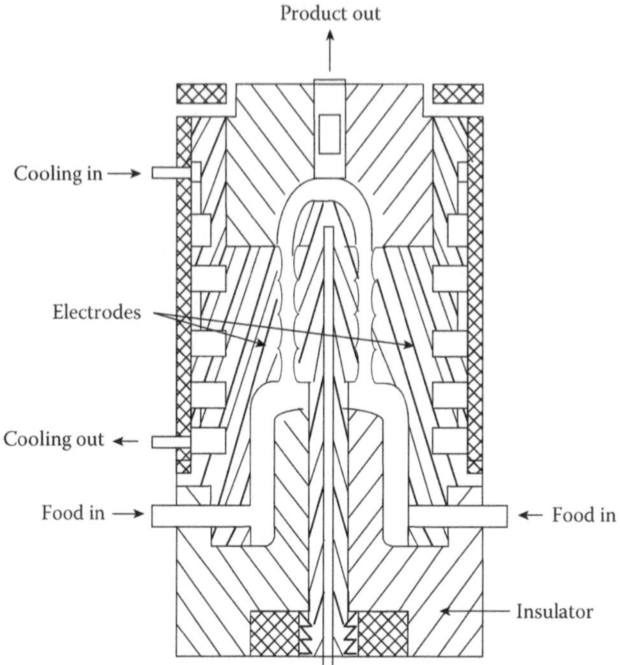

FIGURE 5.16 A longitudinal sectional view of a treatment cell. (Reproduced from Barbosa-Cánovas, G. V., Qin, B.-L, Zhang, H., Olsen, R. G., Swanson, B. G., and Pedrow, P. D., US Patent 6019031, 2000.)

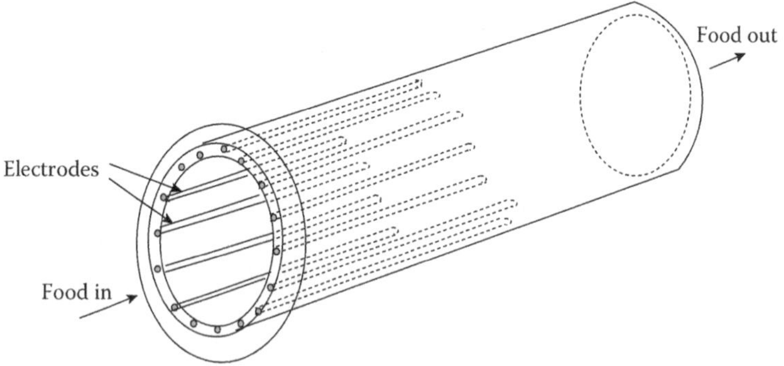

FIGURE 5.17 PEF treatment chamber designed by Morshuis et al. (Reproduced from Morshuis, P. H. F., Van Den Bosch, H. F. M., De Haan, S. W. H., and Ferreira, J. A., EP Patent 1123662, 2001.)

everywhere, a criteria that is hard to fulfilin intersections between electrode and insulator greater than 90°. The reason is that the equipotential lines also tend to be parallel to the electrode surfaces and electric field lines (lines of force) tend to be parallel with the insulator surfaces. So, the electric field strength at the interception of an insulator and an electrode will deviate from the assumed electric field strength as the angle if interception deviates from 90°. As a "rule of thumb" when designing PEF treatment chambers, 90° angles should be used wherever an electrode intersects with an insulator (Lindgren et al. 2002).

5.3.2.8 Designs Based on the Effective Area of Flow

Morshuis et al. (2001) provided a treatment chamber without corners, or alternatively with rounded corners, so that no contaminants can accumulate on the walls. The chamber (Figure 5.17) consists of a number of electrodes having a crescent-shaped cross-section of which one side has a radius of curvature that matches the radius of curvature of the effective area of flow. The electrodes were disposed in such away that the field lines of the electric field run parallel to one another and the potential controller is of such a design that the electric field in the effective area of flow is uniform.

5.3.3 Temperature Control

The change of temperature during PEF processing should be monitored and controlled to maintain the process close to nonthermal operation. The difference between the inlet and outlet temperature is due to heat generation as a result of ohmic heating.

Different cooling systems were reported in the previously discussed designs. To maintain the fluid at the designated temperature during processing, intermediate cooling between chambers must be used. In order to maintain low electrode temperature, cooling water could be circulated in cavities around the electrodes. The temperature can be measured using fiber optic temperature sensors to control the temperature during processing as these sensors are immune to electromagnetic, electrical, microwave, and radio-frequency fields. The PEF system constructed in the University of Auckland consists of fiber optic temperature sensors that have a fast response time (less than 1 msec), very small size (diameter of 150 μm), and are capable of sampling at rates up to 20 Hz.

The fluid food product must be always cooled to refrigeration temperature (5–9°C) using a heat exchanger following the electric field treatment, to prevent cell wall repair of the microorganisms being treated.

5.3.4 Degassing Unit

A PEF system should include a degassing unit to reduce the possibility of dielectric breakdown within the fluid food during processing, by eliminating the gas bubble formation. The presence of gas bubbles in the liquid food causes the electric field magnitude to decrease significantly near the boundary of the bubble, thus intimidating the uniformity of the PEF treatment across the chamber gap. The liquid food may be subjected to vacuum conditions of at least about 20 inches of mercury in order to remove dissolved gases and/or product bubbles (Dunn and Pearlman 1987).

Pressurizing the fluid food during processing can also eliminate bubble formation. A pressure-release device must be included in the treatment chamber design to ensure safety of the operation.

5.4 CONCLUSIONS

The design and construction of a PEF unit for food processing requires both state-of-the-art equipment and common sense. In order to achieve efficient treatment, a number of factors must be considered. One of the main factors is the proper design of the treatment chamber(s), where uniform electric field can be maintained so that the actual applied field strength dose not exceed the dielectric strength of the fluid foods under the test conditions. The energy required for the treatment may be reduced by a proper chamber design that can concentrate the electric field in the treatment region. Proper design of the insulator separating the electrodes and control of the temperature change during processing should be considered in designing the PEF treatment chamber.

ACKNOWLEDGMENTS

The authors would like to acknowledge the funding and support of the Fonterra Research Centre and the Foundation of Research Science and Technology.

REFERENCES

Alkhafaji, S., and Farid, M. 2007. An investigation on pulsed electric fields technology using new treatment design. *Innovative Food Science and Emerging Technologies* 8 (2), 205–12.

Álvarez, I., Raso, J., Sala, F. J., and Condón, S. 2003. Inactivation of Yersinia enterocolitica by pulsed electric fields. *Food Microbiology* 20 (6), 691–700.

Barbosa-Cánovas, G. V., Gongora-Nieto, M. M., Pothakamury, U. R., and Swanson, B. G. 1999. In *Preservation of Foods with Pulsed Electric Fields*, ed. Taylor, S. T. San Diego: Academic Press.

Barbosa-Cánovas, G. V., Qin, B.-L, Zhang, H., Olsen, R. G., Swanson, B. G., and Pedrow, P. D. 2000. Continuous flow electrical treatment of flowable food products. US Patent 6019031.

Barbosa-Cánovas, G. V., and Zhang, Q. H. 2001. In *Pulsed Electric Fields in Food Processing*. Washington, DC: Technomic.

Bartos, F. J. 2000. Medium voltage AC drives: Addendum and applications. *Control Engineering*, CE. International, online feature artical.

Boulart, J. 1983. Process for protecting a fluid product and installations for the realization of that process. French Patent 2,513,087 (in French).

Bushnell, A. H., Dunn, J. E., Clark, R. W., and Pearlman, J. S. 1993. High pulsed voltage systems for extending the shelf life of pumpable food products. US Patent 5235905.

Bushnell, A. H., Dunn, J. E., Lloyd, S. W., and Clark, R. W. 1996. Process for reducing levels of microorganisms in pumpable food products using a high pulsed voltage system. US Patent 5514391.

Calderon-Miranda, M. L., Raso, J., Gongora-Nieto, M., and Barbosa-Cánovas, G. V. 1999. *Minimal Methods of Processing*. Academic Press, USA.

Cornelis, M. H., and Vincent, B. P. 2000. Integrated modular design of a pulsed electric field treatment chamber. EP Patent 1000554.

De Jong, P., and Van Heesch, E. J. M. 1999. Pulsed electric field treatment system. US Patent 2001013467.

Doevenspeck, H. 1960. Verfahren and vorrichtung zur gewinnung der einzeknen phasen nus dispersen systemen. German Patent 1,237,541.

Dunn, J. E., and Pearlman, J. S. 1987. Methods and apparatus for extending the shelf life of fluid food products. US Patent 4695472.

Grahl, T., and Märkl, H. 1996. Killing of microorganisms by pulsed electric fields. *Applied Microbiology and Biotechnology* 45, 148–57.

Hamilton, W. A., and Sale, A. J. H. 1967. Effects of high electric fields on microorganisms. II. Mechanism of action of the lethal effect. *Biochimica et Biophysica Acta* 148, 789–800.

Ho, S. Y., Mittal, G. S., Cross, J. D., and Griffiths, M. N. 1995. Inactivation of *Pseudomonas fluorescens* by high voltage electric field pulse. *Journal of Food Science* 60 (6), 1337–43.

Hofmann, G. A., and Evens, G. A. 1986. Electronic genetic physical and biological aspects of cellular electromanipulation. *IEE Engineering in Medicine and Biology Magazine* 5, 6–25.

Hulsheger, H., Potel, J., and Niemann, E. G. 1983. Electric field effects on bacteria and yeast cells. *Radiation and Environmental Biophysics* 22, 149–62.

Lindgren, M., Aronsson, K., Ohlsson, T., and Galt, S. 2002. Simulation of the temperature increase in pulsed electric field (PEF) continuous flow treatment chambers. *Innovative Food Science & Emerging Technologies* 3, 233–45.

Matsumoto, Y., Satake, T., Shioji, N., and Sakuma, A. 1991. Inactivation of microorganisms by pulsed high voltage applications. *Conference Record of IEEE Industrial Applications Society Annual Meeting*, 28 Sept–4 Oct, Vol. 1, 652–59.

Morshuis, P. H. F., Van Den Bosch, H. F. M., De Haan, S. W. H., and Ferreira, J. A. 2001. Apparatus and method for preserving food products in a pulsed electric field. EP Patent 1123662.

Qin, B.L. 1995. Food pasteurization using high-intensity pulsed electric fields. *Food Technology* 49, 55.

Qin, B., Zhang, Q., Barbosa-Cánovas, G. V., Swanson, B. G., and Pedrow, P. D. 1994. Inactivation of microorganisms by pulsed electric field of different voltage wave-forms. *IEEE Transactions on Dielectrics and Electrical Insulation* 1 (6), 1047–57.

Qiu, X., and Zhang, H. Q. 2001. High voltage pulse generator. US Patent 6214297.

Sale, A. J. H., and Hamilton, W. A. 1967. Effects of high electric fields on microorganisms. 1. Killing of bacteria and yeasts. *Biochemica et Biophysica Acta* 148, 781–88.

Sato, M., and Kawata, H. 1991. Pasteurisation method for liquid foodstuffs. Japanese Patent 3,98,565 (in Japanese).

Sensoy, I., Zhang, Q. H., and Sastry, S. K. 1996. Inactivation kinetics of Salmonella dublin by pulsed electric field. *Journal of Food Process Engineering* 20 (5), 367–81.

Yin, Y. G, Sastry, S. K., and Zhang, Q. H. 1997. High voltage pulsed electric field treatment chambers for the preservation of liquid food products. US Patent 5690978.

Zhang, Q. H., Qui, X., and Sharma, S. K. 1997. Recent developments in pulsed electric processing. In *New Technologies Yearbook*, ed. Chandrana, D. I. Washington, DC: National Food Processors Association, 31–42.

[illegible]

6 Effect of Ultrasound on Food Processing

Jordi Salazar, Juan A. Chávez, Antoni Turó, and Miguel J. García-Hernández
Universitat Politècnica de Catalunya

CONTENTS

6.1 INTRODUCTION

Nowadays, there is growing interest in foods that are minimally preserved and processed. Consumers are demanding healthier, more natural foods that are less processed, and have less added preservatives and a longer shelf life without diminishing their nutritional properties. Consequently, there is currently much emphasis on the development of novel and emerging technologies for minimal preservation and processing methods in contrast to most conventional methods that are by far based on the use of thermal energy, which may alter food properties.

As a nonthermal technology, power ultrasound is attracting considerable interest in the food industry. By means of mechanical vibrations of high enough intensity, power ultrasound can produce changes in food either by disrupting its structure or promoting certain chemical reactions. The range of applications of power ultrasound in the food industry is vast and is growing rapidly. Some examples include emulsification, one of the earliest applications, drying, degassing, and inactivation of microorganisms. The use of this technology provides some valuable benefits such as reduced processing and maintenance costs, and shorter processing times. However, there is very little information about the effects of this technology on sensory and nutritional food properties.

After a review of selected power ultrasound applications for food processing and preservation, the principles of how power ultrasound works, and an update on ultrasonic systems, this chapter focuses on the effects of power ultrasound on food properties such as organoleptic, nutritional, and functional properties. These effects will be illustrated through examples of foods.

6.2 PRINCIPLES OF POWER ULTRASOUND

6.2.1 Ultrasound in the Food Industry

Ultrasound refers to sound waves, mechanical vibrations, which propagate through solids, liquids, or gases with a frequency greater than the upper limit of human hearing. Although this limit can vary from person to person, the ultrasonic frequency range is considered to be at frequencies over 20 kHz. The upper limit of the frequency range of ultrasound is mainly limited by the ability to generate ultrasonic signals.

Although ultrasound has recently attracted considerable interest in the food industry for both the analysis and processing of foods, ultrasonic techniques have been available in the industry since World War II (Povey and McClements 1988). Typically ultrasound applications are divided into two broad categories: high- and low-intensity applications. Low-intensity applications are mainly characterized by frequencies above 100 kHz with energies below 1 W/cm^2. In addition, the ultrasonic wave does not have any significant effect on the material being tested, in contrast to high-intensity applications. In the food industry, low-intensity ultrasound is used as an analytical technique either to control a process or to obtain information about different physicochemical properties such as air bubbles in aerated foods, ratio of fat in meats, vegetable and fruit characterization, quality of eggs, cracks in cheese, texture of biscuits, milk coagulation, wine fermentation control, or dough characterization (Povey and McClements 1989; Contreras et al. 1992; McClements 1997; Benedito et al. 2002; Simal et al. 2003; Salazar et al. 2004; Resa et al. 2007; Álava et al. 2007; Mizrach 2008).

On the other hand, high-intensity or power ultrasound applications tend to use frequencies below 100 kHz with energies above 10 W/cm^2. These applications are intended to have an effect on the material being tested generally by generation of intense cavitation (Mason and Lorimer 1988). Cavitation is an important physical phenomenon of high-intensity ultrasound and involves the formation, growth, and implosive collapse of bubbles in liquid. Some typical examples include disruption of biological cells, emulsifying, drying, mixing materials, or microbial inactivation. In addition, many of the applications of power ultrasound are not exclusively ultrasonic processes but ultrasonically assisted processes.

6.2.2 Cavitation

Several mechanisms can be activated by power ultrasound such as heating, turbulence, agitation, friction, surface instability, and others. Although not all the mechanisms involved in power ultrasound are known or well understood, in the food industry most of them can be attributed to a very complex nonlinear phenomenon known as cavitation. The phenomenon of cavitation was first reported in 1895 by Thornycroft and Barnaby (1895). They postulated that the reason for the poor performance of a new torpedo was the inefficiency and loss of power caused by the formation of cavities in the water.

Power ultrasound enhances chemical and physical changes in a liquid medium through the generation and subsequent destruction of cavitation bubbles. When a liquid is subjected to sufficiently powerful ultrasound, the liquid is alternately compressed and expanded, forming small bubbles or cavities. These bubbles react and grow with the compression and expansion cycles of the ultrasonic wave causing them to expand and when they attain a critical size they eventually collapse. This process is depicted in Figure 6.1. The collapse of bubbles can be so violent that it can cause considerable damage to the surrounding medium. The gas in the bubble has been estimated to reach temperatures around 5000°C and pressures of more than 1000 atm on a nanosecond timescale (Suslick and Price 1999). This intense local heating or hot spots can drive significant gas phase chemical reactions,

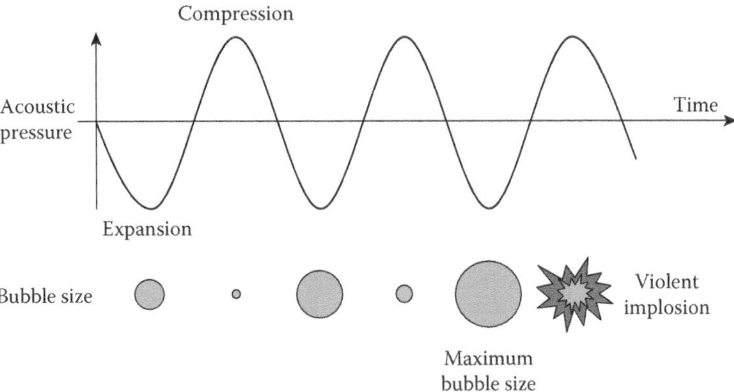

FIGURE 6.1 Cavitation bubble collapse.

which are important in a variety of applications (Gould 2001). However, this heat is transient and very localized, so the overall temperature increase in the matrix is relatively low, around 5°C. When power ultrasound is used in combination with conventional heating, the effect of ultrasound treatment increases.

6.3 POWER ULTRASOUND APPLICATIONS IN FOOD PROCESSING

The use of power ultrasound in the food industry is a promising area. Ultrasound technology has a wide range of current and future applications. Table 6.1 shows a list summarizing current and potential power ultrasound applications and their benefits in different fields of the food industry.

However, despite the large number of potential ultrasonic processes, most of them are still restricted to a research environment and few of them have been introduced in the industry. The main reason is that power ultrasound is not an off-the-shelf technology. Each application needs to be specifically developed and scaled up.

Herein an overview of major applications of power ultrasound in different fields is provided.

6.3.1 Microbial and Enzyme Inactivation

This application is a promising field that relies on the ability of power ultrasound to disrupt biological cell walls through intense cavitation. However, the effectiveness of ultrasound is dependent on the type of microorganism or enzyme being treated, amplitude and frequency of the ultrasonic signal, exposure time, volume of food to be processed, and type of food. In addition, power ultrasound alone is not very effective in the destruction of microorganisms or inactivation of enzymes unless very high intensities are used and, precisely for this reason, generally ultrasound is used in conjunction with another technique such as pressure (manosonication), heat (thermosonication), or both (manothermosonication), to achieve efficient destruction of microorganisms or inactivation of enzymes. Thus, for instance, using pressure during power ultrasound treatment increases the rate of microbial inactivation in a variety of microorganisms, even at temperatures well below the boiling point of the medium. Figure 6.2 shows the effectiveness of this process at a very low temperature when pressure varies from 0 to 400 kPa. All these processes have been studied and described in depth by several researchers (Ordoñez et al. 1984; Raso et al. 1998; Mañas et al. 2000; Mañas and Pagán 2005).

So far, research has been successfully done on a great number of bacteria such as *Escherichia coli*, *Listeria monocytogenes*, *Salmonella*, *Staphylococcus aureus*, *Saccharomyces cerevisiae*, and others (Piyasena, Mohareb, and McKellar 2003). Exposure time, temperature, and intensity

TABLE 6.1
List of Power Ultrasound Applications in the Food Industry

Application	Benefits
Microbial and enzyme inactivation	Destruction of microorganisms or inactivation of enzymes at lower temperatures
Crystallization	Formation of very small and uniform crystals with enhanced rate of seeding
Filtration	Higher filtration rates, filters cleaner and prevents from fouling
Drying	Increased product throughput at lower temperatures avoiding degradation of food
Extraction	Increased yield of extracted components, higher extraction efficiency and processing throughput
Emulsification	Online use for flow processes providing a very effective, stable, and homogenized mixing of two or more immiscible liquids without or with very few additives
Meat tenderization	Increased tenderness by disruption of myofibrillar components in shorter aging periods
Degassing	Rapid removal of unwanted air or gas in liquids
Defoaming	Enhanced degree of defoaming
Freezing	Reduced freezing time with smaller ice crystal size distribution leads to a product of better quality
Oxidation processes	Early maturation of alcoholic beverages

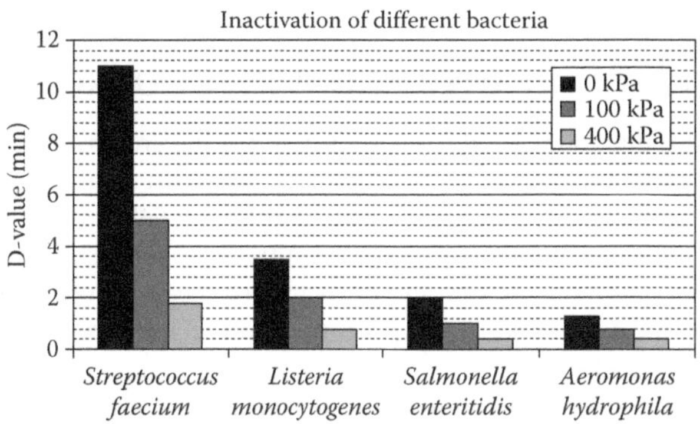

FIGURE 6.2 D-value of different bacteria treated by ultrasound (20 kHz, 117 μm amplitude) at 40°C when pressure varies from 0 to 400 kPa.

and frequency of the ultrasonic wave are key factors for successful destruction of microorganisms. Also, new areas where power ultrasound has not yet been applied show potential use, such as the winemaking industry. Jiranek et al. (2008) explore the application of power ultrasound in modulating microbial activity and load at various stages of the winemaking process in juice, musts, wines, and barrels. Thus, for instance, rather than washing fermentation tanks and barrels with high pressure and hot water, the use of power ultrasound could be more effective and quicker, with a lower water temperature.

Research under different conditions has also been carried out in order to inactivate different enzymes. Thakur and Nelson (1997) showed that pH of the medium was the most important parameter for the inactivation of lipoxygenase. Raviyan, Zhang, and Feng (2005) studied the effect of cavitation intensity and temperature on inactivation of tomato pectinmethylesterase (PME). A comparison between thermal, sonication, and thermosonication treatments under different conditions was conducted. Increasing the temperature increases the inactivation rate, which has been demonstrated in many tomato enzyme thermal inactivation tests (López et al. 1997; Crelier et al. 2001). From Figure 6.3, it is clear that the use of thermosonication increases the inactivation rate. In addition, taking into account that inactivation caused exclusively by thermal treatment at 50°C is

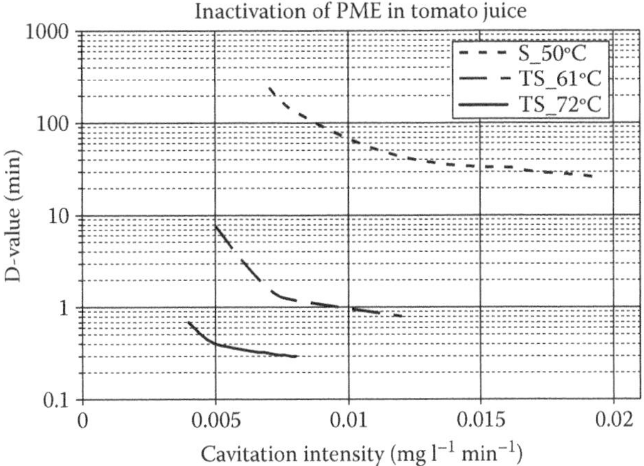

FIGURE 6.3 Inactivation of PME from tomato juice using sonication and thermosonication at different cavitation intensities and temperatures. S, sonicaton; TS, thermosonication.

negligible, inactivation by ultrasonic treatment at 50°C must be due to sonication itself. With reference to the different reported values of reduction in the residual activity at various temperatures by different researchers, the discrepancies might be caused by differences in tomato variety, degree of ripening, heating and temperature measurement techniques, and enzyme preparation and assay method.

Also, experiments under pressure at different temperatures were carried out by Mañas et al. (2006) in order to inactivate lysozyme. Ultrasonic waves at room temperature and pressure hardly inactivate the enzyme. However, the application of an external pressure of 200 kPa and temperatures between 60 and 80°C increased the inactivating effect of ultrasound.

It seems clear that the effect of ultrasound on enzymes is dependent on the type of the enzyme and also influenced by many parameters. In this respect, Kadkhodaee and Povey (2008) studied how and to what extent ultrasonic effects vary under different experimental conditions in order to develop a mathematical model that could quantify the contribution of variable parameters to the overall effect of ultrasound and consequently predict changes in the efficiency of the process. The results show that ultrasound effectively inactivated α-amylase with a minimum overall inactivation rate at 50°C. Also, it was shown that the contribution of the different parameters to the inactivation of the enzyme seems to be dominated by the temperature and the acoustic intensity, although the amino acid composition and the conformational structure of the enzyme also have a great influence.

6.3.2 Crystallization

The application of power ultrasound to the crystallization process was first reported more than 50 years ago, but it has been in the last decade that it has received much attention. The most important effect of ultrasound on crystallization is the induction of nucleation, also mainly based on the cavitation phenomena. During cavitation it is possible to break down crystals and nuclei into a large number of uniform small crystals that will act as nucleation centers. This effect is often referred to as sonocrystallization. A scheme of sonocrystallization by ultrasonic cavitation is shown in Figure 6.4. The crystallization process can be controlled by means of the amplitude and frequency of the ultrasonic signal and the exposure time, thus allowing manipulation of crystal size distribution as well as when the point crystallization occurs (Povey and Mason 1998).

The application of power ultrasound to crystallization has been found particularly appropriate for pharmaceuticals and fine chemicals, which are amongst the hardest materials to crystallize well

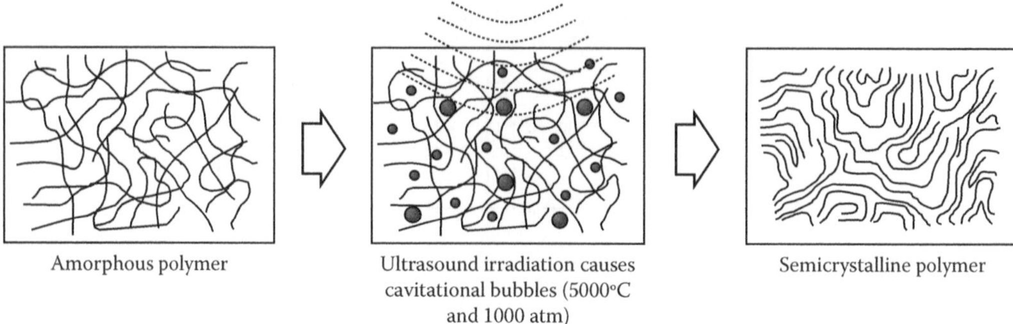

FIGURE 6.4 A scheme of sonocrystallization by ultrasonic cavitation.

TABLE 6.2
Ultrasonic Crystallization of Sugar Solutions

Solute	Quantity Dissolved in 10 ml of Water at 50°C (g)	Temperature at Which Solid Appeared (°C)	
		Control	Ultrasound Treatment
D-xylose	25	36	46
D-sucrose	18	<40	47
D-lactose	5.5	41	43
D-maltose	13	No crystals appeared at 20	40
D-cellubiose	2.0	No crystals appeared at 20	42

because they tend to be high-molecular-weight organics (Ruecroft et al. 2005). There are also some interesting applications in the food industry based on the modification and control of the crystallization process in many food products. One of these applications deals with the control of sugar crystallite size. This application is of particular interest since the texture of food products will be affected by the size of undissolved sugar crystals dispersed in the material (Mason, Paniwnyk, and Lorimer 1996). Table 6.2 gives some measured data on the metastable zone width reduction for a range of solutions cooled from 50°C to 20°C (Ruecroft et al. 2005). In all cases except D-lactose, the zone width was significantly reduced.

The application of power ultrasound in honey has long been studied (Kaloyereas 1955; Liebl 1978; Thrasyvoulou, Manikis, and Tselios 1994). Honey is a supersaturated sugar solution whose natural tendency is to crystallize. This is an undesirable process that must be avoided during the overall steps of the production process of liquid honey. Traditionally, crystallization of honey is delayed by applying a heat treatment. Heat helps to dissolve D-glucose monohydrate crystals but also negatively affects the delicate flavors of honey. Alternatively, the application of power ultrasound eliminates existing crystals in honey and also retards the crystallization process, resulting in a cost-effective technology. Analysis of crystallization shows that honey treated by power ultrasound remains in the liquid state for a much longer period than does heat-treated honey. In addition, no significant effects on honey's quality parameters, such as moisture content, electrical conductivity, or pH, are observed.

Crystallization of fats is another area that is gaining considerable interest in the chocolate, butter, and ice-cream industries. Thus, for instance, the effects of power ultrasound (20 kHz and 100–300 W) on tripalmitoylglycerol and cocoa butter have been studied (Higaki et al. 2001; Ueno et al. 2003). Results indicate that ultrasound irradiation is an efficient tool for controlling polymorphic crystallization of fats and induction time is shortened.

In order to explain the relationship between cavitation and nucleation, several theories have been proposed but the contribution of ultrasound to crystallization is still not fully understood. Therefore, further development of the model is still required.

6.3.3 Filtration

Solid–liquid filtration is a common process in many industries today. Filtration can be applied either to clarify a liquid from solids or to remove solids from a liquid. Ultrasound-assisted filtration has been widely studied in the past. Besides substantially increasing the rate of liquid flow through the filter, application of ultrasound also keeps the filter cleaner and prevents fouling (Fairbanks 1973; Semmelink 1973). Filtration enhancement relies on two effects. First, sonication causes agglomeration of fine particles, and secondly, sonication supplies enough vibratory energy to the system to keep the particles partly suspended and therefore leaves more free channels for solvent elution (Mason et al. 1996).

Ultrasound filtration has been applied to several sectors of the food and beverage industry. When applied to fruit extracts and drinks, this technique is used to increase the fruit juice extracted from pulp (Mason et al. 1996). Other potential applications include sugar, beer, wine, and edible oils (Brennan, Grandison, and Lewis 2006). Thus, for instance, the juice produced by extraction from sugar beet is treated with lime and filtered to produce a clear juice for further processing. Beer is clarified by filtration to remove deposits of yeast and trub formed on the bottom of the maturation tank during the maturation stage. In addition, wine and edible oils are filtered at different stages of production.

In an extensive review, Kyllönen, Pirkonen, and Nyström (2005) studied several parameters that influence the effectiveness of ultrasound filtration such as frequency, power intensity, feed properties, cross-flow velocity, and temperature. In general, higher filtration efficiencies are achieved with lower ultrasound frequencies and high-power intensities using intermittent ultrasonic sonication. However, ultrasound is less effective for highly concentrated suspensions such as slurries. In this sense, one of the current requirements in many industrial applications is the processing of highly concentrated suspensions of fine particles. Conventional filtration techniques are not satisfactory because fouling often occurs, resulting in slow filtration rates. Gallego-Juárez, Elvira-Segura, and Rodríguez-Corral (2003) have been working on the development and application at pilot scale of a power ultrasonic technology to assist cake deliquoring in the filtration of concentrated fine particle slurries in rotary vacuum disk filters. One or various power transducers are placed parallel to the filter surface and very close to it, obtaining an increase in the filtration efficiency with two extremely fine highly concentrated suspensions of particulate materials. The experimental ultrasonic system is schematically depicted in Figure 6.5.

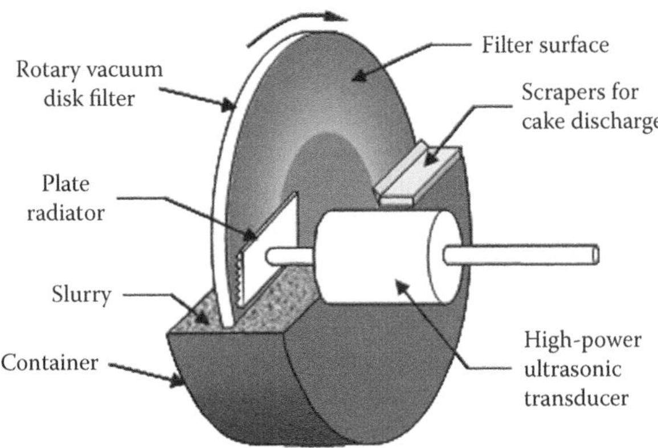

FIGURE 6.5 Schematic diagram of the ultrasonic assisted filtration system.

6.3.4 Drying

Conventional systems for food dehydration are mainly based on either hot air drying or freeze-drying. Hot air drying is a widely used method but the food product can be deteriorated. In freeze-drying product deterioration is negligible but the process is expensive. In contrast to conventional dehydration methods, power ultrasound reduces the costs of the process and the probability of food degradation since the drying process can be carried out at lower temperatures.

Ultrasonic dehydration is still under development, the low drying rates of the ultrasonic process being one of its major drawbacks. However, there has been a lot of interest in recent years in contributing to the development of this potential application.

When power ultrasound is directly coupled to the food material to be dried, a rapid series of alternative compressions and expansions produce a kind of sponge effect and the quick migration of moisture from the product (Floros and Liang 1994; Gallego-Juárez et al. 2007). In addition, ultrasound produces cavitation, which may be beneficial for the removal of moisture that is strongly attached (Tarleton and Wakeman 1998; Fuente-Blanco et al. 2006).

Power ultrasound has been extensively applied as a pretreatment to dry fruits and vegetables such as apples, melons, bananas, papayas, cauliflowers, or carrots. In all cases, the drying time after ultrasound treatment is shortened as compared to untreated products. The effect of pretreatments and drying conditions on the structure of products such as mushrooms, Brussels sprouts, and cauliflowers has been investigated in order to see how later rehydration is influenced (Jambrak et al. 2007a). When compared with the freeze-drying method (reference method), results indicate that the rehydration properties (weight gain, %) are found to be the best for freeze-dried samples. However, the rehydration properties for ultrasound-treated samples are higher than for untreated samples, so the rehydration behavior of plant food can also be enhanced by the use of power ultrasound.

Fernandes and Rodrigues (2007), Fernandes, Oliveira, and Rodrigues (2007), and Fernandes, Gallão, and Rodrigues (2008) have investigated the effect of ultrasonic pretreatment prior to air-drying on dehydration of bananas, papayas, and melons. In all cases, the integrated process (power ultrasound and air-drying) was optimized searching for the operating condition that minimizes total processing time. Time reductions of between 10 and 16% are achieved when ultrasound pretreatment is applied for approximately 20 min.

With the aim to scale up the drying process, during the last decade Gallego-Juárez et al. (1999) have focused their efforts on developing a new technology of airborne power ultrasound for vegetable dehydration that applies ultrasonic vibration directly coupled to the product, avoiding the usual ultrasonic bath. Two experimental procedures have been developed: forced-air drying assisted by airborne ultrasound and ultrasonic dehydration by applying ultrasound in direct contact with the material. In addition, a parametric study of the relative influence of the main physical parameters involved in the process has been carried out with a stepped plate transducer working at 20 kHz and a power capacity of about 100 W (García-Pérez et al. 2006) in order to establish the starting points needed for the development of the system at a preindustrial stage. Experimental results in carrots, apples, and potatoes as well as an extensive review of this new technology can be found in Fuente-Blanco et al. (2006) and Gallego-Juárez et al. (2007). Results obtained with direct contact show that the drying effect is a very effective method for food dehydration. However, the improvement obtained when applying high-intensity airborne ultrasound is less than that for direct contact application. The main reason for this is the low penetration of ultrasonic energy in the vegetable material produced by the mismatch between the acoustic impedance of transducer-air-food.

6.3.5 Extraction

Ultrasonic extraction is based on cavitation phenomena and has become an efficient alternative to traditional methods for solvent extraction. Cavitation breaks down biological cell walls and releases cell contents into the solvent (Mason et al. 1996). Thus, the extraction of sugar from sugar beets has been found to be improved with the use of power ultrasound (Chendke and Fogler 1975). Wang (1975) studied protein extraction from defatted soybeans using a 550 W probe operating at a 20 kHz frequency, which resulted in a more efficient way to extract than any other previously technology. Later, the experiment was scaled up to pilot plant level for the extraction of soybean protein (Moulton and Wang 1982).

More recently, an ultrasonic extraction procedure for Ca, K, and Mg from in vitro embryogenic and nonembryogenic *Citrus* sp. cultures was proposed by Arruda, Rodriguez, and Arruda (2003). During the experiments, different extracting media and sonication times of 5, 10, 15, and 30 min were used to optimize the ultrasonic extraction procedure, which resulted in an excellent alternative with an important reduction of sample handling and operational costs. Albu et al. (2004) used power ultrasound to increase the extraction efficiency of carnosic acid from the herb *Rosmarinus officinalis* using butanone, ethyl acetate, and ethanol as solvents. Power ultrasound reduces the dependence on the extraction solvent and greatly enhances the performance of ethanol, a poor conventional extraction solvent. In addition, sonication appears to have great potential as a method for the extraction of antioxidant materials with comparable levels of extracted carnosic acid obtained, employing an ultrasonic bath or probe system thus indicating a potential for scale up of the extraction process. Another interesting application is the extraction of phenolic compounds from coconut shell powder (Rodrigues and Pinto 2007). The effects of toasting time, toasting temperature, and extraction time are evaluated. The results indicate that high amounts of phenolics can be extracted from coconut shell by ultrasound-assisted extraction technology, the extraction time being the most significant parameter for the process. The best condition to obtain high phenolic contents (406.93 mg/l) in the extracts was found to be toasting time of 60 min at 100°C and ultrasound extraction time of 60 min.

Table 6.3 describes a list of ultrasound-assisted extraction applications. However, a more extensive review of power ultrasound extraction applications is done by Vilkhu et al. (2008), including herbs, oils, proteins, and bioactives.

6.3.6 Emulsification

One of the first applications of power ultrasound was in emulsification. It was first reported by Wood and Loomis (1927). This application, also based on cavitation, provides a very effective, stable, and homogenized mixing of two or more immiscible liquids without or with very few additives (Shoh 1975; Mason et al. 1996). Emulsification's great advantage is that it can be used online for flow processes. Thus, volumes up to 12,000 l/h can be processed, as is the case in the manufacture of fruit juices, tomato ketchup, and mayonnaise (Mason et al. 1996).

The effects of power ultrasound on milk homogenization have been studied by various authors (Villamiel and de Jong 2000; Wu, Hulbert, and Mount 2001; Ertugay, Şengül, and Şengül 2004). It was found that sonication of fresh cow milk at 20 kHz resulted in a reduction in the size of fat globules. They have found that power ultrasound is very effective in reducing fat globule size. An average size of the fat globules of less than 1 μm was achieved. Also, the application of power ultrasound results in more uniformly distributed fat particles than the conventional homogenization method. Thus, milk homogenized by the conventional homogenizer has smaller fat globules than nonhomogenized milk and milk homogenized by power ultrasound also has smaller fat globules than milk homogenized by a conventional homogenizer. Figure 6.6 compares the average fat globule diameter of nonhomogenized milk with the diameter achieved after conventional homogenization and under different ultrasonic treatment conditions. The fat globule diameter obtained in conventional homogenization could be achieved by means of ultrasonic emulsification at a power level of 180 W for 10 min (Villamiel and de Jong 2000). In general, a substantial reduction in fat globule size, up to 81.5%, is achieved after all treatments.

TABLE 6.3
Ultrasound-Assisted Extraction Applications and Their Experimental Results

Product	Ultrasound System	Solvent	Results	Reference
Phenolic compounds from coconut shell powder	Batch, 25 kHz	Water and ethanol	The extraction of phenolic content was mainly affected by toasting time and extraction time	Rodrigues and Pinto (2008)
Hesperidin from Penggan peel	Batch, 20 kHz, 60 kHz and 100 kHz	Methanol. Ethanol and isopropanol	Shorter extraction time (up to 8 times) with lower temperatures over conventional method	Ma et al. (2008)
Pungent compounds from ginger	Batch, 20 kHz	Supercritical CO_2	Up to 30% increased yield	Balachandran et al. (2006)
Almond oils	Batch, 20 kHz	Supercritical CO_2	Up to 20% increased yield and similar yields in about 30% shorter time	Riera et al. (2004)
Carnosic acid from rosemary	Batch, 20 kHz and 40 kHz	Butanone and ethyl acetate	Reduction in extraction time	Albu et al. (2004)
Soy isoflavones	Batch, 24 kHz	Ethanol, methyl cyanide and methanol	Up to 15% increase	Rostagno et al. (2003)
Ca, K and Mg from in vitro citrus culture	Batch, 47 kHz	Water	Remarkable efficiency, handless manipulation and reduced operational costs	Arruda et al. (2003)
Rutin from flower buds of *Sophora japonica*	Batch, 20 kHz	Water and methanol	Up to 20% increase in 30 min when using methanol over conventional method	Paniwynk et al. (2001)
Herbal extracts	Stirred batch, 20 to 1000 kHz	Water and ethanol	Up to 34% increased yield over stirred	Vinatoru (2001)
Soy protein	Continuous, 20 kHz	Water and alkali	53% and 23% yield increase over equivalent ultrasonic batch conditions	Moulton and Wang (1982)

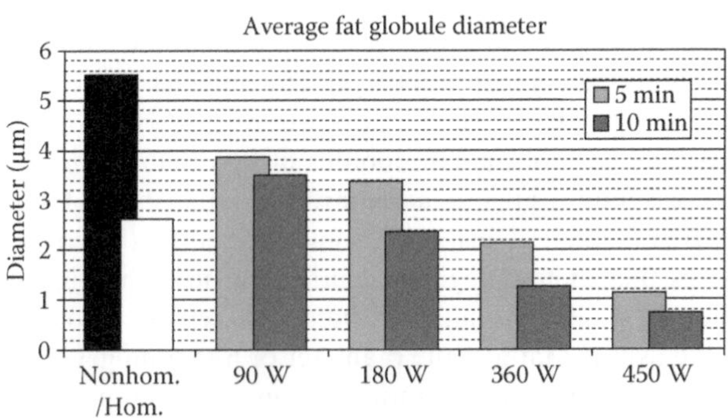

FIGURE 6.6 Effect of ultrasonic homogenization on fat globule diameter of milk (20 kHz ultrasonic probe). As a function of power level (W) and exposure time (min). First two bars refer to nonhomogenized milk (Nonhom.) and milk homogenized by conventional homogenizer (Hom.).

Wu et al. (2001) also studied the effect of power ultrasound on yogurt fermentation. Ultrasound treatment before inoculation of yogurt starter results in an increase of water-holding capacity. In addition, when ultrasonic treatment is applied after inoculation, the total fermentation time is reduced by 0.5 h, which entails clear industrial benefits.

In a more recent work, the effect of irradiation time, irradiation power, and physicochemical properties of oil on the dispersed phase volume and dispersed phase droplet size have been studied in typical emulsions consisting of oil and water (Gaikwad and Pandit 2008). Oils with different viscosity are used in the experiments. In particular, two edible oils and two mineral oils are used. Results indicate that with an increase in the irradiation time, the dispersed phase volume increases while the dispersed phase droplet size decreases. Likewise, as the irradiation power is increased, there is an increase in the hold-up of the dispersed phase while the droplet size of the dispersed phase decreases. In addition, the higher viscosity of the oil is related to a larger droplet size in the emulsion.

6.3.7 Meat Tenderization

Tenderness is one of the most important features of meat quality perceived by the consumer. Meat tenderness varies considerably among species, among animals within a species, and among different muscles, but is highly affected by the way it is cooked as well. In general, meat is tender just after slaughter, however, tenderness is negatively affected by changes that occur during *rigor mortis* and *post-rigor* ageing. Traditionally, ageing has been regarded as the failsafe method of producing tender meat. Carcasses are hung for several days at a temperature just above freezing to allow meat to tenderize, but this process adds a considerable cost to production because of the requirements for space and controlled temperatures.

When using power ultrasound, by means of cavitation effects, muscle and cell structure of the tissue can be disrupted, leading to tenderization of meat. However, existing studies are somehow contradictory since ultrasound treatment conditions vary widely and, consequently, effects on meat tenderization vary as well. In some cases power ultrasound proved effective although in a number of other studies the treatment conditions were found to have no effect or even a negative effect on tenderness. Table 6.4 summarizes experimental results from some studies on meat tenderization.

Smith et al. (1991) suggested that tenderization of meat samples (semitendinosus muscle from cattle) by the application of power ultrasound is of sufficient magnitude to be detected by the average consumer. They found a significant increase in tenderness after 2–4 min of sonication. Nevertheless, opposite effects were observed when meat was sonicated for more than 8 min, but the reasons for this behavior were not explained by the authors. In two consecutive studies published by Lyng et al. (1997) and Lyng, Allen, and McKenna (1998a) using longissimus thoracis et lumborum and semimembranosus muscles, no significant effects on tenderness due to power ultrasound treatment have been found compared to untreated samples. According to the authors the reason could be the method of treatment as well as the ultrasound intensity levels and exposure times. These factors, as well as the use of different frequencies and type of muscles used, might have caused such inconsistent results.

The influence of high-frequency power ultrasound was also investigated (Got et al. 1999). The authors applied a 2.6 MHz transducer to semimembranosus muscles but no conclusive effects on meat tenderization were observed. One of the reasons could be that in the study the benefits from cavitation were not achieved since cavitation is unlikely at such high frequencies.

Recently, Jayasooriya et al. (2004) reviewed the effects and potential benefits of high-power ultrasound on physical, biochemical, and microbial properties of meat. The specific effects on meat tenderness were discussed. Power ultrasound appears to be a very attractive alternative technique for modifying properties of meat and meat products, although further research is needed. In this sense, Jayasooriya et al. (2007) studied the effect of power ultrasound and ageing on the physical properties of bovine semitendinosus and longissimus muscles in order to determine the exposure

TABLE 6.4
Experimental Results From Some Studies on Meat Tenderization by Power Ultrasound

Muscle	Ultrasound Conditions	Results	Reference
Beef semitendinosus	25.9 kHz, 1000 W and 0, 2, 4, 8, 16 min treatment	Significant decrease in shear force at 2 and 4 min of exposure, while shear force increases after 8 min of exposure.	Smith et al. (1991)
Chicken muscle	40 kHz, 2400 W and 15 min treatment	Significant reduction of shear force.	Dickens et al. (1991)
Lamb longissimus thoracis et lumborum	20 kHz, 56–62 W/cm² and 10–180 sec treatment	Ultrasonic treatment of fibers enhanced proteolytic activity. Cell damage was not very extensive, although it depended upon ultrasonication and fiber conditions.	Roncalés et al. (1993)
Beef longissimus thoracis et lumborum, semimembranosus, and biceps femoris	30–47 kHz, 0.62–0.29 W/cm² and 30 min per side	Peak force values were not affected by ultrasound treatment.	Lyng et al. (1997)
Beef longissimus thoracis et lumborum and semimembranosus	20 kHz, 62 W/cm² and 15 sec treatment	No significant effects on tenderness were found.	Lyng et al. (1998a)
Lamb longissimus thoracis et lumborum	20 kHz, 63 W/cm² and 15 sec treatment	Bite-force tenderometry, sensory analysis, and collagen solubility all showed no effect due to ultrasound treatment	Lyng et al. (1998b)
Beef semimembranosus	2.6 MHz, 10 W/cm² and two consecutive periods of 15 sec	No conclusive effects on the post mortem rate of tenderization of meat.	Got et al. (1999)
Beef longissimus thoracis et lumborum and semimembranosus	24 kHz, 12 W/cm² and 0, 30, 60, 120, 240 sec treatment	Improvement in tenderness without detrimental effects on drip, cook or total losses, and color. Benefits obtained decrease with increasing aging time.	Jayasooriya et al. (2007)
Beef semimembranosus	45 kHz, 2 W/cm² and 120 sec treatment	High water-holding capacity of sonicated samples suggests an acceleration of rigor mortis followed by a fragmentation of proteins structures of cell.	Stadnik et al. (2008)
Pork longissimus dorsi	20 kHz, 2–4 W/cm² and 30–180 min treatment	Water-holding capacity and textural properties were improved by ultrasonic treatment. However, higher intensities and/or longer treatment times caused denaturation of proteins.	Siró et al. (2009)

time needed to tenderize different type of muscles and its impact on ageing as well as to determine if power ultrasound negatively affects other meat characteristics such as color or pH. After 60 sec of power ultrasound treatment (24 kHz and 12 W/cm^2), reduction of Warner-Bratzler shear force and hardness is comparable to that of meat aged for three to five days. The benefits of ultrasound are reduced as ageing time increases. In addition, the improvement in tenderness is achieved without detrimental effects on drip, cook, or total losses, and color.

Similarly, Stadnik, Dolatowski, and Baranowska (2008) studied whether low-frequency and relatively low-intensity ultrasound (45 kHz, 2 W/cm^2) treatment at 24 h post mortem affected the pH of meat and its water-holding capacity during 96 h of meat ageing. Ultrasound treatment does not have a statistically significant effect on the pH of the meat. In addition, the high water-holding capacity of sonicated samples, typical of meat in an advanced post mortem stage, suggests that due to ultrasound treatment acceleration of rigor mortis followed by a fragmentation of protein structures of cells occurred.

6.4 UNDESIRED EFFECTS OF POWER ULTRASOUND ON FOOD PROPERTIES

In the previous section, the benefits of power ultrasound have been reviewed through numerous different applications such as homogenization of milk, destruction of microorganisms, extraction of fruit juice from pulp, control of the crystallization process, and meat tenderization. In all these applications, mainly based on the cavitation phenomena, power ultrasound has contributed positively either to the efficiency of the process, generally resulting in higher product yields, reduced costs, and shorter processing times, or enhanced food properties.

Although extensive research have been carried out over many years, current applications of power ultrasound alone or in combination with another technology have been developed under the premise that ultrasound is a minimal processing and nonthermal technology that helps in modification of functional properties of food while keeping food quality attributes, so flavors, essential nutrients, and vitamins undergo minimal or no changes. Also, few studies concluded that under the conditions of treatment no beneficial effects on the food product were observed. Until now, research efforts have been focused on developing this new technology and seeing whether its application can be extended to new products without paying much attention to identifying, if any, the negative effects associated with the application of power ultrasound to food products. However, limited research has reported some negative effects. This is now a matter of study and therefore there is little data on these undesirable effects.

6.4.1 Formation of Hydroxyl Radicals

Despite the many benefits of power ultrasound, the high enough temperatures and pressures achieved during cavitation are thought to be responsible for the generation of hydrogen atoms and hydroxyl radicals, which can also bring about undesired chemical reactions that negatively affect the nutritional food properties.

Ashokkumar et al. (2008) observed, in their experiments at different frequencies, that the amount of hydroxyl radicals could be minimized by selecting very low frequencies. Thus, the amount of hydroxyl radicals generated was minimal at 20 kHz. At this frequency the cavitation bubbles are primarily transient, whereas stable bubbles are generated at higher frequencies. Both stable cavitation and an increase in the number of active bubbles can be expected to increase the amount of hydroxyl radicals generated with an increase in the ultrasound frequency. This could explain the observed increase in the hydroxyl radical yield when the frequency was increased from 20 to 358 kHz. However, it could not explain the decrease in the hydroxyl radical yield when the frequency was further increased to 1062 kHz. This decrease may be due to the relatively less time available during the expansion phase of the bubble growth. At relatively high frequencies, the acoustic cycle becomes very short, and the time available for nucleation and growth decreases and lower free radical yields are produced (Riesz, Berdahl, and Christman 1985). In addition, generally higher intensities lead to higher temperatures and, thus, to higher free radical yields.

6.4.2 Inappropriate Use of Power Ultrasound

From the few data on negative effects, it should be noted that most of the negative effects reported in the literature could be due to experiments being carried out with an inappropriate combination of the ultrasonic parameters for the given application such as the intensity and frequency of the power ultrasound, the exposure time, the type of treatment, temperature, type of food, and volume of food to be processed. Thus, for instance, comparing the bath and probe treatments, Jambrak et al. (2007b) showed in their experiments that when using an ultrasonic probe, a more aggressive effect on the vegetable surface was observed. Otherwise, the effect of ultrasonic bath was more superficial and less aggressive for the tissue, and the damaged areas appeared on the surface more randomly. In addition, the temperature of the surrounding water of vegetables was raised after ultrasonic treatments by only 1°C in the bath to 25°C with the probe. This is an important result because no damage is caused by high temperature in either case. Similarly, Jambrak et al. (2008), studying the effect of ultrasound treatment on solubility and foaming properties of whey protein suspensions, concluded that although using power ultrasound in food processing can lead to several advantages like increased protein solubility, foaming ability, and others, disadvantages may arise when using power ultrasound, for a given frequency, without testing the right power for the treatment time that may lead to a destructive effect of ultrasound like protein denaturation. Indeed, as mentioned earlier (Ashokkumar et al. 2008), unwanted reactions between ultrasonically generated radicals and food ingredients could be minimized by selecting lower frequencies.

With regard to the extraction process of pectin from apple pomace, Panchev, Kirtchev, and Kratchanov (1988, 1994) reported that the overall gel strength was not affected by the sonication procedure. However, no small deformation oscillatory measurements were performed to study the viscoelastic properties of pectin dispersions during the course of gelation. In contrast, Seshadri et al. (2003), using a ½ in. ultrasonic probe, concluded that the rheological properties of pectin can be negatively influenced if treated with ultrasound. As sonication time and intensity is increased, the gel strength is reduced and time of gelation is increased. On the other hand, measurements of UV-visible absorption spectra illustrated that the optical properties of pectin gels may be improved through application of power ultrasound. Gels formed from ultrasonically pretreated pectin were significantly less turbid.

Milk is one of the food products more widely studied in order to analyze its functional properties after processing with power ultrasound (Villamiel and de Jong 2000; Wu et al. 2001; Vercet et al. 2002; Ertugay et al. 2004; Muthukumaran et al. 2005; Krešić et al. 2008; Bermúdez-Aguirre and Barbosa-Cánovas 2008). Vitamin and color changes are of great importance. For instance, color is a key factor influencing consumer sensory acceptance. Muthukumaran et al. (2005) sonicated various whey solutions for up to 4 h and then the soluble protein contents of the resulting solutions were analyzed using high-performance liquid chromatography, showing identical concentration profiles in all samples. This result is in accordance with the findings of other workers. However, Villamiel and de Jong (2000) do find some evidence for the denaturation of whey solutions in excess of 60°C. Muthukumaran et al. (2005), concluded that sonication does not appear to change the protein concentration profile within the whey solution when using low temperatures (22–25°C) and power levels (2 W/l). Similar observations were carried out by Bermúdez-Aguirre and Barbosa-Cánovas (2008), who reported minor changes in physicochemical and nutritional properties, and no degradation of protein content or color variation was observed after treatments.

Once again, results are contradictory among different authors. Vercet, Burgos, and López-Buesa (2001) studied the effect of manothermosonication on nutrient content and nonenzymatic browning reactions in milk and orange juice. Thiamin, riboflavin, ascorbic acid, and carotenoids are rather sensitive to oxidation. No effect was found on thiamin and riboflavin vitamins in milk. With reference to ascorbic acid and carotenoid content in orange juice, only small losses of these nutrients were found, but the average content was within the range of normal values. Thus, vitamins seem to be not especially affected by the combined action of heat and ultrasound under moderate pressure. However, ultrasound clearly increased brown pigments since thermal conditions applied in that study had almost no effect on brown pigments. In another study, Valero et al. (2007) studied the influence of ultrasound under different processing conditions on the inactivation of microorganisms in orange juice. They also showed that the sensory properties associated with limonin content, brown pigment production in nonenzymatic browning reactions, and color were not negatively affected by several ultrasound treatments. Similarly, Tiwari et al. (2008) investigated the effect of power ultrasound on orange juice as a function of exposure time and intensity. Although there were no significant pH differences for pH between the untreated juice and the treated ones, they found that power ultrasound significantly influences key orange juice quality parameters such as color values. Probable reasons for these discrepancies could be that the experiments were not carried out under the same conditions. Firstly, ultrasound was used in combination with heat and pressure in Vercet et al. (2001). Secondly, although all authors used similar frequencies, Tiwari et al. (2008) used intensity levels over 600 W up to 1500 W while, for example, Valero et al. (2007) used values from 120 to 300 W in most of their experiments, and the value of 600 W was not exceeded. However, they did not state the measured number of watts per centimeter squared. In addition, treated volume of orange juice, pulp content, and exposure time were also different in the experiments.

With regard to tomato paste, the most important quality parameter is, together with color, high viscosity. Inactivation of pectic enzyme is therefore, necessary to obtain high-viscosity tomato products. Vercet et al. (2002) have demonstrated that the rheological properties of tomato paste are not only not negatively affected, but improved by manothermosonication. Indeed, the improvement was related to the cavitation phenomena. After manothermosonication, tomato paste presents higher viscosity, higher consistency index, higher yield stress values, and less serum liberated from tomato paste.

Similar oxidation reactions were reported by Chemat et al. (2004). During food emulsification and processing of sunflower oil, a metallic and rancid odor was detected. Some off-flavor compounds (hexanal and hept-2-nal) resulting from the sonodegradation of sunflower oil were identified. These volatile compounds contribute to grassy, fishy, pungent, and oxidized notes, and result from the degradation of linoleic acid and sterols. In addition, results from experiments carried out by Schneider et al. (2006) with a 40 kHz cutting assembly sonicating 50 ml of sunflower oil for 10 sec extend the findings of Chemat et al. (2004). Also a significant increase of the content of free fatty acids after treating the oil in an ultrasonic bath (1 h at 60°C) was observed.

The effects of power ultrasound depend on too many parameters that produce physical, chemical, or mechanical effects. These parameters, individually or in combination, play an important role in the efficiency of the process as well as in determining the benefits and innocuousness of power ultrasound in food products. As a general conclusion, further research is still needed in order to better understand how and to what extent ultrasonic effects vary under different experimental conditions in order to know and quantify the contribution of each variable parameter to the power ultrasound effects.

6.5 CONCLUSIONS

Most studies reported in the literature provide evidence of the benefits of power ultrasound in the different applications, either enhancing the product quality or the process efficiency, and no detrimental effects on the food quality attributes were observed or at least they seem to be not especially affected. Generally, variable parameters are the ultrasonic frequency, intensity, and exposure time. Likewise, when power ultrasound is used in combination with another technique, other parameters such as temperature of the process, air flow in drying processes, or pressure in microorganism inactivation are varied.

Also, but much less numerous, there are some studies that show no beneficial effects or contradictory results on the food product. It is important to note that most of these undesired effects could be due to the performance of the experiments with an inappropriate combination of the ultrasonic parameters for the given application such as the intensity and frequency of the power ultrasound, the exposure time, the type of treatment, temperature, or volume of food to be processed. Moreover, it is usually rather difficult to establish comparisons between experiments carried out among different authors since one of the key parameters that strongly influences results is the type of food.

Although extensive research has been carried out over many years on the application of power ultrasound, a great deal of research on the development of ultrasonic technology and also on the effects of power ultrasound on the properties of foods is still required. On one hand, power ultrasound technology needs to be specifically developed and scaled up for each application. On the other hand, a better understanding of how and to what extent ultrasonic effects vary under different experimental conditions is required in order to know and quantify the contribution of each variable parameter to the power ultrasound effects.

REFERENCES

Álava, J. M., Sahi, S. S., García-Álvarez, J., Turó, A., Chávez, J. A., García, M. J., and Salazar, J. 2007. Use of ultrasound for the determination of flour quality. *Ultrasonics*, 46, 270–76.

Albu, S., Joyce, E., Paniwnyk, L., Lorimer, J. P., and Mason, T.J. 2004. Potential for the use of ultrasound in the extraction of antioxidants from *Rosmarinus officinalis* for the food and pharmaceutical industry. *Ultrasonics Sonochemistry*, 11, 261–65.

Arruda, S. C. C., Rodriguez, A. P. M., and Arruda, M. A. Z. 2003. Ultrasound-assisted extraction of Ca, K and Mg from *in vitro* citrus culture. *Journal of the Brazilian Chemical Society*, 14, 470–74.

Ashokkumar, M., Sunartio, D., Kentish, S., Mawson, R., Simons, Ll., Vilkhu, K., and Versteeg, C. 2008. Modification of food ingredients by ultrasound to improve functionality: A preliminary study on a model system. *Innovative Food Science and Emerging Technologies*, 9, 155–60.

Balachandran, S., Kentish, S. E., Mawson, R., and Ashokkumar, M. 2006. Ultrasonic enhancement of the supercritical extraction from ginger. *Ultrasonics Sonochemistry*, 13, 471–79.

Benedito, J., Carcel, J. A., González, R., and Mulet, A. 2002. Application of low intensity ultrasonics to cheese manufacturing processes. *Ultrasonics*, 40, 19–23.

Bermúdez-Aguirre, D., and Barbosa-Cánovas, G. V. 2008. Study of butter fat content in milk on the inactivation of *Listeria innocua* ATCC 51742 by thermo-sonication. *Innovative Food Science and Emerging Technologies*, 9, 176–85.

Brennan, J. G., Grandison, A. S., and Lewis, M. 2006. Separations in food processing. In: Brennan, J. G. (Ed.), *Food Processing Handbook*. Weinheim: Wiley. 429–512.

Chemat, F., Grondin, I., Shum Cheong Sing, A., and Smadja, J. 2004. Deterioration of edible oils during food processing by ultrasound. *Ultrasonics Sonochemistry*, 11, 13–15.

Chendke, P. K., and Fogler, H. S. 1975. Macrosonics in industry. Part 4: Chemical processing. *Ultrasonics*, 13, 31–37.

Contreras, N. I., Fairley, P., McClements, D. J., and Povey, M. J. W. 1992. Analysis of the sugar content of fruit juices and drinks using ultrasound velocity measurements, *International Journal of Food Science and Technology*, 27, 515–29.

Crelier, S., Robert, M.-C., Claude, J., and Juillerat, M.-A. 2001. Tomato (*Lycopersicon esculentum*) pectinmethylesterase and polygalacturonase behaviors regarding heat- and pressure-induced inactivation. *Journal of Agricultural and Food Chemistry*, 49, 5566–75.

Dickens, J. A., Lyon, C. E., and Wilson, R. L. 1991. Effect of ultrasonic radiation on some physical characteristics of broiler breast muscle and cooked meat. *Poultry Science*, 70, 389–96.

Ertugay, M. F., Şengül, M., and Şengül, M. 2004. Effect of ultrasound treatment on milk homogenisation and particle size distribution of fat. *Turkish Journal of Veterinary and Animal Sciences*, 28, 303–308.

Fairbanks, H. V. 1973. Use of ultrasound to increase filtration rate. *Proceedings of the Ultrasonics International Conference*, 11–15.

Fernandes, F. A. N., Gallão, M. I., and Rodrigues, S. 2008. Effect of osmotic dehydration and ultrasound pre-treatment on cell structure: Melon dehydration. *Lebensmittel-Wissenschaft und-Technologie*, 41, 604–10.

Fernandes, F. A. N., Oliveira, F. I. P., and Rodrigues, S. 2007. Use of ultrasound for dehydration of papayas. *Food and Bioprocess Technology*, 1, 339–45.

Fernandes, F. A. N., and Rodrigues, S. 2007. Ultrasound as pre-treatment for drying of fruits: Dehydration of banana. *Journal of Food Engineering*, 82, 261–67.

Floros, J. D., and Liang, H. 1994. Acoustically assisted diffusion through membranes and biomaterials. *Food Technology*, 48, 79–84.

Fuente-Blanco, S., Riera-Franco, E., Acosta-Aparicio, V. M., Blanco-Blanco, A., and Gallego-Juárez, J. A. 2006. Food drying process by power ultrasound. *Ultrasonics*, 44, e523–27.

Gaikwad, S. G., and Pandit, A. B. 2008. Ultrasound emulsification: Effect of ultrasonic and physicochemical properties on dispersed phase volume and droplet size. *Ultrasonics Sonochemistry*, 15, 554–63.

Gallego-Juárez, J. A., Elvira-Segura, L., and Rodríguez-Corral, G. 2003. A power ultrasonic technology for deliquoring. *Ultrasonics*, 41, 255–59.

Gallego-Juárez, J. A., Riera, E., Fuente-Blanco, S., Rodríguez-Corral, G., Acosta-Aparicio, V. M., and Blanco, A. 2007. Application of high-power ultrasound for dehydration of vegetables: Processes and devices. *Drying Technology*, 25, 1893–901.

Gallego-Juárez, J. A., Rodríguez-Corral, G., Gálvez-Moraleda, J.C., and Yang, T. S. 1999. A new high intensity ultrasonic technology for food dehydration. *Drying Technology*, 17, 597–608.

García-Pérez, J. V., Cárcel, J. A., Fuente-Blanco, S., and Riera-Franco, E. 2006. Ultrasonic drying of foodstuff in a fluidized bed: Parametric study. *Ultrasonics*, 44, e539–43.

Got, F., Culioli, J., Berge, P., Vignon, X., Astruc, T., Quideau, J. M., and Lethiecq, M. 1999. Effects of high intensity high frequency ultrasound on ageing rate, ultrastructure and some physico-chemical properties of beef. *Meat Science*, 51, 35–42.

Gould, G. W. 2001. New processing technologies: an overview. *Proceedings of the Nutritional Society*, 60, 463–74.

Higaki, K., Ueno, S., Koyano, T., and Sato, K. 2001. Effects of ultrasonic irradiation on crystallization behaviour of tripalmitoylglycerol and cocoa butter. *Journal of the American Oil Chemists' Society*, 78, 513–18.

Jambrak, A. R., Mason, T. J., Lelas, V., Herceg, Z., and Herceg, I. L. 2008. Effect of ultrasound treatment on solubility and foaming properties of whey protein suspensions. *Journal of Food Engineering*, 86, 281–87.

Jambrak, A. R., Mason, T. J., Paniwnyk, L., and Lelas, V. 2007a. Accelerated drying of button mushrooms, Brussels sprouts and cauliflower by applying power ultrasound and its rehydration properties. *Journal of Food Engineering*, 81, 88–97.

———. 2007b. Ultrasonic effect on pH, electric conductivity, and tissue surface of button mushrooms, Brussels sprouts and cauliflower. *Czech Journal of Food Science*, 25, 90–99.

Jayasooriya, S. D., Bhandari, B. R., Torley, P., and D'Arcy, B. R. 2004. Effect of high power ultrasound waves on properties of meat: A review. *International Journal of Food Properties*, 7, 301–19.

Jayasooriya, S. D., Torley, P. J., D'Arcy, B. R., and Bhandari, B. R. 2007. Effect of high power ultrasound and ageing on the physical properties of bovine *Semitendinosus* and *Longissimus* muscles. *Meat Science*, 75, 628–39.

Jiranek, V., Grbin, P., Yap, A., Barnes, M., and Bates, D. 2008. High power ultrasonics as a novel tool offering new opportunities for managing wine microbiology. *Biotechnology Letters*, 30, 1–6.

Kadkhodaee, R., and Povey, M. J. W. 2008. Ultrasonic inactivation of Bacillus ⊠-amylase. I. Effect of gas content and emitting face of probe. *Ultrasonics Sonochemistry*, 15, 133–42.

Kaloyereas, S. A. 1955. Preliminary report on the effect of ultrasonic waves on the crystallisation of honey. *Science*, 121, 339–40.

Krešić, G., Lelas, V., Jambrak, A. N., Herceg, Z., and Brnčić, S. R. 2008. Influence of novel food processing technologies on the rheological and thermophysical properties of whey proteins. *Journal of Food Engineering*, 87, 64–73.

Kyllönen, H. M., Pirkonen, P., and Nyström, M. 2005. Membrane filtration enhanced by ultrasound: a review. *Desalination*, 181, 319–35.

Liebl, D. E. 1978. Ultrasound and granulation of honey. *American Bee Journal*, 2, 107.

López, P., Sánchez, A. C., Vercet, A., and Burgos, J. 1997. Thermal resistance of tomato polygalacturonase and pectinmethylesterase at physiological pH. *Zeitschrift für Lebensmittel-Untersuchung und -Forschung*, 204, 146–50.

Lyng, J. G., and Allen, P. 1997. The influence of high intensity ultrasound bath on aspects of beef tenderness. *Journal of Muscle Foods*, 8, 237–49.

Lyng, J. G., Allen, P., and McKenna, B. M. 1998a. The effect on aspects of beef tenderness of pre- and post-rigor exposure to a high intensity ultrasound probe. *Journal of the Science of Food and Agriculture*, 78, 308–14.

———. 1998b. The effects of *Pre-* and *Post-rigor* high-intensity ultrasound treatment on aspects of lamb tenderness. *Lebensmittel-Wissenschaft und-Technologie*, 31, 334–38.

Ma, Y., Ye, X., Hao, Y., Xu, G., Xu, G., and Liu, D. 2008. Ultrasound-assisted extraction of heperidin from penggan (*Citrus reticulata*) peel. *Ultrasonics Sonochemistry*, 15, 227–32.

Mañas, P., Pagan, R., Raso, J., Sala, F. J., and Condón, S. 2000. Inactivation of *Salmonella Typhimurium* and *Salmonella Senftenberg* by ultrasonic waves under pressure. *Journal of Food Protection*, 63, 451–56.

Mañas, P., Muñoz, B., Sanz, D., and Condón, S. 2006. Inactivation of lysozyme by ultrasonic waves under pressure at different temperatures. *Enzyme and Microbial Technology*, 39, 1177–82.

Mañas, P., and Pagán, R. 2005. Microbial inactivation by new technologies of food preservation. *Journal of Applied Microbiology*, 98, 1387–99.

Mason, T. J., and Lorimer, J. P. 1988. *Sonochemistry: Theory, Applications and Uses of Ultrasound in Chemistry*. New York: John Wiley.

Mason, T. J., Paniwnyk, L., and Lorimer, J. P. 1996. The uses of ultrasound in food technology. *Ultrasonics Sonochemistry*, 3, S253–60.

McClements, D. J. 1997. Ultrasonic characterization of foods and drinks: Principles, methods, and applications. *Critical Reviews in Food Science and Nutrition*, 37, 1–46.

Mizrach, A. 2008. Ultrasonic technology for quality evaluation of fresh fruit and vegetables in pre- and post-harvest processes. *Postharvest Biology and Technology*, 48, 315–30.

Moulton, K. J., and Wang, L. C. 1982. A pilot-plant study of continuous ultrasonic extraction of soybean protein. *Journal of Food Science*, 47, 1127–29.

Muthukumaran, S., Kentish, S. E., Ashokkumar, M., and Stevens, G. 2005. Mechanisms for the ultrasonic enhancement of dairy whey ultrafiltration. *Journal of Membrane Science*, 258, 106–14.

Ordoñez, J. A., Sanz, B., Hernández, P. E., and López-Lorenzo, P. 1984. A note on the effect of combined ultrasonic and heat treatments on the survival of thermoduric streptococci. *Journal of Applied Bacteriology*, 54, 175–77.

Panchev, I. N., Kirtchev, N. A., and Kratchanov, G. 1988. Improving pectin technology II. Extraction using ultrasonic treatment. *International Journal of Food Science and Technology*, 23, 337–41.

———. 1994. On the production of low esterified pectins by acid maceration of pectic raw materials with ultrasound treatment. *Food Hydrocolloids*, 8, 9–17.

Paniwynk, L., Beaufoy, E., Lorimer, J. P., and Mason, T. J. 2001. The extraction of rutin from flower buds of *Sophora japonica*. *Ultrasonics Sonochemistry*, 8, 299–301.

Piyasena, P., Mohareb, E., and McKellar, R. C. 2003. Inactivation of microbes using ultrasound: A review. *International Journal of Food Microbiology*, 87, 207–16.

Povey, M. J. W., and Mason, T. J. 1998. *Ultrasound in Food Processing*. London: Blackie Academic and Professional.

Povey, M. J. W., and McClements, D. J. 1988. Ultrasonics in food engineering. Part I: Introduction and experimental methods. *Journal of Food Engineering*, 8, 217–45.

———. 1989. Ultrasonics in food engineering. Part II: Applications. *Journal of Food Engineering*, 9, 1–20.

Raso, J., Palo, A., Pagan, R., and Condón, S. 1998. Inactivation of *Bacillus subtilis* spores by combining ultrasonic waves under pressure and mild heat treatment. *Journal of Applied Microbiology*, 85, 849–54.

Raviyan, P., Zhang, Z., and Feng, H. 2005. Ultrasonication for tomato pectinmethylesterase inactivation: Effect of cavitation intensity and temperature on inactivation. *Journal of Food Engineering*, 70, 189–96.

Resa, P., Bolumar, T., Elvira, L., Pérez, G., and Montero de Espinosa, F. 2007. Monitoring lactic acid fermentation in culture broth using ultrasonic velocity. *Journal of Food Engineering*, 78, 1083–91.

Riera, E., Golás, Y., Blanco, A., Gallego, J. A., Blasco, M., and Mulet, A. 2004. Mass transfer enhancement in supercritical fluids extraction by means of power ultrasound. *Ultrasonics Sonochemistry*, 11, 241–44.

Riesz, P., Berdahl, D., and Christman, C. L. 1985. Free radical generation by ultrasound in aqueous and non-aqueous solutions. *Environmental Health Perspectives*, 64, 233–52.

Rodrigues, S., and Pinto, G. A. S. 2007. Ultrasound extraction of phenolic compounds from coconut (*Cocos nucifera*) shell powder. *Journal of Food Engineering*, 80, 869–72.

Roncalés, P., Ceña, P., Beltrán, J. A., and Jaime, L. 1993. Ultrasonication of Lamb skeletal muscle fibres enhances *post mortem* proteolysis. *Zeitschrift für Lebensmittel-Untersuchung und -Forschung*, 196, 339–42.

Rostagno, M. A., Palma, M., and Barroso, C. G. 2003. Ultrasound-assisted extraction of soy isoflavones. *Journal of Chromatography A*, 1012, 119–28.

Ruecroft, G., Hipkiss, D., Ly, T., Maxted, N., and Cains, P. W. 2005. Sonocrystallization: The use of ultrasound for improved industrial crystallization. *Organic Process Research & Development*, 9, 923–32.

Salazar, J., Turó, A., Chávez, J. A., and García, M. J. 2004. Ultrasonic inspection of batters for on-line process monitoring. *Ultrasonics*, 42, 155–59.

Schneider, Y., Zahn, S., Hofmann, J., Wecks, M., and Rohm, H. 2006. Acoustic cavitation induced by ultrasonic cutting devices: A preliminary study. *Ultrasonics Sonochemistry*, 13, 117–20.

Semmelink, A. 1973. Ultrasonically enhanced liquid filtering. *Proceedings of the Ultrasonics International Conference*, Guilford (England), 7–10.

Seshadri, R., Weiss, J., Hulbert, G. J., and Mount, J. 2003. Ultrasonic processing influences rheological and optical properties of high-methoxyl pectin dispersions. *Food Hydrocolloids*, 17, 191–97.

Shoh, A. 1975. Industrial applications of ultrasound – A review. I. High-power ultrasound. *IEEE Transactions on Sonics and Ultrasonics*, SU-22, 60–71.

Simal, S., Benedito, J., Clemente, G., Femenia, A., and Rosselló, C. 2003. Ultrasonic determination of the composition of a meat-based product. *Journal of Food Engineering*, 58, 253–57.

Siró, I., Vén, Cs., Balla, Cs., Jónás, G., Zeke, I., and Friedrich, L. 2009. Application of an ultrasonic assisted curing technique for improving the diffusion of sodium chloride in porcine meat. *Journal of Food Engineering*, 91, 353–62.

Smith, N. B., Cannon, J. E., Novakofski, J. E., McKeith, F. K., and O'Brien, W. D. 1991. Tenderization of semitendinosus muscle using high intensity ultrasound. *Ultrasonics Symposium*, 1371–73.

Stadnik, J., Dolatowski, Z. J., and Baranowska, H. M. 2008. Effect of ultrasound treatment on water holding properties and microstructure of beef (*m. semimembranosus*) during ageing. *Lebensmittel-Wissenschaft und Technologie*, 41, 2151–58.

Suslick, K. S., and Price, G. J. 1999. Applications of ultrasound to materials chemistry. *Annual Review of Materials Science*, 29, 295–326.

Tarleton, E. S., and Wakeman, R. J. 1998. Ultrasonically assisted separation process. In: Povey, M. J. W., and Mason, T. M. (Eds.), *Ultrasound in Food Processing*, 193–218. London: Blackie Academic and Professional.

Thakur, B. R., and Nelson, P. E. 1997. Inactivation of lipoxygenase in whole soy flour suspension by ultrasonic cavitation. *Die Nahrung*, 41, 299–301.

Thornycroft, J., and Barnaby S. W. 1895. Torpedo boat destroyers. *Proceedings of the Institution of Civil Engineers*, 122, 51–103.

Thrasyvoulou, A., Manikis, J., and Tselios, D. 1994. Liquefying crystallized honey with ultrasonic waves. *Apidologie*, 25, 297–302.

Tiwari, B. K., Muthukumarappan, K., O'Donnell, C. P., and Cullen, P. J. 2008. Colour degradation and quality parameters of sonicated orange juice using response surface methodology. *Lebensmittel-Wissenschaft und Technologie*, 41, 1876–83.

Ueno, S., Ristic, R. I., Higaki, K, and Sato, K. 2003. *In situ* studies of ultrasound-stimulated fat crystallization using synchrotron radiation. *Journal of Physical Chemistry B*, 107, 4927–35.

Valero, M., Recrosio, N., Saura, D., Muñoz, N., Martí, N., and Lizama, V. 2007. Effects of ultrasonic treatments in orange juice processing. *Journal of Food Engineering*, 80, 509–16.

Vercet, A., Burgos, J., and López-Buesa, P. 2001. Manothermosonication of foods and food-resembling systems: Effect on nutrient content and nonenzymatic browning. *Journal of Agriculture and Food Chemistry*, 49, 483–89.

Vercet, A., Sánchez, C., Burgos, J., Montañés, L., and López-Buesa, P. 2002. The effects of manothermosonication on tomato pectic enzymes and tomato paste rheological properties. *Journal of Food Engineering*, 53, 273–78.

Vilkhu, K., Mawson, R., Simons, Ll., and Bates, D. 2008. Applications and opportunities for ultrasound assisted extraction in the food ind ustry – A review. *Innovative Food Science and Emerging Technologies*, 9, 161–69.

Villamiel, M., and de Jong, P. 2000. Influence of high-intensity ultrasound and heat treatment in continuous flow on fat, proteins, and native enzymes of milk. *Journal of Agriculture and Food Chemistry*, 48, 472–78.

Vinatoru, M. 2001. An overview of the ultrasonically assisted extraction of bioactive principles from herbs. *Ultrasonics Sonochemistry*, 8, 303–13.

Wang, L. C. 1975. Ultrasonic extraction of proteins from autoclaved soybean flakes. *Journal of Food Science*, 40, 549–51.

Wood, R. W., and Loomis, A. L. 1927. The physical and biological effects of high frequency sound waves of great intensity. *Philosophical Magazine*, 4, 417–36.

Wu, H., Hulbert, G. J., and Mount, J. R. 2001. Effects of ultrasound on milk homogenization and fermentation with yogurt starter. *Innovative Food Science & Emerging Technologies*, 1, 211–18.

7 Ultrasound Processing: Rheological and Functional Properties of Food

Kasiviswanathan Muthukumarappan
South Dakota State University

B. K. Tiwari and Colm P. O'Donnell
University College Dublin

P. J. Cullen
Dublin Institute of Technology

CONTENTS

7.1 INTRODUCTION

The use of ultrasound in the food industry has been extensively investigated with reported applications in both food analysis and food processing. Ultrasound is a sound wave at a frequency above the threshold of human hearing (>20 MHz). Application of ultrasound in food processing can be classified into three categories. Low-intensity ultrasound uses low power levels, typically less than 1 W/cm^2, at a frequency range of 5–10 MHz (McClements 1995; Mason 1998). Due to the low power levels, low-intensity ultrasound causes no physical or chemical alterations in the properties of the material through which the wave passes. It can be used for diagnostic measurements of food properties including texture, composition, viscosity, or concentration. In contrast, high-intensity ultrasound uses much higher power levels, typically in the range of 10–1000 W/cm^2, at a frequency of 20–100 kHz (McClements 1995; Mason 1998).

Power ultrasound has been recognized as a promising processing technique to replace or complement conventional thermal treatments in the food industry. Advantages of sonication include reduced processing time, higher throughput and lower energy consumption (Zenker et al. 2003; Knorr et al. 2004) while reducing thermal effects. Various research groups have demonstrated the inactivation of pathogenic and spoilage microorganisms (*Escherichia coli*, *Listeria*), and spoilage enzymes (pectin methyl esterase, polyphenol oxidase) with reduced effects on quality or nutritional parameters compared to conventional thermal processing. Although ultrasound is regarded as a nonthermal processing technique, an increase in product macro temperature occurs, which depends on the intrinsic and extrinsic parameters. Most ultrasound applications are limited to liquid foods, mainly fruit juices, smoothies, and milk. Ultrasonic radiation (sonication) is used widely to break cells and organelles, largely because sonication disrupts the larger particles in a suspension, leaving smaller particles unaffected.

Ultrasound alone or in combination with other nonthermal technologies may inactivate microorganisms by physical (cavitation, mechanical effects) and/or chemical (formation of free radicals due to sonochemical reaction) principles. This chapter outlines the impact of ultrasound on the rheological and functional properties of food.

7.2 GENERATION OF POWER ULTRASOUND

An ultrasonic power supply converts 50/60 Hz line voltage to high-frequency electrical energy. This high-frequency electrical energy is transmitted to a piezoelectric transducer within the converter, where it is changed to mechanical vibrations. There are three types of ultrasonic transducers in common usage including liquid-driven transducers, magnetostrictive transducers, and piezoelectric transducers (Mason 1998). Piezoelectric transducers are the most common devices employed for the generation of ultrasound and can be used over the whole range of ultrasonic frequencies (Mason 2000). Piezoelectric material such as barium titanate or lead metaniobate expands and contracts in alternating electrical fields, generating ultrasonic waves. The piezoelectric elements commonly used in ultrasonic transducers are potentially brittle and so it is normal practice to clamp them between metal blocks (front and back drivers). Typically generation of ultrasound is carried out using the electrostrictive transformer principle. This is based on the elastic deformation of ferroelectric materials within a high-frequency electrical field and is caused by the mutual attraction of the molecules polarized in the field (Raichel 2000). As ultrasound propagates through and interacts with a liquid, the energy is attenuated by scattering or absorption and it results in alternate rarefaction and compression. Bubbles or cavities are formed if the amplitude of the wave is sufficiently large. This phenomenon is known as cavitation. The collapse of bubbles creates localized high pressure that disrupts cell membranes and causes cell walls to break down (Scherba, Weigel, and O'Brien 1991).

7.3 APPLICATIONS OF POWER ULTRASOUND IN FOOD PROCESSING

Power ultrasound has been actively investigated for food processing applications. Reported applications of power ultrasound in food processing include inactivation of microorganisms and enzymes, generation of dispersions and emulsions, and the promotion of chemical reactions. Power ultrasound is used in a number of unit operations in fruit juice processing such as cleaning, extraction, homogenization, emulsification, sieving, filtration, crystallization, and pasteurization. High-energy ultrasound has been applied for degassing of liquid foods, for the induction of oxidation/reduction reactions, for extraction of enzymes and proteins, for enzyme inactivation, and for the induction of nucleation for crystallization (Roberts 1993; Thakur and Nelson 1997; Villamiel and de Jong, 2000a). Power ultrasound has been employed for the inactivation of *E. coli* in apple cider (Baumann, Martin, and Feng 2005) and orange juice processing (Valero et al. 2007; Tiwari et al.

2008a). Similarly, enzymes such as peroxidase (De Gennaro et al. 1999), proteases, and lipases (Vercet, Burgos, and Lopez-Buesa 2001) were reported to be inactivated. Application of ultrasound may be classified as sonication, monosonication, themosonication, or manothermosonication (MTS), depending upon whether it is combined with heat or pressure. For example, MTS is the simultaneous application of heat and high-energy ultrasonic waves under moderate pressure. MTS is able to inactivate food-related enzymes and microorganisms at much higher rates than thermal treatments of comparable temperatures (Burgos 1999). Despite the potential of power ultrasound as a nonthermal food processing technology, ultrasound may induce both desirable and undesirable effects on the nutritional and quality parameters of food. This chapter reviews changes in rheological properties of sonicated foods.

7.4 EFFECT OF ULTRASOUND ON FOOD RHEOLOGY

The effect of ultrasound on food rheology is mainly due to the cavitation phenomenon. Cavitation is the formation, growth, and, in some cases, implosion of bubbles within liquids (Figure 7.1). Two different types of cavitation phenomena can be generated by acoustic waves, namely *inertial* and *noninertial* cavitation. Inertial cavitation involves large-scale variations in bubble size (relative to the equilibrium size) over a timescale of a few acoustic cycles, where the rapid growth terminates in bubble collapse with varying degrees of intensity. Noninertial cavitation (stable) involves small-amplitude oscillations (compared to bubble radius) (Atchley and Crum 1988). Table 7.1 lists some of the proposed mechanisms of action for ultrasound. The thermal, mechanical, and chemical effects of high-intensity ultrasound have been attributed to the rapid formation and collapse of cavitational bubbles, generating intense normal and shear stresses (Crum 1995; Stephanis, Hatiris, and Mourmouras 1997). Implosion of cavitation bubbles leads to energy accumulation in hot spots where temperatures of 5000°C and pressures of 100 MPa have been measured (Suslick 1988). As a result of these conditions, water molecules can be broken, generating highly reactive free radicals that can react with other molecules (Riesz and Kondo, 1992). Cavitational thermolysis may produce hydroxyl radicals and hydrogen atoms that can be followed by formation of hydrogen peroxide and, in the absence of oxygen, hydroperoxyl radicals (Portenlaenger and Heusinger 1997; Cains, Martin, and Price 1998; Ashokkumar and Grieser 1999). Production of H_2O_2 during sonication is temperature dependent, decreasing with increase in temperature. Although increasing liquid temperature during sonication allows a reduction in the cavitation threshold, the maximum temperature and pressure during cavitational bubble collapse will be decreased (Mason and Lorimer 2002). Thus the sonochemical reaction that generates H_2O_2 (Reaction 1), would be less intense at elevated temperatures (Suslick 1991).

$$H_2O \rightarrow OH^- + H^+ \rightarrow H_2O_2 + H_2 \quad (7.1)$$

These transient reactive species can subsequently react with carbohydrates. In addition hydrolysis and cleavage due to the strong mechanical forces have been reported for a variety of polysaccharides (Kardos and Luche 2001), reducing molecular weight and changing the rheological

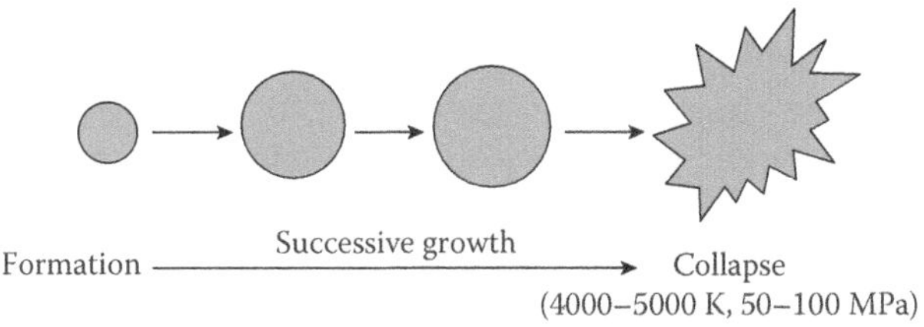

FIGURE 7.1 Formation and collapse of bubbles.

TABLE 7.1
Theory and Mechanism for Power Ultrasound

Theory	Mechanism	Reference
Cavitation	Mechanical removal of attached or entrapped bacteria	Seymour et al. 2002
	Localized hot spots (5500°C) and high pressure (500 MPa)	Suslick 1988; Mason and Luche 1996; Vercet et al. 2001
	Increase in permeability of membranes or loss of selectivity	Lehmann and Krusen 1954
	Disruption (shear stress, localized heating) and chemical reactions within the cell of the microorganism	Piyasena et al. 2003
Formation of free radicals	Sonolysis of water may produce OH^- and H^+ species and hydrogen peroxide	Suslick,1988; Vercet et al. 1997
Intracellular micromechanical shocks	Disruption of cellular structural and functional components up to the point of cell lysis	De Guerrero et al. 2001
	Thinning of cell membranes	Sams and Feria 1991; Butz and Tauscher 2002; Fellows 2000
Generation of mechanical energy	Cleaning action on surfaces	Scherba et al. 1991; Sala et al. 1995
Compressions and rarefactions or compression/expansion cycles	Acoustic microstreaming	Scherba et al. 1991; Floros and Liang 1994

properties of food. Mechanical stress, generated by shock waves derived from bubble implosion or from microstreaming derived from bubble size oscillation, may also break large macromolecules or particles (Basedow and Ebert 1977). Because of the nature of these ultrasound effects on molecules or particles dissolved or suspended in liquids, it is expected that the large multimolecular structures such as casein micelles and fat globules in milk will be affected by ultrasound.

Another important factor that influences the effectiveness of cavitation is the viscosity of the liquid. In highly viscous products, ultrasound diffusion is easily disrupted and this reduces the degree to which cavitation occurs. Low-frequency high-power ultrasound is better at penetrating viscous products than higher-frequency high-power ultrasound, which is more easily dispersed within a viscous liquid. Alternatively, if the liquid is heated, viscosity is reduced and ultrasound penetration improves (Earnshaw, Appleyard, and Hurst 1995).

7.4.1 Rheological Properties of Food

Rheology is defined as the science of deformation and flow, providing insight into material phenomena, which are governed by the microscopic scale at a macroscopic level. Rheology studies the response of materials when subjected to force. A viscous liquid may be defined as a medium where the energy needed to deform it is completely dissipated in the process of deformation. Consequently, it will not recover from the deformation. Conversely an elastic solid is a material that stores the work from the deformation process and returns energy after removal of the deformation forces. As rheology describes how matter responds to an applied stress or strain, it finds application in diverse areas such as product development, process engineering calculations, quality control, stability studies, and correlations to sensory data. Many processed foods are formulated to display the desired rheological behavior under specific stress conditions such as gravity, pouring, mouth feel, etc.

Dispersions are defined as homogeneous or heterogeneous systems where a solid (rigid or deformable) or liquid phase is dispersed in a liquid medium. Dispersions may destabilize or transform over time under a range of forces including interparticle, gravitational, and shear. Shear thinning behavior is observed with structured foods, where viscosity decreases with applied shear. Complex rheological behavior such as shear thinning and time-dependency effects may be associated with the shear-induced breakdown and possible rebuilding of structure. The fluid's microstructure, which may be due to either macromolecule entanglement or particle–particle interaction, is broken down under shear. The exhibition of a yield stress occurs where interparticle attraction forms flocs that in turn develop a three-dimensional network throughout the fluid. Such a network provides additional structure to a fluid, resulting in solid-like behavior under low stresses. A minimum stress is required for sufficient structural breakdown and the initiation of flow. Hydrocolloids are widely used as structuring agents to control suspension stability of large particulates along with organoleptic and processing parameters of foods.

7.4.2 Effect of Power Ultrasound on Viscosity

Viscosity is one of the most important rheological properties of food. Viscosity of tomato juice/puree is highly dependent on pectic substances that form an entanglement where other particles are physically entrapped (Rao and Cooley 1992). In tomato juices, MTS treatments results in a thicker consistency and initial apparent viscosities compared with unprocessed juice (Vercet et al. 2002a). Tomato juices are pseudogels, whose flow properties depend on the interaction or entanglement of cell particles (mostly cell walls), soluble pectin concentration, and the chemical properties of the latter. MTS treatments of pure pectin solutions yielded molecules with lower apparent viscosities due to size reduction (Mason et al. 2005). A similar study conducted by Tiwari et al. (forthcoming) observed a significant reduction in apparent viscosity for 2% w/v pectin dispersion (Figure 7.2). Seshadri et al. (2003) also reported a similar influence of sonication on the gel strength properties of pectin. It is difficult to predict what might be expected from a modification of pectin properties in gels or pseudogels derived from pectin. Longer molecules show a higher resistance to flow however shorter ones can interact in a different way with suspended particles also leading to an increased resistance to flow.

Vercet et al. (2002a) reported that the rheological properties of tomato paste are improved by MTS. In addition they observed that the thixotropic and pseudoplastic behavior of tomato paste was not affected by MTS. These are important properties to maintain the characteristic mouthfeel of the product (Rosenthal 1999). Particle interactions and particle size both play a role in determining tomato consistency. An increase in the number and/or intensity of interactions leads to increased consistency. These interactions can be chemical and/or physical. Pectin molecules are the main agents of physical interactions (Voragen et al. 1995), whereas chemical interactions are dependent on many parameters (Tsai and Zammouri 1988). Consistency is dependent on particle size in a complex manner. Reducing particle size leads to a decrease in viscosity. However, there is a point where

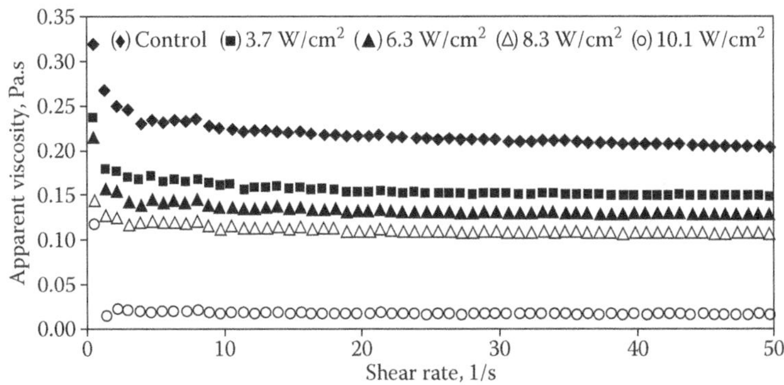

FIGURE 7.2 Effect of ultrasonic intensity (W/cm^2) on apparent viscosity of 2% (w/v) pectin dispersion.

viscosity begins to increase when particle size decreases (Beresovsky, Kopelman, and Mizrahi 1995). This has been ascribed to an increase in the number of interactions between particles or to the fact that smaller particles fit better within the pectin network. The mechanism by which MTS improves the rheological properties of tomato pastes is related to cavitation. This phenomenon results in the breakage of molecules or particles. Particle size reduction and molecule breakage induced by ultrasound has been described in the literature (Price 1990). Application of MTS to break pectin molecules in a purified pectin solution was reported by Mason et al. (2005). It is also possible that ultrasound promotes protein denaturation (Villamiel and de Jong 2000b). Denatured proteins can adhere nonspecifically to tomato particles and facilitate better interaction between particles. Whatever the mechanism, it is worth noting that another emerging technology of food preservation, high pressure, also increases the viscosity of tomato pastes (Porretta et al. 1995).

7.4.3 Juice Clarification

Cloud stability is an important rheological property of fruit juices. Loss of cloudiness and gelation of concentrate is a common problem associated with citrus juices (fresh squeezed, concentrated, and preserved) (Basak and Ramaswamy 1996). Cloud stability is a critical orange juice quality parameter imparting characteristic flavor, color, and mouthfeel. Cloud is attributed to the suspension of particles composed of a complex mixture of protein, pectin, lipids, hemicellulose, cellulose, and other minor components (Klavons, Bennett, and Vannier 1991; Baker and Cameron 1999). Cloud loss arising from the gelation of juice concentrates is primarily attributed to the activity of Pectinmethyl esterase (PME) (Cameron, Baker, and Grohmann 1998; Versteeg et al. 1980). Sequential cleavage of the methyl esters at C6 of galacturonic acid residues in pectin produces free acid groups. When of sufficient size, such blocks on adjacent pectin molecules can be cross-linked by divalent cations, leading to protein precipitation. It has been reported that the degree of esterification of the pectin backbone necessary to cause cloud loss in orange juice is <36% (Baker 1979; Krop and Pilnik 1974). Figure 7.3 shows changes in cloud value and corresponding chord length distributions as a function of ultrasound intensity. This size reduction contributes to improved cloud stability.

Cloud retention after sonication may be attributed to the dispersion stability of macromolecules due to the promotion of certain chemical reactions during sonication (Floros and Liang 1994; Mason 1998; McClements 1995). High-intensity ultrasound enhances protein solubility (Krešić et al. 2008) by changing protein conformation and structure in such a way that hydrophilic parts of amino acids are opened toward water (Morel et al. 2000; Moulton and Wang 1982). Lacroix, Fliss, and Makhlouf (2005) reported that electrostatic interactions between PME and its substrate pectin have

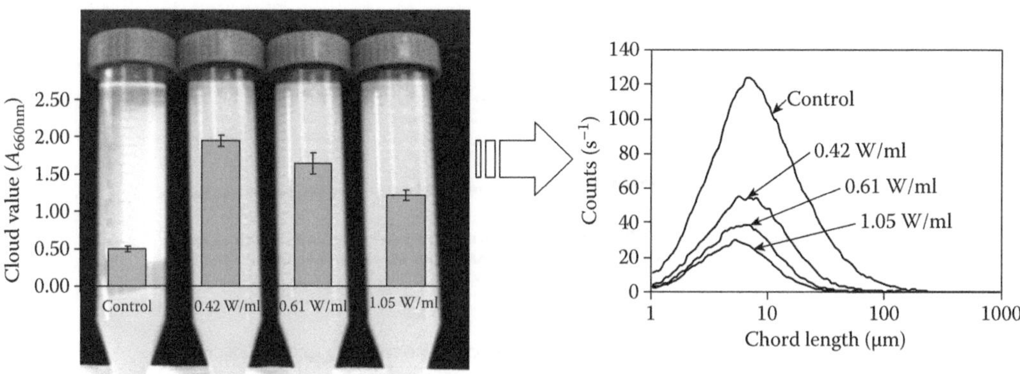

FIGURE 7.3 Effect of acoustic energy density on cloud value and juice stability with corresponding particle size distribution for control and sonicated orange juice.

a major impact on orange juice cloud stability. Lacroix et al. (2005) reported that the particle suspension stability of orange juice depends not only on PME activity but likely also on modifications of pectin. It has been reported that to retain particles in suspension in the juice, the particles must be sufficiently small (0.5–10 μm) (Mizrahi and Berk 1970; Lam Quoc et al. 2006). Seshadri et al. (2003) suggested that the application of ultrasound breaks the linear pectin molecule, reducing its molecular weight, resulting in weaker network formations. Structural damage of pectin may result from microjets of liquid generated by the asymmetrical collapse of the cavitation bubble. The resultant high-pressure gradient may also cause fragmentation of macromolecules or other structural modifications, reducing the accessibility of the pectin molecule to PME.

7.5 EFFECT OF SONICATION ON FOOD HYDROCOLLOIDS

Sonication is reported to induce both reversible and nonreversible structural modifications. Structural changes induced by sonication may be advantageous for meeting processing requirements and differentiated functionality. Iida et al. (2008) studied the effectiveness of ultrasound for the reduction of viscosity after gelatinization of starch. They indicated that the reduction in viscosity during sonication is rapid, does not require any chemical additives and the process will not induce large changes in the chemical structure. Ultrasonic depolymerization has been applied to a variety of homo- and heteropolysaccharides such as dextran (Lorimer et al. 1995), pullulan (Koda et al. 1994), chitosan (Chen, Chang, and Shyur 1997), hyaluronic acid (Miyazaki, Yamamoto, and Okada 2001), xyloglucan (Vodeničarová et al. 2006), starch (Chung et al. 2002), and other food hydrocolloids such as pectin, guar gum, and xanthan.

Camino, Pérez, and Pilosof (2009) studied the effect of high-intensity ultrasound on the particle size of hydroxypropylmethylcellulose (HPMC), which is a surface-active polysaccharide and is a cellulose derivative with methyl and hydroxypropyl groups added to the anhydroglucose backbone. Camino et al. (2009) reported that ultrasound treatment of HPMC induced the formation of concentration-dependent transient clusters that can be regarded as preaggregates which would facilitate the formation of bigger aggregates during the pregel regime, lowering the cloud point. Lowering of cloud point indicates that ultrasound treatment modifies the HPMC performance during the first stage of the gelation process, related to the cloud point. The phenomenon of cavitation during sonication induces the formation of clusters that can be regarded as preaggregates that would facilitate the formation of bigger aggregates during the pregel regime (cloud point). They also observed changes in viscosity and water mobility for high-molecular-weight HPMC, which indicates the structural modifications that were not apparent for the low-molecular-weight HPMC without influencing their emulsifying behavior.

Ultrasound treatment can be employed for partial depolymerization and preparation of medium-sized macromolecules from large ones. For example, Kasaai, Arul, and Charlet (2008) studied the kinetics of chitosan fragmentation by ultrasonic irradiation at a frequency of 20 kHz. They observed that the optimum conditions for preparation of lower-molecular-weight fragments from large macromolecules were a low chitosan concentration, a high ultrasonic power, and a high solution temperature. Polymer chain scission increases with an increase in power level and solution temperature, but a decrease in chitosan concentration. Ultrasound with a frequency of 20 kHz can cause both polymerization and depolymerization (Kruus, Lawrie, and O'Neill 1988), which in turn depends upon the nature of cavitation (Kasaai et al. 2008). Generally, polymerization and degradation processes occur simultaneously and the molecular weight is a complex function of sonication time (Price, Smith, and West 1991).

7.6 EFFECT OF SONICATION ON EMULSIONS

An emulsion is a mixture of two immiscible liquids. Many food products are oil-in-water emulsions, where proteins are generally used to stabilize the oil droplets. Milk is an example of an oil/water emulsion, which is made up of fat globules dispersed in a continuous skim milk phase. The physical stability of the emulsion system is influenced by several factors, such as fat content, particle size of fat globules, type of emulsifier, and the ratio of emulsifier to fat (Bergenstahl and Claesson 1990; Shahidi and Han 1993; Shaker, Jumah, and Jdayil 2000; Xu et al. 1998). Generally, various types of emulsifiers such as monoglycerides, phospholipids, polysaccharides (gum, maltodextrin, corn syrup solids), and proteins (sodium caseinate and skim milk powder) are used to stabilize liquid emulsion systems (Schwartzberg and Hartel 1992; Shahidi and Han 1993; Jena and Das 2006). Emulsion stability is generally affected by the formation of flocs or clumps (flocculation), creaming, or coalescing. Approaches to enhancing emulsion stability or formation include the mechanical mixing of the immiscible liquids and the time for surfactant molecules to organize at the interface of the two phases. Fat globule size plays a predominant role in influencing the stability of an emulsion. Large globules coalesce faster than small ones (Bergenstahl and Claesson 1990).

Ultrasound may be used as an alternative method to produce an emulsion, as well as to improve chemical reactions and surface chemistry (sonochemistry) (Knorr, Ade-Omowaye, and Heinz 2002). Ultrasound assists in the formation of very fine and highly stable emulsions. This occurs due to the effective mixing of the two immiscible liquids when a cavitating bubble collapses near the surface of the phase boundary layer. Power ultrasound involves acoustic cavitation of two main types: "transient" and "stable." The mechanism by which acoustic emulsification occurs is not fully understood but transient cavitation is thought to be responsible for most sonochemical effects, including the acoustic emulsification of oil droplets (Li and Fogler 1978). The advantages of ultrasound include lower energy consumption, reduced surfactant levels, and a more homogeneous emulsion compared to the mechanical process (Abismaïl et al. 1999; Cucheval and Chow 2008).

Cucheval and Chow (2008) investigated the effect of power and duty cycle (% ultrasound applied by time unit) in combination with processing time on the formation of an emulsion for a single layer of oil and water. They employed a high-speed camera to visualize the formation of the emulsion in a model system containing water, soybean oil, and Tween 80. Figure 7.4 shows that during sonication, the interface just below the probe is first pushed downward by the radiation force and forms a large oil droplet. As the interface is pushed down, the radiation forces force the liquid up toward the probe tip. When the interface is directly below the probe tip an emulsion is formed. Li and Fogler (1978) proposed the theoretical mechanism for emulsion formation based on the deformation and breakup of an oil droplet exposed to a cavitation shock wave generated by an acoustic field. The mechanism for acoustic emulsification is that large oil droplets originally formed from the instability of the oil-water interface are disintegrated into smaller ones by cavitation until a critical size is reached.

7.7 FUNCTIONAL PROPERTIES

Functional properties of proteins such as organoleptic, kinesthetic, hydration, interfacial, enzymatic, and rheological properties are of interest to manufacturers of pharmaceutical, food, and cosmetic products (Gülseren et al. 2007). These structural and functional properties are influenced by high-intensity ultrasound mainly due to formation of covalent and/or noncovalent bonds (McClements 1995). The structure–function relationship in proteins may be altered during sonication either by initiating or retarding chemical or enzymatic reactions, which may include oxidation, glycosylation, hydroxylation, phosphorylation, methylation, and acylation. Structural or functional properties of proteins are influenced because of the mechanism of action of high-intensity ultrasound (Table 7.1). The alteration of functional properties by sonication is mainly by direct or indirect effects of sonication on enzymes such as glycosidase (Barton, Bullock, and Weir 1996), invertases (Vargas et al. 2004), and β-glucosidase (Wang and Sakakibara 1997). High-power ultrasound is reported to influence the functional properties of whey proteins including solubility and foaming

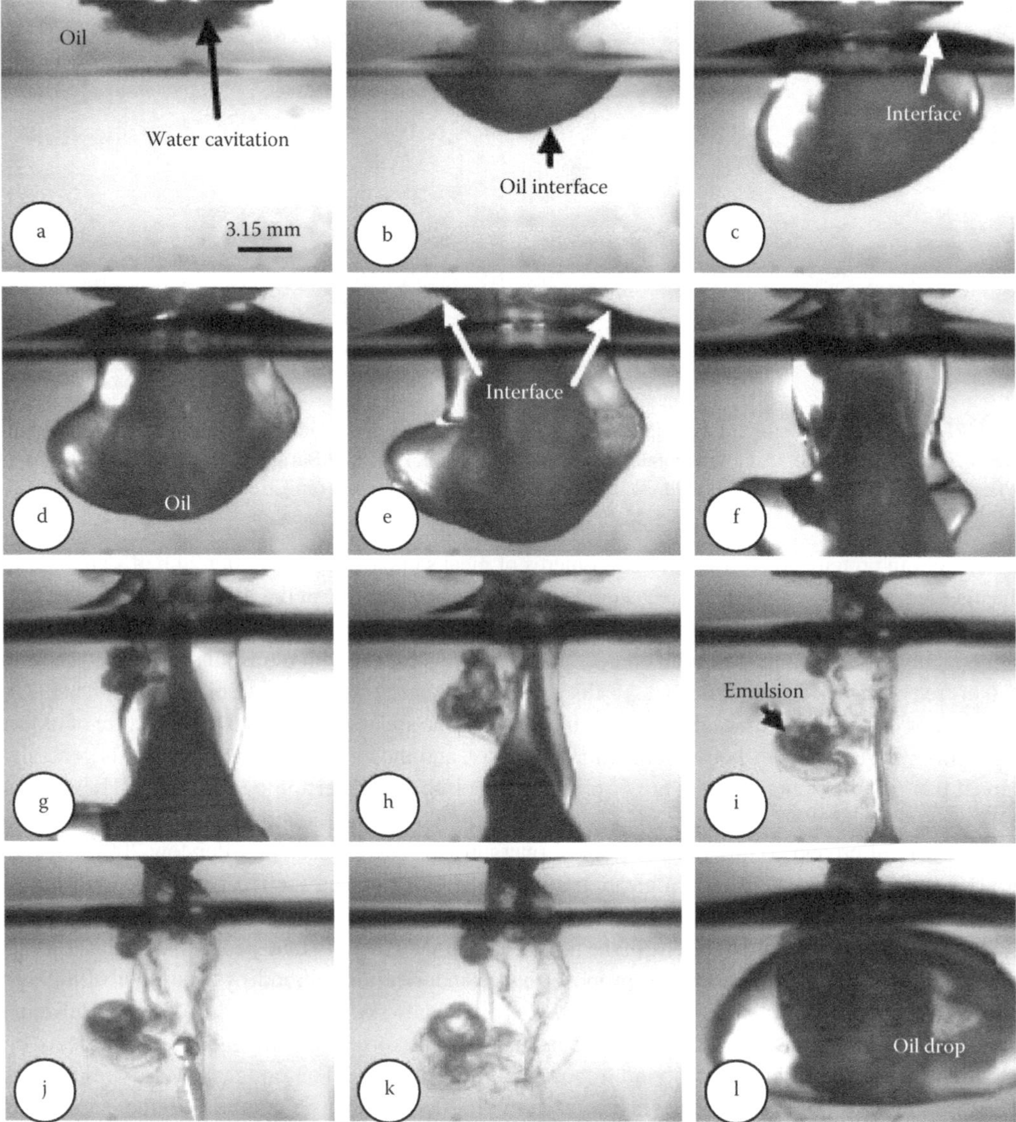

FIGURE 7.4 Formation of emulsion in a model system containing water, soybean oil, and Tween 80. (Adapted from Cucheval, A., and Chow, R. C. Y., *Ultrasonics Sonochemistry*, 15, 916–20, 2008. With permission.)

ability by changing the temperature and conductivity of the surrounding media of the whey protein (Jambrak et al. 2008).

Gülseren et al. (2007) studied the effect of high-intensity ultrasound on the structure and functionality of bovine serum albumin (BSA). They observed that high-intensity ultrasound increases the surface activity, surface hydrophobicity, and surface charge of BSA. Potential applications of sonication may include maneuvering of protein–phospholipid and protein–protein interactions in cells and cell membranes, and alteration of functional properties such as stabilization of lipid and gas interfaces in emulsions, foams, and gels (Gülseren et al. 2007).

Stadnik et al. (2008) studied the effect of low-intensity sonication on the structural, water-holding capacity (WHC) and water compartmentalization changes in beef muscle (m. semimembranosus). They observed that ultrasound treatment can accelerate the aging process. Figures 7.5a and b show the

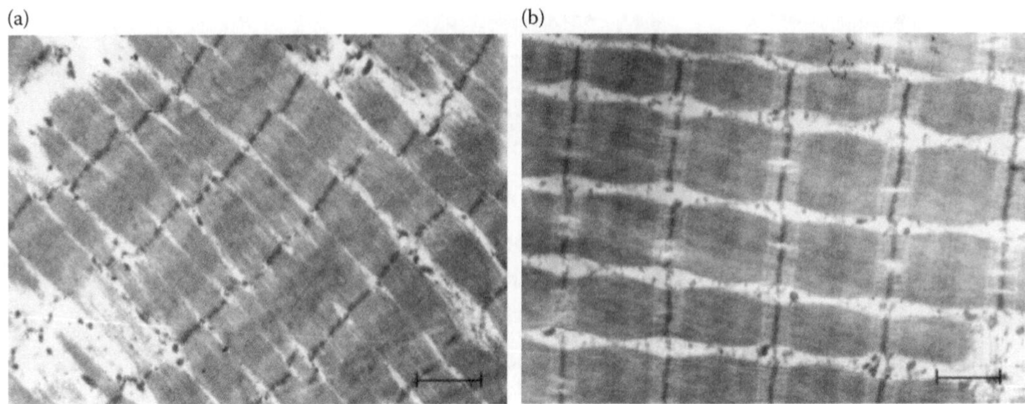

FIGURE 7.5 Microstructure of muscle fibers 24 h after slaughter. (longitudinal section for (a) control sample, (b) sonicated sample) (Adapted from Stadnik, J., Dolatowski, Z. J., and Baranowska, H. M., *Lebensmittel-Wissenschaft und Technologie*, 41 (10), 2151–58, 2008. With permission from Elsevier.)

differences in protein structures of the sarcomeres of meat samples observed after 24 h of slaughter. In the muscle fibers of the sonicated meat sample (Figure 7.5b) a swelling at the A-band region, especially near the H-band and M-line, is shown. Aging has traditionally been regarded as a reliable method for tenderization in beef. Some research on accelerating aging of meat with ultrasound is focused on cavitation-free changes of meat structures produced by low-intensity ultrasound (Dolatowski 1988, 1989). Interactions of proteins and water molecules in muscle tissue are associated with the structure of tissue, especially with interactions between myofibrils, acidity, and the ionic strength of meat. It is generally agreed that post mortem processes are of primary importance in determining the final WHC of meat (Offer and Cousins 1992). WHC increases during aging due to degradation of the cytoskeleton and other structural proteins (Huff-Lonergan and Lonergan 2005; Kristensen and Purslow 2001; Melody et al. 2004; Schäfer et al. 2002). Sonication after slaughter or during the rigor mortis period is one method for modifying the water-protein interactions (McClements 1995).

Tenderness is considered to be one of the most important quality attributes of cooked meat (Lawrie 1985). Ultrasonication has also been proposed as a tenderization method by a number of authors (Zayas and Orlova 1970; Zayas and Gorbatow 1978; Dolatowski 1989; Lyng, Allen, and McKenna 1997). However, sonication treatment of meat has been reported to have inconsistent effects on meat tenderness, with some ultrasound treatments producing no effect on tenderness, while others decreased or increased tenderness (Jayasooriya et al. 2004). Extrinsic control parameters such as ultrasonic frequency, intensity and/or power, treatment time, and temperature determine the extent of the desired result achieved using sonication. For example, low-frequency sonication (22–40 kHz) shows increased tenderness (Dickens, Lyon, and Wilson 1991; Dolatowski 1988, 1989). Sonication in brine solution results in improvement in the tenderization of meat (Zayas and Gorbatow 1978). However, the effect of ultrasound on the brining of meat is not clear. However, Paulsen, Hagen, and Bauer (2001) reported that ultrasound had no effect on brining of meat. Ultrasound can lead to several advantages such as increased protein solubility, foaming ability, etc. However disadvantages such as protein denaturation may arise when using ultrasound without testing the power or treatment time employed (Jambrak et al. 2008).

7.8 EFFECT OF SONICATION ON RHEOLOGICAL PROPERTIES OF DAIRY PRODUCTS

Sonication of milk is reported to result in physicochemical changes in macromolecules, ranging from homogenization (Villamiel and de Jong 2000b) to reduction in fermentation time during yogurt manufacture (Wu, Hulbert, and Mount 2000) and improved rheological properties of yogurt prepared from sonicated milk (Vercet et al. 2002b). Although many pathogenic and spoilage microorganisms are destroyed under standard heat treatments, extracellular lipase and protease may be produced (Stead 1986). These thermoresistant enzymes can reduce the quality and shelf life of heat-treated milk and other dairy products. MTS has been found to be more effective than heat treatment in the inactivation of these heat-resistant proteases and lipases secreted by *Pseudomonas fluorescens* (Vercet, Lopez, and Burgos 1997). The effect of ultrasound on enzymes involved in the coagulation of milk (chymosin, pepsin, and several fungal enzymes) has been studied in model systems using batch processes (Villamiel and de Jong 2000a). Ultrasonic processing of milk may provide comparable microbiological reduction to conventionally thermized milk and, at the same time, reduce the size of the fat globule. More research is needed to improve the processing equipment and elucidate the effect of ultrasound on the main milk components (proteins, vitamins, enzymes, and fat) in order to apply this method for the production of food with improved quality and safety (Villamiel and de Jong 2000b).

7.8.1 Fat Globule Size and Microstructure

Ultrasonic treatment of milk has been considered as an effective system for the reduction of fat globule size (Martinez et al. 1987), inducing a homogenization effect in milk. Conventionally, during homogenization milk is forced to pass through a small gap under high pressure. This breaks down the fat globule size and forms stable oil in water emulsion. Homogenization diminishes the tendency of the fat globules to clump together and coalesce into cream. Villamiel and de Jong (2000b) reported a significant reduction in fat globule particle size during sonication. In a similar study Wu et al. (2000) reported on improved homogenization effect of ultrasound at high amplitude levels compared with conventional homogenization. One of the mechanisms involved in ultrasonic effects is the production of strong eddies within the liquid (Floros and Liang 1994; Earnshaw et al. 1995).

Bermudez-Aguirre, Mawson, and Barbosa-Canovas (2008) observed the structural changes of fat globules in whole milk after heat alone and thermosonication using scanning electron microscopy. They observed that the surface of the fat globule was roughened after thermosonication. Ultrasound waves were responsible for disintegrating the milk fat globule membrane (MFGM) by releasing triacylglycerols. They also observed size reduction of fat globules (<1 μm) and a granular surface. This was due to the interaction between the disrupted MFGM and some casein micelles. Figure 7.6 shows a visible orifice on the surface of each globule, allowing observation of the disruption and cracking of the fat globule membrane, which is composed of cholesterol, phospholipids, proteins, and enzymes (Bermudez-Aguirre et al. 2008).

Power ultrasound can be used as an additional processing tool to modify the crystallization behavior of different systems (sonocrystallization) and therefore obtain the desired physicochemical characteristics in the food. Martini, Suzuki, and Hartel (2008) studied the effect of power ultrasound on the crystallization behavior and microstructure of anhydrous milk fat (AMF) with the objective of using high-intensity ultrasound as a novel processing condition to improve the quality. Figure 7.7 shows the morphology of AMF crystallized at 22, 24, 26, 28, and 30°C as a function of crystallization time without and with the application of sonication. Sonication not only influences the crystallization behavior of AMF but also induces the onset of crystallization and promotes crystal growth.

7.8.2 Effect of Sonication on Dairy Products

Studies have shown that the use of ultrasound as a processing aid can reduce the production time of yogurt by up to 40%. Moreover, sonication may reduce the normal dependence of the process on the origin of the milk as well as improve both the consistency and the texture of the final product

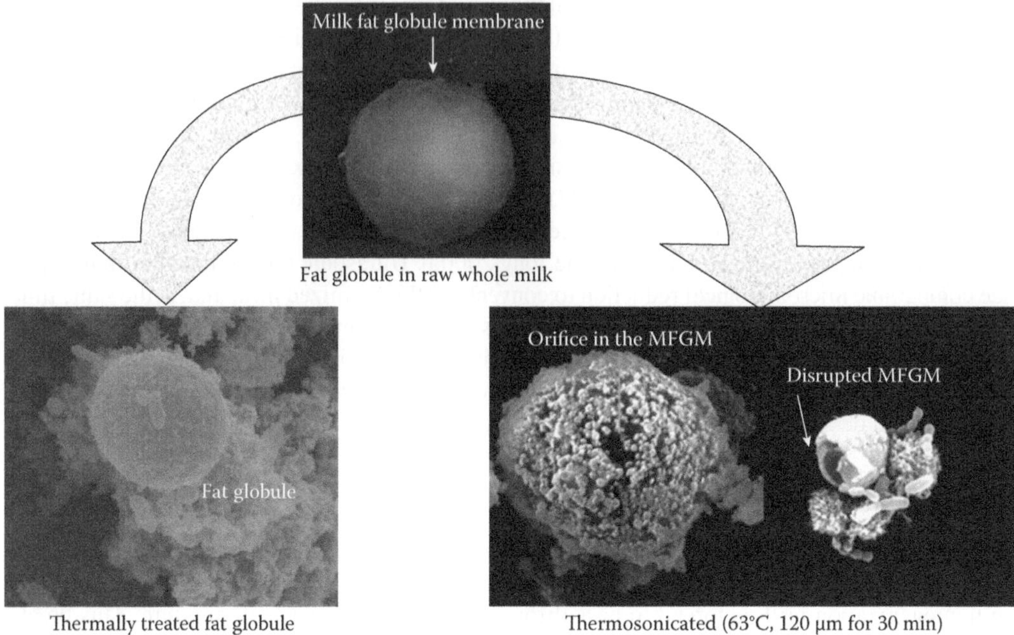

FIGURE 7.6 Changes in microstructure of fat globule during thermal treatment and thermosonication. (Adapted from Bermudez-Aguirre, D., Mawson, R., and Barbosa-Canovas, G. V., *Journal of Food Science,* 73, E325–32, 2008. With permission.)

(Dolatowski, Stadnik, and Stasiak 2007). Texture (high viscosity and consistency) is a very important characteristic of yogurt. It is directly related to yogurt's structure, which is based on the strings or clusters of casein micelles interacting physically with each other and with denatured serum proteins (mostly β-lactoglobulin) entrapping serum and fat globules (Villamiel and de Jong 2000b). Vercet et al. (2002b) observed that the textural characteristics of MTS yogurts have a firmer structure compared to controls. Figure 7.8 shows the texture profile analysis of MTS-treated and four control yogurts. As discussed earlier, sonication has been shown to effect fat globule size (Wu et al. 2000) and denature dairy protein (Villamiel and de Jong 2000b). It is expected that yogurt's structure, which is highly dependent on fat globule size and the denaturation and aggregation state of proteins, will be affected. Müller (1992) suggested that the application of ultrasound for milk homogenization before cheese-making could improve the yield of the cheese due to an increase in binding locations for protein links on the fat globule membrane. Vercet et al. (2002b) demonstrated that MTS substantially improves the homogenizing effects compared to sonication alone. However, sonication at low temperature (<30°C) yields a biphasic distribution of fat globule sizes with two peaks. Conversely, at elevated temperature (~70°C) there is a substantially improved degree of homogenization (Villamiel and de Jong 2000b).

Wu et al. (2000) indicated that ultrasound treatment before inoculation of culture results in an increase in WHC and a decrease in syneresis, without influencing the protein content. This could be due to reduction in fat globule size, leading to an increased total fat membrane surface area. The increased amount of membrane includes a significant amount of new-bonded casein, which would be hydrophilic. An increase in WHC during sonication is due to the ability of the proteins to retain water within the yogurt structure (Kinsella 1984). Ultrasound-induced homogenization causes a change in WHC of the milk proteins, which tends to reduce syneresis (Tamime and Robinson 1985).

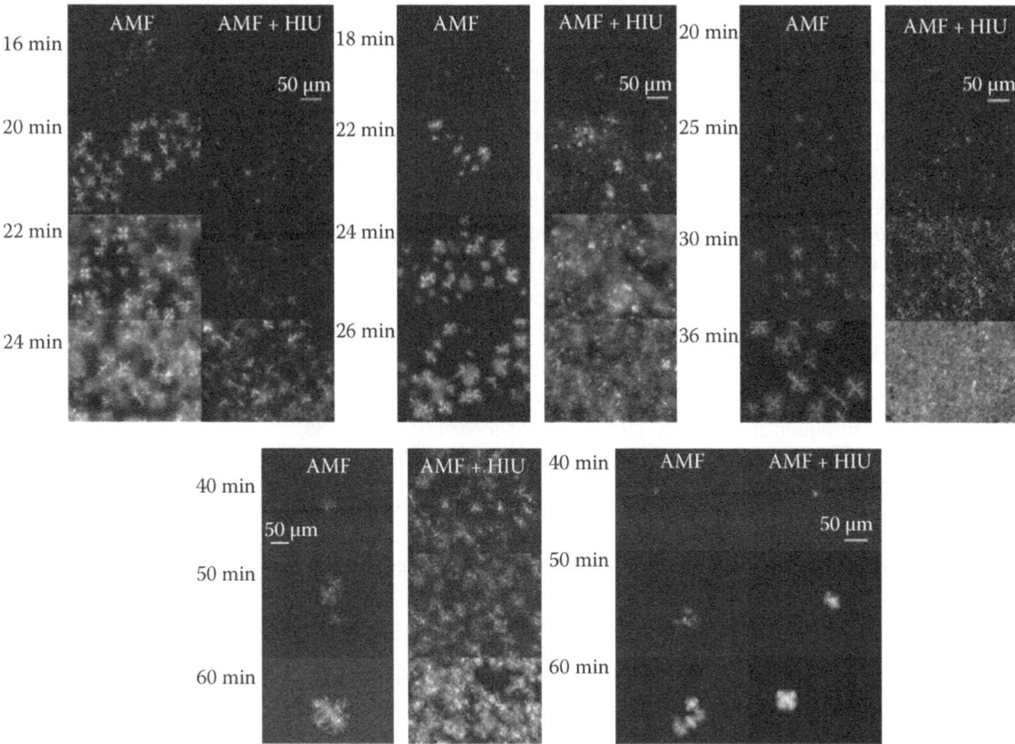

FIGURE 7.7 Microstructure of AMF crystallized at 22, 24, 26, 28 and 30°C with and without the application of high-intensity ultrasound after as a function of crystallization time. (Adapted from Martini, S., Suzuki, A. H., and Hartel, R. W., *Journal of the American Oil Chemists' Society*, 85, 621–28, 2008. With permission from Springer.)

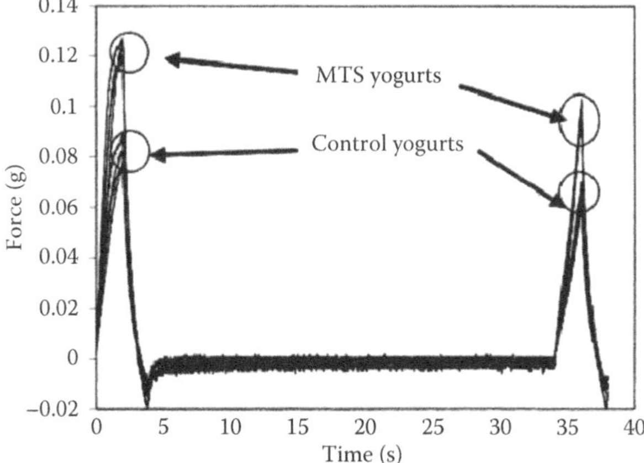

FIGURE 7.8 Texture profile analysis of four MTS-treated and four control yogurts. (Adapted from Vercet, A., Oria, R., Marquina, P., Crelier, S., and Lopez-Buesa, P., *Journal of Agricultural and Food Chemistry*, 50, 6165, 2002.)

7.9 CONCLUSION

The potential of ultrasound either as a stand-alone technique or in combination with mild heat or pressure treatments has been demonstrated. It has been identified as a possible technology to meet the U.S. Federal Drug Administration's (FDA) mandatory 5-log reduction for microbes within juices. Ultrasound finds wide application in an array of foods influencing both their rheological and functional properties. Certain sonication effects such as improved juice cloud stability, enhanced meat tenderization, and improved rheological properties of milk and milk products have been demonstrated. Extrinsic control parameters such as frequency, amplitude, ultrasonic intensity, treatment time, and temperature strongly influence the effects of sonication. Ultrasonic irradiation can also be employed for partial depolymerization and preparation of medium-sized macromolecules from large ones. The studies presented could be used to stimulate further research and development leading to increased industry adoption of ultrasound for food processing operations. To date, industrial application of ultrasound is restricted due to limited data availability for industrial scale up.

REFERENCES

Abismaïl, B., Canselier, J. P., Wilhelm, A. M., Delmas, H., and Gourdon, C. 1999. Emulsification by ultrasound: drop size distribution and stability. *Ultrasonics Sonochemistry*, 6:75–83.

Ashokkumar, M., and Grieser, F. 1999. Ultrasound assisted chemical process. *Reviews in Chemical Engineering*, 15 (1): 41–83.

Atchley, A. A., and Crum, L. A. 1988. Acoustic cavitation and bubble dynamics. In: Suslick, K.S. (ed.), *Ultrasound: Its Chemical, Physical and Biological Effects*, 1–64. New York: VCH.

Baker, R. A. 1979. Clarifying properties of pectin fractions separated by ester content. *Journal of Agriculture and Food Chemistry*, 27:1387–89.

Baker, R. A., and Cameron, R. G. 1999. Clouds of citrus juices and juice drinks. *Food Technology*, 53:64–69.

Barton, S., Bullock, C., and Weir, D. 1996. The effects of ultrasound on the activities of some glucosidase enzymes of industrial importance. *Enzyme and Microbial Technology*, 18:190–94.

Basak, S., and Ramaswamy, H. S. 1996. Ultra high pressure treatment of orange juice: a kinetic study on inactivation of pectin methyl esterase. *Food Research International*, 29 (7): 601–607.

Basedow, A. M., and Ebert, K. H. 1977. Ultrasonic degradation of polymers in solution. *Advances in Polymer Science*, 22:83–148.

Baumann, A. R., Martin, S. E., and Feng, H. 2005. Power ultrasound treatment of *Listeria monocytogenes* in apple cider. *Journal of Food Protection*, 68:2333–40.

Beresovsky, N., Kopelman, I. J., and Mizrahi, S. 1995. The role of pulp interactions in determining tomato juice viscosity. *Journal of Food Processing and Preservation*, 19:133–46.

Bergenstahl, B. A., and Claesson, P. M. 1990. Surface forces in emulsions. In: Larsson, K., and Friberg, S. E. (eds.), *Food Emulsions*, 41–96. New York: Marcel Dekker.

Bermudez-Aguirre, D., Mawson, R., and Barbosa-Canovas, G. V. 2008. Microstructure of fat globules in whole milk after thermosonication treatment. *Journal of Food Science*, 73 (7): E325–32.

Burgos, J. 1999. Manothermosonication. In: Robinson, R. K., Batt, C. A., and Patel, P. D. (eds.), *Encyclopedia of Food Microbiology*, 1462–69. New York: Academic Press.

Butz, P., and Tauscher, B. 2002. Emerging technologies: Chemical aspects. *Food Research International*, 35 (2/3): 279–84.

Cains, P. W., Martin, P. D., and Price, C. J. 1998. The use of ultrasound in industrial chemical synthesis and crystallization. 1. Applications to synthetic chemistry. *Organic Process Research and Development*, 1:234–48.

Cameron, R. G., Baker, R. A., and Grohmann, K. 1998. Multiple forms of pectinmethylesterase from citrus peel and their effects on juice cloud stability. *Journal of Food Science*, 63: 253–56.

Camino, N. A., Pérez, O. E., and Pilosof, A. M. R. 2009. Molecular and functional modification of hydroxypropylmethylcellulose by high-intensity ultrasound. *Food Hydrocolloids*, 23:1089–95.

Chen, R. H., Chang J. R., and Shyur, J. S. 1997. Effects of ultrasound conditions and storage in acidic solutions on changes in molecular weight and polydispersity of treated chitosan. *Carbohydrate Research*, 299:287–94.

Chung, K. M., Moon, T. W., Kim, H., and Chun, J. K. 2002. Physicochemical properties of sonicated mung bean, potato, and rice starches. *Cereal Chemistry*, 79:631–33.

Crum, L. A. 1995. Comments on the evolving field of sonochemistry by a cavitation physicist. *Ultrasonics Sonochemistry*, 2:147–62.

Cucheval, A., and Chow, R. C. Y. 2008. A study on the emulsification of oil by power ultrasound. *Ultrasonics Sonochemistry*, 15 (5): 916–20.

De Gennaro, L., Cavella, S., Romano, R., and Masi, P. 1999. The use of ultrasound in food technology I: Inactivation of peroxidase by thermosonication. *Journal of Food Engineering*, 39:401–407.

De Guerrero, S., López-Malo, A., and Alzamora, S. M. 2001. Effect of ultrasound on the survival of *Saccharomyces cerevisiae*: influence of temperature, pH and amplitude. *Innovative Food Science and Emerging Technologies*, 2:31–39.

Dickens, J. A., Lyon, C. E., and Wilson, R. L. 1991. Effect of ultrasonic radiation on some physical characteristics of broiler breast muscle and cooked meat. *Poultry Science*, 70 (2): 389–96.

Dolatowski, Z. J. 1988. Ultrasonics 2. Influence of ultrasonics on the ultrastructure of muscle tissue during curing. *Fleischwirtschaft*, 68 (10): 1301–303.

———. 1989. Ultrasonics 3. Influence of ultrasonics on the production technology and quality of cooked ham. *Fleischwirtschaft*, 69 (1): 106–11.

Dolatowski, Z. J., Stadnik, J., and Stasiak, D. 2007. Applications of ultrasound in food technology. *Acta Scientiarum Polonorum, Technologia Alimentaria*, 6 (3): 89–99.

Earnshaw, R. G., Appleyard, J., and Hurst, R. M. 1995. Understanding physical inactivation processes: combined preservation opportunities using heat, ultrasound and pressure. *International Journal of Food Microbiology*, 28 (2): 197–219.

Fellows, P. 2000. *Food Processing Technology: Principles and Practice*, 2nd ed. New York: CRC Press.

Floros, J. D., and Liang, H. 1994. Acoustically assisted diffusion through membranes and biomaterials. *Food Technology*, 48:79–84.

Gülseren, I., Güzey, D., Bruce, B. D., and Weiss, J. 2007. Structural and functional changes in ultrasonicated bovine serum albumin solutions. *Ultrasonics Sonochemistry*, 14 (2): 173–83.

Huff-Lonergan, E., and Lonergan, S. M. 2005. Mechanism of water-holding capacity of meat: the role of postmortem biochemical and structural changes. *Meat Science*, 71:194–204.

Iida, Y., Tuziuti, T., Yasui, K., Towata, A., and Kozuka, T. 2008. Control of viscosity in starch and polysaccharide solutions with ultrasound after gelatinization. *Innovative Food Science & Emerging Technologies*, 9 (2): 140–46.

Jambrak, A. R., Mason, T. J., Lelas, V., Herceg, Z., and Ljubić Herceg, I. L. 2008. Effect of ultrasound treatment on solubility and foaming properties of whey protein suspensions. *Journal of Food Engineering*, 86:281–87.

Jayasooriya, S. D., Bhandari, B. R., Torley, P., and D'Arcy, B. R. 2004. Effect of high power ultrasound waves on properties of meat: A review. *International Journal of Food Properties*, 7 (2): 301–19.

Jena, S., and Das, H. 2006. Modeling of particle size distribution of sonicated coconut milk emulsion: effect of emulsifiers and sonication time. *Food Research International*, 39 (5): 606–11.

Kardos, N., and Luche, J. L. 2001. Sonochemistry of carbohydrate compounds. *Carbohydrate Research*, 332: 115–31.

Kasaai, M. R., Arul, J., and Charlet, G. 2008. Fragmentation of chitosan by ultrasonic irradiation. *Ultrasonics Sonochemistry*, 15:1001–1008.

Kinsella, J. E. 1984. Milk proteins: Physico-chemical and functional properties. *CRC Critical Reviews in Food Science and Nutrition*, 21:197–262.

Klavons, J. A., Bennett, R. D., and Vannier, S. H. 1991. Nature of the protein constituent of commercial orange juice cloud. *Journal of Agricultural and Food Chemistry*, 39:1546–48.

Knorr, D., Ade-Omowaye, B. I. O., and Heinz, V. 2002. Nutritional improvement of plant foods by non-thermal processing. *Proceedings of the Nutrition Society*, 61:311–18.

Knorr, D., Zenker, M., Heinz, V., and Lee, D. 2004. Applications and potential of ultrasonics in food processing. *Trends in Food Science & Technology*, 15 (5): 261–66.

Koda, S., Mori, H., Matsumoto, K., and Nomura, H. 1994. Ultrasonic degradation of water-soluble polymers. *Polymer*, 35:30–33.

Krešić, G., Lelas, V., Jambrak, A. R., Herceg, Z., and Brnčić, S. R. 2008. Influence of novel food processing technologies on the rheological and thermophysical properties of whey proteins. *Journal of Food Engineering*, 87 (1): 64–73.

Kristensen, L. and Purslow, P. P. 2001. The effect of ageing on the waterholding capacity of pork: role of cytoskeletal proteins. *Meat Science*, 58:17e23.

Krop, J. J. P., and Pilnik, W. 1974. Effect of pectic acid and bivalent cations on cloud loss of citrus juices. *Lebensmittel-Wissenschaft und Technologie*, 7:62–63.

Kruus, P., Lawrie, J. A. G., and O'Neill, M. L. 1988. Polymerization and depolymerization by ultrasound. *Ultrasonics*, 26:352–55.

Lacroix, N., Fliss, I., and Makhlouf, J. 2005. Inactivation of pectin methylesterase and stabilization of opalescence in orange juice by dynamic high pressure. *Food Research International*, 38 (5): 569–76.

Lam Quoc, A., Mondor, M., Lamarche, F., Ippersiel, D., Bazinet, L., and Makhlouf, J. 2006. Effect of a combination of electrodialysis with bipolar membranes and mild heat treatment on the browning and opalescence stability of cloudy apple juice. *Food Research International*, 39 (7): 755–60.

Lawrie, R. A. 1985. The eating quality of meat. In: *Meat Science*, 159–207. Oxford, UK: Pergamon Press.

Lehmann, J. F., and Krusen, F. H. 1954. Effect of pulsed and continuous application of ultrasound on transport of ions through biologic membranes. *Archives of Physical Medicine and Rehabilitation*, 35:20–23.

Li, M. K., and Fogler, H. S. 1978. Acoustic emulsification. Part 2: Breakup of the large primary oil droplets in a water medium. *Journal of Fluid Mechanics*, 88:513–28.

Lorimer, J. P., Mason, T. J., Cuthbert, T. C., and Brookfield, E. A. 1995. Effect of ultrasound on the degradation of aqueous native dextran. *Ultrasonics Sonochemistry*, 2:s55–57.

Lyng, J. G., Allen, P., and McKenna, B. M. 1997. The influence of high intensity ultrasound baths on aspects of beef tenderness. *Journal of Muscle Foods*, 8:237–49.

Martinez, F. E., Desai, F. D., Davidson, A. G. E., Nakai, S., and Radcliffe, A. J. 1987. Ultrasonic homogenisation of expressed human milk to prevent fat loss during tube feeding. *Journal of Pediatric Gastroenterology and Nutrition*, 6:593–97.

Martini, S., Suzuki, A. H., and Hartel, R. W. 2008. Effect of high intensity ultrasound on crystallization behavior of anhydrous milk fat. *Journal of the American Oil Chemists' Society*, 85:621–28.

Mason, T. J. 1998. Power ultrasound in food processing – The way forward. In: Povey, M. J. W., and Mason, T. J. (eds.), *Ultrasound in Food Processing*, 105–26. London, UK: Thomson Science.

———. 2000. Large scale sonochemical processing: aspiration and actuality. *Ultrasonics Sonochemistry*, 7 (4): 145–49.

Mason, T. J., and Lorimer, J. P. 2002. *Applied Sonochemistry*, 285–90. Weinheim: Wiley.

Mason, T. J., and Luche, J. L. 1996. Ultrasound as a new tool for synthetic chemists. In: van Eldik, R., and Hubbard, C. D. (eds.), *Chemistry Under Extreme or Non Classical Conditions*, 317–80. New York: John Wiley and Spektrum Akademischer Verlag.

Mason, T. J., Riera, E., Vercet, A., and Lopez-Bueza, P. 2005. Application of ultrasound. In: Sun, D.-W. (ed.), *Emerging Technologies for Food Processing*, 323–51. Cambridge, MA: Elsevier.

McClements, J. 1995. Advances in the application of ultrasound in food analysis and processing. *Trends in Food Science and Technology*, 6:293–99.

Melody, J. L., Lonergan, S. M., Rowe, L. J., Huiatt, T. W., Mayes, M. S., and Huff-Lonergan, E. 2004. Early postmortem biochemical factors influence tenderness and water-holding capacity of three porcine muscles. *Journal of Animal Science*, 82:1195–205.

Miyazaki, T., Yamamoto, C., and Okada, S. 2001. Ultrasonic depolymerization of hyaluronic acid. *Polymer Degradation and Stability*, 74:77–85.

Mizrahi, S., and Berk, Z. 1970. Physico-chemical characteristics of orange juice cloud. *Journal of the Science of Food and Agriculture*, 21:250–53.

Morel, M. H., Dehlon, P., Autran, J. C., Leygue, J. P., and Bar-L'Helgouac'h, C. 2000. Effects of temperature, sonication time, and power settings on size distribution and extractability of total wheat flour proteins as determined by size-exclusion high-performance liquid chromatography. *Cereal Chemistry*, 77:685–91.

Moulton, K. J. and Wang, L. C. 1982. A pilot-plant study of continuous ultrasonic extraction of soybean protein. *Journal of Food Science*, 47:1127–29.

Müller, H. 1992. Die Käseausbeute als Kostenfaktor. *Deutshe Milchwirtschaft*, 37:1131–34.

Offer, G., and Cousins, T. 1992. The mechanism of drip production: formation of two compartments of extracellular space in muscle post mortem. *Journal of the Science of Food and Agriculture*, 58:107–16.

Paulsen, P., Hagen, U., and Bauer, F. 2001. Physical and chemical changes of pork loin: Ultrasonic curing compared to conventional pickle curing. *Fleischwirtschaft*, 81 (12): 91–93.

Piyasena, P., Mohareb, E., and McKellar, R. C. 2003. Inactivation of microbes using ultrasound: a review. *International Journal of Food Microbiology*, 87:207–16.

Porretta, S., Birzi, A., Ghizzoni, C., and Vicini, E. 1995. Effects of ultra-high hydrostatic pressure treatments on the quality of tomato juice. *Food Chemistry*, 52:35–41.

Portenlaenger, G., and Heusinger, H. 1997. The influence of frequency on the mechanical and radical effects for the ultrasonic degradation of dextranes. *Ultrasonics Sonochemistry*, 4:127–30.

Price, G. J. 1990. The use of ultrasound for the controlled degradation of polymer solutions. *Advances in Sonochemistry*, 1:231–87.

Price, G. J., Smith, P. F., and West, P. J. 1991. Ultrasonically initiated polymerization of methyl methacrylate. *Ultrasonics*, 29:166.

Raichel, D. R. 2000. *The Science and Applications of Acoustics*. New York: Springer.

Rao, M. A., and Cooley, H. J. 1992. Rheology of tomato pastes in steady dynamic shear. *Journal of Texture Studies*, 12:521–38.

Riesz, P. and Kondo, T. 1992. Free radical formation induced by ultrasound and its biological implications. *Free Radical Biology & Medicine*, 13:247–70.

Roberts, R. T. 1993. High intensity ultrasonics in food processing. *Chemistry and Industry*, 15:119–21.

Rosenthal, A. J. 1999. *Food Texture Measurement and Perception*. Gaithersburg, MD: Aspen.

Sala, F. J., Burgos, J., Condón, S., Lopez, P., and Raso, J. 1995. Effect of heat and ultrasound on microorganisms and enzymes. In: Gould, G. W. (ed.), *New Methods of Food Preservation*, 176–204. London: Blackie Academic & Professional.

Sams, A. R., and Feria, R. 1991. Microbial effects of ultrasonication of broiler drumstick skin. *Journal of Food Science*, 56:247–48.

Schäfer, A., Rosenvold, K., Purslow, P. P., Andersen, H. J., and Henckel, P. 2002. Physiological and structural events post mortem of importance for drip loss in pork. *Meat Science*, 61:355–66.

Scherba, G., Weigel, R. M., and O'Brien, W. D. 1991. Quantitative assessment of the germicidal efficacy of ultrasonic energy. *Applied and Environmental Microbiology*, 57 (7): 2079–84.

Schwartzberg, H. G., and Hartel, R. W. 1992. *Physical Chemistry of Foods*. New York: Marcel Dekker.

Seshadri, R., Weiss, J., Hulbert, G. J., and Mount, J. 2003. Ultrasonic processing influences rheological and optical properties of high-methoxyl pectin dispersions. *Food Hydrocolloids*, 17:191–97.

Seymour, I. J., Burfoot, D., Smith, R. L., Cox, L. A., and Lockwood, A. 2002. Ultrasound decontamination of minimally processed fruits and vegetables. *International Journal of Food Science & Technology*, 37:547–57.

Shahidi, F., and Han, X. Q. 1993. Encapsulation of food ingredients. *Critical Reviews in Food Science and Nutrition*, 33 (6): 501–47.

Shaker, R. R., Jumah, R. Y., and Jdayil, B. A. 2000. Rheological properties of plain yogurt during coagulation process: Impact of fat content and preheat treatment of milk. *Journal of Food Engineering*, 44:175–80.

Stadnik, J., Dolatowski, Z. J., and Baranowska, H. M. 2008. Effect of ultrasound treatment on water holding properties and microstructure of beef (*m. semimembranosus*) during ageing. *Lebensmittel-Wissenschaft und Technologie*, 41 (10): 2151–58.

Stead, D. 1986. Microbial lipases: their characeristics, role in food spoilage and industrial uses. *Journal of Dairy Research*, 53:481.

Stephanis, C., Hatiris, J., and Mourmouras, D. 1997. The process (mechanism) of erosion of soluble brittle materials caused by cavitation. *Ultrasonics Sonochemistry*, 4 (3): 269–71.

Suslick, K. 1988. *Ultrasound: Its Chemical, Physical and Biological Effects*. New York: VCH.

Suslick, K. S. 1991. The temperature of cavitation. *Science*, 253:1397–98.

Tamime, A. Y., and Robinson, R. K. 1985. *Yogurt: Science and Technology*. Oxford: Pergamon Press.

Thakur, B. R., and Nelson, P. E. 1997. Inactivation of lipoxygenase in whole soy flour suspension by ultrasonic cavitation. *Nahrung*, 41:299–301.

Tsai, S. C., and Zammouri, K. 1988. Role of interparticular van der Waals forces in rheology of concentrated suspensions. *Journal of Rheology*, 32:737–50.

Valero, M., Recrosio, N., Saura, D., Munoz, N., Martıc, N., and Lizama, V. 2007. Effects of ultrasonic treatments in orange juice processing. *Journal of Food Engineering*, 80: 509–16.

Vargas, L., Piao, A., Domingos, R., and Carmona, E. 2004. Ultrasound effects invertase from *Aspergillus niger*. *World Journal of Microbiology and Biotechnology*, 20: 137–42.

Vercet, A., Burgos, J., and Lopez-Buesa, P. 2001. Manothermosonication of foods and food-resembling systems: Effect on nutrient content and nonenzymatic browning. *Journal of Agriculture and Food Chemistry*, 49 (1): 483–89.

Vercet, A., Lopez, P., and Burgos, J. 1997. Inactivation of heat resistant lipase and protease from *Pseudomonas fluorescens* by manothermosonication. *Journal of Dairy Science*, 80:29–36.

Vercet, A., Oria, R., Marquina, P., Crelier, S., and Lopez-Buesa, P. 2002b. Rheological properties of yoghurt made with milk submitted to manothermosonication. *Journal of Agricultural and Food Chemistry*, 50 (21): 6165–71.

Vercet, A., Sanchez, C., Burgos, J., Montanes, L., and Lopez-Buesa, P. 2002a. The effects of manothermosonication on tomato pectic enzymes and tomato paste rheological properties. *Journal of Food Engineering*, 53: 273–78.

Versteeg, C., Rombouts, F. M., Spaansen, C. H., and Pilnik, W. 1980. Thermostability and orange juice cloud destabilizing properties of multiple pestinesterases from orange. *Journal of Food Science*, 45:969–72.

Villamiel, M., and de Jong, P. 2000a. Inactivation of *Pseudomonas fluorescens* and *Streptococcus thermophilus* in trypticase soy broth and total bacteria in milk by continuous-flow ultrasonic treatment and conventional heating. *Journal of Food Engineering*, 45:171–79.

———. 2000b. Influence of high-intensity ultrasound and heat treatment in continuous flow on fat, proteins and native enzymes of milk. *Journal of Agriculture and Food Chemistry*, 48:472–78.

Vodeničarová, M., Dřimalová, G., Hromádková, Z., Malovíková, A., and Ebringerová, A. 2006. Xyloglucan degradation using different radiation sources: a comparative study. *Ultrasonics Sonochemistry*, 13:157–64.

Voragen, A. G. J., Pilnik, W., Thibault, J. F., Axelos, M. A.V., and Renard, C. M. G. C. 1995. Pectins. In: Stephen, A. M. (ed.), *Food Polysaccharides and Their Applications*, 287–340. New York: Marcel Dekker.

Wang, D., and Sakakibara, M. 1997. Lactose hydrolysis and β-galactosidase activity in sonicated fermentation with lactobacillus strains. *Ultrasonics Sonochemistry*, 4:255–61.

Wu, H., Hulbert, G. J., and Mount, J. R. 2000. Effects of ultrasound on milk homogenisation and fermentation with yoghurt starter. *Innovative Food Science and Emerging Technologies* 1 (3): 211–18.

Xu, W., Nikolov, A., Wasan, D. T., Gonsalves, A., and Borwankar, R. P. 1998. Fat particle structure and stability of food emulsions. *Journal of Food Science*, 63 (2): 183–88.

Zayas, J. F., and Gorbatow, W. M. 1978. The use of ultrasonics in meat technology. *Fleischwirtsch*, 58: 1009–21.

Zayas, J. F., and Orlova, T. N. 1970. The application of ultrasonic vibrations for tenderisation of meat. *Izvestia vysshykh unhebnykh zavedeniy*, 4:54–56.

Zenker, M., Hienz, V., and Knorr, D. 2003. Application of ultrasound-assisted thermal processing for preservation and quality retention of liquid foods. *Journal of Food Protection*, 66:1642–49.

8 Effect of Irradiation on Food Texture and Rheology

Paramita Bhattacharjee and Rekha S. Singhal
Institute of Chemical Technology

CONTENTS

8.1 INTRODUCTION

Irradiation is the process of exposing food to ionizing radiation in order to destroy food-poisoning bacteria such as *Salmonella*, *Campylobacter*, and *Escherichia coli* and viruses, and for insect disinfestation in foods. Over 40 countries presently allow food irradiation, and volumes are estimated to exceed 500,000 metric tons annually worldwide. Food irradiation exposes food to electron beams, x-rays, or γ-rays, and produces an effect similar to pasteurization, cooking, or other forms of heat treatment, with less detrimental effect on appearance and texture. The most important applications of food irradiation include sprout inhibition, delay of maturation and senescence in fruits and vegetables, ease of juice extraction, shortening of cooking time, improvement of malting properties of grains, and rehydration characteristics of pulses and dried vegetables (Diehl 1995). When subjected to irradiation, cells experience DNA damage that cannot be repaired. These cells undergo apoptosis due to which the potential genetic damage from the larger tissue is eliminated, or they undergo nonlethal DNA mutations that are passed on to subsequent cell divisions.

The specialty of processing food by ionizing radiation is that the energy density per atomic transition is very high; it can cleave molecules and induce ionization, which is not achieved by mere heating. This is the reason for both new effects and new concerns. The treatment of solid food by ionizing radiation can provide an effect similar to heat pasteurization of liquids such as milk. However, the use of the term "cold pasteurization" to describe irradiated foods is controversial, since pasteurization and irradiation are fundamentally different processes.

Because of the wide variation in their structure and composition, both fluid and solid foods exhibit flow behavior ranging from simple Newtonian to time-dependent non-Newtonian and viscoelastic. Further, a given food may exhibit Newtonian or non-Newtonian behavior depending on its origin, concentration, and previous storage and/or processing. The complex nature of foods, their variability, and their diverse behavior necessitates categorizing foods under specific rheological behaviors. The study of rheology of foods is important for understanding its structure and therefore helps in quality control and in process engineering applications. It also allows correlation of sensory properties of food products with its definite rheological properties (Rao and Rizvi 1995).

Irradiation is a well-known technology to preserve shelf life and quality attributes of foods. It fundamentally affects the food structure, and in the process modifies its rheological properties and texture. Voluminous literature is devoted to various aspects of irradiation and irradiated foods with relatively less stress on the rheology and texture.

Low-dose gamma irradiation is known to preserve nutritional and sensory qualities of postharvest fresh-cut produce and several food products (Siddhuraju, Makkar, and Becker 2002; Hu and Jiang 2007). It has been reported to retain vitamin C and overall nutritional quality in fresh vegetables such as spinach, lettuce, tomato, cilantro, green onion, parsley, and carrot (Fan and Sokorai 2008); green gram and garden pea sprouts (Hajare et al. 2007); and bitter gourd (Khattak et al. 2005b). It is also known to retain organic and amino acids in Korean fermented vegetables—more popularly known as Kimchi (Song et al. 2004) and in bean seeds, with simultaneous reduction in the antinutritional behenic acid (Bhat, Sridhar, and Seena 2008); minimize browning and loss of vitamin C and antioxidants in lettuce (Fan et al. 2003c); extend shelf life of whole wheat flour up to six months without any change in nutritional attributes (Marathe et al. 2002); improve sensory quality of fishery products (Venugopal, Doke, and Thomas 1999); and reduce phytic acid and tannin content and increase in vitro protein digestibility of maize and sorghum (Hassan et al. 2009). However, gamma irradiation is reported to significantly lower vitamin C content in minimally processed pineapples without adversely affecting the total carotenoids (Hajare et al. 2006a).

Besides gamma radiation, infrared radiation has also been reported to preserve the nutritional quality of maize, rice, and sorghum and reduce antinutritional factors in beans, maize, and sorghum (Keya and Sherman 1997). Preservation and extension of shelf life of apples (Hemmaty, Moallemi, and Naseri 2007) and improvement of antioxidant capacity of mangoes (González-Aguilar et al. 2007) by UV-radiation has been similarly documented in recent times.

Though there are several reports on the effects of irradiation on microbiological, nutritional, and sensory quality of foods, scientific literature on the effect on rheological profile and texture attributes of different foods subjected to irradiation is relatively scant. In this chapter, we endeavor to focus on these lesser known aspects of food irradiation.

8.2 EFFECT OF IRRADIATION ON RHEOLOGY

Macromolecules in food, mainly starch, proteins, and other biomolecules such as pectins, cellulose, or added hydrocolloids, are well known to influence the rheological and texture profile of foods. Irradiation is known to affect these macromolecules, and hence their related properties.

8.2.1 Effect on Starch and Starch-Based Products

Irradiation results in significant physical modification of starch, causing it to depolymerize. It is assumed that γ-irradiation can alter the stability of the crystallites in starch leading to their partial disruption. γ-Irradiation hydrolyzes chemical bonds, cleaving large molecules of starch into dextrin, which could be either electrically charged or uncharged free radicals (Diehl 1995). The granule structure of starch remains visually undamaged at low doses of irradiation but may suffer severe damage at higher doses of irradiation (100 kGy). Degradation of starch is accompanied by uncoiling of starch chains as well as breaking of hydrogen bonds, thereby shortening the starch chain length. This results in the formation of dense interlaced connecting structure of the starch exudates when starch granules swell as temperature approaches 90°C. These highly swollen starch granules break easily, leading to decreased viscosity of the starch exudates. Irradiation causes depolymerization of molecules by the breakage of glycosidic bonds resulting in smaller polysaccharide units, which give solutions with a lower consistency index (CI). Breakdown values of starch granules generally decrease with an increase in the irradiation dose level. Irradiation also seems to diminish the tendency of starch to retrograde because of starch chain break up, thereby resulting in the inability to produce longer helical structure which and negatively impacts association between starch molecules. It is opined that the increase in viscosity during heating of starch suspension is mainly due to the swelling of starch granules (Wu et al. 2002; Yu and Wang, 2007).

Three major parameters of a typical starch pasting curve are peak viscosity (PKV), cool pasting viscosity or final viscosity (CPV), and hot pasting viscosity or holding viscosity (HPV), which are obtained with a Rapid Visco Analyzer (RVA). A programed heating and cooling cycle is reportedly used in RVA processing, wherein the samples are held at 50°C for 1 min, heated to 95°C at 6°C/min, held at 95°C for 2.7 min, then cooled from 95°C to 50°C at 6°C/min and finally held again at 50°C for 2 min. PKV and HPV are measured at 95°C and CPV at 50°C. From the recorded values of PKV, CPV, and HPV, the values of breakdown viscosity (PKV–HPV) and setback viscosity (CPV–PKV) are calculated. All these parameters considerably decrease with increasing dose levels of irradiation (Wu et al. 2002; Ezekiel et al. 2007; Wang 2007; Sung, Hong, and Chang 2008).

It is generally accepted that the increase in viscosity that occurs during heating of starch suspension is mainly due to the swelling of the starch granules and breakdown of viscosity is caused by the rupture of the swollen starch granules. The swelling degree of starch granules is reported to be in direct proportion to the average size of the starch granules in the rice grain. With increasing dose, an increasing number of starch granules rupture prior to RVA processing and the average size of the starch granules in the rice grains decreases. Therefore, the observed decrease in PKV, HPV, and breakdown values are possibly due to the decreased size of starch granules (commonly known as starch depolymerization), degradation, and uncoiling of starch grains as well as due to breaking of hydrogen bonds within the molecules, caused by gamma irradiation. The parameters of CPV and setback viscosity (SBV) have a significant correlation with the degree of polymerization after RVA processing. With increasing dose of irradiation, the increasing breakage of starch granules causes a reduction in degree of polymerization; in other words, the starch is depolymerized, which further accelerates decrease in CPV and SBV (Wu et al. 2002; Ezekiel et al. 2007; Wang 2007; Sung et al. 2008). Results of studies conducted with different rice cultivars (irradiated with 0–1 kGy doses) with similar apparent amylose content (AAA) on the above-mentioned parameters of starch viscosity are shown in Table 8.1 (Wu et al. 2002). Similar trends were observed with rice starch, irradiated

TABLE 8.1
Mean Values for the Major Parameters of Starch Viscosity Curves in Different Types of Rice With Similar AAA Cultivars after Gamma Irradiation

Type of Rice Cultivars	Dose (kGy)	PKV (RVU)	HPV (RVU)	CPV (RVU)	SBV (RVU)	PKT (min)	PT (°C)
Indica rice	0	156.6	113.3	305.3	148.7	5.233	75.2
(ZHE 852)	0.2	152.3	109.3	276.1	123.8	5.217	76.4
	0.4	147.3	102.1	211.3	64.0	4.873	78.9
	0.6	138.7	98.5	172.5	33.8	4.621	82.3
	0.8	110.3	90.3	143.1	32.8	4.451	84.1
	1.0	61.3	50.1	75.3	14.0	4.398	85.9
Japonica rice	0	215.2	125.1	239.6	24.4	4.7996	70.5
(Xiushui11)	0.2	169.2	102.1	187.2	18	4.7996	72.1
	0.4	149.1	75.2	154.2	5.1	4.533	76.3
	0.6	133.5	64.7	131.4	–2.1	4.5996	83.2
	0.8	114.6	48.2	98.5	–16.1	4.2663	86.2
	1.0	74.4	28.7	49.7	–24.7	4.1996	88.2
Hybrid rice	0	241.3	172.4	293.2	–9.6	5.133	79.2
(Xieyou46)	0.2	231.1	159.2	270.2	–48.6	5.133	80.6
	0.4	212.2	145.4	231.7	–68.5	4.7996	81.5
	0.6	179.4	113.3	182.7	–67.6	4.9996	86.5
	0.8	153.8	87.6	143.7	–153.8	4.4663	87.3
	1.0	126.1	79.2	111.8	–126.1	4.3996	90.7

PKV, peak viscosity; HPV, hot pasting viscosity; CPV, cool pasting viscosity; SBV, setback viscosity; PKT, peak time; PT, pasting temperature, RVU, rapid viscosity units

Source: Wu, D., Shu, Q., Wang, Z., and Xia, Y., *Radiation Physics and Chemistry*, 65, 79–86, 2002. With permission.

TABLE 8.2
The Major Parameters of Rice Starch Viscosity Curves With Different Doses

Dose (kGy)	PKV (RVU)	HPV (RVU)	CPV (RVU)	SBV (RVU)	Breakdown (RVU)	PKT (min)
0	251.33	162.42	271.42	20.09	88.91	6.996
2	213.92	148.83	221.42	7.50	65.09	6.5996
5	186.17	134.25	178.50	–7.67	51.92	6.0663
8	144.17	97.25	122.00	–22.17	46.92	5.7330
10	96.08	72.92	59.25	–36.83	23.16	5.6663

PKV, peak viscosity; HPV, hot pasting viscosity; CPV, cool pasting viscosity; SBV, setback viscosity; PKT, peak time; RVU, rapid viscosity units

Source: Wang, Y. W. J., *Food Research International*, 40, 297–303, 2007. With permission.

at higher doses of 2–10 kGy, as shown in Table 8.2 (Wang 2007). Figure 8.1 shows the effect of irradiation at 2–10 kGy on the structure of starch from inner rice endosperm.

However, reports on the effect of irradiation on the PKV of rice are inconsistent. Most reports suggest the PKV of irradiated rice is reduced with storage time. A significant reduction in PKV over one year of storage has been found at 1 kGy dose of irradiation compared to lower doses. Peak viscosities of the irradiated rice have been observed to shift to lower temperature. This is suggested

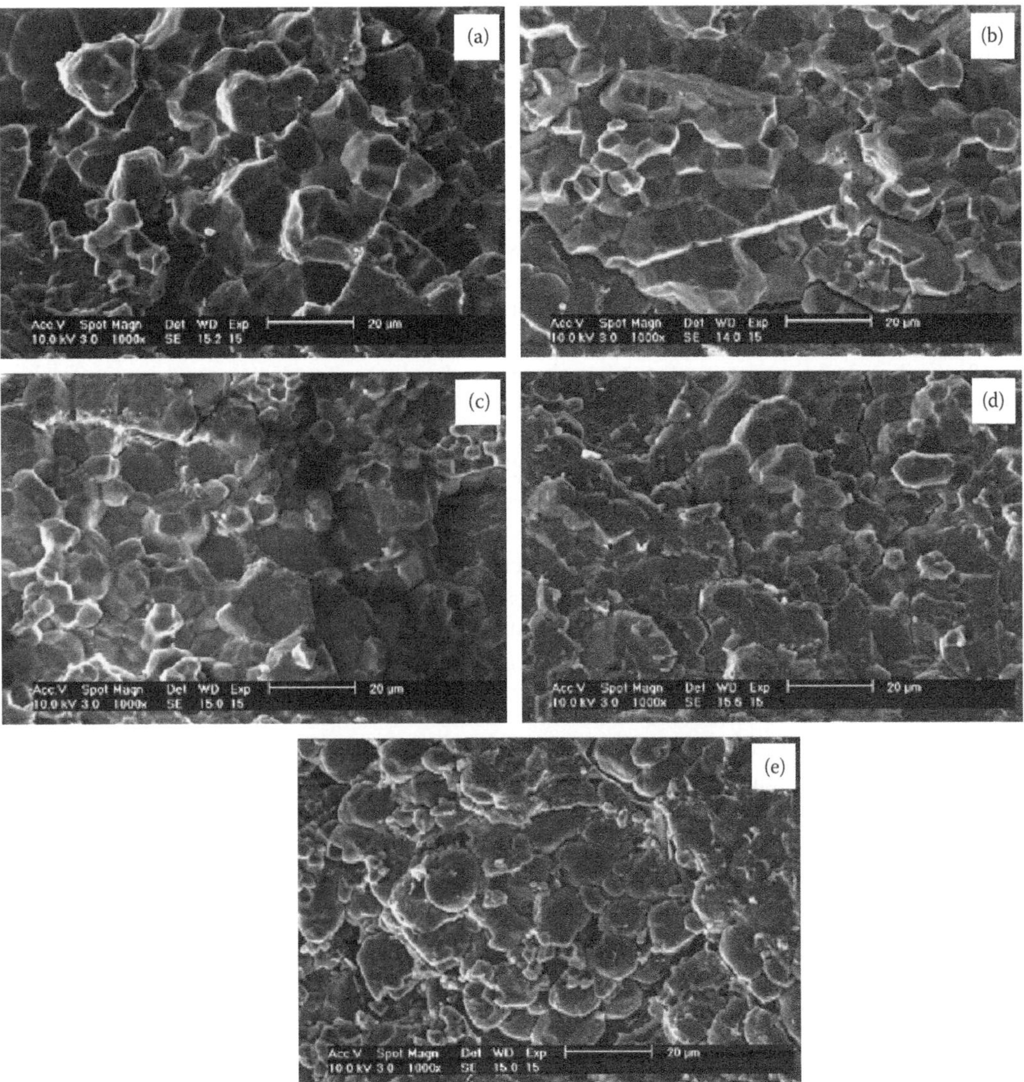

FIGURE 8.1 Effect of irradiation dose on the starch structure of outer rice endosperm: (a) 0 kGy (b) 2 kGy (c) 5 kGy (d) 8 kGy, and (e) 10 kGy. (From Wang, Y. W. J., *Food Research International*, 40, 297–303, 2007. With permission.)

to be principally due to protein, which serves as a barrier to swelling of starch granules. Disulfide linkages between proteins in rice are known to increase during irradiation and subsequent storage, which may be responsible for lower pasting viscosities. The final viscosity and consistency of rice samples remain almost unchanged during storage. CPV shows the lowest reduction with dose between the three parameters. Pasting temperature (PT) of the irradiated rice starch is known to increase with an increase in the irradiation dose. Breakdown and setback are the sensitive indices to differentiate the changes in irradiated and nonirradiated rice during the aging process. The breakdown of nonirradiated and low-dose irradiated rice decreases significantly over two months of storage, after which it remains fairly constant until the end of storage, while that of high-dose irradiated rice (1.0 kGy) decreases throughout the storage period (Sung 2005).

In contrast to rice irradiated at high dose (1.0 kGy) which does not show any changes in SBV, that of the nonirradiated and low-dose irradiated rice significantly increases. This behavior of high-dose irradiated rice is seen throughout the storage period. The lower setback, consistency, and breakdown values of high-dose irradiated rice during storage for one year manifests itself as a softer texture of the cooked irradiated rice. This suggests possible use of γ-irradiation for shortening the aging time of *indica* and *japonica* rice varieties (Sung 2005; Sirisoontaralak and Noomhorm 2006, 2007). Fragments of starch chains caused by γ-irradiation do not produce long helical structures on association, which is typically seen during retrogradation of starch from cooked nonirradiated rice. Thus helix reformation which accelerates retrogradation of starch paste, is diminished during storage of high-dose irradiated rice starch (Ohashi et al. 1980; Sirisoontaralak and Noomhorm 2007).

Irradiation of potatoes at 0.5 kGy also causes a significant effect on pasting properties of starch isulated from the same, resulting in starch with significantly lower peak, trough, and breakdown viscosity than the starches treated with 0.1 kGy of irradiation. Irradiation also reduces the viscosity of starch isolated from potatoes irradiated with 0.05–0.20 kGy and stored at 5 and 20°C, which is chiefly attributed to degradation of starch to simpler molecules such as dextrins and sugars (Ezekiel et al. 2007). Irradiation of potato starch gels at 5 or 10 kGy disrupts the physical state of the gels, as is true for cereal starches. The irradiation-destroyed gels undergo a phase separation by forming a water phase over the denser gel phase. Radiolysis yields unstable reactive agents, which are subsequently converted to stable end products. Since water is the most abundant molecule in starch gels, its radiolysis products (namely hydroxyl radicals, free hydrogen atoms, and solvated electrons) predominantly determine the radiation-induced chemical changes in foodstuffs. The free radicals so formed trigger these radiation-induced modifications in food structure through very complicated free-radical reactions (Kizil and Irudayaraj 2006). A similar effect of reduction in PKV up to 30% consequent to radiolysis has also been observed for irradiated starch from semolina, when irradiated at 0.25 kGy (Rao et al. 1994).

8.2.1.1 Role of Amylose/Amylopectin

The quality traits in starch properties consequent to irradiation have been studied in rice, wheat, and maize. Amylose/amylopectin ratio is a quality parameter for rice cooking and its rheological characteristics. Amylopectin possesses higher crystalline order than amylose and since its content is higher in rice starch, it is more prone to degradation by γ-rays. Contrary to the expectations that the crystallinity of starch would decrease with increasing radiation dose, the crystallinity of irradiated starch has been found to be higher when subjected to higher irradiation dose in rice and maize. It was hypothesized that although the crystallinity of starch increased on irradiation, the quality of the crystallites would be poorer. In nonirradiated starches, there are hydrogen bonds between the starch molecules. On irradiation, these hydrogen bonds are disrupted, which enables hydrogen bond formation between water and the exposed hydroxyl groups on amylose and amylopectin in the starch. Consequently, there is a decrease in the viscosity of the irradiated starches (Bao and Corke 2002).

Addition of acids reduces viscosity of irradiated starches even further. However, these reports are confined to the application of irradiation to acidified (pH-adjusted) starches. Literature on the reverse treatment (i.e., acid to irradiated starches) as it may apply to foods has not been reported. In contrast to the native starches that have lower viscosity at pH 2.5 than at pH 7.0, γ-irradiated starches reportedly have a higher viscosity at pH 2.5 than at pH 7.0. It is opined that although γ-irradiation produces poor-quality crystallites, the crystalline quality is improved when the irradiated starches are placed in acidic solutions. This could possibly be due to the action of the protons on the intra- and/or intermolecular linkages induced by irradiation between starch chains, which in turn gives rise to more perfect crystallites of irradiated starch compared to those in nonirradiated starches. This could explain the higher viscosity of irradiated starches at low pH (Bao and Corke 2002).

The evolution of linear viscoelastic functions, G' as the storage modulus and G'' as the loss modulus, with frequency, shows a similar behavior for the nonirradiated and irradiated samples of white pepper gels. In all cases, the storage and loss moduli curves are parallel, i.e., G' values being higher than G'', the magnitude being less than one order of magnitude, and both being slightly frequency dependent. This is typical of a weak gel spectrum. However, the effect of irradiation on gel structure is expressed by a decrease of the magnitude of the G′ and G″ values. The effect is different for amylose and amylopectin. Irradiation of amylose favors its gel structure due to loss of linearity of the polysaccharide molecules, while amylopectin shows decreased gel strength with increasing irradiation dose. Since amylopectin represents the main polymeric fraction of white pepper, the effect of the ionizing radiation follow the behavior of the amylopectin fraction (Esteves et al. 2002).

8.2.1.2 Role of Starch–Protein Interaction

Several attempts to explain the changes in pasting viscosity associated with aging of rice starch (isolated from irradiated rice) have focused on the properties of starch, lipid, and protein and the interaction of these components during storage. The modification of the proteins in rice principally influences the pasting viscosity of the starch and also that of its products. Proteins influence pasting viscosity through binding water to increase the concentration of the dispersed and viscous phase of gelatinized starch and through the network linked by disulfide bonds. This modification causes proteins to lose the ability to bind water and form a network during starch gelatinization. Degradation of proteins become prominent at relatively high doses of irradiation and can involve deamination, decarboxylation, reduction of disulfide bridges, oxidation of sulfhydryl groups, decomposition of amino acid side groups, and changes in peptide linkages. Irradiation can induce cleavage or a scission of polypeptide chains into lower molecular weight fragments and/or aggregation of proteins, increase protein solubility, and affect enzymatic activities, even at relatively low dose (1.5 kGy). The degradation in the ordered structure of starch as a consequence of changes in its protein networks has been suggested for waxy rice (Urbain 1986; Chrastil 1994; Martin and Fitzgerald 2002; Zhao et al. 2003; Sirisoontaralak and Noomhorm 2007).

8.2.1.3 Role of Starch–Lipid Interactions

Besides protein, interaction of starch with lipid significantly alters viscosity of starch. For nonirradiated rice, it has been observed that formation of the helical structure of the starch molecule with fatty acids, increase during storage which restricts the passage of water, represses swelling of starch granules, and increases viscosity of rice starch paste; whereas in the case of irradiated rice, cleavage of glycosidic bonds of linear chain molecules of starch results in the formation of short-chain polymers and hinders the formation of helical structures with fatty acids during storage. The passage of water through starch is therefore not as restricted as is in naturally aged rice. This facilitates the swelling of the starchy granules and lowers viscosity. Thus starch pastes from irradiated rice flour have low consistency and form soft gels (Sirisoontaralak and Noomhorm 2007).

Besides γ-irradiation, studies have also been carried out on UV irradiation. UV irradiation markedly decreases the paste viscosity of cassava starch, showing a maximal reduction with the mercury vapor lamp. However, UV irradiation does not change the rheological properties of corn starch. The unchanged paste viscosity of irradiated corn starch could result from an inhibition of swelling induced by amylose–lipid complexes. There are, however, contradictory reports on decrease of viscosity of corn starch suspension consequent to UV irradiation. It has been suggested that expansion is related to starch swelling, attributable to amylopectin. Moreover, starch–lipid complexes are considered to inhibit swelling and the lesser degree of disruption of starch granules during gelatinization is supposedly due to lipid surface coating. Thus the decrease in the pasting viscosity of UV-irradiated cassava starch samples can result from a decrease in swelling due to partial depolymerization of amylopectin and/or amylose in amorphous regions (Bertolini, Mestres, and Colonna 2000).

8.2.1.4 Role of Other Constituents

Suppression of amylograph peak viscosities of long and medium grain rice flours by bran hemicelluloses has also been reported. This phenomenon also holds good for wheat starches (Bao and Corke 2002; Wu et al. 2002; MacArthur and D'appolonia 1984; Sabularse et al. 1992; Kang et al. 1999).

8.2.1.5 Role of Additives

The cleavage of starch polymer by γ-irradiation is accompanied by the production of free radicals that reduce the viscosity of starch by chain reaction. The inorganic peroxides used as additives are easily decomposed by γ-irradiation to produce free radicals. Therefore, a combination of γ-irradiation and inorganic peroxides has a synergistic effect on the formation of free radicals within the starch molecules, which can play a decisive role in decreasing the initial viscosity and increasing the viscosity stability of starch. In one study, the effect of γ-irradiation (10 kGy) on viscosity stability of starches consequent to addition of 1–3% ammonium persulfate (APS) to starches has been investigated. It has been found that initial viscosity decreases with an increase in the concentration of added APS. Addition of more that 2% APS dramatically improves the viscosity stability of starch. These investigations suggest that production of modified starches with various levels of viscosity as well as with excellent stability are feasible by control of γ-irradiation dose levels and the addition of APS. Figure 8.2 shows the effects of APS on the viscosity stability of starch paste (Kang et al. 1999).

8.2.2 Effect on Protein and Protein-Based Food Matrices

Irradiation at 1 kGy has shown a decrease in gluten viscosity of commercial Mexican bread-making wheat flour (Arvizu et al. 2006). Studies are reported on the effects of gamma irradiation on the rheological behavior of mixtures of proteins (soy, caseinates, and whey) and glycerol, wherein a decrease in viscosity is observed with irradiation dose. Similar behavior has also been observed for dispersions containing calcium caseinates and glycerol. At a 2:1 ratio of proteins:glycerol, 21, 35, and 40% reductions in viscosity are observed at 5, 15, and 25 kGy, respectively; the corresponding values for

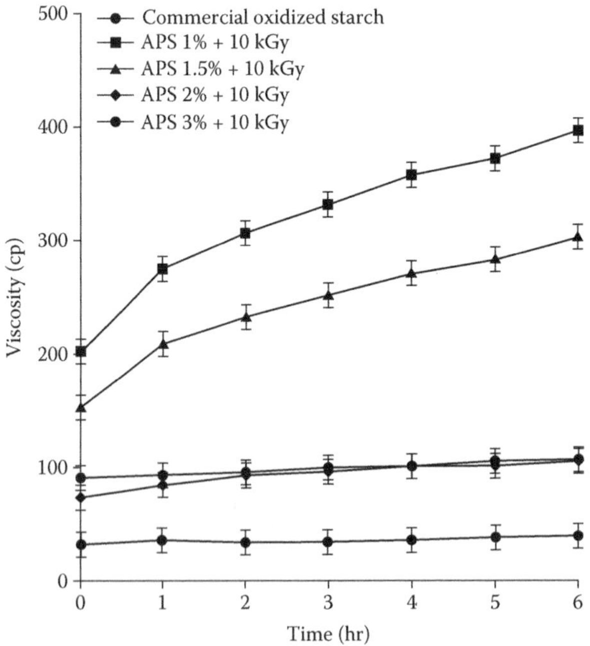

FIGURE 8.2 Effects of ammonium persulfate on the viscosity stability of starch paste. Starch paste (15%) was made from 10 kGy irradiated starch after mixing 1–3% ammonium persulfate by soaking method. (From Kang, I. J., Byun, M. W., Yook, H. S., Bae, C. H., Lee, H. S., Kwon, J. H., and Chung, C. K., *Radiation Physics and Chemistry*, 54, 425–30, 1999. With permission.)

a 1:1 ratio are 23, 33, and 37%. For these non-Newtonian dispersions, the apparent viscosities are, however, unaffected by irradiation. Sodium caseinate shows a trend toward formation of aggregation of macromolecules at 5 kGy. The change in viscosities of solutions of soy protein isolate, whey protein concentrate, calcium caseinate, and sodium caseinate in glycerol with irradiation dose are illustrated in Figures 8.3a, b, c, and d respectively (Sabato and Lacroix 2002).

The viscosity of liquid egg white decreases dramatically on irradiation regardless of irradiation doses used. The egg white becomes watery even after irradiating the shell eggs at 1.0 kGy. Ovomucin is one of the major proteins in egg white, which plays an important role in its gel-like structure. Irradiation causes changes in carbohydrate and protein moieties involved in formation of ovomucin complex, resulting in a loss of gel-like structure. The dramatic decrease in the viscosity of egg white is an important physical change in egg by irradiation, which can be used in egg processing. Watery egg white will facilitate the separation of egg white and yolk and low viscosity can improve the flow of liquid egg white or liquid whole egg in plant facilities that break eggs (Min et al. 2005).

Several studies have suggested that irradiation below 3.5 kGy does not affect the gelation properties of liquid egg white significantly. Hardness, springiness (elasticity), cohesiveness, gumminess, and chewiness of irradiated eggs are not very different from their nonirradiated counter parts. Sensory analysis is also unable to detect any texture differences between irradiated and nonirradiated hard-cooked egg whites. Therefore, it has been suggested that irradiation of shell eggs below

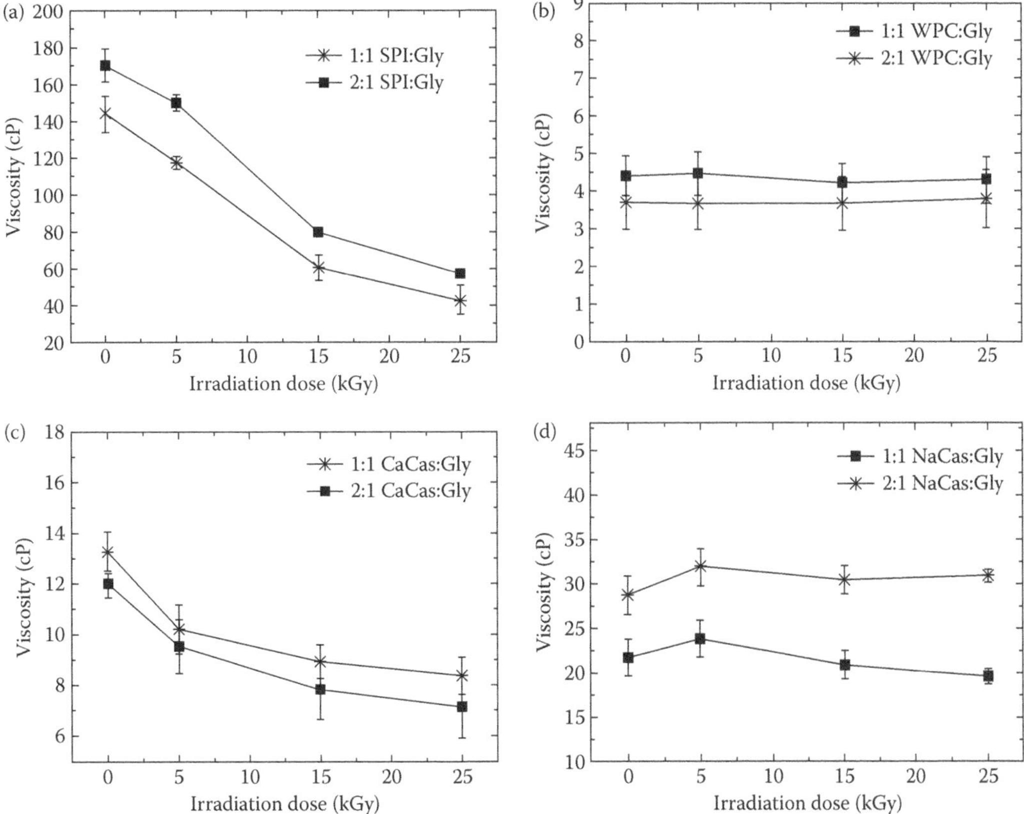

FIGURE 8.3 (a) Viscosity of soy protein isolate and glycerol solution vs irradiation dose. (b) Viscosity of whey protein isolate and glycerol solution vs irradiation dose. (c) Viscosity of calcium caseinate and glycerol solution vs irradiation dose. (d) Viscosity of sodium caseinate and glycerol solution vs irradiation dose. (From Sabato, S. F., and Lacroix, M., *Radiation Physics and Chemistry*, 63, 357–59, 2002. With permission.)

2.0 kGy does not alter the thermal characteristics of egg white proteins. If used for pasteurizing liquid eggs, however, irradiation can improve the efficiency of egg processing steps such as adding or mixing for removal of sugars and for spray drying (Min et al. 2005).

Studies have also found that the storage modulus of the egg yolk as a function of frequency increases during frozen storage for both nonprocessed and irradiated samples. This may be due to the freezing-induced aggregation and gelation of lipoproteins in egg yolk. The storage modulus of the irradiated samples is a little higher than that of nonprocessed samples before storage. This suggests some coagulation of lipoproteins in irradiated egg yolk, thus causing the loss of soluble protein immediately after irradiation. Irradiated samples show aggregation during storage. A significant decrease in the storage modulus is observed in nonprocessed samples, which indicates structural breakdown in proteins and other polymers by enzyme or microbial activities. A corresponding decrease in the irradiated samples is delayed and is insignificant. Electron beam irradiation can therefore be an attractive alternative to other preservation methods for liquid egg proteins (Huang, Herald, and Mueller 1997).

8.2.3 Effect on Other Foods and Food Biopolymers

Irradiation may not always affect rheology of foods at certain doses. The best example is that of honey, when irradiated at 5 and 10 kGy, which does not show any significant change ($p < 0.05$) in viscosity (Table 8.3) and rheology compared to the control nonirradiated samples. In three different temperature regimes studied, both control and irradiated samples display Newtonian behavior. Equations of linearity obtained from plots of shear stress vs shear rate of irradiated honey samples show very high correlation coefficients (Table 8.4), as is true for Newtonian fluids (Sabato 2004).

TABLE 8.3
Averages and Standard Deviation of Viscosity Values for Honey (Parana Region) as a Function of Irradiation Doses, Measured at Three Different Temperatures

	Viscosity (cP)		
Temperature (°C)	0 kGy	5 kGy	10 kGy
30	6142±510	5849±1157	6939±1815
35	3849±239	3594±397	4112±579
40	2433±211	2229±526	2530±428

Source: Sabato, S. F., *Radiation Physics and Chemistry*, 71, 99–102, 2004. With permission.

TABLE 8.4
Equations of Plots of Shear Stress and Shear Rate for Irradiated and Control Honey Samples

	Equations		
Temperature (°C)	0 kGy	5 kGy	10 kGy
30	$y = 69.822x–0.7803$ ($R^2 = 0.9991$)	$y = 51.430x–2.0321$ ($R^2 = 0.9975$)	$y = 65.346x–1.2754$ ($R^2 = 0.9989$)
35	$y = 39.833x–1.8628$ ($R^2 = 0.9999$)	$y = 31.094x–1.744$ ($R^2 = 0.9994$)	$y = 35.868x–1.8315$ ($R^2 = 0.9970$)
40	$y = 22.495x–3.1913$ ($R^2 = 0.9944$)	$y = 19.682x–1.5001$ ($R^2 = 0.9994$)	$y = 21.921x –0.6642$ ($R^2 = 0.9996$)

Source: Sabato, S. F., *Radiation Physics and Chemistry*, 71, 99–102, 2004. With permission.

The rheological properties of hydrocolloids are particularly important when they are used in the formulation of any food, for their effects on their textural attributes. Many factors including the concentration of hydrocolloids, temperature, dissolution, electrical charge, previous thermal and mechanical treatments, and the presence of electrolytes may affect the rheology of the fluid food containing hydrocolloids. Irradiation has an important influence on the flow behavior of hydrocolloid solutions. Viscosity and consistency of the same decreases with increasing radiation dose. Guar gum is very sensitive to irradiation and its solution apparently loses consistency after irradiation. There is a decrease in CI of salep (which principally contains the polysaccharide glucomannan) up to irradiation dose of 6 kGy, beyond which it increases. The change in CI with irradiation dose in salep occurs within a very narrow range (Dogan, Kayacier, and Ic 2007).

In a study on the effect of γ-irradiation (3 kGy) on water-soluble polysaccharides, chiefly pentosans, from hard red spring wheat, the viscosity of gel produced from the same was found to increase on irradiation. Water-soluble pentosans are highly branched polymers that form very viscous gels. Irradiation apparently changes the number and/or sequence of branching within the pentosan molecule, resulting in a very viscous gel. It is hypothesized that the irradiation dose is not high enough to effect hydrolysis of chemical bonds to reduce viscosity but is high enough to alter the structure. Nonirradiated pentosans have a high degree of branching wherein high numbers of L-arabinose units are present on the D-xylose chain. Postirradiation, the bran pentosans appear to be more linear with less arabinose side chains (Grant and D'Appolonia 1991). Carrageenans, agar, and alginate gels also decrease in viscosity with increasing dose of

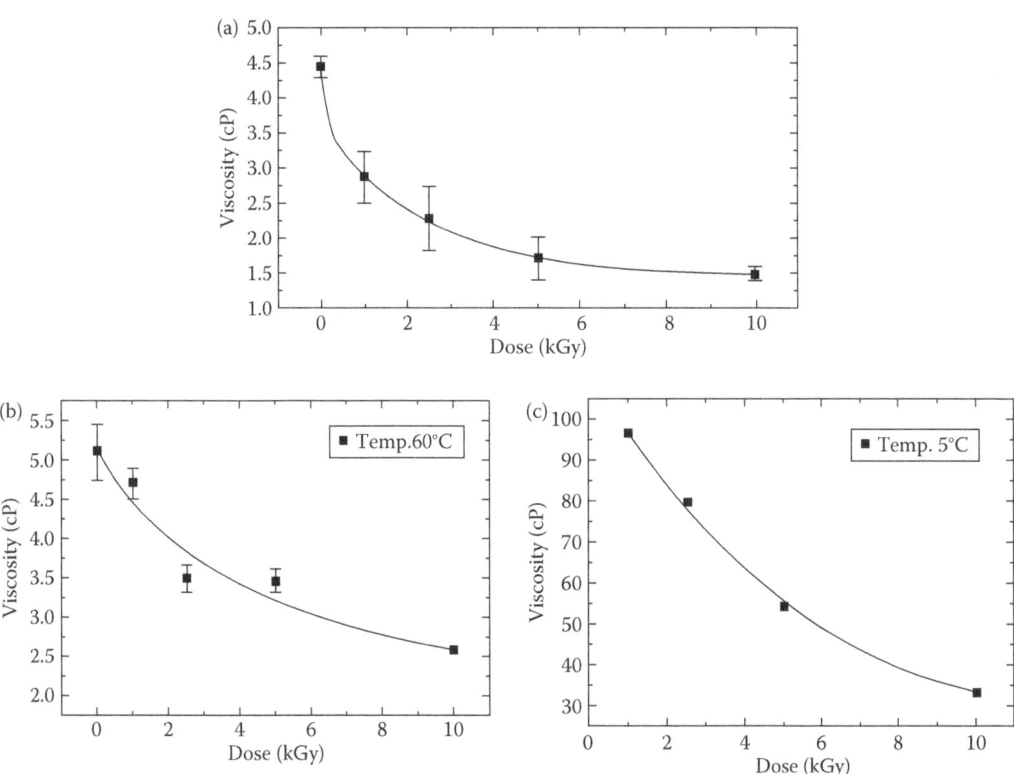

FIGURE 8.4 (a) Viscosity of carrageenan solution vs radiation dose, 250 rpm, 50°C. (b) Viscosity of agar solutions vs radiation dose, 250 rpm, 60°C. (c) Viscosity of alginate solution vs radiation dose, 60 rpm, 5°C. (From Aliste, A. J., Vieira, F. F., and Del Mastro, N. L., *Radiation Physics and Chemistry*, 57, 305–308, 2000. With permission.)

TABLE 8.5
Apparent Viscosities (Mpa·s) of Starch–Xanthan Gum Mixtures

	Shear Rate (s^{-1})					
Xanthan Gum (%)	**0.6**	**1.5**	**3**	**6**	**12**	**30**
25	2628	1275	771	528	311	177
20	1656	877	570	386	263	136
15	1367	613	389	260	169	100
10	533	263	170	135	67	38
5	225	139	103	74	55	19
Control starch	68	50	41	29	20	11

Source: Hanna, M. A., Chinnaswamy, R., Gray, D. R., and Miladinov, V. D., *Journal of Food Science*, 62, 816–20, 1997. With permission.

γ-irradiation (Figures 8.4a, b, and c respectively), principally due to similar reasons of shortening of the polysaccharide chain consequent to radiolysis resulting in softer gels (Aliste, Vieira, and Del Mastro 2000).

The apparent viscosities of nonextruded and irradiated starch–xanthan gum samples are greater than that for the nonextruded irradiated starch in which xanthan gum is added after irradiation (Table 8.5). The water solubility index for starch–gum extrudates increases with an increase in the dose of irradiation. In general, water absorption index exhibits higher values in irradiated samples compared to nonirradiated samples. Similar trends are observed for overall extrudate expansion (Hanna et al. 1997).

8.3 EFFECT OF IRRADIATION ON FOOD TEXTURE

Firmness is an important quality factor for several horticultural products, especially to fetch good market value. Firmness is associated with cell morphology, turgor, and cell-wall middle lamella structure. Loss of firmness may be attributed to the cell injury caused by certain treatments. The texture of several fruits and vegetables has been reported to deteriorate both during and after irradiation. Changes during irradiation are attributed to direct action on substances of cell wall and middle lamella responsible for the mechanical strength of tissues. Irradiation at high doses promotes softening of plant tissue caused by changes of pectic substances as well as degradation of celluloses.

8.3.1 Postirradiation Textural Changes in Fruits

γ-Irradiation causes random hydrolytic breakage of polygalacturonide macromolecules, yielding fragments of lower molecular weight. The breakdown of protopectin to soluble pectin is linear with radiation dose and radiation scissors the 1,4-glycosidic bond. Pectin is degraded at a comparatively lower dose of irradiation. This degradative effect can be reduced by the addition of sugar and can be completely eliminated up to 2.12 kGy when the pH and sugar concentration of the irradiated mixture is in the range suitable for jelly formation. Pasteurization doses of radiation result in leaching of calcium from the plant tissues which causes loss of texture. This radiation-induced softening can be minimized by dipping the fruits in calcium chloride solution after irradiation (Ahmad and Hussain 1973).

8.3.1.1 Mangoes

Exposure of mangoes (*Tommy Atkins* variety) to ionizing radiation of 1.0–3.1 kGy has been shown to induce a significant softening as a function of irradiation dose and storage time. The

TABLE 8.6
Effect of Dose and Storage Time on Texture Attributes of Irradiated and Nonirradiated Mangoes Stored up to 21 Days at 12°C

Texture Parameter	Storage Day	Control (0 kGy)	Low Dose (1 kGy)	Medium Dose (1.5 kGy)	High Dose (3.1 kGy)
Force to rupture (N)	0	177.9	73.50	66.90	24.29
	5	121.4	47.54	16.21	17.13
	10	119.27	82.87	57.78	28.70
	21	140.99	102.87	82.65	31.30
Toughness (J)	0	0.42	0.14	0.12	0.50
	5	0.37	0.06	0.02	0.03
	10	0.30	0.13	0.06	0.04
	21	0.35	0.11	0.08	0.04
Young's modulus or stiffness (MPa @ 3% strain)	0	0.78	0.32	0.29	0.11
	5	0.53	0.21	0.07	0.08
	10	0.53	0.37	0.25	0.13
	21	0.63	0.45	0.36	0.14

Source: Moreno, M., Castell-Perez, E., Gomes, C., Da Silva, P. E., and Moreira, R. G., *Journal of Food Science*, 71, E80–86, 2006. With permission.

effect of dose and storage time on textural attributes of irradiated and nonirradiated mangoes stored at 12°C for 21 days is shown in Table 8.6 (Moreno et al. 2006). Similar results have been reported by Lacroix et al. (1990) and Lacroix, Jobin, and Gagnon (1992) for mango samples irradiated at 0.50 and 0.95 kGy, and by El-Samahy et al. (2000) for mangoes exposed to γ-irradiation between 0.5 and 1.5 kGy. Irradiation affects stiffness of fruits, measured as Young's modulus. Fruits become softer when exposed to the high dose of 3.1 kGy and show the lowest values of Young's modulus at the end of the storage period. Microstructural studies show irradiated fruits to have more collapsed cells than nonirradiated controls. These changes in cell structure are consistent with measured texture characteristics wherein irradiation at high doses significantly reduces the stiffness and firmness of mangoes. Thus it can be rationally concluded that irradiation at 1.0 kGy can retain textural attributes, while that at 1.5 and 3.1 kGy induces undesirable texture or softening. However, exposure up to 3.1 kGy does not negatively affect juiciness of mangoes (Moreno et al. 2006). Investigations on texture of whole mangoes (*Thai* variety) and pulp after γ-irradiation showed it to be hard and slightly undesirable on the first day of storage but it became subsequently softer and more acceptable during storage with the progress of ripening (Lacroix et al. 1993).

Mangoes, being a climacteric fruit, should be at a proper stage of ripeness at the time of irradiation. The use of radiation to offset senescence of mangoes has also been investigated. *Alphanso* and *Desi* mangoes irradiated at 0.25 kGy are firmer than the nonirradiated ones and there is a progressive deterioration of texture with increasing radiation dose. Irradiated *Kent* mangoes are relatively softer immediately after irradiation and contain higher water-soluble and lower insoluble pectins as compared to the control fruits. However, during storage, the increase in water-soluble and decrease in insoluble pectins of control mangoes proceeds at a faster rate than the corresponding irradiated ones. Similar changes are observed in the pectic substances of irradiated *Dusehri* mangoes and the fruits retain better texture when irradiated at 0.30–0.35 kGy, while higher doses have a deleterious effect. Irradiation of mangoes in their ripening stages at 0.64–0.92 kGy and storage at 18°C and 65% RH show a significant weakening of texture compared to the control (Ahmad and Hussain 1973).

8.3.1.2 Apples

An immediate softening of apples has been observed on irradiation above 0.1 kGy. However, the dose threshold is different for different cultivars. The softening may be due to the decrease of protopectin and total pectin or to a change of insoluble pectic materials to soluble forms (Al-Bachir 1999). However, the fruits become firmer during storage compared to the controls. γ-Irradiation over 0.34 kGy causes significant softening of apple slices, which is related to an increase in the content of water-soluble pectin, but not the total pectin content. Total pectin content is unaffected by irradiation. Similarly, softening of minimally processed apple slices is also associated with increased water-soluble pectin and decreased oxalate-soluble pectin content. Both the water-soluble and the oxalate-soluble pectin fractions significantly correlate with the decrease in firmness upon irradiation. Dose lower than 0.4 kGy and higher than 2 kGy have a statistically significant effect on the firmness of irradiated apple slices during storage. However, these levels are far from practical since higher dose cause higher variation in absorbed dose within slices and a lower dose requires excessive treatment periods. Oxygen concentration and storage time do not affect firmness of irradiated slices, suggesting that the softening of irradiated apple slices is probably a direct physical effect rather than that mediated by free radicals. As opposed to the observation on apple slices, protective effects of anoxia against textural changes in irradiated nectarines, peaches, and pears have been reported. Heat treatment of whole fruits yield firmer products compared to nonheated fruits (Ahmad and Hussain 1973; Gunes, Hotchkiss, and Watkins 2001).

Calcium chloride treatment is carried out postradiation to retain texture of apple slices. However, irradiation at 2.5 and 5 kGy softens the slices stored under a controlled atmosphere for 4 weeks irrespective of calcium treatment. Dipping in 0.5% calcium chloride causes a limited but significant improvement on firmness during four-week storage under similar conditions. Slices treated with both calcium and irradiation are comparatively softer than nonirradiated control slices. The inability of calcium chloride to prevent irradiation-induced softening can be due to limited penetration into the tissues. Electron beam is reported to cause less softening, probably due to its lower penetration effect than γ-irradiation. Softening associated with electron beam irradiation may be confined to the surface of the slices and can be eliminated by normal calcium treatment (Gunes et al. 2001).

8.3.1.3 Pears

The optimum dose for radiation of pears has been found to be 2 kGy whereas 3 kGy is high and causes softening of the flesh. Irradiation seems to stimulate ripening in peaches and causes significant softening in some varieties, although in some varieties it shows no significant effect. Tissue softening in peaches and pears consequent to γ-irradiation corresponds to a decrease in protopectin and an increase in pectin and pectate. It has been postulated that increased polymethyl esterase activity may contribute to the initial pectin degradation in irradiated fruits (Ahmad and Hussain 1973).

8.3.1.4 Berries

For the berry family too, similar results as seen in apples are obtained for strawberries wherein 2 kGy is found to be optimum for radiopasteurization of the same. Higher doses cause disagreeable softening of tissues because of formation of a spongy water-soaked structure. For, raspberries, grapes, and black currants, irradiation has a similar adverse effect on their texture. Blueberries exposed to electron beam ionizing radiation endure significant softening throughout storage. Shear force values decrease significantly with irradiation dose. This softening effect induced by irradiation may be associated with the changes in the cell wall structure of the berries and the solubility of its pectin substances. A similar trend is observed with fruit toughness. Samples irradiated at 1.1, 1.6, and 3.2 kGy are reported to be 26, 34, and 49% softer than the control fruit respectively. The ones irradiated at 3.2 kGy are the softest and organoleptically totally unacceptable (Moreno et al. 2007).

8.3.1.5 Cucumbers

The effects of irradiation on texture of cucumbers are contradictory. Khattak et al. (2005a) reported irradiation to have a significant effect on firmness of cucumbers. The firmness of cucumbers decreases gradually from 0–1 kGy, sharply at 1–2 kGy, and immediately after irradiation at 3 kGy. The texture, however, remains within acceptable limits up to a dose of 2.5 kGy after 14 days of storage. Hajare et al. (2006b) did not find any effect on both the central and peripheral regions of cucumber on radiation processing at 2 kGy. However, interestingly during storage, the hardness of the peripheral regions increased significantly until 16 days. This increase in firmness was attributed to the loss of moisture during storage at 8–10°C. The deviations in the studies can be accounted to the varietal differences of cucumbers studied across the Indian subcontinent.

8.3.1.6 Melons

Integrity in texture is the most important shelf-limiting quality factor in melons. Whole cantaloupes irradiated at 1.5 and 3.1 kGy are less firm (lower values of Young's modulus) after the fourth day of storage and less tough after the eighth day of storage at 10°C. Fresh-cut samples of the same irradiated at 3.1 kGy have firmer texture (higher values of Young's modulus) and are harder (higher value of rupture force) and tougher up to the eighth day of storage, after which values begin to decline steadily (Table 8.7). This apparent contradiction in texture improvement with increasing irradiation dose is attributed to the presence of air in polystyrene tray packages, which is believed to reduce the overall density of irradiation target, thus improving penetration characteristics. However, no significant textural changes are obtained for the fresh-cut samples irradiated at lower doses (Castell-Perez et al. 2004). In a study conducted with low-dose electron beam irradiation on melons, "juiciness" was chosen as the texture attribute. This parameter showed a significant effect on irradiation at 0.5–1.0 kGy, with the least significant difference recorded at 0.5 kGy and the highest difference for the high-dose irradiated sample. However, no clear trend of firmness and juiciness on storage of the same in modified atmosphere packages is observed in the said range of irradiation dose (Boynton et al. 2006).

8.3.1.7 Papayas

Some fruits do not show significant changes in texture on irradiation as seen with papayas irradiated at 0.55–0.95 kGy in its ripening stages and subsequent storage at 18°C and 65% RH (Lacroix et al. 1990).

TABLE 8.7
Rupture Force of Cantaloupes up to 12 Days of Storage at 10°C

		Rupture Force (N) of Irradiated Cantaloupes			
	Dose/Day	**Control**	**1.0 kGy**	**1.5 kGy**	**3.1 kGy**
Whole fruit	0	13.98	12.32	10.04	11.74
	4	9.73	5.70	5.50	7.94
	8	9.49	8.86	6.74	5.91
	12	10.79	7.41	5.45	7.55
	Mean	10.79	8.57	6.93	8.29
Fresh-cut	0	4.95	4.75	8.37	8.22
	4	9.22	9.23	9.91	8.50
	8	8.11	4.25	8.27	9.29
	12	6.05	7.64	6.06	10.64
	Mean	7.01	7.72	8.15	9.16

Source: Castell-Perez, E., Moreno, M., Rodrigue, O., and Moreira, R. G., *Food Science and Technology International*, 10, 383–90, 2004. With permission.

8.3.1.8 Pineapple, Lime, and Oranges

Texture analysis of pineapple samples (tissue surrounding the central pith portion which is edible) also reveals no significant changes on irradiation. The texture of the irradiated pineapple samples remain almost unchanged until 12 days of storage at 8°C (Hajare et al. 2006a). Texture of mandarins (*Nagpur* variety) and sweet oranges (*Mosambi*) are not affected by irradiation doses of 0.25–1.5 kGy, as recorded after one week of storage at 20°C. The texture of *Washington Navel* oranges treated at 2 kGy is found to be acceptable after 30 days at 5°C. In case of acid limes however, pulp texture of the treated ones appears slightly melting type and darker yellow-colored as compared with the non-treated fruit, which has normal texture of juice vesicles and normal greenish-yellow color. Lemons, limes, and oranges, however, are very susceptible to transit injury following irradiation (Ladaniya, Singh, and Wadhawan 2003).

8.3.2 Post-Irradiation Textural Changes in Vegetables

Textural attributes such as cohesiveness, springiness, gumminess, and chewiness have been determined for carrot, potato, and beetroot irradiated at 3, 6, 9, and 12 kGy. Texture profile analysis shows significant reduction in hardness of all these vegetables, induced by γ-irradiation. The histological examination of the plant materials further reinforced that γ-irradiation results in breaking up of cell walls, leading to cell damage, loss of turgor pressure, and pectin degradation, and consequently tissue softening. The hardness reduces by 47, 37, and 59% for potato, carrot, and beetroot, respectively, at 12 kGy. Calcium pretreatment results in a significant increase in hardness of the control as well as corresponding irradiated samples from 3 to 12 kGy in all the three vegetables. This can be attributed to the formation of calcium pectate bonds, which increase the rigidity of the middle lamella and cell wall. Cohesiveness, defined as the strength of the internal bonds making up the body of a product, is known to decrease in all of these on account of pectin degradation when irradiation dose is increased to 12 kGy. The springiness, gumminess, and chewiness of calcium-pretreated samples are found to be higher than the corresponding irradiated samples (Nayak et al. 2007). An undesirable change is observed in potato starch beginning at a dose of 0.6 kGy and the degradation follows a linear relationship with dose. Changes in firmness of minimally processed carrots exposed to different doses of γ-irradiation are, however, not significant (Chaudry et al. 2004).

8.3.2.1 Carrots

A linear relationship exists between radiation dose and softening in carrot tissues. The initial effect of irradiation on the texture of raw carrots is that of destruction of the semipermeability of the cell membrane and successive texture deterioration is primarily due to the degradation of pectin and cellulose constituents. A significant increase in ammonium oxalate-oxalic acid soluble pectin and corresponding decrease in insoluble pectin in cell wall of carrots irradiated up to 10 kGy is observed. Radiation does not seem to alter the total amount of hemicellulose and lignin, but does reduce the ∝-cellulose slightly. The texture profile of the central and peripheral regions of irradiated carrots is similar to that observed for cucumbers, as reported above.

8.3.2.2 Tomatoes

Tomatoes are significantly softened by radiation doses from 1.25 to 3.75 kGy. With an increase in the irradiation dose, there is increased damage to texture. Ripe fruits are affected to a greater extent as compared to unripe ones. Firmness of diced *Roma* tomatoes is also reported to decrease with increasing irradiation dose from 0.5 up to 3.7 kGy (Figure 8.5). Treatment of tomatoes with 0.5 and 3.70 kGy results in 30 and 50% loss of firmness, respectively, immediately after irradiation (Table 8.8) (Prakash et al. 2002b).

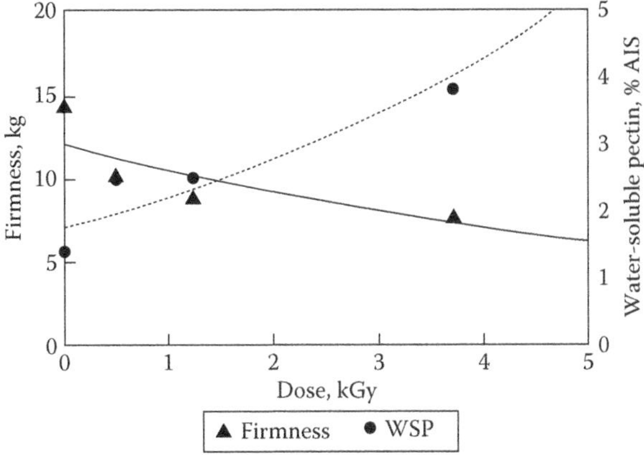

FIGURE 8.5 Correlation between firmness (kg) and WSP content of diced tomatoes as a function of irradiation dose. (From Prakash, A., Manley, J., DeCosta, S., Caporaso, F., and Foley, D. M., *Radiation Physics and Chemistry*, 63, 387–90, 2002b. With permission.)

TABLE 8.8
Effects of Irradiation on the Firmness (kg) of Diced Roma Tomatoes Stored at 4°C for 15 Days

	Irradiation Dose (kGy)			
Day	**0.0**	**0.50**	**1.24**	**3.70**
0	14.53±0.92	10.26±1.12	8.87±1.62	7.60±0.61
3	13.48±1.46	11.33±0.69	10.24±1.06	8.45±0.56
6	13.93±2.42	13.30±1.63	11.03±0.72	8.51±0.23
9	13.13±1.13	11.06±1.25	9.91±1.21	6.99±0.66
12	14.62±1.95	12.91±1.07	8.81±0.32	7.58±0.45
15	13.22±1.56	11.16±1.12	9.33±1.18	6.80±1.30

Source: Prakash, A., Manley, J., DeCosta, S., Caporaso, F., and Foley, D. M., *Radiation Physics and Chemistry*, 63, 387–90, 2002b. With permission.

8.3.2.3 Lettuce

Iceberg lettuce (Hagenmaier and Baker 1997) and Roman lettuce (Prakash et al. 2002a) are amenable to low-dose irradiation (less than 0.5 kGy), with texture and shelf life equal or superior to those of untreated controls. Studies have been conducted with iceberg lettuce by subjecting them to 0–2 kGy irradiation after dipping them in cold water (5°C) or warm water (47°C) and packing them in modified atmosphere film bags. Firmness of the same was found to decrease with increasing irradiation dose after one day of storage. No consistent change in firmness as a function of irradiation dose was observed on other days of storage. Lettuce treated in warm water (47°C) had significantly lower firmness than that treated at 5°C (Fan et al. 2003c). Firmness of fresh cilantro decreases linearly with radiation doses (1–3 kGy) and with storage time (up to seven days) at 3°C. After 14 days of storage, however, there is no significant difference in firmness among the irradiated and nonirradiated samples (Fan, Niemira, and Sokorai 2003a).

8.3.2.4 Other Vegetables

Crispy vegetables that are liked for their desirable textural qualities usually soften or wilt at high radiation dose. γ-Irradiation increases the tenderness of asparagus and doses above 2.5 kGy accelerate softening in onions. In a study on fresh-cut green onion leaves, no consistent change in firmness with doses up to 3 kGy is observed (Fan, Niemira, and Sokorai 2003b). This is especially true for the first eight days of storage in a refrigerator maintained at 4°C. However, there is a slight decrease in firmness of all samples after this time period and this decrease is more pronounced for warm-water-treated samples than for those without warm water treatment. However, studies on fresh-cut green onions show that neither irradiation nor warm-water treatment has any consistent effect on their texture (Kim et al. 2005). Radiation also causes softening of summer squash and peppers at radiopasteurization dose levels (Boynton et al. 2006).

No effect of radiation treatment is observed on the firmness of irradiated sprout samples of green gram (2 kGy), even during storage at 8°C (Hajare et al. 2007). Immediately after irradiation, minimally processed Chinese cabbage does not show a consistent change in firmness in response to irradiation (2 kGy) during refrigerated storage. After 3 weeks of storage, however, there is significant reduction in firmness of the same under aerobic packaging conditions, while modified atmospheric packaging effectively maintains the texture (Table 8.9) (Ahn et al. 2005; Khattak et al. 2005a). Irradiation may not affect the texture of certain vegetables such as bitter gourd. Studies have shown that irradiation doses up to 2 kGy do not significantly affect the texture of minimally processed bitter gourd stored at 5°C for 7 days (Khattak et al. 2005b).

8.3.3 Post-Irradiation Textural Changes in Cereals

The hardness of cooked, irradiated rice is reported to be less than that of cooked, nonirradiated rice. SBV is often used as an indicator of firmness of cooked rice, with higher values indicating firmer texture. SBV of cooked irradiated rice is found to be reduced with increasing dose of irradiation as seen in Table 8.1 (Wu et al. 2002). It increases in all the rice samples with storage time. Changes in the hardness of cooked, nonirradiated rice, and low-dose irradiated

TABLE 8.9
Puncture Strength (N/Mm) of Minimally Processed Salted Chinese Cabbage Treated With Modified Atmosphere Packaging and Irradiation During Refrigerated Storage

		Irradiation Dose (kGy)			
Storage (Week)	**Packaging**	**0**	**0.5**	**1**	**2**
0	Air	0.34	0.40	0.33	0.38
	CO_2	0.35	0.36	0.36	0.36
	CO_2/N_2	0.33	0.36	0.34	0.35
1	Air	0.33	0.31	0.35	0.34
	CO_2	0.34	0.36	0.34	0.34
	CO_2/N_2	0.33	0.34	0.34	0.34
2	Air	0.35	0.43	0.35	0.37
	CO_2	0.34	0.35	0.32	0.33
	CO_2/N_2	0.34	0.35	0.37	0.37
3	Air	0.36	0.31	0.32	0.29
	CO_2	0.35	0.41	0.38	0.36
	CO_2/N_2	0.32	0.38	0.37	0.36

Source: Ahn, H. J., Kim, J. H., Kim, J. K., Kim, D. H., Yook, H. S., and Byun, M. W., *Food Chemistry*, 89, 589–97, 2005. With permission.

rice (0.2 kGy) are noticeable, whereas there are no changes in rice irradiated at 0.5 and 1.0 kGy after four months of storage at 4°C. Significant differences among all samples are observed at the beginning of storage; at the end of storage, hardness of cooked, nonirradiated rice is not significantly different from that of irradiated rice at 0.2 kGy. A greater increase in cooked rice hardness (0.7 kgf) is found during storage of nonirradiated rice for one year whereas the increase is in the range of 0.4–0.5 kgf for rice irradiated at 0.5 and 1.0 kGy. The softer texture of cooked, irradiated rice is due to greater water absorption during cooking. During storage, the texture of cooked rice becomes less sticky and glossy, and more hard and fluffy. Minor changes of texture of cooked, irradiated rice after storage (6–8 months) are observed; more noticeable changes are observed when a higher dose of irradiation is applied. The texture is observed to be slippery with mucus-like material covering the rice kernel. This occurs consistently with leaching of more soluble solids into the cooking water which in turn is due to the disintegration of grains during cooking of irradiated rice. The effect is seen even after storage of the irradiated rice for one year (Sirisoontaralak and Noomhorn 2007).

γ-Irradiation of wheat weakens the dough mixing properties and decreases the amount of insoluble glutenin in flour, both of which tend to be more pronounced at 10 kGy or above. It is highly likely that the direct effect of irradiation is on glutenin, which results in deterioration of rheological properties (Köksel et al. 1998).

8.3.4 Post-Irradiation Textural Changes in Meat

Irradiation affects the quality attributes of meat products including texture due to an increase in lipid oxidation. In a study conducted to determine the effectiveness of electron beam irradiation of 1.0 and 1.8 kGy on the elimination of bacteria in boneless, skinless chicken breast fillets, texture deterioration has been observed with increasing time of storage at 0°C (Lewis et al. 2002). Irradiation of aged white poultry meat has shown minor changes in texture; however, the changes are significant if irradiation is carried out prior to aging. Cooked, irradiated and refrigerated dark poultry meat is found to be more tender than the controls. This study suggests that irradiation has varied effects on poultry meat depending on whether the meat is white or dark. Several studies on irradiation of chicken meat indicates that the conditions of irradiation, the sources of irradiation, the portion of meat that is irradiated, and the objectivity of measurements of the texture characteristics (Yoon 2003) profoundly influence textural variations.

Irradiation of meat under air at high dose rate (20 kGy/h) weakens the texture of aerobically packaged irradiated fresh pork loins, while irradiation under vacuum reportedly has better protection on the texture of meat (Lacroix et al. 2000). Texture of iced refrigerated sea bream on irradiation at 2.5–5 kGy shows decreasing acceptability with respect to texture after 15–17 days of storage *vis-à-vis* 13 days for irradiated control samples. Irradiation slightly increases the resilience of turkey breast rolls. This could be attributed to the cross-linking of amino acid residues by irradiation. In amino acid solutions too, it is observed that irradiation induces cross-linking of amino acids and the solution turns turbid after irradiation. The addition of preservatives such as potassium benzoate, after irradiation to enhance shelf life does not yield satisfactory results since it produces a high amount of benzene after irradiation. This suggests that certain spices or food containing high amounts of phenolic compounds may not be suitable for irradiation (Zhu et al. 2004).

The overall effect of γ-irradiation on texture of frankfurters formulated with potassium lactate and sodium diacetate is found to be minimal. γ-Irradiation has an insignificant influence on springiness and juiciness of the frankfurters. Quality deterioration is minimal for the irradiated samples compared to the aerobically packaged frankfurters (Knight et al. 2007).

Studies on prepared meal products consisting of Salisbury steak, gravy, and mashed potatoes, irradiated at 0.8, 2.9, and 5.7 kGy and stored at 4°C for 3 weeks, do not show any significant difference in the texture of mashed potatoes for hardness over time. Although no change in texture is detected in Salisbury steak (meat patty) on irradiation, a softening is noticed on day 15 of storage. No further significant change is detected with storage time. Overall, the irradiated prepared meals are as acceptable as the nonirradiated ones with respect to texture (Foley et al. 2001).

8.3.5 Post-Irradiation Textural Changes in Nuts, Legumes, and Mushrooms

A loss in texture that is proportional to the applied dose in cooked Brazilian beans has been reported (Villavicencio et al. 2000) In contrast, irradiation dose has an insignificant effect on the texture of pine nut kernels, since texture deformation caused by irradiation is less in foods that have a low water content (Golge and Ova 2008). Also, better retention of texture (firmness) has been found for mushrooms irradiated at 0.5–2.5 kGy when stored in Biaxially Oriented Polypropylene (BOPP) bags at 4, 10, and 20°C for 15 days (Gautam, Sharma, and Thomas 1998).

8.4 CONCLUSION

While irradiation as a technology is attractive for microbial and sprout inhibition and control of insect infestation, changes in the rheological and textural properties of foods do take place. This necessitates formulation alterations to counter the deleterious effects of the same. The effects of combination technologies involving irradiation are emerging and their effect on rheological and texture properties need thorough investigation.

REFERENCES

Ahmad, M., and Hussain, A. M. 1973. Radiation induced textural changes in fruits and vegetables. *Pakistan Journal of Science* 25:194–202.

Ahn, H. J., Kim, J. H., Kim, J. K., Kim, D. H., Yook, H. S., and Byun, M. W. 2005. Combined effects of irradiation and modified atmosphere packaging on minimally processed Chinese cabbage (*Brassica rape* L.). *Food Chemistry* 89:589–97.

Al-Bachir, M. 1999. Effect of gamma irradiation on storability of apples (*Malus domestica* L.). *Plant Foods for Human Nutrition* 54:1–11.

Aliste, A. J., Vieira, F. F., and Del Mastro, N. L. 2000. Radiation effects of on agar, alginates and carrageenan to be used as food additives. *Radiation Physics and Chemistry* 57:305–308.

Arvizu, Z. A., Fernández-Ramírez, M. V., Arce-Corrales, M. E., Cruz-Zaragoza, E., Melédrez, R., Chernov, V., and Barboza-Flores, M. 2006. Gamma irradiation effects on commercial Mexican bread making wheat flour. *Nuclear Instruments and Methods in Physics Research B* 245:455–58.

Bao, J., and Corke, H. 2002. Pasting properties of γ-irradiated rice starches as affected by pH. *Journal of Agricultural and Food Chemistry* 50:336–41.

Bertolini, A. C., Mestres, C., and Colonna, P. 2000. Rheological properties of acidified and UV-irradiated starches. *Starch/Stärke* 52:340–44.

Bhat, R., Sridhar, K. R., and Seena, S. 2008. Nutritional quality evaluation of velvet bean seeds (*Mucuna pruriens*) exposed to gamma irradiation. *International Journal of Food Sciences and Nutrition* 59:261–78.

Boynton, B. B., Welt, B. A., Sims, C. A., Balaban, M. O., Brecht, J. K., and Marshall, M. R. 2006. Effects of low-dose electron beam irradiation on respiration, microbiology, texture, color, and sensory characteristics of fresh-cut cantaloupe stored in modified-atmosphere packages. *Journal of Food Science* 71:387–90.

Castell-Perez, E., Moreno, M., Rodrigue, O., and Moreira, R. G. 2004. Electron beam treatment of cantaloupes: effect on product quality. *Food Science and Technology International* 10:383–90.

Chaudry, M. A., Bibi, N., Khan, M., Khan, M., Badshah, A., and Qureshi, M. J. 2004. Irradiation treatment of minimally processed carrots for ensuring microbiological safety. *Radiation Physics and Chemistry* 71:169–73.

Chrastil, J. 1994. Effect of storage on the physicochemical properties and quality factors of rice. In: Marshall, W. E., and Wadsworth, J. I. (Eds.) *Rice Science and Technology*, 49–81. New York: Marcel-Dekker.

Diehl, J. F. 1995. Potential and current applications of food irradiation. In: *Safety of Irradiated Foods*, 293–321. New York: Marcel-Dekker.

Dogan, M., Kayacier, A., and Ic, E. 2007. Rheological characteristics of some food hydrocolloids processed with gamma irradiation. *Food Hydrocolloids* 21:392–96.

El-Samahy, S. K., Youssef, B. M., Askar, A. A., and Swailam, H. M. M. 2000. Microbiological and chemical properties of irradiated mango. *Journal of Food Safety* 20:139–56.

Esteves, M. P., Raymundo, A., de Sousa, I., Andrade, M. E., and Empis, J. 2002. Rheological behavior of white pepper gels: A new method for studying the effect of irradiation. *Radiation Physics and Chemistry* 64:323–29.

Ezekiel, R., Rana, G., Singh, N., and Singh, S. 2007. Physicochemical, thermal and pasting properties of starch separated from γ-irradiated and stored products. *Food Chemistry* 105:1420–29.

Fan, X., Niemira, B. A., and Sokorai, K. J. B. 2003a. Sensorial, nutritional and microbiological quality if fresh cilantro leaves as influenced by ionizing radiation and storage. *Food Research International* 36:713–19.

———. 2003b. Use of ionizing radiation to improve sensory and microbial quality of fresh cut green onion leaves. *Food Science* 68:1478–83.

Fan, X., and Sokorai, K. J. B. 2008. Retention of quality and nutritional value of 13 fresh-cut vegetables treated with low-dose radiation. *Journal of Food Science* 73:S367–72.

Fan, X., Toivonen, P. M. A., Rajkowski, K. T., and Sokorai, K. J. B. 2003c. Warm water treatment in combination with modified atmosphere packaging reduces undesirable effects of irradiation on the quality of fresh-cut iceberg lettuce. *Journal of Agricultural and Food Chemistry* 51:1231–36.

Foley, D. M., Reher, E., Caporaso, F., Trimboli, S., Musherraf, Z., and Prakash, A. 2001. Elimination of *Listeria monocytogenes* and changes in physical and sensory qualities of a prepared meal following gamma irradiation. *Food Microbiology* 18:193–204.

Gautam, S., Sharma, A., and Thomas P. 1998. Gamma irradiation effect on shelf-life, texture, polyphenol oxidase and microflora of mushroom (*Agaricus bisporus*). *International Journal of Food Sciences & Nutrition* 49:5–10.

Golge, E., and Ova, G. 2008. The effects of food irradiation on quality of pine nut kernels. *Radiation Physics and Chemistry* 77:365–69.

González-Aguilar, G. A., Villeges-Ochoa, M. A., Martinez-Téllez, M. A., Gardea, A. A., and Ayala-Zavala, J. F. 2007. Improving antioxidant capacity of fresh-cut mangoes treated with UV-C. *Journal of Food Science* 72:S197–202.

Grant, L. A., and D'Appolonia, B. L. 1991. Effect of low-level gamma irradiation on water-soluble non-starchy polysaccharides isolated from hard red spring wheat flour and bran. *Cereal Chemistry* 68:651–52.

Gunes, G., Hotchkiss, J. H., and Watkins, C. B. 2001. Effects of gamma irradiation on the texture of minimally processed apple slices. *Journal of Food Science* 66:63–67.

Hagenmaier, R. D., and Baker, R. A. 1997. Low-dose irradiation of cut iceberg lettuce in modified atmosphere packaging. *Journal of Agricultural and Food Chemistry* 45:2864–68.

Hajare, S. N., Dhokane, V. S., Shashidhar, R., Saroj, S., Sharma, A., and Bandekar, J. R. 2006a. Radiation processing of minimally processed pineapple (*Ananas comosus* Merr.): Effect on nutritional and sensory quality. *Journal of Food Science* 71:S501–505.

Hajare, S. N., Dhokane, V. S., Shashidhar, R., Sharma, A., and Bandekar, J. R. 2006b. Radiation processing of minimally processed carrot (*Daucus carota*) and cucumber (*Cucumis sativus*) to ensure safety: Effect on nutritional and sensory quality. *Journal of Food Science* 71:S198–203.

Hajare, S. N., Saroj, S. D., Dhokane, V. S., Shashidhar, R., and Bandekar, J. R. 2007. Effect of radiation processing on nutritional and sensory quality of minimally processed green gram and garden pea sprouts. *Radiation Physics and Chemistry* 76:1642–49.

Hanna, M. A., Chinnaswamy, R., Gray, D. R., and Miladinov, V. D. 1997. Extrudates of starch-xanthan gum mixtures as affected by chemical agents and irradiation. *Journal of Food Science* 62:816–20.

Hassan., A. B., Osman, G. A. M., Rushdi, M. A. H., Eltayeb, M. M., and Diab, E. E. 2009. Effect of gamma irradiation on the nutritional quality of maize cultivars (*Zea mays*) and sorghum (*Sorghum bicolor*) grains. *Pakistan Journal of Nutrition* 8:167–71.

Hemmaty, S., Moallemi, N., and Naseri, L. 2007. Effect of UV-C radiation and hot water on the calcium content and postharvest quality of apples. *Spanish Journal of Agricultural Research* 5:559–68.

Hu, W., and Jiang, Y. 2007. Quality attributes and control of fresh-cut produce. *Stewart Postharvest Review* 3: art. no. 3.

Huang, S., Herald, T. J., and Mueller, D. D. 1997. Effect of electron beam irradiation on physical, physicochemical, and functional properties of liquid egg yolk during frozen storage. *Poultry Science* 76:1607–15.

Kang, I. J., Byun, M. W., Yook, H. S., Bae, C. H., Lee, H. S., Kwon, J. H., and Chung, C. K. 1999. Production of modified starches by gamma irradiation. *Radiation Physics and Chemistry* 54:425–30.

Keya, E. L., and Sherman, U. 1997. Effects of a brief, intense infrareds radiation treatment on the nutritional quality of maize, rice, sorghum and beans. *Food and Nutrition Bulletin* 18:382–87.

Khattak, A. B., Bibi, N., Chaudry, M. A., Khan, M., Khan, M., and Qureshi, M. J. 2005a. Shelf life extension of minimally processed cabbage and cucumber through gamma irradiation. *Journal of Food Protection* 68:105–10.

Khattak, A. B., Bibi, N., Khattack, A. B., and Chaudry, M. A. 2005b. Effect of irradiation on microbial safety and nutritional quality of minimally processed bitter gourd (*Momordica charantia*). *Journal of Food Science* 70:M255–59.

Kim, H. J., Feng, H., Toshkov, S. A., and Fan, X. 2005. Effect of sequential treatment of warm water dip and low-dose gamma irradiation on the quality of fresh-cut green onions. *Journal of Food Science* 70:M179–85.

Kizil, R., and Irudayaraj, J. 2006. Discrimination of irradiated starch gels using FT-Raman spectroscopy and chemometrics. *Journal of Agricultural and Food Chemistry* 54:13–18.

Knight, T. D., Miller, R., Maxim, J., and Keeton, J. T. 2007. Sensory and physicochemical characteristics of frankfurters formulated with potassium lactate and sodium diacetate before and after irradiation. *Journal of Food Science* 72:S112–18.

Köksel, H., Sapirstein, H. D., Celik, S., and Bushuk, W. 1998. Effects of gamma-irradiation of wheat on gluten proteins. *Journal of Cereal Science* 28:243–50.

Lacroix, M., Bernard, L., Jobin, M., Milot, S., and Gagnon, M. 1990. Effect of irradiation on the biochemical and organoleptic changes during ripening of papaya and mango fruits. *Radiation Physics and Chemistry* 35:296–300.

Lacroix, M., Gagnon, M., Pringsulaka, V., Jobin, M., Latreille, B., Nouchpramool, K., Prachasitthisak, Y., Charoen, S., Adulyatham, P., Lettre, J., and Grad. B. 1993. Effect of gamma irradiation with or without hot water dip and transportation from Thailand to Canada on nutritional qualities, ripening index and sensorial characteristics of Thai mangoes (Nahng Glahng Wahn variety). *Radiation Physics and Chemistry* 42:273–77.

Lacroix, M., Jobin, M., and Gagnon, M. 1992. Irradiation and storage effects on sensorial and physical characteristics of Keitt mangoes (*Mangifera indica* L.) quality of irradiated mangoes. *Sciences des Aliments* 12:63–81.

Lacroix, M., Smoragiewicz, W., Jobin, M., Latreille, B., and Krzystyniak, K. 2000. Protein quality and microbiological changes in aerobically or vacuum-packaged, irradiated fresh pork loins. *Meat Science* 56:31–39.

Ladaniya, M. S., Singh, S., and Wadhawan, A. K. 2003. Response of 'Nagpur' mandarin, 'Mosambi' sweet orange and 'Kagzi' acid lime to gamma radiation. *Radiation Physics and Chemistry* 67:665–75.

Lewis, S. J., Velásquez, A., Cuppett, S. L., and McKee, S. R. 2002. Effect of electron beam irradiation on poultry meat safety and quality. *Poultry Science* 81:896–903.

MacArthur, L. A., and D'appolonia, B. L. 1984. Gamma irradiation of wheat. II. Effects of low-dosage radiations on starch properties. *Cereal Chemistry* 61:321–26.

Marathe, S. A., Machaiah, J. P., Rao, B. Y. K., Pednekar, M. D., and Rao, V. S. 2002. Extension of shelf-life of whole-wheat flour by gamma radiation. *International Journal of Food Science and Technology* 37:163–68.

Martin, M., and Fitzgerald, M. A. 2002. Proteins in rice grains influence cooking properties. *Journal of Cereal Science* 36:285–94.

Min, B. R., Nam, K. C., Lee, E. J., Ko, G. Y., Trampel, D. W., and Ahn, D. U. 2005. Effect of irradiating shell eggs on quality attributes and functional properties of yolk and white. *Poultry Science* 84:1791–96.

Moreno, M., Castell-Perez, E., Gomes, C., Da Silva, P. E., and Moreira, R. G. 2006. Effects of electron beam irradiation on physical, textural and microstructural properties of 'Tommy Atkins' mangoes (*Mangifera indica* L.). *Journal of Food Science* 71:E80–86.

———. 2007. Quality of electron beam irradiation of blueberries (*Vaccinium corymbosum* L.) at medium dose levels (1.0-3.2 kGy). *Lebensmittel - Wissenschaft + Technologie* 40:1123–32.

Nayak, C. A., Suguna, K., Narasimhamurthy, K., and Rastogi, N. K. 2007. Effect of gamma irradiation on histological and textural properties of carrot, potato and beetroot. *Journal of Food Engineering* 79:765–70.

Ohashi, K., Goshima, G., Kusuda, H., and Tsuge, H. 1980. Effect of embraced lipid on the gelatinization of rice starch. *Starch* 32:54.

Prakash, A., Guner, A. R., Caporaso, F., and Foley, D. M. 2002a. Effects of low-dose gamma irradiation on the shelf life and quality characteristics of cut romaine lettuce packed under modified atmosphere. *Journal of Food Science* 65:549–53.

Prakash, A., Manley, J., DeCosta, S., Caporaso, F., and Foley, D. M. 2002b. The effects of gamma irradiation on the microbiological, physical and sensory qualities on diced tomatoes. *Radiation Physics and Chemistry* 63:387–90.

Rao, M. A., and Rizvi, S. S. H. 1995. *Engineering Properties of Foods*, 1, 2, 55. New York: Marcel Dekker.

Rao, V. S., Srirangarajan, A. N., Kamat, A. S., Adhikari, H. R., and Nair, P. M. 1994. Studies on extension of shelf-life of *Rawa* by gamma irradiation. *Journal of Food Science and Technology* 31:311–15.

Sabato, S. F. 2004. Rheology of irradiated honey from Parana region. *Radiation Physics and Chemistry* 71:99–102.

Sabato, S. F., and Lacroix, M. 2002. Radiation effects on viscosimetry of protein based solutions. *Radiation Physics and Chemistry* 63:357–59.

Sabularse, V. C., Liuzzo, J. A., Rao, R. M., and Grodner, R. M. 1992. Physicochemical characteristics of brown rice as influenced by gamma irradiation. *Journal of Food Science* 57:143–45.

Siddhuraju, P., Makkar, H. P. S., and Becker, K. 2002. The effect of ionizing radiation on antinutritional factors and the nutritional value of plant materials with reference to human and animal food. *Food Chemistry* 78:187–205.

Sirisoontaralak, P., and Noomhorm, A. 2006. Changes to physicochemical properties and aroma of irradiated rice. *Journal of Stored Products Research* 42:264–76.

Sirisoontaralak, P., and Noomhorm, A. 2007. Changes in physicochemical and sensory-properties of irradiated rice during storage. *Journal of Stored Products Research* 43:282–89.

Song, H. P., Kim, D. H., Yook, H. S., Kim, M. R., Kim, K. S., and Byun, M. W. 2004. Nutritional, physiological, physicochemical and sensory stability of gamma irradiated Kimchi (Korean fermented vegetables). *Radiation Physics and Chemistry* 69:85–90.

Sung, W.C. 2005. Effect of gamma irradiation on rice and its food products. *Radiation Physics and Chemistry* 73:224–28.

Sung, W. C., Hong, M. C., and Chang, T. S. 2008. Effects of storage and gamma irradiation on (*japonica*) waxy rice. *Radiation Physics and Chemistry* 77:92–97.

Urbain, W. M. 1986. *Food Irradiation*, 351. Orlando, Florida: Academic Press.

Venugopal, V., Doke, S. N., and Thomas, P. 1999. Radiation processing to improve the quality of fishery products. *Critical Reviews in Food Science and Nutrition* 30:391–440.

Villavicencio, A. L. C. H., Mancini-Filho, J., Delincee, H., and Greiner, R. 2000. Effect of irradiation on antinutrients (total phenolics, tannins and phytate) in Brazilian beans. *Radiation Physics and Chemistry* 57:289–93.

Wu, D., Shu, Q., Wang, Z., and Xia, Y. 2002. Effect of gamma irradiation on starch viscosity and physicochemical properties of different rice. *Radiation Physics and Chemistry* 65:79–86.

Yoon., K. S. 2003. Effect of gamma irradiation on the texture and microstructure of chicken breast meat. *Meat Science* 63:273–77.

Yu, Y. and Wang, J. 2007. Effect of γ-irradiation on starch granule structure and physicochemical; properties of rice. *Food Research International* 40:297–303.

Zhou, Z., Robards, K., Helliwell, S., and Blanchard, C. 2003. Effect of rice storage on pasting properties of rice flour. *Food Research International* 36:625–34.

Zhu, M. J., Mendonca A., Min, B., Lee, E. J., Nam, K. C., Park, K., Du, M., Ismail, H. A., and Ahn, D. U. 2004. Effects of electron beam irradiation and antimicrobials on the volatiles, color, and texture of ready-to-eat turkey breast roll. *Journal of Food Science* 69:C383–87.

9 Ozone and CO_2 Processing: Rheological and Functional Properties of Food

Kasiviswanathan Muthukumarappan
South Dakota State University

B. K. Tiwari and Colm P. O'Donnell
University College Dublin

P. J. Cullen
Dublin Institute of Technology

CONTENTS

9.1 INTRODUCTION

Ozone has a wide antimicrobial spectrum which, combined with a high oxidation potential, makes it an attractive processing option for the food industry. Relatively small quantities of ozone and short contact times are sufficient for the desired antimicrobial effect and it rapidly decomposes into oxygen, leaving no toxic residues (Muthukumarappan, Halaweish, and Naidu 2000). Ozone decomposes, producing numerous free radicals, predominantly hydroxyl free radicals, which increase with increasing temperature and pH (Graham 1997). The interest in ozone as an antimicrobial agent is based on its high biocidal efficacy and wide antimicrobial spectrum. Ozone is a powerful broad-spectrum antimicrobial agent that is active against bacteria, fungi, viruses, protozoa, and bacterial and fungal spores (Khadre, Yousef, and Kim 2001). Ozone is 50% more effective than chlorine and is active over a wider spectrum of microorganisms than chlorine and other disinfectants. It reacts up to 3000 times faster than chlorine with organic materials and produces no harmful decomposition products (Graham 1997). Excess ozone autodecomposes rapidly to produce oxygen and thus it leaves no residues in food. Such advantages make ozone attractive to the food industry and consequently it has been declared as generally recognized as safe (GRAS) for use in food processing by the U.S. Food and Drug Administration (FDA) in 1997 (Graham 1997). Ozone subsequently gained approved as a direct food additive for the treatment, storage, and processing of foods in the gaseous and aqueous phases in 2001 (Khadre et al. 2001).

Dense-phase carbon dioxide (DPCD) is a food preservation technology often referred to as cold pasteurization. Ever since Fraser (1951) showed the potential of DPCD for inactivation of bacterial cells, the use of DPCD has continued to attract attention as a potent nonthermal preservation technique. The potential of supercritical CO_2 as an antimicrobial agent is appealing as it is nontoxic and easily removed by simple depressurization and outgassing. Supercritical carbon dioxide ($SCCO_2$) is attracting interest in the food industry because of its potential for inactivation of microorganisms and enzymes. It was shown that $SCCO_2$ has significant lethal effects on microorganisms in liquid foods (Ballestra and Cuq 1998; Ballestra, Abreuda, and Cuq 1996; Hong and Pyun 2001; Shimoda et al. 1998; Corwin and Shellhammer 2002; Erkmen and Karaman 2001; Park, Lee, and Park 2002; Shimoda et al. 2001), along with deactivation of enzymes and deodorization of liquid materials, for liquid foodstuffs and medicines with either semicontinuous (Osajima, Shimoda, and Kawano 1996, 1997) or continuous operations (Osajima et al. 1998, 1999, 2003).

This chapter deals with the effect of ozone and DPCD processing on the rheological and functional properties of food.

9.2 OZONE

Ozone is a triatomic molecule consisting of three oxygen atoms. Figure 9.1 shows the structural arrangement of the oxygen atoms in an ozone molecule. Ozone is generated in the upper atmosphere when UV radiation from the sun dissociates or splits oxygen molecules to form oxygen atoms. An unstable oxygen atom quickly combines with an oxygen molecule to form a highly unstable ozone molecule. This ozone is a colorless gas with a pungent odor and is highly corrosive and toxic. Ozone is highly unstable, with a short half-life of 20–30 min in distilled water at 20°C (Khadre et al. 2001). Ozone spontaneously decomposes by a complex mechanism that involves the generation of hydroxyl free radicals (Hoigné and Bader 1983).

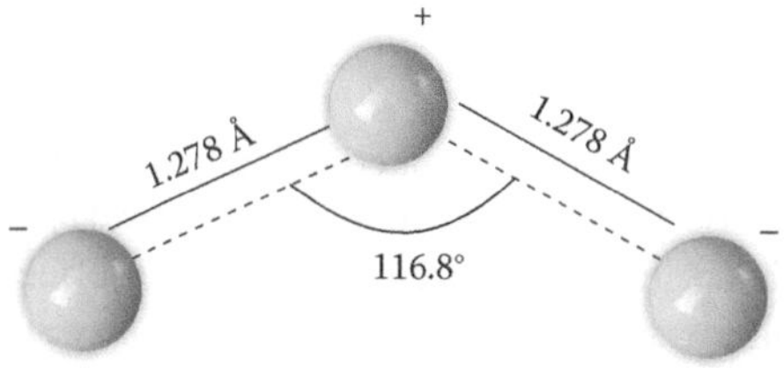

FIGURE 9.1 Ozone molecule.

9.2.1 Ozone Generation

Industrially, ozone is produced either from air or O_2. Ozone (O_3) results from the rearrangement of atoms when oxygen molecules are subjected to high-voltage electric discharge. Generation of the free oxygen radical occurs by breakage of strong O–O bonds, requiring a significant energy input. UV radiation and corona discharge methods can be used to initiate free radical oxygen formation and thereby generate ozone. In addition to photochemical (UV radiation) and electric discharge methods, ozone can be produced by chemical, thermal, chemonuclear, and electrolytic methods (Kim, Yousef, and Dave 1999a). Generally the corona discharge method is used for generation of ozone at a commercial level. Two electrodes, high tension and low tension (ground electrode), separated by a ceramic dielectric medium provide a narrow discharge gap (Muthukumarappan et al. 2000; Mahapatra, Muthukumarappan, and Julson 2005) (Figure 9.2). Electrons with sufficient kinetic energy (around 6–7 eV) to dissociate the oxygen molecule collide and a molecule of ozone can be formed from each oxygen atom (Güzel-Seydim, Bever, and Greene 2004). If atmospheric air is passed through the generator as the feed gas, 1–3% ozone can be produced, however, using pure oxygen up to 6% ozone production can be achieved. Consequently, ozone concentration cannot be increased beyond the point where the rate of formation and destruction are equal. Ozone gas cannot be stored since ozone spontaneously degrades back to oxygen atoms.

9.2.2 Application of Ozone

Ozone has numerous potential applications in the food industry because of its advantages over traditional antimicrobial agents such as chlorine, potassium sorbates, etc. Ozone finds application in the decontamination of a wide array of foods. Applications includes dried foods to reduce *Bacillus* spp. and *Micrococcus* counts in cereal grains, peas, beans, and whole spices by up to 3 log units, depending on ozone concentration, temperature, and relative humidity conditions (Naitoh, Okada, and Sakai 1988), black pepper (Zhao and Cranston 1995), and various fresh fruits and vegetables. Within the food industry, ozone has been used routinely for washing and storage of fruits and vegetables (Liangji 1999) to reduce microbial load and enhance shelf life. Washing of fresh produce by ozonated water is reported to reduce bacterial content in shredded lettuce, blackberries, grapes, black pepper, shrimp, beef, broccoli, carrots, and tomatoes (Kim, Yousef, and Chism 1999b; Chen et al. 1992; Barth et al. 1995; Zhao and Cranston 1995). Ozone has been used in several studies to decontaminate freshly caught fish (Goche and Cox 1999), poultry products (Dave 1999), meat and milk products (Dondo et al. 1992; Gorman et al. 1997), to purify and artificially age wine and spirits (Hill and Rice 1982), to reduce aflatoxin in peanut and cottonseed meals (Dwankanath et al. 1968), to sterilize bacon, beef, bananas, eggs, mushrooms, cheese, and fruit (Kaess and Weidemann 1968; Gammon and Karelak 1973), and to preserve lettuce (Kim et al. 1999b), strawberries (Lyons-Magnus 1999), green peppers (Han et al. 2002) and sprouts (Singh, Singh, and Bhuniab 2003). Most contemporary applications of ozone include treatment of drinking water (Bryant et al. 1992) and municipal wastewater (Rice, Overbeck, and Larson 2000). Ozone was declared as GRAS for use in

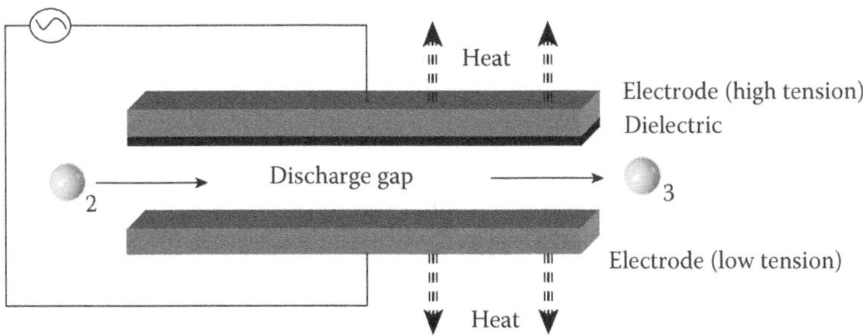

FIGURE 9.2 Ozone generation by corona discharge method.

food processing by the U.S. FDA in 1997 (Graham 1997). Subsequently, ozone has gained approval as a direct food additive for the treatment, storage, and processing of foods both in the gaseous and aqueous phases (Khadre et al. 2001). Use of ozone has been reported for processing of various fruit juices (Tiwari et al. 2008; Steenstrup and Floros 2004) and washing of fruits and vegetables with ozonated water (Zhang et al. 2005).

9.2.3 Rheological Properties of Food Hydrocolloids

Food hydrocolloids such as guar gum, carboxyl methyl cellulose (CMC), pectin, chitosan, xanthan, etc. are high-molecular-weight polysaccharides that find wide application in the food industry, optimizing the rheological and textural characteristics of food systems. These are typically used in food formulations for optimizing viscosity, creating gel structures, and enhancing structural stability (Yaseen et al. 2005; Albert and Mittal 2002; Fagan et al. 2006). Ozonation is reported to decrease molecular weight and viscosity of food hydrocolloids, which could be desirable or undesirable depending on the end use. For example, application of ozone in reducing molecular weight of chitosan has been reported by Yue et al. (2008). This innovative process involves preparation of low-molecular-weight water-soluble chitosan. Chitosan, a natural polysaccharide obtained by the deacetylation of chitin from crustacean shells, has an array of applications in the pharmaceutical and biomedical industries, and in the food industry for food formulations for altering or modifying the rheological or functional properties of food and food products. In the food industry it is generally used as binding, gelling, thickening, and stabilizing agents. Seo, King, and Prinyawiwatkul (2007) investigated the potential of ozone in depolymerization of this polysaccharide, which is generally done by using enzymatic or chemical processes. They reported a decrease in viscosity and molecular weight of chitosan. The molecular weight of ozone-treated chitosan in acidic conditions, i.e., acetic acid solution, caused a significant decrease with increase in ozone treatment compared to ozone treatment in water. A decrease of 92% (104 kDa) was found after a treatment time of 20 min compared to the untreated chitosan (1333 kDa). Since the molecular weight of a polymer is associated with viscosity, a similar decrease in viscosity of chitosan solution with ozone treatment is expected. No et al. (1999) reported a decrease of 63% (206 mPa·s) during ozone treatment of 10 min with ozone concentration of 0.5 ppm compared to untreated chitosan solution (556 mPa·s). Tiwari et al. (2008) studied the effect of ozone treatment on aqueous dispersions of pectin, guar gum, and CMC. A decrease of 95.5%, 81.6%, and 31.7% in apparent viscosity compared to control was reported for guar, CMC, and pectin dispersions respectively at an ozone concentration of 7.8% w/w for 10 min. Figure 9.3 shows the effect of ozone concentration on the apparent viscosity of the dispersions.

The effect of ozone on polymers is mainly dependent on extrinsic parameters such as ozone concentration, flow rate, time, temperature, and the characteristics of the food material under investigation. Degradation of the chitosan polymer is mainly due to the strong oxidative properties of ozone, which selectively degrade β-D-glucoside bonds between units by the electrophilic attack on the C(1)–H bond by ozone molecules (Kabalnova et al. 2001). Degradation of polymers may be due to the formation of other highly reactive species, such as •OH, $HO^{2\bullet}$, $\bullet O_2^-$, and $\bullet O_3^-$, which facilitates degradation (Figure 9.4).

Degradation of organic polymers and subsequent reduction in viscosity in the presence of ozone could be due to either direct reaction with ozone or indirect reaction because of secondary oxidators. Direct reaction is described by the Criegee mechanism (Criegee 1975) where ozone molecules undergo 1–3 dipolar cycloadditions with the double bonds present, leading to the formation of ozonides (1,2,4-trioxolanes) from alkenes and ozone with aldehyde or ketone oxides as decisive intermediates, all of which have finite lifetimes (Criegee 1975). This leads to the oxidative disintegration of ozonide and formation of carbonyl compounds, while oxidative work-up leads to carboxylic acids or ketones. Ozone attacks OH radicals, preferentially to the double bonds in organic compounds

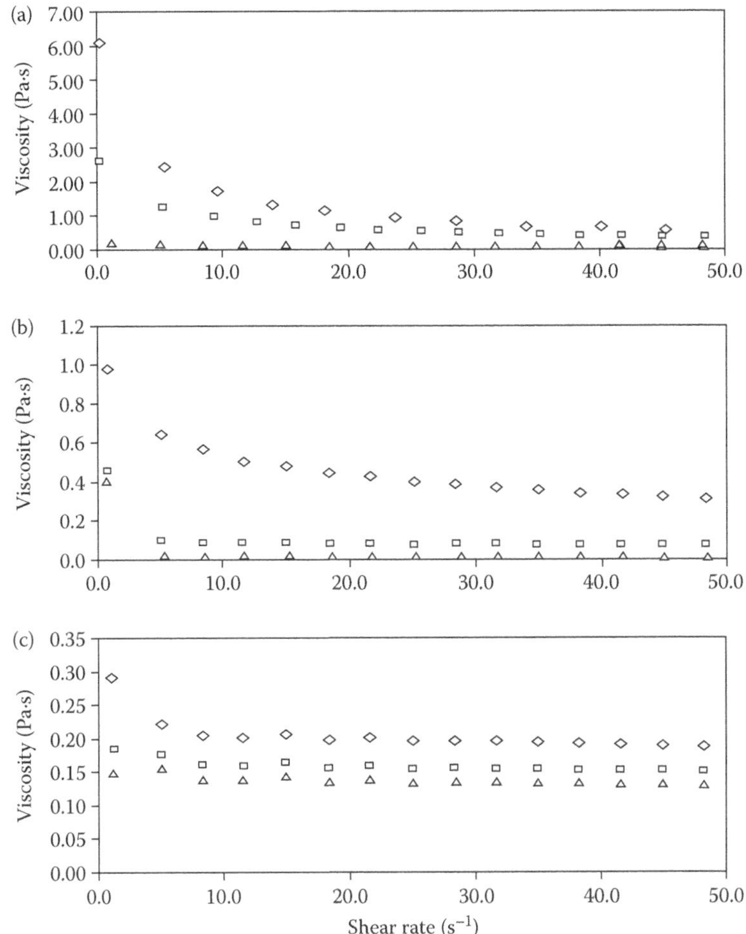

FIGURE 9.3 Effect of ozonation on the apparent viscosity of (a) guar gum (1%), (b) CMC 1%, and (c) pectin (2%). Control (◊), ozone (7.8%w/w) for 5 min (□) and 10 min (Δ) processing times.

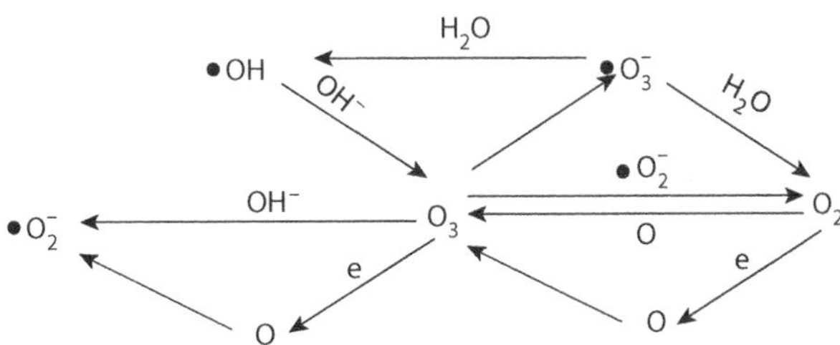

FIGURE 9.4 Formation and transformations of intermediate radicals.

(Figure 9.5). Degradation of organic compounds is also reported to be due to various intermediate radical formations (Figure 9.2) leading to electrophilic and nucleophilic reactions occurring with aromatic compounds that are substituted with an electron donor (e.g., OH^-) having high electron density on the carbon compounds in *ortho* and *para* positions. Chemical reactions involve breakage of old bonds and formation of new bonds and according to bond dissociation energy

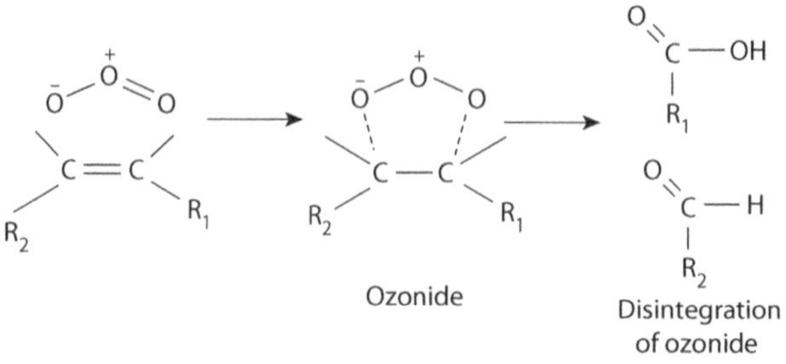

FIGURE 9.5 Direct reaction of ozone with the formation of ozonide.

theory, the lower the bond dissociation energy the more active the bond. Ozone has been reported to degrade organic dye by the formation of highly reactive species, such as •OH, HO^2•, •O_2^-, and •O_3^- (Xue et al. 2008).

9.2.4 Effect of Ozone on Rheological Properties of Cereals and Grains

Ozone application is mainly reported for fresh fruits and vegetables, however a limited number of studies have been reported on ozone treatment of cereals and cereal-based products as an alternative to chlorine treatment. Ibanoglu (2002) studied the effect of ozonation of wheat grains and the subsequent effect on wheat flour properties. He reported that wheat washing with ozonated water (1.5 mg/l) for 30 min significantly reduces microbial load compared to normal water. However, the rheological properties of wheat flour such as extensibility and maximum resistance to extension (flour strength) were significantly affected for soft wheat compared to hard wheat. This could be due to easier penetration of ozone into the endosperm in soft wheat compared to hard wheat. Farinograms, an important rheological property influencing baking properties of wheat flour, were not affected for either soft or hard wheat samples. Studies conducted by Ibanoglu (2002) show that tempering of wheat grains with ozonated water does not have a significant effect on the rheological properties of wheat flour pertinent to baking. Ozonation of soft to medium wheat flour is reported to cause an increase in the resistance to extension of wheat flours and a decrease in extensibility. A comprehensive study conducted by Desvignes et al. (2008) on ozone treatment (10 g/kg) of wheat flour by a patented ozone process, "Oxygreen®," resulted in a reduction of the aleurone layer extensibility and affected the local endosperm resistance to rupture, i.e., an increase in friability. The ozonation on wheat grain causes a significant effect on the maximum viscosity and the setback value of ozonated wheat flour, as indicated by a viscoamylograph.

Storage of grains in an ozone-rich atmosphere does not influence the rheological properties of grains, for example, Mendez et al. (2003) investigated the efficacy of ozone in controlling pests for stored wheat and rice. They reported that ozone treatment does not significantly change the breadmaking properties of hard wheat, including tolerance of the dough to overmixing, absorption of water, dough weight, and proof height. Ozone treatments of rice grain during storage had no significant effect on the adhesiveness of cooked rice. Adhesiveness is one of the more important measures of texture quality of cooked rice which is dependent on amylose content and endosperm proteins (Juliano, Onate, and Del Mundo 1965; Hamaker and Griffin 1990).

9.2.5 Effect of Ozone on Food Texture

Firmness or texture is an important rheological property pertinent to fresh fruits and vegetables. Fruits and vegetables that maintain a firm, crunchy texture are desirable as consumers associate these textures with freshness and wholesomeness. The appearance of a soft or limp product may give rise to consumer rejection prior to consumption (Rico et al. 2007). Textural changes in fruits and vegetables could be due to various enzymatic and nonenzymatic processes. Ozone treatment of fresh fruits and vegetables either by washing or in storage consisting of ozone gas is reported to have a significant effect on texture. Firmness of fresh cilantro leaves was reported to decrease when washed with ozonated water compared to controls. The decrease in firmness was also reported by washing with chlorinated water (Wang et al. 2004). Another study conducted by Selma et al. (2008) reported nonsignificant changes in firmness during storage of fresh-cut cantaloupe, irrespective of gaseous ozone concentration (5000 or 20,000 ppm) for 30 min treatment of compared to controls.

Wei et al. (2007) reported only subtle changes in lettuce firmness throughout 21 days of storage, regardless of ozone concentration. Conversely, chlorine treatment had an adverse impact on lettuce firmness. Figure 9.6 shows the changes in firmness value for lettuce over 21 days of storage. They observed similar trends in the firmness for strawberry samples treated with ozone and chlorine, respectively. Firmness of fresh produce during ozone washing is also dependent on wash-water temperature and pH. Wei et al. (2007) showed that temperatures between 4 and 23°C had little impact on lettuce texture whereas washing of lettuce at higher pH (>7) resulted in lower firmness values. Crispiness or firm texture lettuce can be maintained with improved shelf life by ozone treatment at appropriately controlled conditions (low temperature and pH) and may enhance the texture of fresh produce. Change in texture during ozonation and subsequent storage is possibly due to post-harvest changes in cellulose and hemicellulose contents due to ozone application during modified atmospheric packaging (MAP). This could be due to polymerization and epimerization of cellulose and hemicellulose contents of the cell wall, inducing thickening, causing textural changes in the fresh produce. An et al. (2007) reported an increase in cellulose, hemicellulose, and lignin content of green asparagus during MAP storage after pretreatment with aqueous ozone. Firmness of fresh fruits and vegetables is one of the key quality parameters for commercial value. Ozonation of fruits has been reported to enhance firmness of citrus fruits and cucumbers compared to controls (Nadas, Olmo, and Garcia 2003; Skog and Chu 2001). Ozone is reported to delay softening in strawberries during both cold-room storage and storage at room temperature (Nadas et al. 2003).

9.2.6 Effect of Ozone on Oil

Ozonation of oil or unsaturated triglycerides has gained much attention recently and ozonated vegetable oils have been proposed to be used in a variety of applications. Ozonated vegetable oils have been attributed antibacterial and fungicidal effects, with potential applications in the food, cosmetic, and pharmaceutical industries. Antibacterial activity of ozonated sunflower oil (Sechi et al.

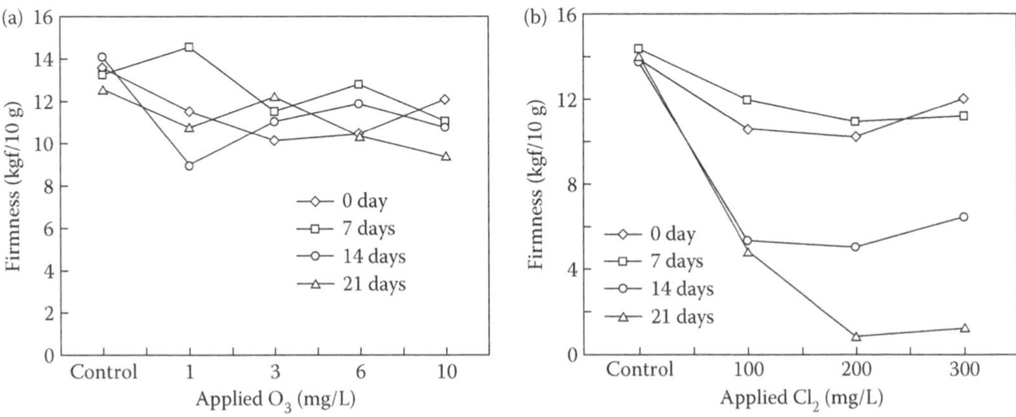

FIGURE 9.6 Changes in firmness during storage for (A) ozone-treated lettuce (4°C, pH 7, and 5 min) and (B) chlorine-treated lettuce (4°C, pH 7, and 5 min). (Reprinted from Wei, K. J., Zhou, H. D., Zhou, T., and Gong, J. H., *Ozone Science & Engineering*, 29, 113–20, 2007. With permission.)

2001), antifungal activity of ozonated olive oil (Geweely 2006), and various other pharmacological applications (Rodríguez et al. 2007; Díaz et al. 2006a, 2006b; Guzel-Seydim et al. 2004; Valacchi et al. 2005; Bocci 2006; Sadowska et al. 2008) have been reported. The therapeutic use of ozonated vegetable oil has been attributed to many favorable effects, however their use has not been widely accepted in orthodox medicine (Bocci 2006).

Sadowska et al. (2008) reported an increase in viscosity and showed that viscosity is a function of molecules' dimension and orientation. The decrease in the degree of unsaturation due to ozonation and the increase in molar mass both contribute to the increase in viscosity of the ozonated oils. Figure 9.7 shows the changes in viscosity of ozonated olive and soybean oil as a function of temperature. Oil viscosity decreases with temperature and increasing degree of unsaturation of triglyceride fatty chains. This is due to reduced intermolecular forces and increased thermal movement among molecules. Ozonolysis of fatty acids, olive oil, and soybean oil was suggested to follow the Criegee mechanism. The reaction of ozone with vegetable oils occurs almost exclusively with carbon-carbon double bonds present in unsaturated fatty chains. The disappearance of unsaturation and the formation of ozonide were almost equal. The ozonide to aldehyde ratio was always above 90%, which indicates that the major product in the early stage of the reaction was ozonide. The ozonation time for complete consumption of the double bonds was ten times longer for the oils than for the pure fatty acids (Sadowska et al. 2008). Conversely, Daiz et al. (2005) reported a decrease in viscosity of ozonized coconut oil with ethanol compared to ozonized coconut oil with water. This behavior might be due to the solubility of oil with ethanol. When water is used in the systems, the viscosity increased due to the poor solubility of oil with water. The reaction of ozone with oil occurs exclusively with carbon-carbon double bonds present in unsaturated fatty chains. The disappearance of unsaturation (double bond) leads to the formation of ozonide.

9.3 DENSE-PHASE CARBON DIOXIDE (DPCD)

Dense-phase carbon dioxide or supercritical CO_2 denotes phases of CO_2 that remain fluid, yet are dense with respect to gaseous CO_2. Moreover, in the supercritical state, CO_2 has low viscosity ($3–7 \times 10^{-5}$ Ns/m^2) and zero surface tension (McHugh and Krukonis 1993), so it can quickly penetrate complex structures and porous materials. Finally, CO_2 is inexpensive and readily available, which makes switching to CO_2-based sterilization economically feasible (Zhang et al. 2006). When

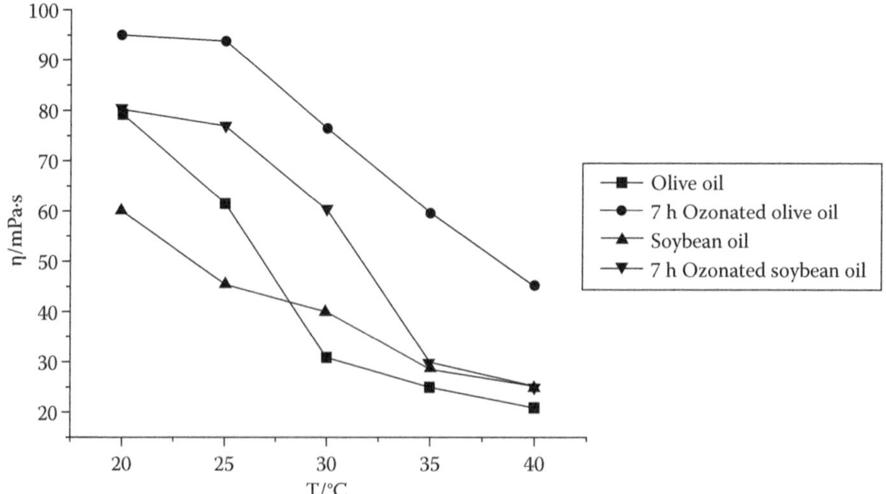

FIGURE 9.7 Changes in the dynamic viscosity of ozonated and unozonated olive oil, and soybean oil as a function of varying temperatures. (From Sadowska, J., Johansson, B., Johanssen, E., Friman, R., Broniary-Press, L., and Rosenholm, J.B., *Chemistry and Physics of Lipids,* 151, 85–91, 2008. With permission.)

gaseous or liquid CO_2 is heated and compressed above the critical temperature (31°C) and pressure (73 atm), it becomes a dense, highly compressible fluid that demonstrates properties of both liquid and gas. Hence, at relatively low pressures and temperatures carbon dioxide transitions to a supercritical state (Figure 9.8). The liquid state of CO_2 conserves some of the solvent properties of the supercritical state, with low viscosity and high diffusion coefficients, therefore the term dense-phase fluid refers to both the supercritical and liquid states. The properties of supercritical CO_2 lend themselves to deep penetration of substrates, which has led to uses in areas ranging from bioremediation to natural product extraction (van der Velde and de Haan 1992; Ge and Yan 2002).

9.3.1 Dense-Phase Carbon Dioxide (DPCD) Treatment System

To date supercritical CO_2 sterilization has not delivered on its promise as a potential preservative, due in part to the inability of existing methodologies to achieve industrial levels of treatment (Spilimbergo and Bertucco 2003). Several systems for possible application of $SCCO_2$ have been developed, including batch, semicontinuous, and continuous. A typical batch-type system mainly comprises of a carbon dioxide cylinder, a pressure regulator, a pressure vessel, a water bath or heater, and a CO_2 release valve (Figure 9.9). CO_2 and the solution under study are stationary in a container during treatment whereas the semicontinuous system allows flow of CO_2 through the treatment chamber, unlike in continuous systems where both CO_2 and the liquid food flows through the system (Balaban and Sibel 2006). Figure 9.9c shows a schematic diagram of continuous DPCD treatment unit.

9.3.2 Effects of Dense-Phase Carbon Dioxide (DPCD) on Food Texture

The use of DPCD as a nonthermal means of food preservation has been reported by various researchers, concentrating mainly on microbial and to some extent on enzymatic inactivation. Table 9.1 shows some of the food applications involving DPCD. There are a limited number of studies available in the literature regarding the effect of DPCD or $SCCO_2$ on product quality and its influence on the rheological and functional properties of foods. Furthermore, most of the work is concentrated on liquid foods as shown in Table 9.1. Haas et al. (1989) treated whole fruits with DPCD to inhibit mould growth but observed tissue damage in some fruits even at low pressures, influencing textural properties of fruits. DPCD as a nonthermal pasteurization technique for beer was found to preserve aroma, flavor, foam capacity, and stability coupled with reducing beer haze and extending shelf life (Dagan and Balaban 2006). Parton et al. (2007) tested a continuous DPCD system for liquid foods such as grape must, orange juice, and tomato paste. No qualitative physical or chemical changes after high-pressure CO_2 treatment were reported with respect to control samples. DPCD treatment of

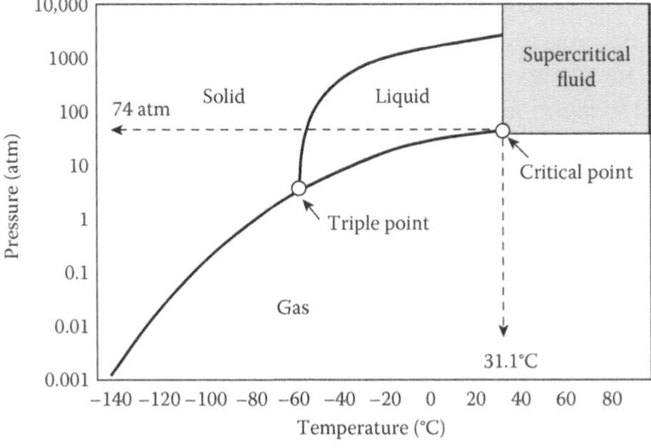

FIGURE 9.8 Phase diagram of carbon dioxide.

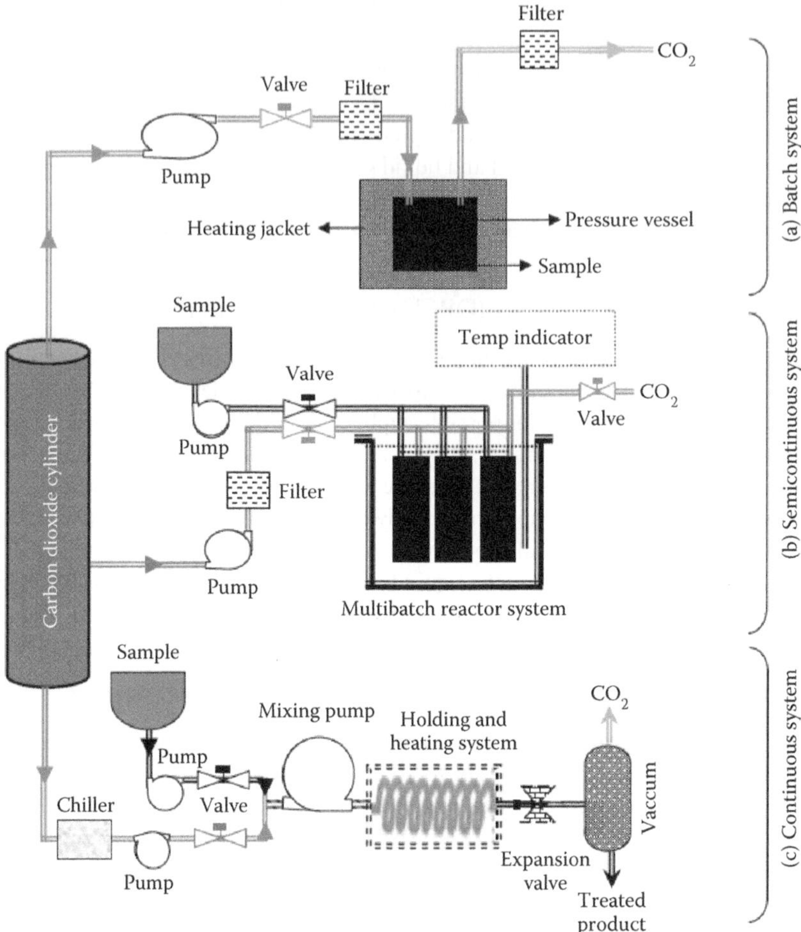

FIGURE 9.9 Schematic diagram of DPCD treatment systems.

orange juice improves some physical and nutritional quality attributes such as cloud formation and stability (Arreola et al. 1991). Moreover, color and cloudiness of DPCD-treated juice was preferred over untreated juice. Dagan and Balaban (2006) employed a continuous DPCD system for pasteurization of beer and reported a reduction in haze by DPCD processing from 146 NTU to 95 NTU compared to fresh beer at processing conditions of 26.5 MPa, 21°C, 9.6% CO_2, and 4.77 min residence time. The aroma and flavor was found to be unaffected, which could be due to the pH-lowering effect of DPCD processing. Haze is an important rheological property of beer that can be defined as the formation of a colloidal suspension that scatters light and makes a beverage appear cloudy. However, they reported a slight decrease in foaming capacity and stability by DPCD. Changes reported for foam characteristics due to DPCD processing may have been caused by the extraction of yeast cell membrane or cell wall parts that may have changed the amount of hydrophobic compounds in the beer, therefore affecting foaming.

9.3.3 Effect of Dense-Phase Carbon Dioxide (DPCD) on Rheological Properties of Dairy Products

DPCD can inactivate pathogenic and spoilage microorganisms and enzymes pertinent to milk and milk products (Hong and Pyun 2001). A major disadvantage for potential applications of DPCD in dairy products is its negative influence on the rheological properties of milk due to changes in

TABLE 9.1
Effect of DPCD on Food Preservation and Quality

Food Product	Treatment System	Quality Attributes	Reference
Apple cider	Continuous	Organoleptic quality (√)	Gunes et al., 2006
Apple Juice	Batch		Liao et al., 2007
Carrot juice	Batch	Color (×), Cloud (×)	Park et al, 2002
Orange juice	Continuous		Kincal et al., 2005
Orange juice	Batch	pH (√), oBrix (√), Color (↑), acidity (↓), AA (↓), organoleptic quality (~)	Arreola et al., 1991
Orange juice	Continuous	pH (~), oBrix (~), acidity (↑), Color (↓), Cloud (↑), organoleptic quality (~)	Kincal et al., 2006
Orange juice	Continuous	pH (~), oBrix (~), acidity (~), Vitamin C (~) and folic acid (~), and aroma profile (~).	Ho, 2003
Orange Juice	Batch	Cloud (↑), pH (~), oBrix (~), Color (↑), organoleptic quality (~)	Balaban, 2005
Orange juice		Color (↓), pH (↓)	Wei et al., 1991
Grape juice	Continuous	pH (~), organoleptic quality (~)	Gunes et al., 2005
Milk	Continuous		Werner and Hotchkiss, 2006
Mandarin juice	Continuous	Cloud (↑), pH (~), oBrix (~), acidity (~), organoleptic quality (~), color (×)	Yagiz et al., 2005
Coconut water	Continuous	organoleptic quality (√), shelf life (↑)	Damar and Balaban, 2005
Beer	Continuous	Haze (↓), aroma (~) and flavor (~) Foam capacity and stability (~)	Dagan and Balaban, 2006
		Solid or Semi Solid	
Kimchi	Batch	pH (↑), TA (↓), organoleptic quality (~), acceptance (↑)	Hong et al., 1997 Hong et al, 1999
Meat (porcine muscle)	Batch	Muscle pH(√), cooking loss (√), protein solubility (√), tenderness (√), water holding capacity (√), protein denaturation (×)	Choi et al., 2008
Chicken		Color (×), cooking quality (×), expulsion of liquid Water Holding Capacity (×)	Wei et al., 1991
Shrimp		Color (×), cooking quality (×), expulsion of liquid (Water Holding Capacity) (×)	Wei et al., 1991

milk protein (casein). As CO_2 dissolves in the aqueous portion of food it undergoes reaction of CO_2 with water to form carbonic acid, thus lowering pH (Damar and Balaban 2006). This further dissociates to yield bicarbonates, carbonates, and H+ ions, lowering extracellular pH and aiding inactivation on microorganisms. This lowering of pH in milk causes precipitation of casein, which has an isoelectric point of pH 4.6. This pH-lowering effect of DPCD can be employed in casein production (Tomasula 1997; Hofland 1999).

$$CO_2 + H_2O \leftrightarrow H_2CO_3$$

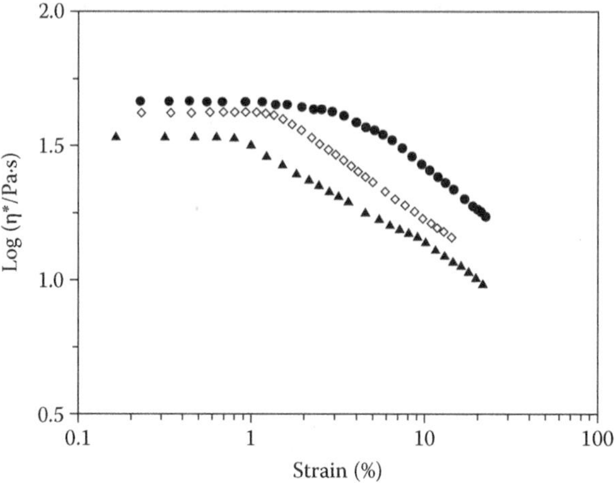

FIGURE 9.10 Changes in complex viscosity as a function of strain for mayonnaise emulsions (25°C; 1 Hz; emulsions aged for 24 h). Key: ▲, spray-dried yolk (D_v = 10.1 μm); ●, yolk extracted with 20:80 (v/v) ethanol/water containing 1.5% (w/v) polysorbate 80 (D_v = 13.2 μm); ◊, yolk extracted with supercritical CO_2 (D_v = 11.7 μm). (From Paraskevopoulou, A., Kiosseoglou, V., Alevisopoulos, S., Kasapis, S., *Colloids and Surfaces B: Biointerfaces,* 12, 107–111, 1999. With permission.)

$$H_2CO_3 \leftrightarrow H^+ + HCO_3^-$$

$$HCO_3^- \leftrightarrow H^+ + CO_3^{-2}$$

Precipitation of casein has a negative influence on consumer sensory perception. Dairy products prepared from DPCD-processed milk such as cheese have resulted in improved rheological parameters of cheese making such as reduced clotting time, augmentation of curd hardness, and whey losses along with a slight increase of cheese yield (Ruas-Madiedo et al. 2002). DPCD treatment of milk is reported to reduce particle size distribution (PSD) compared to controls. The change in the PSD indicates a modification to the fat when milk was treated with CO_2 and pressure (Tisi 2004).

9.3.4 Effect of Dense-Phase Carbon Dioxide (DPCD)-Assisted Extraction on Rheological Properties

Supercritical CO_2 has been proposed as an effective way of removing cholesterol from egg yolk (Froning et al. 1990; Paraskevopoulou et al. 1997). Such low-fat, low-cholesterol egg products are used in various food formulations. Paraskevopoulou et al. (1999) studied the effect on the viscoelastic properties of mayonnaise prepared from reduced-cholesterol yolk after extracting with 20:80 ethanol:water or $SCCO_2$. They found that the method of lipid extraction influenced both the emulsion droplet size and the viscoelastic parameters. The yolk extracted with an ethanol:water mixture, containing 1.5% polysorbate 80, resulted in emulsions that exhibited higher viscoelastic moduli values compared to those prepared with supercritical CO_2-extracted yolk. Figure 9.10 shows the changes in complex viscosity as a function of generated strain for mayonnaise emulsions prepared from reduced cholesterol yolk after extraction.

9.3.5 Dense-Phase Carbon Dioxide (DPCD)-Assisted Extrusion

Extrusion is a well-known commercially adopted cooking process to produce a large variety of expanded food products. The final texture and quality is closely related to the morphology and cell structure of the extrudate (Cho and Rizvi 2008). Supercritical fluid extrusion (SCFX) is an innovative process for continuous generation of microcellular structures in various matrices under low shear and low process temperatures (<100°C) (Rizvi et al. 1995; Mulvaney and Rizvi 1993; Rizvi and Mulvaney 1992). This process offers an attractive extension of the conventional extrusion cooking process to create a new generation of expanded products. It is well established that macro/microstructure formation in extrusion processes is the consequence of several overlapping events including biopolymer structural transformations (starch gelatinization and/or protein denaturation), nucleation, die-swell, cell growth, and cell collapse (Moraru and Kokini 2003).

In an extrusion process when the melt passes through the extruder die, it undergoes a sudden pressure drop resulting in water vapor nuclei generation. These cells grow in size as additional water vapor diffuses into the nuclei. Furthermore, thermal expansion of water vapor causes further expansion. At the maximum level of extrudate expansion, it starts to experience some collapse due to elastic forces in the cell wall (Arhaliass et al. 2003). Conventional steam-based extrusion usually requires low moisture (18–28 wt%), high temperature (120–180°C), and high shear conditions for good expansion. These harsh operating conditions often lead to excessive dextrin formation and limit the application of heat- and shear-sensitive ingredients such as proteins, vitamins, and certain flavors (Cho and Rizvi 2008). Although limited control of the extrudate expansion and the microstructure in conventional steam-based extrusion can be obtained by manipulating parameters such as moisture content, die geometry, die temperature, screw rotation speed, and feed rate, steam-expanded products usually show nonuniform cellular structure and cell sizes in the range of 1–3 mm (Barrett and Peleg 1992). Conversely, use of DPCD as a blowing agent instead of steam uncouples the dual role of water as a blowing agent and a plasticizer during SCFX. In this process, expansion of the melt is achieved by first solubilizing DPCD in the melt and then inducing nucleation due to pressure drop in the die, which is followed by cell growth caused by diffusion of CO_2 into the nucleated cells (Rizvi et al. 1995). A higher moisture content (30–45 wt%) in the extruder barrel is utilized to keep the product temperature low via reduction of viscous dissipation of energy and to maximize $SCCO_2$ solubilization in the melt. Since SCFX extrusion is conducted at temperatures lower than 100°C as well as at lower shear, it enables the use of temperature- and shear-sensitive ingredients in product formulations. Due to the pressure-dependent solubility of DPCD in melts, it can also be used to adjust the pH by formation of carbonic acid to modify the in-barrel rheological properties of the extrudate. The dual role of water, which acts both as a plasticizer and a blowing agent for expansion in conventional extrusion cooking, is decoupled since the expanded structure formation is via puffing (Alavi et al. 2003).

The SCFX process allows control of not only the average cell size of extrudates, which may range from about 50 to 250 μm with a high polydispersity index of ~0.95, but also the volumetric expansion of final products by manipulating process parameters such as die dimension, DPCD concentration, and residence time (Alavi and Rizvi 2005; Winoto 2005). SCFX extrudates show non-porous surfaces and predominantly closed cell structures. High moisture content in the melt reduces the glass transition temperature of the product and the SCFX extrudates are still in a rubbery state upon exiting the die and have a sufficiently low viscosity to cause cell growth or shrinkage. As a result, SCFX extrudates tend to expand further until their structure is set during postextrusion processes such as drying and frying (Alavi et al. 1999; Chen et al. 2002).

Cho and Rizvi (2008) investigated the effects of DPCD injection rate and die dimension on the time-dependent expansion behavior of SCFX extrudates using a visualization technique. Scanning electron microscope images of SCFX extrudates were produced using the formulation 49.5 wt% pregelatinized corn starch, 24 wt% pregelatinized potato starch, 24 wt% sugar, 1 wt% salt, and 1.5 wt% distilled monoglyceride as a dough conditioner with 7 wt% whey protein concentrate to improve the structural stability of the extrudates (Winoto 2005) with barrel moisture content of 43.8 wt%. They observed that, as the die diameter was decreased from 5.9 to 2.9 mm, the cross-sectional expansion and the number of cells increased, whereas the average cell size decreased (Figure 9.11). When the ratio of $SCCO_2$ to the feed was increased from 0.5 to 0.75 wt% $SCCO_2$, the expansion and the cell number density further increased for each die. SCFX extrudates with 0.75 wt% $SCCO_2$ level

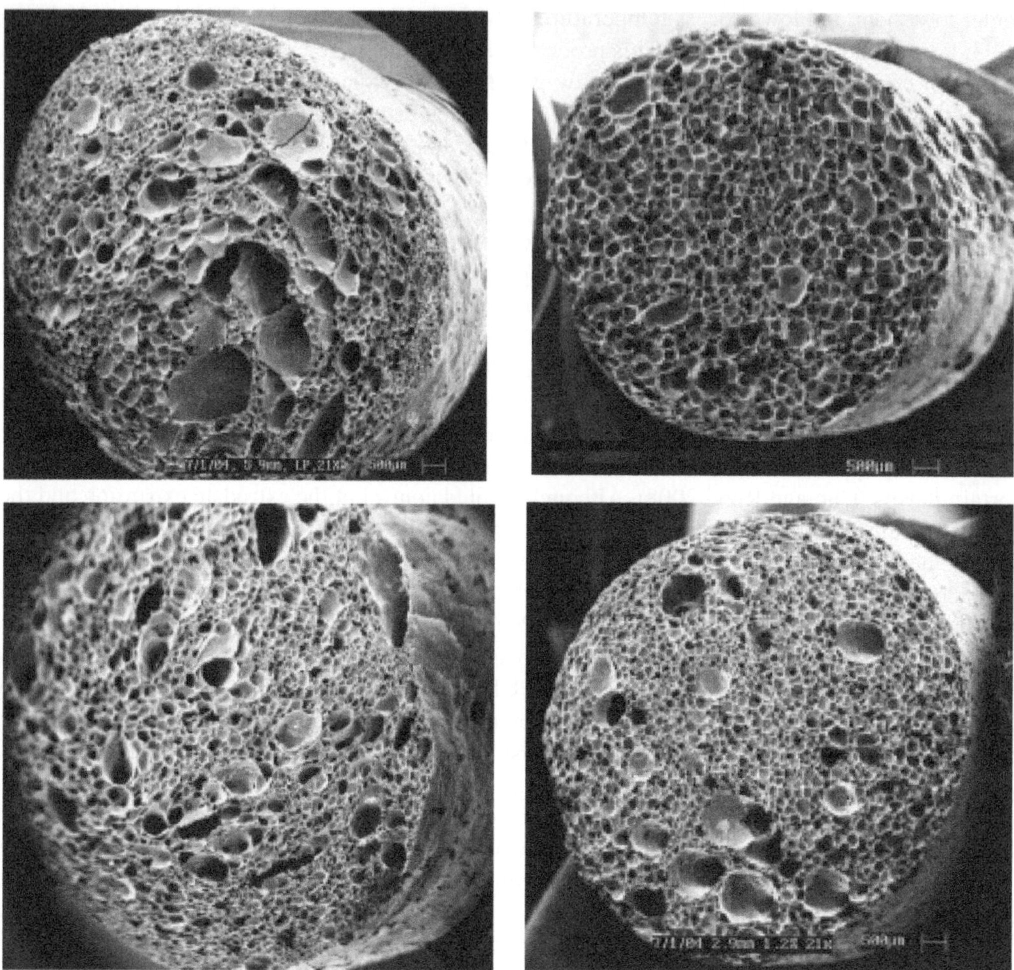

FIGURE 9.11 Effects of $SCCO_2$ and die size on product morphology (21 × magnification). Left: 5.9 mm die; right: 2.9 mm die. Top: 0.5 wt% $SCCO_2$; bottom: 0.75 wt% $SCCO_2$. (Adapted from Winoto, C. W., *Process parameters and their influence on supercritical fluid extrudate properties.* Master thesis, Cornell University, Ithaca, NY, 2005. With permission and Cho, K. Y., and Rizvi, S. S. H., *Food Research International*, 41, 31–42, 2008. With permission.)

and using 5.9 mm die showed significant structure collapse, implying that the prevention of gas loss would be essential to increase the final volumetric expansion.

9.3.6 Effect of Dense-Phase Carbon Dioxide (DPCD) on Food Microstructure

Alavi et al. (1999) observed that SCFX extrudates exhibited the unique characteristic of a nonporous skin surrounding the internal cellular morphology (Figure 9.12). This skin was comprised of unexpanded starch and very small cells. Rapid diffusion of CO_2 out of the sample creates a depletion layer near the edges in which the gas concentration is too low to contribute significantly to cell growth. Moreover, rapid drying and gelation of the proteins near the edges sets the material quickly and inhibits growth of any nucleated sites. A combination of these factors caused the formation of the nonporous skin (Alavi et al. 1999). Conversely, the "skin effect" does not occur for extrudates dried at 100°C because of rupture of any skin that might have developed initially. The presence of skin reduces water penetration and delays onset of sogginess, which is a desirable characteristic for ready-to-eat breakfast cereal. Moreover, the skin provides a composite structure

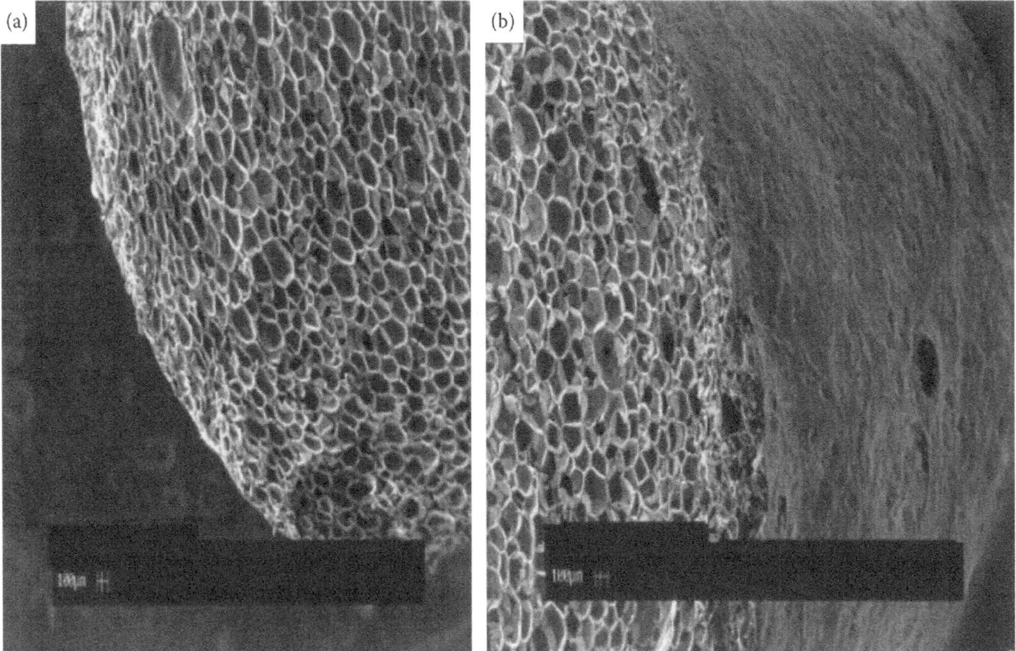

FIGURE 9.12 Scanning electron micrographs of a typical SCFX extrudate showing: (a) cell size distribution near the edge and (b) a nonporous external skin. (From Alavi, S.H., Gogoi, B.K., Khan, M., Bowman, B.J., and Rizvi, S.S.H., *Food Research International,* 32, 107–118, 1999. With permission.)

to the extrudates and variation of its thickness can provide another means for manipulating extrudate mechanical properties and texture (Alavi et al. 1999). The polydispersity index of cell size distribution of SCFX extrudates is reported to be greater compared to steam extrudates (Barrett and Peleg 1992; Alavi et al. 1999). This indicates that distribution of the cell size of SCFX extrudates is uniform. The cell density of the SCFX extrudates can be explained on the basis of the classical nucleation theory. Hence it can be said that extrusion of products with SCFX results in a more uniform microcellular structure.

9.4 CONCLUSIONS AND FUTURE TRENDS

Within the food industry, there is an increasing emphasis and trend toward natural food preservation technologies in response to growing consumer demand for "greener" additives. During the last two decades natural nonthermal technologies have been investigated for practical applications. These technologies have been shown to inactivate microorganisms and enzymes without significant adverse effects on organoleptic or nutritional properties. Reported studies have demonstrated that natural nonthermal food preservation techniques described in this chapter such as ozone and DPCD may offer unique advantages for food processing with minimal or desired effects on the rheological properties of food. More complex considerations arise for combinations of technologies, particularly with respect to optimization of practical application. Intelligent selection of appropriate systems based on detailed, sequential studies is necessary.

REFERENCES

Alavi, S. H., and Rizvi, S. S. H. 2005. Strategies for enhancing expansion in starch-based microcellular foams produced by supercritical fluid extrusion. *International Journal of Food Properties*, 8:23–34.

Alavi, S. H., Rizvi, S. S. H., and Harriot, P. 2003. Process dynamics of starch based microcellular foams produced by supercritical fluid extrusion I: Model development and II. Numerical simulation and experimental evaluation. *Food Research International*, 36:309–30.

Alavi, S. H., Khan, M., Bowman, B.J., and Rizvi, S. S. H. 1999. Structural properties of protein-stabilized starch-based supercritical fluid extrudates. *Food Research International*, 31:107–18.

Albert, S., and Mittal, G. S. 2002. Comparative evaluation of edible coatings to reduce fat uptake in a deep-fried cereal product. *Food Research International*, 35:445–58.

An, J. S., Zhang, M., Lu, Q.R. 2007. Changes in some quality indexes in fresh-cut green asparagus pretreated with aqueous ozone and subsequent modified atmosphere packaging. *Journal of Food Engineering*, 78(1):340–44.

Arhaliass, A., Bouvier, J. M., and Legrand, J. 2003. Melt growth and shrinkage at the exit of the die in the extrusion-cooking process. *Journal of Food Engineering*, 60:185–92.

Arreola, A. G., Balaban, M. O., Marshall, M. R., Peplow, A. J., Wei, C. I., and Cornell, J. A. 1991. Supercritical CO_2 effects on some quality attributes of single strength orange juice. *Journal of Food Science,* 56(4): 1030–1033.

Balaban, M.O. 2005. Effect of dense phase CO_2 on orange juice enzymes and microorganisms in a batch system. In: IFT Annual Meeting Book of Abstracts; No: 50-2, July 15–20; New Orleans, LA.

Balaban, M. O., and Sibel, D. 2006. Review of dense phase CO_2 technology: Microbial and enzyme inactivation, and effects on food quality. *Journal of Food Science*, 71 (1): R1–11.

Ballestra, P., Abreuda, S. A., and Cuq, J. L. 1996. Inactivation of *Escherichia coli* by CO_2 under pressure. *Journal of Food Science*, 61(4): 829–36.

Ballestra, P., and Cuq, J. L. 1998. Influence of pressurized carbon dioxide on the thermal inactivation of bacterial and fungal spores. *Lebensmittel-Wissenschaft und-Technologie*, 31:84–88.

Barrett, A. M., and Peleg, M. 1992. Cell size distributions of puffed corn extrudates. *Journal of Food Science*, 57(1): 146–48, 154.

Barth, M. M., Zhou, C., Mercier, J., and Payne, F. A. 1995. Ozone storage effects on antocyanin content and fungal growth in blackberries. *Journal of Food Science*, 60 (6): 1286–88.

Bocci, A. V. 2006. Scientific and medical aspects of ozone therapy. State of the art. *Archives of Medical Research*, 37:425–35.

Bryant, E. A., Fulton, G. P., and Budd, G. L. 1992. Disinfection alternatives for safe drinking water. Van Nostrand Reinhold, New York.

Chen, K. H., Dogan, E., and Rizvi, S. S. H. 2002. Supercritical fluid extrusion of masa-based snack chips. *Cereal Foods World*, 47:44–51.

Chen, H. C., Huang, S. H., Moody, M. W., and Jiang, S. T. 1992. Bacteriocidal and mutagenic effects of ozone on shrimp (Penaeus-Monodon) meat. *Journal of Food Science*, 57:923–27.

Cho, K. Y. and Rizvi, S. S. H. 2008. The time-delayed expansion profile of supercritical fluid extrudates. *Food Research International*, 41 (1): 31–42.

Choi, Y. M., Ryu, S. H., Lee, S. H., Go, G. W, Shin, H. G., Kim, K. H., Rhee, M. S., and Kim, B.C. 2008. Effects of supercritical carbon dioxide treatment for sterilization purpose on meat quality of porcine longissimus dorsi muscle. *LWT-Food Science and Technology*, 41(2): 317–22

Criegee, R. 1975. Mechanism of ozonolysis. DOI: 10.1002/anie.197507451.

Corwin, H. and Shellhammer, T. H. 2002. Combined carbon dioxide and high pressure inactivation of pectin methylesterase, polyphenol oxidase, Lactobacillus plantarum and Escherichia coli. *Journal of Food Science*, 67(2):697–701.

Dagan, G. F., and Balaban, M. O. 2006. Pasteurization of beer by a continuous dense-phase CO_2 system. *Journal of Food Science*, 71 (3): 715–19.

Damar, S. and Balaban M. O. 2006. Review of dense phase CO_2 technology: microbial and enzyme inactivation, and effects on food quality, *Journal of Food Science*, 71:R1-R11.

Damar, S., and Balaban, M.O. 2005. Cold pasteurization of coconut water with a dense phase CO_2 system. In: IFT Annual Meeting Book of Abstracts 54F-2; 2005 July 15–20 p. 27; New Orleans, LA.

Dave, S. A. 1999. Effect of ozone against *Salmonella enteritidis* in aqueous suspensions and on poultry meat, 26–68. MSc Thesis, Ohio State Univ.

Desvignes, C., Chaurand, M., Dubois, M., Sadoudi, A., Abecassis, J., and Lullien-Pellerin, V. 2008. Changes in common wheat grain milling behavior and tissue mechanical properties following ozone treatment. *Journal of Cereal Science*, 47:245–51.

Diaz, M. F., Gavín Sazatornil, J. A., Ledea, O., Hernandez, F., Alaiz, M., and Garces, R. 2005. Spectroscopic characterization of ozonated sunflower oil. *Ozone Science & Engineering*, 27: 247–53.

Díaz, M. F., Hernandez, R., Martinez, G., Vidal, G., Gomez, M., Fernandez, H., and Garcés, R. 2006a. Comparative study of olive oil and ozonized sunflower oil. *Journal of Brazilian Chemical Society*, 17(2):403–07.

Díaz, M. F., Gavin, J. A., Gomez, M., Curtielles, V., and Hernández, F. 2006b. Study of ozonated sunflower oil using 1H NMR and microbiological analysis. *Ozone Science & Engineering*, 28:59–63.

Dondo, A., Nachman, C., Doglione, L., Rosso, A., and Genetti, A. 1992. Foods: their preservation by combined use of refrigeration and ozone. *Ing. Aliment. Conserve Anim*, 8:16–25.

Dwankanath, C. I., Rayner, E. T., Mann, G. E., and Dollar, F. G. 1968. Reduction of aflatoxin levels in cotton seed and peanut meals by ozonization. *Journal of American Oil Chemical Society*, 45 (2): 93–97.

Erkmen, O., and Karaman, H. 2001. Kinetic studies on the high pressure carbon dioxide inactivation of *Salmonella typhimuriun*. *Journal of Food Engineering*, 50:25–28.

Fagan, C. C., O'Donnell, C. P., Cullen, P. J., and Brennan, C. S. 2006. The effect of dietary fibre inclusion on milk coagulation kinetics. *Journal of Food Engineering*, 77:261–68.

Fraser, D. 1951. Bursting bacteria by release of gas pressure. *Nature*, 167:33–34.

Froning, G. W., Wehling, R. L., Cuppett, S. L., Pierce, M. M., Niemann, L., and Siekman, D. K. 1990. *Journal of Food Science*, 55:95.

Gammon, R., and Karelak, A. 1973. Gaseous sterilization of foods. *American Institute of Chemical Engineering Symposium Series*, 69 (132): 91–102.

Ge, Y., and Yan, H. 2002. Extraction of natural Vitamin E from wheat germ by supercritical carbon dioxide. *Journal of Agricultural Food Chemistry*, 50 (4): 685–89.

Geweely, N. S. I. 2006. Antifungal activity of ozonized olive oil (Oleozone). *International Journal of Agriculture and Biology*, 8 (5):670–75.

Goche, L., and Cox, B. 1999. Ozone treatment of fresh H&G Alaska salmon. Report to Alaska Science and Technology Foundation and Alaska Department of Environmental Conservation, Seattle, WA.

Gorman, B. M., Kuchevar, S. L., Sofos, L. W., Morgan, J. B., Schmidt, G. R., and Smith, G. C. 1997. Changes on beef adipose tissue following decontamination with chemical solutions or water 35 1C or 741C. *Journal of Muscle Foods*, 8:185–97.

Graham, D. M. 1997. Use of ozone for food processing. *Food Technology*, 51:121–37.

Güzel-Seydim, Z., Bever Jr., P. I., and Greene, A. K. 2004. Efficacy of ozone to reduce bacterial populations in the presence of food components. *Food Microbiology*, 21:475–79.

Guzel-Seydim, Z. B., Greene, A. K., and Seydim, A. C. 2004. Use of ozone in food industry. *Lebensmittel-Wissenschaft Und-Technologies*, 37:453–60.

Gunes, G., Blum, L.K., and Hotchkiss, J.H. 2006. Inactivation of *Escherichia coli* (ATCC 4157) in Diluted Apple Cider by Dense-Phase Carbon Dioxide. *Journal of Food Protection*, 69(1): 12–16.

Gunes, G., Lisa, B., and Hotchkiss, J.H. 2005. Inactivation of yeasts in grape juice using a continuous dense phase carbon dioxide processing system. *Journal of the Science of Food Agriculture* 85(14): 2362–68.

Haas, G. J., Prescott, J. R., Dudley, E., Dik, R., Hintlian, C., and Keane, L. 1989. Inactivation of microorganisms by carbon dioxide under pressure. *Journal of Food Safety*, 9:253–65.

Hamaker, B. R. and V. K. Griffin. 1990. Changing the viscoelastic properties of cooked rice through protein disruption. *Cereal Chemistry*, 67:261–64.

Han, Y., Floros, J. D., Linton, R. H., Nielsen, S. S., and Nelson, P. E. 2002. Response surface modeling for the inactivation of *E. coli* O157: H7 on green peppers by ozone gas treatment. *Journal of Food Science*, 67 (3): 3188–93.

Hill, D. G., and Rice, R. G. 1982. Historical background properties and applications. In: Rice, R. G., and Netzer, A. (Eds.), *Handbook of Ozone Technology and Applications*, Vol. 1, 1–37. Ann Arbor, MI: Ann Arbor Science.

Ho, K. L. G. 2003. Dense phase carbon dioxide processing for juice. In: IFT annual meeting book of abstracts; 2003 July 12-16; Chicago Ill.: Inst. of Food Technologists. Abstract no. 50–3.

Hofland, G.W., Van Es, M., Luuk, A. M., Van der, W., Witkamp, G. J. 1999. Isoelectric precipitation of casein using high pressure CO_2. *Industrial & Engineering Chemical Research*, 38(12):4919–27.

Hoigné, J., and Bader, H. R. 1983. Constants of reactions of ozone with organic and inorganic compounds in water. I: Non-dissociating organic compounds. *Water Research*, 17:173–83.

Hong, S. I., and Pyun, Y. R. 2001. Membrane damage and enzyme inactivation of *L. plantarum* by high pressure CO_2 treatment. *International Journal of Food Microbiology*, 63:19–28.

Hong, S. I., Park, W. S., and Pyun, Y. R. 1997. Inactivation of Lactobacillus sp. From kimchi by high pressure carbon dioxide. *LWT-Food Science and Technology,* 30 (7): 681–85.

Hong, S. I., Park, W. S., and Pyun, Y. R. 1999. Non-thermal inactivation of *Lactobacillus plantarum* as influenced by pressure and temperature of pressurized carbon dioxide. *International Journal of Food Science Technology* 34(2): 125–30.

Ibanoglu, S. 2002. Wheat washing with ozonated water: effects on selected flour properties. *International Journal of Food Science & Technology*, 37(5):579–84.

Juliano, B. O., Onate, L. U., and Del Mundo, A. M. 1965. Relation of starch composition, protein content, and gelatinization temperature to cooking and eating qualities of milled rice. *Food Technology*, 119:1006–1008.

Kabalnova, N. N., Murinov, K. Y., Mullagaliev, R., Krasnogorskaya, N. N., Shereshovets, V. V., Monakov, Y. B., and Zaikov, G. E. 2001. Oxidative destruction of chitosan under effect of ozone and hydrogen peroxide. *Journal of Applied Polymer Science*, 81:875–81.

Kaess, G., and Weidemann, J. F. 1968. Ozone treatment of chilled beef. Effect of low concentrations of ozone on microbial spoilage and surface color of beef. *Journal of Food Technology*, 3:325–34.

Khadre, M. A., Yousef, A. E., and Kim, J. 2001. Microbiological aspects of ozone applications in food: a review. *Journal of Food Science*, 6:1242–52.

Kim, J. G., Yousef, A. E., and Chism, G. W. 1999b. Use of ozone to inactivate microorganisms on lettuce. *Journal of Food Safety*, 19:17–34.

Kim, J. G., Yousef, A. E., and Dave, S. 1999a. Application of ozone for enhancing the microbiological safety and quality of foods: a review. *Journal of Food Protection*, 62 (9): 1071–87.

Kincal, D. 2000. A continuous high pressure carbon dioxide system for cloud retention, microbial reduction and quality change in orange juice. Unpublished MSc thesis, Univ. of Florida, Gainesville, FL, USA.

Kincal, D., Hill, W. S., Balaban, M., Portier, K. M., Sims, C. A., Wei, C. I., and Marsahll, M. R. 2006. A continuous high-pressure carbon dioxide system for cluod and quality retention in orange juice. *Journal of Food Science* 71(6): c338–c344.

Liangji, X. 1999. Use of ozone to improve the safety of fresh fruits and vegetables. *Food Technology*, 53:58–61.

Liao, X., Hu, X., Gui, F., Wu, J., Chen, F., Zhang, Z., and Wang, Z. 2007. Inactivation of polyphenol oxidases in cloudy apple juice exposed to supercritical carbon dioxide. *Food Chemistry* 100: 1678–85.

Lyons-Magnus, M. 1999. Ozone Use Survey Data. Ozone Treatment of Fresh Strawberries. Data submitted to EPRI Agriculture and Food Alliance, September 28, 1999, Fresno, CA, USA.

Mahapatra, A. K., Muthukumarappan, K., and Julson, J. L. 2005. Applications of ozone, bacteriocins and irradiation in food processing: A review. *Critical Reviews in Food Science and Nutrition*, 45 (6): 447–61.

McHugh, M. and Krukonis, V. 1993. Introduction. Supercritical Fluid Extraction Butterworth-Heinemann. Newton, MA, 1–16.

Mulvaney, S. J. and Rizvi, S. S. H. 1993. Extrusion processing with supercritical fluids. *Food Technology*, 47: 74–82.

Mendez, F., Maier, D. E., Mason, L. J., and Woloshuk, C. P. 2003. Penetration of ozone into columns of stored grains and effects on chemical composition and processing performance. *Journal of Stored Products Research*, 39(1):33–44.

Moraru, C. I. and Kokini, J. L. 2003. Nucleation and expansion during extrusion and microwave heating of cereal foods. *Comprehensive Reviews in Food Science and Food Safety*, 2:120–38.

Muthukumarappan, K., Halaweish, F., and Naidu, A. S. 2000. Ozone. In: Naidu, A. S. (Eds.), *Natural Food Anti-Microbial Systems*, 783–800. Boca Raton: CRC Press.

Nadas, A., Olmo, M., and Garcia, J. M. 2003. Growth of Botrytis cinerea and strawberry quality in ozone-enriched atmospheres. *Journal of Food Science*, 68 (5): 1798–802.

Naitoh, S., Okada, Y., and Sakai, T. 1988. Studies on utilization of ozone in food preservation: V. Changes microflora of ozone treated cereals, grains, peas, beans and spices during storage. *Journal of Japanese Society of Food Science and Technology*, 35:69–77.

No, H. K., Kim, S. D., Kim, D. S., Kim, S. K., and Meyers, S. P. 1999. Effect of physical and chemical treatment on chitosan viscosity. *Journal of Chitin Chitosan*, 4 (4): 177–83.

Osajima, Y., Shimoda, M., and Kawano, T. 1996. Method for modifying the quality of liquid foodstuff. United States Patent, US 5,520,943.

———. 1997. Method for inactivating enzymes, microorganisms, and spores in a liquid foodstuff. United States Patent, US 5,667,835.

Osajima, Y., Shimoda, M., Kawano, T., and Okubo, K. 1998. System for processing liquid foodstuff or liquid medicine with a supercritical fluid of carbon dioxide. United States Patent, US 5,704,276.

———. 1999. System for processing liquid foodstuff or liquid medicine with a supercritical fluid of carbon dioxide. United States Patent, US 5,869,123.

Osajima, Y., Shimoda, M., Takada, M., and Miyake, M. 2003. Method of and system for continuously processing liquid materials, and the product processed thereby. United States Patent, US 6,616,849.

Paraskevopoulou, A., Panayiotou, K., and Kiosseoglou, V. 1997. *Food Hydrocolloids*, 11, 385.

Paraskevopoulou, A., Kiosseoglou, V., Alevisopoulos, S., and Kasapis, S. 1999. Influence of reduced-cholesterol yolk on the viscoelastic behaviour of concentrated O/W emulsions. *Colloids and Surfaces B: Biointerfaces,* 12 (3–6): 107–111.

Park, S. I., Lee, J. I., and Park, J. 2002. Effects of a combined process of high pressure carbon dioxide and high hydrostatic pressure on the quality of carrot juice. *Journal of Food Science*, 67:1827–34.

Parton, T., Bertucco, A., Elvassore, N., and Grimolizzi, L. 2007. A continuous plant for food preservation by high pressure CO_2. *Journal of Food Engineering*, 79:1410–17.

Rice, R. G., Overbeck, P., and Larson, K. A. 2000. Costs of ozone in small drinking water systems. In: *Proceedings on Small Drinking Water and Wastewater Systems*, 27. Ann Arbor, MI: NSF International.

Rico, D., Martín-Diana, A. B., Barat, J. M., and Barry-Ryan, C. 2007. Extending and measuring the quality of fresh-cut fruit and vegetables: a review. *Trends in Food Science & Technology*, 18 (7): 373–86.

Rodríguez, Z. B. Z., Alvarez, R. G., Guanche, D., Merino, N., Rosales, F. H., Cepero, S. M., Gonzalez, Y. A., and Schulz, S. 2007. Antioxidant mechanism is inlvolved in gastroprotective effects of ozonized sunflower oil in ethanolinduced ulcers in rats. *Mediators of Inflammation*, pages 6.

Ruas-Madiedo P., Alonso L., Delgado T., Bada-Gancedo J. C., de los Reyes-Gavilan, C. G. 2002. Manufacture of Spanish hard cheeses from CO_2-treated milk. *Food Research International* 35 (7):681–90.

Rizvi, S. S. H., Mulvaney, S. J., and Sokhey, A. S. 1995. The combined application of supercritical fluid and extrusion technology. *Trends in Food Science and Technology*, 6:232–40.

Rizvi, S. S. H. and Mulvaney, S. J. 1992. Extrusion processing with supercritical fluids. US Patent 5120559.

Sadowska, J., Johansson, B., Johannessen, E., Friman, R., Broniarz-Press, L., and Rosenholm, J. B. 2008. Characterization of ozonated vegetable oils by spectroscopic and chromatographic methods. *Chemistry and Physics of Lipids* 151 (2):85–91.

Sechi, L. A., Lezcano, I., Nunez, N., Espim, M., Duprè, I., Pinna, A., Molicotti, P., Fadda, G., and Zanetti, S. 2001. Antibacterial activity of ozonated sunflower oil (Oleozon). *Journal of Applied Microbiology*, 90:279–84.

Selma, M. V., Ibanez, A. M., Allende, A., Cantwell, M. and Suslow, T. 2008. Effect of gaseous ozone and hot water on microbial and sensory quality of cantaloupe and potential transference of Escherichia coli O157:H7 during cutting. *Food Microbiology*, 25:162–68.

Shimoda, M., Cocunubo-Castellanos, J., Kago, H., Miyake, M., Osajima, Y., and Hayakawa, I. 2001. The influence of dissolved CO_2 concentration on the death kinetics of Saccharomyces cerevisiae. *Journal of Applied Microbiology*, 91:306–11.

Seo, S., King, J. M., and Prinyawiwatkul, W. 2007. Simultaneous depolymerization and decolorization of chitosan by ozone treatment. *Journal of Food Science* 72 (9): C522–26.

Shimoda, M., Yamamoto, Y., Cocunubo-Castellanos, J., Tonoike, H., Kawano, T., and Ishikawa, H. 1998. Antimicrobial effects of pressured carbon dioxide in a continuous flow system. *Journal of Food Science*, 63:709–12.

Singh, N., Singh, R. K., and Bhuniab, A. K. 2003. Sequential disinfection of E. coli O157: H7 inoculated alfalfa seeds before and during sprouting using aqueous chloride dioxide, ozonated water and thyme essential oil. *Lebensmittel-Wissenschaft und-Technologie*, 36:235–43.

Skog, J. L. and Chu, C. L. 2001. Effect of ozone on qualities of fruits and vegetables in cold storage. *Canadian Journal of Plant Science*, 81:773–78.

Spilimbergo, S., and Bertucco, A. 2003. Non-thermal bacteria inactivation with dense CO_2. *Biotechnology and Bioengineering*, 84 (6): 627–38.

Steenstrup, L. D., and Floros, J. D. 2004. Inactivation of *E. coli* 0157:H7 in apple cider by ozone at various temperatures and concentrations. *Journal of Food Processing Preservation*, 28:103–16.

Tisi, A. D. 2004. Effects of dense phase CO_2 on enzyme activity and casein proteins in raw milk. Ithaca, N.Y.: Cornell Univ. Available from: http://dspace.library.cornell.edu/handle/1813/60. Accessed June 14, 2005.

Tiwari, B. K., Muthukumarappan, K., O'Donnell, C. P., and Cullen, P. J. 2008. Modeling colour degradation of orange juice by ozone treatment using response surface methodology. *Journal of Food Engineering*, 88:553–60.

Tomasula, P. M., Craig, J. C., and Boswell, R. T. 1997. A continuous process for casein production using high pressure CO_2. *Journal of Food Engineering*, 33 (3/4):405–19.

Valacchi, G., Fortino, V., and Bocci, V. 2005. The dual action of ozone on the skin. *British Journal of Dermatology*, 153:1096–1100.

van der Velde, E. G., and de Haan, W. 1992. Supercritical fluid extraction of polychlorinated biphenyls and pesticides from soil. Comparison with other extraction methods. *Journal of Chromatography*, 626 (1): 135–43.

Wang, H., Feng, H., and Luo, Y. 2004. Microbial reduction and storage quality of fresh-cut cilantro washed with acidic electrolyzed water and aqueous ozone. *Food Research International*, 37(10):949–956.

Wei, C. I., Balaban, M. O., Fernando, S. Y., and Peplow, A. J. 1991. Bacterial effect of high-pressure CO_2 treatment on foods spiked with Listeria or Salmonella, *Journal of Food Protection*, 54: 189–94.

Wei, K. J., Zhou, H. D., Zhou, T., and Gong, J. H. 2007. Comparison of aqueous ozone and chlorine as sanitizers in the food processing industry: Impact on fresh agricultural produce quality. *Ozone Science & Engineering*, 29 (2): 113–20.

Werner, B. G., and Hotchkiss, J. H. 2006. Continuous flow non-thermal CO_2 processing: the lethal effects of sub-critical and super-critical CO_2 on total microbial populations and bacterial spores in raw milk. *Journal of Dairy Science* 89:872–81.

Xue, J., Chen, L., and Wang, H. 2008. Degradation mechanism of Alizarin Red in hybrid gas–liquid phase dielectric barrier discharge plasmas: experimental and theoretical examination. *Chemical Engineering Journal*, 138:120–27.

Yagiz, Y., Lim, S. L., and Balaban, M. O. 2005. Continuous high pressure CO_2 processing of mandarin juice. In: IFT annual meeting book of abstracts; 54F-16, 2005 July 15–20; New Orleans, LA.

Yaseen, E. I., Herald, T. J., Aramouni, F. M., and Alavi, S. 2005. Rheological properties of selected gum solutions. *Food Research International*, 38(2):111–19.

Yue, W., Yao, P., Wei, Y., Li, S., Lai, F., and Liu, X. 2008. An innovative method for preparation of acid-free-water-soluble low-molecular-weight chitosan (AFWSLMWC). *Food Chemistry*, 108:1082–87.

Criegee, R. 1975. Mechanism of ozonolysis. DOI: 10.1002/anie.197507451.

Zhang, J., Davis, T. A., Matthews, M. A., Drews, M. J., LaBerge, M., and Yuehuei, H. 2006. Sterilization using high-pressure carbon dioxide. *Journal of Supercritical Fluids*, 12:245–49.

Zhang, L., Lu, Z., Yu, Z., and Gao, X. 2005. Preservation of fresh-cut celery by treatment of ozonated water. *Food Control*, 16:279–83.

Zhao, J., and Cranston, P. M. 1995. Microbial decontamination of black pepper by ozone and effects of treatment on volatile oil constituents of the spice. *Journal of the Science of Food and Agriculture*, 68:11–18.

10 Gelation and Thickening with Globular Proteins at Low Temperatures

Ruben Mercade-Prieto and Sundaram Gunasekaran
University of Wisconsin-Madison

CONTENTS

10.1 WHEY PROTEINS

Whey proteins, not long ago a waste product of the dairy industry, have become an important ingredient for many food products because of their high nutritional value and easy digestibility (de Wit 1998). In the last few decades, technologies have been developed to concentrate, isolate, and fractionate whey proteins and make them available as food ingredients (Zadow 1992). Whey proteins are well known for their functional properties, including emulsification, stabilization, foaming, and gelation (Huffman 1996; Foegeding et al. 2002; Totosaus et al. 2002). They are often used to improve texture in various foods, such as sausages, dairy products, desserts, bakery products, cold sauces, surimi, mayonnaise, and gelatin-like deserts (Barbut 1995a; Elofsson et al. 1997; Britten and Giroux 2001), and they can mimic the role of fats in enhancing textural properties of foods (Clark 1992).

Whey proteins are able to form gels and thicken solutions (Mulvihill and Kinsella 1987; Kinsella and Whitehead 1989). However, solutions have to be heated above 65°C to see an increase in the viscosity, which limits their application to many food products. In addition, heat-set gels formed at acidic pH and high salt concentrations may not have desirable properties, for example being too opaque (McClements and Keogh 1995). Another example is that they may have bad water-holding capacity (WHC), an essential property in food products because most of them contain more than 50% water, and consumers tend do avoid products that have condensed water in the package (Barbut 1996).

The gelation of whey protein solutions is a multistep process. Considering the major whey protein, β-lactoglobulin (βLg), at neutral pH and low salt concentrations, there is first the disruption of the βLg dimer to monomers, followed by the denaturation of the monomers, which lead to the formation of small aggregates, which, upon further cross-linking, will grow and eventually percolate (Pouzot et al. 2004). The last stage, the formation of a gel network from soluble protein aggregates, can be induced by many other parameters in addition to heat, even at room or refrigeration temperatures, thus resulting in a cold gelation (Bryant and McClements 1998). A traditional cold-set gel is formed first by preheating the protein solution in order to form soluble aggregates and second by adding a salt or decreasing the pH, at room temperature, to induce gelation. This last stage is not necessary if a thickened solution—not a gel—is desired.

Traditional cold gelation is a special case of multistep gelation. If the number of steps is two, there are three stages: (native) proteins, soluble protein aggregates, and gel network. Following Ju and Kilara (1998d) (Figure 10.1), the environmental factors *i* and *j* are the initial pretreatment and the final gelation step, respectively. If *i*=*j*=heat, it corresponds to a heat-induced gelation, in one or two steps. If *i* is heat and *j* is salt, a traditional cold-set gel is formed. In the present review, a generic approach will be taken for a two-step gelation, with an emphasis on traditional cold gelation, which is by large the most studied. The *i* and *j* factors are many times interchangeable. If a factor can be used to form aggregates it can certainly be used to form gels. The opposite, even if true, may not result in an acceptable starting solution to form a gel. An example would be to decrease the pH close to the pI, conditions where native βLg aggregate, the *i* factor (Ju, Hettiarachchy, and Kilara 1999; Majhi et al. 2006), followed by the addition of salts (*j* factor) would result in the formation of a coagulum or precipitate, not a gel. An example of the nonspecificity of other factors in the different steps is the study of Ju et al. (1997). The authors formed similar whey protein isolate (WPI) gels from heated solution that where subsequently gelled with a proteolytic enzyme, and from partially hydrolyzed solutions (using the same proteolytic enzyme) followed by a heat treatment. The limitations of heat processing to form gels can be partially solved by adding this extra processing step. As will be discussed later on, cold-set gels can be less opaque, and have better mechanical and water holding properties, extending the range of applications of whey proteins as functional ingredients in the food industry. An example of potential application of cold gelation is reported for meat batters

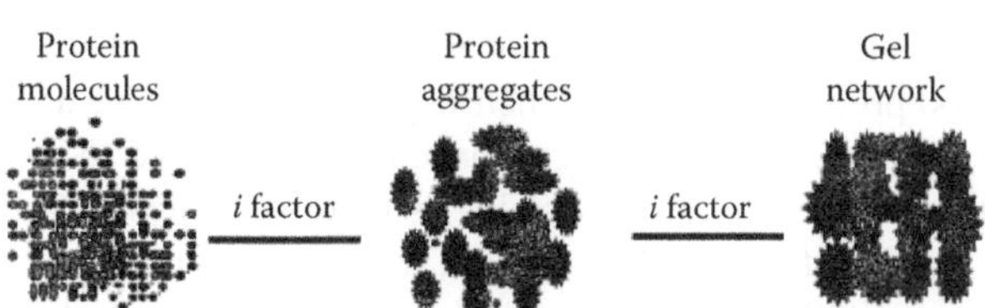

FIGURE 10.1 Schematic representation of a two-step whey protein gelation. (From Ju, Z. Y., and Kilara A., *Journal of Agricultural and Food Chemistry,* 46, 1830–35, 1998.)

with NaCl, resulting in a beneficial increase of the WHC, reduction of cook losses, and an increase in the gel strength of the raw and cooked products (Hongsprabhas and Barbut 1999a, 1999b).

10.1.1 Pretreatment Step

In the first step of cold gelation, the protein solutions are partially or fully denatured, resulting in a significant degree of aggregation, but a 3D gel is not formed. This protein solution is then brought back to ambient conditions if required. The second step involves the addition of a chemical or of a physical treatment to finally induce the gelation of the primary aggregates. The initial aggregation step can theoretically be induced by any process that can also cause gelation in a single step. For example, high pressures are well known to denature and aggregate food proteins (Messens, VanCamp, and Huyghebaert 1997; Lopez-Fandino 2006a, 2006b), resulting in either gels (Vancamp and Huyghebaert 1995; VanCamp et al. 1997b) or soluble aggregates (VanCamp et al. 1997a; Olsen et al. 1999), depending on the conditions. These soluble aggregates would then be susceptible to a second cold gelation step. Another example would be the use of chemicals, such as urea (Xiong and Kinsella 1990), which, of course, is of no use for food applications.

A feasible but rare way to form soluble aggregates is by incubation with $CaCl_2$. Aggregates are formed faster at higher calcium concentrations and incubation temperatures, up to a ratio of $CaCl_2$/whey proteins of 6.7×10^{-3} wt/wt (Ju and Kilara 1998a). Larger calcium ratios result in larger aggregates (Ju et al. 1999). These calcium-induced aggregates can form gels after the addition of enzymes, lowering of pH, heat treatment, and even by the addition of more $CaCl_2$ (Ju and Kilara 1998e).

10.1.1.1 Enzymatic Pretreatment

Other low-temperature treatments include the use of enzymes. Enzymatic modification has long been applied to whey proteins, for example to partially hydrolyze the proteins and improve their functionality close to the pI (Panyam and Kilara 1996). Other types of enzymes can be employed to cross-link proteins, modifying their functional properties (Dickinson 1997). Transglutaminases, which link glutamine and lysine residues, are able to induce the full gelation of whey protein solutions. These enzymes often require the presence of a denaturing or reducing agent (e.g., Dithiothreitol, DDT) to break the globular structure, allowing the enzyme to work (Faergemand, Otte, and Qvist 1997; Eissa and Khan 2006; Eissa et al. 2006). The functionality of proteins can be further enhanced by a limited enzymatic hydrolysis and controlled enzymatic cross-linking (Wilcox et al. 2002).

Transglutaminase cross-linking followed by cold gelation at acidic pH was initially applied to skim milk (Faergemand and Qvist 1997). Eissa, Bisram, and Khan (2004) expanded the method for whey proteins without using reagents not allowed in food application, such as DTT. WPI solutions were incubated at 50°C for 5 h at a slightly alkaline pH of 8, necessary for the enzymatic cross-linking of βLg. While some disulfide bonds are formed at those conditions, their presence, and their effect on the rheology, is negligible. This methodology was later modified to improve the formation of disulfide cross-linking and to extend the life of the enzyme by lowering the pH. The pretreatment was split first in a conventional heat treatment at 80°C for 1 h at pH 7, followed by the incubation with the transglutaminase at 50°C for 10 h (Eissa and Khan 2005), shown in Figure 10.2 (lanes 4 and 6). Cold-set gels with an enzymatic pretreatment at pH 4, a pH where gels are usually weak and brittle (Burke et al. 2002), have significantly higher fracture stress and strain compared with traditional acid cold-set gels and heat-set gels at 80°C. BLP can also transform α-lactalbumin (αLa), the second main whey protein that usually does not gel on its own, into hydrolyzates that self-assemble, forming nonbranching strands ~20 nm in diameter, which can gel after the addition of calcium (Ipsen, Otte, and Qvist 2001).

Proteolytic enzymes transform proteins into smaller peptides by cleaving peptide bonds. This limited hydrolysis usually worsens the gelation functionality of whey proteins, but some enzymes can enhance it (Ju et al. 1995). A specific protease from *Bacillus licheniformis*, usually termed BLP, was found to induce the aggregation and gelation of whey proteins at neutral pH (Otte et al. 1996),

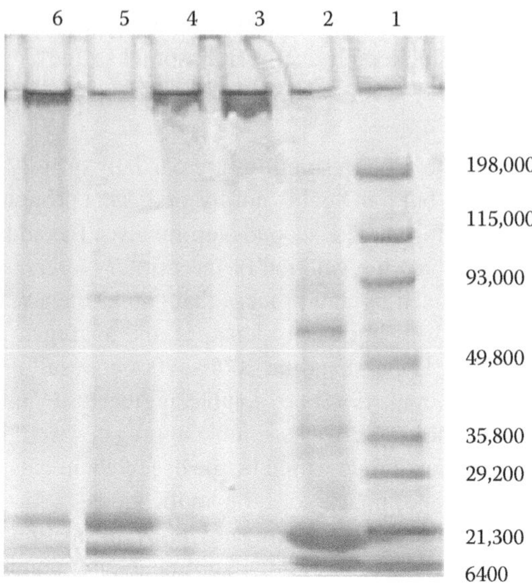

FIGURE 10.2 SDS-PAGE of different samples: M_w marker (lane 1); native WPI (lane 2); preheated WPI (pH 7, 80°C for 1 h) in nonreducing (lane 3) and in reducing conditions (lane 5); preheated WPI which is then incubated with transglutaminase for 10 h at 50°C, under nonreducing (lane 4) and reducing conditions (lane 6). (From Eissa, A. S., and Khan, S. A., *Journal of Agricultural and Food Chemistry,* 53, 5010–17, 2005.)

preferably if there is a previous preheating step, although it is not necessary (Ju et al. 1997; Ipsen et al. 2000). BLP is a glutamyl endopetidase, which cleaves hydrophilic segments while preserving the hydrophobic ones. These hydrophobic peptides, which are initially buried in the native protein, can now interact with other peptides or proteins, resulting in the formation of aggregates, particularly through hydrophobic interactions (Creusot and Gruppen 2007a, 2007b). Whey hydrolyzates using Alcalase, a commercial protease from *subtilisin Carlsberg*, are also capable of gelling through hydrophobic interactions (Doucet et al. 2003a). Extensive hydrolysis is required with this enzyme to induce gelation, compared to BLP, resulting in small peptides <2 kDa (Doucet et al. 2003b; Doucet and Foegeding, 2005). Once soluble aggregates have been formed after incubation with the enzymes, cold-set gels can be normally formed after the addition of NaCl, $CaCl_2$, glucono-δ-lactone (GDL), or more BLP, or by conventional heating (Ju and Kilara 1998c, 1998d).

10.1.1.2 Heat Pretreatment

The most common and convenient way to pretreat whey in the industry is by heating. The formation of soluble aggregates during the heating of whey proteins has long been established (Watanabe and Klostermeyer 1976). Early studies verified that the preheated solutions that were later on used in cold gelation experiments corresponded to highly aggregated whey solutions (Kawamura et al. 1993; Ju and Kilara 1998b; Marangoni et al. 2000). For example, high performance liquid chromatography (HPLC) shows that a typical heat treatment at 80°C for 30 min converts all the native proteins to large aggregates (Ju and Kilara 1998f). Visual confirmation of the formation of large aggregates was provided early on by Nakamura et al. (1995) using transmission electron microscopy (TEM). Heated WPI samples (90°C for 10 min) presented aggregates 30–50 nm large, which were absent in the unheated samples. The soluble aggregates formed are easily observed with nonreducing SDS-polyacrylamide gel electrophoresis (PAGE) (Figure 10.2).

Many heating regimes have been used to form soluble protein aggregates, from 68.5°C for 2 h (Alting et al. 2003a) to 90°C for 30 min (McClements and Keogh 1995). The heating regime must be severe enough, otherwise a gel will not be formed in the second gelation step. For example, a 10 wt% WPI solution heated at 70°C for 10 min does not form a gel in 200 mM NaCl; the heating process must be extended to 20 min or the temperature increased to 75°C (Bryant and McClements 2000d). The minimum heating regime to apply is determined by the denaturation and aggregation kinetics of the proteins. In order to maximize the number of proteins involved in the network formation, which will determine the mechanical properties of the final gels, a denaturation and aggregation degree of >95% is usually considered (Alting et al. 2003a). While one heating step is the norm, the effect of two heating steps has also been studied (Glibowski et al. 2006).

The heat-induced denaturation of βLg is a multistep mechanism (Iametti et al. 1996; Qi et al. 1997; Navea, de Juan, and Tauler 2003). The dimeric form, which accounts for most of the proteins at neutral pH, dissociates first to the monomeric form before denaturation, due to the higher thermal stability of the dimers (Apenten and Galani 2000; Apenten, Khokhar, and Galani 2002). An R-type structure is formed around 40–55°C, which differs from the monomer state in only a few conformational changes on some chains. Above ~70°C, usually termed the denaturation temperature, βLg presents a molten globule state (de Jongh, Groneveld, and de Groot 2001), defined as "a compact protein conformation that has a secondary structure content like that of the native protein, but poorly defined tertiary structure" (Ewbank and Creighton 1991). Thus, the protein can swell more and the hydrophobic structure has greater accessibility to the solvent (de la Fuente, Singh, and Hemar 2002). The secondary structure is reduced at temperatures above ~60°C, irreversibly above ~70°C, but at 90°C about 25–35% of the β-sheets are still present in the protein structure (Iametti et al. 1996; Qi et al. 1997; de Jongh et al. 2001; Fessas et al. 2001). During the early stages of βLg denaturation by heat, >50°C (Owusu-Apenten and Chee 2004), the free cysteine in βLg is exposed to the solvent (Croguennec et al. 2003), which has a crucial role in aggregation by acting as an initiator of thiol–disulfide exchange reactions with other proteins (Hoffmann and van Mil 1997, 1999). Thiol–disulfide exchange reactions are significantly faster at pHs above neutral (Galani and Apenten 1999).

Once the first oligomers are formed through intermolecular disulfide bridges (usually termed "primary aggregates"), they start to assemble into larger aggregates. This leads to a two-step aggregation process (Vardhanabhuti and Foegeding 1999; Ikeda 2003). These primary aggregates only form above a critical association concentration (>5 g/l) (Baussay et al. 2004). The interactions involved in the growth and aggregation of the primary aggregates are not uniquely established as in the formation of the primary aggregates. Many studies assert that large aggregates are formed almost exclusively through intermolecular disulfide bonds (Hoffmann et al. 1997; Hoffmann and van Mil 1997; Schokker et al. 1999), in agreement with the reaction model of Roefs and De Kruif (1994). The critical role of disulfide bonding during aggregation is clearly observed in Figure 10.2; note the difference of polymeric material in nonreducing conditions (lane 3) with reducing conditions (lane 5). Subsequent studies showed that noncovalent interactions are also involved (Anema 2000; Havea, Singh, and Creamer 2001), such as intermolecular β-sheets (Allain, Paquin, and Subirade 1999). The main weak interactions at neutral pH are hydrogen bonding and hydrophobic interactions (Shimada and Cheftel 1988). The formation of hydrophobic interactions requires heating above ~70°C, in order that βLg is sufficiently denatured and the buried hydrophobic amino acids are exposed to the solvent (Galani and Apenten 1999) and high protein concentrations (Havea et al. 1998). Bauer et al. (2000) showed that the formation of primary aggregates was enhanced when increasing the pH from 6.7 to 8.4, as thiol/disulfide exchange reactions are favored, but the formation of larger aggregates was retarded due to the increase of repulsive forces (Hoffmann et al. 1997; Hoffmann and van Mil 1997). A double preheating step has been proposed to maximize the formation and cross-linking of aggregates, by first heating at pH 8 to favor the formation of disulfide bridges, followed by a second heating at pH 6–7 to enhance noncovalent interactions (Mleko et al. 2002).

The preheating should be performed at low salt concentrations and at pH away from the pI, preferably at neutral conditions, in order that these aggregates do not aggregate further, and eventually form a heat-set gel, but remain soluble. Heat-set gels become increasingly easy to form at high protein concentrations, as covalent and noncovalent interactions are facilitated. In practice, there is a maximum protein concentration where a large amount of aggregates can be formed without forming a gel, about 10–12 wt% at neutral pH in the absence of salts, usually referred to as the critical gelation concentration (Kavanagh, Clark, and Ross-Murphy 2000b). The addition of salts and acidification decrease this gelation concentration markedly (Mehalebi, Nicolai, and Durand 2008a, 2008b). Salts reduce the interprotein repulsion and facilitate interactions (Kitabatake, Wada, and Fujita 2001; Croguennec, O'Kennedy, and Mehra 2004), which is a desired event in the final gelation step but not during the pretreatment.

The typical whey protein aggregates formed in the preheating step, at low salt concentrations and pH close to neutral, are formed from primary aggregates of about 100 βLg monomers (Durand, Gimel, and Nicolai 2002), curved strands of 10×50 nm according to TEM (Pouzot et al. 2005), or ~20–30 nm with AFM (Elofsson et al. 1997; Ikeda and Morris 2002). Figure 10.3c and d show the typical morphology of heat-induced βLg aggregates. These globular aggregates subsequently interact, suggested to be head-to-tails (Pouzot et al. 2005), to form elongated aggregates (Durand et al. 2002) with a fractal dimension of 1.7 at low ionic strength (Baussay et al. 2004; Mehalebi et al. 2008b) and of 2 at >30 mM (Le Bon, Durand, and Nicolai 2002; Pouzot et al. 2005). A pH above

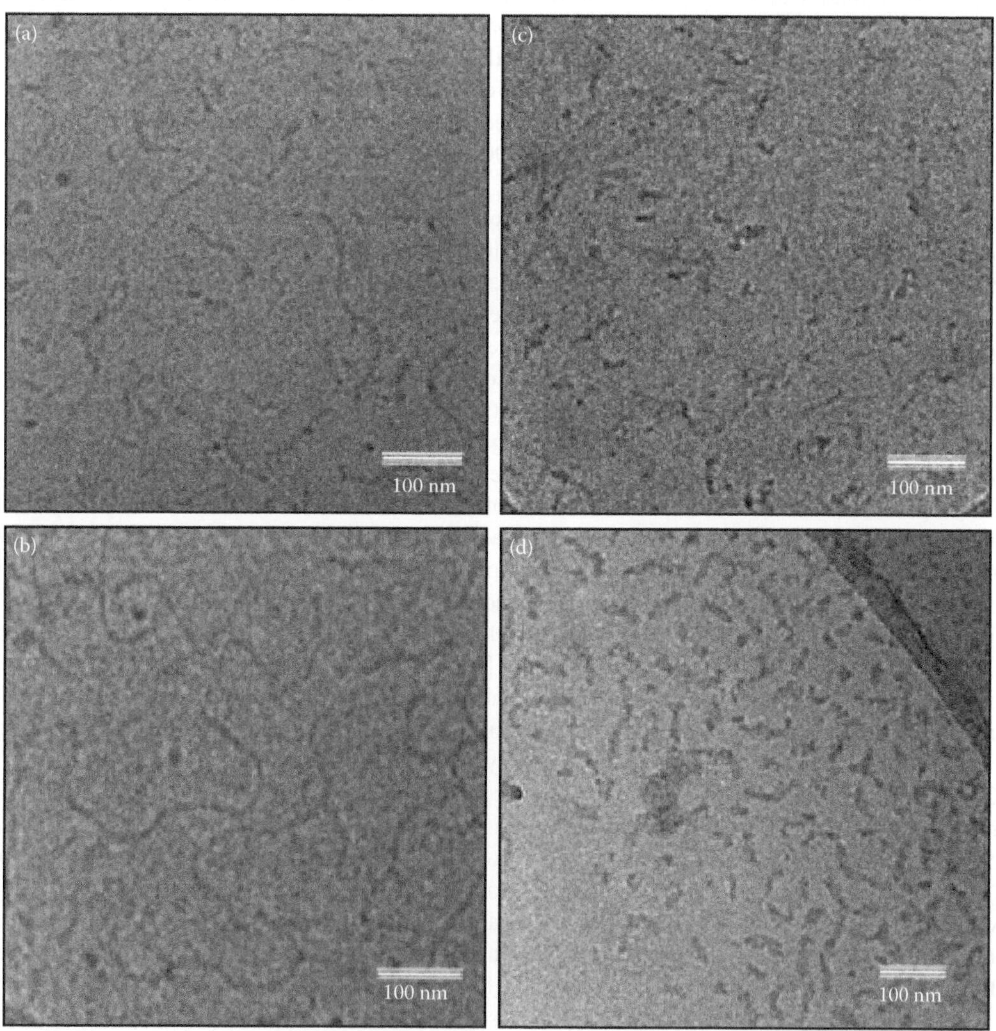

FIGURE 10.3 Cry-TEM micrographs of dispersions of aggregates heated at different protein concentrations. (a) 2.5 wt% ovalbumin, 78°C for 22 h; (b) 5 wt% ovalbumin, 78°C for 22 h; (c) 3 wt% WPI, 68.6°C for 24 h; (d) 9 wt% WPI, 68.5°C for 2 h. (From Alting, A. C., Weijers, M., De Hoog, E. H. A., van de Pijpekamp, A. M., Stuart, M. A. C., Hamer, R. J., De Kruif, C. G., and Visschers, R. W., *Journal of Agricultural and Food Chemistry,* 52, 623–31, 2004.)

5.8 does not affect the structure of the aggregates at small length scales (Mehalebi et al. 2008b), nor does the ionic strength below 100 mM (Pouzot et al. 2005).

The use of pure βLg, preferred in academic studies, is economically unsound for any other applications. Cheaper whey mixtures, such as WPI and whey protein concentrate (WPC), can only be used in the industry. The aggregation behavior of bovine serum albumin (BSA) and αLa, the other major whey proteins, and mixtures with βLg is briefly discussed. BSA can form good heat-induced gels (Tobitani and Ross-Murphy 1997), but not αLa on its own (Matsudomi et al. 1992; Dalgleish, Senaratne, and Francois 1997). Gels formed from BSA/βLg mixtures have intermediate properties depending on the protein ratio (Gezimati, Singh, and Creamer 1996). In typical whey mixtures, the observed behavior will be that of βLg, the most abundant. In heated αLa/βLg mixtures, αLa is incorporated in the βLg aggregates and enhances the gelation process (Dalgleish et al. 1997; Kavanagh et al. 2000a; Schokker, Singh, and Creamer 2000), probably because αLa has four internal disulfide bonds susceptible to interact with the free thiol group in βLg and BSA (Havea et al. 2001). Therefore, all major whey proteins present positive interactions in the formation of aggregates at neutral pH (Hines and Foegeding, 1993). If preheating is conducted at very acidic pH, such as 2, when βLg aggregates in amyloid-like fibrils (Bromley, Krebs, and Donald 2005), αLa and BSA are not incorporated in these fibrils and there are no mixed aggregates (Bolder et al. 2006b), and therefore, would not be part of the stress-bearing gel network (Bolder et al. 2006a). However, such low pH conditions are highly atypical in the formation of soluble aggregates for cold gelation applications; at neutral pH, the information collected from pure βLg studies is usually directly applicable to typical whey mixtures, e.g., microstructure of WPI and βLg aggregates and gels (Langton and Hermansson 1992; Ikeda and Morris 2002; Mahmoudi et al. 2007).

10.1.1.3 Microstructure of Aggregates and Gels

If the protein aggregates discussed previously were allowed to aggregate further they would form a fine-stranded gel, characteristic of gelation pHs different from the pI (<4 and >6) and low salt concentrations (Langton and Hermansson 1992). These gels are transparent, as the strands are too small to scatter light (<100 nm [Doi 1993]), and present a homogeneous network with very small pore sizes, about 20–100 nm (Croguennoc et al. 2001; Colsenet, Soderman, and Mariette 2006). At pH close to the pI or at high salt concentrations, the heated whey solutions form particulate aggregates. As mentioned before, these heating conditions are not desirable for cold gelation applications. However, the final cold gelation step is usually triggered by a decrease in pH close to the pI, which will be termed acid cold gelation, or by the addition of salts (salt cold gelation). Particulates are spherical structures formed by the random association of aggregates. Particulate sizes are in the order of microns and decrease at higher heating rates and temperature (Langton and Hermansson 1996; Bromley, Krebs, and Donald 2006). The large particle size of particulate gels results in high turbidity values and in opaque gels with larger pore sizes (>100 μm) than in fine-stranded gels (Stading, Langton, and Hermansson 1993).

Particulate structures start to form at pH<5.8 in the absence of salts. At higher pH, the same behavior is observed at increasing salt concentrations, e.g., 40 mM NaCl at pH 6, 100 mM at pH 6.5, and 200 mM at pH 7 (Mahmoudi et al. 2007). In fact, the type of aggregate formed, whether particulate or fine stranded, seems to be governed only by the balance of the protein charge and the screening conditions. Both aggregation mechanisms have been found in many proteins under the right conditions, suggesting that it is not related to the amino acid sequence of the proteins, and that they are universal forms of aggregation of proteins (Krebs, Devlin, and Donald 2007).

The particulate aggregation of βLg at pH 5.3 (>pI) has been described using a nucleation-based model, where the nucleation rate depends on the concentration of unfolded protein capable to react (Bromley et al. 2006). Majhi et al. (2006) has suggested that at room temperature native βLg dimer first reacts quickly to form ~200 nm aggregates, without the formation of smaller oligomers, followed by a second aggregation step to form large particulates. Dimeric βLg is also the starting point of particulate aggregation at high temperatures (Lefevre and Subirade 2000).

10.1.1.4 Effect of Preheating in Cold Gelation

The soluble aggregates formed during the pretreatment step are the building blocks of the final gel; therefore, the characteristics of these aggregates will influence the properties of the gel. Higher preheating temperatures result in clearer, stronger gels, with high water-holding capacity (WHC) and extensibility (Hongsprabhas and Barbut 1996). Higher protein concentrations further increase the gel clarity, fracture force, and Young's modulus (Hongsprabhas and Barbut 1997c; Hongsprabhas, Barbut, and Marangoni 1999). A longer preheating step, once all the proteins are aggregated, has no effect on the properties of the cold gel (e.g., increasing from 30 to 60 min at 80°C) (Ju and Kilara 1998b), albeit the gel is a bit clearer (Hongsprabhas and Barbut 1997a). An example of how to determine the aggregation state of a whey protein solution, in order to determine when most of the proteins are aggregated and longer holding times are not necessary, is given in Alting et al. (2003a). From an economic perspective, Bryant and McClements (2000d) suggested the following preheating conditions: 85°C for 15 min at pH 7, when the highest protein concentration can be heated without gelling, about 10 wt% WPI.

The pretreatment pH (6.5–8.5) has little influence on the hydrodynamic radius (R_h) of the aggregates formed in the absence of salts, at ~35 nm (Britten and Giroux 2001). A small addition of calcium (4 mM) can increase R_h to ~60 nm at pH 6.5, but cause a reduction to ~20 nm at pH 8.5. The intrinsic viscosity of the aggregates, which relates to the hydration and/or shape of the aggregates, decreases markedly with pH from 6.5 to 8.5. This suggests that increasing the pH alters the shape to more spherical but not the hydrodynamic size, which could be caused by more branching reactions.

The effect of the whey protein concentration during the preheating experiments, at a constant pH of 7, was studied by Ju and Kilara (1998b). Increasing the WPI concentration resulted in larger aggregates, from 20 nm at 2 wt% to 60 nm at 9 wt%. Cold gelation at a constant gelation concentration of 3 wt% with the different preheated solutions resulted in gels the hardness of which increased with the preheating protein concentrations. Linear relationships were observed relating the particle size and the gel hardness with the preheating whey concentration. The initial protein concentration also has a marked effect increasing the intrinsic viscosity of the soluble aggregates (Vardhanabhuti and Foegeding 1999). The relevance of the size of the aggregates formed during the initial heating step was also observed by Mleko (1999). In this study, the second gelation was the result of another heating step. Nevertheless, it was found again that the hardness and storage modulus of the final gels increased with the protein concentration of the first heating step, when the protein concentration in the second heating step was constant at 3 wt%.

Alting et al. (2003a) questioned the conclusions of the previous studies that related the structural properties of the initial aggregates (intrinsic viscosity, size) with the mechanical properties of the final cold-set gel. The authors corroborated that a higher preheating concentration results in aggregates with larger size, intrinsic viscosity, and voluminosity. However, these differences had a minor effect if the aggregates were previously incubated with a thiol-blocking agent; the modified proteins resulted in cold-set gels much weaker than the unmodified ones. The authors thus, concluded that the formation of new disulfide cross-links between aggregates is more important that the initial size or shape of the aggregates.

On the other hand, the structural properties of the aggregates, as inferred from the apparent viscosity of the preheated solutions, have a deep influence on the gelation time (Bryant and McClements 2000d). This is also observed when large fibrils are formed at pH 2, with contour length about 2–7 μm, about one order of magnitude larger than the fibrils formed at neutral pH. Those long fibrils, after being adjusted to neutral pH, showed a gelation concentration about five times smaller, only 0.12 wt% βLg, compared to a neutral pH preheating treatment (Veerman et al. 2003). Finally, the effect of the preheating can be further seen by mixing heated whey aggregates with unheated whey proteins. A higher percentage of aggregates results in gels with higher fracture stress, fracture modulus, held water, and translucency (Vardhanabhuti et al. 2001), all of them desired properties in cold gelation.

10.1.1.5 Rheological Properties of Aggregated Whey Solutions

Higher heating temperature and longer holding times increase the viscosity of the preheated solution (Bryant and McClements 2000d). The apparent viscosity plateaus with time once the aggregation process is finished. A lower solution pH, in the range of 6–8, results in more viscous solutions (Mleko and Foegeding 2000). Shear thinning is observed in the most viscous solutions, those heated at higher temperature and longer times. The flow behavior is usually described by a power-law model,

$$\eta = k\dot{\gamma}^{n-1}, \tag{10.1}$$

where η is the apparent viscosity, $\dot{\gamma}$ is the shear rate, n is the flow index, and k is the consistency. Increasing the protein concentration in the heating process results in more viscous solutions and higher consistencies, but up to 10 wt% WPI they are still very Newtonian (n~1) (Vardhanabhuti and Foegeding 1999). Shear thinning behavior is clearly observed at 11 wt% WPI, with n~0.45–0.75. The overall rheology is similar to that observed for other hydrocolloids at lower concentrations (Mleko and Foegeding 1999; Vardhanabhuti and Foegeding 1999). The large aggregates formed when heating at high protein concentrations result in more viscous solutions after being diluted than heated solutions at the diluted concentration (Mleko 1999; Mleko and Foegeding 1999).

10.1.2 Thickening

There are many hydrocolloids that are used as instant thickeners for food applications, such as pregelatinized starch, but they lack the nutritional benefits of whey proteins (Regester et al. 1996). Whey proteins can also be used as food thickeners at room temperature as opposed to high temperatures (Kinsella and Whitehead 1989), resulting in weak cold-set gels capable of holding water (Hudson et al. 2000, 2001). Protein-based thickening agents could succeed in applications where it is desirable to reduce the consumption of carbohydrates while increasing the consumption of proteins. Such an example was presented by Hudson, Daubert, and Foegeding (2001) for dysphagia patients (difficulty swallowing). For thickening applications, there is no need for a secondary gelation step, only an aggregation step. This aggregation step however can be more severe than that discussed previously. For example, aggregates can be produced from the destruction of gelled systems. For thickening applications, these aggregates should be in a powder form.

A few patents have been granted for modified whey protein powders for food applications (Tamaki, Nishiya, and Tatsumi 1991; Holst et al. 1993; Kawachi, Takeachi, and Nishiya 1993). Studies have been performed to study their improved functionality as emulsifiers (Firebaugh and Daubert 2005) or as thickeners that improve the WHC in yogurts (Li and Guo 2006). The elementary steps to form a derivatized whey protein thickening powder are as follows: protein hydration, (thermal) gelation, drying, and milling. In principle, any initial conditions capable of resulting in a heat-set gel (e.g., pH, salts) could be used during heat gelation. However, these powders will be mixed with other foods, thus they must be highly soluble in order to have any application. Soluble powers (>60%) are only formed at acid conditions, using, for example, HCl, at low enough pH to obtain fine-stranded gels (pH<4)(Hudson et al. 2000). Particulate gels formed close to the pI or fine-stranded gels formed above the pI yield powders with very low solubility values and therefore of little use. Only low pH conditions, particularly pH 3.35, have been used in subsequent studies (Resch, Daubert, and Foegeding 2005a, 2005b). Therefore, heating conditions for thickening applications are very different than for cold gelation applications.

The derivatized proteins also present good stability once they are reconstituted in water. Like many food thickeners, reconstituted powders show shear thinning behavior. Using the right gelation conditions (e.g., pH 3.35, no salt, 80°C for 3 h), the rheological characteristics of the solutions can be fairly independent of the temperature and of the solution pH (4 or 8) (Hudson et al. 2000). Increasing the protein concentration of the reconstituted solution obviously increases the viscosity greatly.

The consistency of a 5.6 wt% solution is one order of magnitude higher than of a 3 wt% solution, and that of a 7 wt% is two orders of magnitude higher than of a 5.6 wt% solution. Increasing the protein solution also accentuates the sear thinning behavior; the flow index in Equation 10.1 decreases from 0.86 at 3 wt% to 0.75 at 5.6 wt% and to 0.24 at 7 wt% (Clare et al. 2007). The WHC of the reconstituted solutions is around 8–10 g water/g dry protein (Resch and Daubert 2002; Resch et al. 2005b; Clare et al. 2007).

The addition of NaCl prior to heating under acidic conditions has a detrimental effect on the derivatized whey proteins; it decreases the solubility of the powders and reduces the viscosity of the reconstituted powder solutions. A prolonged heating time during thermal gelation results in an improvement in the solubility and thickening capability. It has been suggested that the beneficial effect of a prolonged heating and of a low salt concentration is caused by the higher degree of acid hydrolysis under those conditions. Good derivitized powders showed significant amounts of low-molecular-weight peptides (<10 kDa) (Hudson et al. 2000).

At constant pH conditions (3.35), the type of acid used to acidify the protein solution has a marked effect on the initial gel formed, and in the final derived powders. Gels formed with lactic acid were stronger and the reconstituted solutions more viscous than using the usual HCl (Resch et al. 2005b). Very weak heat-set gels are achieved using phosphoric acid and this result in powders with little thickening capability and worse WHC. The addition of citric acid speeds up the heat gelation kinetics compared to lactic acid, although the final complex modulus is very similar. However, the powders made from citric acid are very difficult to solubilize, resulting in a solution with very low viscosity and no thickening ability. Resch et al. (2005b) argued that the differences among the different acid anions follow the Hofmeister series, which already occurs in the thermal stability of βLg (Damodaran 1989). Citric acid, and the reciprocal citrate anion, destabilize whey proteins—a strong salting out effect—causing rapid aggregation, resulting in particulate-like gels. As discussed previously at different pH, particulate gels are bad candidates to form thickening powders. On the other hand, phosphoric acid has a stabilizing effect on whey proteins, increasing the denaturation temperature, reported as the peak transition temperature with DSC, from 85 to 89°C. This explains why good gels are not formed with phosphoric acid when heating at 80°C.

Increasing the gelation temperature using HCl and lactic acid improved the thickening functionality up to 85°C; higher temperatures did not show further advantages. Phosphoric acid, which yields bad gels and powders at low heating temperatures, can equal the HCl and lactic acid powders by increasing the gelation temperature to 90°C, because the denaturation temperature is higher in the presence of phosphoric acid. The rheological differences observed using different acids are nevertheless too subtle to be observed under scanning electron microscope (SEM) (Resch et al. 2005a). In summary, the thickening and water-holding functionality of reconstituted derivatized powders parallels the rheological properties of the gels from which the powders were formed. Both parameters are affected by the type of acid used to decrease the pH before the heating process.

While lower heating rates allow gelation to be observed at lower gelation temperatures, in nonisothermal experiments, the final rheological properties, once the modulus plateaus, appear very similar between 0.2 and 2°C/min. Only at very low values (e.g., 0.1°C/min) is an increase in the gel strength observed. However, this small benefit will be offset by higher processing costs, making the heating rate an impractical parameter to manipulate for improving gelation and derivatized powder functionality in a commercial setting (Resch et al. 2005a).

The final production of the whey powder is commonly carried out in the lab by freezing 12 wt% heat-set gels (80°C for 3 h) at –5°C for 16–18 h, followed by freeze-drying up to a moisture content <5%, and finally milling. This procedure is economically unfeasible on an industrial scale. Resch, Daubert, and Foegeding (2004) developed a friendlier protocol for the food industry, which is as follows: an 8 wt% WPC solution is heated under constant agitation at 80°C for 1 h, resulting in a semisolid gel which is transferred to a laboratory spray-dryer, using an inlet temperature of 190°C and an outlet temperature of 90°C. The derivatized powders using both methods presented comparable thickening capabilities, despite the different heating protocols, which can lead to the development of a (semi-)continuous manufacturing process. The functionality of the derivatized powders depends on the initial composition of the whey protein powder used. In addition to pure βLg and WPI, commercial WPC powders are also capable to from good thickening agents (Resch and Daubert 2002; Firebaugh and Daubert 2005).

Whey-protein-based thickening agents at room temperature show great promise in food applications, but following the previous procedures a higher concentration—more than double—is so far required compared to traditional hydrocolloids (e.g., carrageenan, xanthan gum, or starch) to achieve the same viscosity (Resch et al. 2004). Nevertheless, the thickening ability of derivatized whey powders can be greatly improved in the presence of calcium during hydration (after heating) (Clare et al. 2007). Increasing the calcium concentration to 75 mM in a 5.6 wt% solution greatly increases the consistency (two orders of magnitude), while the flow index decreases (from 0.75 to 0.17). This leads to a significant increase of the apparent viscosity, particularly at low shear rates, comparable with a reconstituted solution of 10 wt% without calcium (Hudson, Daubert, and Foegeding 2000). Calcium also induces significant gel strengthening during cold storage in a time frame of hours (Clare et al. 2007).

10.1.3 Gelation Step

Once soluble denatured proteins and aggregates are obtained, the final gelation step can be achieved by many means. For example, the final gelation can be achieved after another heat treatment, when the pH (Mleko 1999) or the presence of salts (McClements and Keogh 1995) can be adjusted before the second heat treatment. However, these examples do no constitute a cold gelation protocol, although they fall within the generalized two-step gelation discussed previously (Figure 10.1). In this section the cold gelation after the addition of salts and acids will be considered in detail.

10.1.3.1 $CaCl_2$ Cold Gelation

Divalent salts are capable of inducing the cold gelation of preheated whey protein solutions, being in fact much more effective than monovalent salts. The most widely studied divalent salt is calcium, owing to its presence in milk and its health benefits. Calcium has a strong effect on gelation even if large quantities of monovalent cations are present (120 mM) (Kuhn and Foegeding 1991). Calcium cold-set gels are usually manufactured in the laboratory by dialyzing a preheated whey solution against a salt solution at the desired concentration (Barbut and Foegeding 1993). This allows a slow increase in the salt concentration and a more homogeneous network. Long times—about 12 h—are required for the cold-set gels to reach constant mechanical properties (Roff and Foegeding 1996).

Calcium binds strongly with βLg, the amount of which depends on the net negative charge of the protein, and it is well known to promote the isoelectric aggregation of βLg (Zittle and Dellamonica 1956; Zittle et al. 1957), even at low temperatures (Sherwin and Foegeding 1997). Calcium has a strong salting-out effect at low concentrations in native whey proteins, with a maximum aggregation rate around 20–40 mM Ca^{2+}, while at higher concentrations there is resolubilization of the aggregates (salting-in) (Zhu and Damodaran 1994). Calcium has long been employed in heat-induced gelation, resulting in good mechanical properties, such as hardness (Schmidt et al. 1979), compressive strength (Mulvihill and Kinsella 1988), and maximum elastic modulus (Gault and Fauquant 1992), at low concentrations, around 10 mM Ca^{2+}; about 10–20 times lower than with monovalent salts (Mulvihill and Kinsella 1988). The addition of calcium does not significantly alter the circular dichroism spectrum of heated whey protein aggregates in the near-UV; the tertiary structure is already lost during heating (Marangoni et al. 2000). The initial heating step causes a redshift and a significant decrease in intensity in the tryptophan fluorescence as proteins unfold. The addition of calcium causes a small blueshift and a dramatic increase of the fluorescence intensity, to a value 50% higher than for native WPI. Marangoni et al. (2000) argued that the effect of calcium could be caused by an increase of the quantum yield (the efficient fluorescence of the tryptophan) due to charge neutralization after calcium binding and due to aggregation via hydrophobic groups.

The first use of calcium to induce cold gelation was reported by Barbut and Foegeding (1993) in a comparison with heat gelation. Calcium cold-set gels result in different mechanical properties from equivalent calcium heat-set gels; however, the most remarkable is the different microstructure at a constant calcium concentration of 10 mM: fine-stranded in cold gelation against particulate in heat gelation. In fact, calcium heat-set gels are always opaque because of their particulate nature (Mulvihill and Kinsella 1988; Kuhn and Foegeding 1991). Cold-set gels have a much finer structure, even at high calcium concentrations (360 mM), compared to a heat-set gel in 10 mM Ca^{2+} (Barbut 1995c). This finer structure is due to the fine-stranded nature of the primary aggregates previously formed. The calcium concentration required to induce cold gelation is of the same order as that required to induce heat gelation (~<10 mM), and a bit lower for cold gelation (Kuhn and Foegeding 1991; Barbut and Foegeding 1993; Roff and Foegeding 1996).

Increasing the calcium concentration in cold gelation increases the opacity due to the formation of larger aggregates (Barbut 1995c; Hongsprabhas and Barbut 1996). Above a calcium concentration of 15–30 mM Ca^{2+}, cold-set gels cease to be transparent and become increasingly white (Barbut 1995c; Hongsprabhas and Barbut 1996; Roff and Foegeding 1996; Bryant and McClements 2000a); above 100 nM there are no visual differences in the gels' clarity (Barbut 1997). The opacity of the gels, caused by the light scatter of large aggregates, is given by the calcium-driven aggregation of the protein aggregates. Figure 10.4 shows a schematic representation of the two types of gels that can be formed depending on the concentration of salt; this representation is independent of the salt used. The rate constant of aggregation increases at 10–30 mM Ca^{2+} but decreases if the concentration is increased further (Hongsprabhas and Barbut 1997c; Hongsprabhas et al. 1999; Marangoni et al. 2000). At the molecular level, a fast aggregation of the large aggregates of the preheated protein solution is first observed, followed by a slower aggregation of smaller protein aggregates (Marangoni et al. 2000). Aggregation with calcium is so quick that intermediates are not observed (Croguennec et al. 2004). The quick aggregation at >10 mM Ca^{2+} results in a sharp decrease of the gelation time to almost instantaneous values above 25 mM if the calcium and the whey solution are directly mixed (Bryant and McClements 2000a; Wu, Xie, and Morbidelli 2005). In addition, an increase of the gelation temperature from 1 to 24°C results in larger aggregate formation, which decreases the gel strength and water-holding properties (Hongsprabhas and Barbut 1997b). Reheating the cold-set gel at high temperatures (80°C for 30 min) induces rearrangements that result in a harder and less cohesive structure, with higher opacity and larger pore sizes (Hongsprabhas and Barbut 1997a). At low calcium concentrations (10 mM), the increase in aggregate size and network connectivity over

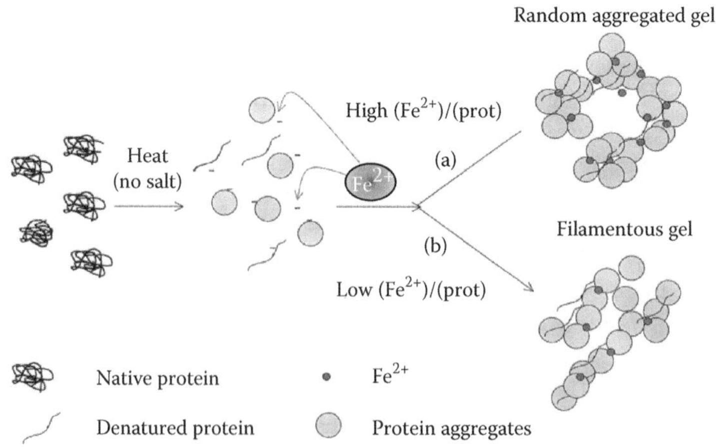

FIGURE 10.4 Mechanisms of aggregation and gelation of βLg cold-set gels: (a) random-aggregated gels at high salt concentration and (b) filamentous gels at low salt concentrations. (From Remondetto, G. E., and Subirade, M., *Biopolymers,* 69, 461–69, 2003.)

time is achieved by clustering of adjacent aggregates, while at high calcium concentrations (120 mM) the increase in aggregate size and connectivity is by enlargement of the aggregates which formed connected paths and filled up interstitial spaces (Hongsprabhas and Barbut 1997d). This also results in an increase of the pore sizes at higher calcium concentrations as a more particulate microstructure is formed, from 60 to 150 nm at 10 mM to 180–250 nm at 120 mM (Hongsprabhas and Barbut 1997a).

The microstructure change from fine-stranded at 10 mM to a particulate-like mixture >30 mM Ca^{2+} results in deep mechanical differences. From large deformation tests, this transition correlates with a significant increase of the Young's modulus—gels become more elastic, while there is a decrease of the fracture strain (Barbut 1997; Hongsprabhas and Barbut 1997c; Bryant and McClements 2000a; Marangoni et al. 2000). Higher calcium concentration slightly decreases the Young's modulus and increases fracture strain, thus at ~30 mM Ca^{2+} gels are the most rigid and brittle (Hongsprabhas and Barbut 1997c). Increasing the whey protein concentration of the gels increases the fracture stress and the Young's modulus at all calcium concentrations (10–150 mM), while the fracture strain is only increased at high concentrations (10 wt%) (Hongsprabhas and Barbut 1997c). The hardness of the gels also increases greatly between 10 and 20 mM Ca^{2+}, and plateaus at higher concentrations (Ju and Kilara 1998f). The adhesiveness increases between 10 and 40 mM, while the cohesiveness slightly decreases (Ju and Kilara 1998f). Comparing the fracture properties of heat- and cold-set gels in the presence of calcium, heat-set gels present a higher increase of the shear stress with the calcium concentration. At low calcium concentrations (10 mM) the shear stress is higher in cold-set than in heat-set gels (Barbut 1995c), but the maximum values found with both methods are comparable, between 30 and 100 mM (Roff and Foegeding 1996). On the other hand, the shear strain at fracture presents different behavior. In cold gelation the shear strain decreases with the salt concentration (e.g., from 2.5 at <5 mM to 1.5 above 50 mM $CaCl_2$), but increases in heat gelation, becoming constant above 30 mM (Roff and Foegeding 1996). The opposite evolution of the shear stress and strain with the calcium concentration is also observed, surprisingly, in the heat gelation in the presence of NaCl (Kuhn and Foegeding 1991). Finally, the storage modulus increase during calcium cold gelation is much slower than a comparable heat-set gel with calcium (Barbut and Foegeding 1993).

Another major physical difference between heat-set and cold-set gels in the presence of calcium is the WHC. High WHC is observed for fine and homogeneous microstructures, because they are able to retain water better with capillary forces (Barbut 1995c). The different microstructure of both types of gels, particulate vs fine-stranded/mix, results in heat-set gels syneresing at 50 mM $CaCl_2$ while comparable cold-set gels are able to retain water, despite both gels being opaque. This translates into a WHC of ~4.8 g water held/g protein in the latter compared to only ~1.7 in the former (Roff and Foegeding 1996). The microstructure change in cold gelation at higher salt concentrations also causes a decrease in WHC (Barbut 1997; Hongsprabhas and Barbut 1997c). However, this decrease is mild compared to that observed in heat gelation, e.g., ~7 g water/g protein at 1 mM $CaCl_2$ compared to ~5 at 100 mM $CaCl_2$ (Roff and Foegeding 1996).

10.1.3.2 NaCl Cold Gelation

Gelation by calcium and by sodium salts is based on the same principle, reduction of the electrostatic repulsion between the proteins, but there are marked differences. As discussed previously, the heat-induced gelation in the presence of calcium results in particulate structures and gels; fine-stranded structures are only observed in cold gelation at low concentrations. On the other hand, fine-stranded transparent gels are formed at low NaCl concentrations (e.g., 25 mM) in heat-set gels, with protein strands between 20 and 50 nm (Barbut 1995b). Denser structure is observed at higher NaCl concentrations: ~100 nm at 50 mM and ~200 nm at 100 mM (semiclear gels). Correspondingly, the pore size also increases, from 0.1 to 0.2 μm at 50 mM to 0.5 μm at 100 mM. At higher NaCl concentrations, the gels adopt a particulate nature, becoming opaque, with very large aggregates and pore sizes, and weaker gels. These gels resemble those formed at low calcium concentrations, and therefore, present similarly low WHCs. A second difference between the two salts is the amount required for a similar behavior: gel hardness is maximum in heat-induced gels at ~100 mM NaCl compared to ~10 mM $CaCl_2$ (Schmidt et al. 1978, 1979), although the actual values vary depending on the salt already present in the whey proteins (Schmidt et al. 1978; Kuhn and Foegeding 1991). Higher NaCl concentrations fasten the gelation process, the gelation time is reduced considerably above 100 mM, and the gels formed present higher modulus (Bryant and McClements 2000a; Mleko 2000). Marangoni et al. (2000) described the NaCl-induced cold gelation as a one-step process involving the slow and simultaneous aggregation of very large aggregates and smaller aggregated species to form a percolating network.

Cold gelation with NaCl is an exception to the nonspecificity of the factors used for aggregation and gelation discussed for Figure 10.1. In this case, a *j* factor that can induce the gelation of protein aggregates cannot be used as an *i* factor to form large aggregates. In other words, NaCl alone does not induce gelation of native whey proteins, as preheating or similar pretreatment is required (Barbut and Drake 1997). The minimum NaCl concentration to induce cold gelation is reported to be higher than for an equivalent heat-set gel, 75 and 25 mM NaCl, respectively (Barbut and Drake 1997). Cold gelation also forms less opaque gels, as previously observed for calcium (Mleko 2000). The opacity increases at higher NaCl concentration, up to 250 mM. The gel hardness and adhesiveness also increase up to 200 mM, and further increases have little effect (Barbut and Drake 1997; Ju and Kilara 1998f). The gels become less elastic at higher concentrations (Marangoni et al. 2000), although the Young's modulus increases continuously between 100 and 400 mM NaCl (Bryant and McClements 2000a). NaCl cold-set gels present much lower WHC than equivalent heat-set gels at low salt concentrations. The WHC increases with the NaCl concentrations in cold-set gels up to ~50% of the initial water content (Barbut and Drake 1997), similar to that for particulate heat-set gels at high NaCl concentrations (Barbut 1995b). Therefore, if high WHCs are desired, sodium salts are worse than calcium salts.

10.1.3.3 Calcium vs. Sodium Salts

The reasons behind the effectiveness of calcium compared to sodium in heat or cold gelation of whey proteins, requiring about 10–20 times lower concentrations (Mulvihill and Kinsella 1988; Gault and Fauquant 1992; Barbut 1995b; Barbut and Drake 1997; Bryant and McClements 2000a), have been an area of extensive speculation in the past. Firstly, notice that the ionic strength of $CaCl_2$ solution is only three times higher than NaCl solution at the same concentration. Therefore, at equal ionic strength concentration, calcium seems to be more effective than sodium (Gault and Fauquant 1992). Nevertheless, there are some gelation parameters that are equivalent, with either salt at the same ionic strength concentration. For example, the ionic strength required to achieve heat gelation (10 wt% WPI pH 7, 80°C 30 min) is independent of the type of salt, 7.5 mM Ca and 25 mM Na (Kuhn and Foegeding 1991). The aggregation rate in cold gelation peaks at an ionic strength of 100 mM, also independent of the salt type. However, the aggregation rate is about three times higher in the presence of calcium than with sodium (Marangoni et al. 2000). The role of calcium in enhancing aggregation has been proposed to be caused by:

1. Interprotein ion bridges, involving carboxylic groups and the calcium ion (Mulvihill and Kinsella 1988; Bryant and McClements 2000a).
2. Electrostatic shielding of negative charges.

One of the key differences of calcium compared to sodium is that the former binds extensively with βLg. The number of moles of calcium ions bound to βLg is directly related to the net charge of the protein and is independent of both protein concentration and the denatured state (Zittle and Custer 1957; Zittle et al. 1957), although the binding affinity is higher after preheating (Jeyarajah and Allen 1994). This suggests that calcium may not be involved in the formation of ionic bridges between the protein molecules. Simons et al. (2002) evaluated this hypothesis by modifying the amount of carboxylates in βLg. They observed that increasing the amount of carboxylates in the protein required much higher calcium concentrations to induce aggregation, as the number of calcium ions bound to the protein also increased. This observation is at odds with hypothesis (1), because if calcium forms ion bridges between proteins, when increasing the amount of carboxylates these crosslinks would be easier to form, and aggregation would then be facilitated. However, the contrary was observed. The authors concluded that the site-specific calcium binding screens intermolecular carboxylates, thereby reducing interprotein repulsion and facilitating hydrophobic aggregation. Hence, calcium is observed to mainly act like sodium by screening charges, hypothesis (2), but because it binds specifically to the carboxylates (the main contributors of negative charge in the protein, i.e., Asp and Glu) much lower concentrations than with sodium are required for a similar screening effect. Interprotein ion bridges using calcium have also not been observed using sedimentation flield-flow fractionation (Saeseaw, Shiowatana, and Siripinyanond 2006), strengthening hypothesis (2). Nevertheless, Veerman et al. (2003) claimed that the formation of salt bridges with calcium is essential for the cold gelation of long βLg fibrils formed at pH 2, as equivalent NaCl concentrations (0.03–0.15 M) do not induce gelation. In addition, calcium is reported to facilitate the denaturation of βLg (Li, Hardin, and Foegeding 1994; Law and Leaver 2000), while high NaCl concentrations tend to stabilize the protein (Boye et al. 1995; Verheul, Roefs, and de Kruif 1998a). NaCl has little effect on the secondary and tertiary structure of βLg, either in its native form or after being heated (Matsuura and Manning 1994). Thus, conformational changes induced by calcium may lead to more hydrophobic interactions (Jeyarajah and Allen 1994), facilitating aggregation and gelation.

10.1.3.4 Cold Gelation With Other Salts

Few other cations have been used in the cold gelation of whey proteins. According to heat gelation studies, monovalent cations (Li, K, Rb, and Cs) have similar effects to those described for sodium, while calcium behaves like Mg and Ba (Kuhn and Foegeding 1991). In cold gelation studies, barium induces similar gels to calcium, concluding that the size of the cation is not important, as in the gelation of alginate, and the gelation mechanism is generic for all divalent cations (Roff and Foegeding 1996). For instance, cold gelation has also been observed in the presence of 1,6-hexanediamine, an organic salt. The effects of other ions, such as zinc and copper, have only been studied while heating (Navarra, Leone, and Militello 2007).

It has been of recent interest to add cations that have a health benefit to protein gels, as nutraceuticals, where these cations are responsible for the final gelation step. An example is the use of zinc, the deficiency of which is associated with many health problems. Zinc-led whey protein aggregation is faster than with calcium, but unlike the latter, it seems to be involved in intermolecular ionic bridges (Saeseaw et al. 2006). Iron, deficient in a large part of the worldwide population, has been used successfully in cold gelation. At low iron concentrations (e.g., 10 mM) the gels formed present a fine-stranded structure, while at higher concentrations they present a particulate nature (Remondetto, Paquin, and Subirade 2002; Remondetto and Subirade 2003), as discussed previously for calcium. The elastic behavior, the WHC, and the strength of rupture decrease at higher iron concentrations (Remondetto et al. 2002). It is suggested that iron is not involved in interprotein ion bridges either, following its negative effect in the gel elasticity.

10.1.3.5 Acid Cold Gelation

During the same years of the 1990s when salt cold gelation was discovered, Kawamura et al. (1993) found that preheated WPI solutions could also form gels at room temperatures when the pH was lowered to less than 5.8. Low pH is desirable to the food industry as it increases shelf stability (Errington and Foegeding 1998) and requires less stringent sterilization processes (Potter and Hotchkiss 1995). An example of acid-induced cold gelation with whey proteins would be to increase the viscosity and WHC of yogurts, instead of using milk powder (Britten and Giroux 2001).

Acidification is usually performed using GDL. GDL is an internal ester that slowly hydrolyzes in the presence of water to gluconic acid, which is a weak acid (pK_a 3.9) that further dissociates (de

Kruif 1997). These equilibrium reactions allow GDL to decrease the pH of the solution gradually, resulting in more regular gels. Inorganic or other organic acids, such as ascorbic or citric acids, cause an instantaneous decrease of the pH and therefore an irregular gelation.

Acid-induced whey protein gels using GDL have better mechanical properties, such as hardness, than salt-induced gels (Ju and Kilara 1998f), and the network formed is finer (Nakamura et al. 1995). Gel harness is maximum at pH 4.7, the adhesiveness at pH 4.4, while the cohesiveness does not change significantly between pH 5.3 and 3.5 (Ju and Kilara 1998f). The WHC is better at pH 5.2 and 3.9 than in between, while the elastic modulus and the stress at rupture tend to decrease at lower pH, and the strain at rupture remains unchanged (Cavallieri et al. 2007; Cavallieri and da Cunha 2008). The hardness of the gels increases with the final protein concentration in a power-law manner (Ju and Kilara 1998f). Higher incubation temperatures (20–50°C) result in gels with finer networks, and larger fracture stress and strain (Kawamura et al. 1993). Once a gel is formed, the modulus is minimum at ~20°C and it increases at lower and higher temperatures (Britten and Giroux 2001).

Dynamic studies of acid cold gelation have been performed by Cavallieri and da Cunha. (2008) by following the formation of the mechanical properties of the gels with time as the pH decreases. Previous studies only took into account the final pH of the system, not the acidification rate. A higher acidification rate, which leads to a lower final pH, reduces the time required for the system to gel and increases the rate of growth of the complex modulus after gelation, but the pH at the gelation point is very similar (5.8 and 5.6 at low and high acidification rates, respectively). Once a self-supported gel is formed, the stress at rupture and the elasticity value increase quickly in the first 30 h at a gelation temperature of 10°C (not at room temperature as in many other studies), increasing slowly thereafter until reaching an equilibrium value. These values are very similar at a final pH of 5.1, 4.9, and 4.7, about 20 kPa for the stress at rupture and 15 kPa for the elastic modulus. Much lower values are found at a pH of 4.2, ~15, and ~5 kPa, respectively. The time profiles at pH 4.2 are different from those at higher pH: the elastic modulus increases quickly in the early ~15 h but it decreases thereafter and the rupture stress does not increase further. This corresponds when the pH decreases below pH ~4.3. The strain at rupture of the gels increases quickly in the first 17 h, regardless of the acidification rate, followed by a slight decrease during a prolonged storage. If the evolution of the mechanical properties is followed with the pH, not with time, the data do not superimpose. This suggests that the acidification rate is relevant in the formation of the gel network, resulting in weaker gels at higher rates. Gels at equal pH, but at higher acidification rates, have lower rupture stress and elasticity modulus. The network deformability is also dependent on the rate, more than on the pH. Dynamic analysis of the cold gelation process also shows that the time required to reach the equilibrium pH is much shorter than to reach the equilibrium stress rupture. Increasing the acidification rate shortens the former but increases the latter. This results in a greater amount of molecular rearrangements occurring after the pH is set at higher acidification rates. This aging process is particularly important at high rates because a significant increase in the mechanical properties occurs once the pH of the gel does not change further.

In addition to chemical acidification using GDL, acid cold gelation can also be achieved with acid-producing bacteria. Bacterial acidification is often preferred in the food industry. In addition, the acidification method can vary the acidification rate and the final pH by varying the inoculum size (concentration) and glucose concentration (feed stock). Alting et al. (2004a) used a lactid acid bacteria (*Lactobacillus plantarum*) to induce the acid cold gelation of WPI solutions below pH 5.9. Increasing the concentration of glucose decreases the final acidification pH. The gel hardness does not change significantly between 5.2 and 4.5, while it decreased at lower values, similar to that in chemical acidification (Ju and Kilara 1998f). Changing the amount of bacteria affects the acidification profile, while the final pH is the same if the glucose concentration is kept constant, e.g., the time to reach pH 5.5 changes form 220 min to 750 min from an inoculum size of 10–0.5%, respectively. A fast acidification rate resulted in harder gels, probably because these gels have been in the gel state for longer.

10.1.4 Interactions between Protein Aggregates

There are several intermolecular interactions that lead to the formation and stabilization of protein gels: van der Waals (0.1–1 kJ/mol), hydrophobic (5–10 kJ/mol), hydrogen bonds (10–40 kJ/mol), electrostatic interactions (25–80 kJ/mol), and covalent bonds (200–400 kJ/mol) (Dickinson 1997). In this section the key interactions involved in the different types of cold-set gels reviewed previously will be discussed, particularly salt and acid cold-set gels.

10.1.4.1 Interactions in Salt Cold Gelation

Salt (and acid) cold gelation is driven by the reduction of the electrostatic repulsion between the aggregates (Kawamura et al. 1993); divalent salts are just more effective in screening charges than their monovalent counterparts, as discussed previously. This screening allows charged proteins, like whey proteins at neutral pH, to overcome the initial repulsion and to engage in more interparticle contacts, which facilitates the formation of short-ranged and weak interactions. Initial attempts to study the interactions involved in cold gelation were performed by following the evolution of a structural parameter, like the storage modulus of the gel, with the gelation temperature. McClements and Keogh (1995) reported that the rate of gel formation increased on cooling, which could suggest a hydrophobic effect in the initial stages of gel formation. Hydrophobic interactions are greatly facilitated after a preheating step due the increased protein hydrophobicity (Sato et al. 1995; Sava et al. 2005) and accessibility of hydrophobic groups to the solvent after denaturation. Weak interactions, in general, are reported to control the way salt cold-set gels are formed (Hongsprabhas and Barbut 1997d). Hydrogen bonding could be involved in this process, together with hydrophobic interactions, but studies suggest otherwise. Fourier transform infrared spectroscopy (FTIR) is a spectroscopic technique that allows following the evolution of intra- and interprotein hydrogen bonds during aggregation and gelation. Heated proteins solutions at neutral pH present two characteristic peaks in the amide I region (1600–1700 cm^{-1}), at 1614 and 1682 cm^{-1}, characteristic of intermolecular β-sheets between proteins (Lefevre and Subirade 2000). The FTIR spectrum of preheated solutions does not change after the formation of a salt cold-set gel. This implies that hydrogen bonding is not significant during the final gelation step. This is confirmed when the addition of urea does not change the modulus of the final gel formed (Beaulieu et al. 2002; Remondetto and Subirade 2003).

On the other hand, once the gels are fully formed, they became less rigid at lower temperatures, suggesting that nonhydrophobic forces are more important once a gel is fully formed (McClements and Keogh 1995). These interactions are primarily disulfide cross-links, well known to stabilize fine-stranded gels (Errington and Foegeding 1998), and which will be discussed extensively for acid cold-set gels. The addition of a thiol-blocking agent to a preheating solution, inhibiting the formation of new disulfide cross-links, results in very weak salt cold-set gels (Sato et al. 1995). In addition, the presence of β-mercaptoethanol during cold gelation, a chemical agent known to cleave disulfide bonds, yields weaker gels (Remondetto and Subirade 2003). The relevance of disulfide cross-linking, and the absence of hydrogen bonding, is true at different salt concentrations, independent of whether a fine-stranded or particulate gel is formed. On the other hand, hydrophobic interactions are very important at low salt concentrations, where the presence of sodium dodecylsufate (SDS) can inhibit the formation of a gel at all, but SDS has little effect at high salt concentrations. Figure 10.5 shows the remarkable effect of SDS inhibiting cold gelation at low salt concentrations (line d) and the weakening role of β-mercaptoethanol in the final gel (line c). At high salt concentrations, Remondetto and Subirade (2003) found that typical disrupting chemicals had no major significant effect in the gel formed, concluding that van der Waals interactions, the only interactions left, were the key interaction stabilizing the gel network. They further proposed a gelation model hypothesizing that the fine-stranded structure at low salt concentration is caused by the linear protein aggregation through exposed hydrophobic patches. The particulate nature is caused, they suggested, by the random aggregation—at complete screening conditions—controlled by van der Waals interactions, which favor the formation of clusters and growth in all directions.

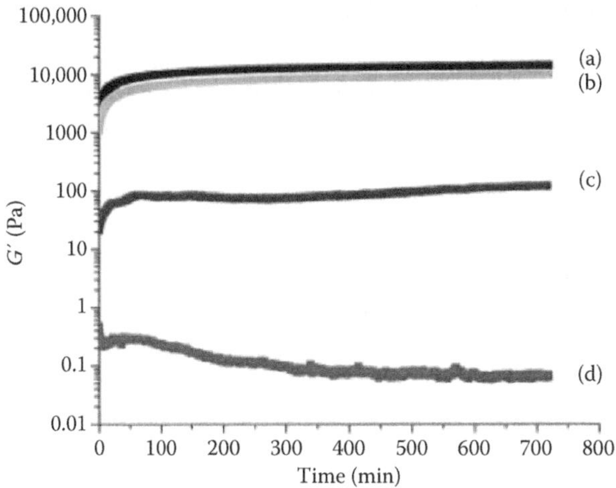

FIGURE 10.5 Storage modulus (*G*') evolution for cold-set gels of βLg (6% w/v) with 10 mM Fe^{2+} in: (a) water, (b) 2 M urea, (c) 0.2 M 2-mercaptoethanol, and (d) 1% SDS. (From Remondetto, G. E., and Subirade, M., *Biopolymers,* 69, 461–69, 2003.)

10.1.4.2 Interactions in Acid Cold Gelation

When the pH is decreased to values close to the pI, electrostatic interactions start to become important. Aggregation may not have to be maximum at the pI, where the net charge of the protein is zero. For example, dimeric βLg associates faster at pH 4.6 in nondenaturing conditions, much lower than the pI ~5.2, the same pH where βLg solubility is minimum (Majhi et al. 2006). At this pH, the positive and negative electrostatic potential contours are highly asymmetric, the negative much smaller than the positive because pH<pI, which allows branching aggregation. This reversible aggregation is eliminated at high salt concentrations, proving the electrostatic nature of the interactions.

Electrostatic interactions are present in acid cold gelation, specially considering the high dipole moment of βLg (Arakawa and Timasheff 1987), but the details are likely to be different than the above-mentioned reversible aggregation with native proteins. Under normal conditions, preheated βLg and WPI solution show a high turbidity between pH ~5.3 and 3.8, about the same pH range where acid cold-set gels are formed (Alting et al. 2002). This high-turbidity pH range can be decreased by modifying the net charge of the proteins. Succinylation of primary amino groups reduces the amount of positively charged amino groups, thus decreasing the pI of the proteins. Particulate aggregation (e.g., high turbidity) and gelation of succinylated proteins is shifted to lower pH, in good agreement with the predicted shift to lower values of the pI (Alting et al. 2002), clearly demonstrating the importance of the net electric charge of the aggregates during acid cold gelation.

At pH above the isoelectric point, spectroscopic measurements (small-angle x-ray and neutron scattering, SAXS and SANS) of WPI aggregates show a peak in the scattering wave vector, correlating to an interparticle distance of ~25 nm for 2 wt% solution. As the pH is decreased closer to the pI the peak disappears, indicating that this interpaticle distance also disappears, and no other characteristic length scale could be observed in the range of nanometers to micrometers (Alting et al. 2004b).

The nature of the interactions involved during gelation is usually investigated by using chemicals that disrupt specific interactions. Therefore, the percentage of a gel that is soluble in a given chemical solution can be used as an approximation of the number of protein aggregates that were stabilized, incorporated to the gel network, by the interaction known to be disrupted by the chemical used. Solubility measurements in the presence of disruptive agents have long been used to study heat-induced gels (Shimada and Cheftel 1988; Yamul and Lupano 2003). This approach was extended to the acid cold gelation of WPI solutions by Cavallieri et al. (2007). The authors reported that if gels were solubilized in just water, the solubility was ~10% at pH 5.2 and 3.9, and ~5% at pH in between. Such low solubility values in water are also typical for heat-induced gels (Mercadé-Prieto et al. 2006) and imply that most of the proteins are attached to the gel network, remaining insoluble. BSA is the least soluble of the whey proteins, while βLg is the most. In a pH 8 Tris-Glycine-EDTA buffer, the solubility of acid cold-set gels is very high, >85%, except for gels made at pH 4.2 (~72%). The addition of β-mercaptoethanol to the pH 8 buffer does not increase the solubility. Gels are completely soluble in a 6 M urea solution. The fact that acid cold-set gels are so soluble in a buffer with electrolytes is direct evidence that the interactions in the gel matrix are mainly electrostatic in nature (Morr and Ha 1993; Cavallieri et al. 2007). Limited hydrogen bonding and/or hydrophobic interactions could be responsible for the insoluble proteins in a pH 8 buffer that are soluble in urea.

In addition to the formation of noncovalent interactions during acid cold gelation, Alting et al. (2000, 2002) showed that covalent disulfide cross-links are also formed between the primary aggregates. The size of the initial heated aggregates (9 wt% WPI, 68.5°C for 2 h) was ~80 nm. Their size remained unchanged when adding thiol-blocking agents (N-ethylmaleimide, iodoacetamide, and *p*-chloromercuribenzoic acid) after acid gelation with GDL and resolubilization in a SDS solution. However, the resolubilized aggregates of gels without thiol-blocking agents were substantially larger than the initial aggregates (~290 nm); reducing electrophoresis confirmed the disulfide nature of the cross-links. It may be surprising that disulfide cross-linking occurs at acidic pH, as the reactive form is the anionic thiolate. Alting et al. (2000) suggested that these interactions occurred due to the large increase of the local protein concentration, a result of the noncovalent gelation, which could bring thiols and disulfide bonds from other proteins much closer. Other studies have shown that extensive disulfide cross-linking occurs in traditional heat-induced gelation (Otte, Zakora, and Qvist 2000) and in acid cold-set gels induced by bacteria (Alting et al. 2004a).

Disulfide cross-linking has a profound effect on stabilizing the gel network. Blocking the reactive thiol groups markedly decreases the hardness of the acid gels formed, as observed by the filled points in Figure 10.6 (Alting et al. 2000). Similar results have been reported in the high-pressure gelation of whey proteins (Keim and Hinrichs 2004). Nevertheless, it is important to realize these disulfide cross-links, while important ion any whey protein gels, are not sufficient to induce acid cold gelation on their own. As discussed previously, gels are highly soluble in a pH 8 buffer, and the presence of β-mercaptoethanol has no beneficial effect, meaning that despite larger disulfide cross-linked aggregates being formed, they are still small and soluble (Cavallieri et al. 2007). If too much GDL is added such that the pH is reduced below 4, the end state will be a redissolved solution of aggregates.

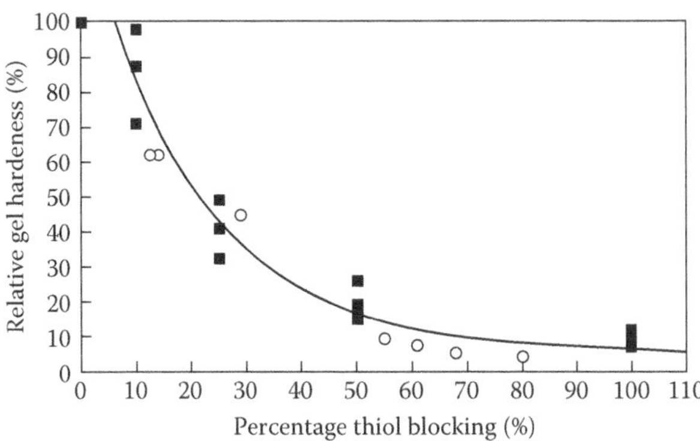

FIGURE 10.6 Dependence of the relative gel hardness on the relative number of thiol groups present in the soluble aggregates. The closed squares represent the hardness of gels prepared from chemically thiol-blocked WPI aggregates. The open circles represent the variation in the number of thiol groups caused by differences in the heat treatment. (From Alting, A. C., Hamer, R. J., de Kruif, C. G., Paques, M., and Visschers, R. W., *Food Hydrocolloids,* 17, 469–79, 2003a.)

The reduction of the gel modulus at pH<pI is attributed to the development of electrostatic repulsion between the protein aggregates (Eissa et al. 2004; Eissa and Khan 2005). In addition, at such low pH disulfide bridges are not formed (Otte et al. 2000; Surroca, Haverkamp, and Heck 2002).

The formation of covalent and noncovalent interactions during acid cold gelation is suggested to occur in two different steps (Alting et al. 2003b, 2004b). In the initial step noncovalent interactions are first formed. Unmodified and thiol-blocked WPI solutions have the same initial aggregation and gelation kinetics as measured by the increase of turbidity and the storage modulus. This initial step also determines the gel microstructure of the gel; no differences are observed in the micrometer scale between modified or unmodified gels between 1 and 9 wt% (Alting et al. 2003b). The gel permeability, which decreases at higher protein concentrations as the pore become smaller, also presents little difference between modified and unmodified gels (Alting et al. 2000, 2003b). Therefore, the initial morphology of the gel is established by noncovalent interactions.

The second step involves the formation of novel intermolecular disulfide cross-links. This occurs in a time frame of many hours. The storage modulus increase of thiol-blocked gels plateaus earlier than in the unmodified ones. This gel strengthening is followed by a continuous increase in the size of disulfide cross-linked aggregates with time, from the initial 80 nm to >500 nm after 48 h. Unmodified gels present higher maximum linear strain and higher plateau modulus (above 2 wt%) than the thiol-blocked gels, suggesting a mechanical strengthening of the network due to the formation of disulfide cross-links (Alting et al. 2003b).

The addition of thiol-blocking agents, and how they can greatly reduce the gel hardness, has been discussed previously. Similar effect can be achieved by modifying the heating conditions in a way that decreases the amount of free thiol groups in the aggregates, for example by using different preheating whey protein concentrations and heating times; see open points in Figure 10.6 (Alting et al. 2003a). Low protein solutions require extensive heating in order to achieve a high aggregation degree. However, after a prolonged heating the number of free thiol groups decreases markedly, which has a negative effect on the hardness of the cold-set gels. Alting et al. (2003a) reported that for a constant aggregation rate of 95%, a protein concentration higher than 4.5 wt% (68.5°C for 21 h) was required in order to have extensive disulfide cross-linking between the aggregates and hard gels.

10.1.5 Gel Models

It has been of great interest to understand the macroscopic properties of protein gels with physical models and how microstructural changes (e.g., by the addition of salts) affect these properties. Physical models are usually tested by comparing the evolution with the protein concentration of a physical gel property, like the elastic modulus. Experiments show that a power-law relationship can usually be used to fit the data, specially for particulate-like heat-set gels in the presence of salts (Verheul et al. 1998b; Ikeda et al. 1999; Pouzot et al. 2004). Criticism has been raised by pointing out that if low enough protein concentrations are used, close to the minimum required to form a gel, a single power-law is not observed (Gosal, Clark, and Ross-Murphy 2004b; Ross-Murphy 2007). This is particularly true for fine-stranded gels that have a much higher minimum gelation concentration than particulate-like gels (Mehalebi et al. 2008a). A power-law relationship has often been interpreted using the so-called fractal models (Bremer, Vanvliet, and Walstra 1989; Shih et al. 1990; Bremer et al. 1993), which assume that the gels can be described as a collection of fractal blobs that act as springs. These models predict that the plateau elastic modulus G' relates with the protein volume ϕ in a power-law relationship

$$G' \sim \varphi^{\alpha/3-D_f}, \tag{10.2}$$

where D_f is the fractal dimension of the clusters; the higher the value, the more compact is the aggregate structure. The parameter α depends on the nature of the gel strands, ranging from one for rigid strands in stretching deformation mode up to five for randomly curved flexible strands in bending deformation. Similar relationships have been developed for the maximum linear strain and yield stress (Mellema, van Opheusden, and van Vliet 2002). The model that has been commonly applied for cold-set gels is the weak-link regime of the model developed by Shih et al. (1990), which assumes that the interactions within a cluster are stronger and more elastic than the interactions between clusters. This regime considers that α in Equation 10.2 equals 1. Values of D_f for calcium cold gelation have been reported to increase from 2.3 to 2.5 at low concentrations (<10 mM Ca^{2+}) up to ~2.6 above 20 mM (Hongsprabhas et al. 1999; Marangoni et al. 2000). These D_f values agree with estimations from the microscopic analysis of the aggregates. Cold-set gel formed at NaCl concentrations higher than 200 mM have a calculated D_f of ~2.45 using rheological and microscopic methods (Marangoni et al. 2000). These values are much higher than would be expected by conventional random reaction or diffusion-limited aggregation (DLA) of particulates, D_f~1.7–2.2 (Meakin 1987; Ikeda et al. 1999). In fact, the assumption of a weak-link regime is ill chosen, as this implies that the fracture strain increases with the protein concentration, while the opposite is usually observed in heat-set gels (Ikeda et al. 1999) and in cold-set gels (Alting et al. 2003b; Eissa and Khan 2005). The strong-link regime of the Shih model should be used to calculate D_f, where α~4 in Equation 10.2, resulting in lower values, about 2–2.2 in acid cold gelation (Alting et al. 2003b; Eissa and Khan 2005). However, the strong-link regime is hard to justify because it considers that the links between the neighboring clusters have a higher elasticity than those in the cluster. In cold-set gels the opposite is true: protein aggregates are formed primarily from strong and elastic disulfide cross-links during the preheating step, while the major forces between these aggregates after the addition of salts or acids are weak noncovalent interactions.

Another source of confusion arises from the reported increase of D_f with the salt concentration, which suggests that there is a transition from DLA to reaction-limited aggregation (RLA) (Hongsprabhas and Barbut 1997d; Hongsprabhas et al. 1999). However, RLA is slower than DLA due to the presence of a repulsive electrostatic energy barrier between the approaching particles (Meakin 1987; Vreeker et al. 1992). Therefore, considering the effect of the salts on the gelation time, RLA must occur at low salt concentrations and DLA at high concentrations (Wu et al. 2005), as observed in heat-set gels (Ikeda, Foegeding, and Hagiwara 1999). The fractal dimension of clusters during cold gelation in 100 mM Ca^{2+} is ~1.85 using light scattering (Wu et al. 2005), a reasonable value for DLA conditions, but it is at odds with the D_f~2.6 found in previous studies using rheological and microscopic methods. D_f values typical of RLA are found away from the pI (e.g., ~2.1 at pH 5.4 and 3.2), but they decrease to ~1.8 at pH 4.9. The addition of NaCl at pH 5.4 can also induce such transition (Vreeker et al. 1992). However, the applicability of fractal models in cold gelation is still not fully solved. Fine-stranded aggregates and gels are usually not well characterized with fractal models (Kavanagh et al. 2000b; Gosal et al. 2004b). The former are the starting material of most cold-set gels, and the latter are formed at low salt concentrations. Therefore, comparisons of fractal dimensions calculated at different salt concentrations do not have a solid theoretical background.

10.1.6 Cold Gelation of Whey Protein Emulsions

The consumer preference for foods containing bioactive compounds such as nutraceuticals and functional ingredients is rapidly growing. The effectiveness of nutraceuticals in providing healthful benefits depends on preserving the bioavailability of the active ingredients. However, only a small proportion of the molecules remain available following consumption because of various reasons: insufficient gastric residence time, low permeability and/or solubility within the gut, as well as instability under conditions encountered in food processing (temperature, oxygen, light) or in the gastrointestinal tract (pH, enzymes, presence of other nutrients), all of which limit the activity and potential health benefits (Bell 2001). Therefore, safe and effective delivery of these molecules is a formidable challenge. The development of good nutraceutical delivery systems that do not compromise food safety and quality should (McClements, Decker, and Weiss 2007): (1) use safe and approved ingredients and methods; (2) protect encapsulated bioactive compounds against degradation; (3) ensure adequate loading capacity and retention properties for bioactive compounds; (4) possess release properties in response to a specific environmental stimulus; and (5) enhance the bioavailability/bioactivity of the encapsulated components.

Food emulsions and emulsion-filled hydrogels have been utilized to encapsulate functional ingredients (see next Section) (Appelqvist et al. 2007). Emulsions, consisting of two immiscible phases, exist widely in food products such as milk, cream, butter, margarine, etc. In addition to imparting desirable mouthfeel characteristics to the food and acting as key ingredients in the formation of structure in certain foods, food emulsions also possess a number of potential advantages as delivery systems for encapsulation, protection, delivery, and release of both hydrophilic and lipophilic bioactive compounds. Since emulsions contain a nonpolar region (the oil phase), a polar region (the aqueous phase), and an amphiphilic region (the interfacial layer), it is possible to incorporate multiple components within the same delivery system (McClements 1999). Conventional o/w emulsions have been studied extensively to encapsulate and deliver a variety of different bioactive lipids for the purpose of incorporating into foods products such as milk, yogurts, ice-cream, and meat patties, and to prevent oxidation of the polyunsaturated fats within the lipid droplets (Heinzelmann and Franke 1999; McClements and Decker 2000; Osborn and Akoh 2004; Klinkesorn et al. 2005). These applications usually take advantage of some biopolymers such as proteins and polysaccharides, which possess the encapsulation properties to form a matrix in which the fat globules are trapped or with which they interact.

Whey proteins are often used as emulsifiers to stabilize droplets; they adsorb at the emulsion droplet surface, creating an electrostatic barrier against flocculation and coalescence (Moreau et al. 2003; Dalgleish 2006; Ye and Singh 2006). Whey proteins, and in particular βLg, unfold and self-aggregate after adsorption, forming continuous and homogeneous membranes around the oil droplets through intermolecular β-sheets interactions (Lefevre and Subirade 2003). Whey proteins can form a multilayer on the droplet surface with a limiting surface concentration of approximately 3.2 mg/m^2 (Hunt and Dalgleish 1994). Both protein adsorbed on the droplet surfaces and unadsorbed in the aqueous phase may inhibit lipid oxidation of emulsions (Donnelly, Decker, and McClements 1998; Tong et al. 2000; Park et al. 2005). All major whey proteins—βLg, αLa, and BSA—contain cysteine and cystine residues that can scavenge free radicals and inhibit the formation of lipid hydroperoxides, potentially acting as antioxidant systems.

Once an oil-whey emulsion is formed with a conventional homogenization method, the solution can be gelled, for example by heating (Monahan, McClements, and Kinsella 1993; Dickinson, Hong, and Yamamoto 1996). The role of the oil droplet—the filler—in the gel properties can be classified as being "active" or "inactive," depending on whether there are strong interactions between the gel matrix and the filler or there is little chemical affinity, respectively (Ring and Stainsby 1982). WPI-covered oil droplets in emulsion gels behave as active filler: the storage modulus of the heat-set emulsion gels is significantly higher than that of pure WPI protein gels; the higher the oil concentration the higher the modulus (Chen and Dickinson 1998; Dickinson and Chen 1999). The improved elasticity is suggested to occur from disulfide cross-linking between the absorbed and nonabsorbed proteins (Chen and Dickinson 1999). The addition of oil markedly reduces the minimum concentration to form a gel, from ~5 wt% in a pure WPI gel to ~0.13 wt% in 45 vol% oil emulsion gels (Chen and Dickinson 1998). Note that whey proteins can be used in both the emulsification process and in the gelation process, or only in one of them and a polysaccharide in the other (Sala et al. 2007).

Line, Remondetto, and Subirade (2005) first characterized emulsion cold-set gels formed in the presence of calcium. The WPI solution is first pretreated to form soluble aggregates (e.g., by heating—9.5 wt% WPI at 85°C for 45 min), followed by the addition of oil and homogenization. At low oil concentrations the emulsions exhibit Newtonian behavior, but a shear thinning behavior is developed at higher concentrations due to depletion flocculation (Boutin et al. 2007). Aggregated WPI solutions yield more stable emulsions than using native WPI. The adsorbed layer is thicker using aggregated proteins; this increases the density of the emulsion droplets and reduces the creaming rate. In addition, higher oil and WPI concentrations result in lower creaming rates (Rosa et al. 2006).

Once the emulsion is prepared, cold gelation can be induced by just mixing with a calcium salt (Line et al. 2005). Higher calcium concentrations reduce the WHC of the cold-set gel, as expected for a salt-induced gelation, while the opposite effect is observed with the oil concentration. A higher oil concentration also has a beneficial effect by increasing the storage modulus, as expected for an active filler, but calcium concentrations higher than 20 mM have a detrimental effect. Similar results have been reported in the cold gelation of emulsions by acidification using GDL (Boutin et al. 2007). While the previous studies reported a strong enhancement of the gel stiffness at high oil concentrations (e.g., G' increase from 5 kPa in a pure WPI gel to 40 kPa in a 30% oil emulsion gel [Line et al. 2005]), Rosa et al. (2006) only observed a slight improvement of the storage modulus and of the fracture stress in GDL cold gelation, which could be caused by the much milder heating conditions used (68.5°C instead of 85°C). The fracture strain, not reported previously, decreases slightly at higher oil concentrations, resulting in more brittle gels.

The van der Poel theory (van der Poel 1958; Smith 1974, 1975) is commonly used to analyze the rheological properties of emulsion gels where the filler particles are assumed not to interact with each other. The theory states that shear modulus of a composite filled gel (G') can be predicted knowing the shear modulus of the matrix gel (G'_m) and filler material (G'_f):

$$\frac{G'}{G'_m} = 1 + \frac{15(1-v_m)(M-1)\phi}{Q-(8-10v_m)(M-1)\phi}, \tag{10.3}$$

where φ is the volume fraction of the filler (i.e., oil), $M = G'_f / G'_m$, $Q = (8-10v_m)M + 7 - 5v_m$, and v_m is the Poisson's ratio of the matrix, which can be taken as 0.5. This theory was quite successful in explaining the G' data of Rosa et al. (2006): G' increases with the oil concentration at low WPI concentrations (3 wt%) but not at high concentrations (8.9 wt%). However, the G' increase in some studies is too large to be explained with the van der Poel theory (Chen and Dickinson 1998; Line et al. 2005); the reason could be related to the aggregation of the emulsion droplets that the theory does not consider.

10.1.7 Encapsulation With Cold-Set Gels

Whey proteins are increasingly being used as oral nutraceutical delivery systems, recently reviewed by Chen, Remondetto, and Subirade (2006), following their great encapsulation capability, reduced cost, extensive availability, nutritional value, and because they are generally recognized as safe (GRAS). During encapsulation, the gelation functionality of whey proteins is used to trap the active ingredients. The size of the encapsulated system will depend on the delivery system.

The emulsification properties of whey proteins have long been used in microencapsulation. Heelan and Corrigan (1998) reported a method to form WPI microspheres loaded with drugs using organic solvents and glutaraldehyde as a cross-linking agent. WPI can form similar microspheres, with similar release profiles, as those using methylcellulose or sodium caseinate (Corrigan and Heelan 2001). Lee and Rosenberg proposed another encapsulation method consisting of a double emulsification, with glutaraldehyde cross-linking, in order to yield water-insoluble microspheres (Rosenberg 1997; Lee and Rosenberg 1999, 2000a). Insoluble microspheres are required in order to avoid burst-type core release and to achieve the desired controlled release. However, the use of organic solvents and cross-linking agents limits the applicability of those methods. Cross-linking agents can be avoided using the heat gelation properties of whey proteins (Lee and Rosenberg 2000b) and an "all-aqueous" solution was finally developed using a calcium alginate coating (Rosenberg and Lee 2004a, 2004b).

These latter microencapsulation methods still present the limitation of a final heat treatment, which limits its applicability to the encapsulation of heat-insensitive compounds. To be able to encapsulate heat-sensitive chemicals, the heat gelation step can be changed to a cold gelation step, first developed by Beaulieu et al. (2002). This encapsulation method again benefits from the emulsification properties of whey proteins. It is discussed briefly as a typical example of cold gelation for encapsulation purposes. First, the whey protein solution (8 wt% WPI, pH 8) is heated extensively (80°C for 30 min) under mechanical agitation. Once the whey solution is equilibrated to room temperature, an oil/water emulsion is formed with a whey protein concentration of 5.6 wt% and 30 wt% of vegetable oil. The mixture is first pre-homogenized in an Ultra-Turrax, and then it is passed through a high-pressure homogenizer, first at 100 MPa and then at 3 MPa. The resulting emulsion is added dropwise to a concentrated $CaCl_2$ solution. The beads formed are finally rinsed and dried. The drug chemical or nutraceutrical is mixed with the oil before emulsification. This method, therefore, is valid for hydrophobic bioactive components soluble in oils. The calcium concentration in the solution at the final gelation step (10–20 wt%) has little effect on the mechanical properties of the beads. Nevertheless, a 20 wt% $CaCl_2$ concentration is preferred because the beads formed are more regular and spherical, with a smooth surface and desirable characteristics for controlled release, and the fat globules are smaller and the network is more homogeneous. The size of the beads was about 2 mm (e.g., Figure 10.7).

Whey protein beads with a cold gelation protocol can also be formed in all aqueous conditions, without the presence of an oil-based emulsion. In this case, the preheated whey solution is directly dispersed dropwise to the calcium solution. This methodology has been applied for the encapsulation of recombinant yeasts (Hebrard et al. 2006); the resulting beads are shown in Figure 10.7. The viability and the growth capability of yeasts are not altered by entrapment, highlighting a major benefit of the mild conditions required for cold gelation.

In addition these whey protein beads are highly resistant to gastric incubation (pH<2 in the presence of pepsin) (Chen et al. 2006), an important result as the survival of the active bacteria in gastric conditions is essential for any potential application. Beaulieu et al. (2002) suggested that pepsin, which preferentially attacks hydrophobic amino acids, is not capable of hydrolyzing the whey proteins in emulsified beads because the hydrophobic groups of the denatured proteins are adsorbed in the oil droplets and are thus, masked from the solvent. The beads are destroyed under pancreatic conditions, when the bioactive ingredient is released. However, the gastric resistance could be an intrinsic property of whey proteins as it is also observed in gels when oil emulsions are not used (Guerin, Vuillemard, and Subirade 2003; Hebrard et al. 2006) and even when gels are not formed (Dave and Shah 1998; Kos et al. 2000).

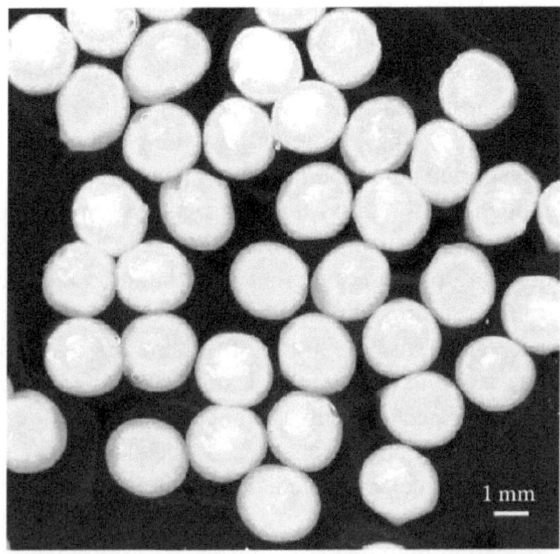

FIGURE 10.7 Whey protein beads formed by cold gelation in a 0.1 M $CaCl_2$ solution. (From Hebrard, G., Blanquet, S., Beyssac, E., Remondetto, G., Subirade, M., and Alric, M., *Journal of Biotechnology,* 127, 151–60, 2006.)

The beneficial effects of whey proteins in protecting bacteria have been extended to probiotic products, of great interest to the food industry. Probiotic bacteria must reach the gut at high concentrations in order to confer health benefits to the host. They must survive the acidic conditions of the stomach, digestive enzymes and bile salts of the small intestine (Marteau et al. 1997), and the food processing and storage conditions (Champagne et al. 2005). Microencapsulation is seen as a suitable technique to solve both problems (Siuta-Cruce and Goulet 2001). In vitro studies using an aqueous cold gelation method showed that microencapsulation had a positive effect increasing the survival of the bacteria (Reid et al. 2005). However, it was found that the gelation step was highly damaging, although it could not be concluded if it was due to the calcium (16.7 wt%) or due to low pH (4.6). Microencapsulation with cold-set whey beads has also been shown to have a protective effect to a short heat-treatment and to rehydration conditions at extreme pH (Reid et al. 2007). Probiotics, once encapsulated in a whey bead, can increase significantly after incubation in milk, offering a plausible alternative process to manufacturing beads with high-cell densities (Reid et al. 2005).

Alginate beads have also been shown to have a protective effect on probiotics against adverse conditions in food processing (Sheu, Marshall, and Heymann 1993; Khalil and Mansour 1998). However, whey protein beads in similar applications would have the added nutritional value of whey proteins, and may release greater numbers of cells in the gastrointestinal system due to protein hydrolysis during digestion. Whey-based microspheres, following a cold gelation procedure, can be formed mixed with polysaccharides, such as alginate. The gelation of microspheres occurs while the whey-alginate-calcium-bioactive ingredient solution is dispersed in oil under agitation followed by acetic acid addition (Chen and Subirade 2006, 2007). Core-shell nanoparticles with chitosan/βLg (Chen and Subirade 2005) or beet pectin/βLg (Santipanichwong et al. 2008) and alginate/pectin/whey beads (Guerin et al. 2003) have also been successfully developed.

It is interesting to discuss the role of a fine-stranded or a particulate morphology in the release of nutraceuticals or drugs; we consider the case of iron-induced whey gels (Remondetto, Beyssac, and Subirade 2004). A high iron concentration in the gels is desirable. This will result in a particulate morphology with large pore sizes (Remondetto et al. 2002), which may facilitate the iron release. However, iron in particulate gels is trapped within the large aggregates, while in fine-stranded gels it is located at their outer surface. This results in more iron being released and made available for adsorption in the intestinal wall in fine-stranded than in particulate gels, despite the lower initial iron concentration.

10.1.8 Cold Gelation With Saccharides

One of the major potential applications of cold gelation is in food products. Foods are composed of many different chemical compounds that could affect the functionality of whey proteins (Baier and McClements 2005). Polysaccharides are of particular interest because they are commonly used as thickening agents. The interactions between protein and polysaccharides will not be discussed here; for recent reviews see Doublier et al. (2000), Nishinari, Zhang, and Ikeda (2000), Benichou, Aserin, and Garti (2002), and Ye (2008). Bryant and McClements (2000c) studied NaCl cold gelation in the presence of xanthan gum. The presence of polysaccharide increases the gelation rate, shear modulus, and opacity of the cold-set gels formed. The authors suggested that the thermal incompatibility of xanthan and the whey protein aggregates form a water-in-water emulsion, which, once a gel is formed in the presence of a salt, results in xanthan-rich regions embedded in the continuous protein gel.

The final microstructure is the result of the balance between gel formation and phase separation, which depends on the polysaccharide charge density, mole of negative charge per mole of monosaccharide; a schematic representation is shown in Figure 10.8 (de Jong and van de Velde 2007). At low charge densities (<0.3, e.g., guar gum, locust bean gum, xanthan gum, and gellan gum), there is microphase separation in a continuous protein matrix at low polysaccharide concentrations. At higher concentrations, the serum phase increases and no gel is formed. The gel modulus reaches a maximum with the polysaccharide concentration. At low polysaccharide concentrations, more

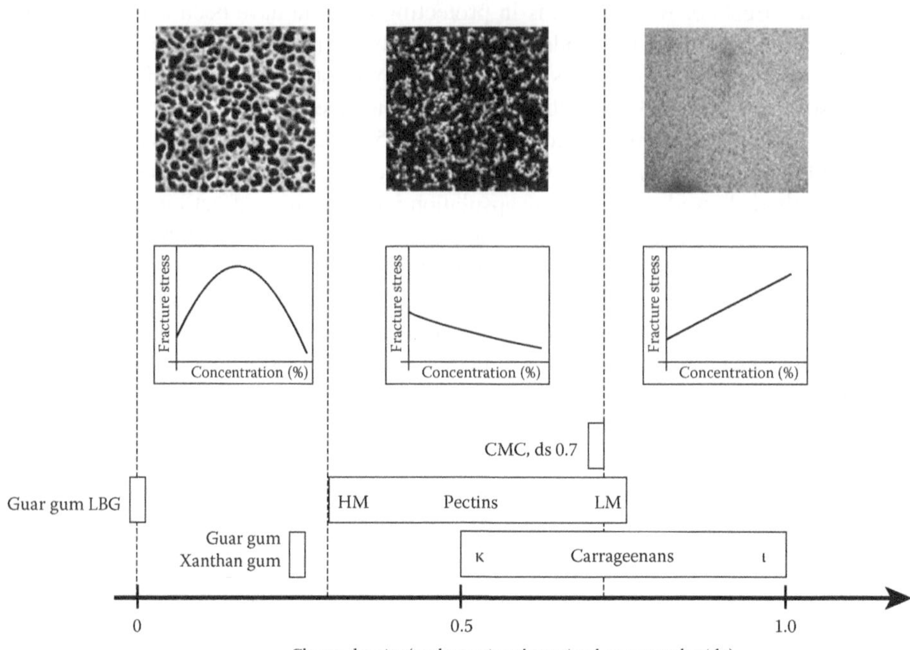

FIGURE 10.8 Schematic representation of the effect of the polysaccharide charge density on microstructure and fracture properties of mixed whey protein-polysaccharide cold-set gels. (From de Jong, S., and van de Velde, F., *Food Hydrocolloids,* 21, 1172–87, 2007.)

microphase separation increases the protein concentration of the protein aggregate strands, strengthening the gel. If there is extensive phase separation, or even phase inversion to a polysaccharide continuum phase, the number of effective protein strands decreases and so does the gel modulus. At medium charge densities (0.3–0.6, e.g., κ-carrageans, pectins, and CMC), the gels formed become increasingly coarse at higher polysaccharide concentrations, with lower modulus and fracture stress. On the other hand, the addition of highly charged polysaccharides (>0.6, e.g., ι-carragean and low-molecular-weight pectin) does not change the microstructure at the micrometer scale; phase separation is actually prevented. Understanding the final microstructure of the mixed gels is important because it affects their sensorial perception by modifying the breakdown and serum-release properties (van den Berg et al. 2007, 2008a, 2008b).

Cold-gelling ingredients can also be used as a replacement for gelation or polysaccharides in sweetened foods such as desserts (Kulmyrzaev, Cancelliere, and McClements 2000b). Sucrose is known to stabilize whey proteins during heating (Kulmyrzaev, Bryant, and McClements 2000a). However, when used during cold gelation, sucrose enhances gelation up to 8 wt% because of the increase of viscosity of the aqueous phase. Higher concentrations favor gel formation by promoting protein interactions (Bryant and McClements 2000b; Kulmyrzaev et al. 2000b). Gels formed with sucrose are less opaque than those formed without.

10.2 SOY PROTEINS

Soy proteins are also used in many food products. Heat-induced gelation of soy protein has been studied extensively (Utsumi and Kinsella 1985; Renkema, Knabben, and van Vliet 2001), showing many similarities with whey protein gels, such as forming fine-stranded gels at neutral pHs and particulate-like gels close to the pI (~4.8). Soy proteins gels can also be formed with transglutaminases (Tang et al. 2006) or high pressures (Alvarez, Ramaswamy, and Ismail 2008). Soy protein isolate (SPI) is composed of two protein fractions: β-conglycinin (~200 kDa) and glycinin (~350 kDa), with denaturation temperatures of ~70 and 85°C, respectively (Renkema et al. 2000). Complete denaturation is achieved after heating at 85°C for 1 h at neutral pH (Renkema and van Vliet 2002). Soeda and coworkers have long reported that concentrated solutions of freeze-dried aggregated soy proteins, formed after a heat treatment (Soeda 1994b, 1994a) or using transglutaminases (Soeda 2003), can form gels when stored for several days at refrigeration temperatures (~5°C). The procedure, despite being performed at low temperatures, cannot be considered a cold gelation method as described in the introduction. The final gelation step occurs because of the high protein concentrations of the reconstituted solutions (~20 wt%); there is not a new stimulus that drives gelation.

Maltais et al. (2005) first reported calcium cold gelation of SPI solutions that were previously heated at 105°C for 30 min, at neutral pH and 9.5 wt%. Lower heating temperatures (95°C) can also be used (Kuipers et al. 2005). At the end of the preheating step soy proteins are highly aggregated (Cramp, Kwanyuen, and Daubert 2008); these aggregates constitute the building blocks of the cold-set gels as observed with FTIR (Maltais, Remondetto, and Subirade 2008). Heating higher protein concentrations also results in larger aggregates, with result in more viscous reconstituted solutions than comparable heat-treated diluted solutions (Cramp et al. 2008). The interactions involved in the formation of aggregates are reported to be primarily intermolecular disulfide bonds, which allow the subsequent formation of noncovalent interactions, like hydrophobic interactions and hydrogen bonding.

The effect of the calcium concentration and the SPI concentration upon gelation follows that previously discussed for whey proteins. At 10 mM Ca^{2+} the gels formed are transparent and soft and at 20 mM the gels have become white, granular, and brittle. Higher calcium concentrations (10–20 mM) increase the light reflectance (larger particulates), decrease the WHC, and increase the storage modulus. Gels at low calcium concentrations are homogeneous and have small pores, while a coarse microstructure is observed at 20 mM Ca^{2+} using SEM and TEM, which is consistent with the decreased WHC (Maltais et al. 2005, 2008).

Soy protein hydrolyzates also present novel funtionalities during cold gelation. Kuipers et al. (2005) reported that SPI hydrolyzates form gels during acid cold gelation at higher pH, e.g., pH ~6 for untreated SPI to ~7.6 for a degree of hydrolysis of 5%, which could expand its use to neutral foods. Lower pH and higher degrees of hydrolysis result in coarser gels, and larger aggregates and larger pores. Soy protein hydrolyzates have a higher tendency to aggregate at neutral pH than comparable whey protein hydrolyzates (Kuipers, Alting, and Gruppen 2007), caused by an increase in hydrophobicity during hydrolysis that drives aggregation (Kuipers and Gruppen 2008).

Heat-set gels of SPI, or enriched fractions of glycinin and β-conglycinin, have similar protein concentration dependence of the storage modulus (Renkema and van Vliet 2004). Fractal models can be applicable at some gelation conditions, such as at pH 3.8 and 0.2 M NaCl yielding a D_f of 2.3, but not when fine-stranded-like gels are formed, as discussed previously for whey protein gels, because the minimum gelation concentration is high (Renkema and van Vliet 2004). Similar analysis in calcium cold-set gels shows an increase of D_f from 2 to 2.65 from 10 to 20 mM Ca^{2+} (Maltais et al. 2008); the former is described with the Wu and Morbidelli (2001) model to be in the transition regime (between strong and weak links) while the latter falls in the weak-link regime.

Finally, we have considered the cold gelation of the purified soy protein fractions, which are expected to be used mixed with different chemicals or food ingredients in their applications. Nevertheless, if cold gelation is performed with pure soymilk then tofu is formed. Heated soymilk (e.g., at 95°C) forms gels after the addition of salts, GDL, or transglutaminases (Liu and Chang 2004; Tang et al. 2007), in agreement with the two-step gelation process depicted in Figure 10.1.

10.3 OVALBUMIN AND EGG WHITE

Initial studies on two-step gelation of egg white were those that considered an initial preheating at 60°C to simulate pasteurization, in a subsequent heat gelation step (Xu, Shimoyamada, and Watanabe 1997). The morphology of the soluble ovalbumin aggregates formed during the preheating step is different from that discussed previously for WPI. Whey protein aggregates formed at neutral pH are curved and stranded-like, but ovalbumin aggregates are fibrillar (Koseki et al. 1989a; Koseki, Kitabatake, and Doi 1989b), like those formed at pH 2 with βLg (Gosal, Clark, and Ross-Murphy 2004a). These fibrils are very linear and show little branching, and can be considered a semiflexible string of monomers (Pouzot et al. 2005). Figure 10.3a and b show the cryo-TEM of ovalbumin aggregates; note that the fibrils are much larger than in typical WPI aggregates at neutral pH (Figure 10.3c and d). The fractal dimension of these self-similar structures is 1.7 and 2 at low and high salt concentrations, respectively; the same values were reported previously for WPI (Weijers, Visschers, and Nicolai 2002). Typical heating conditions to form soluble ovalbumin aggregates are 78°C for 22 h (Alting et al. 2004b). Prolonged heating is required because the aggregation reactions are slow, despite the protein being fully denatured (Weijers et al. 2003). Noncovalent interactions occur first, followed by disulfide bond stabilization, the opposite to that for whey proteins. The length of the fibrils formed depends on the heating conditions, such as the protein concentration, but they are significantly larger than those formed with WPI. Preheating is also done at low salt concentrations to improve the solubility of the aggregates resulting from strong repulsive interactions (Weijers et al. 2002).

The longer ovalbumin fibrils result in acid cold-set gels that are less opaque and harder than WPI gels (Alting et al. 2004b). Gel hardness, storage modulus, and turbidity profiles are not affected by the presence of thiol-blocking agents, contrary to WPI acid cold gelation. Agarose gel electrophoresis does not show novel disulfide cross-linked aggregates during cold gelation, despite the presence of reactive thiol groups in ovalbumin. This proves that disulfide cross-linking is not a critical interaction in the cold gelation of ovalbumin, either with GDL or bacterial acidification (Alting et al. 2004a). Alting et al. (2004b) suggested that upon the reduction of electrostatic repulsion, during acidification, the long ovalbumin fibrils could directly form a percolating network from physical entanglements and hydrophobic and van der Waals interactions. A small addition of NaCl (~10 mM) during the preheating step is reported to result in more transparent and stronger cold-set gels (Choi, Lee, and Moon 2008).

Food-grade purified ovalbumin is not available. For this reason Weijers et al. (2006) studied how to form good transparent cold-set gels using industrial egg white powder (EWP). EWP has a high salt concentration, which inhibits the formation of fibrils or strands upon heating; a desalting step is first required (e.g., with ultrafiltration). Nevertheless, desalted EWP still forms turbid gels and much shorter aggregates than those observed for pure ovalbumin. The authors showed that this was caused primarily by the presence of ovotransferrin, which only accounts for ~12% of EWP but has 15 cystine bonds per protein, compared to only one in ovalbumin. Ovotransferrin can be eliminated from egg white by a preliminary heating at 60°C for 30 min, to fully denature the protein, followed by precipitation at pH 4.7 and centrifugation. The remaining solution is adjusted to pH 7 and desalted with ultrafiltration. The ovotransferrin-free solution will then be heated at 78°C for 22 h to form the soluble aggregates to be used in cold gelation. The authors suggested that the large amount of disulfide bonds in ovotransferrin allow the formation of disulfide bridges with ovalbumin and that disrupts the formation of large fibrils.

10.4 CONCLUSIONS

Cold gelation procedures further extend the range of possible protein gelation mechanisms. Gels can be made efficiently with many other procedures other than by heat. Moreover, the gels formed usually have better desirable properties than traditional protein gels, like those formed from only heat treatment: gels can be less opaque and have better WHCs. The mechanical and failure properties of the gels vary greatly depending on the pretreatment and on the final gelation step.

The mixed nature of many cold-set gels, fine-stranded primary aggregates randomly aggregated to form particulate-like systems, is a challenge to relate microstructure with the physical parameters of the gels, such as the elastic modulus. New models have to be developed to accommodate the unique mixed characteristics of cold-set gels.

Recent studies have shown that whey proteins in particular are suitable to form delivery systems due to their great emulsifying and gelling properties. Nevertheless, more research is required using real active ingredients, particularly using sensible live systems like bacteria. While promising results are available, extensive optimization of the encapsulation process is required to increase the load and the "survival" of the active ingredients. Long-term stability studies will also be required before developing commercial products. Finally, good understanding of how these protein-based carriers behave under gastrointestinal conditions will be required to prove that not only do they release the active ingredients at the desired tract, but that they improve the bioavailability and the adsorption of the active ingredients. In addition, new experimental protocols should be developed in order to produce smaller micron-size capsules, which should still be free of non-GRAS chemical cross-linkers and organic solvents.

The work currently being performed on mixtures of proteins with polysaccharides should be extended in the future to accommodate more complex mixtures with other food ingredients, such as fats and other proteins.

While cold gelation is a fairly universal mechanism, in such a way as heat-induced gelation, different globular proteins result in significantly different gels due to the different way proteins aggregate. A clear example has been presented between whey proteins, and βLg in particular, and ovalbumin. Therefore, different proteins are going to be more suitable for different applications. More proteins, and combinations of them, should be investigated for cold gelation.

REFERENCES

Allain, A. F., Paquin, P., and Subirade, M. 1999. Relationships between conformation of beta-lactoglobulin in solution and gel states as revealed by attenuated total reflection Fourier transform infrared spectroscopy. *International Journal of Biological Macromolecules* 26 (5): 337–44.

Alting, A. C., de Jongh, H. H. J., Visschers, R. W., and Simons, J. 2002. Physical and chemical interactions in cold gelation of food proteins. *Journal of Agricultural and Food Chemistry* 50 (16): 4682–89.

Alting, A. C., Hamer, R. J., de Kruif, C. G., Paques, M., and Visschers, R. W. 2003a. Number of thiol groups rather than the size of the aggregates determines the hardness of cold set whey protein gels. *Food Hydrocolloids* 17 (4): 469–79.

Alting, A. C., Hamer, R. J., de Kruif, C. G., and Visschers, R. W. 2003b. Cold-set globular protein gels: Interactions, structure and rheology as a function of protein concentration. *Journal of Agricultural and Food Chemistry* 51 (10): 3150–56.

———. 2000. Formation of disulfide bonds in acid-induced gels of preheated whey protein isolate. *Journal of Agricultural and Food Chemistry* 48 (10): 5001–5007.

Alting, A. C., van der Meulena, E. T., Hugenholtz, J., and Visschers, R. W. 2004a. Control of texture of cold-set gels through programmed bacterial acidification. *International Dairy Journal* 14 (4): 323–29.

Alting, A. C., Weijers, M., De Hoog, E. H. A., van de Pijpekamp, A. M., Stuart, M. A. C., Hamer, R. J., De Kruif, C. G., and Visschers, R. W. 2004b. Acid-induced cold gelation of globular proteins: Effects of protein aggregate characteristics and disulfide bonding on rheological properties. *Journal of Agricultural and Food Chemistry* 52 (3): 623–31.

Alvarez, P. A., Ramaswamy, H. S., and Ismail, A. A. 2008. High pressure gelation of soy proteins: Effect of concentration, pH and additives. *Journal of Food Engineering* 88 (3): 331–40.

Anema, S. G. 2000. Effect of milk concentration on the irreversible thermal denaturation and disulfide aggregation of beta-lactoglobulin. *Journal of Agricultural and Food Chemistry* 48 (9): 4168–75.

Apenten, R. K. O., and Galani, D. 2000. Thermodynamic parameters for beta-lactoglobulin dissociation over a broad temperature range at pH 2.6 and 7.0. *Thermochimica Acta* 359 (2): 181–88.

Apenten, R. K. O., Khokhar, S., and Galani, D. 2002. Stability parameters for beta-lactoglobulin thermal dissociation and unfolding in phosphate buffer at pH 7.0. *Food Hydrocolloids* 16 (2): 95–103.

Appelqvist, I., Golding, M., Vreeker, R., and Zuidam, N. J. 2007. Emulsions as delivery systems in foods. In: J. M. Lakkis (ed.) *Encapsulation and Controlled Release Technologies in Food Systems*. Oxford, UK: Blackwell Publishing Ltd.

Arakawa, T., and Timasheff, S. N. 1987. Abnormal solubility behavior of beta-lactoglobulin: Salting-in by glycine and NaCl. *Biochemistry* 26 (16): 5147–53.

Baier, S. K., and McClements, D. J. 2005. Influence of cosolvent systems on the gelation mechanism of globular protein: Thermodynamic, kinetic, and structural aspects of globular protein gelation. *Comprehensive Reviews in Food Science and Food Safety* 4 (3): 43–54.

Barbut, S. 1995a. Cold gelation of whey proteins. *Scandinavian Dairy Information* 2:20.

———. 1995b. Effect of sodium level on the microstructure and texture of whey protein isolate gels. *Food Research International* 28 (5): 437–43.

———. 1995c. Effects of calcium level on the structure of pre-heated whey protein isolate gels. *Food Science and Technology - Lebensmittel-Wissenschaft & Technologie* 28 (6): 598–603.

———. 1996. Determining water and fat holding. In: G. M. Hall (ed.) *Methods of Testing Protein Functionality*, 187–225. London: Chapman & Hall.

———. 1997. Relationships between optical and textural properties of cold-set whey protein gels. *Food Science and Technology - Lebensmittel-Wissenschaft & Technologie* 29 (6): 590–93.

Barbut, S., and Drake, D. 1997. Effect of reheating on sodium-induced cold gelation of whey proteins. *Food Research International* 30 (2): 153–57.

Barbut, S., and Foegeding, E. A. 1993. Ca^{2+}-induced gelation of pre-heated whey-protein isolate. *Journal of Food Science* 58 (4): 867–71.

Bauer, R., Carrotta, R., Rischel, C., and Ogendal, L. 2000. Characterization and isolation of intermediates in beta-lactoglobulin heat aggregation at high pH. *Biophysical Journal* 79 (2): 1030–38.

Baussay, K., Le Bon, C., Nicolai, T., Durand, D., and Busnel, J. P. 2004. Influence of the ionic strength on the heat-induced aggregation of the globular protein beta-lactoglobulin at pH 7. *International Journal of Biological Macromolecules* 34 (1–2): 21–28.

Beaulieu, L., Savoie, L., Paquin, P., and Subirade, M. 2002. Elaboration and characterization of whey protein beads by an emulsification/cold gelation process: Application for the protection of retinol. *Biomacromolecules* 3 (2): 239–48.

Bell, L. N. 2001. Stability testing of nutraceuticals and functional foods. In: R. E. C. Wildman (ed.) *Handbook of Nutraceuticals and Functional Foods*. New York: CRC Press.

Benichou, A., Aserin, A., and Garti, N. 2002. Protein-polysaccharide interactions for stabilization of food emulsions. *Journal of Dispersion Science and Technology* 23 (1–3): 93–123.

Bolder, S. G., Hendrickx, H., Sagis, L. M. C., and Van der Linden, E. 2006a. Ca^{2+}-induced cold-set gelation of whey protein isolate fibrils. *Applied Rheology* 16 (5): 258–64.

———. 2006b. Fibril assemblies in aqueous whey protein mixtures. *Journal of Agricultural and Food Chemistry* 54 (12): 4229–34.

Boutin, C., Giroux, H. J., Paquin, P., and Britten, M. 2007. Characterization and acid-induced gelation of butter oil emulsions produced from heated whey protein dispersions. *International Dairy Journal* 17 (6): 696–703.

Boye, J. I., Alli, I., Ismail, A. A., Gibbs, B. F., and Konishi, Y. 1995. Factors affecting molecular characteristics of whey-protein gelation. *International Dairy Journal* 5 (4): 337–53.

Bremer, L. G. B., Bijsterbosch, B. H., Walstra, P., and van Vliet, T. 1993. Formation, properties and fractal structure of particle gels. *Advances in Colloid and Interface Science* 46:117–28.

Bremer, L. G. B., Vanvliet, T., and Walstra, P. 1989. Theoretical and experimental-study of the fractal nature of the structure of casein gels. *Journal of the Chemical Society - Faraday Transactions I* 85:3359–72.

Britten, M., and Giroux, H. J. 2001. Acid-induced gelation of whey protein polymers: Effects of pH and calcium concentration during polymerization. *Food Hydrocolloids* 15 (4–6): 609–17.

Bromley, E. H. C., Krebs, M. R. H., and Donald, A. M. 2005. Aggregation across the length-scales in beta-lactoglobulin. *Faraday Discussions* 128:13–27.

———. 2006. Mechanisms of structure formation in particulate gels of beta-lactoglobulin formed near the isoelectric point. *European Physical Journal E* 21 (2): 145–52.

Bryant, C. M., and McClements, D. J. 1998. Molecular basis of protein functionality with special consideration of cold-set gels derived from heat-denatured whey. *Trends in Food Science & Technology* 9 (4): 143–51.

———. 2000a. Influence of NaCl and $CaCl_2$ on cold-set gelation of heat-denatured whey protein. *Journal of Food Science* 65 (5): 801–804.

———. 2000b. Influence of sucrose on nacl-induced gelation of heat denatured whey protein solutions. *Food Research International* 33 (8): 649–53.

———. 2000c. Influence of xanthan gum on physical characteristics of heat-denatured whey protein solutions and gels. *Food Hydrocolloids* 14 (4): 383–90.

———. 2000d. Optimizing preparation conditions for heat-denatured whey protein solutions to be used as cold-gelling ingredients. *Journal of Food Science* 65 (2): 259–63.

Burke, M. D., Ha, S. Y., Pysz, M. A., and Khan, S. A. 2002. Rheology of protein gels synthesized through a combined enzymatic and heat treatment method. *International Journal of Biological Macromolecules* 31 (1–3): 37–44.

Cavallieri, A. L. F., Costa-Netto, A. P., Menossi, M., and Da Cunha, R. L. 2007. Whey protein interactions in acidic cold-set gels at different ph values. *Lait* 87 (6): 535–54.

Cavallieri, A. L. F., and da Cunha, R. L. 2008. The effects of acidification rate, pH and ageing time on the acidic cold set gelation of whey proteins. *Food Hydrocolloids* 22 (3): 439–48.

Clare, D. A., Lillard, S. J., Ramsey, S. R., Amato, P. M., and Daubert, C. R. 2007. Calcium effects on the functionality of a modified whey protein ingredient. *Journal of Agricultural and Food Chemistry* 55 (26): 10932–40.

Clark, A. 1992. Gels and gelling. In: H. Schwartzberg, and R. Hartel (eds) *Physical Chemistry of Foods*, 263–305. New York: Marcel Dekker.

Colsenet, R., Soderman, O., and Mariette, F. O. 2006. Pulsed field gradient nmr study of poly(ethylene glycol) diffusion in whey protein solutions and gels. *Macromolecules* 39 (3): 1053–59.

Corrigan, O. I., and Heelan, B. A. 2001. Characterization of drug release from diltiazem-loaded polylactide microspheres prepared using sodium caseinate and whey protein as emulsifying agents. *Journal of Microencapsulation* 18 (3): 335–45.

Cramp, G. L., Kwanyuen, P., and Daubert, C. R. 2008. Molecular interactions and functionality of a cold-gelling soy protein isolate. *Journal of Food Science* 73 (1): E16–24.

Creusot, N., and Gruppen, H. 2007a. Enzyme-induced aggregation and gelation of proteins. *Biotechnology Advances* 25 (6): 597–601.

———. 2007b. Protein-peptide interactions in mixtures of whey peptides and whey proteins. *Journal of Agricultural and Food Chemistry* 55 (6): 2474–81.

Croguennec, T., Bouhallab, S., Molle, D., O'Kennedy, B. T., and Mehra, R. 2003. Stable monomerie intermediate with exposed cys-119 is formed during heat denaturation of beta-lactoglobulin. *Biochemical and Biophysical Research Communications* 301 (2): 465–71.

Croguennec, T., O'Kennedy, B. T., and Mehra, R. 2004. Heat-induced denaturation/aggregation of beta-lactoglobulin a and b: Kinetics of the first intermediates formed. *International Dairy Journal* 14 (5): 399–409.

Croguennoc, P., Nicolai, T., Kuil, M. E., and Hollander, J. G. 2001. Self-diffusion of native proteins and dextran in heat-set globular protein gels. *Journal of Physical Chemistry B* 105 (24): 5782–88.

Champagne, C. P., Gardner, N. J., and Roy, D. 2005. Challenges in the addition of probiotic cultures to foods. *Critical Reviews in Food Science and Nutrition* 45 (1): 61–84.

Chen, J. S., and Dickinson, E. 1998. Viscoelastic properties of heat-set whey protein emulsion gels. *Journal of Texture Studies* 29 (3): 285–304.

———. 1999. Effect of surface character of filler particles on rheology of heat-set whey protein emulsion gels. *Colloids and Surfaces B - Biointerfaces* 12 (3–6): 373–81.

Chen, L. Y., Remondetto, G. E., and Subirade, M. 2006. Food protein-based materials as nutraceutical delivery systems. *Trends in Food Science & Technology* 17 (5): 272–83.

Chen, L. Y., and Subirade, M. 2005. Chitosan/beta-lactoglobulin core-shell nanoparticles as nutraceutical carriers. *Biomaterials* 26 (30): 6041–53.

———. 2006. Alginate-whey protein granular microspheres as oral delivery vehicles for bioactive compounds. *Biomaterials* 27 (26): 4646–54.

———. 2007. Effect of preparation conditions on the nutrient release properties of alginate-whey protein granular minicrospheres. *European Journal of Pharmaceutics and Biopharmaceutics* 65 (3): 354–62.

Choi, S. J., Lee, S. E., and Moon, T. W. 2008. Influence of sodium chloride and glucose on acid-induced gelation of heat-denatured ovalbumin. *Journal of Food Science* 73 (5): C313–22.

Dalgleish, D. G. 2006. Food emulsions: Their structures and structure-forming properties. *Food Hydrocolloids* 20 (4): 415–22.

Dalgleish, D. G., Senaratne, V., and Francois, S. 1997. Interactions between alpha-lactalbumin and beta-lactoglobulin in the early stages of heat denaturation. *Journal of Agricultural and Food Chemistry* 45 (9): 3459–64.

Damodaran, S. 1989. Influence of protein conformation on its adaptability under chaotropic conditions. *International Journal of Biological Macromolecules* 11 (1): 2–8.

Dave, R. I., and Shah, N. P. 1998. Ingredient supplementation effects on viability of probiotic bacteria in yogurt. *Journal of Dairy Science* 81 (11): 2804–16.

de Jong, S., and van de Velde, F. 2007. Charge density of polysaccharide controls microstructure and large deformation properties of mixed gels. *Food Hydrocolloids* 21 (7): 1172–87.

de Jongh, H. H. J., Groneveld, T., and de Groot, J. 2001. Mild isolation procedure discloses new protein structural properties of beta-lactoglobulin. *Journal of Dairy Science* 84 (3): 562–71.

de Kruif, C. G. 1997. Skim milk acidification. *Journal of Colloid and Interface Science* 185 (1): 19–25.

de la Fuente, M. A., Singh, H., and Hemar, Y. 2002. Recent advances in the characterisation of heat-induced aggregates and intermediates of whey proteins. *Trends in Food Science & Technology* 13 (8): 262–74.

de Wit, J. N. 1998. Nutritional and functional characteristics of whey proteins in food products. *Journal of Dairy Science* 81 (3): 597–608.

Dickinson, E. 1997. Enzymic crosslinking as a tool for food colloid rheology control and interfacial stabilization. *Trends in Food Science & Technology* 8 (10): 334–39.

Dickinson, E., and Chen, J. S. 1999. Heat-set whey protein emulsion gels: Role of active and inactive filler particles. *Journal of Dispersion Science and Technology* 20 (1–2): 197–213.

Dickinson, E., Hong, S. T., and Yamamoto, Y. 1996. Rheology of heat-set emulsion gels containing beta-lactoglobulin and small-molecule surfactants. *Netherlands Milk and Dairy Journal* 50 (2): 199–207.

Doi, E. 1993. Gels and gelling of globular proteins. *Trends in Food Science & Technology* 4:1–5.

Donnelly, J. L., Decker, E. A., and McClements, D. J. 1998. Iron-catalyzed oxidation of menhaden oil as affected by emulsifiers. *Journal of Food Science* 63 (6): 997–1000.

Doublier, J. L., Garnier, C., Renard, D., and Sanchez, C. 2000. Protein-polysaccharide interactions. *Current Opinion in Colloid & Interface Science* 5 (3–4): 202–14.

Doucet, D., and Foegeding, E. A. 2005. Gel formation of peptides produced by extensive enzymatic hydrolysis of beta-lactoglobulin. *Biomacromolecules* 6 (2): 1140–48.

Doucet, D., Gauthier, S. F., Otter, D. E., and Foegeding, E. A. 2003a. Enzyme-induced gelation of extensively hydrolyzed whey proteins by alcalase: Comparison with the plastein reaction and characterization of interactions. *Journal of Agricultural and Food Chemistry* 51 (20): 6036–42.

Doucet, D., Otter, D. E., Gauthier, S. F., and Foegeding, E. A. 2003b. Enzyme-induced gelation of extensively hydrolyzed whey proteins by alcalase: Peptide identification and determination of enzyme specificity. *Journal of Agricultural and Food Chemistry* 51 (21): 6300–308.

Durand, D., Gimel, J. C., and Nicolai, T. 2002. Aggregation, gelation and phase separation of heat denatured globular proteins. *Physica A - Statistical Mechanics and Its Applications* 304 (1–2): 253–65.

Eissa, A. S., Bisram, S., and Khan, S. A. 2004. Polymerization and gelation of whey protein isolates at low pH using transglutaminase enzyme. *Journal of Agricultural and Food Chemistry* 52 (14): 4456–64.

Eissa, A. S., and Khan, S. A. 2005. Acid-induced gelation of enzymatically modified, preheated whey proteins. *Journal of Agricultural and Food Chemistry* 53 (12): 5010–17.

———. 2006. Modulation of hydrophobic interactions in denatured whey proteins by transglutaminase enzyme. *Food Hydrocolloids* 20 (4): 543–47.

Eissa, A. S., Puhl, C., Kadla, J. F., and Khan, S. A. 2006. Enzymatic cross-linking of beta-lactoglobulin: Conformational properties using ftir spectroscopy. *Biomacromolecules* 7 (6): 1707–13.

Elofsson, C., Dejmek, P., Paulsson, M., and Burling, H. 1997. Characterization of a cold-gelling whey protein concentrate. *International Dairy Journal* 7 (8–9): 601–608.

Errington, A. D., and Foegeding, E. A. 1998. Factors determining fracture stress and strain of fine-stranded whey protein gels. *Journal of Agricultural and Food Chemistry* 46 (8): 2963–67.

Ewbank, J. J., and Creighton, T. E. 1991. The molten globule protein conformation probed by disulfide bonds. *Nature* 350 (6318): 518–20.

Faergemand, M., Otte, J., and Qvist, K. B. 1997. Enzymatic cross-linking of whey proteins by a Ca^{2+}-independent microbial transglutaminase from *Streptomyces lydicus*. *Food Hydrocolloids* 11 (1): 19–25.

Faergemand, M., and Qvist, K. B. 1997. Transglutaminase: Effect on rheological properties, microstructure and permeability of set style acid skim milk gel. *Food Hydrocolloids* 11 (3): 287–92.

Fessas, D., Iametti, S., Schiraldi, A., and Bonomi, F. 2001. Thermal unfolding of monomeric and dimeric beta-lactoglobulins. *European Journal of Biochemistry* 268 (20): 5439–48.

Firebaugh, J. D., and Daubert, C. R. 2005. Emulsifying and foaming properties of a derivatized whey protein ingredient. *International Journal of Food Properties* 8 (2): 243–53.

Foegeding, E. A., Davis, J. P., Doucet, D., and McGuffey, M. K. 2002. Advances in modifying and understanding whey protein functionality. *Trends in Food Science & Technology* 13 (5): 151–59.

Galani, D., and Apenten, R. K. O. 1999. Heat-induced denaturation and aggregation of beta-lactoglobulin: Kinetics of formation of hydrophobic and disulphide-linked aggregates. *International Journal of Food Science and Technology* 34 (5–6): 467–76.

Gault, P., and Fauquant, J. 1992. Heat-induced gelation of beta-lactoglobulin: Influence of pH, ionic-strength and presence of other whey proteins. *Lait* 72 (6): 491–510.

Gezimati, J., Singh, H., and Creamer, L. K. 1996. Heat-induced interactions and gelation of mixtures of bovine beta-lactoglobulin and serum albumin. *Journal of Agricultural and Food Chemistry* 44 (3): 804–10.

Glibowski, P., Mleko, S., Wasko, A., and Kristinsson, H. G. 2006. Effect of two-stage heating on Na^{+}-induced gelation of whey protein isolate. *Milchwissenschaft - Milk Science International* 61 (3): 252–55.

Gosal, W. S., Clark, A. H., and Ross-Murphy, S. B. 2004a. Fibrillar beta-lactoglobulin gels: Part 1. Fibril formation and structure. *Biomacromolecules* 5 (6): 2408–19.

———. 2004b. Fibrillar beta-lactoglobulin gels: Part 2. Dynamic mechanical characterization of heat-set systems. *Biomacromolecules* 5 (6): 2420–29.

Guerin, D., Vuillemard, J. C., and Subirade, M. 2003. Protection of bifidobacteria encapsulated in polysaccharide-protein gel beads against gastric juice and bile. *Journal of Food Protection* 66 (11): 2076–84.

Havea, P., Singh, H., and Creamer, L. K. 2001. Characterization of heat-induced aggregates of beta-lactoglobulin, alpha-lactalbumin and bovine serum albumin in a whey protein concentrate environment. *Journal of Dairy Research* 68 (3): 483–97.

Havea, P., Singh, H., Creamer, L. K., and Campanella, O. H. 1998. Electrophoretic characterization of the protein products formed during heat treatment of whey protein concentrate solutions. *Journal of Dairy Research* 65 (1): 79–91.

Hebrard, G., Blanquet, S., Beyssac, E., Remondetto, G., Subirade, M., and Alric, M. 2006. Use of whey protein beads as a new carrier system for recombinant yeasts in human digestive tract. *Journal of Biotechnology* 127 (1): 151–60.

Heelan, B. A., and Corrigan, O. I. 1998. Preparation and evaluation of microspheres prepared from whey protein isolate. *Journal of Microencapsulation* 15 (1): 93–105.

Heinzelmann, K., and Franke, K. 1999. Using freezing and drying techniques of emulsions for the microencapsulation of fish oil to improve oxidation stability. *Colloids and Surfaces B: Biointerfaces* 12:223–29.

Hines, M. E., and Foegeding, E. A. 1993. Interactions of alpha-lactalbumin and bovine serum-albumin with beta-lactoglobulin in thermally induced gelation. *Journal of Agricultural and Food Chemistry* 41 (3): 341–46.

Hoffmann, M. A. M., Sala, G., Olieman, C., and de Kruif, K. G. 1997. Molecular mass distributions of heat-induced beta-lactoglobulin aggregates. *Journal of Agricultural and Food Chemistry* 45 (8): 2949–57.

Hoffmann, M. A. M., and van Mil, P. 1999. Heat-induced aggregation of beta-lactoglobulin as a function of pH. *Journal of Agricultural and Food Chemistry* 47 (5): 1898–905.

Hoffmann, M. A. M., and van Mil, P. J. J. M. 1997. Heat-induced aggregation of beta-lactoglobulin: Role of the free thiol group and disulfide bonds. *Journal of Agricultural and Food Chemistry* 45 (8): 2942–48.

Holst, H. H., Albertsen, K., Clausen, P. M., Thomsen, B., and Hartmann, U. 1993. Partly denatured whey protein product. PCT/EP93/1093.

Hongsprabhas, P., and Barbut, S. 1996. Ca^{2+}-induced gelation of whey protein isolate: Effects of pre-heating. *Food Research International* 29 (2): 135–39.

———. 1997a. Ca^{2+}-induced cold gelation of whey protein isolate: Effect of two-stage gelation. *Food Research International* 30 (7): 523–27.

———. 1997b. Effect of gelation temperature on Ca^{2+}-induced gelation of whey protein isolate. *Food Science and Technology - Lebensmittel-Wissenschaft & Technologie* 30 (1): 45–49.

———. 1997c. Protein and salt effects on Ca^{2+}-induced cold gelation of whey protein isolate. *Journal of Food Science* 62 (2): 382–85.

———. 1997d. Structure-forming processes in Ca^{2+}-induced whey protein isolate cold gelation. *International Dairy Journal* 7 (12): 827–34.

———. 1999a. Effect of pre-heated whey protein level and salt on texture development of poultry meat batters. *Food Research International* 32 (2): 145–49.

———. 1999b. Use of cold-set whey protein gelation to improve poultry meat batters. *Poultry Science* 78 (7): 1074–78.

Hongsprabhas, P., Barbut, S., and Marangoni, A. G. 1999. The structure of cold-set whey protein isolate gels prepared with Ca^{++}. *Food Science and Technology - Lebensmittel-Wissenschaft & Technologie* 32 (4): 196–202.

Hudson, H. M., Daubert, C. R., and Foegeding, E. A. 2000. Rheological and physical properties of derivitized whey protein isolate powders. *Journal of Agricultural and Food Chemistry* 48 (8): 3112–19.

———. 2001. Thermal and pH stable protein thickening agent and method of making the same. US Patent, 6,261,624.

Huffman, L. M. 1996. Processing whey protein for use as a food ingredient. *Food Technology* 50 (2): 49–52.

Hunt, J. A., and Dalgleish, D. G. 1994. Effect of pH on the stability and surface-composition of emulsions made with whey-protein isolate. *Journal of Agricultural and Food Chemistry* 42 (10): 2131–35.

Iametti, S., DeGregori, B., Vecchio, G., and Bonomi, F. 1996. Modifications occur at different structural levels during the heat denaturation of beta-lactoglobulin. *European Journal of Biochemistry* 237 (1): 106–12.

Ikeda, S. 2003. Heat-induced gelation of whey proteins observed by rheology, atomic force microscopy, and Raman scattering spectroscopy. *Food Hydrocolloids* 17 (4): 399–406.

Ikeda, S., Foegeding, E. A., and Hagiwara, T. 1999. Rheological study on the fractal nature of the protein gel structure. *Langmuir* 15 (25): 8584–89.

Ikeda, S., and Morris, V. J. 2002. Fine-stranded and particulate aggregates of heat-denatured whey proteins visualized by atomic force microscopy. *Biomacromolecules* 3 (2): 382–89.

Ipsen, R., Otte, J., Lomholt, S. B., and Qvist, K. B. 2000. Standardized reaction times used to describe the mechanism of enzyme-induced gelation in whey protein systems. *Journal of Dairy Research* 67 (3): 403–13.

Ipsen, R., Otte, J., and Qvist, K. B. 2001. Molecular self-assembly of partially hydrolysed alpha-lactalbumin resulting in strong gels with a novel microstructure. *Journal of Dairy Research* 68 (2): 277–86.

Jeyarajah, S., and Allen, J. C. 1994. Calcium-binding and salt-induced structural-changes of native and preheated beta-lactoglobulin. *Journal of Agricultural and Food Chemistry* 42 (1): 80–85.

Ju, Z. Y., Hettiarachchy, N., and Kilara, A. 1999. Thermal properties of whey protein aggregates. *Journal of Dairy Science* 82 (9): 1882–89.

Ju, Z. Y., and Kilara, A. 1998a. Aggregation induced by calcium chloride and subsequent thermal gelation of whey protein isolate. *Journal of Dairy Science* 81 (4): 925–31.

———. 1998b. Effects of preheating on properties of aggregates and of cold-set gels of whey protein isolate. *Journal of Agricultural and Food Chemistry* 46 (9): 3604–608.

———. 1998c. Gelation of hydrolysates of a whey protein isolate induced by heat, protease, salts and acid. *International Dairy Journal* 8 (4): 303–309.

———. 1998d. Gelation of ph-aggregated whey protein isolate solution induced by heat, protease, calcium salt, and acidulant. *Journal of Agricultural and Food Chemistry* 46 (5): 1830–35.

———. 1998e. Properties of gels induced by heat, protease, calcium salt, and acidulant from calcium ion-aggregated whey protein isolate. *Journal of Dairy Science* 81 (5): 1236–43.

———. 1998f. Textural properties of cold-set gels induced from heat-denatured whey protein isolates. *Journal of Food Science* 63 (2): 288–92.

Ju, Z. Y., Otte, J., Madsen, J. S., and Qvist, K. B. 1995. Effects of limited proteolysis on gelation and gel properties of whey-protein isolate. *Journal of Dairy Science* 78 (10): 2119–28.

Ju, Z. Y., Otte, J., Zakora, M., and Qvist, K. B. 1997. Enzyme-induced gelation of whey proteins: Effect of protein denaturation. *International Dairy Journal* 7 (1): 71–78.

Kavanagh, G. M., Clark, A. H., Gosal, W. S., and Ross-Murphy, S. B. 2000a. Heat-induced gelation of beta-lactoglobulin/alpha-lactalbumin blends at pH 3 and pH 7. *Macromolecules* 33 (19): 7029–37.

Kavanagh, G. M., Clark, A. H., and Ross-Murphy, S. B. 2000b. Heat-induced gelation of globular proteins: 4. Gelation kinetics of low pH beta-lactoglobulin gels. *Langmuir* 16 (24): 9584–94.

Kawachi, K., Takeachi, M., and Nishiya, T. 1993. Solution containing whey protein, whey protein gel, whey protein powder and processed food product produced by using the same. US Patent, 5,217, 741.

Kawamura, F., Mayuzumi, A., Nakamura, M., Koizumi, S., Kimura, T., and Nishiya, T. 1993. Preparation and properties of acid-induced gel of whey-protein. *Journal of the Japanese Society for Food Science and Technology - Nippon Shokuhin Kagaku Kogaku Kaishi* 40 (11): 776–82.

Keim, S., and Hinrichs, J. 2004. Influence of stabilizing bonds on the texture properties of high-pressure-induced whey protein gels. *International Dairy Journal* 14 (4): 355–63.

Khalil, A. H., and Mansour, E. H. 1998. Alginate encapsulated bifidobacteria survival in mayonnaise. *Journal of Food Science* 63 (4): 702–705.

Kinsella, J. E., and Whitehead, D. M. 1989. Proteins in whey: Chemical, physical, and functional properties. *Advances in Food and Nutrition Research* 33:343–438.

Kitabatake, N., Wada, R., and Fujita, Y. 2001. Reversible conformational change in beta-lactoglobulin a modified with n-ethylmaleimide and resistance to molecular aggregation on heating. *Journal of Agricultural and Food Chemistry* 49 (8): 4011–18.

Klinkesorn, U., Sophanodora, P., Chinachoti, P., Decker, E. A., and McClements, D. J. 2005. Encapsulation of emulsified tuna oil in two-layered interfacial membranes prepared using electrostatic layer-by-layer deposition. *Food Hydrocolloids* 19 (6): 1044–53.

Kos, B., Suskovic, J., Goreta, J., and Matosic, S. 2000. Effect of protectors on the viability of lactobacillus acidophilus M92 in simulated gastrointestinal conditions. *Food Technology and Biotechnology* 38:121–27.

Koseki, T., Fukuda, T., Kitabatake, N., and Doi, E. 1989a. Characterization of linear polymers induced by thermal denaturation of ovalbumin. *Food Hydrocolloids* 3 (2): 135–48.

Koseki, T., Kitabatake, N., and Doi, E. 1989b. Irreversible thermal denaturation and formation of linear aggregates of ovalbumin. *Food Hydrocolloids* 3 (2): 123–34.

Krebs, M. R. H., Devlin, G. L., and Donald, A. M. 2007. Protein particulates: Another generic form of protein aggregation? *Biophysical Journal* 92 (4): 1336–42.

Kuhn, P. R., and Foegeding, E. A. 1991. Mineral salt effects on whey-protein gelation. *Journal of Agricultural and Food Chemistry* 39 (6): 1013–16.

Kuipers, B. J. H., Alting, A. C., and Gruppen, H. 2007. Comparison of the aggregation behavior of soy and bovine whey protein hydrolysates. *Biotechnology Advances* 25 (6): 606–10.

Kuipers, B. J. H., and Gruppen, H. 2008. Identification of strong aggregating regions in soy glycinin upon enzymatic hydrolysis. *Journal of Agricultural and Food Chemistry* 56 (10): 3818–27.

Kuipers, B. J. H., van Koningsveld, G. A., Alting, A. C., Driehuis, F., Gruppen, H., and Voragen, A. G. J. 2005. Enzymatic hydrolysis as a means of expanding the cold gelation conditions of soy proteins. *Journal of Agricultural and Food Chemistry* 53 (4): 1031–38.

Kulmyrzaev, A., Bryant, C., and McClements, D. J. 2000a. Influence of sucrose on the thermal denaturation, gelation, and emulsion stabilization of whey proteins. *Journal of Agricultural and Food Chemistry* 48 (5): 1593–97.

Kulmyrzaev, A., Cancelliere, C., and McClements, D. J. 2000b. Influence of sucrose on cold gelation of heat-denatured whey protein isolate. *Journal of the Science of Food and Agriculture* 80 (9): 1314–18.

Langton, M., and Hermansson, A. M. 1992. Fine-stranded and particulate gels of beta-lactoglobulin and whey-protein at varying pH. *Food Hydrocolloids* 5 (6): 523–39.

Langton, M., and Hermansson, A. M. 1996. Image analysis of particulate whey protein gels. *Food Hydrocolloids* 10 (2): 179–91.

Law, A. J. R., and Leaver, J. 2000. Effect of pH on the thermal denaturation of whey proteins in milk. *Journal of Agricultural and Food Chemistry* 48 (3): 672–79.

Le Bon, C., Durand, D., and Nicolai, T. 2002. Influence of genetic variation on the aggregation of heat-denatured beta-lactoglobulin. *International Dairy Journal* 12 (8): 671–78.

Lee, S. J., and Rosenberg, M. 1999. Preparation and properties of glutaraldehyde cross-linked whey protein-based microcapsules containing theophylline. *Journal of Controlled Release* 61 (1–2): 123–36.

———. 2000a. Preparation and some properties of water-insoluble, whey protein-based microcapsules. *Journal of Microencapsulation* 17:29–44.

———. 2000b. Whey protein-based microcapsules prepared by double emulsification and heat gelation. *Lebensmittel-Wissenschaft Und-Technologie - Food Science and Technology* 33 (2): 80–88.

Lefevre, T., and Subirade, M. 2000. Molecular differences in the formation and structure of fine-stranded and particulate beta-lactoglobulin gels. *Biopolymers* 54 (7): 578–86.

———. 2003. Formation of intermolecular beta-sheet structures: A phenomenon relevant to protein film structure at oil-water interfaces of emulsions. *Journal of Colloid and Interface Science* 263 (1): 59–67.

Li, H., Hardin, C. C., and Foegeding, E. A. 1994. NMR-studies of thermal-denaturation and cation-mediated aggregation of beta-lactoglobulin. *Journal of Agricultural and Food Chemistry* 42 (11): 2411–20.

Li, J., and Guo, M. 2006. Effects of polymerized whey proteins on consistency and water-holding properties of goat's milk yogurt. *Journal of Food Science* 71 (1): C34–38.

Line, V. L. S., Remondetto, G. E., and Subirade, M. 2005. Cold gelation of beta-lactoglobulin oil-in-water emulsions. *Food Hydrocolloids* 19 (2): 269–78.

Liu, Z. S., and Chang, S. K. C. 2004. Effect of soy milk characteristics and cooking conditions on coagulant requirements for making filled tofu. *Journal of Agricultural and Food Chemistry* 52 (11): 3405–11.

Lopez-Fandino, R. 2006a. Functional improvement of milk whey proteins induced by high hydrostatic pressure. *Critical Reviews in Food Science and Nutrition* 46 (4): 351–63.

———. 2006b. High pressure-induced changes in milk proteins and possible applications in dairy technology. *International Dairy Journal* 16 (10): 1119–31.

Mahmoudi, N., Mehalebi, S., Nicolai, T., Durand, D., and Riaublanc, A. 2007. Light-scattering study of the structure of aggregates and gels formed by heat-denatured whey protein isolate and b-lactoglobulin at neutral pH. *Journal of Agricultural and Food Chemistry* 55 (8): 3104–11.

Majhi, P. R., Ganta, R. R., Vanam, R. P., Seyrek, E., Giger, K., and Dubin, P. L. 2006. Electrostatically driven protein aggregation: Beta-lactoglobulin at low ionic strength. *Langmuir* 22 (22): 9150–59.

Maltais, A., Remondetto, G. E., Gonzalez, R., and Subirade, M. 2005. Formation of soy protein isolate cold-set gels: Protein and salt effects. *Journal of Food Science* 70 (1): C67–73.

Maltais, A., Remondetto, G. E., and Subirade, M. 2008. Mechanisms involved in the formation and structure of soya protein cold-set gels: A molecular and supramolecular investigation. *Food Hydrocolloids* 22 (4): 550–59.

Marangoni, A. G., Barbut, S., McGauley, S. E., Marcone, M., and Narine, S. S. 2000. On the structure of particulate gels: The case of salt-induced cold gelation of heat-denatured whey protein isolate. *Food Hydrocolloids* 14 (1): 61–74.

Marteau, P., Minekus, M., Havenaar, R., and Huis in't Veld, J. H. 1997. Survival of lactic acid bacteria in a dynamic model of the stomach and small intestine: Validation and the effects of bile. *Journal of Dairy Science* 80 (6): 1031–37.

Matsudomi, N., Oshita, T., Sasaki, E., and Kobayashi, K. 1992. Enhanced heat-induced gelation of beta-lactoglobulin by alpha-lactalbumin. *Bioscience Biotechnology and Biochemistry* 56 (11): 1697–700.

Matsuura, J. E., and Manning, M. C. 1994. Heat-induced gel formation of beta-lactoglobulin: A study on the secondary and tertiary structure as followed by circular-dichroism spectroscopy. *Journal of Agricultural and Food Chemistry* 42 (8): 1650–56.

McClements, D. J. 1999. *Food Emulsions: Principles, Practice, and Techniques*. Boca Raton, FL: CRC Press.

McClements, D. J., and Decker, A. E. 2000. Lipid oxidation in oil-in-water emulsions: Impact of molecular environment on chemical reactions in heterogeneous food systems. *Journal of Food Science* 65 (8): 1270–82.

McClements, D. J., Decker, E. A., and Weiss, J. 2007. Emulsion-based delivery systems for lipophilic bioactive components. *Journal of Food Science* 72 (8): R109–24.

McClements, D. J., and Keogh, M. K. 1995. Physical-properties of cold-setting gels formed from heat-denatured whey-protein isolate. *Journal of the Science of Food and Agriculture* 69 (1): 7–14.

Meakin, P. 1987. Fractal aggregates. *Advances in Colloid and Interface Science* 28:249–331.

Mehalebi, S., Nicolai, T., and Durand, D. 2008a. The influence of electrostatic interaction on the structure and the shear modulus of heat-set globular protein gels. *Soft Matter* 4 (4): 893–900.

———. 2008b. Light scattering study of heat-denatured globular protein aggregates. *International Journal of Biological Macromolecules* 43 (2): 129–35.

Mellema, M., van Opheusden, J. H. J., and van Vliet, T. 2002. Categorization of rheological scaling models for particle gels applied to casein gels. *Journal of Rheology* 46 (1): 11–29.

Mercadé-Prieto, R., Falconer, R. J., Paterson, W. R., and Wilson, D. I. 2006. Effect of gel structure on the dissolution of heat-induced beta-lactoglobulin gels in alkali. *Journal of Agricultural and Food Chemistry* 54 (15): 5437–44.

Messens, W., VanCamp, J., and Huyghebaert, A. 1997. The use of high pressure to modify the functionality of food proteins. *Trends in Food Science & Technology* 8 (4): 107–12.

Mleko, S. 1999. Effect of protein concentration on whey protein gels obtained by a two-stage heating process. *European Food Research and Technology* 209 (6): 389–92.

———. 2000. Studies on permeability and rheology of heat and sodium ions-induced whey protein gels. *Journal of Food Science and Technology - Mysore* 37 (3): 307–10.

Mleko, S., and Foegeding, E. A. 1999. Formation of whey protein polymers: Effects of a two-step heating process on rheological properties. *Journal of Texture Studies* 30 (2): 137–49.

———. 2000. pH induced aggregation and weak gel formation of whey protein polymers. *Journal of Food Science* 65 (1): 139–43.

Mleko, S., Glibowski, P., Gustaw, W., and Janas, P. 2002. Calcium ions induced gelation of double-heated whey protein isolate. *Journal of Food Science and Technology - Mysore* 39 (5): 563–65.

Monahan, F. J., McClements, D. J., and Kinsella, J. E. 1993. Polymerization of whey proteins in whey protein-stabilized emulsions. *Journal of Agricultural and Food Chemistry* 41 (11): 1826–29.

Moreau, L., Kim, H. J., Decker, E. A., and McClements, D. J. 2003. Production and characterization of oil-in-water emulsions containing droplets stabilized by beta-lactoglobulin-pectin membranes. *Journal of Agricultural and Food Chemistry* 51 (22): 6612–17.

Morr, C. V., and Ha, E. Y. W. 1993. Whey-protein concentrates and isolates - processing and functional-properties. *Critical Reviews in Food Science and Nutrition* 33 (6): 431–76.

Mulvihill, D. M., and Kinsella, J. E. 1987. Gelation characteristics of whey proteins and beta-lactoglobulin. *Food Technology* 41 (9): 103–11.

———. 1988. Gelation of beta-lactoglobulin: Effects of sodium-chloride and calcium-chloride on the rheological and structural-properties of gels. *Journal of Food Science* 53 (1): 231–36.

Nakamura, M., Sato, K., Koizumi, S., Kawachi, K., Nishiya, T., and Nakajima, I. 1995. Preparation and properties of salt-induced gel of whey-protein. *Journal of the Japanese Society for Food Science and Technology - Nippon Shokuhin Kagaku Kogaku Kaishi* 42 (1): 1–6.

Navarra, G., Leone, M., and Militello, V. 2007. Thermal aggregation of [beta]-lactoglobulin in presence of metal ions. *Biophysical Chemistry* 131 (1–3): 52–61.

Navea, S., de Juan, A., and Tauler, R. 2003. Modeling temperature-dependent protein structural transitions by combined near-IR and mid-IR spectroscopies and multivariate curve resolution. *Analytical Chemistry* 75 (20): 5592–601.

Nishinari, K., Zhang, H., and Ikeda, S. 2000. Hydrocolloid gels of polysaccharides and proteins. *Current Opinion in Colloid & Interface Science* 5 (3–4): 195–201.

Olsen, K., Ipsen, R., Otte, J., and Skibsted, L. H. 1999. Effect of high pressure on aggregation and thermal gelation of beta-lactoglobulin. *Milchwissenschaft - Milk Science International* 54 (10): 543–46.

Osborn, H. T., and Akoh, C. C. 2004. Effect of emulsifier type, droplet size, and oil concentration on lipid oxidation in structured lipid-based oil-in-water emulsions. *Food Chemistry* 84 (3): 451–56.

Otte, J., Ju, Z. Y., Faergemand, M., Lomholt, S. B., and Qvist, K. B. 1996. Protease-induced aggregation and gelation of whey proteins. *Journal of Food Science* 61 (5): 911–15.

Otte, J., Zakora, M., and Qvist, K. B. 2000. Involvement of disulfide bands in bovine beta-lactoglobulin b gels set thermally at various pH. *Journal of Food Science* 65 (3): 384–89.

Owusu-Apenten, R., and Chee, C. 2004. Sulfhydryl group activation for commercial beta-lactoglobulin measured using kappa-casein 2-thio, 5′ nitrobenzoic acid. *International Dairy Journal* 14 (3): 195–200.

Panyam, D., and Kilara, A. 1996. Enhancing the functionality of food proteins by enzymatic modification. *Trends in Food Science & Technology* 7 (4): 120–25.

Park, E. Y., Murakami, H., Mori, T., and Matsumura, Y. 2005. Effects of protein and peptide addition on lipid oxidation in powder model system. *Journal of Agricultural and Food Chemistry* 53 (1): 137–44.

Potter, N. N., and Hotchkiss, J. J. 1995. *Food Science*. New York: Chapman & Hall.

Pouzot, M., Nicolai, T., Durand, D., and Benyahia, L. 2004. Structure factor and elasticity of a heat-set globular protein gel. *Macromolecules* 37 (2): 614–20.

Pouzot, M., Nicolai, T., Visschers, R. W., and Weijers, M. 2005. X-ray and light scattering study of the structure of large protein aggregates at neutral pH. *Food Hydrocolloids* 19 (2): 231–38.

Qi, X. L., Holt, C., McNulty, D., Clarke, D. T., Brownlow, S., and Jones, G. R. 1997. Effect of temperature on the secondary structure of beta-lactoglobulin at pH 6.7, as determined by CD and IR spectroscopy: A test of the molten globule hypothesis. *Biochemical Journal* 324:341–46.

Regester, G. O., McIntosh, G. H., Lee, V. W. K., and Smithers, G. W. 1996. Whey proteins as nutritional and functional food ingredients. *Food Australia* 48 (3): 123–27.

Reid, A. A., Champagne, C. P., Gardner, N., Fustier, P., and Vuillemard, J. C. 2007. Survival in food systems of *Lactobacillus rhamnosus* R011 microentrapped in whey protein gel particles. *Journal of Food Science* 72 (1): M31–37.

Reid, A. A., Vuillemard, J. C., Britten, M., Arcand, Y., Farnworth, E., and Champagne, C. P. 2005. Microentrapment of probiotic bacteria in a Ca^{2+}-induced whey protein gel and effects on their viability in a dynamic gastro-intestinal model. *Journal of Microencapsulation* 22 (6): 603–19.

Remondetto, G. E., Beyssac, E., and Subirade, M. 2004. Iron availability from whey protein hydrogels: An in vitro study. *Journal of Agricultural and Food Chemistry* 52 (26): 8137–43.

Remondetto, G. E., Paquin, P., and Subirade, M. 2002. Cold gelation of beta-lactoglobulin in the presence of iron. *Journal of Food Science* 67 (2): 586–95.

Remondetto, G. E., and Subirade, M. 2003. Molecular mechanisms of Fe^{2+}-induced beta-lactoglobulin cold gelation. *Biopolymers* 69 (4): 461–69.

Renkema, J. M. S., Knabben, J. H. M., and van Vliet, T. 2001. Gel formation by [beta]-conglycinin and glycinin and their mixtures. *Food Hydrocolloids* 15 (4–6): 407–14.

Renkema, J. M. S., Lakemond, C. M. M., de Jongh, H. H. J., Gruppen, H., and van Vliet, T. 2000. The effect of pH on heat denaturation and gel forming properties of soy proteins. *Journal of Biotechnology* 79 (3): 223–30.

Renkema, J. M. S., and van Vliet, T. 2002. Heat-induced gel formation by soy proteins at neutral pH. *Journal of Agricultural and Food Chemistry* 50 (6): 1569–73.

______. 2004. Concentration dependence of dynamic moduli of heat-induced soy protein gels. *Food Hydrocolloids* 18 (3): 483–87.

Resch, J. J., and Daubert, C. R. 2002. Rheological and physicochemical properties of derivatized whey protein concentrate powders. *International Journal of Food Properties* 5 (2): 419–34.

Resch, J. J., Daubert, C. R., and Foegeding, E. A. 2004. A comparison of drying operations on the rheological properties of whey protein thickening ingredients. *International Journal of Food Science and Technology* 39 (10): 1023–31.

______. 2005a. Beta-lactoglobulin gelation and modification: Effect of selected acidulants and heating conditions. *Journal of Food Science* 70 (1): C79–86.

______. 2005b. The effects of acidulant type on the rheological properties of beta-lactoglobulin gels and powders derived from these gels. *Food Hydrocolloids* 19 (5): 851–60.

Ring, S., and Stainsby, G. 1982. Filler reinforcement of gels. *Progress in Food Nutrition and Science* 6:323–29.

Roefs, S., and De Kruif, K. G. 1994. A model for the denaturation and aggregation of beta-lactoglobulin. *European Journal of Biochemistry* 226 (3): 883–89.

Roff, C. F., and Foegeding, E. A. 1996. Dicationic-induced gelation of pre-denatured whey protein isolate. *Food Hydrocolloids* 10 (2): 193–98.

Rosa, P., Sala, G., Van Vliet, T., and Van De Velde, F. 2006. Cold gelation of whey protein emulsions. *Journal of Texture Studies* 37 (5): 516–37.

Rosenberg, M. 1997. Milk derived whey protein-based microencapsulating agents and a method of use. US Patent, 5,601,760.

Rosenberg, M., and Lee, S. J. 2004a. Calcium-alginate coated, whey protein-based microspheres: Preparation, some properties and opportunities. *Journal of Microencapsulation* 21 (3): 263–81.

———. 2004b. Water-insoluble, whey protein-based microspheres prepared by an all-aqueous process. *Journal of Food Science* 69 (1): E50–58.

Ross-Murphy, S. B. 2007. Biopolymer gelation: Exponents and critical exponents. *Polymer Bulletin* 58 (1): 119–26.

Saeseaw, S., Shiowatana, J., and Siripinyanond, A. 2006. Observation of salt-induced beta-lactoglobulin aggregation using sedimentation field-flow fractionation. *Analytical and Bioanalytical Chemistry* 386 (6): 1681–88.

Sala, G., Van Aken, G. A., Stuart, M. A. C., and Van De Velde, F. 2007. Effect of droplet-matrix interactions on large deformation properties of emulsion-filled gels. *Journal of Texture Studies* 38 (4): 511–35.

Santipanichwong, R., Suphantharika, M., Weiss, J., and McClements, D. J. 2008. Core-shell biopolymer nanoparticles produced by electrostatic deposition of beet pectin onto heat-denatured beta-lactoglobulin aggregates. *Journal of Food Science* 73 (6): N23–30.

Sato, K., Nakamura, M., Koizumi, S., Kawachi, K., Nishiya, T., and Nakajima, I. 1995. Changes in hydrophobicity and sh content on salt-induced gelation of whey-protein. *Journal of the Japanese Society for Food Science and Technology - Nippon Shokuhin Kagaku Kogaku Kaishi* 42 (1): 7–13.

Sava, N., Van der Plancken, I., Claeys, W., and Hendrickx, M. 2005. The kinetics of heat-induced structural changes of beta-lactoglobulin. *Journal of Dairy Science* 88 (5): 1646–53.

Schmidt, R. H., Illingworth, B. L., Ahmed, E. M., and Richter, R. L. 1978. The effect of dialysis on heat induced gelation of whey protein concentrate. *Journal of Food Processing and Preservation* 2 (2): 111–22.

Schmidt, R. H., Illingworth, B. L., Deng, J. C., and Cornell, J. A. 1979. Multiple-regression and response surface-analysis of the effects of calcium-chloride and cysteine on heat-induced whey-protein gelation. *Journal of Agricultural and Food Chemistry* 27 (3): 529–32.

Schokker, E. P., Singh, H., and Creamer, L. K. 2000. Heat-induced aggregation of beta-lactoglobulin a and b with alpha-lactalbumin. *International Dairy Journal* 10 (12): 843–53.

Schokker, E. P., Singh, H., Pinder, D. N., Norris, G. E., and Creamer, L. K. 1999. Characterization of intermediates formed during heat-induced aggregation of beta-lactoglobulin ab at neutral pH. *International Dairy Journal* 9 (11): 791–800.

Sherwin, C. P., and Foegeding, E. A. 1997. The effects of $CaCl_2$ on aggregation of whey proteins. *Milchwissenschaft - Milk Science International* 52 (2): 93–96.

Sheu, T. Y., Marshall, R. T., and Heymann, H. 1993. Improving survival of culture bacteria in frozen desserts by microentrapment. *Journal of Dairy Science* 76 (7): 1902–907.

Shih, W.-H., Shih, W. Y., Kim, S.-I., Liu, J., and Aksay, I. A. 1990. Scaling behavior of the elastic properties of colloidal gels. *Physical Review A* 42 (8): 4772.

Shimada, K., and Cheftel, J. C. 1988. Texture characteristics, protein solubility, and sulfhydryl-group disulfide bond contents of heat-induced gels of whey-protein isolate. *Journal of Agricultural and Food Chemistry* 36 (5): 1018–25.

Simons, J., Kosters, H. A., Visschers, R. W., and de Jongh, H. H. J. 2002. Role of calcium as trigger in thermal beta-lactoglobulin aggregation. *Archives of Biochemistry and Biophysics* 406 (2): 143–52.

Siuta-Cruce, P., and Goulet, J. 2001. Improving probiotic survival rates. *Food Technology* 55 (10): 36–42.

Smith, J. C. 1974. Correction and extension of vanderpoels method for calculating shear modulus of a particulate composite. *Journal of Research of the National Bureau of Standards Section A – Physics and Chemistry* 78 (3): 355–61.

———. 1975. Simplification of Van Der Poel's formula for shear modulus of a particulate composite. *Journal of Research of the National Bureau of Standards Section A – Physics and Chemistry* 79 (2): 419–23.

Soeda, T. 1994a. Effect of heating on the molecular-structure of soy protein studies on gelation of soy protein during cold-storage. 2. *Journal of the Japanese Society for Food Science and Technology - Nippon Shokuhin Kagaku Kogaku Kaishi* 41 (10): 676–81.

———. 1994b. Gelation of soy protein isolate during cold-storage studies on gelation of soy protein during cold-storage. 1. *Journal of the Japanese Society for Food Science and Technology - Nippon Shokuhin Kagaku Kogaku Kaishi* 41 (10): 670–75.

———. 2003. Effects of microbial transglutaminase for gelation of soy protein isolate during cold storage. *Food Science and Technology Research* 9 (2): 165–69.

Stading, M., Langton, M., and Hermansson, A. M. 1993. Microstructure and rheological behavior of particulate beta-lactoglobulin gels. *Food Hydrocolloids* 7 (3): 195–212.

Surroca, Y., Haverkamp, J., and Heck, A. J. R. 2002. Towards the understanding of molecular mechanisms in the early stages of heat-induced aggregation of beta-lactoglobulin ab. *Journal of Chromatography A* 970 (1–2): 275–85.

Tamaki, K., Nishiya, T., and Tatsumi, K. 1991. Process for producing textured protein food materials. US Patent, 5,011, 702.

Tang, C.-H., Li, L., Wang, J.-L., and Yang, X.-Q. 2007. Formation and rheological properties of 'cold-set' tofu induced by microbial transglutaminase. *Lebensmittel-Wissenschaft und Technologie* 40 (4): 579–86.

Tang, C.-H., Wu, H., Yu, H. P., Li, L., Chen, Z., and Yang, X. Q. 2006. Coagulation and gelation of soy protein isolates induced by microbial transglutaminase. *Journal of Food Biochemistry* 30 (1): 35–55.

Tobitani, A., and Ross-Murphy, S. B. 1997. Heat-induced gelation of globular proteins. 1. Model for the effects of time and temperature on the gelation time of BSA gels. *Macromolecules* 30 (17): 4845–54.

Tong, L. M., Sasaki, S., McClements, D. J., and Decker, E. A. 2000. Antioxidant activity of whey in a salmon oil emulsion. *Journal of Food Science* 65 (8): 1325–29.

Totosaus, A., Montejano, J. G., Salazar, J. A., and Guerrero, I. 2002. A review of physical and chemical protein-gel induction. *International Journal of Food Science and Technology* 37 (6): 589–601.

Utsumi, S., and Kinsella, J. E. 1985. Forces involved in soy protein gelation: Effects of various reagents on the formation, hardness and solubility of heat-induced gels made from 7S, 11S, and soy isolate. *Journal of Food Science* 50 (5): 1278–82.

van den Berg, L., Carolas, A. L., van Vliet, T., van der Linden, E., van Boekel, M. A. J. S., and van de Velde, F. 2008a. Energy storage controls crumbly perception in whey proteins/polysaccharide mixed gels. *Food Hydrocolloids* 22 (7): 1404–17.

van den Berg, L., van Vliet, T., van der Linden, E., van Boekel, M., and van de Velde, F. 2008b. Physical properties giving the sensory perception of whey proteins/polysaccharide gels. *Food Biophysics* 3 (2): 198–206.

———. 2007. Breakdown properties and sensory perception of whey proteins/polysaccharide mixed gels as a function of microstructure. *Food Hydrocolloids* 21 (5–6): 961–76.

van der Poel, C. 1958. On the rheology of concentrated dispersion. *Rheologica Acta* 1:198–205.

Vancamp, J., and Huyghebaert, A. 1995. High pressure-induced gel formation of a whey-protein and hemoglobin protein-concentrate. *Food Science and Technology - Lebensmittel-Wissenschaft & Technologie* 28 (1): 111–17.

VanCamp, J., Messens, W., Clement, J., and Huyghebaert, A. 1997a. Influence of pH and calcium chloride on the high-pressure-induced aggregation of a whey protein concentrate. *Journal of Agricultural and Food Chemistry* 45 (5): 1600–607.

———. 1997b. Influence of pH and sodium chloride on the high pressure-induced gel formation of a whey protein concentrate. *Food Chemistry* 60 (3): 417–24.

Vardhanabhuti, B., and Foegeding, E. A. 1999. Rheological properties and characterization of polymerized whey protein isolates. *Journal of Agricultural and Food Chemistry* 47 (9): 3649–55.

Vardhanabhuti, B., Foegeding, E. A., McGuffey, M. K., Daubert, C. R., and Swaisgood, H. E. 2001. Gelation properties of dispersions containing polymerized and native whey protein isolate. *Food Hydrocolloids* 15 (2): 165–75.

Veerman, C., Baptist, H., Sagis, L. M. C., and Van der Linden, E. 2003. A new multistep Ca^{2+}-induced cold gelation process for beta-lactoglobulin. *Journal of Agricultural and Food Chemistry* 51 (13): 3880–85.

Verheul, M., Roefs, S., and de Kruif, K. G. 1998a. Kinetics of heat-induced aggregation of beta-lactoglobulin. *Journal of Agricultural and Food Chemistry* 46 (3): 896–903.

Verheul, M., Roefs, S., Mellema, J., and de Kruif, K. G. 1998b. Power law behavior of structural properties of protein gels. *Langmuir* 14 (9): 2263–68.

Vreeker, R., Hoekstra, L. L., Denboer, D. C., and Agterof, W. G. M. 1992. Fractal aggregation of whey proteins. *Food Hydrocolloids* 6 (5): 423–35.

Watanabe, K., and Klostermeyer, H. 1976. Heat-induced changes in sulfhydryl and disulfide levels of beta-lactoglobulin-a and formation of polymers. *Journal of Dairy Research* 43 (3): 411–18.

Weijers, M., Barneveld, P. A., Stuart, M. A. C., and Visschers, R. W. 2003. Heat-induced denaturation and aggregation of ovalbumin at neutral pH described by irreversible first-order kinetics. *Protein Science* 12 (12): 2693–703.

Weijers, M., van de Velde, F., Stijnman, A., van de Pijpekamp, A., and Visschers, R. W. 2006. Structure and rheological properties of acid-induced egg white protein gels. *Food Hydrocolloids* 20 (2–3): 146–59.

Weijers, M., Visschers, R. W., and Nicolai, T. 2002. Light scattering study of heat-induced aggregation and gelation of ovalbumin. *Macromolecules* 35 (12): 4753–62.

Wilcox, C. P., Clare, D. A., Valentine, V. W., and Swaisgood, H. E. 2002. Immobilization and utilization of the recombinant fusion proteins trypsin-streptavidin and streptavidin-transglutaminase for modification of whey protein isolate functionality. *Journal of Agricultural and Food Chemistry* 50 (13): 3723–30.

Wu, H., and Morbidelli, M. 2001. A model relating structure of colloidal gels to their elastic properties. *Langmuir* 17 (4): 1030–36.

Wu, H., Xie, J. J., and Morbidelli, M. 2005. Kinetics of cold-set diffusion-limited aggregations of denatured whey protein isolate colloids. *Biomacromolecules* 6 (6): 3189–97.

Xiong, Y. L., and Kinsella, J. E. 1990. Mechanism of urea-induced whey-protein gelation. *Journal of Agricultural and Food Chemistry* 38 (10): 1887–91.

Xu, J., Shimoyamada, M., and Watanabe, K. 1997. Gelation of egg white proteins as affected by combined heating and freezing. *Journal of Food Science* 62 (5): 963–66.

Yamul, D. K., and Lupano, C. E. 2003. Properties of gels from whey protein concentrate and honey at different pHs. *Food Research International* 36 (1): 25–33.

Ye, A., and Singh, H. 2006. Heat stability of oil-in-water emulsions formed with intact or hydrolysed whey proteins: Influence of polysaccharides. *Food Hydrocolloids* 20 (2–3): 269–76.

Ye, A. Q. 2008. Complexation between milk proteins and polysaccharides via electrostatic interaction: Principles and applications – a review. *International Journal of Food Science and Technology* 43 (3): 406–15.

Zadow, J. G. 1992. *Whey and Lactose Processing*. London: Elsevier Applied Science.

Zhu, H., and Damodaran, S. 1994. Effects of calcium and magnesium ions on aggregation of whey protein isolate and its effect on foaming properties. *Journal of Agricultural and Food Chemistry* 42 (4): 856–62.

Zittle, C. A., and Custer, J. H. 1957. Electrophoresis of b-lactoglobulin in the presence of calcium chloride. *Archives of Biochemistry and Biophysics* 71:229–34.

Zittle, C. A., DellaMonica, E. S., Rudd, R. K., and Custer, J. H. 1957. Binding of calcium ions by β-lactoglobulin both before and after aggregation by heating in the presence of calcium ions. *Journal of the American Chemical Society* 79:4661–66.

11 Fundamental Considerations in the Comparison between Thermal and Nonthermal Characterization of Bioglasses

Stefan Kasapis
Royal Melbourne Institute of Technology

CONTENTS

High pressure technology is the outcome of demand for better quality control of biomaterials. It is hoped that the technology will demonstrate advantages over conventional thermal treatments in terms of processing time, end-product acceptability attributes, ice-crystal tissue damage, etc. The present treatise deals with the effect of high hydrostatic pressure on the rubber-to-glass transformation of biomaterials with potential industrial interest in confectionery formulations. Early studies in pressure-induced functionality were based on the classical free volume theory which in conjunction with the Ferry-Stratton equation was unable to rationalize the pressure dependence of mechanical properties at constant temperature. Incorporation of the correct pressure dependence of the compressibility coefficient led to the Fillers-Moonan-Tschoegl equation, which followed the combined thermorheologically and piezorheologically simple behaviour of amorphous synthetic polymers. Work in this laboratory demonstrated the utility of the sophisticated "synthetic polymer approach" in the time-temperature superposition of mechanical functions of biomaterials undergoing vitrification. Nevertheless, the time-temperature-pressure equivalence of synthetic materials is not operational in the glass-like behaviour of high sugar systems in the presence of gelatin or gelling polysaccharides. This deviation from the "normal" course is discussed in terms of the irreversible destabilization of intermolecular aggregates (mainly) in polysaccharide networks following pressure-induced vitrification.

11.1 INTRODUCTION

The effect of processing conditions, storage temperature, and aging on the vitrification properties of various biopolymers and their synthetic counterparts has been a matter of interest for decades considering the multitude of related industrial applications. Inevitably, these are linked with the concept of the glass transition temperature (T_g), at which the consistency of materials with a considerable amorphous component changes from that of a soft rubber to that of a hard glass (Slade and Franks 2002). Above this temperature, and for a variety of model systems (e.g., amorphous synthetic polymers or biopolymer/sugar mixtures), it was shown that temperature may have an equivalence in time (Zimeri and Kokini 2002). Thus controlled cooling of a biomaterial may yield a consistency which is also noted for the effect of short timescales of observation on relaxation processes, as monitored mechanically, electrically, volumetrically, etc.

Traditionally, reduction in free volume with cooling is considered by the "synthetic polymer approach" as the mechanism responsible for the universal behavior of the rates of molecular relaxations, which appear to remain unaffected by the chemical nature of materials (Ferry 1991). The approach will be discussed in some detail in this treatise. The free volume theory is not, however, the only concept proposed in an effort to address the molecular origins of vitrification phenomena. Recently, "the coupling model" or concept of thermorheological complexity (TC) has been put forward to overcome potential oversimplifications associated with the application of the free volume theory to the entirety of the broad glass transition region (Ngai 2000). The coupling model considers much of the chemical detail of the material and its effect on the intermolecular cooperative dynamics at the vicinity of the glass transition temperature. It is anticipated that researchers will pay attention to the contrasting outcomes of these two schools of thought, especially for the rationalization of cases of "anomalous" experimental facts (Kasapis 2008).

The above work dealt primarily with the combined effect of time and temperature on structural properties and the sophisticated synthetic polymer approach epitomized in the concept of free volume was extended to the effect of changing pressure on vitrification (Mpoukouvalas et al. 2005). Today, there is no data in the literature in pursuit of the applicability of the framework of the coupling model to pressure-induced molecular processes. Based on free volume as a primary mechanistic variable, work on synthetic polymers demonstrated an increase in small-deformation viscoelasticity and accompanying relaxation time with applied pressure. It was then feasible to add the dimension of pressure to those of time and temperature in the characterization of rubber-to-glass transformations (Ferry 1980). This is the second area of coverage within the scope of the present book chapter.

In biomaterials, aspects of pressure-induced functionality have been investigated in the solubilization of galactomannans and xyloglucans (guar, locust bean, tara, tamarind, and detarium gum) in order to produce true "molecular" systems with a reduced rate of subsequent aggregation (Picout et al. 2001, 2002, 2003). Reduction in the degree of crystallinity and gelatinization temperature of corn and potato starch with increasing pressure treatment has also been demonstrated (Blaszczak, Valverde, and Fornal 2005; Blaszczak et al. 2005a). Pressure shift freezing in aqueous- and sucrose-containing gelatin, agar, and deacylated gellan preparations yielded a large number of small ice crystals that helped to retain an acceptable texture in the frozen product (Zhu, Ramaswamy, and Le Bail 2005; Fuchigami and Teramoto 2003; Fuchigami, Teramoto, and Jibu 2006). In addition, alteration of the gelation or flow characteristics of gelatin, xanthan gum, soy, and muscle proteins, and reduction in microbial population, digestibility, and inactivation of antinutritional factors in cereal grains and legumes have been discussed as a function of variable levels of pressure (Montero, Fernandez-Diaz, and Gomez-Guillen 2002; Ahmed and Ramaswamy 2004; Molina, Defaye, and Ledward 2002; Jimenez Colmenero 2002; Estrada-Giron, Swanson, and Barbosa-Canovas 2005).

Increasingly, fundamental information is being made available about the effect of pressure on the vitrification of biomacromolecules, especially at high solid systems that relate to soft

FIGURE 11.1 Picture of the commercial fruit leather referred to in this work. (From Torley, P. J., de Boer, J., Bhandari, B. R., Kasapis, S., Shrinivas, P., and Jiang, B., *Journal of Food Engineering*, 86, 243–50, 2008. With permission.)

confections or boiled-down hard candy. Figure 11.1 depicts such a material, a commercial "fruit leather" with a solids content of about 70% w/w (Torley et al. 2008). High-pressure vitrification work on these biomaterials should be contrasted with corresponding data available from their synthetic counterparts. That would be the final objective of this work, which aims to examine the application of high pressure to vitrification of gelling biopolymer/co-solute samples and then discuss it, using the synthetic polymer approach, in conjunction with the effect of temperature.

11.2 THE COMBINED EFFECT OF TIME AND TEMPERATURE ON THE VITRIFICATION OF SYNTHETIC AND BIOMATERIALS

11.2.1 Theoretical Considerations

The first breakthrough in this area was achieved back in 1957 when Doolittle and Doolittle were interested in the Newtonian viscosity of the liquid *n*-alkanes (C_5 to C_{64}) over a wide range of temperatures. According to custom, the simple Andrade temperature equation (Allen 1993):

$$\eta = A \exp(B/T), \tag{11.1}$$

where η is the steady-shear viscosity, T is the absolute temperature, and A, B are constants for non-associated substances, was used to interpolate or extrapolate viscosity data over moderate ranges of temperature (equation of Arrhenius if B is taken to be the activation energy, E_a, of a molecular process). However, the equation failed to accurately reproduce viscosity values over an extended temperature range required, for example, in petroleum research and product development.

Recognizing this, Doolittle and Doolittle advanced an alternative equation which not only offered a good representation of data but, through the years, proved to serve as a basis for the development of an altogether different theory of viscosity and, consequently, a glass-transition related theory (Angell 1988):

$$\eta = A \exp[B(u_o/u_f)]. \tag{11.2}$$

Equation 11.2 is based on the concept of free volume (u_f), which is the difference between the total (u) and the occupied (u_o) volumes of a molecule. The latter is considered to be the "limiting specific volume" to which a liquid will contract without undergoing a change in phase during cooling to absolute zero (Ward and Hadley 1993).

The next step was to define the fractional increase in free volume (f) as the dimensionless ratio u_f/u and work out from Equation 11.2 a relationship that encompasses data of two distinct experimental temperatures (normally 3–5°C temperature interval). Horizontal shifting of the data to produce a continuum in the viscoelastic spectrum produces the so-called "shift factor" (a_T). The logarithmic form of the equation of factor a_T is (Hutchinson 1995):

$$\log a_T = B/2.303\left(\frac{1}{f}-\frac{1}{f_o}\right) + \log(T_o\rho_o/T\rho), \tag{11.3}$$

where ρ_o and f_o are the density and fractional free volume of the sample at the reference temperature, T_o. In practice, this equation is almost always used without the second term due to the slow temperature variation of the denominator $T\rho$.

An assumption that further advances this way of thinking is of a rapid and linear development of the fractional free volume at temperatures above the glass transition. The thermal expansion coefficient (α_f in deg^{-1}) thus generated can be considered at T_o as follows (Slade and Levine 1991):

$$f = f_o + \alpha_f(T - T_o). \tag{11.4}$$

Use of this relation for the fractional free volume in Equation 11.3 provides a mathematical expression applicable to a variety of rheological functions, and in the case of stress relaxation modulus $G(t)$ we obtain (Arridge 1975):

$$\log a_T = \log[G(t)(T)/G(t)(T_o)] = -\frac{(B/2.303f_o)(T - T_o)}{(f_o/\alpha_f) + T - T_o}. \tag{11.5}$$

In the form of $C_1^o = B/2.303f_o$ and $C_2^o = f_o/\alpha_f$, this is the equation proposed by Williams, Landel, and Ferry (WLF) to describe the temperature dependence of viscoelasticity throughout the glass transition region. The WLF equation can be recast in the following form:

$$\log a_T = C_1^o\ (T - T_o)/(T_\infty - T), \tag{11.6}$$

where the temperature T_∞ has been introduced by Vogel and is equal to $T_o - C_2^o$.

For a number of different synthetic polymers, T_∞ was found to be about 50°C below the glass transition temperature (Rahman 1999). An expedient way to determine the Vogel temperature is by minimizing the calculated shift factors against the temperature fraction of Equation 11.6 until the best straight line is obtained. This will also allow estimation of the parameters C_1^o and C_2^o at the reference temperature T_o. The WLF theory becomes inappropriate at temperatures below T_g or higher than T_g + 100°C when the temperature dependence of relaxation processes is heavily controlled by specific features, for example, the chemical structure of molecules in the melt.

The temperature bracket of 100°C is not absolute but depends on the ratio of amorphous to "crystalline" components in the matrix, with increasing levels of amorphicity in the system approaching the aforementioned limit of phenomenological glassy rheology (Kasapis, Al-Marhoobi, and Mitchell 2003b). Within these limits, however, Equation 11.5 holds for any reference temperature (including the T_g) and a straightforward algorithm of simultaneous equations can be devised to determine parameters (C_1^g, C_2^g, f_g, α_f) related to the glassy state of a polymer (Dannhauser, Child, and Ferry 1958):

$$C_1^o = C_1^g\ C_2^g/(C_2^g + T_o - T_g), \tag{11.7a}$$

$$C_2^o = C_2^g + T_o - T_g. \tag{11.7b}$$

11.2.2 Application of the Time/Temperature Effect to the Vitrification of High-Solid Biomaterials

The combined theoretical framework of WLF/free volume (Equations 11.1 through 11.7) described in the previous section was found to be applicable to the vitrification of amorphous synthetic polymers (Mansfield 1993). Then it was extended to low-molecular-weight organic liquids and inorganic materials (Angell 2002; Tobolsky 1956). This has prompted calls for the universality of the approach in glass forming systems where changes in the free volume appear to be independent of chemical features, as it may be the case in the mechanical manifestation of the glass transition region.

Recently, a start has been made in discussing the structural properties of biopolymer/co-solute that relate directly to industrial formulations, an example of which is shown in Figure 11.1. Understanding of structure–function relationships is sought in relation to the experimentally accessible temperature range and time of observation. The relaxation transition of a typical system within the scope of this treatise is discussed presently. This is made of 15% gelatin, 31.5% glucose syrup (dry solids of dextrose equivalent of 42), and 31.5% sucrose. The mixture was composed as it was in order to avoid ice formation or sucrose crystallization, thus focusing efforts on vitrification phenomena. The rubber-to-glass transition was studied using the stress relaxation modulus due to the fundamental and applied importance of this phenomenon lying partially in the associated changes of that mechanical function.

Research was carried out by implementing a series of time runs at fixed temperatures covering the range of 5 to −70°C and monitoring the stress relaxation modulus shown in Figure 11.2 (Kasapis 2006b). Then, horizontal shifts along the logarithmic time axis were implemented, which centered on the reference temperature of −25°C. The arbitrary choice of T_0 is inconsequential as long as it is confined within the glass transition region. The outcome of such data shifting is the so-called "master or composite curve" of viscoelasticity, and the scheme of developing a composite curve is known as the method of reduced variables or time/temperature superposition (TTS). Through the years, TTS was applied successfully to a plethora of measurements and systems (Roland et al. 1993;

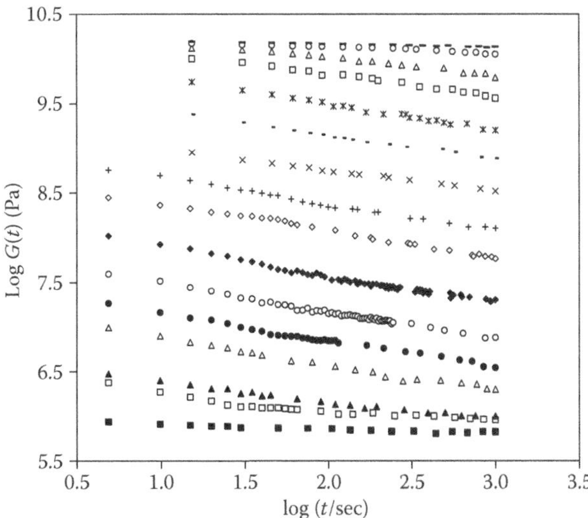

FIGURE 11.2 Variation of the stress relaxation modulus at fixed temperatures for a sample containing 15% gelatin, 31.5% glucose syrup, and 31.5% sucrose. Bottom curve is taken at 5°C (■); other curves successively upward, 0 (□), −5 (▲), −10 (△), −15 (●), −20 (○), −25 (◆), −30 (◇), −35 (+), −40 (×), −45 (–), −50 (✱), −55 (□), −60 (△), −65 (○), −70 (—) °C, respectively. (From Kasapis, S., *Handbook of Food Science, Technology and Engineering*, CRC Press, Boca Raton, 2006. With permission.)

Robertson and Palade 2006). Good matching of the shapes of adjacent curves is a critical criterion for the applicability of the method of reduced variables.

Figure 11.3 illustrates the master curve produced by an empirical superposition of the stress relaxation data obtained at different temperatures in Figure 11.2. The transition zone from a "rubbery" state at long times (>10^8 s) to a "glassy" state at short times (<10^{-7} s) is clearly discernable, thus demarcating a spectacular dependence of mechanical properties on time. Values of the reduced stress relaxation modulus, $G_p(t)$, do not change much with time in the rubbery plateau and glassy state, but they do so rapidly in the glass transition region (10^6–10^{10} Pa).

To move from a qualitative description to a quantitative treatment, we note that the horizontal superposition of the gelatin/co-solute data yields a factor a_T discussed, for example, in Equation 11.3. This is utilized in the reduction of data in Figure 11.2 by plotting in Figure 11.3 as follows (Alves et al. 2004):

$$G_p(t) = G(t)T_o\rho_o/T\rho \text{ vs } t/a_T. \tag{11.8}$$

It should be mentioned that in the case of dynamic oscillation data, this fundamental descriptor (i.e., the shift factor, a_T) of the temperature dependence is plotted as: $\omega\, a_T$, where ω is the frequency of oscillation. In practice, satisfactory matching of adjacent curves is achieved without the vertical shift of the temperature and density factors in Equation 11.8, since their logarithms are relatively small compared to the rapid changes in viscoelastic functions with experimental temperature (Angell 1997).

Figure 11.4 reproduces the effect of temperature on the molecular mobility of the gelatin/co-solute system from the glassy state to the rubbery plateau. Clearly, the set of horizontal shift factors yields a plot that is not a straight line in its entirety, so the temperature dependence of a_T is not a simple exponential function. Two discontinuities appear at −10 and −30°C, creating three log a_T vs temperature curves, which fan out but do not superpose onto a single line. Qualitatively, it is known that molecular motions are greatly inhibited as the biopolymer solution is cooled through the gel to the glassy state (Shirke and Ludescher 2005), but the question is of identifying a description which is close to a molecular interpretation of the observed phenomena.

It was mentioned earlier that Equation 11.1 was unable to follow progress in viscoelasticity within the glass transition region, an outcome that forced researchers to develop the concept of free volume epitomized mathematically in Equation 11.3. In the gelatin/co-solute case, it was verified that above −10°C and below −30°C log a_T is a linear function of reciprocal absolute temperature. These results are plotted against degrees centigrade in Figure 11.4 and clearly do not obey the postulates of Equation 11.3. By analogy with the theory of rate processes, and extending Equation 11.1 to include two experimental temperatures, the following expression emerges for the interplay of relaxation time and temperature within the glassy state and rubbery plateau (Plazek 1996):

$$\log a_T = Ea/2.303R\left(\frac{1}{T}-\frac{1}{T_o}\right), \tag{11.9}$$

where R is the gas constant. Equation 11.9 relates to the constant activation energy of a particular molecular mechanism within a relevant temperature range (i.e., glassy or rubbery state), which is obtained from the gradient of a linear relationship between log a_T and $1/T$.

Results thus far indicate that the amount of shift on the logarithmic scale being, of course, equal to log a_T is followed by Equation 11.9 within the rubbery plateau and glassy state. However, at the intermediate regime in Figure 11.4 (between −10 and −30°C) a curve is obtained which argues strongly for a non-Arrhenius process. The inability of predictions of the reaction rate theory to hold at the glass transition region brought into play the Williams, Landel, and Ferry approach in the form of Equations 11.5 or 11.6. The approach was able to follow the development of shift factor as a function of temperature, thus making free volume the overriding parameter determining molecular processes within the glass transition region.

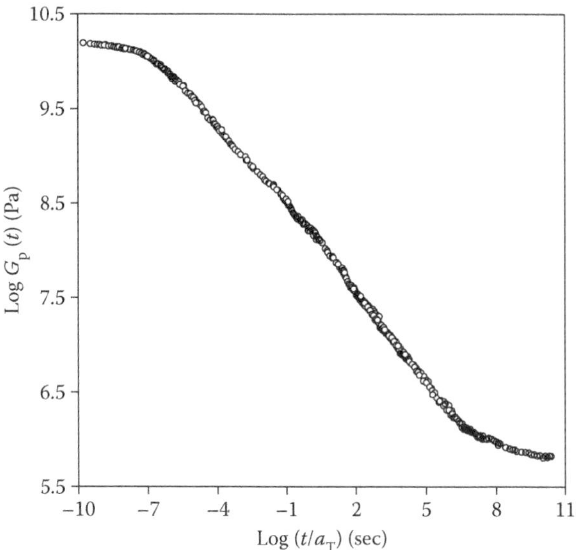

FIGURE 11.3 Master curve of stress relaxation modulus for the preparation in Figure 11.2 reduced to −25°C and plotted logarithmically against reduced time ($\tilde{t}/a_T$) utilizing the experimental isothermal runs of Figure 11.2. (From Kasapis, S., *Handbook of Food Science, Technology and Engineering*, CRC Press, Boca Raton, 2006. With permission.)

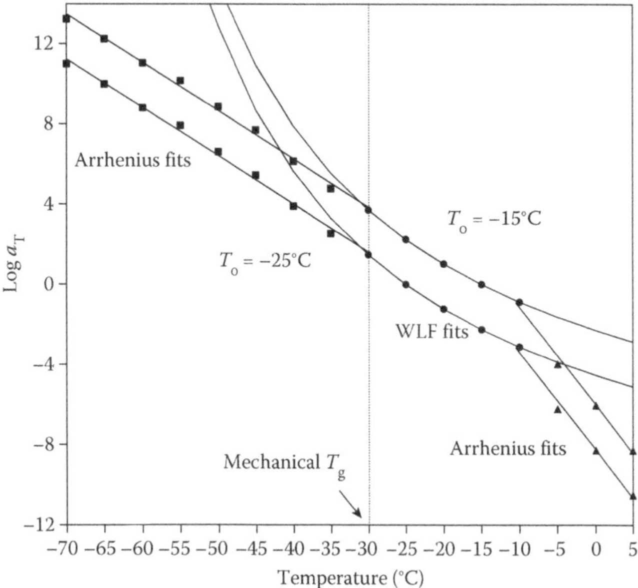

FIGURE 11.4 Logarithm of the reduction factor, a_T, for the sample in Figure 11.2 plotted against temperature from the data of the master curve in Figure 11.3. Reference temperatures, mechanistic modeling and the mechanical glass transition temperature are also shown. (From Kasapis, S., *Handbook of Food Science, Technology and Engineering*, CRC Press, Boca Raton, 2006. With permission.)

Clearly, the analysis amounts to more than curve fitting since it defines at −30°C in Figure 11.4 a turning point where large configurational vibrations requiring free volume in the glass transition region cease to be of the utmost importance (Kasapis et al. 2003a). At lower temperatures, the need to overcome an energetic barrier for the occurrence of local rearrangements from one state to the other becomes the primary consideration according to Equation 11.9. This framework of thought affords a fundamental definition of the mechanical glass transition temperature for the gelatin/co-solute system. This is an improvement on several empirical indices of the mechanical T_g, which recorded an array of discontinuities in the slope of storage (G') or loss modulus (G'') traces on shear and large peaks in the damping factor, tan δ (ratio of G'' to G') in the glassy state or glass transition region (Rieger 2001; Peleg 1995; Marshall and Petrie 1980). In the absence of a fundamental criterion, however, it seems barely credible to pinpoint glassy phenomena, or any other molecular events (thermal relaxation, molecular degradation, etc.) to empirical indices of pictorial rheology (Kasapis and Sablani 2005). Nonetheless, the glass transition temperature can be identified in terms of fundamental features of the molecular vitrification of gelatin discussed presently.

Interpretation of results in terms of the framework of the preceding paragraph yields a value of mechanical T_g (–30°C) which coincides with the completion of the glass transition region upon cooling in Figure 11.4. As mentioned earlier, within the limits of the transition region, Equation 11.5 holds for any reference temperature (including the T_g), thus, facilitating derivation of Equations 11.7a and 11.7b. Using the latter set of equations, reduction of the gelatin/co-solute data produces free volume indicators, which are congruent with those for synthetic polymers and diluted mixtures (Plazek et al. 1995; Hahn and Hillmyer 2003); at the reference temperature, $T_o = -25$°C: $C_1^o = 13.91$, $C_2^o = 52$ deg, $f_o = 0.031$, and at the glass transition temperature, $T_g = -30$°C: $C_1^g = 14.47$, $C_2^g = 50$ deg, $f_g = 0.030$; $\alpha_f = 6.0 \times 10^{-4}$ deg^{-1}. Finally, Figure 11.4 demonstrates that regardless of the choice of reference temperature (for example, –25 or –15°C), pinpointing T_g, and thus, the WLF-related parameters, remains unaffected, an outcome that further validates the strength of reasoning of this analysis.

11.3 ADDING THE DIMENSION OF HIGH PRESSURE TO THE GLASS TRANSITION OF SYNTHETIC MATERIALS

Once the effect of time and temperature on the vitrification of amorphous solid materials was fairly well understood, researchers turned their attention to a new problem, i.e., the mechanical manifestation of the application of high levels of pressure. It was already known that in the melt the viscosities of polymers and their viscoelastic relaxation times would increase under static confining pressure (Keshtiban, Belblidia, and Webster 2004). Qualitatively that behavior could be considered as the reverse effect of increasing the temperature of ordinary liquids. Based on this observation, it was hypothesized that free volume during vitrification should decrease with increasing pressure just as it does with decreasing temperature. Therefore, the concept of free volume was the starting point for the derivation of quantitative relationships between pressure and mechanical properties in the glass transition region (Danch, Osoba, and Wawryszczuk 2007).

As mentioned in Section 11.2.1, Equation 11.3 is applicable to two distinct experimental temperatures (states 1 and 2), which represent particular relaxation times. If, instead of temperature, states 1 and 2 refer to atmospheric pressure (P_o) and another (higher) pressure (P), then at constant temperature this mathematical expression can be generalized to give the shift factor a_T for pressure dependence (Ferry and Stratton 1960):

$$\log a_T = B/2.303 \left(\frac{1}{f_p} - \frac{1}{f_{P0}}\right), \tag{11.10}$$

where, f_p and f_{P0} are the fractional free volumes in these two states. It should be noted that besides the reduction in free volume with increasing pressure, there would be a small diminution in the occupied volume of the macromolecule. Both free and occupied volumes contribute to the parameter of total volume whose reduction can be measured experimentally.

The combined theoretical framework of WLF/free volume described as a function of temperature via Equations 11.2 through 11.7) assumed that the change of volume with temperature is linear, thus yielding a constant value of thermal expansion coefficient above T_g. Assuming that the change in volume is also linear with pressure, a compressibility coefficient, β_f in centimeters square per dyne, can be incorporated into the following relationship (Dlubek et al. 2005):

$$f_P = f_{P0} + \beta_f(P - P_o). \tag{11.11}$$

By analogy with temperature, substitution of the expression of fractional free volume into Equation 11.10 produces the pressure analog of the WLF equation, which is known as the Ferry and Stratton equation (Moonan and Tschoegl 1985):

$$\log a_p = \frac{(B/2.303 f_0)(P - P_0)}{(f_0/\beta_f) - (P - P_0)}, \tag{11.12}$$

where the factor a_p follows the pressure dependence of mechanical properties at constant temperature.

Analysis of data with Equation 11.12 following procedures described for temperature in Section 11.2.2 worked for the pressure dependence of viscosity for some ordinary liquids and of bulk relaxation times of polyvinyl acetate (Moonan and Tschoegl 1984). However, the preliminary success on limited data could not hide the fact that compressibility is markedly dependent on pressure as defined by the following:

$$\beta_f = -(1/u)(du_f/dP)_T. \tag{11.13}$$

Equation 11.13 implies an exponential dependence of volume, with β_f being a decreasing function of P. This is not surprising since, intuitively, it is expected that compression of a material of a certain volume would become progressively more difficult.

It was found that incorporation of the correct pressure dependence of the compressibility coefficient was required to allow satisfactory description of the mechanical response of single and filled elastomers throughout the glass transition region. The free volume approach was followed to derive a new equation (Fillers–Moonan–Tschoegl) for the general case of combined temperature and pressure effects. Thus, the final expression for the shift factor $a_{T,P}$ is (Moonan and Tschoegl 1983):

$$\log a_{T,P} = -\frac{(B/2.303 f_0)\,[T - T_0 - \Theta(P)]}{f_0/\alpha_f(P) + T - T_0 - \Theta(P)}, \tag{11.14}$$

where, $\Theta(P)$ is a rather complicated function, but can be summarized as:

$$\Theta(P) = f_{T_0}(P)/\alpha_f(P). \tag{11.15}$$

At the reference pressure, $\Theta(P) = 0$, thus preserving the form of the WLF equation for isobaric measurements at any pressure.

In conclusion, the shift factor a_p follows the progress in viscoelasticity as a function of time or frequency measured at distinct pressures, exactly as a_T is employed for a range of temperatures, and the factor $a_{T,P}$ combines all data at different pressures and temperatures. A large number of synthetic materials has been analyzed in this way, an example of which is given in Figure 11.5. This involves stress relaxation studies in simple elongation of a chlorosulfonated polyethylene strip lightly filled with 4% carbon black known as Hypalon 40 (Fillers and Tschoegl 1977). Data were converted

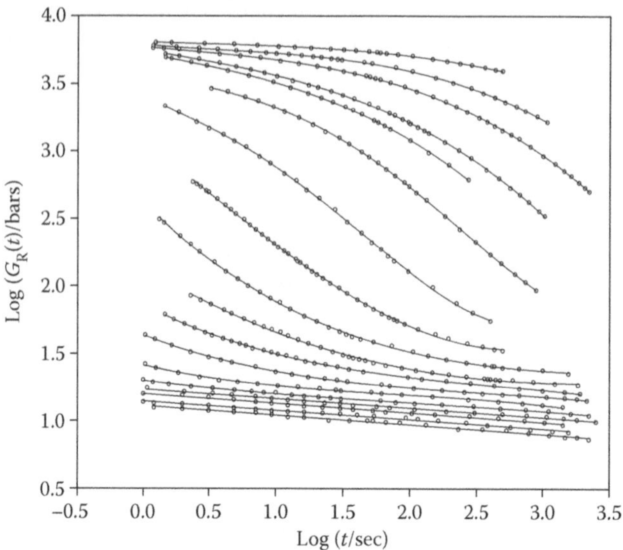

FIGURE 11.5 Logarithmic plot of shear relaxation modulus against time at 25°C and pressures from bottom to top: 1, 400, 800, 1200, 1600, 2000, 2400, 2600, 2800, 3000, 3200, 3400, 3600, 3800, 4000, 4200, 4400, and 4600 bars for chlorosulfonated polyethylene. (From Fillers, R. W., and Tschoegl, N. W., *Transactions of the Society of Rheology*, 21, 51–100, 1977. With permission.)

to relaxation modulus on shear and covered a range of pressures from atmospheric to 4600 bars. Temperature remained constant during experimentation at 25°C.

Shifting the mechanical spectra along the logarithmic time axis produces a smooth master curve of viscoelasticity (A) in Figure 11.6. Over the timescale covered by the horizontal superposition of the experimental traces, the polymer exhibits rubbery consistency at low applied pressures with the stress relaxation modulus being of the order of 10^6 Pa. In contrast, Hypalon 40 is a hard glass at high pressures reaching $G_R(t)$ values of about 6×10^8 Pa. At intermediate experimental pressures, the modulus trace goes through the sigmoidal transition zone as a function of time. An analogous pattern of shift factor as a function of temperature was obtained for mechanical spectra from 25 to −25°C at 1 bar, shown as curve B in Figure 11.6 (Fillers and Tschoegl 1977).

This is a very successful treatment, with the coincidence in traces at the rubbery and most of the transition region arguing that the material is piezo- and thermorheologically simple. The slender difference in the glassy state is due to a dense glass formed by the application of pressure (as compared to temperature) that displays greater shear modulus. This type of analysis allows determination of molecular parameters during vitrification from a combination of isobaric measurements at atmospheric pressure as a function of temperature, and isothermal measurements as a function of pressure. The outcome is a pressure/time/temperature equation-of-state that predicts the behavior of amorphous polymers when subjected to combined pressure and temperature during industrial processing (Ngai and Fytas 1986; Tribone and O'Reilly 1989; Utracki et al. 2003; Roland and Casalini 2003; Zhang et al. 2003).

11.4 HIGH-PRESSURE EFFECTS ON THE GLASS TRANSITION OF BIOMATERIALS

11.4.1 The Example of Gelatin in a High Sugar Environment

Besides man-made polymers used in a bewildering array of applications: packaging, films, fibers, tubing, pipes, etc., the pressure dependence of mechanical properties is of significant practical

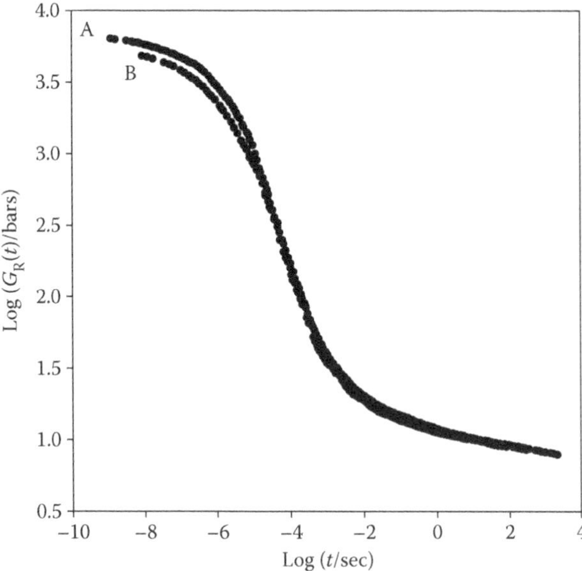

FIGURE 11.6 Isothermal data of Figure 11.5 from 1 to 4600 bars (A), and others (isobaric) (B) for Hypalon 40 at different temperatures (from 25 to −25°C) reduced to 1 bar and 25°C by shift factors $a_{T,P}$. (From Fillers, R. W., and Tschoegl, N. W., *Transactions of the Society of Rheology*, 21, 51–100, 1977. With permission.)

importance for biomaterials. The latter is of animal, plant, or bacterial origin, and the use of hydrostatic pressure is encountered in various food extrusion applications or the intelligent design and molding of implants in three-dimensional shapes that actively induce bone and cartilage regeneration (Fernandez et al. 2006; Liu and Wang 2007). Food technologists keep an inventory of biomaterials among which gelatin has proved to be a versatile ingredient in formulations ranging from soft confectionery to low-fat dairy products (Rougier, Bonazzi, and Daudin 2007). Furthermore, advances in site-specific drug delivery have considered using cross-linked gelatin as a biocompatible matrix encapsulating "small" bioactive compounds (Saxena et al. 2005).

From extruded products to molded tablets and capsules, gelatin alone or in mixture with bulking agents (sugars, sequestrants, etc.) delivers the required functionality via manipulation of the mechanical manifestation of the rubber-to-glass transition. The following constitutes an example of making high-solid materials in order to investigate the effect of pressure on vitrification behavior: Gelatin is prepared by soaking the granules in distilled water overnight and then heating to 65°C. Appropriate amounts of co-solute, i.e., glucose syrup and sucrose 50:50, are added to the gelatin solution. Glucose syrup should be added first, followed by sucrose, and excess water is evaporated by mild heating up to 70°C with stirring, leading to preparations with the required composition. Cylindrical gels of gelatin/co-solute are airtight sealed and left to equilibrate for a few days at ambient temperature (below 25°C). The aged gels are then sealed under vacuum in plastic pouches and placed in a high-pressure vessel apparatus. The pressure medium can be demineralized water and pressure buildup records values of up to 700 MPa or even higher (Bauer and Knorr 2005).

Figure 11.7 reproduces illustrative mechanical data obtained during heating or cooling of 15% gelatin plus 63% co-solute at a scan rate of 1°C/min (Kasapis 2007). Temperature-dependent changes are depicted for three distinct applications of pressure, namely 0.1, 400, and 700 MPa for 30 min. Small-deformation dynamic-oscillation measurements were taken that utilize the parameters of storage (G') and loss modulus (G'') (Richardson and Kasapis 1998). The transition from the rubbery plateau to the glassy state shows perfect thermal reversibility, i.e., absence of thermal hysteresis. At temperatures above −15°C, the mechanical response is predominantly solid-like, with $G' > G''$ at the experimental frequency of 1 rad/s. On further reduction in temperature, there

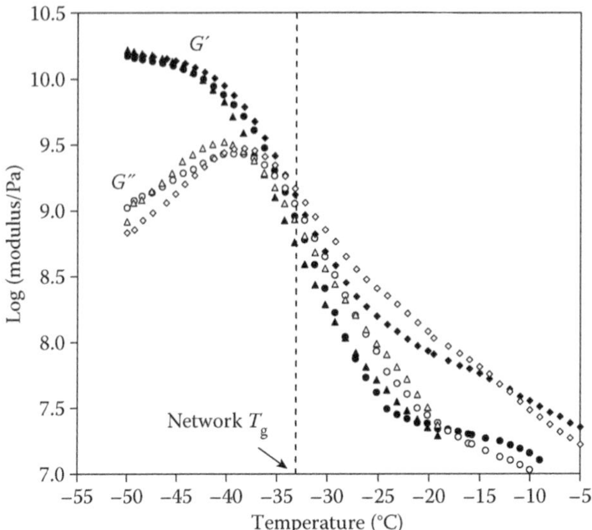

FIGURE 11.7 Temperature variation of oscillatory moduli for 15% gelatin plus 63% co-solute at a scan rate of 1°C/min and a strain range of 0.00071 to 1%. Prior to mechanical analysis gels were pressurized at 0.1 [G' (●); G'' (O)], 400 [G' (▲); G'' (△)], and 700 MPa [G' (◆); G'' (◇)]. (From Kasapis, S., *International Journal of Biological Macromolecules*, 40, 491–97, 2007. With permission.)

is considerable reinforcement in mechanical properties, with the liquid-like component becoming more pronounced, thus, producing a good measure of separation between the G'' and G' traces. As discussed earlier in this chapter, this is known as the glass transition region and reflects the diminishing transverse string-like vibrations of polymeric segments upon cooling (Ronan et al. 2007).

At the lower range of experimental temperatures in Figure 11.7, there is yet another development. The solid-like character becomes dominant once more and exceeds readings of 10^{10} Pa at −50°C. This part of the master curve of viscoelasticity is known as the glassy state, also seen for stress relaxation studies of gelatin in Figure 11.3. In the glassy state, there is little variation of storage modulus with temperature, and an increasing separation between the G' and G'' traces. It is expected that the solid-like character of residual contraction will dramatically reduce the diffusion rate of water molecules and compounds needed to support chemical reactions and biological processes (Rahman 1995). The absence of a first-order phase transition as a function of changing temperature or pressure in Figure 11.7 allows implementation of the analytical framework of WLF/free volume for the gelatin/co-solute system.

In doing so, gels were pressurized for 30 min at 400 MPa and standard mechanical spectra were taken at regular temperature intervals of 5°C between 5 and −50°C. Following the TTS principle demonstrated in Figures 11.2 and 11.3 (Kasapis and Al-Marhoobi 2005), data were processed arbitrarily choosing a point within the glass transition region as the reference temperature (T_o = −20°C) and shifting the remaining data along the log frequency axis until a uniform curve was obtained. Construction of the master curve of viscoelasticity yields a set of shift factors that are plotted against temperature in Figure 11.8. The WLF algorithm (Equation 11.5) follows the progress in viscoelasticity within the glass transition region well, thus making free volume the molecular mechanism dictating diffusional mobility. However, the WLF/free volume approach does not hold for the temperature range of the rubbery plateau (> −5°C) and the glassy state (< −35°C), at which the alternative plot of Equation 11.9, i.e., the modified Arrhenius equation, achieves good linear relationships for the horizontal shift factors.

The appearance of an Arrhenius-related linear superposition delimits a discontinuity in the development of shift factors at the end of the WLF curvature upon cooling (Figure 11.8). This change in operational kinetics can be considered to be a measure of the fundamental glass transition

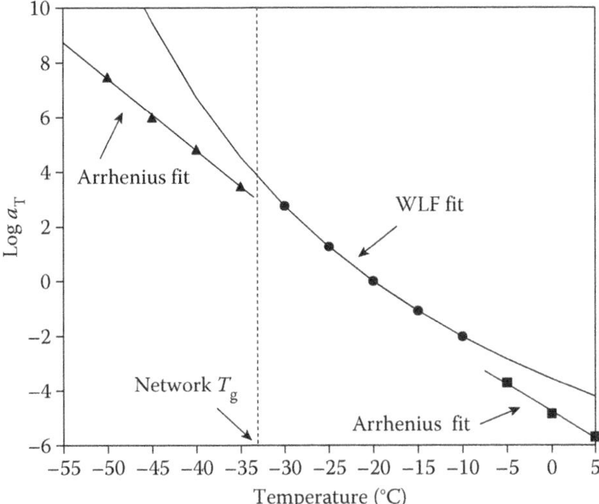

FIGURE 11.8 Temperature variation of the factor a_T from the rubbery (■), and the glass transition region (●) to the glassy state (▲) for 15% gelatin plus 63% co-solute. The solid lines reflect the WLF fit in the glass transition region, and the modified Arrhenius fits in the rubbery and glassy states. The dashed line pinpoints the prediction of the network glass transition temperature. Prior to mechanical analysis gels were pressurized at 400 MPa. (From Kasapis, S., *International Journal of Biological Macromolecules*, 40, 491–97, 2007. With permission.)

temperature at which free volume considerations become secondary to an energetic barrier of molecular rearrangements from one state to another (Levi and Karel 1995). Mechanical studies were extended to atmospheric pressure and 700 MPa, with frequency sweeps being reduced to master curves via the set of calculated shift factors illustrated in Figure 11.8 (data are not shown to avoid clutter). This exercise gave a clear confirmation that the temperature-course of vitrification phenomena remains unaffected within the experimentally accessible range of pressure, thus indicating a glass transition temperature of −33±1°C.

Rheological experiments were supported using differential scanning calorimetry (DSC) (Verdonck, Schaap, and Thomas 1999). The latter recorded a well-defined change in heat flow (heat capacity) of gelatin/co-solute, which can be associated with the "endothermic" glass transition region upon heating. The midpoint of this thermal event at subzero temperatures was readily detectable and was considered as the DSC T_g (Lopes and Felisberti 2004). Further heating of the gelatin/co-solute mixture recorded an extensive but linear drift in heat flow followed by a well-defined endotherm owing to melting of the gelatin network. Table 11.1 summarizes results at the point of vitrification and the melting transition obtained for gels pressurized between 0.1 and 700 MPa prior to thermal analysis (Kasapis 2007). Clearly, the progressive increase in the intensity of applied pressure is not manifest in an irreversible (compaction) relaxation of a densified network morphology. It appears that gelatin/co-solute possesses a high degree of network reversibility (recovery) following application of pressure, with the glass transition temperature remaining about −52°C. Similarly, the high temperature transition owing to the melting of the gelatin network is recorded to be 42.3±1.0°C.

It is known that calorimetrically determined glass transition temperatures are affected by the heating rate, whereas there is a frequency dependence on the vitrification parameters obtained with mechanical spectroscopy (Kasapis, Al-Marhoobi, and Mitchell 2003c). Keeping in mind that the measuring principles of the two techniques may set hurdles for meaningful comparisons, it is clear from the data in Table 11.1 and Figure 11.8 that rheological measurements are appropriate implements for the characterization of the vitrification behavior of the gelatin network (three-dimensional structure), thus, producing estimates for the network glass

TABLE 11.1
Glass Transition (T_g) and Melting (T_m) Temperatures of 15% Gelatin Gels in the Presence of 63% Co-Solute Determined Using Modulated Differential Scanning Calorimetry

Pressure (MPa)	T_g Replicates (°C)			Means	T_m Replicates (°C)			Means
0.1	−51.8	−51.4	−51.0	−51.4	42.0	43.5	44.5	43.3
120	−52.1	−53.0	−52.3	−52.5	42.5	44.5	43.0	43.3
210	−50.9	−52.0	−52.8	−51.9	42.0	42.0	44.0	42.7
290	−51.3	−52.9	−52.7	−52.3	42.5	42.0	42.5	42.3
400	−52.7	−53.3	−51.7	−52.6	42.0	42.0	41.5	41.8
500	−50.8	−52.2	−51.5	−51.5	43.5	41.0	41.5	42.0
600	−50.7	−49.4	−53.7	−51.3	39.5	41.0	42.5	41.0
700	−55.0	−48.8	−50.5	−51.4	41.0	43.0	41.0	41.7

Source: From Kasapis, S., *International Journal of Biological Macromolecules*, 40, 491–97, 2007. With permission.

transition temperature (T_g = −33°C). On the other hand, calorimetry monitors primarily the mobility of the sugar molecules yielding a lower prediction (T_g = −52°C).

11.4.2 The Example of Gelling Polysaccharides in a High Sugar Environment

In the last fifteen years or so, gelatin has been tarred (in our view unjustifiably) with the brush of uncertainty that surrounds BSE in terms of safety of products utilizing the protein as a structuring ingredient (Anand et al. 2005). In these intervening years, plant/marine and bacterial polysaccharides have come to prominence as gelatin replacers, of which agarose, carrageenans, and deacylated gellan gum are the most notable examples in high sugar preparations (Fonkwe, Narsimhan, and Cha 2003). This section will address the dependence of relaxation processes, as manifest in changes of the glass transition temperature under pressure (0.1–700 MPa) for the gelling polysaccharide/co-solute system, thus contrasting results with those in gelatin-containing formulations.

Figure 11.9 reproduces typical polysaccharide/glucose syrup data illustrated for 2% deacylated gellan plus 76% glucose syrup (6.7 mM $CaCl_2$ added) during heating or cooling at as scan rate of 1°C/min. Furthermore, temperature-dependent changes are depicted for two distinct applications of pressure, namely 0.1 and 700 MPa. As for gelatin (Figure 11.7), the transition from the rubbery region to the glassy state is perfectly thermoreversible and at temperatures above −5°C the mechanical response is predominantly solid-like (some G' data are not shown to avoid clutter) (Kasapis and Sablani 2008). On further reduction in temperature, there is considerable reinforcement in mechanical properties, signifying the glass transition region. At the lower range of experimental temperatures (below −25 and −31°C at 0.1 and 700 MPa, respectively) the solid-like character of the glassy state becomes dominant and reaches readings of 10^{10} Pa at −45°C. Further elucidation of the pressure dependence of relaxation processes requires development of quantitative relations which include parameters that affect the molecular free volume. Thus treatments analogous to those in Figure 11.8 produced T_g values of about −24.0 and −30.5 at atmospheric pressure and 700 MPa, respectively.

Regarding κ-carrageenan and agarose (Kasapis 2006b), experimental evidence combined with WLF/free volume analysis yielded T_g estimates of about −15.2°C (0.1 MPa) and −19.0 (700 MPa) for 1.5% κ-carrageenan + 76.5% glucose syrup (30 mM KCl added), and −32.3°C (0.1 MPa) and −36.7°C (700 MPa) for 3% agarose + 75% glucose syrup. DSC studies were carried out next to examine the effect of high hydrostatic pressure on the micromolecular properties of polysaccharide/sugar mixtures.

Figure 11.10 reproduces the shapes of typical DSC curves obtained for samples of 3% agarose in the presence of 75% glucose syrup solids that have been treated at 0.1 (ambient pressure), 300,

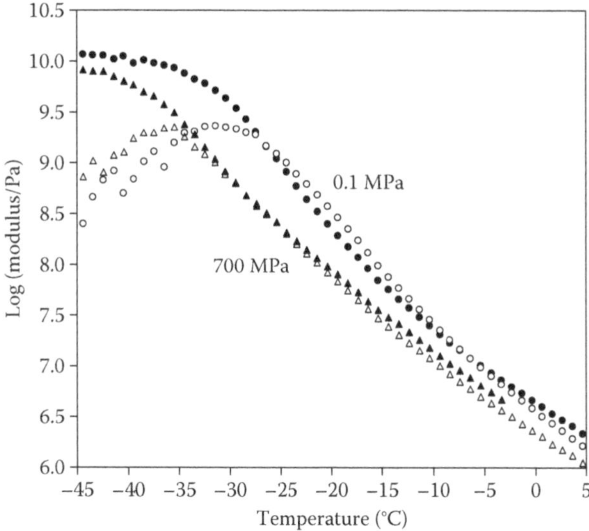

FIGURE 11.9 Temperature variation of shear moduli for 2% gellan plus 76% glucose syrup (6.7 mM $CaCl_2$ added) at a scan rate of 1°C/min and a strain range of 0.00071 to 1%. Prior to mechanical analysis, gels were pressurized at 0.1 [G' (●); G'' (○)] and 700 [G' (▲); G'' (△)] MPa. (From Kasapis, S., and Sablani, S. S., *Carbohydrate Polymers*, 72, 537–44, 2008. With permission.)

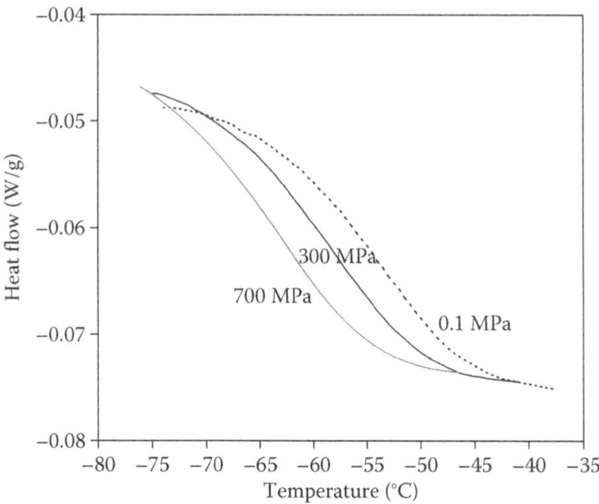

FIGURE 11.10 Heat flow variation as a function of temperature for 3% agarose plus 75% glucose syrup obtained with DSC at a heating rate of 1°C/min. Gels were prepared at 0.1 (atmospheric pressure), 300, and 700 MPa. (From Kasapis, S., and Sablani, S. S., *Carbohydrate Polymers*, 72, 537–44, 2008. With permission.)

and 700 MPa (Kasapis and Sablani 2008). Materials were cooled slowly (1°C/min) from ambient temperature (24°C) and then heated at the same scan rate from temperatures well below the glass transition temperature of the mixture, thus exhibiting a "pseudoequilibrium" relaxation response to the changing thermal regime. Several replicates for each experimental pressure were recorded. These reproduced a sigmoidal variation in heat flow (heat capacity), which is related to the "endothermic" devitrification upon controlled heating (Chung and Lim 2003).

TABLE 11.2
Glass Transition Temperatures (T_g) of Polysaccharide/Co-Solute Preparations (78% Total Level of Solids) Determined Using Modulated Differential Scanning Calorimetry

	0.1 MPa	150 MPa	300 MPa	500 MPa	700 MPa
3% agarose + 75% glucose syrup					
T_g	–56.3	–57.0	–60.4	–59.4	–61.3
Replications	–56.8	–56.4	–58.1	–63.5	–63.9
(°C)	–55.0	–59.4	–60.5	–61.2	–59.9
2% gellan + 76% glucose syrup (6.7 mM $CaCl_2$)					
T_g	–56.5	–56.4	–57.3	–57.4	–62.2
Replications	–57.0	–55.5	–56.6	–59.2	–62.3
(°C)	–56.2	–53.5	–56.6	–60.6	–61.5
1.5% κ-carrageenan + 76.5% glucose syrup (30 mM KCl)					
T_g	–52.9	–54.3	–53.5	–59.5	–60.3
Replications	–53.2	–55.6	–60.6	–62.9	–63.6
(°C)	–52.6	–57.4	–57.0	–59.0	–60.2
Average T_g	–55.2±1.8	–56.2±1.2	–57.8±2.0	–60.3±1.5	–61.7±1.9

Source: From Kasapis, S. and Sablani, S. S., *Carbohydrate Polymers*, 72, 537–44, 2008. With permission.

As for the DSC work on gelatin, the midpoint of the heat capacity change was considered as the T_g of the glassy consistency. Figure 11.10 attempted to detect possible variations in the thermal profile of agarose/glucose syrup gels by increasing the applied hydrostatic pressure at intervals of about 300 MPa. Table 11.2 summarizes results at the glass transition region obtained between 0.1 and 700 MPa for agarose, deacylated gellan gum ($CaCl_2$ added), and κ-carrageenan (KCl added) in the presence of glucose syrup as co-solute (Kasapis and Sablani 2008). Clearly, the progressive increase in the intensity of applied pressure is manifest in irreversible relaxation phenomena, which appear to be of comparable magnitude for the three polysaccharide/co-solute preparations.

In general, there is a progressive decrease in T_g values from about –55 to –62°C with increasing experimental hydrostatic pressure in Table 11.2. It appears that the treatment of gelling polysaccharide/sugar with high pressure resulted in the disruption of the three-dimensional structure. At higher pressure (700 MPa), vitrification points approach those for the single sugar systems. For example, values between –62 and –65°C have been reported in the glass transition curve of the state diagram of glucose preparations at 78% solids (Roos 1995).

11.5 CONCLUDING REMARKS

This chapter deals with the molecular mechanisms underlying thermal and nonthermal treatments of high-solid biomaterials with relevance to soft confectionery products, boiled-down hard candy, ice-cream, etc. Distinct effects of the mechanical and calorimetric manifestation of structural properties were documented for the vitrification phenomena of gelatin from animal collagen and gelling polysaccharides of marine/bacterial origin in mixture with sugar (sucrose and/or glucose syrup). Thus the effect of static confining pressure (hydrostatic in this treatise) on vitrification phenomena was found to be variable according to ingredient source. Conclusions were facilitated via experimental observations in combination with the theoretical treatment of WLF/free volume that pinpointed the network glass transition temperature from mechanical measurements.

Quantification of the effect of pressure on the structural properties of individual polysaccharide/co-solute samples documents a break in the time/temperature/pressure superposition, which is operational in the vitrification properties of amorphous synthetic materials. Thus destabilization of thermal and viscoelastic relaxation processes is monitored under pressure for gelling polysaccharides, as opposed to the positive temperature or pressure effect on the corresponding behavior of elastomers with hydrophobic interactions (e.g., styrene or butadiene rubbers). Furthermore, the synthetic-polymer behavior is in itself dissimilar to the high-pressure response obtained for the molecular interplay between gelatin and co-solute under an identical experimental treatment, although the protein is considered to be the archetype of a "biological rubber."

Critical examination of the data available in the literature supports the concept of nonaggregated hydrogen-bonding stabilized associations in gelatin, although in the presence of sugar the level of order in the protein network seems to increase somehow (Gupta, Mohanty, and Bohidar 2005; Kasapis 2003a). The intermolecular associations exhibit a degree of reversibility once the pressure-generating experiment is complete. Thus values of T_g remain constant throughout the range of applied hydrostatic pressure (0.1–700 MPa), an external stimulus which is comparable to that reported earlier in the literature of synthetic polymers.

In the case of gelling polysaccharides, recent work demonstrated that these form rubbery gels that maintain a degree of intermolecular aggregation in the high co-solute environment of glucose syrup/sucrose (Al-Marhoobi and Kasapis 2005). The degree of aggregation is diminished in comparison with the extensive volume of enthalpic associations of the aqueous counterparts. We can now state with a certain degree of confidence that pressure-induced vitrification patterns appear to be largely irreversible due to the destabilization of the remaining aggregated assemblies of the high co-solute polysaccharide networks. The outcome of such an effect is lower values in the micro (DSC) and macro (rheology) manifestation of the glass transition temperature.

REFERENCES

Ahmed, J., and Ramaswamy, H. S. 2004. Effect of high-hydrostatic pressure and concentration on rheological characteristics of xanthan gum. *Food Hydrocolloids*, 18:367–73.

Allen, G. 1993. A history of the glassy state. In: Blanshard, J. M. V., and Lillford, P. J. (eds.), *The Glassy State in Foods*, 1–12. Nottingham: Nottingham University Press.

Al-Marhoobi, I. M., and Kasapis, S. 2005. Further evidence of the changing nature of biopolymer networks in the presence of sugar. *Carbohydrate Research*, 340:771–74.

Alves, N. M., Mano, J. F., Gomez Ribelles, J. L., and Gomez Tejedor J. A. 2004. Departure from the Vogel behaviour in the glass transition-thermally stimulated recovery, creep and dynamic mechanical analysis studies. *Polymer*, 45:1007–17.

Anand, A., Moreira, R., Henry, J. Chowdhury, M., Cote, G., and Good, T. 2005. A bio-sensing strategy for the detection of prions in foods. *Lebensmittel-Wissenschaft und Technologie*, 38:849–58.

Angell, C. A. 1988. Perspective on the glass transition. *Journal of Physics and Chemistry of Solids*, 49:863–71.

———. 1997. Why C_1 = 16–17 in the WLF equation is physical-and the fragility of polymers. *Polymer Papers*, 38:6261–66.

———. 2002. Liquid fragility and the glass transition in water and aqueous solutions. *Chemistry Reviews*, 102:2627–50.

Arridge, R. G. C. 1975. The glass transition. In: *Mechanics of Polymers*, 24–50. Oxford: Clarendon Press.

Bauer, B. A., and Knorr, D. 2005. The impact of pressure, temperature and treatment time on starches: Pressure-induced starch gelatinisation as pressure time temperature indicator for high hydrostatic pressure processing. *Journal of Food Engineering*, 68:329–34.

Blaszczak, W., Fornal, J., Valverde, S., and Garrido, L. 2005a. Pressure-induced changes in the structure of corn starches with different amylose content. *Carbohydrate Polymers*, 61:132–40.

Blaszczak, W., Valverde, S., and Fornal. J. 2005. Effect of high pressure on the structure of potato starch. *Carbohydrate Polymers*, 59:377–83.

Chung, H.-J., and Lim, S.-T. 2003. Physical aging of glassy normal and waxy rice starches: Effect of aging time on glass transition and enthalpy relaxation. *Food Hydrocolloids*, 17:855–61.

Danch, A., Osoba, W., and Wawryszczuk, J. 2007. Comparison of the influence of low temperature and high pressure on the free volume in polymethylpentene. *Radiation Physics and Chemistry*, 76:150–52.

Dannhauser, W., Child, W. C. Jr., and Ferry, J. D. 1958. Dynamic mechanical properties of poly-*n*-octylmethacrylate. *Journal of Colloid Science*, 13:103–13.

Dlubek, G., Sen Gupta, A., Pionteck, J., Habler, R., Krause-Rehberg, R., Kaspar, H., and Lochhaas, K. H. 2005. Glass transition and free volume in the mobile (MAF) and rigid (RAF) amorphous fractions of semicrystalline PTFE: A positron lifetime and PVT study. *Polymer*, 46:6075–89.

Estrada-Giron, Y., Swanson, B. G., and Barbosa-Canovas, G. V. 2005. Advances in the use of high hydrostatic pressure for processing cereal grains and legumes. *Trends in Food Science & Technology*, 16:194–203.

Fernandez, P. P., Otero, L., Guignon, B., and Sanz, P. D. 2006. High-pressure shift freezing versus high-pressure assisted freezing: Effects on the microstructure of a food model. *Food Hydrocolloids*, 20:510–22.

Ferry, J. D. 1980. Dependence of viscoelastic behaviour on temperature and pressure. In: *Viscoelastic Properties of Polymers*, 264–320. New York: John Wiley.

———. 1991. Some reflections on the early development of polymer dynamics: Viscoelasticity, dielectric dispersion, and self-diffusion. *Macromolecules*, 24:5237–45.

Ferry, J. D., and Stratton, R. A. 1960. The free volume interpretation of the dependence of viscosities and viscoelastic relaxation times on concentration, pressure, and tensile strain. *Kolloid-Zeitscherift*, 171:107–11.

Fillers, R. W., and Tschoegl, N. W. 1977. The effect of pressure on the mechanical properties of polymers. *Transactions of the Society of Rheology*, 21:51–100.

Fonkwe, L. G., Narsimhan, G., and Cha, A. S. 2003. Characterisation of gelation time and texture of gelatin and gelatin-polysaccharide mixed gels. *Food Hydrocolloids*, 17:871–83.

Fuchigami, M., and Teramoto, A. 2003. Texture and structure of high-pressure-frozen gellan gum gel. *Food Hydrocolloids*, 17:895–99.

Fuchigami, M., Teramoto, A., and Jibu, Y. 2006. Texture and structure of pressure-shift-frozen agar gel with high visco-elasticity. *Food Hydrocolloids*, 20:160–69.

Gupta, A., Mohanty, B., and Bohidar, H. B. 2005. Flory temperature and upper critical solution temperature of gelatin solutions. *Biomacromolecules*, 6:1623–27.

Hahn, S. F., and Hillmyer, M. A. 2003. High glass transition temperature polyolefins obtained by the catalytic hydrogenation of polyindene. *Macromolecules*, 36:71–76.

Hutchinson, J. M. 1995. Physical aging of polymers. *Progress in Polymer Science*, 20:703–60.

Jimenez Colmenero, F. 2002. Muscle protein gelation by combined use of high pressure/temperature. *Trends in Food Science & Technology*, 13:22–30.

Kasapis, S. 2006a. Building on the WLF / free volume framework: Utilisation of the coupling model in the relaxation dynamics of the gelatin / co-solute system. *Biomacromolecules*, 7:1671–1678.

Kasapis, S. 2006b. Composition and structure-function relationships in gums. In: Y. H. Hui (ed.) *Handbook of Food Science, Technology and Engineering*, 92-1 to 92-19. Boca Raton: CRC Press.

Kasapis, S. 2007. The effect of pressure on the glass transition of biopolymer/co-solute. Part I: The example of gelatin. *International Journal of Biological Macromolecules*, 40:491–97.

Kasapis, S. 2008. Beyond the free volume theory: Introduction of the concept of cooperativity to the chain dynamics of biopolymers during vitrification. *Food Hydrocolloids*, 22:84–90.

Kasapis, S., and Al-Marhoobi, I. M. 2005. Bridging the divide between the high- and low-solid analyses in the gelatin/κ-carrageenan mixture. *Biomacromolecules*, 6:14–23.

Kasapis, S., Al-Marhoobi, I. M., Deszczynski, M., Mitchell, J. R., and Abeysekera, R. 2003a. Gelatin vs. polysaccharide in mixture with sugar. *Biomacromolecules*, 4:1142–49.

Kasapis, S., Al-Marhoobi, I. M., and Mitchell, J. R. 2003b. Molecular weight effects on the glass transition of gelatin/co-solute mixtures. *Biopolymers*, 70:169–85.

———. 2003c. Testing the validity of comparisons between the rheological and the calorimetric glass transition temperatures. *Carbohydrate Research*, 338:787–94.

Kasapis, S., and Sablani, S. S. 2005. A fundamental approach for the estimation of the mechanical glass transition temperature in gelatin. *International Journal of Biological Macromolecules*, 36:71–78.

———. 2008. The effect of pressure on the structural properties of biopolymer/co-solute. Part II: The example of gelling polysaccharides. *Carbohydrate Polymers*, 72:537–44.

Keshtiban, I. J., Belblidia, F., and Webster, M. F. 2004. Numerical simulation of compressible viscoelastic liquids. *Journal of Non-Newtonian Fluid Mechanics*, 122:131–46.

Levi, G., and Karel, M. 1995. Volumetric shrinkage (collapse) in freeze-dried carbohydrates above their glass transition temperature. *Food Research International*, 28:145–51.

Liu, Y., and Wang, M. 2007. Developing a composite material for bone tissue repair. *Current Applied Physics*, 7:547–54.

Lopes, C. M. A., and Felisberti, M. I. 2004. Thermal conductivity of PET/(LDPE/AI) composites determined by MDSC. *Polymer Testing*, 23:637–43.

Mansfield, M. L. 1993. An overview of theories of the glass transition. In: J. M. V. Blanshard, and P. J. Lillford (eds) *The Glassy State in Foods*, 103–122. Nottingham: Nottingham University Press.

Marshall, A. S., and Petrie, S. E. B. 1980. Thermal transitions in gelatin and aqueous gelatin solutions. *The Journal of Photographic Science*, 28:128–34.

Molina, E., Defaye, A. B., and Ledward, D. A. 2002. Soy protein pressure-induced gels. *Food Hydrocolloids*, 16:625–32.

Montero, P., Fernandez-Diaz, M. D., and Gomez-Guillen, M. C. 2002. Characterisation of gelatin gels induced by high pressure. *Food Hydrocolloids*, 16:197–205.

Moonan, W. K., and Tschoegl, N. W. 1983. Effect of pressure on the mechanical properties of polymers. 2. Expansivity and compressibility measurements. *Macromolecules*, 16:55–59.

———. 1984. Effect of pressure on the mechanical properties of polymers. 3. Substitution of the glassy parameters for those of the occupied volume. *International Journal of Polymeric Materials*, 10:199–211.

———. 1985. The effect of pressure on the mechanical properties of polymers. IV. Measurements in torsion. *Journal of Polymer Science: Polymer Physics Edition*, 23:623–51.

Mpoukouvalas, K., Floudas, G., Zhang, S. H., and Runt, J. 2005. Effect of temperature and pressure on the dynamic miscibility of hydrogen-bonded polymer blends. *Macromolecules*, 38:552–60.

Ngai, K. L. 2000. Dynamic and thermodynamic properties of glass-forming substances. *Journal of Non-Crystalline Solids*, 275:7–51.

Ngai, K. L., and Fytas, G. 1986. Interpretation of differences in temperature and pressure dependences of density and concentration fluctuations in amorphous poly(phenylmethyl siloxane). *Journal of Polymer Science: Part B: Polymer Physics*, 24:1683–94.

Peleg, M. 1995. A note on the tanδ(T) peak as a glass transition indicator in biosolids. *Rheological Acta*, 34:215–20.

Picout, D. R., Ross-Murphy, S. B., Errington, N., and Harding. S. E. 2001. Pressure cell assisted solution characterisation of polysaccharides. 1. Guar gum. *Biomacromolecules*, 2:1301–309.

———. 2003. Pressure cell assisted solubilization of Xylglucans: Tamarind seed polysaccharide and detarium gum. *Biomacromolecules*, 4:799–807.

Picout, D. R., Ross-Murphy, S. B., Jumel, K., and Harding, S. E. 2002. Pressure cell assisted solution characterisation of polysaccharides. 2. Locust bean gum and tara gum. *Biomacromolecules*, 3:761–67.

Plazek, D. J. 1996. 1995. Bingham medal address: Oh, thermorheological simplicity, wherefore art thou? *Journal of Rheology*, 40:987–1015.

Plazek, D. J., Chay, I.-C., Ngai, K. L., and Roland, C. M. 1995. Viscoelastic properties of polymers. 4. Thermorheological complexity of the softening dispersion in polyisobutylene. *Macromolecules*, 28:6432–36.

Rahman, M. S. 1999. Glass transition and other structural changes in foods. In: *Handbook of Food Preservation*, 75–93. New York: Marcel Dekker.

Rahman, S. 1995. Phase transitions in foods. In: *Food Properties Handbook*, 87–177. Boca Raton: CRC Press.

Richardson, R. K., and Kasapis, S. 1998. Rheological methods in the characterisation of food biopolymers. In: D. L. B. Wetzel, and G. Charalambous (eds.), *Instrumental Methods in Food and Beverage Analysis*, 1–48. Amsterdam: Elsevier.

Rieger, J. 2001. The glass transition temperature T_g of polymers: Comparison of the values from differential thermal analysis (DTA, DSC) and dynamic mechanical measurements (torsion pendulum). *Polymer Testing*, 20:199–204.

Robertson, C. G., and Palade, L. I. 2006. Unified application of the coupling model to segmental, Rouse, and terminal dynamics of entangled polymers. *Journal of Non-Crystalline Solids*, 352:342–48.

Roland, C. M., and Casalini, R. 2003. Temperature and volume effects on local segmental relaxation in poly(vinyl acetate). *Macromolecules*, 36:1361–67.

Roland, C. M., Santangelo, P. G., Ngai, K. L., and Meier, G. 1993. Relaxation dynamics in poly (methylphenylsiloxane), 1,1-bis(*p*-methoxyphenyl)cyclohexane, and their mixtures. *Macromolecules*, 26:6164–70.

Ronan, S., Alshuth, T., Jerrams, S., and Murphy, N. 2007. Long-term stress relaxation prediction for elastomers using the time-temperature superposition method. *Materials and Design*, 28:1513–23.

Roos, Y. H. 1995. Prediction of the physical state. In: *Phase Transitions in Foods*, 157–92. San Diego: Academic Press.

Rougier, T., Bonazzi, C., and Daudin, J.-D. 2007. Modeling incidence of lipid and sodium chloride contents on sorption curves of gelatin in the high humidity range. *Lebensmittel-Wissenschaft und Technologie*, 40:1798–807.

Saxena, A., Sachin, K., Bohidar, H. B., and Verma, A.-K. 2005. Effect of molecular weight heterogeneity on drug encapsulation efficiency of gelatin nano-particles. *Colloids and Surfaces B: Biointerfaces*, 45:42–48.

Shirke, S., and Ludescher, R. D. 2005. Molecular mobility and the glass transition in amorphous glucose, maltose, and maltotriose. *Carbohydrate Research*, 340:2654–60.

Slade, L., and Franks, F. 2002. Appendix I: Summary report of the discussion symposium on chemistry and application technology of amorphous carbohydrates. In: H. Levine (ed.), *Amorphous Food and Pharmaceutical Systems*, x–xxvi. Cambridge: The Royal Society of Chemistry.

Slade, L., and Levine, H. 1991. Beyond water activity: Recent advances based on an alternative approach to the assessment of food quality and safety. In: F. M. Clydesdale (ed.), *Critical Reviews in Food Science and Nutrition*, Vol. 30, 115–360. Boca Raton: CRC Press.

Tobolsky, A. V. 1956. Stress relaxation studies of the viscoelastic properties of polymers. *Journal of Applied Physics*, 27:673–85.

Torley, P. J., de Boer, J., Bhandari, B. R., Kasapis, S., Shrinivas, P., and Jiang, B. 2008. Application of the synthetic polymer approach to the glass transition of fruit leathers. *Journal of Food Engineering*, 86:243–50.

Tribone, J. J., and O'Reilly, J. M. 1989. Pressure-jump volume-relaxation studies of polystyrene in the glass transition region. *Journal of Polymer Science: Part B: Polymer Physics*, 27:837–57.

Utracki, L. A., Simha, R., and Garcia-Rejon, A. 2003. Pressure-volume-temperature dependence of poly-ε-caprolactam/clay nanocomposites. *Macromolecules*, 36:2114–21.

Verdonck, E., Schaap, K., and Thomas, L. C. 1999. A discussion of the principles and applications of modulated temperature DSC (MTDSC). *International Journal of Pharmaceutics*, 192:3–20.

Ward, I. M., and Hadley, D. W. 1993. Experimental studies of linear viscoelastic behaviour as a function of frequency and temperature: Time-temperature equivalence. In: *An Introduction to the Mechanical Properties of Solid Polymers*, 84–108. Chichester: John Wiley.

Zhang, S. H., Casalini, R., Runt, J., and Roland, C. M. 2003. Pressure effects on the segmental dynamics of hydrogen-bonded polymer blends. *Macromolecules*, 36:9917–23.

Zhu, S., Ramaswamy, H. S., and Le Bail, A. 2005. Ice-crystal formation in gelatin gel during pressure shift versus conventional freezing. *Journal of Food Engineering*, 66:69–76.

Zimeri, J. E., and Kokini, J. L. 2002. The effect of moisture content on the crystallinity and glass transition temperature of inulin. *Carbohydrate Polymers*, 48:299–304.

12 Effect of High-Pressure and Ultrasonic Processing of Foods on Rheological Properties

Jirarat Tattiyakul
Chulalongkorn University

M. A. Rao
Cornell University

CONTENTS

12.1 HIGH-PRESSURE (HP) PROCESSING OF FOODS

Several novel or alternative food processing techniques have been used mainly to replace the long-used heat treatment. Among the alternative technologies, high-pressure (HP) processing technology has found wide acceptance as a preservation technique. HP technology is also an effective and safe method of modifying protein structure, and self-assembly properties. Pressure processing can lead to protein denaturation and different states of aggregation or gelation depending on the protein system, the treatment temperature, the protein solution conditions (e.g., pH and ionic strength), and the magnitude and duration of the applied pressure (Galazka, Dickinson, and Ledward 2000). A comprehensive review of HP processing of foods can be found in Rastogi et al. (2007). However, the corresponding changes in rheological properties and structural characteristics have not been covered in depth.

HP appears to be the only processing technology alternative to heat treatment that has reached the consumer with a variety of products that include fruit jams, jellies, sauces, juices, avocado pulp, guacamole, and cooked ham. However, there do not seem to be, as yet, commercial HP-treated dairy products, even though many research studies have been published on them. Most of the applications take advantage of the fact that HP exerts antimicrobial effects without impairing nutritional quality (López-Fandiño 2006). Therefore, the effect of HP on the rheological and functional properties of several proteins, especially milk proteins, and foods containing them has been studied.

Specifically, HP treatments influence the functional properties of proteins through the disruption and reformation of hydrogen bonds, as well as hydrophobic interactions and the separation of ion pairs. These changes have been found to depend on protein structure, pressure level, temperature, pH, ionic strength, solvent composition, and protein concentration. Here, the significant changes in structure and rheological properties of foods containing proteins due to the application of HP are reviewed.

12.2 VISCOELASTIC AND TEXTURAL PARAMETERS OF FOODS

Small amplitude oscillatory shear (SAOS), also called dynamic rheological experiment, can be used to determine viscoelastic properties of foods without altering the material's structure (Rao 2007). In an SAOS experiment, a sinusoidal oscillating stress or strain with a frequency ω is applied to the material and the phase difference between the oscillating stress and strain as well as the amplitude ratio are measured. For shear deformation within the linear viscoelastic range, the generated stress (σ_0) is expressed in terms of the storage modulus G' (Pa) and a loss modulus G'' (Pa); G' is a measure of the magnitude of the energy that is stored in the material and G'' is a measure of the energy which is lost as viscous dissipation per cycle of deformation, respectively. For a viscoelastic material the resultant stress is also sinusoidal but shows a phase lag of δ radians when compared with the strain. The phase angle δ covers the range of 0 to $\pi/2$ as the viscous component increases. The dependence of the storage modulus, G', and the loss modulus, G'', on the oscillatory frequency (ω) in the linear viscoelastic region is one set of useful information. If $G'' > G'$, the material is behaving predominantly as viscous liquid; if $G' > G''$, the material is behaving predominantly as a solid.

Often, empirical tests, using very large deformations, are used to characterize solid foods. These tests can provide a simple measure of a food, such as firmness, or a number of defined parameters, such as the texture profile analysis (TPA) parameters (Rao and Quintero 2005). The following TPA parameters, based on a two-cycle compression, have been used to characterize the effect of HP processing of foods:

Hardness is defined as the peak force during the first compression cycle;

Adhesiveness is defined as the negative force area for the first cycle, and

Springiness is defined as the height to which the food recovers during the time between the first and second compression cycles.

12.3 HIGH-PRESSURE (HP) PROCESSING OF MILK AND MILK PROTEINS

Because numerous studies have been conducted on HP processing of milk and its proteins (López-Fandiño 2006), some of the quantitative studies are reviewed first. Milk processing at 150–400 MPa caused a certain amount of irreversible fragmentation of casein micelles, together with calcium release, increased milk viscosity, decreased milk turbidity, and decreased noncasein nitrogen (Cheftel and Dumay 1996). The exact changes in individual caseins and whey proteins (WPs) are not known, but milk treatment accelerates subsequent casein coagulation by rennet or glucono-δ-lactone, and enhances the strength and water-holding capacity (WHC) of acid-set gels. Processing solutions of β-lactoglobulin (β-Lg) in the same HP range causes partial structure denaturation. β-Lg unfolding is more extensive and irreversible at neutral than at acid pH. In addition to the reduction of particle size, micelles in HP-treated milk are irregularly shaped and of enhanced voluminosity, due to higher hydration, and all these factors contribute to an increased viscosity in skimmed milk with increasing pressure and treatment time (López-Fandiño 2006).

As a consequence of micellar disruption and protein solubilization, the amount of milk protein associated with the milk fat globules also increases by HP treatments and this could, at least partially, influence the creaming properties of HP-treated milk. The milk fat globules also increase by HP treatment and this could, at least partially, influence the creaming properties of HP-treated milk (López-Fandiño 2006).

12.3.1 Kinetics of Thermal and Pressure-Induced Changes

Hinrichs and Rademacher (2005) studied the kinetics of the combined effects of thermal and pressure-induced WP denaturation in bovine skim milk. Estimation of kinetic parameters for pressure-induced denaturation was based on isobaric experiments at different pressures. In addition, the pressure treatment was carried out at different temperatures, because the rate of pressure-induced reactions could be accelerated or decelerated due to synergistic or antagonistic effects of temperature. Specifically, skim milk was treated under isobaric conditions (200–800 MPa) in combination with isothermal conditions (1–70°C) and the levels of denaturation of β-lactoglobulin A and B, and α-lactalbumin (α-La) were analyzed.

The kinetic equations applicable for the change in a protein are (Hinrichs and Rademacher 2005):

$$\frac{dC}{dt} = -k_{\mathrm{p,T}} C^{n}; \tag{12.1}$$

$$\text{for } n = 1,\ \frac{C_{\mathrm{t}}}{C_0} = \exp(-k_{\mathrm{p,T}} t). \tag{12.2}$$

The rate constant, $k_{\mathrm{p,T}}$, in Equation 12.2 depends on both temperature and pressure. The isothermal activation volume, $\Delta V^{\#}$, can be calculated from experiments at a constant temperature and using a reference pressure, p_{ref}, say 500 MPa. For isothermal experiments, Equations 12.3 and 12.4 are applicable when $n = 1$ and when $n \neq 1$, respectively:

$$C_{\mathrm{t}} = C_0 \exp\left(k_{\mathrm{ref,T}} \exp\left\{ -\frac{\Delta V^{\#}}{RT}(p - p_{\mathrm{ref}}) \right\} t \right), \tag{12.3}$$

$$C_{\mathrm{t}} = C_0 \left(1 + (n-1) k_{\mathrm{ref,T}} \exp\left\{ -\frac{\Delta V^{\#}}{RT}(p - p_{\mathrm{ref}}) \right\} C_0^{n-1} t \right)^{(1/1-n)}. \tag{12.4}$$

The kinetic parameters (order of reaction, rate constant at reference pressure and temperature, and activation volume) in Equations 12.3 and 12.4 were estimated by means of nonlinear regression of the experimental data with an overall fit. Depending on the temperature, the determined formal reaction order, n, decreased from about $n = 3$ at 1–10°C to a value of 2 at 60–70°C, for both β-Lg fractions (β-Lg A, β-Lg B). An order of about 2.5 was estimated in the measured temperature range for the denaturation of α-La.

The isobaric activation energy, E_{a}/RT, was incorporated to obtain the equation:

$$C_{\mathrm{t}} = C_0 \left(1 + (n-1) k_{\mathrm{ref,T}} \exp\left\{ -\frac{E_{\mathrm{a}}}{RT}\left(\frac{1}{T} - \frac{1}{T_{\mathrm{ref}}} \right) \right\} C_0^{n-1} t \right)^{(1/1-n)}. \tag{12.5}$$

The activation volume, $\Delta V^{\#}$, for the β-Lg fraction decreased from around 0 at 1°C to less than −80 ml/mol at 70°C. A similar behavior was found for the temperature dependence of the activation volume of α-La. The activation volume, $\Delta V^{\#}$, of about 0 at 1°C indicated that denaturation showed no pressure dependence at this temperature. In addition, the rate constant, $k_{500,\mathrm{T}}$, increased with increasing temperature, resulting in an accelerated denaturation reaction whereby pressure and temperature act synergistically.

Thus, the higher the applied temperature the higher the pressure dependence of the denaturation reaction. The results suggest a synergistic mechanism, where the molecule unfolds more easily under pressure at elevated temperatures. However, α-La showed a much higher pressure resistance than β-Lg. The former molecule is much more stable, as α-La is stabilized by four intramolecular disulfide bonds compared to two stabilizing β-Lg. Additionally, α-La possesses no free SH groups as β-Lg does. In analogy to thermal treatment, it can be assumed that α-La is only denatured if free SH-groups are available. Due to the fast pressure-induced aggregation of β-Lg among each other and with αS2-caseins and k-caseins, not enough free SH-groups are available for the oligomerization with α-La, resulting in the observed low rate of denaturation.

12.3.2 Mechanisms of Heat- and Pressure-Induced Changes

Considine et al. (2007) noted that in general terms heat treatment and pressure treatment have similar effects: denaturing and aggregating the WPs and diminishing the number of viable microorganisms. However, it was pointed out that there are significant differences between the effects of the two treatments on protein unfolding and the subsequent thiol-catalyzed disulfide-bond interchanges that lead to different structures and product characteristics.

It was also noted that β-Lg is one of the most pressure-sensitive proteins and α-La is one of the most pressure resistant. In a heated whey protein concentrate (WPC) system, bovine serum albumin (BSA) is very sensitive and β-Lg is more resistant. In a heated milk system, β-Lg reacts with κ-casein (κ-CN) and not with αS2-CN, but in pressure-treated milk β-Lg forms adducts with either κ-CN or αS2-CN. In both treatments, the role of β-Lg is central to the ongoing reactions, involving α-La and κ-CN in heated systems but involving κ-CN, αS2-CN, and α-La in pressurized systems.

12.3.3 Structural Changes in Milk Proteins

Kanno, Mu, and Rikimaru (2002) studied the pressure-induced denaturation and gelation of β-Lg, α-La, and BSA at various concentrations at 200–800 MPa and 30°C for 10 min. The α-helix content of β-Lg was found to decrease with increasing pressure and the random structure was increased; however, the secondary structure of α-La and of BSA was not affected. The microstructure of the β-Lg gels showed a porous network, while the BSA gels had a plate-like structure. The addition of β-Lg or cysteine to α-La induced a stronger gel at 800 MPa with a similar microstructure to that of the gel from BSA.

12.3.4 Effect of Pressure Release Rates

Fertsch, Müller, and Hinrichs (2003) studied the effects of HP on the structure and firmness of whey protein isolate (WPI) and micellar casein (MC) at 600 MPa and 30°C, using a pressure buildup rate of 200 MPa/min and holding times of 0, 15, or 30 min. In addition, pressure release rate was varied between 20, 200, and 600 MPa/min. After 3 min of pressure buildup and a holding time of 0 min, gelling was only detected with a release rate of 20 MPa/min for WPI. Increasing holding time to 15 or 30 min increased gel firmness. Firmness for samples with 15 or 30 min holding times and different rates of pressure release did not differ significantly. As holding time increased, so did the number of disulfide bonds and therefore the strength of cross-linking in the gel structure. MC solution gelled under all treatment conditions, but firmness was less than for WPI gels with the same protein content. Firmness was not significantly affected by holding time but was determined by pressure release rate; the faster the pressure release, the firmer the gel. This was attributed to changes in microstructure; electron micrographs showed a fine microstructure of small particles with a pressure release time of 1 min, but several irregular aggregates of ~1 μm diameter with a release time of 3 min. Thus the firmness of pressure-induced gels depends on the type of protein present as well as the protein content.

12.4 ULTRA-HIGH-PRESSURE HOMOGENIZATION (UHPH) OF MILK

Milk acid gels are widely consumed as a healthy food and there is growing interest in producing gels that do not show syneresis during storage without added stabilizers (Lucey 2001). This is reflected in the great effort dedicated to the study of novel technologies for the processing of yogurt milk base including: enzymatic cross-linking by transglutaminase, extending shelf life involving the use of carbon dioxide, and improving several characteristics of yogurt such as fermentation time, WHC, and postacidification, which may include the use of high hydrostatic pressure.

Thus, in addition to using static HP, the effect of ultra-high-pressure homogenization (UHPH) is of interest. UHPH is based on the same principle as conventional homogenization but works at significantly higher pressures (up to 350 MPa) (Serra et al. 2007).

UHPH samples were compared to heat-treated (HT) milk both with and without skim milk powder (SMP) added, since the increase in the nonfat solid content (to E14%) of the milk base using SMP is the most common current practice in the dairy industry (Serra et al. 2009). Set-type yogurts prepared from milk UHPH treated at 200 or 300 MPa presented higher gel firmness in texture analysis, less syneresis, and lower titratable acidity compared with conventionally treated milk, even fortified with 3% SMP. In view of these previous results and with the objective of giving a practical approach to this new technology, the study of the evolution of UHPH samples during storage was made by comparison with a product manufactured following the normal process applied in industry.

To obtain stirred yogurt, the coagulum is mechanically broken before cooling and packaging, which induces considerable changes in the rheological properties, although the physical properties of stirred yogurts are also affected by the original gel characteristics. However, in order to maintain the stability of the product during the shelf life, some of the approaches adopted by manufacturers include the use of exopolysaccharide (EPS)-producing cultures, which is a common practice (Tamime and Robinson 2007) for improving the final texture of the product, because EPS contributes to a polymer-like behavior of the serum phase, which might have the ability to bind water and increase yogurt viscosity. The addition of pectin, gelatin, starch, and/or blends of these stabilizers (Tamime and Robinson 2007) is also a common way to proceed due to the ease of use in the already disrupted coagulum.

12.4.1 Structural Changes due to Ultra-High-Pressure Homogenization (UHPH) Treatment

During the UHPH treatment, partial disintegration of casein micelles has been reported (Sandra and Dalgleish 2005), resulting in an increase in the number of casein fragments in solution and hence in a higher solvation of these proteins. In addition, the partial disintegration of casein micelles is accompanied by the solubilization of colloidal calcium-phosphate (CCP) (Serra et al. 2008). As a result, it is likely that the structure of UHPH gels is dominated by casein-casein interactions rather than by WP-casein interactions, which is in line with the thicker strands observed in the microstructure of the protein matrix and with the G' values obtained. Dynamic moduli (G' and G'') are related to the number and strength of interactions responsible for the network structure (Lucey et al. 1998). Therefore, the higher values of G' obtained in HT + SMP gels could be due to the branched microstructure composed of WP and casein via disulfide bonds (Lucey, Munro, and Singh 1999).

In UHPH gels the backbone of the matrix is likely to be composed mainly of casein–casein unions by combinations of electrostatic and hydrophobic interactions. Hence, in UHPH gels the intensity of the linkages present in the network was considerably lower than those covalent bonds that constituted the matrix of HT + SMP yogurts. The solid-like behavior of the gel (G') not only depends on the magnitude of interacting forces but also on the number of them. In both types of yogurt, UHPH and HT + SMP, there was an increase in the number of interacting particles compared with nonenriched milk. In UHPH gels, this increase was due to the reduction in particle size, whereas in HT + SMP gels it was because of the higher concentration of protein (Serra et al. 2009). In this way, the reduction in fat droplet size demonstrated in earlier studies (Serra et al. 2008), due to the UHPH treatment, could also explain some of the differences observed. In HT + SMP gels, the lower number of fat droplets resulted in the formation of a continuous protein network occasionally interrupted by embedded fat globules, whereas in UHPH gels the high number of fat droplets resulted in a discontinuous protein–fat droplets network. Hence, considering the same surface in contact with the rheometer probe, less protein-protein interactions were detected in UHPH gels compared with HT + SMP gels, due to the different dispersion of fat; this fact could also explain the lower G' values of UHPH gels.

Textural analysis showed higher values of firmness of UHPH gels (Table 12.1) (Serra et al. 2009). It is important to note that in puncture tests, not only the linkages in the structure are important but the global distribution of particles also has a big effect. Microstructure of UHPH gels was much more homogeneous and compact, with large protein clusters, and less porous than that of HT + SMP gels, in agreement with results of gel density obtained in previous work (Serra et al. 2007). Therefore, the network of UHPH yogurts is more resistant to breakage than conventional yogurts that have a more branched matrix with a large number of pores. Firmness values hardly increased during cold storage, probably because the test applied was not sensitive enough to reflect the rearrangements that occurred at the particle-particle level.

Stirred gels from HT + SMP milk gave finer and less grainy structures than those from UHPH-treated milk; the main structure of the latter is hypothesized based on casein-casein unions. The

TABLE 12.1
Storage Modulus and Firmness of Set-Type and Stirred Yogurts Obtained from Ultra-High-Pressure Homogenization or Heat-Treated (HT) Milk

	Storage Time (days)				
Parameters	**1**	**7**	**14**	**21**	**28**
		Modulus, Set-Type Yogurt (G', Pa)			
HT + SMP[1]	$1481^{a/E}$	$1633^{a/C}$	$1754.5^{a/A}$	$1699^{a/B}$	$1552.1^{a/D}$
200 MPa	$1121^{b/C}$	$1293^{b/B}$	$1342^{b/B}$	$1312.3^{b/B}$	$1417^{b/A}$
300 MPa	$992^{c/C}$	$1181^{c/B}$	$1212^{c/B}$	$1179.3^{c/B}$	$1255.3^{c/A}$
		Modulus, Stirred Yogurt (G', Pa)			
HT + SMP[1]	$177.1^{c/C}$	$189.85^{c/B}$	$202.96^{c/A}$	$180.91^{c/BC}$	$184.5^{c/BC}$
200 MPa	$246.48^{b/B}$	$264.36^{b/AB}$	$274.13^{b/A}$	$264.16^{b/}$AB	$260.16^{b/AB}$
300 MPa	$295.95^{a/C}$	$337.85^{a/A}$	$315.73^{a/B}$	$339.4^{a/A}$	$356.31^{a/A}$
		Firmness, Set-Type Yogurt (N)			
HT + SMP[1]	$1.42^{a/A}$	$1.42^{a/A}$	$1.43^{a/A}$	$1.37^{a/B}$	$1.42^{a/A}$
200 MPa	$1.83^{b/AB}$	$1.79^{b/B}$	$1.81^{b/B}$	$1.86^{b/A}$	$1.87^{b/A}$
300 MPa	$1.99^{c/D}$	$2.13^{c/C}$	$2.27^{c/AB}$	$2.33^{c/A}$	$2.23^{c/B}$
		Strain (Stirred Yogurt) (N)			
HT + SMP	$3.37^{a/B}$	$3.76^{a/B}$	$3.69^{a/B}$	$3.98^{a/A}$	$3.58^{a/B}$
200 MPa	$4.92^{b/A}$	$4.52^{b/B}$	$4.52^{b/B}$	$4.47^{b/B}$	$4.87^{b/A}$
300 MPa	$5.73^{c/A}$	$5.18^{c/BC}$	$4.87^{c/C}$	$5.41^{c/AB}$	$5.43^{c/AB}$

[a–c] Different superscripts denote differences ($p < .05$) between different treatments within the same parameter
[A–D] Different superscripts denote differences ($p < .05$) between different days of storage within the same treatment
[1] Heat-treated milk at 90°C, 90 s, homogenized at 15 MPa and fortified with 3% SMP
Source: From Serra, M., Trujillo, A. J., Guamis, B., and Ferragut, V., *Food Hydrocolloids*, 23, 82, 2009. With permission.

heterogeneity in the structure of stirred gels from UHPH-treated milk could explain the higher G' values obtained, since the presence of microclusters implies more junctions between larger particles in dispersion. During cold storage, G' values increased until day 14 in all cases (Table 12.1), probably due to overacidification and a tendency to recover the gel structure. It is noted that acidification during cold storage enhances casein-casein unions by means of the intensification of hydrophobic interactions and solubilization of CCP. In addition, in stirred gels, one can expect the gels to regain their structure after destructive treatments

12.5 MYOFIBRILLAR PROTEINS

Iwasaki et al. (2006) reported that the microstructure of combined pressure-heat-induced chicken myofibrillar gel was composed of three-dimensional fine strands. In addition, pressurization at 200 MPa prior to heating increased the apparent elasticity of chicken myofibrillar gel and pork patty; however, pressure treatment above 200 MPa decreased it. The apparent elasticity of the pressure-treated (200 MPa) thermal myofibrillar gel was three times higher, and that of pork patty was twice as high as those of the unpressurized ones. The rheological properties of the low-salt (1% NaCl) pork sausage can be improved by pressure treatment at 200 MPa prior to heating.

Zamri, Ledward, and Frazier (2006) studied simultaneous heat and pressure treatments ranging from ambient temperature to 70°C and from 0.1 to 800 MPa, respectively, in various combinations. TPA of treated samples was performed to determine changes in hardness. At treatment temperatures up to and including 50°C, heat and pressure acted synergistically to increase meat hardness. However, at 60 and 70°C, hardness decreased following pressure treatments of > 200 MPa.

TPA performed on extracted myofibrillar protein gels showed similar effects of heat and pressure after treatment under similar conditions. DSC analysis of whole chicken meat samples revealed that at ambient pressure the unfolding of myosin was completed at 60°C, unlike actin, which completely denatured only above 70°C (Zamri, Ledward, and Frazier 2006).

Effects of HP processing on isolated myofibrillar proteins and myosin from cod were investigated, and results were compared with those using myofibrillar proteins from turkey meat (breast) (Angsupanich, Edde, and Ledward 1999). When turkey breast muscle and isolated myofibrillar protein and myosin of cod or turkey (pH~7) were subjected to pressures up to 800 MPa for 20 min, DSC and SDS-PAGE indicated that HP-induced denaturation of myosin led to formation of structures that contained hydrogen bonds and were additionally stabilized by disulfide bonds. Disulfide bonds were also important in heat-induced myosin gels. Hardness of whole cod muscle, estimated by TPA, showed pressure-treated samples (400 MPa) to be harder than cooked (50°C), or cooked and then pressure-treated, or pressure-treated and then cooked samples. The results support the suggestion that pressure induces formation of heat-labile hydrogen-bonded structures while heat treatment gives rise to structures that are primarily stabilized by disulfide bonds and hydrophobic interactions. As expected, turkey myosin was more stable than that of cod; however, the pressure-induced gelation mechanisms of both myosins were similar.

In the review of Rastogi et al. (2007), it was noted that HP treatment of freshwater fish (carp, *Cyprinus carpio*) resulted in the gelling of fish paste, which is useful for product development. Breaking strength of pressure-induced carp gels was much lower than that of heat-induced carp gels or lizardfish gels. The gel-forming ability of myofibrillar proteins was increased by addition of transglutaminase. The gel strength and brightness of the pork paste gel was found to increase with an increase in pressure. The higher the temperature or the sodium chloride concentration used for gelation, the lower was the pressure needed to denature the protein. Changes in rheological properties of bovine myofibrillar proteins in solution were influenced by structural changes caused by HP treatment. Further, increase in process pressure and holding time resulted in decrease in viscosity and shift toward the Newtonian flow behavior.

Chapleau and Lamballerie-Anton (2003) used response surface methodology to study the effect of pressure (0–600 MPa) and time (0–1800 s) on the surface hydrophobicity, reactive sulfhydryl groups content, and the flowing properties of bovine myofibrillar proteins in 10 g/l solution. High-pressure treatment induced a threefold increase in the surface hydrophobicity of myofibrillar proteins between 0 and 450 MPa. The same upward trend was observed with the reactive sulfhydryl groups, whose content increased from 40 to 69%. Concerning rheological properties of solutions, the flow behavior index tended toward a maximum value close to Newtonian behavior ($n = 1$), whereas the viscosity decreased with the increase of pressure.

12.6 SOY PROTEINS

The minimum pressure required for inducing gelation of soy proteins was reported to be 300 MPa for 10–30 min and the gels formed were softer with lower elastic modulus in comparison with HT gels (Okamoto, Kawamura, and Hayashi 1990). The treatment of soymilk at 500 MPa for 30 min changed it from a liquid state to a solid state, whereas at lower pressures and at 500 MPa for 10 min, the milk remained in a liquid state, but indicated improved emulsifying activity and stability (Kajiyama et al. 1995). The hardness of tofu gels produced by HP treatment at 300 MPa for 10 min was comparable to heat-induced gels.

Dynamic viscoelastic behavior of soy protein isolate (SPI) dispersions with concentrations of 10, 15, and 20% that were subjected to HP treatment at (350, 450, 550, and 650 MPa) for 15 min at 23±1.5°C was studied (Ahmed et al. 2007). The frequency sweep rheological data (0.1–10.0 Hz showed that the elastic modulus (G') predominated over the viscous component (G'') for all concentrations. Gel rigidity of pressurized samples showed no systematic pattern with applied pressure. However, concentration significantly increased the mechanical strength of the gels. Calorimetric studies confirmed denaturation of the SPI dispersions at 350 MPa irrespective of concentration. Electrophoresis results (both native and SDS) showed insignificant conformational changes in the protein subunits of the processed samples. At similar concentrations, the firmness of pressure-treated gels was lower than that of thermally induced gels.

Molina and Ledward (2003) studied the textural properties: adhesiveness, springiness (elasticity), and hardness, as well as the water-binding capacity of gels from 12% dispersions of SPI and 11S and 7S globulin fractions subjected to heat (90°C for 15 min) before or after pressurization at 300–700 MPa. Gelation only occurred for the 11S fraction after either heating or pressurization. Pressurization followed by heating (PHT) produced gels in all the dispersions; however, heating followed by pressurization (HP) produced a gel only in the 11S fraction. Adhesiveness of 11S PHT gels was not much affected by pressure, whereas elasticity tended to decrease and hardness to increase as pressure increased. Water loss increased with pressure applied to 11S PHT gels. Adhesiveness of SPI gels followed the trends of 11S gels, but values were higher; the same trend was observed for water-binding capacity. The 7S PHT gel had no adhesiveness at any applied pressure, while the elasticity decreased as pressure increased to 600 MPa and then 700 MPa; hardness decreased as pressure increased from 300 to 500 MPa, then remained stable. Water-binding capacity increased with pressure applied in 7S PHT gels.

Apichartsrangkoon (2003) studied the effects of HP processing (at 200–800 MPa at 20 or 60°C) on the rheological properties, protein solubility, and SDS-PAGE profiles of hydrated (moisture content of 80%) commercial soy protein concentrates (SPC). Shapes of the storage (G') and loss (G'') modulus curves changed only slightly with temperature/pressure treatments. Overall, increasing temperature gave rise to greater rheological changes than did pressure. Chemical analysis of gel systems suggested limited disulfide bonding occurred as a result of pressure and temperature.

Puppo et al. (2004) demonstrated that the application of HP (200–600 MPa) on SPI at pH 8.0 resulted in an increase in protein hydrophobicity and aggregation, a reduction in free sulfhydryl content, and a partial unfolding of the 7S and 11S fractions. A change in the secondary structure leading to a more disordered structure was also reported. At pH 3.0, the protein was partially denatured and insoluble aggregates were formed, the major molecular unfolding resulted in decreased thermal stability, and increased protein solubility and hydrophobicity.

Puppo et al. (2008) studied the effect of the combined T/HP (T 20–60°C/HP 0.1–600 MPa) treatment on physicochemical and rheological properties of emulsions prepared with native SPIs at 7% (w/v). The size and aggregation of oil droplets of emulsions prepared with SPI 7% (w/v) solutions were not altered by the combined T/HP treatment and the emulsions did not flocculate or coalesce. Simultaneously, a significant increase in the apparent viscosity with increase in pressure was observed, which was reinforced by temperature. The phenomenon was attributed to the gelation of nonadsorbed soybean proteins. Temperature seemed to improve the gelation process but only up to 400 MPa. At higher pressures, the combined effect of temperature and pressure resulted in dissociation of the protein aggregates, decreasing gelation.

12.7 EGG PROTEINS

Rastogi et al. (2007) summarized the effect of HP treatment on egg proteins. The conformation of the main protein component, ovalbumin, of egg white remained fairly stable when pressurized at 400 MPa, probably due to the four disulfide bonds and noncovalent interactions stabilizing the three-dimensional structure of ovalbumin. Egg yolk formed a gel when subjected to a pressure of 400 MPa for 30 min at 25°C, kept its original color, and was soft and adhesive. The hardness of the pressure-treated gel increased and adhesiveness decreased with an increase in pressure.

12.8 EFFECT OF ULTRASOUND ON RHEOLOGICAL PROPERTIES OF SOME FOOD MATERIALS

High-frequency and low-energy (~100 mW) ultrasounds* have long been employed for nondestructive testing (Gunasekaran and Chiyung 1994). The properties of low-intensity ultrasounds can be found in McCarthy, Wang, and McCarthy (2005). Since the 1980s, there have been many reports on the relationship between ultrasonic parameters and the composition of food products as well as their texture (Andersen et al. 1983; Benedito et al. 2002; Benedito et al. 2001; Benedito et al. 2000; Ghaedian et al. 1998; Mizrach 2007; Mizrach et al. 1994, 1999; Nielsen, Martens, and Kaack 1998). Power ultrasonic process, employing 20–100 kHz higher-energy ultrasounds, has been employed in the food industry (Patist and Bates 2008). High-energy (10–1000 W/cm^2) ultrasound has been reported to have been used for cleaning surfaces and in some food processing unit operations (Ohlsson and Bengtsson 2002). The effects of power ultrasound on chemical reactions have been reported since it was discovered in the late 1800s (Suslick 1994; Ashokkumar et al. 2008). Transmission of ultrasound through a medium causes cavitation to occur. According to Suslick (1988), cavitation is the formation, growth, and, sometimes, implosion of microbubbles created in a liquid when ultrasound waves propagate through it. The collapse of the bubbles results in hot spots where temperatures above 5000°C and pressures of about 500 MPa have been observed. As a result of cavitation, three mechanisms, which are thermal energy dissipation, water sonolysis, which can cause generation of free radicals, and mechanical action (shear) occur. The three mechanisms can act alone or in combination. Ultrasonic treatment has been reported to aid, for example, separation (Xia et al. 2008; Zhang et al. 2005), extraction (Rodrigues, Pinto, and Fernandes 2008; Hielscher 2006; Vilkhu et al. 2008), homogenization and emulsification (Kentish et al. 2008; Hielscher 2006; Jafari, He, and Bhandari 2007), drying (de la Fuente-Blanco et al. 2006), freezing (Zheng and Sun 2006), sterilization (Piyasena, Mohareb, and McKellar 2003), and filtration (Muthukumaran et al. 2007). This review article focuses on the effect of power ultrasonic treatment on rheological properties of some food materials.

* Ultrasounds are sound waves having frequencies above the audible frequency range, that is, greater than 20 kHz. The use of ultrasound in the food industry can be classified into low-frequency (10–100 kHz) applications, where the ultrasonic beam is transmitted through air, and high-frequency (500 kHz–5 MHz) applications, where the beam is transmitted through liquid or, in some cases, solid wall (Denbow 2001).

12.8.1 Effect of Ultrasonic Treatment on Starch

Starch appears in nature in the form of granules consisting of amylose and amylopectin, which align to form amorphous and crystalline regions. When dispersed in water, starch granules can imbibe a limited amount of water and swell. The swelling is reversible and no noticeable change occurs to either the dispersion or the granules. If one subjects the starch dispersion to heat, the granules imbibe more water and swell to a greater extent. When heating up to a temperature called the "onset gelatinization" temperature, melting of crystalline region takes place. The swelling and consequent changes from this point on would be irreversible. The dispersion becomes more viscous and the opaque dispersion turns somewhat transparent. The viscosity of gelatinized starch dispersions depends on many factors including starch concentration, the amount of amylose and amylopectin in the starch, the presence of other solutes or molecules in the system, and the heat and/or other mechanical force applied to the dispersion during gelatinization.

Sonication at sufficiently large amplitude leads to cavitations with consequent bubble formation (Hielscher 2006; Suslick 1994). The collapse and implosion of cavitation bubbles results in the erosion and, possibly, modification of starch granule structure (Degrois et al. 1974). This was evidenced by scanning electron micrographs of starch granules subjected to ultrasonic treatments. Azhar and Hamdy (1979) reported that degradation of potato and sweet potato starch that caused a reduction in starch paste viscosity was due to the effect of ultrasound on granules, not starch polymers; this was demonstrated by the unchanged enzymatic hydrolysis rate of starch by beta-amylase. Chung et al. (2002) noted a similar finding on the effect of sonication on mung bean, potato, and rice starches. Lipatova, Losev, and Yusova (2002) treated 1–8% starch hydrogels with 22 kHz ultrasound, with acoustic energy of about 1.34 W/cm^3, for 5–60 s. Their result (Figure 12.1) showed that the relative viscosity of the starch hydrogels decreased with increasing sonication time. Concomitant with the reduction in relative viscosity (η/η_0), they found an increase in starch grain cleavage (DC), starch solubility (A), and a reduction in the suspension optical density (D). In comparison with treating starch hydrogel with heat at 105°C and 1.1 atm, sonication at 22 kHz and 40°C caused a greater reduction in starch hydrogel relative viscosity (Figure 12.2). Recent work of Huang, Li, and Fu (2007) showed that ultrasonic treatment of cornstarch caused a decrease in breakdown viscosity but an increase in setback viscosity determined with an Rapid Visco Analyzer (RVA). They noted that the crystalline structure of cornstarch did not change but the amorphous area was slightly destroyed. According to Lipatova, Losev, and Yusova (2002), "The disintegration of the primary structure of starch hydrogels

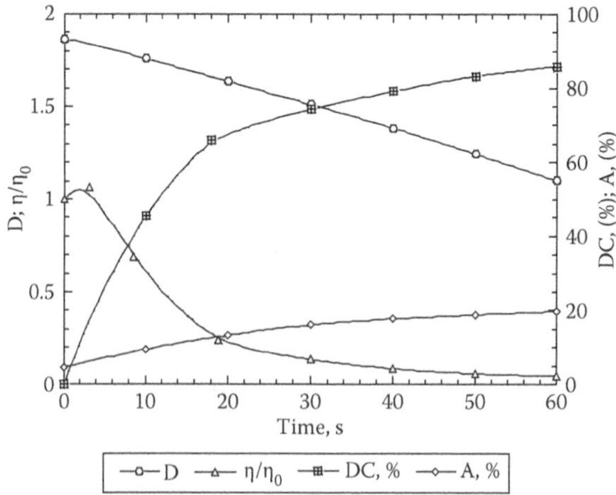

FIGURE 12.1 Optical density D (λ = 400 nm), relative viscosity η/η_0, degree of cleavage DC, and content of water-soluble fraction, A, of starch hydrogels vs time of ultrasonic treatment (22 kHz, 1.34 W/cm^3). (Adapted from Lipatova, I. M., Losev, N. V., and Yusova, A. A., *Russian Journal of Applied Chemistry*, 75, 526–30, 2002. With permission.)

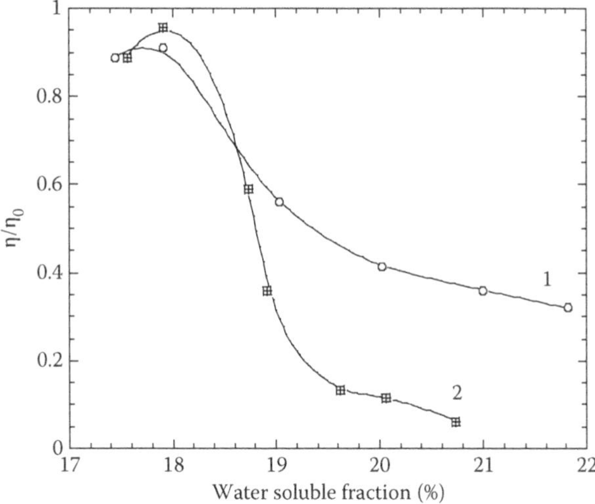

FIGURE 12.2 Relative viscosity (η/η_0) versus water-soluble fraction in cleavage of 5% starch hydrogel by (1) thermal treatment at 105°C and 1.1 atm and (2) ultrasonication at 22 Hz and 40°C. (From Lipatova, I.M., Losev, N.V., and Yusova, A.A., *Russian Journal of Applied Chemistry*, 75, 526–30, 2002. With permission.)

under the action of ultrasonic vibrations is, in fact, the process of dispersion. The term [dispersion] is referred to two different processes: dispersion proper, i.e., disintegration of monolithic particles, and deagglomeration, i.e., rupture of bonds between particles in the agglomerate." However, an earlier work of Szent-Györgyi (1933) showed that ultrasonic treatment caused depolymerization of starch molecules, which was shown by results of an iodine test: a change from blue to red color.

Zhang et al. (2005) employed ultrasonic treatment to enhance separation of starch from corn endosperm; they also found that the ultrasound-treated starches gave higher paste viscosity. In an investigation by Iida et al. (2008), the researchers found that 20–30% starch gel was liquidized and the viscosity of 5–10% gelatinized starch dispersion was reduced after sonication. Sonication (100 W/500 g sample) at 60°C for 30 min caused the viscosity of gelatinized 5% and 10% waxy maize, potato, tapioca, sweet potato, and corn starch dispersion to reduce about two orders of magnitude. A lower sonication frequency caused more reduction in starch dispersion viscosity. Sonication at higher temperature was also more effective in reducing the dispersion viscosity. It was reported that the molecular weight peak of waxy maize starch was reduced by about one order of magnitude after sonication for 30 min and continued to reduce with increasing sonication time. Selected data on the viscosity of ultrasonicated (US) and untreated (silent) 5% waxy maize, potato, tapioca, sweet potato, pectin, and 1% glucomannan dispersion heated at 90°C for 60 min are shown in Table 12.2.

12.8.2 Effect of Ultrasonic Treatment on Other Polymeric Food Materials

In other food polymeric materials also, ultrasonic treatment imparts similar effect on viscosity as it does to starch polymers. It is known that power ultrasound could be employed to cleave polymeric molecules (Price, West, and Smith 1994), which causes a reduction in solution viscosity. The polymer cleavage is a result of shear force involving the movement of solvent molecules around collapsing ultrasonic cavitation bubbles. It was reported that, below a certain molecular weight, degradation by ultrasonic treatment does not take place (Grönroos, Pentti, and Hanna 2008). Earlier studies stated that the rate of polymer degradation decreases with decreasing polymer molecular weight (Mason and Peters 2002; Suslick 1988; Grönroos, Pirkonen, and Kyllonen 2008) and increasing polymer concentration (Mason and Lorimer 2002). Portenlänger and Heusinger (1997) reported that ultrasonic degradation of dextran did not occur when the molecular weight was lower than 40 kDa. They also showed that lower ultrasonic frequency, i.e., 35 kHz, had a greater degradation effect and resulted in dextran with lower

TABLE 12.2
Changes in the Viscosity of Heated (90°C for 60 min) Starches and Polysaccharides by Sonication

Samples		Viscosity (mPa·s)				
		60°C	50°C	40°C	30°C	20°C
5% Waxy maize	Silent	254	300	372	440	586
	US[a]	8	6	8	10	14
5% Potato	Silent	528	782	2100	3110	5410
	US[a]	8	8	10	16	20
5% Tapioca	Silent	574	900	1420	2110	3200
	US[a]	6	8	10	14	16
5% Sweet potato	Silent	656	760	912	1140	1310
	US[a]	4	6	6	8	12
1% Glucomannan	Silent	3480	4480	4400	6520	7600
	US[a]	64	78	104	178	250
5% Pectin	Silent	230	206	480	742	1110
	US[a]	134	190	296	424	700

[a] Sonication at 120 W/100 g was carried out at 60°C for 30 min and the viscosity was measured at 60°C, 50°C, 40°C, 30°C, and 20°C with the same sample but without additional agitation. The data on the table are the average of two replications. "US" and "Silent" denote ultrasonicated and untreated samples, respectively.

Source: Adapted from Iida, Y., Tuziuti, T., Yasui, K., Towata, A., and Kozuka, T., *Innovative Food Science & Emerging Technologies*, 9, 140–46, 2008. With permission.

molecular weight distribution compared with higher ultrasonic frequency, i.e., 0.5, 0.8, and 1.6 MHz (Figure 12.3). Szu et al. (1986) noted that the ultrasonic depolymerization rate of some bacterial polysaccharides depended on solvent viscosity and polysaccharide concentration. Ultrasonic treatment of all studied polysaccharides resulted in depolymerization to a similar finite molecular weight around 50 kDa. Seshadri et al. (2003) treated pectin solutions (1.15 wt% pectin, 41.4 wt% sucrose) with high-intensity ultrasound (0–40 W/cm^2) for various times (0–60 min). Their result showed that the flow behavior index (n) of the dispersion increased from 0.6 for untreated dispersion to 0.97 for ultrasonically pretreated (40 W/cm^2, 30 min) dispersion while the consistency coefficient (K) decreased from 1.93 Pa·s^n to 0.09 Pa·s^n, respectively. The rate of pectin gelation decreased as ultrasonic intensity and application time increased. Results were attributed to an overall reduction in the average molecular weight of pectin due to ultrasonic cavitation. Mohammadifar et al. (2006) reported that treating tragacanthin solution with 20 kHz 100 W ultrasound for 6 min reduced the polymer molecular weight by about one order of magnitude. The intrinsic viscosity of the solution reduced from 19.6 dl/g to 4.42 dl/g. Baxter, Zivanovic, and Weiss (2005) noted that treating chitosan in acetic acid solutions with high-energy ultrasound resulted in a reduction of chitosan molecular weight (Table 12.3) but not its degree of deacetylation. They also reported that the intrinsic viscosity of samples decreased exponentially with increasing sonication time and the rates of intrinsic viscosity decrease increased linearly with ultrasonic intensity. Grönroos, Pentti, and Hanna (2008) showed that ultrasonic degradation of carboxymethylcellulose (CMC) caused a permanent reduction in CMC solution viscosity. The rate of degradation was reported to be dependent on CMC's initial molecular weight and concentration.

12.8.3 Effect of Ultrasonic Treatment on Food Containing Suspended Particles

In a food containing suspended particles, the effect of ultrasound depends upon the extent of particle degradation. Generally, large size particles give rise to high viscosity. Particle size reduction results in a decrease in viscosity. However, further particle size reduction could result

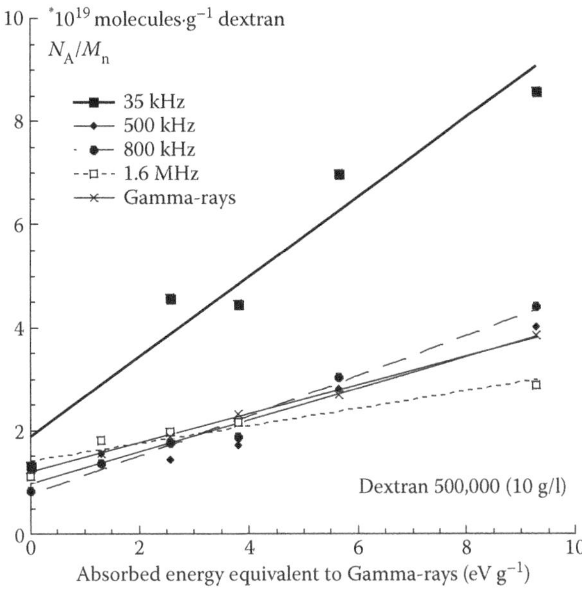

FIGURE 12.3 Number of bond breaks; N_A/M_n (N_A is Avogadro's number, M_n is number average) in dependence of frequency and absorbed energy that is proportional to irradiation time for the degradation of dextran 500,000 ($c = 10$ g/l, dose rate = 0.6±0.05 kGy/h). (From Portenlänger, G., and Heusinger, H., *Ultrasonics Sonochemistry*, 4, 127–30, 1997. With permission.)

TABLE 12.3
Average Molecular Weight of Chitosan Dispersions Calculated from Intrinsic Viscosity Using the Mark–Houwink Parameters $a = 1.1$ and $K = 2.14 \times 10^{-3}$ ml/g

	16.5 W/cm²		28.0 W/cm²		35.2 W/cm²	
Sonication Time	M_w	ΔM_w	M_w	ΔM_w	M_w	ΔM_w
0	867,191	61,117	867,191	61,117	867,191	61,117
0.5	817,339	69,561	815,117	55,220	741,614	62,921
1	803,932	79,496	768,425	35,806	584,547	65,037
5	486,764	39,679	360,799	11,698	249,640	12,057
15	368,853	15,437	241,220	26,696	167,566	29,344
30	325,469	9364	181,141	22,189	140,983	8589

Source: Adapted from Baxter, S., Zivanovic, S., and Weiss, J., *Food Hydrocolloids*, 19, 821–30, 2005. With permission.

in an increase in viscosity (Beresovsky, Kopelman, and Mizrahi 1995) because the number of particle-particle interactions increases. Vercet et al. (2002) showed that tomato paste treated with manothermosonication (MTS), that is, high energy sonication under heat and moderate pressure, exhibited higher apparent viscosity, higher consistency index, higher yield stress values, and less serum liberated compared to that thermally treated. Figure 12.4 shows flow curves of tomato juice thermally treated (a) and manosonicated (b). Despite having observed that MTS caused pectin molecules to breakdown, the researchers noted that the disintegration of other suspended tomato components by ultrasonic cavitation could be responsible for the increase in paste viscosity. Apart from resulting in an increase in particle interactions, smaller particles fit better in the pectin gel networks, causing an increase in viscosity (Beresovsky, Kopelman, and Mizrahi 1995).

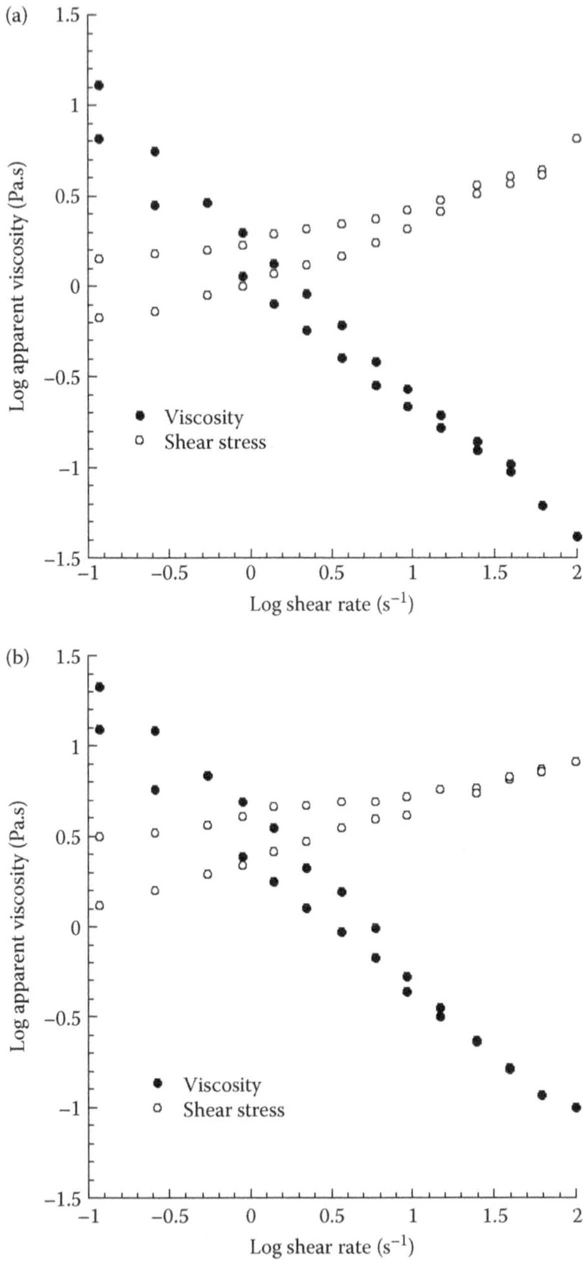

FIGURE 12.4 Typical flow curves of thermal (a) and MTS-treated (b) tomato juice. (From Vercet, A., Sanchez, C., Burgos, J., Montanes, L., and Lopez Buesa, P., *Journal of Food Engineering*, 53, 273–78, 2002. With permission.)

12.9 CONCLUSIONS

HP treatments influence the functional properties of proteins through the disruption and reformation of hydrogen bonds, as well as hydrophobic interactions and the separation of ion pairs. The rheological properties studied include viscosity and dynamic viscoelasticity. In addition, textural characteristics of solid foods have been studied. The changes in those properties have been found to depend on a number of factors. Numerous studies have been conducted on high-pressure processing of milk and its proteins, including the kinetics and mechanisms of thermal- and pressure-induced changes. The effect of UHPH of milk on its acid gels was studied. The rheological and textural properties of myofibrillar proteins, soy protein, and egg proteins have also been studied.

Sonication at sufficiently large amplitudes leads to cavitation and consequent formation of bubbles. The collapse and implosion of those bubbles results in the degradation of starch. The consequent decrease in dispersion viscosity was shown with sweet potato, corn, mung bean, potato, and rice starches. Ultrasonic treatment of several bacterial polysaccharides resulted in depolymerization to a finite molecular weight of about 50 kDa. However, tomato paste treated with high-energy sonication under heat and moderate pressure exhibited higher apparent viscosity and higher yield stress values, and less serum was released; the disintegration of the suspended tomato particles by ultrasonic cavitation might have been responsible for the increase in paste viscosity.

REFERENCES

Ahmed, J., A. Ayad, H. S. Ramaswamy, I. Alli, and Y. Shao. 2007. Dynamic viscoelastic behavior of high pressure treated soybean protein isolate dispersions. *International Journal of Food Properties* 10 (2): 397–411.

Andersen, B. B., H. Busk, J. P. Chadwick, A. Cuthbertson, G. A. J. Fursey, D. W. Jones, P. Lewin, C. A. Miles, and M. G. Owen. 1983. Comparison of ultrasonic equipment for describing beef carcass characteristics in live cattle (Report on a joint ultrasonic trial carried out in the U.K. and Denmark). *Livestock Production Science* 10 (2): 133–47.

Angsupanich, K., M. Edde, and D. A. Ledward. 1999. Effects of high pressure on the myofibrillar proteins of cod and turkey muscle. *Journal of Agricultural and Food Chemistry* 47 (1): 92–99.

Apichartsrangkoon, A. 2003. Effects of high pressure on rheological properties of soy protein gels. *Food Chemistry* 80 (1): 55–60.

Ashokkumar, M., D. Sunartio, S. Kentish, R. Mawson, L. Simons, K. Vilkhu, and C. Versteeg. 2008. Modification of food ingredients by ultrasound to improve functionality: A preliminary study on a model system. *Innovative Food Science & Emerging Technologies* 9 (2): 155–60.

Azhar, A., and M. K. Hamdy. 1979. Sonication effect on potato starch and sweet potato powder. *Journal of Food Science* 44 (3): 801–804.

Baxter, S., S. Zivanovic, and J. Weiss. 2005. Molecular weight and degree of acetylation of high-intensity ultrasonicated chitosan. *Food Hydrocolloids* 19 (5): 821–30.

Benedito, J., J. A. Carcel, R. Gonzalez, and A. Mulet. 2002. Application of low intensity ultrasonics to cheese manufacturing processes. *Ultrasonics* 40 (1–8): 19–23.

Benedito, J., J. A. Carcel, C. Rossello, and A. Mulet. 2001. Composition assessment of raw meat mixtures using ultrasonics. *Meat Science* 57 (4): 365–70.

Benedito, J., J. A. Carcel, N. Sanjuan, and A. Mulet. 2000. Use of ultrasound to assess Cheddar cheese characteristics. *Ultrasonics* 38 (1–8): 727–30.

Beresovsky, N., I. J. Kopelman, and S. Mizrahi. 1995. The role of pulp interparticle interaction in determining tomato juice viscosity. *Journal of Food Processing and Preservation* 19 (2): 133–46.

Chapleau, N. J., and M. I. Lamballerie-Anton. 2003. Changes in myofibrillar proteins interactions and rheological properties induced by high-pressure processing. *European Food Research and Technology* 216 (6): 470–76.

Cheftel, J. C., and E. Dumay. 1996. Effects of high pressure on dairy proteins: A review. In *Progress in Biotechnology*, edited by R. Hayashi and C. Balny. pp. 299–308, UK: Elsevier.

Chung, K. M., T. W. Moon, H. Kim, and J. K. Chun. 2002. Physicochemical properties of sonicated mung bean, potato, and rice starches. *Cereal Chemistry* 79 (5): 631–33.

Considine, T., H. A. Patel, S. G. Anema, H. Singh, and L. K. Creamer. 2007. Interactions of milk proteins during heat and high hydrostatic pressure treatments: A review. *Innovative Food Science & Emerging Technologies* 8 (1): 1–23.

de la Fuente-Blanco, S., E. Riera-Franco de Sarabia, V. M. Acosta-Aparicio, A. Blanco-Blanco, and J. A. Gallego-Juarez. 2006. Food drying process by power ultrasound. *Ultrasonics* 44 (Supplement 1): e523–27.

Degrois, M., D. Gallant, P. Baldo, and A. Guilbot. 1974. The effects of ultrasound on starch grains. *Ultrasonics* 12 (3): 129–31.

Denbow, N. 2001. Ultrasonic instrumentation in the food industry. In *Instrumentation and Sensors for the Food Industry*, edited by E. Kress-Rogers and C. J. B. Brimelow. pp. 326–402, Cambridge, UK: Woodhead Publishing.

Fertsch, B., M. Müller, and J. Hinrichs. 2003. Firmness of pressure-induced casein and whey protein gels modulated by holding time and rate of pressure release. *Innovative Food Science & Emerging Technologies* 4 (2): 143–50.

Galazka, V. B., E. Dickinson, and D. A. Ledward. 2000. Influence of high pressure processing on protein solutions and emulsions. *Current Opinion in Colloid & Interface Science* 5 (3–4): 182–87.

Ghaedian, R., J. N. Coupland, E. A. Decker, and D. J. McClements. 1998. Ultrasonic determination of fish composition. *Journal of Food Engineering* 35 (3): 323–37.

Grönroos, A., Pirakonen, P., and Kyllonen, H. 2008. Ultrasonic degradation of aqueous carboxymethylcellulose: Effect of viscosity, molecular mass, and concentration. *Ultrasonics Sonochemistry* 15 (4): 644–48.

Gunasekaran, S., and A. Chiyung. 1994. Evaluating milk coagulation with ultrasonics. *Food Technology* 48 (12): 74–78.

Hielscher. 2006. *Ultrasound in the Food Industry*. Hielscher Ultrasonics gmbh 2006 [cited December 17 2007]. Available from http://www.hielscher.com/ultrasonics/food_01.htm.

Hinrichs, J., and B. Rademacher. 2005. Kinetics of combined thermal and pressure-induced whey protein denaturation in bovine skim milk. *International Dairy Journal* 15 (4): 315–23.

Huang, Q., L. Li, and X. Fu. 2007. Ultrasound effects on the structure and chemical reactivity of cornstarch granules. *Starch/Starke* 59 (8): 371–78.

Iida, Y., T. Tuziuti, K. Yasui, A. Towata, and T. Kozuka. 2008. Control of viscosity in starch and polysaccharide solutions with ultrasound after gelatinization. *Innovative Food Science & Emerging Technologies* 9 (2): 140–46.

Iwasaki, T., K. Noshiroya, N. Saitoh, K. Okano, and K. Yamamoto. 2006. Studies of the effect of hydrostatic pressure pretreatment on thermal gelation of chicken myofibrils and pork meat patty. *Food Chemistry* 95 (3): 474–83.

Jafari, S. M., Y. He, and B. Bhandari. 2007. Production of sub-micron emulsions by ultrasound and microfluidization techniques. *Journal of Food Engineering* 82 (4): 478–88.

Kajiyama, N., S. Isobe, K. Uemura, and A. Noguchi. 1995. Changes of soya protein under ultra high hydraulic pressure. *International Journal of Food Science and Technology* 30 (2): 147–58.

Kanno, C., T.-H. Mu, and H. Rikimaru. 2002. Gel formation of individual milk whey proteins under hydrostatic pressure. In *Progress in Biotechnology*, edited by R. Hayashi and C. Balny. pp. 453–460, Elsevier Science, Kyoto, Japan.

Kentish, S., T. J. Wooster, M. Ashokkumar, S. Balachandran, R. Mawson, and L. Simons. 2008. The use of ultrasonics for nanoemulsion preparation. *Innovative Food Science & Emerging Technologies* 9 (2): 170–75.

Lipatova, I. M., N. V. Losev, and A. A. Yusova. 2002. Effect of ultrasonic field on the state of starch hydrogels. *Russian Journal of Applied Chemistry* 75 (4): 526–30.

López-Fandiño, R. 2006. High pressure-induced changes in milk proteins and possible applications in dairy technology. *International Dairy Journal* 16 (10): 1119–31.

Lucey, J. A. 2001. The relationship between rheological parameters and whey separation in milk gels. *Food Hydrocolloids* 15 (4–6): 603–608.

Lucey, J. A., P. A. Munro, and H. Singh. 1999. Effects of heat treatment and whey protein addition on the rheological properties and structure of acid skim milk gels. *International Dairy Journal* 9 (3–6): 275–79.

Lucey, J. A., C. T. Teo, P. A. Munro, and H. Singh. 1998. Microstructure, permeability and appearance of acid gels made from heated skim milk. *Food Hydrocolloids* 12 (2): 159–65.

Mason, T. J., and J. P. Lorimer. 2002. *Applied Sonochemistry. The Uses of Poer Ultrasound for Chemistry and Processing.* Wiley-VCH Verlag GmbH. Weinheim, Germany.

Mason, T.J., and D. Peters. 2002. *Practical Sonochemistry, Power Ultrasound Uses and Applications*, 2nd ed. Ellis Horwood. Publishing Limited, Chichester, UK.

McCarthy, M.J., L. Wang, and K. L. McCarthy. 2005. Ultrasound properties. In *Engineering Properties of Foods*, edited by M. A. Rao, S. S. H. Rizvi and A. K. Datta. Boca Raton, pp. 567–609, FL: Marcel Dekker.

Mizrach, A. 2007. Nondestructive ultrasonic monitoring of tomato quality during shelf-life storage. *Postharvest Biology and Technology* 46 (3): 271–74.

Mizrach, A., U. Flitsanov, Z. Schmilovitch, and Y. Fuchs. 1999. Determination of mango physiological indices by mechanical wave analysis. *Postharvest Biology and Technology* 16 (2): 179–86.

Mizrach, A., N. Galili, D. C. Teitel, and G. Rosenhouse. 1994. Ultrasonic evaluation of some ripening parameters of autumn and winter-grown 'Galia' melons. *Scientia Horticulturae* 56 (4): 291–97.

Mohammadifar, M. A., S. M. Musavi, A. Kiumarsi, and P. A. Williams. 2006. Solution properties of targacanthin (water-soluble part of gum tragacanth exudate from *Astragalus gossypinus*). *International Journal of Biological Macromolecules* 38 (1): 31–39.

Molina, E., and D. A. Ledward. 2003. Effects of combined high-pressure and heat treatment on the textural properties of soya gels. *Food Chemistry* 80 (3): 367–70.

Muthukumaran, S., S. E. Kentish, G. W. Stevens, M. Ashokkumar, and R. Mawson. 2007. The application of ultrasound to dairy ultrafiltration: The influence of operating conditions. *Journal of Food Engineering* 81 (2): 364–73.

Nielsen, M., H. J. Martens, and K. Kaack. 1998. Low frequency ultrasonics for texture measurements in carrots (Daucus carota L.) in relation to water loss and storage. *Postharvest Biology and Technology* 14 (3): 297–308.

Ohlsson, T., and N. Bengtsson. 2002. Minimal processing of foods with non-thermal methods. In *Minimal Processing Technologies in the Food Industry*, edited by T. Ohlsson and N. Bengtsson. pp. 34–60, Cambridge, UK: Woodhead Publishing.

Okamoto, M., Y. Kawamura, and R. Hayashi. 1990. Application of high pressure to food processing: textural comparison of pressure- and heat-induced gels of food proteins. *Agricultural and Biological Chemistry* 54 (1): 183–89.

Patist, A., and D. Bates. 2008. Ultrasonic innovations in the food industry: From the laboratory to commercial production. *Innovative Food Science & Emerging Technologies* 9 (2): 147–54.

Piyasena, P., E. Mohareb, and R. C. McKellar. 2003. Inactivation of microbes using ultrasound: A review. *International Journal of Food Microbiology* 87 (3): 207–16.

Portenlänger, G., and H. Heusinger. 1997. The influence of frequency on the mechanical and radical effects for the ultrasonic degradation of dextranes. *Ultrasonics Sonochemistry* 4 (2): 127–30.

Price, G. J., P. J. West, and P. F. Smith. 1994. Control of polymer structure using power ultrasound. *Ultrasonics Sonochemistry* 1 (1): S51–57.

Puppo, C., N. Chapleau, F. Speroni, M. deLamballerie-Anton, F. Michel, C. Anon, and M. Anton. 2004. Physicochemical modifications of high-pressure-treated soybean protein isolates. *Journal of Agricultural and Food Chemistry* 52 (6): 1564–71.

Puppo, M. C., V. Beaumal, N. Chapleau, F. Speroni, M. de Lamballerie, M. C. Anon, and M. Anton. 2008. Physicochemical and rheological properties of soybean protein emulsions processed with a combined temperature/high-pressure treatment. *Food Hydrocolloids* 22 (6): 1079–89.

Rao, M. A. 2007. *Rheology of Fluid and Semisolid Foods: Principles and Applications*, 2nd ed. New York: Springer.

Rao, V. N. M., and X. Quintero. 2005. Rheological properties of solid foods. In *Engineering Properties of Foods*, edited by M. A. Rao, S. S. H. Rizvi and A. K. Datta. pp. 101–147, Boca Raton and New York: CRC Press.

Rastogi, N. K., K. S. M. S. Raghavarao, V. M. Balasubramaniam, K. Niranjan, and D. Knorr. 2007. Opportunities and challenges in high pressure processing of foods. *Critical Reviews in Food Science and Nutrition* 47:69–112.

Rodrigues, S., G. A. S. Pinto, and F. A. N. Fernandes. 2008. Optimization of ultrasound extraction of phenolic compounds from coconut (*Cocos nucifera*) shell powder by response surface methodology. *Ultrasonics Sonochemistry* 15 (1): 95–100.

Sandra, S., and D. G. Dalgleish. 2005. Effects of ultra-high-pressure homogenization and heating on structural properties of casein micelles in reconstituted skim milk powder. *International Dairy Journal* 15 (11): 1095–104.

Serra, M., A. J. Trujillo, B. Guamis, and V. Ferragut. 2009. Evaluation of physical properties during storage of set and stirred yogurts made from ultra-high pressure homogenization-treated milk. *Food Hydrocolloids* 23 (1): 82–91.

Serra, M., A. J. Trujillo, P. D. Jaramillo, B. Guamis, and V. Ferragut. 2008. Ultra-high pressure homogenization-induced changes in skim milk: Impact on acid coagulation properties. *Journal of Dairy Research* 75:69–75.

Serra, M., A. J. Trujillo, J. M. Quevedo, B. Guamis, and V. Ferragut. 2007. Acid coagulation properties and suitability for yogurt production of cows' milk treated by high-pressure homogenisation. *International Dairy Journal* 17 (7): 782–90.

Seshadri, R., J. Weiss, G. J. Hulbert, and J. Mount. 2003. Ultrasonic processing influences rheological and optical properties of high-methoxyl pectin dispersions. *Food Hydrocolloids* 17 (2): 191–97.

Suslick, K. S. 1988. *Ultrasound, its Chemical, Physical, and Biological Effects*. New York: VCH.

———. 1994. The chemistry of ultrasound. In The *Yearbook of Science and the Future*. Encyclopedia Britannica, Chicago, USA.

Szent-Györgyi, A. 1933. Chemical and biological effects of ultra-sonic radiation. *Nature* 131:278.

Szu, S. C., G. Zon, R. Schneerson, and J. B. Robbins. 1986. Ultrasonic irradiation of bacterial polysaccharides. Characterization of the depolymerized products and some applications of the process. *Carbohydrate Research* 152:7–20.

Tamime, A. Y., and R. K. Robinson. 2007. *Yoghurt Science and Technology*. Cambridge, UK: Woodhead Publishing.

Vercet, A., C. Sanchez, J. Burgos, L. Montanes, and P. Lopez Buesa. 2002. The effects of manothermosonication on tomato pectic enzymes and tomato paste rheological properties. *Journal of Food Engineering* 53 (3): 273–78.

Vilkhu, K., R. Mawson, L. Simons, and D. Bates. 2008. Applications and opportunities for ultrasound assisted extraction in the food industry: A review. *Innovative Food Science & Emerging Technologies* 9 (2): 161–69.

Xia, Y., M. E. Rivero-Huguet, B. H. Hughes, and W. D. Marshall. 2008. Isolation of the sweet components from Siraitia grosvenorii. *Food Chemistry* 107 (3): 1022–28.

Zamri, A. I., D. A. Ledward, and R. A. Frazier. 2006. Effect of combined heat and high-pressure treatments on the texture of chicken breast muscle (pectoralis fundus). *Journal of Agricultural and Food Chemistry* 54 (8): 2992–96.

Zhang, Z., Yuxian N., Eckhoff, S. R., and Feng, H. 2005. Sonication enhanced cornstarch separation. *Starch/Starke* 57 (6): 240–45.

Zheng, L., and D.-W. Sun. 2006. Innovative applications of power ultrasound during food freezing processes: A review. *Trends in Food Science & Technology* 17 (1): 16–23.

13 Effect of High Pressure on Structural and Rheological Properties of Cereals and Legume Proteins

Jasim Ahmed
Polymer Source Inc.

CONTENTS

13.1 INTRODUCTION

High-pressure processing of biological macromolecules has received tremendous research interest for its potential benefits in functionality leading to newer product development with desirable structure (Ledward et al. 1995). In 1914, Bridgman reported that proteins (egg white) could be coagulated under certain conditions, demonstrating that high pressure, despite having a lethal affect on some microorganisms, could affect protein reactivity. The denaturation of proteins under high pressure is now well established. Pressure affects protein structure at the secondary, tertiary, and quaternary levels, leading in general to protein denaturation, aggregation, and gelation (Silva and Weber 1993; Balny and Masson 1993; Balny et al. 2002; Silva et al. 2001). The native structure of a protein (conformation) is the result of a delicate balance between stabilizing and destabilizing interactions within the polypeptide chains and with a solvent (Lullien-Pellerin and Balny 2002). Pressure destabilizes the balance of intramolecular and solvent–protein interactions. Water is commonly used as a high-pressure medium and it has a significant effect on protein structure under high pressure (Heremans et al. 2000). Thermal denaturation of proteins occurs with protein unfolding by breaking of covalent bonds. Contrary to thermal denaturation, pressure-induced denaturation mechanisms are somewhat different (Knorr, Heinz, and Buckow 2006). The pressure causes change/breakage in ionic bonds of protein structure (Hayakawa et al. 1996) and also pressure is able to affect the protein structure, at the secondary, tertiary, and quaternary levels, leading in general to protein denaturation (Silva and Weber 1993).

The native protein structure is a delicate balance between stabilizing and destabilizing interactions, within the polypeptide chains and with a solvent. Native proteins are stable in a constricted zone depending on the environmental factors like solvents, pH, temperature, and pressure, which act as physicochemical parameters that disturb the balance of intramolecular and solvent–protein interactions (Lullien-Pellerina and Balny 2002). The presence of water is, therefore, a prerequisite to observe changes in protein structure under high pressure.

The extent of protein dissociation appears to be dependent upon protein structure, temperature, magnitude of the applied pressure, and nature of the solvent (Masson 1992). Generally, the quaternary structure is the most sensitive to pressure. The oligomeric protein (two or more subunits) dissociates at moderate pressure (150–200 MPa) (Silva and Weber 1993). Pressure treatment at above 150 MPa induces unfolding of proteins and reassociation of subunits from dissociated oligomers. The tertiary structure has been significantly affected at above 200 MPa. Secondary structure changes take place at very high pressure (300–700 MPa), leading to nonreversible denaturation, depending on the rate of compression and on the extent of secondary structure rearrangements (Balny and Masson 1993).

Cereal and legume grains are important components in a healthy diet and provide most of the calories in the daily diet. Grains supply about 50% of the protein intake. The most important cereal crops used for human consumption are wheat, maize, and rice. Among the grain legumes, soybeans exhibit exceptional functional properties due to their high protein content compared to cereal grains. Generally, grain legumes are rich in lysine but poor in methionine content and, therefore, complement the reverse amino acid pattern found in cereals. Other important legumes for protein production with better functional properties are peanuts, beans, lentils, and peas. The protein content in cereal grains ranges between 5 and 12% and for legumes the range goes higher. At present cereal and legume protein isolates have been commercially utilized as functional ingredients in food formulations for their high nutritional value and low cost. Protein isolates are also of special interest to processors and consumers due to their low fat content and potential use as a substitute for meat, fish, and dairy products. The major functional properties of cereal and legume proteins are hydration, gelation, emulsification, foaming, and adhesion (Morr 1990). Apart from these, several potential health benefits have been attributed to the consumption of cereal- and legume-based proteinaceous foods. Application areas for protein isolates in the food processing industry include baked goods, extruded high-protein foods along with cereals, nutritional bars, cured meats, and meat analogs. Presently, commercial manufacturers produce different types of isolate according to industry demand and therefore, proper characterization of the functional properties of these products are required.

High-pressure studies on protein foods revealed that pressure treatment appreciably affects food materials: in a few cases high-pressure-assisted changes are insignificant whereas in other cases the extent of modifications becomes noticeable. Protein isolates form excellent gel with desirable water-holding capacity under high pressure. The gel structure, gel point, and its rigidity can be well characterized by rheometric measurements. Small amplitude oscillatory shear (SAOS) measurement is a useful method to study the gelation phenomenon, because it can be carried out at small strain within the linear viscoelastic regime and gelation curves can be monitored as a function of time (Nishinari et al. 1991; Clark and Ross-Murphy 1987).

The degree of protein denaturation increases with increasing pressure level and it depends upon the nature of protein, concentration, pH, and some other factors (Torrezan et al. 2007). A relatively high pressure (>700 MPa) affects protein structure and produces irreversible denaturation. To determine the structural changes of these proteins upon pressurization, different techniques are employed including Fourier transform infrared (FTIR) spectroscopy, circular dichroism (CD) spectroscopy, differential scanning calorimetry (DSC), nuclear magnetic resonance (NMR), X-ray crystallography, and electrophoresis. The information obtained by DSC is on a macroscopic level and enables the assessment of the overall structure of the protein molecule. FTIR spectroscopy—a vibrational spectroscopic technique—could provide better understanding of secondary structure changes of proteins after the influence of high pressure. FTIR spectroscopy is sensitive to protein secondary structure. It is a complementary approach to DSC because it provides direct structural information about the protein at a submolecular level (Brandes et al. 2006).

This chapter describes the effect of high-pressure processing on the rheological and structural properties of cereal and legume proteins in recent years and puts various related aspects into perspective.

13.2 EFFECT OF HIGH PRESSURE ON RHEOLOGY AND STRUCTURE OF PROTEINS

Although the main thrust of the use of high pressure initially was for food preservation, it was recently identified for its potential to change the functional properties of food biopolymers, particularly proteins. The effect of high hydrostatic pressure on the rheology and structure of proteins in aqueous solution has received extensive attention (Hayakawa et al. 1992; Cheftel 1992; Silva and Weber 1993; Heremans 1995; Johnston and Murphy 1995; Galazka et al. 1996; Ahmed et al. 2008).

13.2.1 High-Pressure Effect on Rheology

High pressure-induced protein gels have received incredible research interest among food processors and researchers, and it is worthwhile to understand their gelation properties. A gel is defined as a substantially diluted system that exhibits no flow (Ferry 1980). It is established now that pressure affects functional, rheological, and physical properties of proteins, depending on the extent of applied pressure, pH, ionic strength, concentration, etc. Gelation is an association or cross-linking of protein molecules to form a three-dimensional continuous network that traps and immobilizes water to form a rigid structure that is resistant to flow under pressure (Glicksman 1982). Protein denaturation is a prerequisite for an ordered gel formation. Pressure-induced gelation of the cereal and legume proteins leads to the formation of a three-dimensional network, which exhibits both elastic and viscous properties.

Amongst the several small strain experiments used in rheometry, including stress relaxation and creep, SAOS measurement is often used to characterize the frequency dependence of food materials including gels. SAOS can be used to examine irreversible structure development and destruction of gels. In this measurement, a sample is deformed in an oscillatory shearing flow while placed in a cone-and-plate, parallel plate, or couette geometry. Before making comprehensive dynamic measurements to probe the sample's microstructure, the linear viscoelastic region (LVER) must first be identified. The amplitude sweep is commonly used to locate the LVER of any viscous materials. The length of the LVER of the elastic modulas (G') is used as a measurement of the stability of a sample's structure, as structural properties are best related to elasticity. If the strain/stress is too large, the material will be deformed beyond its LVER, where the measured component becomes dependent on the extent of the deformation.

The elastic and viscous components of a viscoelastic material are generally represented by G' (stored energy inside a material) and G'' (energy dissipation), respectively. An increase in G' during gelation is considered as gel rigidity or stiffness for various protein foods (Clark and Lee-Tuffnell 1986). Viscoeleactic characteristics of gels can also be well described by either phase angle or complex viscosity (Ahmed 2009).

13.2.2 High-Pressure Mechanism and Effect on Structure

The native structure of proteins changes under high pressure analogous to the changes occurring at high temperatures. High pressure may affect native protein structure either reversibly or irreversibly. Thermal properties (thermal expansion, compressibility, and heat capacity) of proteins change during unfolding or denaturation. The enthalpies (ΔH) and entropies (ΔS) of unfolding are very temperature dependent, because the heat capacity of the unfolded state is significantly higher than that of the folded state (Privalov 1979). The native conformational states of proteins may usually be unfolded irreversibly due to breakage of covalent bonds and further aggregation of the molecule. However, with high-pressure treatment parts of the molecular structure remain unchanged and thus, the denaturation mechanisms are largely different. Studies on various small molecule model systems have provided information on the effect of high pressure on noncovalent intermolecular interactions (Royer 2002). One of the important observations is that hydrogen bonds are stabilized by high pressures. Hydrophobic interactions that play a substantial role in the stabilization of the tertiary structure and in protein-protein interactions are destabilized under high pressure (Balny et al. 2002).

The partial molar volume (V_i) of a protein in solution is contributed by three major constituents: (i) the volume of individual atoms; (ii) the void volume of internal cavities due to the imperfect packing of amino acid residues; and (iii) a contribution due to hydration of peptide bonds and amino acid side chains (Kauzmann 1959). It is evident that the effects of pressure and temperature both give rise to changes in the cavities and the hydration. Independent data on the cavities and the hydration are rare and most researchers are forced to make assumptions on the contributions of these factors.

The internal energy of the system at constant temperature under high pressures is almost independent of pressure, and internal interactions are affected solely by the changes in the volumes of the protein molecule and water structure (Silva and Weber 1993). Generally, denaturation by pressure can be described with a simple two-state thermodynamic formalism suggesting a transition between two states of a protein. However, the existence of pressure-induced predenaturation transitions direct to stepwise processes (Jonas and Jonas 1994).

Protein unfolding occurs with a negative molar volume change (ΔV_d) of denaturation, which is about less than 2% of the volume of the protein molecule (Silva and Weber 1993). The net value of change of molar volume includes the effects of formation and disruption of ionic bonds, changes in protein hydration and in conformation. The size of the protein hydration shell increases by attraction of new water molecules by the newly surface-exposed amino acid residues but this increase is more than compensated for by the negative contribution to ΔV_d from the disruption of electrostatic and hydrophobic interactions and disappearance of voids in the protein not accessible to solvent molecules.

The properties of water surrounding proteins are significantly affected by high-pressure treatment. A molecular dynamics simulation supported considerable changes in the protein hydration shell at 1000 MPa (Kitchen et al. 1992). The solvation shell becomes more ordered after high-pressure treatment. At a definite distance (3.5 Å), the number of water molecules increases 1.5 times per protein atom and simultaneously the number of protein to water hydrogen bonds increased. Proteins were denatured without unfolding at 1000 MPa where the simulation was studied. The absence of unfolding is most probably due to the short simulation time (100 ps); however, it is established that an increase in protein–water interactions may be a characteristic feature of pressure-induced denaturation.

X-ray structure analysis of high-pressure-treated protein molecules revealed that the extents of compressibility at different regions in the protein molecule are not same (Kundrot and Richards 1987). It was observed that one domain is almost pressure resistant whereas another domain and the interdomain region are well compressed by pressure. At a pressure level of 100 MPa, the root mean square shift for protein atoms is 0.2 Å with a few atoms moving more than 1 Å. The main-chain angles of α-helix and β-sheet are rarely affected by high pressure. Deformation of β-sheet regions is relatively lower than that of α-helices.

A quantitative description of the effects of temperature and pressure on biomaterials, particularly proteins, is available in the literature and there is little scope to discuss them here.

Table 13.1 presents major techniques used to study changes in protein structure after high-pressure treatment. The effects of high pressure on rheology and structure of individual cereal and legumes proteins are discussed individually in the following sections.

13.3 SOYBEAN PROTEIN

Among various plant proteins, soybean is the most studied for its gelation and functional properties in addition to its health benefits. Soybean proteins consist of two major components, 7S and 11S globulins (Osborne 1924; Wolf et al. 1961), which have different structures and gel properties (Nielsen 1985; Saio and Watanabe 1978; Morr 1990). Soybean protein forms an excellent gel under pressure treatment. The influence of high pressure on rheology and conformational changes

TABLE 13.1
Techniques Used To Study High-Pressure-Treated Protein Structure Change

Protein Structure	Major Techniques/Instrumentation
Primary	Not affected by pressure
Secondary	Vibrational spectroscopy
	Fourier Transform Infrared (FTIR) spectroscopy
	Circular dichrosim (CD) spectroscopy
Tertiary	NMR spectroscopy
	X-ray analysis
	UV-Vis fluorescence spectroscopy
Quaternary	Electrophoresis
	Fluorescence spectroscopy
	NMR spectroscopy

Source: Adapted from Mozhaev, V. V., Heremans, K., Frank, J., Masson, P., and Balny, C., Proteins: Structure, Function, and Genetics, 24, 81–91, 1996. With permission.

of soybean proteins (β-conglycinin and glycinin) was studied by rheometric and textural analysis, infrared spectroscopy, sulfhydryl group detection, spectrofluorimetry, CD, DSC, and electrophoresis. Some of those results are discussed below.

13.3.1 Rheology

Frequency sweep of pressure-treated soy protein isolate (SPI) dispersions (> 10% w/w) showed the elastic behavior at 20°C (Ahmed et al. 2007). Figure 13.1 is a typical dynamic rheological pattern of a pressurized SPI gel. The G' of pressure-treated SPI dispersions predominated over G'' in the frequency range of 0.1–10 Hz. A slow change of G' with frequency indicates a perfect elastic nature of the pressurized gel (Ferry 1980).

Application of high pressure (350–650 MPa) did not result in any significant change in the mechanical strength (G') of SPI dispersions (10–20% w/w), whereas protein concentration played a very important role in gel strength (Ahmed et al. 2007). A 5% (w/w) pressure-induced SPI sample could not form a gel and G' of the dispersion was found to be two orders higher in magnitude than that of pure water (Alvarez, Ramaswamy, and Ismail 2008). It has been observed that a high-pressure-induced gel can be detected well by rheometry at concentration levels higher than 10% (w/w). The G' of 15% (w/w) SPI dispersion increased after pressure treatment at 450 MPa, reached a peak, and decreased on further rise in pressure to 650 MPa (Ahmed et al. 2007). Interestingly, control and pressure-treated samples at 650 MPa (treatment time of 15 min) showed almost similar magnitude of G', particularly at higher frequency (1–10 HZ). The G' of 20% (w/w) SPI after pressure treatment (450–650 MPa) decreased significantly. These observations indicate that the material started to undergo stress on its structure. Pressure resistance of soy proteins containing solely 11S protein has been reported (Ahmed et al. 2006; Zhang et al. 2003) and it is believed that stronger mechanical strength of pressurized soy gel was contributed by the 7S fraction.

The viscous modulus, G'', provides viscous flow characteristics of SPI dispersion after high-pressure treatment. At lower concentration (0.32–3.68%) and selected pH range (2.66–4.34 and 5.16–6.84) pressurized (200–700 MPa) SPI dispersion behaved as a viscous fluid (Torrezan et al. 2007). The G'' of 5% pressure-induced SPI sample was similar to that of distilled water (Alvarez et al. 2008) whereas the G'' of 10% SPI decreased as pressure was increased from 350 to 650 MPa and the magnitude was found to be lower than that of untreated sample. A sudden drop of G'' at 350 MPa irrespective of concentration led to the conclusion that the protein had been denatured. A complicated G''-P profile was noticed for 15 and 20% SPI concentrations (Ahmed et al. 2007).

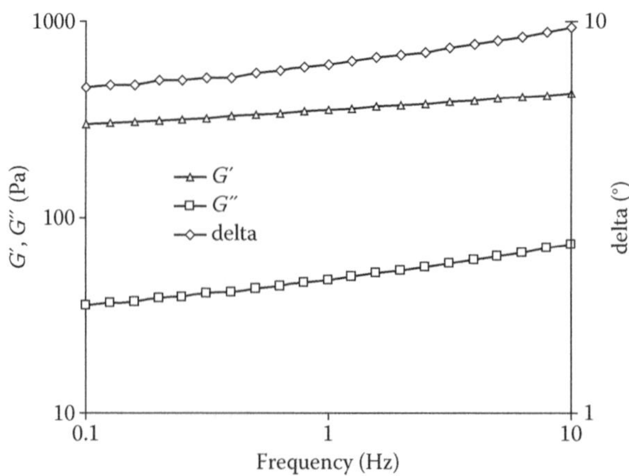

FIGURE 13.1 Typical dynamic rheological pattern of pressurized SPI dispersion.

The viscoelasticity of pressure-treated SPI gels was mathematically calculated from slopes of the linear regressions of ln ω (angular frequency) vs ln G' (elastic modulus) and ln G'' (viscous modulus) following the power-type relationship (Equation 13.1); the magnitudes of slopes are shown in Table 13.1.

$$G' \text{or } G'' = K\omega^n, \tag{13.1}$$

where K (intercept) and n (slope) are regression coefficients for the above equation. The slopes (0.07 and 0.01) indicated that G' is relatively independent of frequency in the concentration range of 15–20%. A true gel is characterized by a zero slope of the power law model and viscous materials demonstrate positive slopes (Ross-Murphy 1984). Pressure treatment decreased the G' of gel in comparison to untreated sample whereas elasticity of the pressurized gel is almost independent of pressure. Similarly, slopes of ln ω vs ln G'' increased with pressure except at 20% concentration and 650 MPa.

The complex viscosity (η*) and phase angle (δ) of pressure-treated samples further confirmed the viscoelastic behavior of SPI gels (Ahmed et al. 2007). There were no significant differences in δ (7.79 to 8.99) observed between pressure-treated and control sample in the concentration range of 15–20% whereas the 10% concentration sample behaved as a viscous fluid with a relatively higher magnitude of δ. The η* values of pressure-treated SPI increased significantly as a function of concentration (C) and followed a power-type relationship at constant frequency of 1 Hz:

$$\eta = AC^n, \tag{13.2}$$

where A and n are constants and the regression coefficients varied between 5×10^{-8} and 1×10^{-6}, and 6.99 and 8.13, respectively.

13.3.2 Textural Properties

The texture profile analysis (TPA) of pressure-treated soybean protein gels showed that the springiness and cohesiveness at similar concentration were marginally lower than thermally treated gels (Molina, Defaye, and Ledward 2002). Pressure-assisted soybean protein showed an increase in hardness and decrease in adhesiveness (Okamoto, Kawamura, and Hayashi 1990). However, no significant differences in textural properties of soybean protein isolate and its subunits (7S, 11S) were observed after pressure treatment (300–700 MPa), as shown in Figure 13.2 (Molina et al. 2002). The adhesiveness of pressure-treated 7S and SPI samples possessed higher magnitudes compared to thermally treated samples. Similar to rheological observation, the hardness of both 7S and 11S gels increased with increasing pressure and concentration. The pressure-induced hardness is found to be significantly lower than for the heat-treated gel. It is worth mentioning that relatively low protein concentrations (3.75% for 11S; 8% for 7S and SPI) could produce a pressure-induced gel (Nakamura, Utsumi, and Mori 1986), whereas a threshold protein concentration of 10% and above is prerequisite to obtain a firm gel under high pressure (Ahmed et al. 2007). The hardness during SPI gel formation does not follow a phase diagram but is influenced by the proportion of 7S and 11S present in the SPI sample. The proportions of each fraction could possibly interact with each other and enhance gel rigidity.

13.3.3 Comparison Between Pressure- and Heat-Induced Soy Protein Isolate (SPI) Gel

Most of the high-pressure-assisted SPI gels have been compared with thermally treated gels and it has been found that pressure-assisted gels are relatively softer than gels obtained by thermal treatment. Gels formed by pressure behaved as a true viscoelastic material. Ahmed et al. (2007)

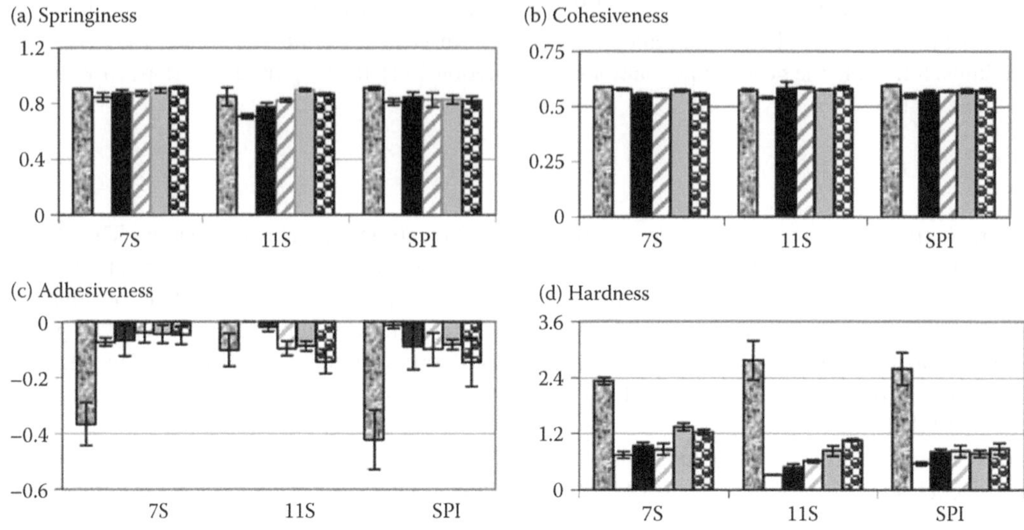

FIGURE 13.2 Textural properties of 20% (w/v) pressure-induced (300–700 MPa) 7S, 11S and SPI gel. (a) Springiness, (b) cohesiveness, (c) adhesiveness, and (d) hardness. (From Molina, E., Defaye, A. B., and Ledward, D. A., *Food Hydrocolloids*, 16, 625–32, 2002. With permission.)

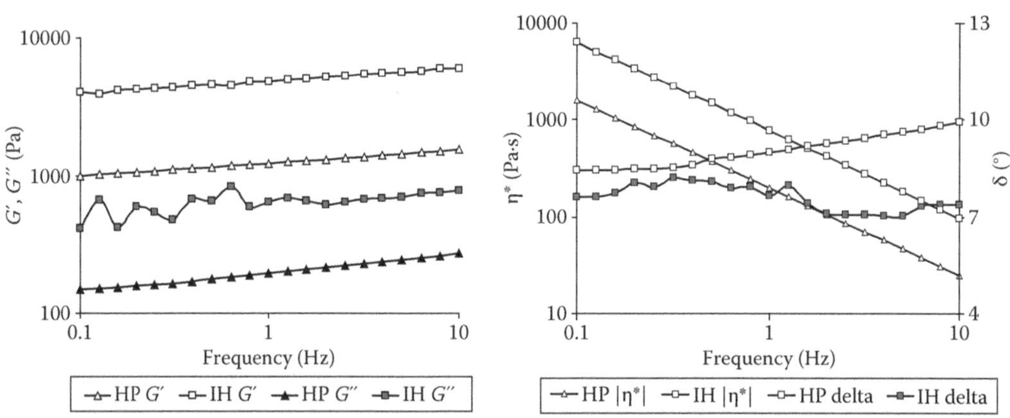

FIGURE 13.3 Comparison of thermal (IH: isothermal heating, 90°C for 30 min) and pressure effect (HP: 650 MPa for 15 min) on dynamic rheological parameters of 20% SPI dispersion. (From Ahmed, J., Ayad, A., Ramaswamy, H. S., Alli, I., and Shao, Y., *International Journal of Food Properties*, 10, 397–411, 2007. With permission.)

have compared the dynamic rheological behavior between thermal- and pressure-treated 20% SPI gels (Figure 13.3). It was found that an isothermally heated (90°C for 30 min) gel attained higher mechanical strength ($G' \approx$ 4–6 times higher), a viscous component ($G'' \approx$ 3–4 times higher), and complex viscosity (≈4 times higher) compared to the pressure-induced gel (650 MPa for 15 min). A similar increase in hardness of thermally treated gels compared to pressure-induced gel has been reported (Molina et al. 2002). A relatively higher protein content and temperature (90°C) resulted in the formation of stronger gels which required four times the rupture force compared to those formed at lower temperature (Kang et al. 1991). In addition, high heating temperature generated more reaction sites for gel networks on the surface of protein molecules (mainly hydrophobic sites) and resulted in the formation of a disordered gel (Nagano et al. 1996).

13.3.4 Differential Scanning Calorimetry (DSC) Analysis

DSC scan of untreated soy protein exhibited two distinct endothermic peaks at about 70–80°C and 90–98°C, corresponding to the thermal denaturation of the β-conglycinin (7S) and glycinin (11S) fractions, respectively (Puppo et al. 2004; Ahmed et al. 2006; Molina et al. 2001; Zhang et al. 2003; Wang et al. 2008). The peak thermal denaturation temperature (T_d) of glycinin increased gradually although the T_d was not detected at pressures higher than 350 MPa due to complete denaturation of the protein fraction (Ahmed et al. 2007; Zhang et al. 2003; Wang et al. 2008) and consequently the native structure of the glycinin was lost. A typical DSC thermogram for untreated and pressure-treated glycinin is shown in Figure 13.4. A pressure level of 200 MPa did not affect the denaturation temperature of glycinin but it decreased enthalpy by 27%. A significant drop in the enthalpy (≈90%) was detected after treatment at 400 and 600 MPa (Puppo et al. 2004). The glycinin transitions caused by high pressure were found to be irreversible (Zhang et al. 2003).

The T_d of β-conglycinin remained unaffected or insignificantly dropped in the pressure range of 0.1–600 MPa (Wang et al. 2008) although the 7S fraction of soybean protein in soymilk was completely denatured after treatment at 300 MPa (Zhang et al. 2005). The enthalpies of pressurized β-conglycinin sample were lowered by 11, 78, and 84% relative to control sample after pressure treatment at 200, 400, and 600 MPa (Puppo et al. 2004). Pressure sensitivity of 7S fraction is explained on the basis of trimer structure without any disulfide bonds, in which the subunits are associated mainly via hydrophobic interactions (Yamauchi et al. 1991).

Different structural behavior of soybean protein fractions has been observed after pressure treatment. Pressure treatments in the range of 200–600 MPa had resulted in the unfolding of 7S and 11S fractions and also the dissociation of glycinin (Puppo et al. 2004). At 400 MPa, the 7S fraction dissociated into partially or totally denatured monomers which enhanced surface activity, but the unfolding of the 11S polypeptides within the hexamer led to aggregation, negatively affecting the surface hydrophobicity of the SPI (Molina et al. 2002).

The effect of pH (3.0 and 8.0) on SPI denaturation under high pressure has been investigated (Puppo et al. 2004). Pressure treatment at 400 and 600 MPa resulted in denaturation and thermal destabilization of both 7S and 11S fractions at pH 8.0. The denaturation enthalpies were dropped to 73 and 79% after pressure treatment at 400 and 600 MPa with respect to the control sample. Denaturation enthalpy of SPI at pH 3.0 was found to be lower than the control sample at pH 8.0. It inferred that soybean proteins at acidic pH are more sensitive to thermal treatments and are partially unfolded and dissociated. Pressure treatment at 200 MPa provided an additional denaturation

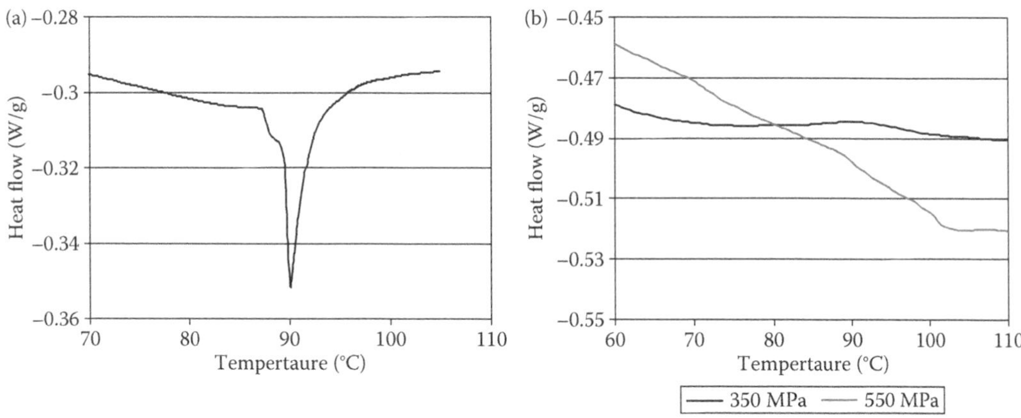

FIGURE 13.4 DSC thermogram for untreated and pressure-treated soybean glycinin (a, untreated; b, pressure treated). (From Ahmed, J., Ayad, A., Ramaswamy, H. S., Alli, I., and Shao, Y., *International Journal of Food Properties*, 10, 397–411, 2007. With permission.)

(57% compared to the control) without changes in the thermal stability of the proteins. Treatment at 400 and 600 MPa produced a total denaturation of the proteins. The control 7S fraction at pH 3.0 presented a unique endotherm at 63°C and a denaturation enthalpy about 40% lower than that corresponding to the control 7S fraction at pH 8.0. The HP treatment induced the unfolding of the protein, reaching complete denaturation at 400 MPa.

13.3.5 Circular Dichroism (CD) Analysis

CD spectroscopy is the most widely used technique to investigate secondary structures directly. It can measure any protein at a low concentration level under various solvent conditions and because it is sensitive to the local secondary structures (Matsuo et al. 2007). The secondary-structure analysis of native proteins by CD spectroscopy has been considerably improved by incorporation of various software programs that are capable of extracting the secondary structures from atomic coordinates and analyzing the CD spectra. However, limited information is available on pressure effects using CD spectroscopy to investigate the detailed secondary structures of denatured soybean proteins.

CD spectra (far UV-CD) of SPI dispersions (soluble fraction) indicated the presence of α-helix, β-sheets, β-turns, and random coils (Puppo et al. 2004), whereas the secondary structure of the native glycinin was found to be rich in β-structure (Zhang et al. 2003). The zero-crossing of the CD spectrum of native glycinin was about 201 nm, which dropped to 196 nm after high pressure treatment (550 MPa) and simultaneously the ellipticity also decreased (Zhang et al. 2003). The trough at about 208 nm in the native glycinin shifted to 204 nm after pressure treatment. CD spectra calculation indicated a significant increase in the random coil of pressure-denatured glycinin while the ordered structures α-helix and β-structure decreased simultaneously (Zhang et al. 2003). This indicated that some of the ordered structures of α-helix and β-structure were destroyed and converted to random coil after pressurization.

Pressure treatment of pH-adjusted (pH 3.0 and 8.0) SPI dispersions exhibited differences in the secondary structure by measuring the absorbance of polarized light in the 180–250 far-UV range (Puppo et al. 2004). The control sample at pH 8.0 showed a positive band near 190 nm with a zero crossing at 200 nm, and a negative band near 210 nm. After high-pressure treatment at 400 and 600 MPa, the intensity of the positive band decreased and the minimum of the negative band was shifted toward lower wavelength. The α-helix content decreased by 24%, while the percentage of β-sheets increased by 24% concurrently after pressure treatment of 200 MPa. However, the α-helix content decreased to about 15% and β-sheets remained constant at higher pressures (400 and 600 MPa). Random coil increased from 41 to 48% while pressure was increased from 200 to 600 MPa and the corresponding β-turn increased up to 13% under similar conditions.

Changes in secondary structure of pressure-treated SPI at pH 3.0 were obvious; however the differences in CD spectra as a function of pressure were less prominent than those observed at pH 8.0. The α-helix content was found to decrease from 27 to 22% while pressure level increased from 200 to 600 MPa. Similarly, the β-sheets and the random coil content increased from 19 to 22%; and 44 to 46% while β-turns remained constant (10%). From a structural point of view, the changes provoked by high-pressure treatments are more important at pH 8 than at pH 3.

13.3.6 Fourier Transform Infrared (FTIR)

Infrared spectroscopy has made progress recently for analyzing high-pressure effects on protein conformation, particularly FTIR spectroscopy (Dzwolak et al. 2002). In infrared spectroscopy, hydrogen bonding with C=O bonds and coupling of C=O vibrations are the predominant factors that determine the features of the Amide *I'* band of proteins (Jackson and Mantsch 1995).

The secondary structure of soybean protein is very complex because of the diversity of the protein components (Tang and Ma 2009). The changes in secondary structure of the soybean proteins are well characterized by FTIR spectroscopy. The amide *I′* absorption region (1700–1600 cm^{-1}) in the infrared spectrum of a protein is one of the most useful for secondary structure elucidation (Byler and Susi 1986; Susi and Byler 1988). The amide *I′* band assignments of soybean proteins are mainly composed of β-type turn and bend, β-strands, β-turns, α-helix, antiparallel β-sheet, and unordered and parallel β-sheet (Abbott et al. 1996; Byler and Susi 1986; Susi and Byler 1988; Tang and Ma 2009). The secondary structure is predominantly β-sheet and unordered, as is obvious by the bands at 1632 and 1646 cm^{-1}, respectively. Pressure treatment (250 MPa for 0.1 min) of 15% (w/w) soy protein concentrate indicated minor changes in the amide *I′* region (Alvarez et al. 2008); on the contrary the unfolding of protein conformation was obvious at 200 MPa (15 min) with a significant increase in integrated intensity of bands (Tang and Ma 2009). Changes in intensity were observed in most of the bands (1692, 1660, 1645, 1632, and 1622 cm^{-1}) (Figure 13.5). These band shifts were accompanied by an increase in relative random coil and β-type turns and a decrease in α-helix content. A shift of anti-parallel β-sheet at above 400 MPa indicated an increase in hydrogen bond strength of β-sheet structure initially buried in the interior of protein structure and pressure-induced unfolding of protein structure (Tang and Ma 2009). At and above 600 MPa, turns and bends increased compared to lower pressure levels and it is believed that the unfolded proteins underwent restoration or reassociation of secondary structure (Tang and Ma 2009).

HP treatment resulted in gradual and significant increase in absolute integrated intensity of amide II bands, in a pressure-level-dependent manner, signifying that unfolding extent of tertiary structure gradually increased with the increase in applied pressure (Tang and Ma 2008). Moreover, the wavenumbers of most of the bands in this amide II region were nearly constant by the HP treatment (at 200–600 MPa), reflecting that tertiary structure of SPI was mainly associated with hydrophobic interactions and unrelated to hydrogen bonds within the molecules (Tang and Ma 2009).

13.3.7 Change in Sulfhydryl Group

Functional properties of soybean proteins are significantly influenced by the free sulfhydryl (SH) content and disulfide (S–S) bonds; both SH groups and SS bonds undergo changes during high-pressure application. The SH content of glycinin was increased modestly at 100 MPa (Kajiyama et al. 1995) and it increased significantly when the pressure was increased to 300 MPa. A further increase in pressure (>500 MPa) lowered the SH content (Zhang et al. 2003). These observations indicate that pressure treatment favors formation of a new sulfhydryl group. The mechanism of

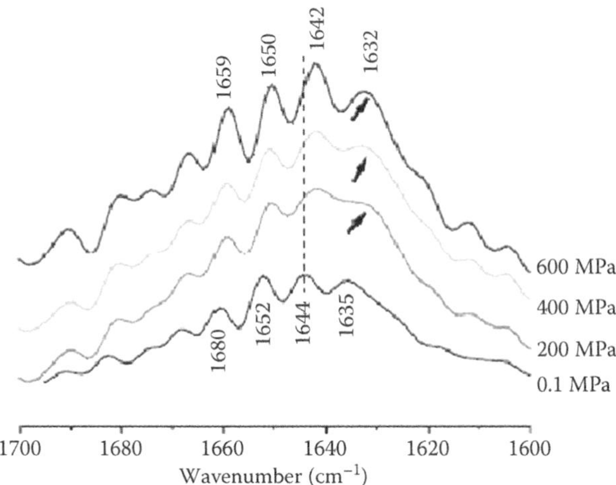

FIGURE 13.5 FTIR spectra of untreated and pressure-treated 10% soybean protein isolate (SPI) (w/w). (From Tang, C.-H., and Ma, C.-Y., *Lebensmittel-Wissenschaft und Technologie*, 42, 606–11, 2009. With permission.)

denaturation might involve the rupture of S–S bonding. Glycinin was dissociated into subunits after pressurization, creating new free SH residues, some of which recombined to give S–S exchange reactions or new S–S bonding by oxidation.

13.3.8 Solubility

The solubility of soybean seed protein was enhanced during pressure treatment. When soybean seeds were immersed in distilled water and treated at 300 MPa and 20°C for 0–180 min, the solubilized proteins accounted for 0.5–2.5% of the total seed proteins. The solubility of protein in water increased as pressure was increased, reaching a maximum value at 400 MPa. Sodium dodecyl sulfate polyacrylamide gel electrophoresis (SDS-PAGE) patterns of high-pressure-treated seeds demonstrated solubilization of 7S globulin, consisting of 27 and 16 kD bands with staining intensity increasing as pressure increased to 400 MPa. The increase in staining intensity is an indication of the amount of released protein; therefore, at higher intensity larger amounts of protein are released. At 700 MPa, 11S glycinin and 2S β-conglycinin also improved their staining intensity (Omi et al. 1996).

13.3.9 Size-Exclusion Chromatography (SEC)

Size-exclusion chromatography (SEC) in combination with multiangle laser light scattering (MALLS) has enabled great advances in characterizing protein aggregation behavior, molar mass (M_w), and hydrodynamic radius (r_h) of soluble aggregate (SA) in solution, without dependence on column calibration or reference standards (Wyatt 1993). This technique has been successfully applied to investigate heat-induced aggregates of oat and buckwheat globulin (Choi and Ma 2006; Zhao, Mine, and Ma 2004) and is also currently explored for high-pressure-assisted aggregates.

A typical SEC for pressure-treated SPI is illustrated in Figure 13.6. The molecular weight (MW) distribution of SPI increased after pressure treatment and also increased as a function of pressure (200–600 MPa) (Tang and Ma 2009). Untreated SPI sample at pH 8.0 showed a distinct peak of 813 kD, most likely corresponding to native 7S and 11S subunits, with a shoulder (50 kD), and a second minor peak of molecular mass higher than 1×10^6 kD (Puppo et al. 2004). Molina and Anon (2000) also reported an SEC peak at 917 kD for native soybean protein isolate and suggested the presence of aggregated forms constituted by 7S and 11S fractions. At 200 MPa, no changes in chromatographic profiles were noticed. At 400 MPa, the first peak was divided into two peaks and moved to

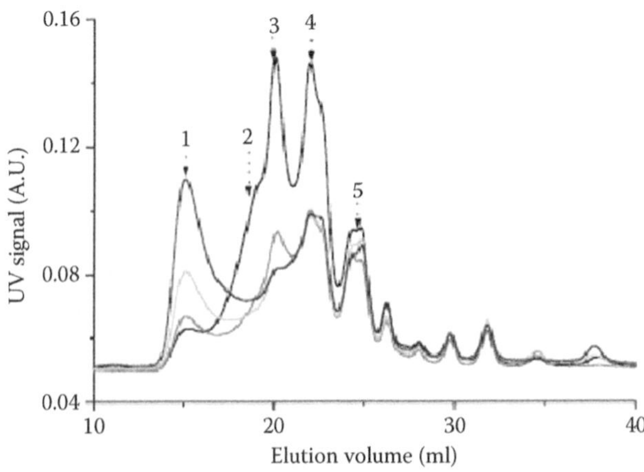

FIGURE 13.6 Typical SEC UV elution profiles of untreated and pressure-treated (200–600 MPa) SPI. Black: control; red: 200 MPa; green: 400 MPa and blue: 600 MPa. (From Tang, C.-H., and Ma, C.-Y., *Lebensmittel-Wissenschaft und Technologie*, 42, 606–11, 2009. With permission.)

low elution volumes (MM 11,343 kD). This process was more obvious at 600 MPa (MM 29,618 kD). Aggregation of SPI significantly increased after pressure treatment due to pressure-induced unfolding and denaturation of proteins, since unfolded or denatured proteins would associate to form higher MW protein products, via hydrophobic and other weak interactions (Tang and Ma 2009).

Pressure treatment at 200 MPa led to transformation of about 38% total protein constituents to insoluble aggregation (IA). The SAs increased significantly while the pressure level increased further with simultaneous decrease in IA. It infers that at a specific protein concentration, more SA would be formed at the expense of insoluble aggregate at higher pressure levels (Tang and Ma 2009).

The chromatographic profile of SPI at pH 8.0 did not change significantly after pressure treatment at 200 MPa. At 400 MPa, the first peak was divided into two peaks and moved to low elution volumes (MM 11,343 kD) and the process was more evident (MM 29,618 kD) at 600 MPa (Puppo et al. 2004). The phenomenon can be described by the relationship between aggregation and SH/S-S interchange of protein subunits. Acidic dispersion of SPI sample (pH 3.0) exhibited two major peaks at elution volumes around 190 and 220 mL, which corresponded to molecular masses lower than 66 kD. These results indicated the presence of a large amount of aggregates of high MW that remained in the insoluble fraction, so that the soluble fraction only contained dissociated polypeptides (Puppo et al. 2004).

The particle size (average diameter) distribution of SPI (pH 8.0) was not influenced by pressure treatment at 200 MPa whereas protein aggregation was detected at higher pressure (400–600 MPa) (Puppo et al. 2004). SPI at pH 3.0 showed two defined populations of proteins with variable average diameters (40 and 139 nm). The average sizes confirmed the protein aggregation as evidenced from solubility and SEC measurements. Pressure treatments at 200–600 MPa induced both aggregation and deaggregation of proteins.

The mean molar mass (Mw) of SAs significantly increased from 7.63×106 (control) to 1.63×107 g/mol at 200 MPa (Tang and Ma 2009). However, with pressure treatment above 400 MPa the mean Mw marginally decreased (5.17×107 g/mol) compared to that of the control, suggesting the occurrence of different kinds of aggregates. The initial aggregates formed with large Mw at low pressure levels (≈200 MPa) and further dissociated into smaller aggregates at higher pressure (>400 MPa). The radius-average (rh) increased from 27.4 nm (control) to 42.8 nm (200 MPa) and dropped marginally to 33.7 nm at 600 MPa. The decrease of polydispersity index (Mw/Mn where Mn is number average MW) of SA from 1.58 (control) to 1.00 (600 MPa) clearly suggested more homogeneity in molar mass of protein components after pressure treatment (Tang and Ma 2009).

13.3.10 Electrophoresis

Both native and SDS electrophoresis have been employed to study pressure effects on SPI protein fractions and subunits. Native PAGE of SPI protein is carried out in the absence of SDS and reducing reagent at ambient temperature. Electrophoresis under these conditions has provided information regarding the relative charge for molecules with the same size and shape or regarding the relative size of molecules with the same charge (Zhang et al. 2005).

Native PAGE patterns detect change of glycinin subunits without interference from SDS, reducing reagent, and heating. Native PAGE electrophoresis, before and after high-pressure treatment of the SPI dispersion, is shown in Figure 13.7. Samples appeared to have changed in the native PAGE profiles after high-pressure treatment. Disappearance and reappearance of the distinctive bands in the pressure range of 300–650 MPa have been observed (Ahmed et al. 2007; Zhang et al. 2005; Torrezan et al. 2007). It seems that the pressure treatment could have dissociated the proteins, breaking the aggregates into smaller units, some of which may associate to aggregate and become insoluble. However, at pH of 2.7, the loss of the larger unit appears to be minor (Torrezan et al. 2007; Puppo et al. 2004). Samples treated at 400 and 600 MPa showed a great amount of polypeptides of very low mobility. This could suggest that, at extremely low pH, the effect of pH is stronger than the effect of high pressure (Torrezan et al. 2007). This result reinforces the lower efficacy of

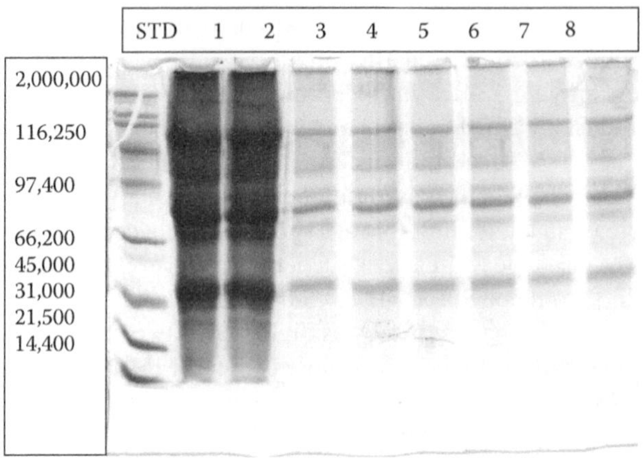

FIGURE 13.7 SDS-PAGE profile of SPI. Lanes 1 and 2, control; 3 and 4, 350 MPa; 5 and 6, 550 MPa; 7 and 8, 650 MPa, (STD) standard marker. (From Ahmed, J., Ayad, A., Ramaswamy, H. S., Alli, I., and Shao, Y., *International Journal of Food Properties*, 10, 397–411, 2007. With permission.)

high-pressure treatment at lower pH. The native PAGE of SPI sample at pH 8.0 showed no change in electrophoretic profile at 200 MPa. HP processing (400 and 600 MPa) produced aggregation of the 11S fraction, yielding large aggregates that could not enter the gel, while no changes in 7S bands were observed.

SDS-PAGE profiles of the soybean protein isolates of the control and pressurized samples demonstrated that SPI has a heterogeneous protein structure (Ahmed et al. 2007). At 100 MPa, there was in fact no shift of bands and treated samples resembled the control. A gradual decrease in the mobility and intensity of protein subunits was observed as a function of pressure (350–650 MPa) due to shifting of bands. This observation advocated high-pressure-assisted SH/SS interchange of globular proteins. However, few authors have observed no visible changes before and after high-pressure treatment while studying SDS-PAGE (Apichartsrangkoon 2003; Torrezan et al. 2007). The soy protein is completely solubilized in SDS and 2-mercaptoethanol, and some limited disulfide bonding may occur in all samples. However, additional disulfide bonding does not happen during the pressure/temperature treatments, although interchange may occur (Apichartsrangkoon 2003).

13.4 WHEAT PROTEINS

Wheat proteins are the best known proteins of cereal grains and the gluten complex plays a major role in determining the technological quality of wheat (MacRitchie 1992, 1994; Shewry 1995; Shewry et al. 1994, 1995; Shewry and Tatham 1997). Wheat flour dough is a composite system that comprises two dispersed filler phases (gas cells and starch granules) and a gluten matrix (Smith et al. 1970). Gluten mainly consists of two major storage protein fractions, namely gliadin (low MW fraction) and glutenin (high MW fraction) (about 65/45 w/w) (Redl et al. 2003), and they both contribute to rheological, structural, and baking characteristics of wheat. The ratio of the rheological properties of the matrix to those of the fillers determines the behavior of the composite (Nielsen 1974). Wheat gluten proteins have unique viscoelastic properties and are capable of forming three-dimensional structures in aqueous systems.

Traditionally, wheat dough properties have been determined using dough mixing instruments such as the mixograph, or descriptive rheological methods such as the alveograph. Dough is a nonlinear viscoelastic fluid that is shear thinning and work hardening, and exhibits a small yield stress. The rheology of bread making is a vast and complicated subject and well reviewed by numerous authors. Dough is an example of complex rheology. Dough rheology is mostly studied based on empirical measurements due to its complexity and the fact that dough rheology changes during different stages. The linear viscoelastic properties (LVPs) of dough are related to dough mixing strength, baking characteristics, and pasta quality (Peressini et al. 1999; Edwards et al. 2001), and large deformation creep tests provide information on dough extensibility properties (Edwards et al. 1999). Measurement of the fundamental LVPs of gluten and its building blocks, gliadin and glutenin, help to better understand the underlying macromolecular structures in gluten that are responsible for its unique physical properties. Dynamic oscillatory rheometers are powerful tools for examining the fundamental viscoelastic characteristics of dough (Petrofsky and Hoseney 1995).

13.4.1 Effect of High Pressure on Rheology and Structure of Wheat Protein

Temperature effects on viscoelastic properties of wheat dough and gluten subfractions (gliadin or glutenin) are abundant in the literature; however, limited information is available on the effect of high pressure on wheat protein characterization.

13.4.1.1 Rheology and Texture

The rheological characteristics of gluten and gliadin treated under various conditions (pressure, temperature, and holding time) are available in the literature (Apichartsrangkoon et al. 1998; Kieffer et al. 2007; Ahmed 2009). The gels formed were very unlike heat-processed gluten, having a more distinct elastic character (Apichartsrangkoon et al. 1998). There were good correlations between the hardness and their moduli of elasticity, which advocated that either hardness or elasticity could be used as an index of textural modification. However, hardness appeared to be a more sensitive parameter for studying pressure-induced structural modification (Apichartsrangkoon et al. 1998). Microextension tests of pressure-treated gluten showed more resistance to extension (RE) and low extensibility (EX) at 200 MPa (Kieffer et al. 2007). This is because of an increase in ethanol-soluble proteins (Timmermann and Belitz 1993) or the drop in viscosity of the gliadin fraction. However, at 500 MPa, RE of wet gluten increased significantly and EX dropped.

The interaction between pressure and temperature (0.1–800 MPa and 20–60°C) at two holding times (20 and 50 min) on the texture of gluten was studied (Apichartsrangkoon et al. 1998). It is noticeable that the hardness values of the gluten samples treated at ambient temperature (20°C) for 20 min were marginally changed, whereas they were significantly affected at the same pressures when treated for 50 min. The effect was more pronounced at higher pressure with significant modification of structure. Thus at 20°C pressure only significantly modified the gluten structure at the higher pressure levels and the effect was time dependent. It is worth mentioning that several combinations (P, T, and t) have similar effects and produce similar magnitudes of hardness. A combination of pressure level higher than 200 MPa and temperature of 60°C is required to induce textural changes (Apichartsrangkoon et al. 1998). Holding time has a significant effect on the mechanical strength of pressure-treated gluten and an extended treatment time enhanced the hardness abruptly. In another study, a significant increase in RE or firmness was recorded from 1.58 to 3.51 N and EX decreased from 78 to 51 mm when pressure holding time was increased from 2 to 20 min at 400 MPa and 40°C. At 600 MPa, the gluten strength started to increase and reached 1.93N RE and 44 mm EX at 800 MPa while the temperature remained constant at 30°C. A combination of moderate pressure (200 MPa) and high temperature (60°C) showed synergistic effects resulting in strong gluten firmness; however, the strength decreased with the loss in gluten extensibility and cohesivity at 600 MPa and 60°C. The gluten lost its cohesiveness under extreme conditions (600 MPa and 80°C or 800 MPa and 60°C) and the measurement of RE and extension was no longer possible. Extensibility was found to be more pressure sensitive than firmness.

The effect of high pressure on rehydrated gliadin is well characterized by rheometry. The dynamic rheology of pressure-treated (350–650 MPa; 25–30°C) commercial gliadin-rich dough (46% moisture content, wet basis) was recently studied by Ahmed (2008). The untreated dough exhibited a predominantly viscous character ($G'' > G'$) which retained its liquid-like character after high-pressure treatment. A typical rheogram is illustrated in Figure 13.8. Both G' and G'' increased

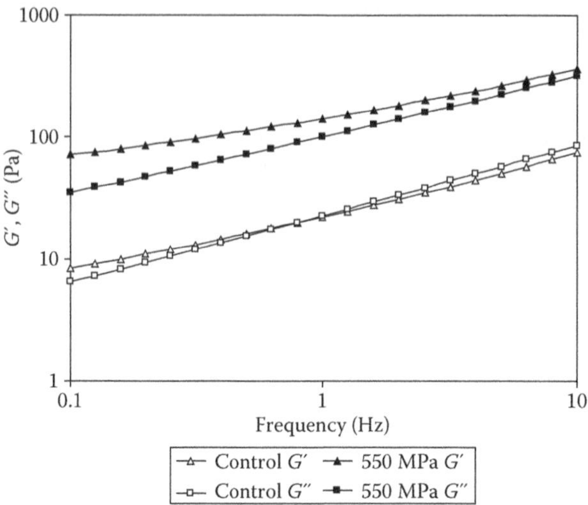

FIGURE 13.8 Typical rheograms for pressure-treated glutenin-rich protein dough.

as pressure was increased from 0.1 to 450 MPa and, thereafter, decreased (Ahmed 2008). The liquid-like characteristics (G'') increased nonsystematically as pressure was increased, as supported by the slope of power-type relationship (Table 13.2). Nevertheless, a combination of high pressure and temperature (500 MPa and 60°C) reduced both G'' and G' of gliadin by around 50% at a holding time of 10 min as compared to the control sample (22°C) (Kieffer et al. 2007). Such a significant drop in dynamic modulii could be supplemented by high temperature.

Commercial glutenin-rich dough (46% m.c., w.b.) behaved as a true viscoelastic material with an equal magnitude of G' and G'' at 0.101 MPa and 20°C (Ahmed 2008). With applied pressure, the elastic modulus predominated over the viscous modulus ($G' > G''$) and significantly increased from 0.101 MPa to 550 MPa (25–30°C). However, both G'' and G' dropped when the pressure level was increased to 650 MPa (Figure 13.9). The drop of dynamic modulii above 550 MPa may be attributed to cleavage and rearrangement of disulfide bonds. However, the solid-like characteristics of glutenin dough increased linearly as a function of pressure (Table 13.3). At a higher pressure and temperature combination (800 MPa and 70°C), the G'' and G' of glutenin were increased by a factor of 2 while the phase angle remained unchanged (Kieffer et al. 2007).

The viscoelasticity of pressure-treated gliadin-rich and glutenin-rich dough has been well characterized by phase angle (δ) and complex viscosity (η^*) (Ahmed 2008; Kieffer et al. 2007). The η^* increased from 0.101 to 450 MPa and decreased thereafter. Intensity of pressure did not show any systematic pattern with η^* as shown in Figure 13.10 (Ahmed 2008).

13.4.1.2 Rheology of Pressurized Soy–Gluten Mix

The effects of pressure on individual soy protein and gluten rheology have already been discussed. The effect of protein combination on rheology under high pressure is also interesting from the product development point of view. Rheological behavior of hydrated gluten and soy mixture (gluten:soy = 20:80; 40:60; 60:40, and 80:20) exhibited solid-like character ($G' > G''$) after pressure treatment at 700 MPa for 50 min at 20 and 60°C (Apichartsrangkoon and Ledward 2002). The temperature had a significant effect on dynamic moduli. At 20°C, the G' of most of the mixed samples resembled the G' of the 100% gluten samples except for 100% SPI. The G'' showed mixed behavior: the 100% gluten, and the mixture of 80:20 and 60:40 gluten soy had similar rheograms, while the mixture of 40:60 and 20:80 gluten–soy had a similar G'' pattern to the 100% SPI. The phase angle values indicated more elastic behavior of SPI-rich samples while gluten-rich samples had more liquid-like characteristics.

TABLE 13.2
Slope of Power-Type Relationship ln G = ln K + n ln ω of Pressure-Treated SPI Dispersion

	G'			G''		
Sample Type	**Slope**	R^2	**SE**	**Slope**	R^2	**SE**
10%						
Untreated	0.117	0.987	0.019	0.208	0.988	0.034
350 MPa	0.145	0.840	0.093	0.266	0.974	0.061
450 MPa	0.218	0.867	0.125	0.295	0.975	0.068
550 MPa	0.294	0.884	0.156	0.312	0.971	0.076
650 MPa	0.296	0.988	0.048	0.392	0.985	0.072
15%						
Untreated	0.070	0.997	0.006	0.127	0.970	0.033
350 MPa	0.080	0.998	0.005	0.145	0.991	0.020
450 MPa	0.079	0.999	0.003	0.143	0.990	0.021
550 MPa	0.090	0.997	0.007	0.141	0.988	0.022
650 MPa	0.089	0.996	0.008	0.153	0.992	0.020
20%						
Untreated	0.078	0.998	0.005	0.128	0.985	0.023
350 MPa	0.094	0.999	0.002	0.115	0.995	0.012
450 MPa	0.093	0.999	0.003	0.116	0.994	0.013
550 MPa	0.092	0.999	0.004	0.117	0.994	0.013
650 MPa	0.099	0.999	0.004	0.130	0.993	0.014

Source: From Ahmed, J., Ayad, A., Ramaswamy, H. S., Alli, I., and Shao, Y., *International Journal of Food Properties*, 10, 397–411, 2007. With permission.

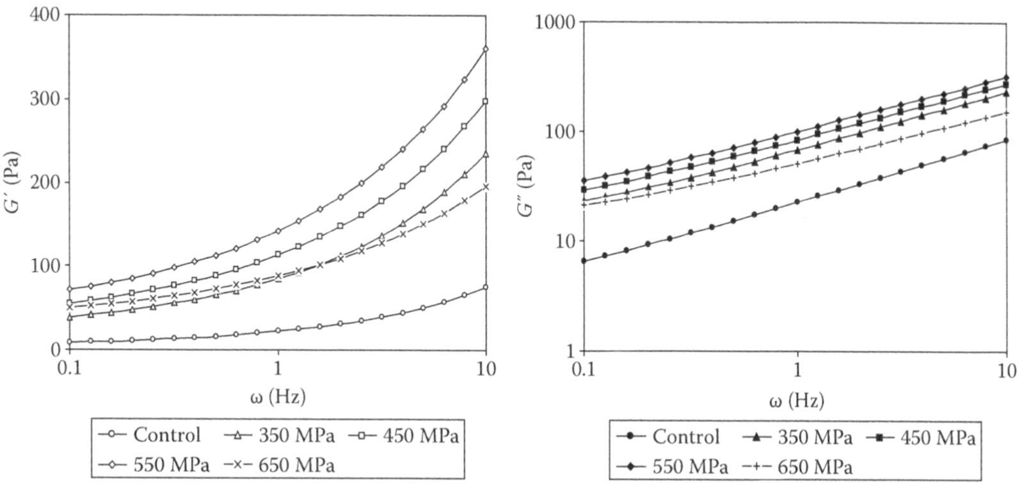

FIGURE 13.9 Effect of high pressure on elastic and viscous modulus of glutenin-rich dough.

At higher temperatures (60°C), both G' and G'' increased significantly at constant pressure of 700 MPa and the pressure-induced soy–gluten mix showed predominating solid-like characteristics with appreciably lower phase angle values. Gels having a high concentration of gluten (80:20 and 60:40) showed an abrupt increase in G' and G'' with increasing temperature, however, in the gels having high concentration of SPI both dynamic moduli appeared to increase marginally

TABLE 13.3
Slope of Power-Type Relationship ln $G'' = \ln K + n \ln \omega$ of Gliadin-Rich Dough at Selected Pressure Levels

Pressure (MPa)	Slope	R^2	SE
0.101	0.92	0.99	0.096
350	0.84	0.99	0.035
450	0.60	0.99	0.017
550	0.77	0.99	0.020
650	0.74	0.99	0.123

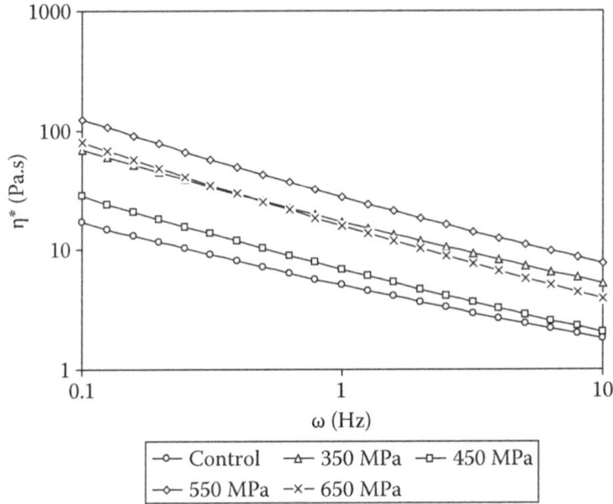

FIGURE 13.10 Viscoelasticity of glutenin-rich dough as a function of applied pressure.

with an increasing temperature. This leads to the conclusion that the combined effect of pressure and temperature on complex gluten molecule is more evident compared to smaller soy globulins. Furthermore, there is no interaction between the two proteins; the soy works as a plasticizer to the gluten network. The structures of the gluten-soy mix (80:20) were very firm due to high permanent cross-link density and, therefore, the rigidity was the maximum among combinations.

13.4.2 Effect of High Pressure on Structure

The solubility of gluten proteins decreased as pressure intensity was increased in some selected solvents (60% ethanol, SDS, ethyldiamine) (Kieffer et al. 2007). Treatment of gluten at lower pressure and temperature (200 MPa and 30°C) increased the proportion of the ethanol-soluble fraction (ESF). Increasing the intensity of treatment reduced the solubility of cysteine containing α- and γ-gliadins in 60% ethanol, but not that of cysteine-free ω-gliadins. This indicated the cleavage of intrachain disulfide bonds and their rearrangement into interchain bonds. The structures of two disulfide peptides isolated from treated gluten supported this mechanism. Because all gluten proteins were soluble in aqueous alcohol after reduction of disulfide bonds, even under extreme treatment conditions, disulfide bonds appear to be the only type of covalent bonds that are affected by HP treatment.

Rehydrated gliadin samples (0.1–800 MPa and 30–70°C) were completely soluble in 60% ethanol and their Reversed Phase-High Performance Liquid Chromatography (RP-HPLC) patterns were unchanged. Thus, cleavage and rearrangement of disulfide bonds did not occur. This result was confirmed by SDS-PAGE: neither band positions nor band intensities were changed (Kieffer and Wieser 2004). However, Lullien-Pellerin et al. (2001) observed that a pressure range above 400 MPa induced a change in conformation that resulted in a decrease of the polarity of the environment of aromatic amino acids combined with an increase in the hydrophobicity of the gliadin surface. The changes observed were found to be reversible.

The CD spectrum of pressure-treated gliadin dissolved in 50% ethanol showed an increase in the maximum intensity at 207 nm, which indicated an increase of regular conformations such as α-helices. The structural changes were reversible, when the probes were kept in a solvent at high dilution (Lullien-Pellerin et al. 2001). The combination of high HP with temperatures up to 60°C did not affect the disulfide structure of the gliadins, but their secondary structures were affected, leading to a decrease of the intrinsic viscosity. Thermal analysis (DSC) of pressure-treated gliadin-rich dough showed that pressure level ranging between 350 and 650 MPa did not denature gliadin completely (Ahmed 2008).

The solubility of treated glutenin in SDS buffer markedly decreased as pressure and temperature increased. Extreme effects were observed at 800 MPa and 70°C. The proportion of the SDS-soluble fraction dropped to 1.9%. The properties of isolated glutenin having relatively high thiol content were strongly influenced by high pressure and temperature. Glutenin having a relatively high SH content was more affected by pressure than total gluten; its solubility in SDS buffer was strongly reduced. The results indicate that SH groups are primarily responsible for the effects of HP treatment.

13.5 RICE PROTEINS

Whole rice grain contains many types of proteins, which have been isolated and characterized, mainly according to their solubility properties, using the Osborne extraction method (Marshall and Wadsworth 1994). The major proteins are mainly prolamins (5–10%), globulins (4–10%), and glutelins (80–90%) (Juliano 1972). In the Osborne extraction process, ground rice is defatted and extracted with water to obtain the albumin fraction, followed by sequential extraction with dilute salt solution, dilute alkali, and 70% ethanol to obtain globulin, glutelin, and prolamin fractions, respectively (Padhye and Salunkhe 1979). Rice protein contains a high quantity of essential amino acids such as methionine, threonine, lysine, and tryptophan; however, lysine is the first limiting amino acid followed by threonine.

The consumption of rice is often associated with allergic disorders such as asthma and dermatitis (Baldo and Wrigley 1984). These disorders are related to the ingestion of rice proteins, particularly 16 kDa albumin and 26 kDa α-globuglobulin, which have been identified as major rice allergens (Limas et al. 1990; Shibasaki et al. 1979). Kato et al. (2000) have applied HP to remove allergenic proteins (mainly 16 kDa albumin and 33 kDa globulin) from rice and they observed solubility and subsequent release of rice allergenic proteins in the pressure range of 100 to 400 MPa. Significant amounts of proteins were released in the pressure range of 300–400 MPa. Therefore, studies on high-pressure processing of rice indicated that post-process rice may have considerable health benefits in addition to functional properties.

Effect of high pressure on gelatinization and rheological characteristics of rice starch have been studied extensively (Ojeda, Tolaba, and Suarez 2000; Yamamoto et al. 2006; Ahmed et al. 2008). Little is known about the pressure effect on protein fractions of rice in excess water.

13.5.1 Rheology of Rice Protein

Rheology of rice protein isolate under high pressure is not available in the literature. However, rheological properties of pressure-treated 8% protein-containing rice flour dispersion differed significantly from those of the untreated sample (Ahmed et al. 2006). The G' predominates over G'' in the selected frequency range (0.1–10 Hz) and a small change in G' with frequency supports the elastic nature of the gel network (Ferry 1980; Ross-Murphy 1984). Both G' and G'' values of the slurry significantly ($p < .05$) increased after the pressure treatment. Phase angle decreased significantly for pressure-treated samples, indicating again the more solid-like characteristic of the gel as compared to the untreated sample.

Mechanical strength of rice slurry gels increased with an increase in pressure level. The slope coefficient of the logarithmic elastic modulus-frequency decreased from 0.14 to 0.06 as the pressure levels increased from 0.1 to 650 MPa (Table 13.4), demonstrating greater pressure sensitivity at both ends. The samples exhibited increasingly solid-like characteristics with pressure.

Elasticity of rice gel increased with pressure holding time at a constant pressure. The slope coefficients of rice gels slightly decreased with longer holding periods (for example, 0.089 to 0.086 at 550 MPa for 7.5 and 15 min). However, the increase in G' at 650 MPa was found to be independent of holding time.

13.5.2 Structure of Rice Protein

13.5.2.1 Differential Scanning Calorimetry (DSC)

The DSC thermograms of rice slurry and isolated protein fraction slurry without pressure treatment are presented in Figure 13.11a and b, respectively. Two distinctive endothermic peaks (T_d), associated with gelatinization/denaturation, were observed in the DSC scan for the untreated rice flour dispersion sample. The first peak (T_{d1}) at 70.6°C resembled the gelatinization peak of rice starch and is in agreement with the observation (70.9°C for milled white rice) of Ellepola and Ma (2006). In addition to starch, the thermal denaturation of albumin also occurs in that range (≈75°C), however the high solubility of starch in water is likely responsible for the insignificant contribution of albumin to the total thermal transitions of flour dispersions. The second endotherm peak (T_{d2}) of slurry at 105.2°C is likely attributed by thermal denaturation of globulin. The extracted isolated rice globulin appeared to show lower thermal transition temperature (T_d) of 103.8°C (Figure 13.11b). The lower T_d of globulin could be influenced by the presence of starch and fat of the rice flour. Seed globulins have been found to possess T_d in the range of 83.8–107.8°C (Marcone et al. 1998; Ellepola and Ma 2006).

In order to investigate the effect of high pressure on the suspensions, samples were scanned immediately after pressurization. The temperature of the peak maximum (T_d) and ΔH of pressure-treated samples are presented in Table 13.4. Results indicated that the rice slurries were significantly influenced by the high-pressure treatment though complete gelatinization/denaturation did not take place. The T_d of the first endotherm decreased systematically from 350 to 550 MPa after pressurization with respect to the untreated sample. The disappearance of the endothermic peak at 650 MPa

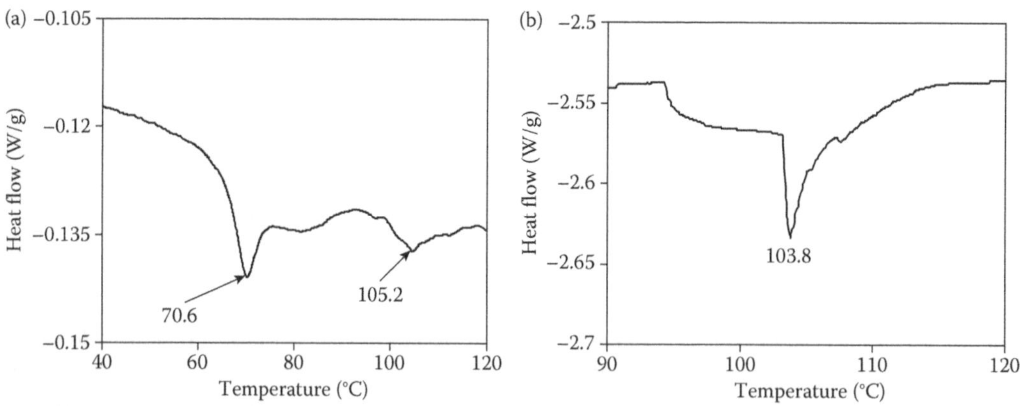

FIGURE 13.11 DSC thermograms for untreated 20% rice slurry and isolated protein fraction (a, rice slurry; b, isolated protein).

TABLE 13.4
Slope of Power-Type Relationship $\ln G'' = \ln K + n \ln \omega$ of Glutenin-Rich Dough at Selected Pressure Levels

Pressure (MPa)	Slope	R^2	SE
0.101	0.47	0.99	0.044
350	0.39	0.99	0.042
450	0.37	0.99	0.042
550	0.35	0.99	0.042
650	0.30	0.99	0.039

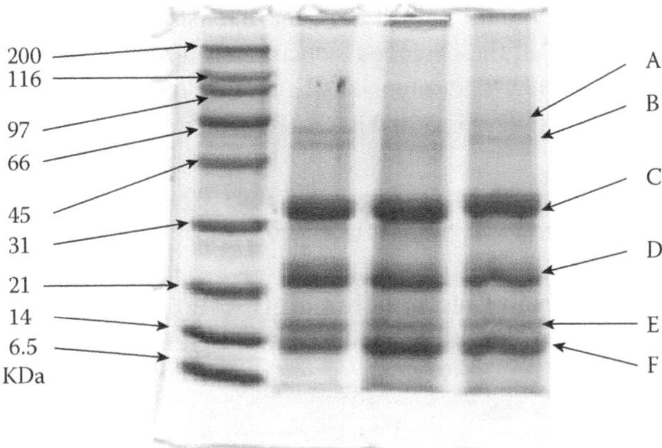

FIGURE 13.12 SDS-PAGE of Basmati rice slurry (M, marker; 1, control; 2, 350 MPa for 15 min; 3, 650 MPa for 15 min). (From Ahmed, J., Ramaswamy, H. S., Ayad, A., Alli, I., and Alvarez, P., *Journal of Cereal Science*, 46, 148–56, 2007. With permission.)

indicated complete gelatinization and/or denaturation of starch/protein in the rice flour dispersion. The T_d of the second peak decreased insignificantly from 350 to 450 MPa whereas an increase was noticed at higher pressure levels (550–650 MPa) (Table 13.5). The partial change of protein structure during pressure treatment can be well explained on the basis of increase in water density (Hayakawa et al. 1996). Water density increased with pressure levels and subsequently the number of water molecules per lattice increased significantly from 4.3 at 0.101 MPa and 4°C to 10 molecules per lattice at 1000 MPa and 25°C. These changes could be attributed to the cleavage of hydrogen bonds between the surface of the protein and the surrounding protecting water molecules.

The residual denaturation enthalpy provides a net value from a combination of endothermic reactions like the disruption of hydrogen bonds, and exothermic processes, including the breakup of hydrophobic interactions and protein aggregation (Privalov and Khechinashvili 1974; Ma and Harwalkar 1991). As a result, DSC thermograms of the pressure-treated rice slurry exhibit the subsequent thermal denaturation of the proteins remaining in a native-like conformation. A reduction in residual enthalpy could be an indication for a partial loss of protein structure during high pressure and the observation is in close agreement with work reported by earlier researchers (Taniguchi et al. 1994; Hayakawa et al. 1996, Van der Plancken et al. 2006) where rearrangement of secondary structure was suggested during high-pressure treatment.

13.5.2.2 Electrophoresis

SDS-PAGE analysis (Figure 13.12) showed a gradual decrease in the intensity of A and B subunits (MW ≈ 52.2 and 35.0) with increasing pressure application. Those major subunits identified in the rice sample are 16.2 (F), 17.5 (E), 20.4 (D), and 22.2 (C) kD, respectively. The results suggest that subunits C, D, and F were affected by the pressure alone. The lower molecular weights (MWs) (16.2–22.2 kDa) are likely globulin subunits (Krishnan, White, and Pueppke 1992) while the higher MW (≈35 kDa) could be glutelin subunits (Komatsu and Hirano 1992).

13.5.2.3 Fourier Transform Infrared (FTIR)

The amide *I′* band assignments of rice proteins are presented in Table 13.6 (Ahmed et al. 2007). Although the secondary structure features shown are the average of features of all the proteins in rice, it still resembles the rice globulin spectrum, since this is the major storage protein in the cereal. FTIR spectra demonstrate an irreversible change of the secondary structures of high-pressure-treated rice proteins at 650 MPa for 15 min (Figure 13.13) (Ahmed et al. 2007). A major change in the secondary structure (loss of intensity) was observed at about 1690 cm^{-1}, which corresponds to β-structure. It was observed that β-structure was significantly influenced by high-pressure treatment. In addition, slight increases in the intensities of two other bands at 1650 and 1641 cm^{-1} suggests a transition of the extended structure (1690 cm^{-1}) to α-helix and random coil. Some differences in band positions and intensities have been reported for rice proteins. These differences could be explained on the basis of the nature of protein, isolation techniques, and presence of lipid and starch.

TABLE 13.5
Thermal Changes in Pressure-Treated (15 Min Holding Time) Rice Proteins Obtained from Differential Scanning Calorimetry

Pressure (MPa)	T_p (°C)	Denaturation (%)
0.101	108.6±2.3	0
350	107.6±1.2	37.6
450	108.1±2.2	47.9
550	111.1±1.6	49.1
650	112.0±0.7	98.4

Source: From Ahmed, J., Ramaswamy, H. S., Ayad, A., Alli, I., and Alvarez, P., *Journal of Cereal Science*, 46, 148–56, 2007. With permission.

TABLE 13.6
Band Assignments of the Deconvoluted Amide *I′* Spectral Region of Rice Protein

Assignment	Band Position (cm^{-1})
β-structure	1690
Antiparallel β-sheet (aggregation)	1680
β-turns	1658
α-helix	1650
Unordered structure (random coil)	1641
β-strand	1632
Antiparallel β-sheet	1623
Side-chain vibrations	1611

Source: From Ahmed, J., Ramaswamy, H. S., Ayad, A., Alli, I., and Alvarez, P., *Journal of Cereal Science*, 46, 148–56, 2007. With permission.

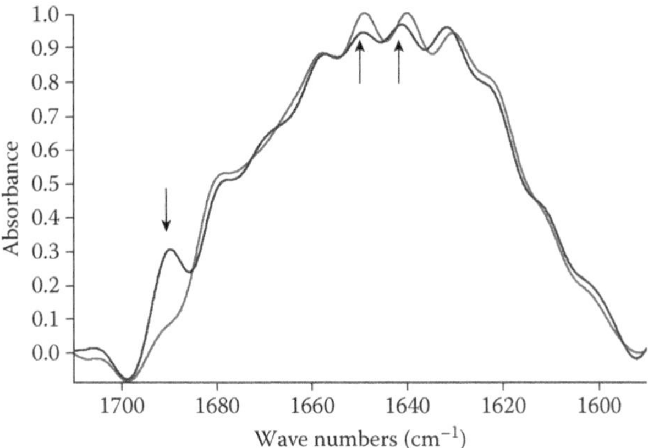

FIGURE 13.13 Amide *I′* region of the FTIR spectra from basmati rice. (Blue line represents control sample and red line is for pressure-treated sample at 650 MPa for 15 min.) (From Ahmed, J., Ramaswamy, H. S., Ayad, A., Alli, I., and Alvarez, P., *Journal of Cereal Science*, 46, 148–56, 2007. With permission.)

13.6 LENTIL PROTEINS

Crop legumes like lentils are highly nutritious and provide a good source of minerals. Lentils contain about 25% protein, 56% carbohydrate, and 1% fat and are one of the best and cheapest sources of vegetable protein (Adsule, Kadam, and Leung 1989). Lentils, which are botanically classified as *Lens culinaris* (Adsule et al. 1989), are an important crop in many developing countries. Legume proteins are primarily storage proteins comprised of two principle globulins, legumin and vicilin (Swanson 1990). Lentil flour has potential for traditional and newer product developments with health benefits since it contains a higher amount of protein, is gluten free, and has a low glycemic index. Lentil protein isolate produced a milk of intermediate quality equivalent to milk prepared from SPIs (Swanson 1990). Both the starch and protein fractions of lentils offer a new source of novel ingredients. New sources of cheaper protein provide new alternatives for the dairy industry, where cheaper protein is required to replace existing proteins (Lee et al. 2007). Extensive research has been carried out on cereal proteins due to their ready availability and wide usage in food and nonfood applications. However, there is little information available on the structure-property relationship of lentil flour and proteins.

13.6.1 Rheology

The rheological characteristics of the pressure-treated lentil dispersions differed significantly over the control sample. A typical rheogram of pressure-treated lentil dispersions at 20°C is illustrated in Figure 13.14. It demonstrates that the G' predominates over G'' in the frequency range of 0.1–10 Hz for control and pressure-treated samples, indicating a gel with more solid-like properties. Over the frequency range measured, the G' and G'' were relatively frequency independent, due to extensive molecular entanglement. The complex viscosity increased with increasing pressure, supporting more viscoelastic characteristics of a pressurized gel.

Pressure treatment significantly increased both dynamic modulii, however the increase of G'' was higher than the incease of G', irrespective of moisture content. The viscoelasticity of the lentil dispersion was adequately supported by the calculated slopes of the linear regression of the power-type relationship of ln ω vs ln G' and ln G'' (Equation 13.1) (Table 13.7). The magnitude of the slope clearly indicates a loss of solid-like character and an increase in liquid-like character with an increase in moisture content. In addition, samples behaved as true viscoelastic fluids with equal magnitude of slope. High pressure converted lentil flour dispersion into a viscous gel by

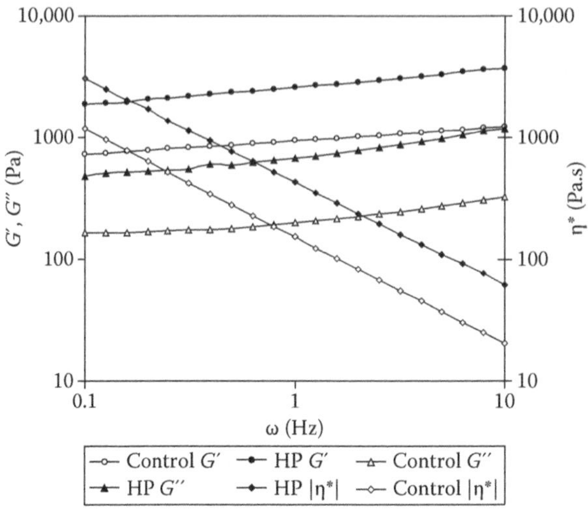

FIGURE 13.14 Typical rheogram for pressure-treated lentil flour dispersion at 20°C.

TABLE 13.7
Slope of Power-Type Relationship ln $G' = \ln K + n \ln \omega$ of Lentil Slurry at Selected Pressure and Moisture Levels

		$\ln G' = \ln K + n \ln \omega$			$\ln G'' = \ln K + n \ln \omega$		
M.C. (%)	Pressure (MPa)	Slope	R^2	SE	Slope	R^2	SE
37	0.101	0.09	0.99	0.00	0.11	0.94	0.06
	450	0.12	0.99	0.01	0.15	0.98	0.05
46	0.101	0.10	0.99	0.01	0.15	0.93	0.07
	450	0.13	0.99	0.01	0.17	0.94	0.077
57	0.101	0.11	0.99	0.00	0.15	0.94	0.047
	450	0.13	0.99	0.00	0.17	0.96	0.047
	650	0.15	0.97	0.04	0.18	0.76	0.14

gelatinization and/or denaturation of starch and protein components. The slopes increased with increasing pressure level. For example, the slope for the 57% moisture-containing sample increased from 0.10 to 0.15 and from 0.15 to 0.18 for solid-like and liquid-like properties, respectively, after pressure treatment at 650 MPa for 15 min. The effect of high pressure on complex viscosity clearly demonstrates an increase in viscoelasticity with increasing pressure intensity and consequently the viscoelastic solid converts to a viscoelastic fluid.

13.6.2 Differential Scanning Calorimetry (DSC) Study

During DSC scans one distinctive endothermic peak (T_d), assigned to lentil protein denaturation, was identified. The peak (T_d) ranged between 135 and 110°C as moisture content increased from 6.5 to 57% (Ahmed et al. 2009). The endothermic peak (T_d) of dispersions is most likely attributed to the thermal denaturation of globulin. 11S globulin of SPI dispersion showed denaturation at 90°C (Ahmed et al. 2006). A higher T_d for lentil could be associated with variation in the legume, the nature of the protein, and the presence of starch and fat.

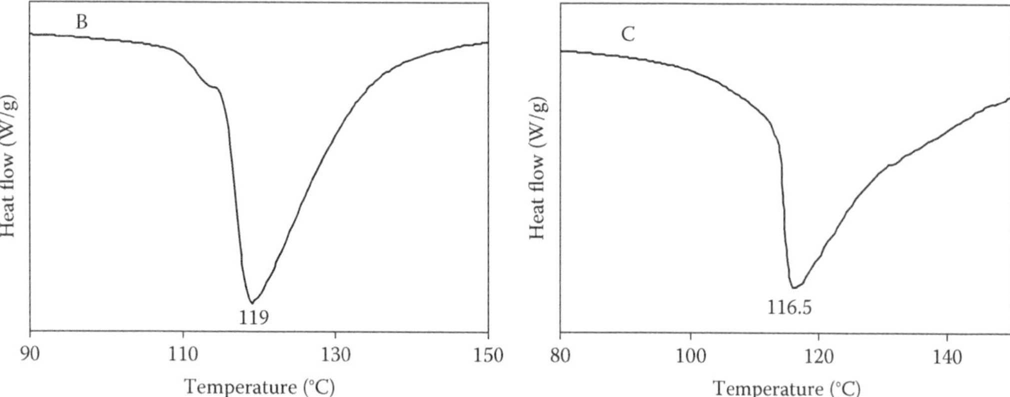

FIGURE 13.15 Typical DSC thermograms of lentil slurry before and after pressure treatment (B, untreated; C, pressure treated).

TABLE 13.8
Effect of High Pressure on Protein Denaturation Temperature as Function of Moisture Content of Lentil Dispersion

		T_d (°C)				
Sample ID	**Moisture Content (% w.b.)**	**Control**	**350 MPa**	**450 MPa**	**550 MPa**	**650 MPa**
Dried flour	6.5	135.5	–	–	–	–
S1	14.9	125.9	115.2	118.2	117.4	137.2
S2	24.6	125.0	124.3	121	117	124
S3	29.1	119	126	129.5	130.8	133
S4	37.2	126.8	115	131.0	126	137
S5	46.7	125	125	117	111	126
S6	57.6	111	118	110	113	120

Typical DSC thermograms of lentil slurries before and after pressure treatment are illustrated in Figure 13.15. It clearly demonstrates the incomplete denaturation of protein and a minor shift of denaturation temperature when pressure was applied. The residual denaturation enthalpy (difference between actual and pressure treated) provides a net value from a combination of endothermic reactions like the disruption of hydrogen bonds, and exothermic processes, including the breakup of hydrophobic interactions and protein aggregation (Privalov and Khechinashvili 1974; Ma and Harwalkar 1991). The shifting of dentauration temperature (T_d) of pressurized lentil slurry as a function of moisture content is presented in Table 13.8. It was observed that T_d varied with pressure level nonsystematically.

13.6.3 Fourier Transform Infrared (FTIR) Spectroscopy

The spectral intense bands for legumes are around 1520 (amide-II; N–H bending) and 1660 (amide-I; C=O stretching) cm^{-1} (Carbonaro et al. 2008). Figure 13.16 illustrates the spectrum of the lentil flour in the spectral region 1000–1700 cm^{-1}. The main spectral features consist of four intense bands located around 1163, 1408, 1545, and 1652 cm^{-1}. The spectra of the control and pressurized samples in the spectral range of the amide *I′* band are shown in Figure 13.17. The spectral shape at 1652 cm^{-1} (amide-I) does not change considerably after pressure treatment although an insignificant

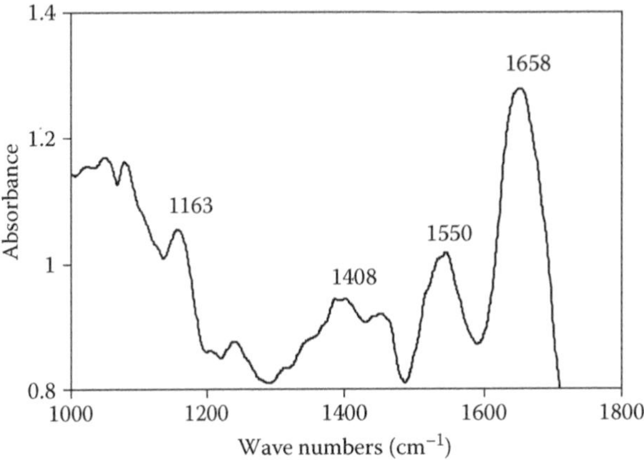

FIGURE 13.16 FTIR spectra for untreated lentil flour.

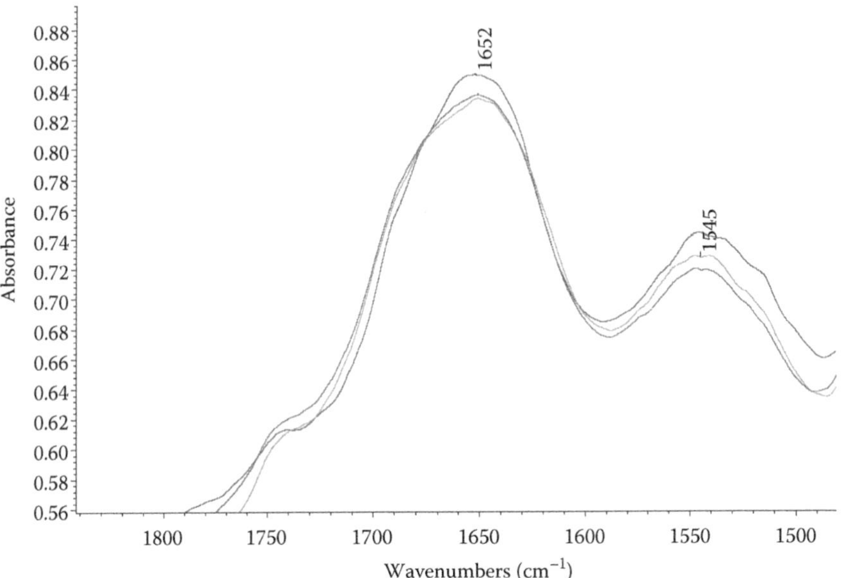

FIGURE 13.17 FTIR spectra for pressure-treated and control lentil flour samples (Top, control; bottom, 350 MPa; centre curve, 650 MPa).

change was noticed at 1545 cm^{-1}. The lentil flour contains a high amount of β-sheet structures, which was retained after dry thermal treatment with only a slight destabilization of the protein structure (Carbonaro et al. 2008).

13.7 CONCLUSION

High-pressure studies have shown that the technique has the potential to influence the rheological and structural properties of cereal proteins. The rheological study of various cereal proteins treated with high pressure indicated gradual transformation of sol to gel and a predominant viscoelastic nature of the treated gel. Increased pressure intensity and heating has strong effects on the rheological properties of cereal proteins and led to an increase in gel strength and finally loss of cohesivity. Concentration has a significant effect in producing pressure-induced gel. The chemical changes of cereal proteins after pressure treatment are well characterized by various techniques. The changes in the amide *I′* band of secondary structure are studied by FTIR and CD analysis. Functional properties of soybean proteins are significantly influenced by the free SH content and disulfide (S–S) bonds; both SH groups and SS bonds undergo changes during high-pressure application. The MW distribution of SPI is increased by pressure treatment and increases as a function of pressure (200–600 MPa). Pressure-assisted protein dissociation and association has been observed from electrophoresis results.

High-pressure processing of food is being used today on an ever-increasing commercial basis. Opportunities clearly exist for innovative applications and new food product development. Still, the effects of high pressure on cereal and legume proteins have not been well investigated and there are ample opportunities to utilize high pressure to produce novel foods with desirable functional properties and health benefits.

REFERENCES

Abbott, T. P., Nabetani, H., Sessa, D. J., Wolf, W. J., Liebman, M. N., and Kor, R. K. 1996. Effects of bound water on FTIR spectra of glycinin. *Journal of Agricultural and Food Chemistry*, 44:2220–24.

Adsule, R. N., Kadam, S. S., and Leung, H. K. 1989. Lentil. In: Salunkhe, D. K., and Kadam, S. S., Eds, *Handbook of World Food Legumes: Nutritional Chemistry, Processing Technology, and Utilization*, 11, 131–52. Boca Raton: CRC Press.

Ahmed, J. 2009. Rheological properties of food. In: M. Farid, Ed. *Mathematical Modeling of Food Processing*, 27–63, Boca Raton: CRC Press.

Ahmed, J. 2008. Effect of high pressure treatment on gliadin and glutenin frction. *Unpublished.*

Ahmed, J., Ayad, A., Ramaswamy, H. S., Alli, I., and Shao, Y. 2007. Dynamic viscoelastic behavior of high pressure treated soybean protein isolate dispersions. *International Journal of Food Properties*, 10:397–411.

Ahmed, J., Ramaswamy, H. S., and Alli, I. 2006. Thermorheological characteristics of soybean protein isolate. *Journal of Food Science*, 71:158–63.

Ahmed, J., Ramaswamy, H.S., Alli, I., and Ngadi, M. 2003. Effect of high pressure on rheological characteristics of liquid egg. *Lebensmittel-Wissenschaft und Technologie*, 36:517–24

Ahmed, J., Ramaswamy, H. S., Ayad, A, and Alli, I. 2008. Thermal and dynamic rheology of insoluble starch from Basmati rice. *Food Hydrocolloids*, 22:278–87.

Ahmed, J., Ramaswamy, H. S., Ayad, A., Alli, I., and Alvarez, P. 2007. Effect of high-pressure treatment on rheological, thermal and structural changes of Basmati rice flour slurry. *Journal of Cereal Science*, 46:148–56.

Ahmed, J., Varshney, S. K., and Ramaswamy, H. S. 2009. Effect of high pressure treatment on thermal and rheological properties of lentil flour slurry LWT. *Food Science and Technology*, 42:1538–44.

Alvarez, P. A., Ramaswamy, H. S., and Ismail, A. A. 2008. High pressure gelation of soy proteins: Effect of concentration, pH and additives. *Journal of Food Engineering*, 88:331–40.

Apichartsrangkoon, A. 2003. Effects of high pressure on rheological properties of soy protein gels. *Food Chemistry*, 80: 55–60.

Apichartsrangkoon, A., Ledward, D. A., Bell, A. E., and Brennan, J. G. 1998. Physicochemical properties of high-pressure treated wheat gluten. *Food Chemistry*, 63:215–20.

Apichartsrangkoon, A., and Ledward, D. A. 2002. Dynamic viscoelastic behaviour of high pressure treated gluten–soy mixtures. *Food Chemistry*, 77:317–23.

Baldo, B. A., and Wrigley, C. W. 1984. Allergies to cereals. *Advance Cereal Science & Technology*, 6:289–356.

Balny, C., and Masson, P. 1993. Effects of high pressure on proteins. *Food Review International*, 9:611–28.

Balny, C., Masson, P., and Heremans, K. (Eds). 2002. *Frontiers in High Pressure Biochemistry and Biophysics.* Amsterdam: Elsevier, 3–10.

Brandes, N., Welzel, P. B., Werner, C., and Kroh, L. W. 2006. Adsorption-induced conformational changes of proteins onto ceramic particles: Differential scanning calorimetry and FTIR analysis. *Journal of Colloid and Interface Science*, 299:56–69.

Bridgeman, P. W. 1914. The coagulation of albumen by pressure. *Journal of Biological Chemistry*, 19:511–12.

Byler, D. M., and Susi, H. 1986. Amide I region of resolution-enhanced Raman spectra of proteins. In *Proceedings of the 10th International Conference on Raman Spectroscopy*, 1–58.

Carbonaro, M., Maselli, P., Dore, P., and Nucara, A. 2008. Application of Fourier transform infrared spectroscopy to legume seed flour analysis. *Food Chemistry*, 108:361–68.

Cheftel, J. C. 1992. Effects of high hydrostatic pressure on food constituents: An overview. In: Balny, C., Hayashi, R., Heremans, K., and Masson, P., Eds, *High Pressure and Biotechnology*, 195–209, 887. Montrouge: John Libbey Eurotext.

Choi, S.-W., and Ma, C.-Y. 2006. Study of thermal aggregation of globulin from common buckwheat (*Fagopyrum esculentum* Moench) by size-exclusion chromatography and laser light scattering. *Journal of Agricultural and Food Chemistry* 54:554–61.

Clark, A. H., and Lee-Tuffnell, C. D. 1986. Gelation of globular proteins. In: Mitchell, J. R., and Ledward, D. A., Eds, *Functional Properties of Food Macromolecules*, 203–72. London: Elsevier Applied Science.

Clark, A. H., and Ross-Murphy, S. B. 1987. Structural and mechanical properties of biopolymer gels. *Advances in Polymer Science*, 83:57–92.

Dzwolak, W., Katoa, M., and Taniguchi, Y. 2002. Fourier transform infrared spectroscopy in high-pressure studies on proteins. *Biochimica et Biophysica Acta (BBA) - Protein Structure and Molecular Enzymology*, 1595:131–44.

Edwards, N. M., Dexter, J. E., Scanlon, M. G., and Cenkowski, S. 1999. Relationship of creep-recovery and dynamic oscillatory measurements to Durum wheat physical dough properties. *Cereal Chemistry*, 76: 638–45.

Edwards, N. M., Peressini, D., Dexter, J. E., and Mulvaney, S. J. 2001. Viscoelastic properties of durum wheat and common wheat dough of different strengths. *Rheologica Acta*, 40:142–53.

Ellepola, S. W., Choi, S. M., and Ma, C. Y. 2005. Conformational study of globulin from rice (Oryza sativa) seeds by Fourier-transform infrared spectroscopy. *International Journal of Biological Macromolecules*, 37:12–20.

Ellepola, S. W., and Ma, C.-Y. 2006. Thermal properties of globulin from rice (Oryza sativa) seeds. *Food Research International*, 39:257–64.

Ferry, J. D. 1980. *Viscoelastic Properties of Polymer*, 3rd ed. New York: John Wiley.

Galazka, V. B., Sumner, I. G., and Ledward, D. A. 1996. Changes in protein-protein and proteinpolysaccharide interactions induced by high pressure. *Food Chemisrry*, 57:393–98.

Glicksman, M. 1982. *Food Hydrocolloids*, Vol. I. Boca Raton, FL: CRC Press.

Hayakawa, I., Kajihara, J., Morikawa, K., Oda, M., and Fujio, Y. 1992. Denaturation of bovine serum albumin (BSA) and ovalbumin by high pressure, heat and chemicals. *Journal of Food Science*, 57:288–92.

Hayakawa, I., Linko, Y., and Linko, P. 1996. Mechanism of high pressure denaturation of proteins. *Lebensmittel-Wissenschaft und Technologie* 29:756–62.

Heremans, K. 1995. High pressure effects on biomolecules. In: Ledward, D. A., Johnston, D. E., Earnshaw, R. G., Hasting, A. P. M., Eds, *High Pressure Processing of Foods*, 81–97. Nottingham: Nottingham University Press.

Heremans, K., Meersman, F., Pfeiffer, H., Rubens, P., and Smeller, L. 2000. Pressure effects on biopolymer structure and dynamics. *High Pressure Research*, 19:233–40.

Hsu, S. 1999. Rheological studies on gelling behavior of soy protein isolates. *Journal of Food Science*, 64:136–40.

Jackson, M., and Mantsch, H. H. 1995. The use and misuse of FTIR spectroscopy in the determination of protein structure. *Critical Review in Biochemistry and Molecular Biology*, 30:95–120.

Jonas, J., and Jonas, A. 1994. High-pressure NMR spectroscopy of proteins and membranes. *Annual Review of Biophysics and Biomolecular Structure*, 23:287–318.

Johnston, D. E., and Murphy, R. J. 1995. Some changes to the properties of milk proteins caused by high-pressure treatment. In: Dickinson, E., Lorient, D., Eds; *Food Macromolecules and Colloids*, 134–140. Cambridge, UK: Royal Society of Chemistry.

Juliano, B. O. 1972. Physicochemical properties of starch and protein in relation to grain quality and nutritional value of rice. In IRRI Rice Breeding. IRRI, Los Baños, Philippines, 389–405.

Kajiyama, N., Isobe, S., Uemura, K., and Noguchi, A. 1995. Changes of soy protein under ultra-high hydraulic pressure. *International Journal of Food Science and Technology*, 30:147–58.

Kang, I. J., Matsumura, Y., and Mori, T. 1991. Characterization of texture and mechanical properties of heat-induced soy protein gels. *Journal of the American Oil Chemists' Society*, 68 (5): 338–45.

Kato, T., Katayama, E., Matsubara, S., Omi, Y., and Matsuda, T. 2000. Release of allergic proteins from rice grains induced by high hydrostatic pressure. *Journal of Agricultural and Food Chemistry*, 48:3124–29.

Kauzmann, W. 1959. Some factors in the interpretation of protein denaturation. In: Anfinsen, C. B., Ed., *Advances in Protein Chemistry*, Vol. 14, 1–63. Academic Press, New York.

Kieffer, R., Schurera, F., Köhlera, P., and Wieser, H. 2007. Effect of hydrostatic pressure and temperature on the chemical and functional properties of wheat gluten: Studies on gluten, gliadin and glutenin. *Journal of Cereal Science*, 45:285–92.

Kieffer, R., and Wieser, H., 2004. Effect of high pressure and temperature on the functional and chemical properties of gluten. In: Lafiandra, D., Masci, S., and D' Ovidio, R., Eds, *The Gluten Proteins*, 235–38. Cambridge, UK: The Royal Society of Chemistry.

Kitchen, D. B., Reed, L. H., and Levv, R. M. 1992. Molecular dvnamics simulation of solvated protein at high pressure. *Biochemistry*, 31:10083–93.

Knorr, D., Heinz, V., and Buckow, R., 2006. High pressure application for food biopolymers. *Biochimica et Biophysica Acta*, 1764:619–31.

Komatsu, S., and Hirano, H. 1992. Rice seed globulin: A protein similar to wheat seed glutelin. *Phytochemistry*, 3 (1): 3455–3459.

Krishnan, H. B., White, J. A., and Pueppke, S. G. 1992. Characterization and localization of rice *(Oryza sativa* L.) seed globulins. *Plant Science*, 81:1–11.

Kundrot, C. E., and Richards, F. M. 1987. Effect of hydrostatic pressure on the solvent in crystals of hen egg-white lysozyme. *Journal of Molecular Biology*, 193:157–170.

Ledward, D. A., Johnston, D. A., Earnshaw, R. G., and Hasting, A. P. M. (Eds.) 1995. *High Pressure Processing of Foods*. Nottingham: Nottingham University Press. 65–79.

Lee, H. C., Htoon, A. K., Uthayakumaran, S., and Paterson J. L. 2007. Chemical and functional quality of protein isolated from alkaline extraction of Australian lentil cultivars: Matilda and Digger. *Food Chemistry*, 102:1199–207.

Limas, G. G., Salinas, M., Moneo, I., Fischer, S., Wittmann-Liebold, B., and Mendez, E. 1990. Purification and characterization of ten new rice NaCl-soluble proteins: identification of four proteinsynthesis inhibitors and two immunoglobulin-binding proteins. *Planta*, 181:1–9.

Lullien-Pellerin, V., Popineau, Y., Meersman, F., Morel, M. H., Heremans, K., Lange, R., and Balny, C. 2001. Reversible changes of the wheat gamma-46 gliadin conformation submitted to high pressures and temperatures. *European Journal of Biochemistry*, 268:5705–12.

Lullien-Pellerina, V., and Balny, C. 2002. High-pressure as a tool to study some proteins' properties: conformational modification, activity and oligomeric dissociation. *Innovative Food Science & Emerging Technologies*, 3:209–21.

Ma, C.-Y., and Harwalkar, V. R. 1991. Effects of medium and chemical modification on thermal characteristics of β-lactoglobulin. *Journal of Thermal Analysis and Calorimetry*, 47:1513–25.

MacRitchie, F. 1994. Role of polymeric proteins in flour functionality. In: Wheat kernel proteins: molecular and functional aspects. Bitervo, Italy, Universita degli studi Della Tuscia. 145–50.

MacRitchie, F. 1992. Physiochemical properties of wheat proteins in relation to functionality. *Advances in Food and Nutrition Research*, 36:1–87.

MacRitchie, F. 1987. Evaluation of contributions from wheat fractions to dough mixing and bread making. *Journal of Cereal Science*, 6: 259–68.

Marcone, M. F., Kakuda, Y., and Yada, R. Y. 1998. Salt-soluble seed globulins of dicotyledonous and monocotyledonous plants. II. Structural characterization. *Food Chemistry*, 63:265–74.

Marshall, W. E., and Wadsworth, J. I. 1994. *Rice Science and Technology*, Marcel Dekker, Inc., New York, NY.

Masson, P. 1992. Pressure denaturation of proteins. In: Balny, C., Hayashi, R., Heremans, K., and Masson, P., Eds, *High Pressure and Biotechnology*, Vol. 224, 89–99. Montrouge: Colloque INSERM/John Libbey, Eurotext.

Matsuo, K., Sakurada, Y., Yonehara, R., Kataoka, M., and Gekko, K. 2007. Secondary-structure analysis of denatured proteins by vacuum-ultraviolet circular dichroism spectroscopy. *Biophysical Journal*, 92: 4088–96.

Molina, S. E., and Anon, M. C. 2000. Analysis of products, mechanisms of reaction, and some functional properties of soy protein hydrolysates. *Journal of the American Oil Chemists' Society*, 77:1293–1301.

Molina, E., Defaye, A. B., and Ledward, D. A. 2002. Soy protein pressure-induced gels. *Food Hydrocolloids*, 16:625–32.

Molina, E., Papadopoulou, A., and Ledward, D. A. 2001. Emulsifying properties of high pressure treated soy protein isolates and 7S and 11S globulins. *Food Hydrocolloids*, 15:263–69.

Morr, C. V. 1990. Current status of soy protein functionality in food systems. *Journal of the American Oil Chemists' Society*, 67 (5): 265–71.

Mozhaev, V. V., Heremans, K., Frank, J., Masson, P., and Balny, C. 1996. High Pressure effects on protein structure and function, *Proteins: Structure, Function, and Genetics*, 24:81–91.

Nakamura, T., Utsumi, S., and Mori, T. 1986. Interactions during heat induced gelation in a mixed system of soybeans 7S and 11S globulins. *Agricultural and Biological Chemistry*, 50 (10): 2429–35.

Nagano, T., Fukuda, Y., and Akasaka, T. 1996. Dynamic viscoelastic study on the gelation properties of beta-conglycininrich and glycinin-rich soybean protein isolates. *Journal of Agricultural and Food Chemistry*, 44:3484–88.

Nielsen, N. C. 1985. The structure and complexity of the 11S polypeptides in soybeans. *Journal of the American Oil Chemists' Society*, 62:1680–86.

Nielson, L. E. 1974. Mechanical Properties of Polymers and Composites – Volume 2. New York, Marcel Dekker.

Nishinari, K., Kohyama, K., Zhang, Y., Kitamura, K., Sugimoto, T., Saio, K., and Kawamura, Y. 1991. Rheological study on the effect of the A5 subunit on the gelation characteristics of soybean proteins. *Agricultural Biological Chemistry*, 55:351–55.

Ojeda, C. A., Tolaba, M. P., and Suarez, C. 2000. Modeling starch gelatinization kinetics of milled rice flour. *Cereal Chemistry*, 77:145–47.

Okamoto, M., Kawamura, Y., and Hayashi, R. 1990. Application of high pressure to food processing: Textural comparison of pressureand heat-induced gels of food proteins. *Agricultural Biological Chemistry*, 54:183–89.

Omi, Y., Kato, T., Ishida, K.-I., Kato, H., and Matsuda, T. 1996. Pressure-induced release of basic 7S globulin from cotyledon dermal tissue of soybean seeds. *Journal of Agricultural Food Chemistry*, 44:3763–67.

Osborne, T. B. 1924. In: *Vegetable Proteins*, 2nd ed. London: Longmans-Green. 13–23.

Padhye, V. W., and Salunkhe, D. K. 1979. Extraction and characterization of rice proteins. *Cereal Chemistry*, 56:389–93.

Peressini, D., Sensidoni, A., Pollini, C. M., and de Cindio, B. 1999. Improvement of wheat fresh pasta-making quality. *Cereal Chemistry*, 76: 452–58.

Peterofsky, K. E., and Hoseney, R. C. 1995. Rheological properties of dough made with starch and gluten from several cereal sources. *Cereal Chemistry*, 72:53–58.

Privalov, P. L. 1979. Stability of proteins. Small globular proteins. *Advanced Protein Chemistry*, 33:167–204.

Privalov, P. L., and Khechinashvili, N. N. 1974. A thermodyanamic approach to the problem of stabilization of globular protein structure: A calorimetry study. *Journal of Molecular Biology*, 86:665–84.

Puppo, C., Chapleau, N., Speroni, F., De Lamballerie-Anton, M., Michel, F., Anoan, C., and Anton, M. 2004. Physicochemical modifications of high-pressure-treated soybean protein isolates. *Journal of Agricultural Food Chemistry*, 52:1564–71.

Redl, A., Guilbert, S., and Morel, M. H. 2003. Heat and shear mediated polymerization of plasticized wheat gluten protein upon mixing. *Journal of Cereal Science*, 38:105–14.

Ross-Murphy, S. B. 1984. Rheological methods. In: Chan, H. W. S., Ed., *Bio-physical Methods in Food Research*, 138–99. London: Blackwell Scientific

Royer, C. A. 2002. Revisiting volume changes in pressure-induced protein unfolding *Biochimica Biophysica Acta*, 1595:201–209.

Saio, K., and Watanabe, T. 1978. Differences in functional properties of 7S and 11S soybean proteins. *Journal of Texture Studies*, 9:135–157.

Shewry, P. R. 1995. Cereal seed storage proteins. In: H Kigel, G Galili, Eds, *Seed Development and Germination*, 45–77. Marcel Dekker, NY.

Shewry, P. R., Miles, M. J., and Tatham, A. S. 1994. The prolamin storage proteins of wheat and related cereals. *Progress in Biophysics and Molecular Biology*, 16:37–59.

Shewry, P. R., and Tatham A. S., 1997. Disulphide bonds in wheat gluten proteins. *Journal of Cereal Science*, 25 (21): 207–27.

Shewry, P. R., Tatham, A. S., Barro, F., Barcelo, P. and Lazzeri, P., 1995. Biotechnology of bread making: Unravelling and manipulating the multi-protein gluten complex. *BioTechnology*, 13:1185–90.

Shibasaki, M., Suzuki, S., Nemoto, H., and Kuroume, T. 1979. Allergenicity and lymphocyte-stimulating property of rice protein. *Journal of Allergy and Clinical Immunology*, 64:259–65.

Silva, J. L., Fogel, D., and Royer, C. A. 2001. Pressure provides new insights into protein folding, dynamics and structure. *Trends in Biochemical Sciences*, 26:612–18.

Silva, J. L., and Weber, G. 1993. Pressure stability of proteins. *Annual Review of Physical Chemistry*, 44:89–113.

Smith, J. R., Smith, T. L., and Tschoegl, N. W. 1970. Rheological properties of wheat flour doughs. III. Dynamic shear modulus and its dependence on amplitude, frequency, and dough composition. *Rheologica Acta*, 9:239–52.

Susi, H., and Byler, D. M. 1988. Fourier deconvolution of the amide I Raman band of proteins as related to conformation. *Applied Spectroscopy*, 42:819.

Swanson, B. G. 1990. Pea and Lentil Protein Extraction and Functionality, *Journal of American Oil Chemists' Society*, 67:276–80.

Tang, C.-H., and Ma, C.-Y. 2009. Effect of high pressure treatment on aggregation and structural properties of soy protein isolate. *Lebensmittel-Wissenschaft und Technologie*, 42:606–11.

Taniguchi, Y., Taked, N., and Kato, M., 1994. Pressure denaturation mechanism for proteins. In: Taniguchi, Y., Sendo, M., Hara, K. Eds, *High Pressure Liquids and Solutions*, 101–117. Amsterdam: Elsevier Science.

Timmermann, F., and Belitz, H.-D. 1993. Aspartic proteinases in wheat flour. III. Effect on dough and gluten. *Zeitschrift für Lebensmittel-Untersuchung und –Forschung*, 196:17–21.

Torrezan, R., Tham, W. P., Bell, A. E., Frazier, R. A., and Cristianini, M. 2007. Effects of high pressure on functional properties of soy protein. *Food Chemistry*, 104:140–47.

Van der Planchen, I., Van Loey, A., and Hendrickx, M. E., 2007. Kinetic study on the combined effect of high pressure and temperature on the physico-chemical properties of egg white proteins. *Journal of Food Engineering*, 78:206–16.

Wang, X. S., Tang, C. H., Li, B. S., Yang, X. Q., Li, L., and Ma, C. Y. 2008. Effects of high pressure treatment on some physicochemical and functional properties of soy protein isolates. *Food Hydrocolloids*, 22:560–67.

Wolf, W. J., Babcock, G. E., and Smith, A. K. 1961. Ultracentrifugal differences in soybean protein composition. *Nature*, 191:1395–96.

Wyatt, P. J. 1993. Light scattering and absolute characterization of macromolecules. *Analytica Chimica Acta*, 272:1–40.

Yamamoto, H., Makita, E., Oki, Y., and Otani, M. 2006. Flow characteristics and gelatinization kinetics of rice starch under strong alkali conditions. *Food Hydrocolloids*, 20:9–20.

Yamauchi, F., Yamagishi, T., and Iwabuchi, S. 1991. Molecular understanding of heat induced phenomenon of soybean protein. *Food Reviews International*, 7 (3): 282–323.

Zhang, H., Li, L., Tatsumi, E., and Kotwal, S. 2003. Influence of high pressure on conformational changes of soybean glycinin. *Innovative Food Science and Emerging Technologies*, 4:269–75.

Zhang, H., Li, L., Tatsumi, E., and Isobe, S. 2005. High-pressure treatment effects on proteins in soy milk. *Lebensmittel-Wissenschaft und Technologie*, 38 (1): 7–14.

Zhao, Y., Mine, Y., and Ma, C. Y. 2004. Study of thermal aggregation of oat globulin by laser light scattering. *Journal of Agricultural and Food Chemistry*, 52:3089–96.

14 High-Pressure Treatment Effects on Food Proteins of Animal Origin

Pedro A. Alvarez, Hosahalli S.Ramaswamy, and Ashraf A. Ismail
McGill University

CONTENTS

14.1 INTRODUCTION

High-pressure (HP) treatment is an emerging alternative to the more traditional thermal processing of foods. HP can kill spoilage and pathogenic bacteria and can also be used for improving functionality of food ingredients and finished foods. The technique consists in placing a prepackaged food in a flexible container/bag into a fluid containing vessel and raising the pressure in the vessel by injecting more fluid using a hydraulic/HP pump. The pressure is created instantaneously and acts uniformly through the vessel without pressure gradients; this property offers a direct advantage over heat processing, where there is always a temperature gradient and a cold spot in the geometry of the packaged food to be processed.

The HP treatment of foods under different product- and process-specific conditions can cause partial unfolding of proteins that can lead to the reversible/irreversible gelation of the product. These changes can be desirable or undesirable depending on the product and processing scenarios. The amount of protein unfolding is dependent on many parameters and can be modulated by altering process-related parameters like pressure level, pressure holding time, pressure cycles, processing temperature, and depressurization time. Other food intrinsic parameters also play a key role: protein nature, protein concentration, food pH, and ionic strength.

Some researchers have worked on the effect of HP on proteins at the structural level (e.g., Smith et al. 2000) and others have evaluated HP effects on protein functionality (e.g., Ahmed and Ramaswamy 2003a; Ibanoglu and Karatas 2001), but few studies have related the two aspects under commercial HP processing conditions. The published information is limited by protein or technique and is usually restricted to a maximum of 400 MPa of pressure (e.g., López-Fandiño, Carrascosa, and Olano 1996). Studies dealing with changes in protein structure under heat-induced gel formation are more abundant but special care of the choice of methods to apply is of utmost importance. In the case of Liu et al. (2007) a practical application was the foundation of their work on gelation of fish and pork meat mixes for surimi production. This article is a good example of a good attempt to correlate product functionality and changes in molecular structure, but some methodological uncertainties linger, like discrepancies in the protein concentration used for rheological determinations and for circular dichroism studies, and also differences in heating rates.

14.2 BRIEF HISTORY OF HIGH-PRESSURE (HP) FOOD APPLICATIONS

The food processing and preservation industry boomed in 1809 with the first canned products and it has evolved over two centuries into a mature industry that still relies heavily on thermal processing to feed a growing global population. The novel nonthermal alternatives include pulsed electric field (PEF) and HP preservation. While PEF is mainly applied to liquids or pumpable products using a flow-through treatment chamber, HP can be applied to either liquid or solid foods in flexible containers, and it has proved valuable for food processing and preservation.

The recent upsurge in demand for fresh and healthy food has presented challenges to the food industry, primarily regarding the need to implement techniques that will allow food to be kept fresher for longer, along with a reasonable shelf life and food safety assurance. Thanks to these changes in consumer preferences, a new impetus has been given to the development of concept-driven technologies that could potentially provide the required processing using non- or mildly thermal means. Therefore, much of the recent scientific research for the food industry has focused on nonthermal processing, with HP processing being one of the few that shows vast potential in commercial aspects of the industry (Alvarez, Ramaswamy, and Ismail 2008; Knorr 1999; Rastogi et al. 2007; San Martín, Barbosa-Cánovas, and Swanson 2002).

In 1899, Japanese researcher Hite attempted HP treatment of food for the first time. His publication represents the first reported case of reduction of spoilage by means of high hydrostatic pressure. Hite and co-workers further expanded their experiments from treating meat and milk products to fruit and vegetables, successfully preserving peaches and pears treated at about 400 MPa at room temperature for 30 min (Hite 1899; Hite, Giddings, and Weakly 1914). While working on a project that involved the generation of HP in 1905, Percy W. Bridgman, a doctoral student at Harvard University, frustrated by the inadequacy of the available equipment, approached the pressure sealing by designing a leak-proof sealing that ensured that the sealing always remained at a higher pressure than the materials under study. This first major change led to other changes in the pumping system, then the development of the opposed-anvil device and further technical innovations, opening a whole new field of study of materials behavior under unprecedented high pressures. Bridgman experimented with organic compounds and biological materials, noticing the denaturation of enzymes submitted to HP (McMillan 2005).

In 1960, Keizo Suzuki reported on the combined effect of pressure and temperature on the kinetics of protein denaturation deduced by turbidity. Suzuki calculated the thermodynamic functions in the activation process of denaturation from the relations of absolute reaction rates obtained from ovalbumin and carbonylhemoglobin (Suzuki 1960). Hawley (1971) determined the pressure–temperature-reversible transition surface for chymotrypsinogen at pH 2.07 by ultraviolet difference spectra (temperature interval 8.5–70°C) between atmospheric pressure and 7000 atm (709 MPa), using the apparatus designed by Bridgman, and was able to fit the surface to a relatively simple equation of state, found to be compatible with known denaturation phenomena associated with pressure–temperature interaction. In further exploration of reversible denaturation of chymotrypsinogen under HP, the conformational relaxation occurred slowly, which allowed for the main states of the equilibrium mixture to separate by electrophoresis; he reported experimental concentration distribution patterns obtained at pH 2.03 and 20.5°C consistent with the expected behavior from a simple two-state isomerism (Hawley and Mitchell 1975).

Taniguchi and Suzuki (1983) examined the pressure effects on the inactivation of α-chymotrypsin (α-CHT) at different pressures up to 5 kbar (500 MPa) from 20 to 55°C; the apparent inactivation rate increased when the temperature increased and then suddenly dropped to zero above a certain temperature, reflecting the temperature dependence of the deacylation and the heat inactivation of α-CHT. Considering the volume alterations and hydrophobic interactions, the results seemed to support a scheme where pressure inactivation is caused by the rupture of hydrophobic interactions amid nonpolar groups of α-CHT. Also in 1983, Weber and Drickamer reviewed their groundbreaking work on the use of fluorescence as a probe of protein conformation; they established the nature of changes in protein conformation induced by temperature or pH, which were invariably irreversible. Further studies have shown that, under a variety of circumstances, pressure can introduce reversible changes (Drickamer 1999). In 1985, Heremans and Wong indicated the role of intermolecular interactions in pressure-induced protein denaturation; and later, in 1987, Kauzmann singled out the importance of HP in the understanding of the hydrophobic effect. Hayashi et al. (1989) applied high hydrostatic pressure to fresh egg white and yolk, further analyzing the resulting gels regarding their texture, protease susceptibility, and nutrients. The egg white set to a stiff gel at above 6000 kg/cm^2 (585 MPa), while the egg yolk reached the same state at a lower pressure of 4000 kg/cm^2 (390 MPa). In both cases, the gels had a natural taste, preserved their vitamins, and showed no amino acid residue or formation of unusual compounds. The authors proposed the use of HP to minimize adverse effects in food processing and preservation.

In the early 1990s the first commercial food applications of HP technology were seen, when the Japanese company Medi-Ya launched and marketed the first high-pressure processed (HPP) food, a high-acid jam (Mozhaev et al. 1994). Due to the commercial success of jams, other products have since been marketed, such as HPP jellies and shellfish in Japan, oysters and guacamole in the United States, and fruit juices in France, Mexico, and the United Kingdom (Smelt 1998; Torres and Velazquez 2005). In 1992, the first HP conference in bioscience and biotechnology took place, organized by Claude Balny. The same year, Groß and Ludwig (1992) reported pressure-temperature phase diagrams for the inactivation of microorganisms and later, in 1994, Jonas and Jonas carried out the first nuclear magnetic resonance (NMR) study of the temperature and pressure of a protein. Also, phase diagrams were developed for synthetic polymers (Kunugi et al. 1997), for the gelation of starch (Rubens et al. 1999) and for staphylococcal nuclease denaturation using fluorescence and NMR (Royer et al. 1993).

HP treatment of milk is currently a popular research topic and it has celebrated a centenary since the original work of Hite in 1899. Hite succeeded in prolonging milk's shelf life, but the technical difficulties of obtaining the necessary pressures meant his work was largely dormant and only reappeared when Japanese researchers teamed up with heavy industry who had the necessary engineering know-how. Clearly, much more is now known about the mechanism of action of HP. A few selected applications of HP in foods have been a commercial success, among them avocado paste and raw ham, and in some dairy applications HP processing has succeeded in eliminating yeasts in fermented products, although the high resistance of spores has proved to be a hard problem to solve (Patazca, Koutchma, and Ramaswamy 2006; Shao, Ramaswamy, and Zhu 2007; Zhang and Mittal 2008).

Mussa and Ramaswamy (1997) studied the kinetics of microbial destruction and changes in physicochemical characteristics of fresh raw milk caused by HP treatment, which was conducted at 200–400 MPa for various holding times (5–120 min). The treatment led to an efficient destruction of microorganisms and a prolonged shelf life of milk up to 18 days at 5°C and 12 days at 10°C. It was concluded that HP processing of milk may be a useful alternative for extending the shelf life with quality advantages.

It becomes obvious that after a century of utilizing HP for food processing, it is only over the last 20 years that significant advances in HPP technology have been made, in the form of semi-continuous systems, to the scaling up of pilot units, to successful commercially viable processes. Present industrial HPP treatment of food is carried out using a batch or semicontinuous process; solid food can only be treated in a batch mode whereas liquid products can also be treated using a continuous or semicontinuous process (Hogan, Kelly, and Sun 2005). An important example of the commercial significance of HPP food equipment at an industrial level is Avomex Inc., which began HP-treating avocado using a 25 l batch processing unit in 1996 and then decided to invest in another 25 l as well as a 50 l vessel as product demand grew. By the twenty-first century, the company had undergone another expansion, investing in a semicontinuous unit and a larger 215 l batch processing vessel (Torres and Velazquez 2005). More recently, HPP has extended to include food products such as salsa, rice products, fish, meal kits (containing HP-treated cooked meats and vegetables), poultry products, and sliced ready-to-eat meats (Murchie et al. 2005). HPP treatment of such foods has enabled the consumer to access foods with distinct advantages over thermally processed foods, such as minimally processed, fresh-tasting, high-quality convenient products with an extended shelf life.

In the past two decades, a lot of scientific material has been published on HP-related topics and interest continues to grow. The following peer-reviewed scientific articles have been published by a single research group in Canada over the past decade:

1. HP destruction of spoilage and pathogenic bacteria in different foods (Basak, Ramaswamy, and Piette 2002; Basak, Ramaswamy, and Smith 2003; Mussa and Ramaswamy 1997; Mussa, Ramaswamy, and Smith 1999a, 1999b; Pandey, Ramaswamy, and Idziak 2003; Ramaswamy and Riahi 2003; Shao et al. 2007)
2. HP-shifted freezing and thawing (Chen et al. 2007; Ousegui et al. 2006; Rouille et al. 2002; Sequeira-Muñoz et al. 2005; Zhu, LeBail, and Ramaswamy 2003, 2006a; Zhu, Ramaswamy, and LeBail 2004d, 2005a, 2005b, 2006b; Zhu, Ramaswamy, and Simpson 2004e; Zhu et al. 2004a, 2004b, 2004c)
3. Changes in quality of foods, including changes derived from HP inactivation of enzymes (Ahmed and Ramaswamy 2006a, 2006b; Ashie, Simpson, and Ramaswamy 1996b; Basak and Ramaswamy 1997, 2001; Basak, Ramaswamy, and Simpson 2001; Pandey and Ramaswamy 2004; Pandey, Ramaswamy, and St-Gelais 2003a, 2003b; Ramaswamy and Riahi 2003; Riahi and Ramaswamy 2003, 2004; Sequeira-Muñoz et al. 2006)
4. HP-induced changes in functional properties of foods, including texture, viscoelasticity, and digestibility (Ahmed et al. 2003, 2007a; Ahmed, Ramaswamy, and Hiremath 2005; Ahmed and Ramaswamy 2003a, 2003b, 2004, 2005, 2007; Ashie, Simpson, and Ramaswamy 1996a; Basak and Ramaswamy 1998; Izquierdo et al. 2005; Pandey, Ramaswamy, and St-Gelais 2000; Sareevoravitkul, Ramaswamy, and Simpson 1996)
5. HP at elevated temperature causing inactivation of bacterial spores (Patazca et al. 2006; Shao et al. 2008; Zhu et al. 2008)
6. HP-induced structural changes of proteins (Ahmed et al. 2007b; Alvarez, Ramaswamy, and Ismail 2007, 2008)

14.3 APPLICATION AREAS OF HIGH-PRESSURE (HP) PROCESSING

Treatment of food with HP has primarily been studied as a means to control bacterial populations with improvement of the food's shelf life and/or increased product safety. However, there are other areas that make use of HP technology, including HP-shifted freezing and thawing; HP-assisted thermal sterilization; inactivation of quality-affecting enzymes; enhancement of activity of enzymes, so as to accelerate processes like cheese ripening; enhancement of functionality of foods with minimal impact on nutritional/quality aspects; improvement of protein digestibility; minimizing allergens and antinutritional factors; and enhanced release of nutraceuticals from foods.

Before the most important applications of HP technology are discussed, the governing principles of this novel process should be reviewed. The pressure applied to foods is transmitted isostatically and instantaneously; thus the process is not dependent on the shape or size of the food. This is a major improvement, since the food is treated evenly throughout its geometry, which has regularly been problematic in thermal processing of large or bulky food products.

HPP acts instantaneously and uniformly throughout a mass of food independent of size, shape, and food composition. Consequently, package size, shape, and composition are not factors in process determination. The work of compression during HPP treatment will increase the temperature of foods through adiabatic heating approximately 3°C per 100 MPa, depending on the composition of the food (Balasubramanian and Balasubramaniam 2003). For example, if the food contains a significant amount of fat, such as butter or cream, the temperature rise can be larger. Foods cool down to their original temperature on decompression if no heat is lost to or gained from the walls of the pressure vessel during the hold time at pressure (Hogan et al. 2005). A uniform initial temperature is required to achieve a uniform temperature increase in a homogenous system during compression (Zhu et al. 2008).

Characteristically, food preservation techniques are for the most part evaluated based on their ability to eradicate pathogenic microorganisms, in doing so improving food safety and extending product shelf life through the inactivation of spoilage microorganisms. HP treatment has a distinct advantage in this respect, producing foods of better quality and nutritional value than thermally processed products (Smelt 1998).

HPP is potentially able to produce high-quality extended-shelf-life foods that exhibit characteristics of fresh products and are microbiologically safe. For today's market, HP-treated foods are novel foods under the criteria of a new manufacturing process being employed in their production, and their history of human consumption, to date, being minimal. However, consumer perception of food quality depends not only on microbial quality but mainly on other more apparent factors such as biochemical and enzymatic reactions, structural changes, taste, and visual appeal. HP treatment of food can have an effect on food yield and on sensory qualities such as color and texture, and it is common knowledge that the appearance and color of food significantly influence consumer sales (Hogan et al. 2005).

During HP treatment of certain high-protein foods, some degree of protein denaturation can take place, but there are significantly fewer changes in physical functionality and/or raw product color than in foods treated with conventional thermal processing techniques (Hogan et al. 2005). Further, the reversibility of HP-induced protein denaturation is determined by treatment conditions such as temperature, time, and pressure, and also depends on the type of protein (Rastogi et al. 2007). By using HP conditions of 100–300 MPa, proteins usually, though not always, denature, dissolve, or precipitate reversibly (Thakur and Nelson 1998).

HPP treatment could extend the shelf life of fresh meat and poultry by controlling the growth of both spoilage and pathogenic bacteria. Hugas, Garriga, and Monfort (2002) report that from both a physicochemical and microbiological point of view, cooked pork ham, dry cured pork ham and, marinated beef loin, vacuum-packed and HP treated at 600 MPa for 10 min at 30°C, are substantially equivalent to the same untreated products. In fresh meat and poultry, pressure-induced color changes are due to changes in myoglobin, heme displacement/release, or ferrous atom oxidation, which can result in a cooked-like appearance that prevents the product from being sold as fresh meat. On the other hand, HP treatment of white or cured meats hardly ever causes major color changes (Cheftel and Culioli 1997).

According to the United Kingdom's Agri-Food and Biosciences Institute, HP processing can be used to give additional microbiological safety assurance to meats already treated with another preservation method without further adversely affecting eating quality. The foods are not sterile and must be refrigerated to ensure optimum quality. HPP is being used commercially in the United States and Europe to treat cooked poultry, cooked beef, vacuum-packaged ham, fermented sausage, and salami (Hayman et al. 2004; Marcos, Aymerich, and Garriga 2005; Patterson 2005). In the light of the unlucky events of sliced ready-to-eat meats in Canada involving contamination with *Listeria monocitogenes*, HP processing should become one of the alternatives to prevent such events in the future.

Along with color, food texture has a vast impact on product sales, as even the most appealing food will be perceived as decaying or just "not good" if its textural properties are not similar to those of fresh food, e.g., soft/spongy texture. Consequently, a detailed understanding of the rheological properties of different kinds of foods is essential for product development and quality control. The physical structure of most high-moisture food products remains unchanged after HPP exposure as the pressure exerted does not generate shear forces; however, color and texture may change in gas-containing products after HP treatment due to gas displacement and liquid infiltration into the collapsed gas pockets from the surrounding food structure. Shape distortion and physical shrinkage can occur due to the collapse of air pockets, generally causing irreversible compression of whole foods such as fruit (Hogan et al. 2005); however, fruit fragments versus whole fruit are better suited for such treatment. Regardless of some of the objectionable textural changes, HP treatment can also be used to induce desirable changes in product texture and structure such as melting of mozzarella cheese during processing (O'Reilly et al. 2002).

Among the main benefits of HP processing of food is the extension of shelf life while retaining the sensory characteristics of fresh food products (Lakshmanan, Patterson, and Piggott 2005). One of the most successful cases is that of fresh avocado. Palou et al. (2000) reported that avocado's delicate sensory attributes could be preserved using HPP while also conferring a sensibly safe and stable shelf life. In general, HPP treatment gives a higher food product yield compared with heat treatment, with effects depending on product type and treatment intensity. This has enormous economic significance to food manufacturers (Hugas et al. 2002).

Whereas conventional thermal sterilization processes are the most frequently used methods for food preservation, and they are effective mechanisms for microbial inactivation, these techniques often lead to undesirable changes in the product quality, flavor, texture, color, and vitamin contents. In contrast, microbial inactivation provided by HPP primarily targets cell membranes of treated cells, even though in some cases the process can cause solute loss, protein denaturation, and key enzyme inactivation. However, in these cases, when the sole use of HP yields unsatisfactory results, the food treatment benefits when HP is combined with other processing techniques, such as acidification and PEFs. In addition, when HP has been used concurrently with mildly thermal processes, it has been found to increase the inactivation of bacterial spores (Raso and Barbosa-Cánovas 2003).

Most of the HP equipment currently used is operated in batch mode. Since pressurizing and depressurizing steps can now be rapidly accomplished, the low efficiency associated with batch processing may be minimized. Rapid pressurizing and depressurizing cycles also can cause metal fatigue and reduce the life of equipment. Above 400 MPa, the weight of equipment increases significantly, as does its cost (Torres and Velazquez 2005).

Foods produced by HPP are already available on both the Japanese and U.S. markets. These foods generally cannot be processed by conventional methods. Guacamole (prepared from avocado) is traditionally prepared immediately prior to consumption owing to its sensitivity to heat and other preservation methods. HP-processed guacamole, having a refrigerated shelf life of several weeks, is now produced by Avomex (Avomex Inc., Keller, TX) for the US market. Pressure-processed fruit juices with sensory qualities similar to those of the freshly prepared products are scheduled to appear on the U.S. market in the near future. The recent application of HPP to fresh seafood allows the sale of minimally processed oysters with a level of contamination that otherwise could not be reached without affecting the sensory properties of the product. Pressure-processed fresh oysters will soon be available on the U.S. market. The novelty of pressure-processed products justifies their relatively high price and the high processing start-up costs (Palou et al. 2000; Kural and Chen 2008).

As for the Canadian market, in 2004 Health Canada approved the sale of applesauce and applesauce/fruit blends treated by high hydrostatic pressure by Leahy Orchard Inc., Franklin Centre, Quebec. The processing consists of a 550 MPa treatment with a 1 min hold.

14.4 HIGH-PRESSURE (HP)-INDUCED CHANGES IN FUNCTIONAL PROPERTIES

HP processing can modify functionality of food products to a certain extent and this can be controlled by varying the pressure level, pressure holding time, and depressurization time, as well as other parameters noninherent to the HP process such as pH, ionic strength, and food composition (Messens, Van Camp, and Huyghebaert 1997).

14.4.1 Food Rheology and Texture

Modern rheology is defined as the deformation and flow of matter and food rheology is the deformation and flow of the raw materials, the intermediate products, and the final products of the food industry. Rheology focuses on the relationship between a force acting on a substance and its resulting deformation; simply put, it is the study of how systems behave when work is applied (liquids might flow, solids may bend) (Walstra 2003).

Rheological tests in food are of great value for engineering process design, since the knowledge about food flow and deformation properties makes possible the design of equipment for handling foods, such as conveyor belts, pumps, pipelines, storage containers, etc. Rheological tests allow the acquisition of information on the structure of the food and its macromolecular elements, as well as assessing the textural attributes of the product prior to its processing. Based on these predictions, the food processing will be oriented toward achieving a final product with the characteristics that have proven desirable for the consumer. The deformation of the food under the influence of a force is frequently used as a measure of quality, so foods that deform to a large extent are categorized as soft, flaccid, or spongy, whereas those that deform to a small extent are classified as firm, hard, or rigid. This does not mean that either group is better than the other, because the association of these characteristics with food quality depends on what the consumer wants regarding that specific type of food (Borwankar 1992).

In order to obtain reproducible results in deformation tests, the geometry of the samples needs to be controlled, so it is customary to use specimens of standard dimensions and predefined geometry to perform the tests, where the force per unit area is called stress and the deformation per dimensional unit is called strain. The geometry determination is especially important in the type of foods that cannot be cut into standard-size/shape pieces (i.e., lettuce and eggs). Only by understanding how size and shape affect the deformability of the food is it possible to separate the deformability measurements into differences due to changes in textural quality from those due to changes in size (Bourne 1967).

14.4.1.1 Dairy Foods

There are a number of studies regarding the rheological properties of HP-treated foods. Regarding the HP treatment of dairy products, HPP can affect the protein structure in milk, giving the potential to produce cheese with unique, desirable characteristics. This has been used successfully in the manufacture of mozzarella cheese with improved "meltability" and stringiness, both desirable characteristics in pizza production. Changes in the protein structure can also lead to novel gelling properties, which could be of value in the manufacture of yogurts and other dairy desserts (Sheehan et al. 2005; Walker et al. 2006).

Arora, Chism, and Shellhammer (2003) examined the effects of surfactant type and concentration, dispersed phase concentration, and the presence of a stabilizer on the rheological properties as well as physical stability of pressure-treated acidic emulsions, using materials and conditions found in commercial salad dressings. This research confirmed that pressure treatment had no significant detrimental effects on the rheological behavior or physical stability of acidified emulsions stabilized by whey protein isolate and polysorbate-60. Differences in the flow behavior, viscoelasticity (G' and G''), particle size distribution, and physical stability of emulsions were influenced largely by the lipid content and the type of surfactant. Emulsions stabilized by soy lecithin were inherently unstable and this instability was further aggravated under pressure. Pressure-stable oil-in-water emulsions can be formed using hydrophilic surfactants such as polysorbate-60 or whey protein. The addition of xanthan gum improved stability in systems emulsified with polysorbate-60 and this stability was further improved by the application of pressure treatment.

Molina et al. (2000) evaluated the suitability of HP-treated milk for reduced-fat cheese production. For this purpose, pasteurized, pressurized, and pasteurized-pressurized milks were compared regarding cheese-making properties, cheese composition, proteolysis during ripening, and development of cheese flavor and texture. The results obtained showed that pressurization (400 MPa, 22°C, 15 min) of semiskim milk prior to cheese-making increased the yield of reduced-fat cheese through an enhanced protein and moisture retention. The sequential application of pasteurization (65°C, 30 min) and pressurization increased this effect over the pressurization of raw milk and, in addition, pressurization of pasteurized skim milk improved its coagulation properties. The cheeses made from pasteurized-pressurized and pressurized milks underwent a faster rate of protein breakdown than the cheeses made from pasteurized milk, which could have been responsible for a more rapid development of texture and flavor. The cheeses with the highest moisture content and proteolytic degradation presented the lowest hardness, as determined by both the sensory panel and instrumental analyses. In addition, the results suggested a relationship between the lower hardness and cohesiveness of the cheeses made from pasteurized-pressurized and pressurized milks, as determined instrumentally, and their higher texture scores determined by sensory analysis. Pressurization of low-fat milk prior to cheese-making improved cheese texture and thus accounted for a higher overall acceptability, except for the cheese made from pasteurized-pressurized milk at 60 days of ripening, whose acceptability score was adversely affected by bitterness. Since pressurized milk coagulates faster and produces firmer curds, the amount of coagulant used can be reduced so the amount of residual rennet can be optimized.

The development of food ingredients with novel functional properties offers the dairy industry an opportunity to revitalize existing markets and develop new ones. HPP can lead to modifications in the structure of milk components, particularly protein, which may provide interesting possibilities for the development of high-value nutritional and functional ingredients. O'Reilly et al. (2000) reported that the application of HPP to cheddar cheese resulted in increased levels of proteolysis, a key measure of maturity, indicating that HPP has the potential to accelerate the ripening of cheddar. However, the level of acceleration observed was not sufficient to support the commercial exploitation of the technology.

Walker et al. (2006) used HPP to treat fruit yogurt in an attempt to develop a shelf-stable fruit yogurt and monitor changes in color, texture, and microbial content over 60 days' storage at 4.4°C and room temperature (25°C). HP-treated yogurt had a noticeably smoother and thicker appearance and maintained higher viscosity at both storage temperatures over the 60-day storage period, compared with the non-pressure-treated control yogurt. In addition, there were no significant effects on color or pH. The HPP inactivated yeasts and moulds and no growth was observed following prolonged storage at room temperature, but the HPP also inactivated the lactic acid bacteria after 40 days, therefore not allowing the "live culture" claim to be stated on the yogurt container.

14.4.1.2 Egg Proteins

Ahmed et al. (2003) investigated the effect of HPP on rheological characteristics of whole liquid egg (WLE), albumen, and yolk, finding that all egg samples behaved as thixotropic fluids. The egg protein structure breakdown was enhanced with pressure and it was complete at 300 MPa for 30 min at 20°C. HP affected the protein structure of albumin and WLE; however, electrophoresis results showed that the protein coagulation was irreversible. The yolk behaved differently with pressure treatment. The study concludes by stating that further work is needed on pressure-temperature combination and microbiological aspects of postprocessed samples before implementation in the industrial sector. Ahmed et al. (2005) observed improved rheological characteristics of mango pulp after HP treatment, with significant differences between canned (thermally pretreated) and fresh mango pulp. No significant changes in color were detected after HP treatment.

The HP treatment also inactivates microorganisms and, therefore, this technique provides an alternative to heat treatments. In contrast, there are considerable differences in protein denaturation and aggregation induced by HP compared to those induced by heat. The use of HP to modify the functionality of food proteins was reviewed by Heremans and Smeller (1997) and Messens et al. (1997), finding that while blood plasma and egg white proteins are known to be sensitive to heat and readily form gel networks at moderate operating temperatures, at 80°C (30 min) no gelation occurs if these proteins are pressurized for 30 min at a pressure of 400 MPa. This stability may be positively correlated with the large number of disulfide bonds stabilizing the three-dimensional structure of both proteins. At the same time, β-lactoglobulin appears far more sensitive to pressure than ovalbumin and bovine serum albumin (BSA). HP processing has been shown to destabilize casein micelles in reconstituted skim milk. The size distribution of the spherical casein micelles changed from ~200 to 120 nm after pressurization, whereas subsequent heating of skim milk at 30°C and at atmospheric pressure restored the original size distribution.

14.4.2 Protein Foaming

Foams are an array of bubbles dispersed within a continuous phase arranged in the form of a thin film called a lamella. In the case of water-based foams, the air bubbles are separated by an hourglass-shaped film containing dissolved solutes such as sugars, fat, ice crystals, low-molecular-weight emulsifiers, thickeners (nongelling), starch, protein, or a combination of these. Foaming capacity (FC) is the volume of gas-phase and foam stability (FS) is the degree to which fluid can be retained within the foam, requiring a cohesive protein film that entraps air bubbles (Bos and van Vliet 2001).

Foam is caused by the protein film lowering surface tension. An example would be cohesive force of water molecules that tend to collapse a bubble and give resistance to shearing/tearing with a high level of stretching capacity. So when a protein solution is whipped or stirred vigorously, air is pulled down into solution, and when it tries to escape up, the flexible protein surface forms a bubble. Not all proteins foam equally well and not all of the foams are particularly stable. So BSA is slightly below average on foaming, while casein forms allow a good foam to be obtained, although it is not stable. Egg protein does not form much foam, but it is quite stable, and gelatin provides a good amount of foam that is also stable (Campbell and Mougeot 1999; Foegeding, Luck, and Davis 2006).

Proteins are without doubt the most employed foaming agents in the food industry. They adsorb sturdily at the air–water interface, where the protein unfolds. Hydrophobic amino acid residues within the protein orient themselves toward the gas phase while hydrophilic portions of amino acids place themselves in the aqueous phase. Since proteins typically have multiple hydrophobic sites, they adsorb strongly at the interface with little likelihood of spontaneous desorption. Unfolded proteins can also interact with other protein molecules, forming a film at the interface. Measuring the strength of the film using surface rheology is helpful in predicting stability against foam destabilization mechanisms such as coalescence and disproportionation (Murray 2007).

14.4.2.1 Dairy Proteins

Rodiles-López et al. (2008) found an increased FC and stability with HPP of α-lactalbumin (600 MPa at 55°C for 10 min). Ibanoglu and Karatas (2001) studied the FC of whey protein isolate as affected by HP treatment; they found a positive correlation between total foam volume (FV) and pressure level applied, whereas pressure holding time did not have a significant effect on FV or FS.

Lim, Swanson, and Clark (2008) studied the foaming properties of fresh and commercial whey protein concentrate (WPC) after HP treatment. Solutions of WPC were treated with 300 and 400 MPa (0 and 15 min holding time) and 600 MPa (0 min holding time) pressure. After HP, the solubility of the WPC was determined at both pH 4.6 and 7.0 using UDY and BioRad protein assay methods. Overrun and FS were determined after protein dispersions were whipped for 15 min. The protein solubility was greater at pH 7.0 than at pH 4.6, but there were no significant differences at different HP treatment conditions. The maintenance of protein solubility after HP indicates that HP-treated WPC might be appropriate for applications to food systems. Untreated WPC exhibited the smallest overrun percentage, whereas the largest percentage for overrun and FS was obtained for WPC treated at 300 MPa for 15 min. Additionally, HP-WPC treated at 300 MPa for 15 min acquired larger overrun than commercial WPC 35. The HP treatment of 300 MPa for 0 min did not improve FS of WPC. However, WPC treated at 300 or 400 MPa for 15 min and 600 MPa for 0 min exhibited significantly greater FS than commercial WPC 35. The HP treatment was beneficial to enhance overrun and FS of WPC, showing promise for ice-cream and whipping cream applications.

14.4.2.2 Egg Proteins

van der Plancken, van Loey, and Hendrickx (2007) studied the foaming properties of egg white proteins affected by HP treatment; they found foams with high volume and average stability and density obtained by pressure treatment at pH 8.8 (above 500 MPa). The processing-induced changes in the foaming properties could not be attributed to the changes in a single physicochemical property; the foaming ability was in part determined by the sulfhydryl content and protein flexibility. Improved protein-protein interactions (solubility and exposed SH groups) contributed to increased FS of treated egg white solutions.

14.4.3 Protein-Emulsifying Capacity/Behavior

An emulsion is a mixture of two immiscible substances, where one of them or dispersed phase exists as discrete droplets suspended in the other or continuous phase. There is an interfacial layer between the two phases formed by some indispensable surfactant material (Schramm 2005).

There are three main types of emulsions of high relevance to the food industry: oil-in-water emulsions (o/w), where droplets of oil are suspended in an aqueous continuous phase; water-in-oil emulsions (w/o), which depend more for their stability on the properties of the oil and the surfactant used than in the properties of the aqueous phase; and water-in-oil-in-water emulsions (w/o/w), an o/w emulsion where the oil droplets are in fact a w/o emulsion (they contain water droplets). This last type of emulsion is the most complex to produce and control, because the stability of the water droplets contained in the oil droplets, as well as the stability of the oil droplets contained in the continuous aqueous phase, must be achieved (McClements 2005).

Emulsions play an important part among the structure units within foods, as not only do they impart a desirable mouthfeel, but they are also essential ingredients in the formation of structures of certain products such as whipped toppings, ice-creams, and even processed cheeses (McClements 2005).

The interfacial layer of most o/w emulsions contains proteins, often mixed with other surfactant materials. Proteins are frequently present in the raw food and they are excellent emulsifiers, but the composition of the interfacial layer if determined by what is present at the moment the emulsion is formed. If proteins are the only emulsifier, they will adsorb to the oil interfaces in proportion to their concentration in the aqueous phase, except for some mixtures of α- and β-caseins, where there will be a preferential adsorption of α-casein (Dickinson 1999).

Several different methods have been used to estimate the dimensions of adsorbed protein monolayers, such as dynamic light scattering, ellipsometry, and neutron reflectance, showing that adsorbed layers of protein may be thick compared to molecular dimensions, one of the reasons why adsorbed proteins can stabilize emulsion droplets either by steric repulsion and electrostatic mechanisms. Most adsorbed proteins exist in conformations unlike their original state, as a result of the tendency of hydrophobic parts of the molecule to be adsorbed by the hydrophobic interface with the disruption of their secondary and tertiary structures. That is the main reason why the properties of the emulsion are not the same of those of the parent protein (Dalgleish 1997).

14.4.3.1 Dairy Proteins

In the food industry, protein macromolecules are usually the universal stabilizer of w/o emulsions and foams. Emulsions in ice-creams, toppings, and creams are consistently stabilized by milk proteins, casein in aggregated form usually being the main stabilizer (Dickinson 1999).

Regarding β-lactoglobulin emulsions, Dickinson and James (1998) reported that moderate thermal processing was shown to have a far greater effect than HPP on the state of flocculation of β-lactoglobulin-coated emulsion droplets, therefore regarding HPP as a milder processing operation than thermal treatment. It was found that at neutral pH, strong emulsion gels were produced from concentrated emulsion samples (40 vol% oil + 1.0 wt% protein) following heat treatment at 70°C for 5 min. The increase in viscoelasticity caused by mild heating (i.e., 65°C for 5 min) resembled the change in rheological behavior induced by rather severe pressure processing (800 MPa for 60 min). Even though both treatments induced droplet flocculation, the rheological properties of moderately dilute β-lactoglobulin-stabilized emulsions were unaffected by HP or thermal treatment, given the insensitivity of small-deformation rheological analysis to the state of flocculation of droplets in a dilute emulsion such as the one used. Under conditions of lower pH or higher ionic strength, β-lactoglobulin-stabilized emulsions became more flocculated following temperature or pressure processing. The extent of emulsion flocculation following HP appeared greater when there was a larger proportion of free protein present in the aqueous continuous phase. The observation of significant levels of droplet flocculation, even in the presence of minimal unadsorbed protein, suggested that changes in adsorbed protein structure also contributed significantly to the overall effect of pressure treatment. Rodiles-López et al. (2008) found improved emulsion activity index and emulsion stability with HPP α-lactalbumin after 600 MPa at 55°C for 10 min at different pH (3.0–9.0).

14.4.3.2 Blood Plasma Proteins

Parés and Ledward (2001) studied the emulsifying and gelling properties of porcine blood plasma as influenced by HPP, finding that while HP treatments up to 300 MPa did not affect the functionality of blood plasma proteins, regardless of the pH, the effects of treatments at pressures above 400 MPa caused pH-dependent changes. At acidic pH (5.5), increasing pressures led to a decrease in the emulsifying properties of plasma solutions and provoked changes in the textures of heat-induced gels, which adversely affected their water-holding capacity (WHC). Both properties were seriously affected by the changes in structure that the proteins underwent following HPP at this pH. At pH 6.5, HP treatments at 400 MPa improved the emulsifying properties of plasma solutions without negatively affecting the characteristics of heat-induced gels. Treatments above 400 MPa produced solutions with reduced emulsifying activity and stability. At pH 6.5, heat-induced gels were always weaker and less elastic but, in some cases (pH 6.5), they showed improved WHC.

14.4.4 Protein Solubility

Solubility is a crucial function of proteins for the food industry. The Osborne classification of proteins such as albumins, globulins, glutelins, and prolamins in based on their solubility. Albumins are those proteins that are readily soluble in water, while globulins are those that may require salt solution for solubilization. Glutelins are soluble in dilute acid or base, and prolamins require alcohol-based media as solvents (Osborne and Campbell 1898).

At the surfaces of proteins are amino acid residues that interact with water. The amino acids are referred to as hydrophilic amino acids and include arginine, lysine, aspartic acid, and glutamic acid. At pH 7 the side chains of these amino acids carry charges—positive for arginine and lysine, negative for aspartic acid and glutamic acid. As the pH increases, lysine and arginine begin to lose their positive charge, and at pHs greater than about 12 they are mainly neutral. In contrast, as pH decreases, aspartic acid and glutamic acid begin to lose their negative charges, and at pHs less than 4 they are mainly neutral (Arakawa and Timasheff 1985).

The surface of a protein has a net charge that depends on the number and identities of the charged amino acids and on pH. At a specific pH the positive and negative charges will balance and the net charge will be zero. This pH is called the isoelectric point (p*I*) and for most proteins it occurs in the pH range of 5.5–8. A protein has its lowest solubility at its isoelectric point. If there is a charge at the protein surface, the protein prefers to interact with water, rather than with other protein molecules. This charge makes it more soluble. Without a net charge, protein-protein interactions and precipitation are more likely (Arakawa and Timasheff 1985).

Rodiles-López et al. (2008) studied the solubility of α-lactalbumin treated with 600 MPa at 55°C for 10 min and found a positive effect on protein solubility after HPP, in particular at low pH values. Additionally, Yin et al. (2008) found the solubility of kidney-bean protein isolate significantly improved at pressures of 400 MPa or higher, possibly due to formation of soluble aggregates from insoluble precipitates.

14.4.5 Protein Digestibility

Digestion typically begins in the stomach when pepsinogen is converted to pepsin by the action of hydrochloric acid, and continued by trypsin and CHT in the intestine. The amino acids and their derivatives into which dietary protein is degraded are then absorbed by the gastrointestinal tract. The absorption rates of individual amino acids are highly reliant on the protein source; for example, the digestibility of many amino acids in humans differs between soy and milk proteins and between individual milk proteins, β-lactoglobulin and casein. For milk proteins, about 50% of the ingested protein is absorbed between the stomach and the jejunum and 90% is absorbed by the time the digested food reaches the ileum (Matthews and Laster 1965).

The digestibility of proteins depends on, among other things, their structure. Compacting of the protein structure by cross-linking prevents unfolding of the peptide chains, which may decrease the digestibility by making some of the peptide bonds unreachable to the enzymes. Cross-linking may be caused by a decrease in the rate of protein hydrolysis due to heating, as well as by interactions of amino groups with reduced saccharides or other compounds containing carbonyl groups. In food processing, these are secondary products of lipid oxidation and aldehydes contained in smoked products. As a result of several further reactions, different unsaturated compounds are generated and interact with the amines, possibly involving different cross-linking reactions (Matthews and Laster 1965).

The protein digestibility-corrected amino acid score (PDCAAS) method has been considered to be a straightforward and scientifically sound approach for routine assessment of dietary protein quality for humans. Major questions have, however, been raised about the validity of the PDCAAS relative to its inability to credit the extra nutritional values of proteins having scores higher than that of the reference protein, its failure to fully account for the possible adverse effects of antinutritional factors, and its assumption about complete biological efficiency of supplemental amino acids in improving quality of proteins, which may not be true in the case of poorly digestible, low-quality proteins (Schaafsma 2005).

Throughout food processing, protein sources are treated with heat, oxidizing agents such as hydrogen peroxide, organic solvents, alkalis, and acids for a variety of reasons such as to sterilize or pasteurize, to improve flavor, texture, and other functional properties, to deactivate antinutritional factors, and to prepare concentrated protein products. These processing treatments may cause the formation of Maillard compounds, oxidized forms of sulfur amino acids, D-amino acids, and cross-linked peptide chains, resulting in lower amino acid bioavailability and a decrease in protein quality (Bender 1972).

In the journey toward a more accurate method to determine protein and amino acid digestibility of food products, several in vitro methods have been developed, simulating the physiological process of digestion. The Agri-Food Agriculture Canada center in St-Hyacinthe (Quebec) has acquired a complete system that simulates the digestive system from mastication to absorption of nutrients; this instrument is intended to give a more holistic approach to food research.

Yin et al. (2008) studied in vitro trypsin digestibility of kidney-bean protein isolate and found that the protein digestibility decreased only at pressures above 200 MPa and for long holding times (up to 120 min). Chicón et al. (2006) previously noted an improved in vitro trypsin proteolysis of β-lactoglobulin as affected by pressures up to 400 MPa, regardless of the digestion carried out under pressure or over a pre-HP-treated protein. They extended their work to the pepsin digestibility of β-lactoglobulin and whey protein isolate after HP treatment and found that HP treatment at 400 MPa promoted the hydrolysis of β-lactoglobulin by pepsin, but this increased susceptibility of β-lactoglobulin to proteolysis was progressively lost during refrigerated storage (Chicón et al. 2008).

Izquierdo et al. (2005) reported complete in vitro hydrolysis of β-lactoglobulin by α-CHT or pronase when using intermediate pressures (100–300 MPa), compared to partial hydrolysis by microwave or conventional heating.

14.4.6 Protein Water-Holding Capacity (WHC)

Water-holding capacity is an important property of gelled foods, particularly in dehydrated and rehydrated products, in which the strength of the water binding to the solid matrix can be weaker than that in the original food. In some cases, water will not leak out of a solid-like food, but it is present in closed cells, in open pores in a solid matrix, or between chains of coiled polymers. Binding sites for water need not be present for the water to be held, and this amount of held water can be large: in some gels, 1 g of polymer can hold 100 g of water, usually in the form of an aqueous solution rather than as pure water (Kneifel et al. 1991).

A simple explanation for the way that water is held would be that of water as a space filler. A coiled polymer molecule has a certain equilibrium conformation and there are water molecules in the space between its segments in such a way that removal of water will cause shrinking of the coil, which costs free energy, since it implies a decrease in conformational entropy. This is also true for gels and gel-like structures, when deformation of the network needs a force and consequently, energy. The amount of held water in a gel-like system would vary with all the factors that affect the equilibrium state of swelling of the gel, such as concentration of cross-links, solvent quality, pH, and ionic strength (Foegeding 2006).

Pandey et al. (2000) studied the WHC of rennet curds obtained from HP-treated milk. They found that in general, with a decrease in pressure level, temperature, and holding time, there was a decrease in WHC and an increase in the gel-strength of the produced rennet curds.

14.5 CHANGES IN MOLECULAR STRUCTURE OF PROTEINS

Proteins are linear condensation products of various α-L-amino acids that differ in molecular mass, charge, and polar character, bound by transpeptide linkages. Proteins differ also in the number and distribution of amino acid residues in the molecule, that is, the secondary structure containing helical regions, β-pleated sheets and β-turns, the tertiary structure or the spatial arrangement of the chains, and the quaternary structure or assembly of polypeptide chains (Richardson 1981).

The structure of protein is an important determinant for its functionality. Therefore, changing the protein structure may improve functionality. The protein structure can unintentionally be changed during processing as in Maillard reactions and oxidation of cysteine or deliberately by modification reactions (Kester and Richardson 1984).

Chemical modification methods that are commercially used include extensive hydrolysis and deamidation. Extensive hydrolysis by boiling with acids results in free amino acids and very small peptides, and is mainly applied for the preparation of hydrolyzed vegetable protein, frequently used to fortify and enhance the aroma of soups. Deamidation, which is the conversion of the amino acids glutamine and asparagine to their acidic counterparts, is meant to change the charge, achieving alteration and extension of the range of functional properties of wheat gluten (Vojdani and Whitaker 1994).

Most studies on modification of protein molecular structure have focused on enzymatic modification, especially on the use of proteases and transglutaminases. There are also a number of publications on the enhancement of solubility, foaming, and emulsifying properties by limited proteolysis leading to changes in charge, hydrophobicity, and molecular mass in the transition from protein to peptide mixtures (Panyam and Kilara 1996).

HPP can affect protein conformation and produce protein denaturation, aggregation, or gelation, depending on the protein system, the applied pressure, the temperature, and the duration of the pressure treatment. Low pressures usually induce reversible changes such as dissociation of protein-protein complexes, the binding of ligands, and conformational changes, while pressures higher than 500 MPa induce, in most cases, irreversible denaturation to HP-sensitive proteins (Gross and Jaenicke 1994; Knorr, Heinz, and Buckow 2006).

14.5.1 Protein Secondary Structure

The secondary structure of a segment of polypeptide chain is the local spatial arrangement of its main-chain atoms without regard to the conformation of its side chains or to its relationship with other segments. There are three common secondary structures in proteins, namely α-helices, β-sheets, and turns. Those which cannot be classified as one of the standard three classes is usually grouped into a category called "random coil" (Richardson 1981).

The α-helix and β-structure conformations for polypeptide chains are generally the most thermodynamically stable of the regular secondary structures. However, particular amino acid sequences of a primary structure in a protein may support regular conformations of the polypeptide chain other than α-helical or β-structure. Thus, whereas α-helical or β-structure are found most commonly, the actual conformation is dependent on the particular physical properties generated by the sequence present in the polypeptide chain and the solution conditions in which the protein is dissolved. In addition, in most proteins there are significant regions of unordered structure in which the angles are not equal. A large proportion of helices are distorted in some way, i.e., radius of curvature greater than 90 Å and deviation of axis from straight line is equal to or greater than 0.25 Å. These may be due to a number of reasons (Richards 1977):

- CO groups form hydrogen bonds with NH groups three residues along the chain forming a 3_{10} helix. A substantial amount of all 3_{10} helices occur at the ends of α-helices. They are called 3_{10} because there are three residues per turn and ten atoms enclosed in a ring formed by each hydrogen bond. Dipoles are not aligned as in a normal right-handed α-helix.
- Packing of buried helices against other secondary structural elements in the core of a protein can lead to distortions since the side chains are on the surface of helices.
- Proline residues induce distortions of around 20° in the direction of a helix. This is because proline cannot form a regular α-helix due to steric hindrance arising from its cyclic side chain, which blocks the main chain NH group. Proline causes two hydrogen bonds in the helix to be broken. Helices containing proline are usually long because shorter helices would be destabilized.

- Exposed helices are often bent away from the solvent. This is because the exposed C=O groups tend to point toward solvent to maximize their hydrogen bonding capacity, i.e., tend to form hydrogen bonds to solvent as well as N-H groups. This gives rise to a bend in the helix axis.
- The π-helix is an extremely rare secondary structural element in proteins. Like the 3_{10} helix, one turn of π-helix is sometimes found at the ends of regular alpha helices. The infrequency of this particular form of secondary structure stems from the following properties: (i) the φ and ψ angles of the pure π-helix (–57.1, –69.7) lie at the very edge of an allowed minimum energy region of the Ramachandran map, (ii) the π helix requires that the angle τ (N-Cα-C′) be larger (114.9) than the standard tetrahedral angle of 109.5°, (iii) the large radius of the π-helix means the polypeptide backbone is no longer in van der Waals contact across the helical axis forming an axial hole too small for solvent water to fill, and (iv) side chains are more staggered than the ideal 3_{10} helix but not as much as the α-helix (Low and Grenville-Wells 1953; Schulz and Schirmer 1990).

Besides the α-helix, β-sheets are another major structural element in globular proteins. The basic unit of a β-sheet is a β-strand (which can be thought of as a helix with $n = 2$ residues/turn) with approximate backbone dihedral angles $\varphi = -120$ and $\psi = +120$ producing a translation of 3.2–3.4 Å/residue for residues in antiparallel and parallel strands, respectively. The β-strand is then like the α-helix, a repeating secondary structure. However, since there are no intrasegment hydrogen bonds and van der Waals interactions between atoms of neighboring residues are not significant due to the extended nature of the chain, this extended conformation is only stable as part of a β-sheet where contributions from hydrogen bonds and van der Waals interactions between aligned strands exert a stabilizing influence. The β-sheet is sometimes called the β-pleated sheet since sequential neighboring C-α atoms are alternately above and below the plane of the sheet, giving a pleated appearance. β-Sheets are found in two forms, designated as "antiparallel" or "parallel," based on the relative directions of two interacting β-strands (Darby and Creighton 1993; Kabsch and Sander 1983).

Like α-helices, β-sheets have the potential for amphiphilicity with one face polar and the other apolar. However, unlike α-helices, which are composed of residues from a continuous polypeptide segment (i.e., hydrogen bonds between CO of residue i and NH of residue $i + 3$), β-sheets are formed from strands that are very often from distant portions of the polypeptide sequence. Hydrogen bonds in β-sheets are on average 0.1 Å shorter than those found in α-helices. The classical β-sheets originally proposed are planar but most sheets observed in globular proteins are twisted (0–30° per residue) (Baker and Hubbard 1984).

Antiparallel β-sheets are more often twisted than parallel sheets. Another irregularity found in antiparallel β-sheets is the hydrogen-bonding of two residues from one strand with one residue from another called a β bulge. Bulges are most often found in antiparallel sheets, with ~5% of bulges occurring in parallel strands (Chan et al. 1993; Richardson 1981).

Turns are the third of the three "classical" secondary structures that serve to reverse the direction of the polypeptide chain. They are located primarily on the protein surface and accordingly contain polar and charged residues. Antibody recognition, phosphorylation, glycosylation, and hydroxylation sites are found frequently at or adjacent to turns. Turns were first recognized from a theoretical conformational analysis by Venkatachalam (1968). He considered what conformations were available to a system of three linked peptide units (or four successive residues) that could be stabilized by a backbone hydrogen bond between the CO of residue n and the NH of residue $n + 3$. He found three general types, one of which (type III) actually has repeating values of –60, –30° and is identical with the 3_{10}-helix. The three types each contain a hydrogen bond between the carbonyl oxygen of residue i and the amide nitrogen of $i + 3$. These three types of turns are designated I, II, and III. Many have speculated on the role of this type of secondary structure in globular proteins. Turns may be viewed as a weak link in the polypeptide chain, allowing the other secondary structures (helix and sheet) to determine the conformational outcome. In contrast, based on the recent experimental finding of "turn-like" structures in short peptides in aqueous solutions, turns are considered to be structure-nucleating segments, formed early in the folding process. Type I turns occur two to three times more frequently than type II. There are position-dependent amino acid preferences for residues in turn conformations. Type I can tolerate all residues in position i to $i + 3$ with the exception of Pro at position $i + 2$. Proline is favored at position $i + 1$ and Gly is favored at $i + 3$ in type I and type II turns. The polar side chains of Asn, Asp, Ser, and Cys often populate position i where they can hydrogen bond to the backbone NH of residue $i + 2$ (Dyson et al. 1988).

The practical limitations and complexities encountered in high-resolution structural studies of proteins stimulated the development of low-resolution techniques such as Fourier transform infrared (FTIR) spectroscopy, which can be utilized for estimating the secondary structure contents of proteins very rapidly. For the analysis of secondary structure of proteins from FTIR spectra, commonly, the amide I region (1700–1600 cm^{-1}) is utilized. Different conformational types, such as helix, sheet, turns, etc. result in different discrete bands in the amide I region, which are usually broad and overlapping. Therefore, to identify the bands from FTIR spectra, mathematical resolution enhancement techniques have to be applied. Different techniques, such as curve fitting, partial least squares analysis, factor analysis, and artificial neural networks (ANN), have been explored to predict the secondary structure of proteins from FTIR spectra by using the correlation between the FTIR spectral bands and the crystallographic data for proteins whose x-ray data is available (Byler and Susi 1986; Jackson and Mantsch 1995).

Native BSA undergoes significant unfolding and aggregation following pressure treatment (Galazka, Sumner, and Ledward 1996), and a loss of secondary structure, depending on the magnitude of applied pressure (Hayakawa et al. 1992). Ahmed et al. (2007b) found some limited HP-induced modification of the secondary structure of Basmati rice proteins exposed to 650 MPa of pressure. Alvarez et al. (2008) also found relatively small changes in secondary structure of soy proteins after a treatment of 250 MPa.

14.5.2 Protein Tertiary Structure

Zhang et al. (2003) investigated the influence of HP on conformational changes of soybean proteins by means of detection of sulfhydryl groups detection, spectrofluorimetry, ultraviolet difference spectra, circular dichroism, differential scanning calorimetry, and electrophoresis. Alvarez et al. (2008) found that soy proteins became loosely folded (packed) after increasing pressure level treatments. Walker et al. (2004) studied β-lactoglobulin tertiary structure by near-UV CD, intrinsic protein fluorescence spectroscopy, hydrophobic fluorescent probe binding, and thiol group reactivity; as affected by HP/low-temperature processing, they found structural changes at all processing conditions, but larger changes at increased temperature (24°C). Yin et al. (2008) found that HP treatment resulted in gradual unfolding of protein structure, as evidenced by gradual increases in fluorescence strength and SS formation from SH groups, and decrease in denaturation enthalpy change.

14.6 HIGH-PRESSURE (HP) EFFECTS ON BIOACTIVE PEPTIDES

Bioactive peptides are extensively distributed among milk proteins. Numerous studies have shown in vitro formation of bioactive peptides from milk proteins and in some of them in vivo formation has also been found. In addition to liberation during in vivo digestion, bioactive peptides may be liberated during the manufacture of milk products. For example, hydrolyzed milk proteins used for hypoallergenic infant formulas, clinical diets, and as food ingredients comprise exclusively peptides. Proteolysis during milk fermentation and cheese ripening leads to the formation of various peptides. Indeed, casomorphins, ACE-inhibitory peptides, and phosphopeptides have been found in fermented milk products (Séverin and Wenshui 2005; Shah 2000).

On the other side of the spectrum we find antinutritional factors that may take place naturally or may be formed during heat processing. Some examples of naturally occurring antinutritional factors include glucosinolates in mustard and rapeseed protein products, trypsin inhibitors and hemagglutinins in legumes, phytates in cereals and oilseeds, and gossypol in cottonseed protein preparations, which could adversely affect nutrient utilization and may contribute to growth supression in animals (Francis, Makkar, and Becker 2001).

The recent evolution of HP technology has allowed the development of new processes, where an enhancement of enzymatic proteolysis can be obtained by combining with HP treatment, which modifies the tertiary and quaternary structures of proteins; when the structure modification is reached in the presence of active proteases, the hydrolysis can be improved. Furthermore, by combining proteolysis and HP treatment, it is possible to produce hydrolysates with lower residual antigenicity (Chicón et al. 2006; Quirós et al. 2007).

There are few published reports available on potential risks of HPP, but it is necessary to collect data to clarify the role of HPP toward allergenicity and nutritional quality of pressurized foods. Allergenicity is a concern in the safety assessment of novel foods, and one of the major challenges of molecular allergy and food sciences is to predict the allergenic potential of a protein. Kato et al. (2000) reported specific protein release from rice grains during pressurization and identified the released proteins as 16 kDa albumin, α-globulin, and 33 kDa globulin, which are known as major rice allergens.

Methods and possibilities for reduction of food allergy on the level of the proteins are as follows: to avoid allergens, production of hypoallergenic infant formula/food, heat treatment, other processes to avoid the most allergenic fractions, HP, enzymatic modifications, proteolysis/fermentation, modification of epitopes (EPM), masking epitopes (transglutaminase), production of oligoantigenic peptide mixture for prevention, proposal for tolerogenic antigen peptides, among others. This is why new studies on the allergenic character of high hydrostatic pressure-treated foods are considered, in a bid to find out if HPP can become a technology to obtain foods with reduced allergenicity or that proteins in foods create or unmask new immunoreactive structures (Meyer-Pittroff, Behrendt, and Ring 2007).

Hajós, Polgár, & Farkas (2004) found HP treatment at 600 MPa for 20 Min to result in a 3-log count reduction in aerobic counts and a 5-log count reduction in *Listeria monocytogenes* in of Hungarian fermented sausage raw batter. They found the pressure treatment to effectively modified the IgE immunoreactivity of the proteins in the sausage raw batter. HP-induced denaturation/aggregation of proteins caused a decrease in the solubility of proteins also in the urea-soluble fractions of sausage batter; however, the patterns suggested that pressure-induced protein aggregations were mostly reversible. Pressurization altered the immunoreactive profiles of the urea-soluble protein fraction of sausage batter, causing complete disappearance of several proteins and also the appearance of other proteins with isoelectric points greater than 8.2. According to these results, HP induced conformational changes in the pork batter's proteins with alteration of some of the epitope structures. HP might form new protein aggregates with weak immunoreactivity. Curiously, high hydrostatic pressure at 600 MPa significantly reduced the potential allergenic character of a number of protein spots, and modified the structure/conformation of some proteins without altering the epitopes. In a study performed by Nakamura, Sado, and Syukunobe (1993), the antigenicity of whey protein hydrolysates treated with HP was found to be lower than that of heat-treated hydrolysates.

14.7 OTHER USES OF HIGH-PRESSURE (HP) PROCESSING

Other potential applications of HP treatment on milk include low-temperature inactivation of enzymes and stabilization of fermented dairy products, improved coagulation of milk, and the manufacture of dairy gels and emulsions with novel textures. Furthermore, studies have been undertaken on the effects of HP treatment on meat proteins myosin and metmyoglobin, egg white, ovalbumin, and soy proteins. Additional experimental research on protein model systems and real food products is required to understand the potential of this technology in the restructuring of food proteins and stabilizing their biological activities (Alvarez et al. 2007, 2008).

HP treatments have been used commercially in the United States to facilitate the shucking of raw oysters for several years, with the additional advantage of inactivation of *Vibrio parahemolyticus* and *V. vulnificus*. He et al. (2002) studied the use of HP for opening of oysters and reported that HP treatment at 310 MPa, with immediate pressure release, resulted in 100% efficiency of shucking or opening of the shell; changes in oyster body color and other visual characteristics were still observed at higher pressures. Cruz-Romero, Kelly, and Kerry (2007) reported that HP processing of oysters is potentially a more suitable postharvest treatment than heat treatment, causing less negative effects on the quality attributes and resulting in significantly higher yield than untreated samples.

Another use for HP treatment is the acceleration of cheese ripening. Yokoyama, Sawamura, and Motobayashi (1992) applied pressure of 10–250 MPa at 25°C for 3 days to cheddar cheese, made using a 10-fold higher level of proteolytic starter, reporting that the flavor was similar to that of six-month-old Cheddar cheese. The authors also treated cheese at 50 MPa and 25°C for 3 days in combination with addition of lipase, protease, and salting, reporting that the resultant cheese developed a Parmesan-like flavor equivalent to a commercial control in terms of flavor scores and levels of free fatty acids. O'Reilly et al. (2000, 2002, 2003) studied the effects of a range of HP treatment conditions on key ripening characteristics of both cheddar and mozzarella cheeses; overall, they found that HP treatment at low pressures (~50–200 MPa) combined with long processing times (up to 82 h) impacted mainly on proteolysis, while HP treatment at higher pressures (200–400 MPa) for relatively shorter processing times (~20 min) caused changes in the protein structure, which in turn improved cheese functional properties. However, the increases in proteolysis following HP treatment at low pressures were not considered to be adequate to warrant the commercial exploitation of HP treatment for acceleration of cheese ripening.

Similar to thermal treatment, HP causes destruction of microorganisms but, unlike thermal treatment, it does not inactivate certain enzymes that play an important role in cheese ripening, such as lipoprotein lipase. Trujillo et al. (1999) found higher production of free fatty acids in HP-treated goat's milk compared to conventionally pasteurized milk, indicating higher lipoprotein lipase activity. The use of HP on cheese results in an increase in moisture content and pH, generating modifications to the cheese matrix and lysis of cells, which assists ripening. HPP of cheese affects the pattern of proteolysis during ripening, with an overall effect that depends on the type of cheese as well as the magnitude, duration, and temperature of pressure treatment.

Other studies concerning goat's cheese by Saldo, Sendra, and Guamis (2000) involved pressure treatment at 400 MPa for 5 min, which led to a reduction in starter counts by ~6 log cycles, while treatment at 50 MPa for 72 h only slightly decreased the starter count. Lysis of starter culture by HP is important in cheese ripening, given that it releases intracellular enzymes into the cheese matrix, which play an important role in the breakdown of large and intermediate size peptides to small peptides and amino acids. The HP treatment resulted in an increase in pH and levels of proteolysis.

REFERENCES

Ahmed, J., Ayad, A., Ramaswamy, H. S., Alli, I., and Shao, Y. 2007a. Dynamic viscoelastic behavior of high pressure treated soybean protein isolate dispersions. *International Journal of Food Properties*, 10 (2), 397–411.

Ahmed, J., and Ramaswamy, H. S. 2006a. Changes in colour during high pressure processing of fruits and vegetables. *Stewart Postharvest Review*, 5 (9), 1–8.

———. 2007. Dynamic rheology and thermal transitions in meat based strained baby foods. *Journal of Food Engineering*, 78 (4), 1274–84.

——— 2003a. Effect of high-hydrostatic pressure and temperature on rheological characteristics of glycomacropeptide. *Journal of Dairy Science*, 86 (5), 1535–40.

———. 2004. Effect of high-hydrostatic pressure and concentration on rheological characteristics of xanthan gum. *Food Hydrocolloids*, 18 (3), 367–73.

———. 2003b. Effect of hydrostatic pressure and temperature on rheological characteristics of alpha-lactalbumin. *Australian Journal of Dairy Technology*, 58 (3), 233–37.

———. 2006b. High pressure processing of fruits and vegetables. *Stewart Postharvest Review*, 1 (8), 1–10.

———. 2005. Rheology of xanthan gum: Effect of concentration, temperature and high-pressure. *Journal of Food Science, and Technology - Mysore*, 42 (4), 355–58.

Ahmed, J., Ramaswamy, H. S., Alli, I., and Ngadi, M. 2003. Effect of high pressure on rheological characteristics of liquid egg. *Lebensmittel-Wissenschaft und Technologie*, 36 (5), 517–24.

Ahmed, J., Ramaswamy, H. S., Ayad, A., Alli, I., and Alvarez, P. 2007b. Effect of high-pressure treatment on rheological, thermal and structural changes in Basmati rice flour slurry. *Journal of Cereal Science*, 46 (2), 148–56.

Ahmed, J., Ramaswamy, H. S., and Hiremath, N. 2005. The effect of high pressure treatment on rheological characteristics and colour of mango pulp. *International Journal of Food Science, and Technology*, 40 (8), 885–95.

Alvarez, P. A., Ramaswamy, H. S., and Ismail, A. A. 2007. Effect of high pressure treatment on the electrospray ionization mass spectrometry (ESI-MS) profiles of whey proteins. *International Dairy Journal*, 17, 881–88.

———. 2008. High pressure gelation of soy proteins: Effect of concentration, pH and additives. *Journal of Food Engineering*, 88, 331–40.

Arakawa, T., and Timasheff, S. N. 1985. Theory of protein solubility. *Methods in Enzymology*, 114, 49–77.

Arora, A., Chism, G. W., and Shellhammer, T. H. 2003. Rheology and stability of acidified food emulsions treated with high pressure. *Journal of Agricultural Food Chemistry*, 51 (9), 2591–96.

Ashie, I. N. A., Simpson, B. K., and Ramaswamy, H. S. 1996a. Changes in texture and microstructure of pressure-treated fish muscle during chilled storage. *Journal of Muscle Foods*, 8 (1), 13–32.

———. 1996b. Control of endogenous enzyme activity in fish muscle by inhibitors and hydrostatic pressure using RSM. *Journal of Food Science*, 61(2), 350–56.

Baker, E. N., and Hubbard, R. E. 1984. Hydrogen bonding in globular proteins. *Progress in Biophysics and Molecular Biology*, 44 (2), 97–179.

Balasubramanian, S., and Balasubramaniam, V. M. 2003. Compression heating influence of pressure transmitting fluids on bacteria inactivation during high pressure processing. *Food Research International*, 36 (7), 661–68.

Basak, S., and Ramaswamy, H. S . 1998. Effect of high hydrostatic pressure processing (HPP) on the texture of selected fruits and vegetables. *Journal of Texture Studies*, 29, 587–601.

———. 2001. Pulsed high pressure inactivation of pectin methyl esterase in single strength and concentrated orange juices. *Canadian Biosystems Engineering*, 43 (3), 25–9.

———. 1997. Ultra high pressure treatment of orange juice: A kinetic study on inactivation of pectin methyl esterase. *Food Research International*, 29 (7), 601–08.

Basak, S., Ramaswamy, H. S., and Piette, G. 2002. High pressure destruction kinetics of *Leucononostoc mesenteroides* and *Sachharomyces cerevisiae* in single strength and concentrated orange juice. *Innovative Food Science and Emerging Technologies*, 3, 223–45.

Basak, S., Ramaswamy, H. S., and Simpson, B. K. 2001. High pressure inactivation of pectin methyl esterase in orange juice using combination treatments. *Journal of Food Biochemistry*, 25 (6), 509–26.

Basak, S., Ramaswamy, H. S., and Smith, J. P. 2003. High-pressure kinetics of microorganisms in selected fruits and vegetables. *Journal of Food Technology*, 1 (3), 142–49.

Bender, A. E. 1972. Processing damage to protein food. A review. *Journal of Food Technology*, 7, 239–50.

Borwankar, R. P. 1992. Food texture and rheology: A tutorial review. *Journal of Food Engineering*, 16 (1–2), 1–16.

Bos, M. A., and van Vliet, T. 2001. Interfacial rheological properties of adsorbed protein layers and surfactants: a review. *Advances in Colloid and Interface Science*, 91 (3), 437–71.

Bourne, M. C. 1967. Deformation testing of foods. *Journal of Food Science*, 32 (5), 601–05.

Byler, D. M., and Susi, H. 1986. Examination of the secondary structure of protein by deconvolved FTIR spectra. *Biopolymers*, 25 (3), 469–87.

Campbell, G. M., and Mougeot, E. 1999. Creation and characterisation of aerated food products. *Trends in Food Science, and Technology*, 10 (9), 283–96.

Chan, A. W., Hutchinson, E. G., Harris, D., and Thornton, J. M. 1993. Identification, classification, and analysis of beta-bulges in proteins. *Protein Science*, 2 (10), 1574–90.

Cheftel, J. C., and Culioli, J. 1997. Effects of high pressure on meat: A review. *Meat Science*, 46 (3), 211–36.

Chen, C. R., Zhu, S. M., Ramaswamy, H. S., Marcotte, M., and Le Bail, A. 2007. Computer simulation of high pressure cooling of pork. *Journal of Food Engineering*, 79 (2), 401–09.

Chicón, R., Belloque, J., Alonso, E., and López-Fandiño, R. 2008. Immunoreactivity and digestibility of high-pressure-treated whey proteins. *International Dairy Journal*, 18 (4), 367–76.

Chicón, R., Belloque, J., Recio, I., and López-Fandiño, R. 2006. Influence of high hydrostatic pressure on the proteolysis of β-lactoglobulin A by trypsin. *Journal of Dairy Research*, 73, 121–28.

Cruz-Romero, M., Kelly, A. L., and Kerry, J. P. 2007. Effects of high-pressure and heat treatments on physical and biochemical characteristics of oysters *Crassostrea gigas*. *Innovative Food Science and Emerging Technologies*, 8, 30–38.

Dalgleish, D. G. 1997. Adsorption of protein and the stability of emulsions. *Trends in Food Science and Technology*, 8 (1), 1–6.

Darby, N. J., and Creighton, T. E. 1993. *Protein Structure*. Oxford: IRL Press at Oxford University Press.

Dickinson, E. 1999. Caseins in emulsions: Interfacial properties and interactions. *International Dairy Journal*, 9 (3–6), 305–12.

Dickinson, E., and James, J. D. 1998. Rheology and flocculation of high-pressure-treated β-lactoglobulin-stabilized emulsions: Comparison with thermal treatment. *Journal of Agricultural and Food Chemistry*, 46 (7), 2565–71.

Drickamer, H. G. 1999. Two applications of pressure tuning spectroscopy to processes in liquids. *Catalysis Today*, 53 (3), 317–23.

Dyson, H. J., Rance, M., Houghten, R. A., Lerner, R. A., and Wright, P. E. 1988. Folding of immunogenic peptide fragments of proteins in water solution I. Sequence requirements for the formation of a reverse turn. *Journal of Molecular Biology*, 201, 161–200.

Foegeding, E. A. 2006. Food biophysics of protein gels: A challenge of nano and macroscopic proportions. *Food Biophysics*, 1 (1), 41–50.

Foegeding, E. A., Luck, P. J., and Davis, J. P. 2006. Factors determining the physical properties of protein foams. *Food Hydrocolloids*, 20 (2–3), 284–92.

Francis, G., Makkar, H. P. S., and Becker, K. 2001. Antinutritional factors present in plant-derived alternate fish feed ingredients and their effects in fish. *Aquaculture*, 199 (3–4), 197–227.

Galazka, V. B., Sumner, I. G., and Ledward, D. A. 1996. Changes in protein-protein and protein-polysaccharide interactions induced by high pressure. *Food Chemistry*, 57 (3), 393–98.

Gross, M., and Jaenicke, R. 1994. Proteins under pressure. The influence of high hydrostatic pressure on structure, function and assembly of proteins and protein complexes. *European Journal of Biochemistry*, 15 (2), 617–30.

Groß, P., and Ludwig, H. 1992. Pressure-temperature-phase diagram for the stability of bacteriophage T4. *High Pressure and Biotechnology*, 224, 57–9.

Hajós, G., Polgár, M., and Farkas, J. 2004. High-pressure effects on IgE immunoreactivity of proteins in a sausage batter. *Innovative Food Science, and Emerging Technologies*, 5 (4), 443–49.

Hawley, S. A. 1971. Reversible pressure-temperature denaturation of chymotrypsinogen. *Biochemistry*, 13 (10), 2436–42.

Hawley, S. A., and Mitchell, R. M. 1975. An electrophoretic study of reversible protein denaturation: Chymotrypsinogen at high pressures. *Biochemistry*, 14 (14), 3257–64.

Hayakawa, I., Kajihara, J., Morikawa, K., Oda, M., and Fujio, Y. 1992. Denaturation of bovine serum albumin (BSA) and ovalbumin by high pressure, heat and chemicals. *Journal of Food Science*, 57 (2), 288–92.

Hayashi, R., Kawamura, Y., Nakasa, T., and Okinaka, O. 1989. Application of high pressure to food processing: Presurization of egg white and yolk and properties of gels formed. *Agricultural and Biological Chemistry*, 53 (11), 2935–39.

Hayman, M. M., Baxter, I., O'Riordan, P. J., and Stewart, C. M. 2004. Effects of high-pressure processing on the safety, quality, and shelf-life of ready-to-eat meats. *Journal of Food Protection*, 67, 1709–18.

He, H., Adams, R. M., Farkas, D. F., and Morrissey, M. T. 2002. Use of high-pressure processing for oyster shucking and shelf-life extension. *Journal of Food Science*, 67 (2), 640–45.

Heremans, K., and Smeller, L. 1997. Pressure versus temperature behaviour of proteins. *European Journal of Solid State Inorganic Chemistry*, 34, 745–58.

Heremans, K., and Wong, P. T. T. 1985. Pressure effects on the Raman spectra of proteins: Pressure induced changes in the conformation of lysozyme in aqueous solutions. *Chemical Physics Letters*, 118 (1), 101–04.

Hite, B. H. 1899. The effect of pressure in the preservation of milk. *Bulletin of West Virginia University of Agriculture Experimental Station Morgantown*, 58, 15–35.

Hite, B. H., Giddings, N. J., and Weakly, C. E. 1914. The effect of pressure on certain microorganisms encountered in the preservation of fruits and vegetables. *Bulletin of West Virginia University of Agriculture Experimental Station Morgantown*, 146, 1–67.

Hogan, E., Kelly, A. L., and Sun, D.-W. 2005. High pressure processing of foods: An overview. In Sun, D.-W. (Ed.) *Emerging Technologies for Food Processing*, pp. 3–32. San Diego, CA: Elsevier Academic Press.

Hugas, M., Garriga, M., and Monfort, J. M. 2002. New mild technologies in meat processing: High pressure as a model technology. *Meat Science*, 62 (3), 359–71.

Ibanoglu, E., and Karatas, S. 2001. High pressure effect on foaming behaviour of whey protein isolate. *Journal of Food Engineering*, 47 (1), 31–6.

Izquierdo, F. J., Alli, I., Gómez, R., Ramaswamy, H. S., and Yaylayan, V. 2005. Effects of high pressure and microwave on pronase and α-chymotrypsin hydrolysis of β-lactoglobulin. *Food Chemistry*, 92 (4), 713–19.

Jackson, M., and Mantsch, H. H. 1995. The use and misuse of FTIR spectroscopy in the determination of protein structure. *Critical Reviews in Biochemistry and Molecular Biology*, 30 (2), 95–120.

Jonas, J., and Jonas, A. 1994. High-pressure NMR spectroscopy of proteins and membranes. *Annual Reviews in Biophysical and Biomolecular Structures*, 23, 287–318.

Kabsch, W., and Sander, C. 1983. Dictionary of protein secondary structure: Pattern recognition of hydrogen-bonded and geometrical features. *Biopolymers*, 22 (12), 2577–637.

Kato, T., Katayama, E., Matsubara, S., Omi, Y., and Matsuda, T. 2000. Release of allergenic proteins from rice grains induced by high hydrostatic pressure. *Journal of Agricultural Food Chemistry*, 48, 3124–29.

Kauzmann, W. 1987. Thermodynamics of unfolding. *Nature*, 325, 763–64.

Kester, J. J., and Richardson, T. 1984. Modification of whey proteins to improve functionality. *Journal of Dairy Science*, 67 (11), 2757–74.

Kneifel, W., Paquin, P., Abert, T., and Richard, J.-P. 1991. Water-holding capacity of proteins with special regard to milk proteins and methodological aspects: A review. *Journal of Dairy Science*, 74, 2027–41.

Knorr, D. 1999. Novel approaches in food-processing technology: New technologies for preserving foods and modifying function. *Current Opinion in Biotechnology*, 10 (5), 485–91.

Knorr, D., Heinz, V., and Buckow, R. 2006. High pressure application for food biopolymers. *BBA - Proteins, and Proteomics*, 1764 (3), 619–31.

Kunugi, S., Takano, K., Tanaka, N., Suwa, K., and Akashi, M. 1997. Effects of pressure on the behavior of the thermoresponsive polymer poly(*N*-vinylisobutyramide) (PNVIBA). *Macromolecules*, 30 (15), 4499–501.

Kural, A. G., and Chen, H. 2008. Conditions for a 5-log reduction of *Vibrio vulnificus* in oysters through high hydrostatic pressure treatment. *International Journal of Food Microbiology*, 122 (1–2), 180–87.

Lakshmanan, R., Patterson, M. F., and Piggott, J. R. 2005. Effects of high-pressure processing on proteolytic enzymes and proteins in cold-smoked salmon during refrigerated storage. *Food Chemistry*, 90 (4), 541–48.

Lim, S.-Y., Swanson, B. G., and Clark, S. 2008. High hydrostatic pressure modification of whey protein concentrate for improved functional properties. *Journal of Dairy Science*, 91, 1299–307.

Liu, R., Zhao, S., Xiong, S., Xie, B., and Liu, H. 2007. Studies on fish and pork paste gelation by dynamic rheology and circular dichroism. *Journal of Food Science*, 72 (7), E399–403.

López-Fandiño, R., Carrascosa, A. V., and Olano, A. 1996. The effects of high pressure on whey protein denaturation and cheese-making properties of raw milk. *Journal of Dairy Science*, 79, 929–36.

Low, B. W., and Grenville-Wells, H. J. 1953. Generalized mathematical relationships for polypeptide chain helices. The coordinates of the Π helix. *Proceedings of the National Academy of Sciences of the United States of America*, 39 (8), 785–802.

Marcos, B., Aymerich, T., and Garriga, M. 2005. Evaluation of high pressure processing as an additional hurdle to control *Listeria monocytogenes* and Salmonella enterica in low-acid fermented sausages. *Journal of Food Science*, 70 (7), M339–44.

Matthews, D. M., and Laster, L. 1965. Absorption of protein digestion products: a review. *Gut*, 6 (5), 411–26.

McClements, D. J. 2005. *Food Emulsions Principles Practices and Techniques*. Boca Raton, FL: CRC Press.

McMillan, P. 2005. Pressing on: The legacy of Percy W. Bridgman. *Nature Materials*, 4, 715–18.

Messens, W., Van Camp, J., and Huyghebaert, A. 1997. The use of high pressure to modify the functionality of food proteins. *Trends in Food Science and Technology*, 8 (4), 107–12.

Meyer-Pittroff, R., Behrendt, H., and Ring, J. 2007. Specific immuno-modulation and therapy by means of high pressure treated allergens. *International Journal of High Pressure Research*, 27 (1), 63–7.

Molina, E., Alvarez, M. D., Ramos, M., Olano, A., and López-Fandiño, R. 2000. Use of high-pressure-treated milk for the production of reduced-fat cheese. *International Dairy Journal*, 10, 467–75.

Mozhaev, V. V., Heremans, K., Frank, J., Masson, P., and Balny, C. 1994. Exploiting the effects of high hydrostatic pressure in biotechnological applications. *Trends in Biotechnology*, 12, 493–501.

Murchie, L. W., Cruz-Romero, M., Kerry, J. P., Linton, M., Patterson, M. F., Smiddy, M., and Kelly, A. L. 2005. High pressure processing of shellfish: A review of microbiological and other quality aspects. *Innovative Food Science and Emerging Technologies*, 6, 257–70.

Murray, B. S. 2007. Stabilization of bubbles and foams. *Current Opinion in Colloid, and Interface Science*, 12 (4–5), 232–41.

Mussa, D. M., and Ramaswamy, H. S. 1997. Ultra high pressure pasteurization of milk: Kinetics of microbial destruction and changes in physico-chemical characteristics. *Lebensmittel-Wissenschaft und Technologie*, 30 (6), 551–57.

Mussa, D. M., Ramaswamy, H. S., and Smith, J. P. 1999a. High pressure destruction kinetics of *Listeria monocytogenes* in milk. *Food Research International*, 31 (5), 343–50.

———. 1999b. Ultra high pressure destruction kinetics of *Listeria monocytogenes* in pork. *Journal of Food Protection*, 62 (1), 165–70.

Nakamura, T., Sado, H., and Syukunobe, Y. 1993. Production of low antigenic whey protein hydrolysates by enzymatic hydrolysis and denaturation with high pressure. *Milchwissenschaft*, 48 (3), 141–44.

O'Reilly, C., O'Connor, P., Murphy, P., Kelly, A., and Beresford, T. 2000. The effect of exposure to pressure of 50 MPa on Cheddar cheese ripening. *Innovative Food Science and Emerging Technologies*, 1 (2), 109–17.

O'Reilly, C. E., Kelly, A. L., Oliveira, J. C., Murphy, P. M., Auty, M. A. E., and Beresford, T. P. 2003. Effect of varying high-pressure treatment conditions on acceleration of ripening of cheddar cheese. *Innovative Food Science and Emerging Technologies*, 4, 277–84.

O'Reilly, C. E., Murphy, P. M., Kelly, A. L., Guinee, T. P., Auty, M. A. E., and Beresford, T. P. 2002. The effect of high pressure treatment on the functional and rheological properties of Mozzarella cheese. *Innovative Food Science, and Emerging Technologies*, 3 (1), 3–9.

Osborne, T. B., and Campbell, G. F. 1898. Proteids of the soy bean (*Glycine hispida*). *Journal of American Chemical Society*, 20, 419–28.

Ousegui, A., Zhu, S., Ramaswamy, H. S., and LeBail, A. 2006. Modelling of a high pressure calorimeter: Application to the measurement of the latent heat of a model food (tylose). *Journal of Thermal Analysis and Calorimetry*, 86 (1), 243–48.

Palou, E., Hernández-Salgado, C., López-Malo, A., Barbosa-Cánovas, G. V., Swanson, B. G., and Welti-Chanes, J. 2000. High pressure-processed guacamole. *Innovative Food Science and Emerging Technologies*, 1 (1), 69–75.

Pandey, P. K., and Ramaswamy, H. S. 2004. Effect of high pressure treatment of milk on lipase and γ-glutamyl transferase activity. *Journal of Food Biochemistry*, 28, 449–62.

Pandey, P. K., Ramaswamy, H. S., and Idziak, E. 2003. High pressure destruction kinetics of indigenous microflora and *Escherichia coli* in raw milk at two temperatures. *Journal of Food Process Engineering*, 26 (3), 265–83.

Pandey, P. K., Ramaswamy, H. S., and St-Gelais, D. 2000. Water-holding capacity and gel strength of rennet curd as affected by high-pressure treatment of milk. *Food Research International*, 33 (8), 655–63.

———. 2003a. Effect of high-pressure processing on rennet coagulation properties of milk. *Innovative Food Science and Emerging Technologies*, 4, 245–56.

———. 2003b. Evaluation of pH change kinetics during various stages of Cheddar cheese-making from raw, pasteurized, micro-filtered and high-pressure-treated milk. *Lebensmittel-Wissenchft und Technologie*, 36 (5), 497–506.

Panyam, D., and Kilara, A. 1996. Enhancing the functionality of food proteins by enzymatic modification. *Trends in Food Science, and Technology*, 7 (4), 120–25.

Parés, D., and Ledward, D. A. 2001. Emulsifying and gelling properties of porcine blood plasma as influenced by high-pressure processing. *Food Chemistry*, 74 (2), 139–45.

Patazca, E., Koutchma, T., and Ramaswamy, H. S. 2006. Inactivation kinetics of *Geobacillus stearothermophilus* spores in water using high-pressure processing at elevated temperatures. *Journal of Food Science*, 71, M110–16.

Patterson, M. F. 2005. Microbiology of pressure treated foods. *Journal of Applied Microbiology*, 98 (4), 541–48.

Quirós, A., Chichón, R., Recio, I., and López-Fandiño, R. 2007. The use of high hydrostatic pressure to promote the proteolysis and release of bioactive peptides from ovalbumin. *Food Chemistry*, 104 (4), 1734–39.

Ramaswamy, H. S., and Riahi, E. 2003. High-pressure inactivation kinetics of polyphenoloxidase in apple juice. *Applied Biotechnology, Food Science and Policy*, 1 (3), 189–97.

Raso, J., and Barbosa-Cánovas, G. V. 2003. Nonthermal preservation of foods using combined processing techniques. *Critical Reviews in Food Science and Nutrition*, 43 (3), 265–85.

Rastogi, N., Raghavarao, K., Balasubramaniam, V., Niranjan, K., and Knorr, D. 2007. Opportunities and challenges in high pressure processing of foods. *Critical Reviews in Food Science, and Nutrition*, 47 (1), 69–112.

Riahi, E., and Ramaswamy, H. S. 2003. High-pressure processing of apple juice: Kinetics of pectin methyl esterase inactivation. *Biotechnology Progress*, 19 (3), 908–14.

———. 2004. High pressure inactivation kinetics of amylase in apple juice. *Journal of Food Engineering*, 64 (2), 151–60.

Richards, F. M. 1977. Areas, volumes, packing, and protein structure. *Annual Review of Biophysics and Bioengineering*, 6, 151–76.

Richardson, J. S. 1981. The anatomy and taxonomy of protein structure. *Advanced Protein Chemistry*, 34, 167–339.

Rodiles-López, J. O., Jaramillo-Flores, M. E., Gutiérrez-López, G. F., Hernández-Arana, A., Fosado-Quiroz, R. E., Barbosa-Cánovas, G. V., and Hernández-Sánchez, H. 2008. Effect of high hydrostatic pressure on bovine α-lactalbumin functional properties. *Journal of Food Engineering*, 87 (3), 363–70.

Rouille, J., LeBail, A., Ramaswamy, H. S., and LeClerc, L. 2002. High pressure thawing of fish and shellfish. *Journal of Food Engineering*, 53 (1), 83–8.

Royer, C. A., Hinck, A. P., Loh, S. N., Prehoda, K. E., Peng, X., Jonas, J., and Markley, J. L. 1993. Effects of amino acid substitutions on the pressure denaturation of staphylococcal nuclease as monitored by fluorescence and nuclear magnetic resonance spectroscopy. *Biochemistry*, 32, 5222–32.

Rubens, P., Snauwaert, J., Heremans, K., and Stute, R. 1999. In situ observation of pressure-induced gelation of starches studied with FTIR in the diamond anvil cell. *Carbohydrate Polymers*, 39, 231–35.

Saldo, J., Sendra, E., and Guamis, B. 2000. High hydrostatic pressure for accelerating ripening of goat's milk cheese: Proteolysis and texture. *Journal of Food Science*, 65 (4), 636–40.

San Martín, M. F., Barbosa-Cánovas, G. V., and Swanson, B. G. 2002. Food processing by high hydrostatic pressure. *Critical Reviews in Food Science and Nutrition*, 42 (6), 627–45.

Sareevoravitkul, R., Ramaswamy, H. S., and Simpson, B. K. 1996. Comparative properties of bluefish (*Pomatomus saltatrix*) gels formulated by high hydrostatic pressure and heat. *Journal of Aquatic Food Product Technology*, 5, 65–79.

Schaafsma, G. 2005. The protein digestibility-corrected amino acid score (PDCAAS): A concept for describing protein quality in foods and food ingredients: A critical review. *Journal of AOAC International*, 88 (3), 988–94.

Schramm, L. L. 2005. *Emulsions, Foams, and Suspensions: Fundamentals and Applications*. Weinheim: Wiley.

Schulz, G. E., and Schirmer, R. H. 1990. *Principles of Protein Structure*, 2nd edn. New York: Springer.

Sequeira-Muñoz, A., Chevalier, D., LeBail, A., Ramaswamy, H. S., and Simpson, B. K. 2006. Physicochemical changes induced in carp (*Cyprinus carpio*) fillets by high pressure processing at low temperature. *Innovative Food Science, and Emerging Technologies*, 7 (1–2), 13–18.

Sequeira-Muñoz, A., Chevalier, D., Simpson, B. K., Le Bail, A., and Ramaswamy, H. S. 2005. Effect of pressure-shift freezing versus air-blast freezing of carp (*Cyprinus carpio*) fillets: A storage study. *Journal of Food Biochemistry*, 29 (5), 504–16.

Séverin, S., and Wenshui, X. 2005. Milk biologically active components as nutraceuticals: Review. *Critical Reviews in Food Science and Nutrition*, 45 (7), 645–56.

Shah, N. P. 2000. Effects of milk-derived bioactives: An overview. *British Journal of Nutrition*, 84 (1), S3–10.

Shao, Y., Ramaswamy, H. S., and Zhu, S. 2007. High pressure destruction kinetics of spoilage and pathogenic bacteria in raw milk cheese. *Journal of Food Process Engineering*, 30 (3), 357–74.

Shao, Y., Zhu, S., Ramaswamy, H., and Marcotte, M. 2008 online. Compression heating and temperature control for high-pressure destruction of bacterial spores: An experimental method for kinetics evaluation. *Food and Bioprocess Technology*, DOI: 10.1007/s11947-008-0057-y.

Sheehan, J. J., Huppertz, T., Hayes, M. G., Kelly, A. L., Beresford, T. P., and Guinee, T. P. 2005. High pressure treatment of reduced-fat Mozzarella cheese: Effects on functional and rheological properties. *Innovative Food Science and Emerging Technologies*, 6 (1), 73–81.

Smelt, J. P. P. M. 1998. Recent advances in the microbiology of high pressure processing. *Trends in Food Science and Technology*, 9 (4), 152–58.

Smith, D., Galazka, V. B., Wellner, N., and Sumner, I. G. 2000. High pressure unfolding of ovalbumin. *International Journal of Food Science and Technology*, 35 (4), 361–70.

Suzuki, K. 1960. Studies on the kinetics of protein denaturation under high pressure. *The Review of Physical Chemistry of Japan*, 29 (2), 91–8.

Taniguchi, Y., and Suzuki, K. 1983. Pressure inactivation of α-chymotrypsin. *Journal of Physical Chemistry*, 87, 5185–93.

Thakur, B. R., and Nelson, P. E. 1998. High pressure processing and preservation of food. *Food Reviews International*, 14 (4), 427–47.

Torres, J. A., and Velazquez, G. 2005. Commercial opportunities and research challenges in the high pressure processing of foods. *Journal of Food Engineering*, 67 (1–2), 95–112.

Trujillo, A. J., Royo, C., Ferragut, V., and Guamis, B. 1999. Ripening profiles of goat cheese produced from milk treated with high pressure. *Journal of Food Science*, 64 (5), 833–37.

van der Plancken, L., van Loey, A., and Hendrickx, M. E. 2007. Foaming properties of egg white proteins affected by heat or high pressure treatment. *Journal of Food Engineering*, 78 (4), 1410–26.

Venkatachalam, C. M. 1968. Stereochemical criteria for polypeptides and proteins. V. Conformation of a system of three linked peptide units. *Biopolymers*, 6, 1425–36.

Vojdani F., and Whitaker, J. R. 1994. Chemical and enzymatic modification of proteins for improved functionality. In Hettiarachchy, N. S., and Ziegler, G. R. (Eds) *Protein Functionality in Food Systems*, 261–310. New York: Marcel Dekker.

Walker, M. K., Farkas, D. F., Anderson, S. R., and Meunier-Goddik, L. 2004. Effects of high-pressure processing at low temperature on the molecular structure and surface properties of β-lactoglobulin. *Journal Agricultural Food Chemistry*, 52 (26), 8230–35.

Walker, M. K., Farkas, D. F., Loveridge, V., and Meunier-Goddik, L. 2006. Fruit yogurt processed with high pressure. *International Journal of Food Science, and Technology*, 41 (4), 464–67.

Walstra, P. 2003. *Physical Chemistry of Foods*. New York: Marcel Dekker.

Weber, G., and Drickamer, H. G. 1983. The effect of high pressure upon proteins and other biomolecules. *Quarterly Reviews in Biophysics*, 16, 89–112.

Yin, S. W., Tang, C. H., Wen, Q. B., Yang, X. Q., and Li, L. 2008. Functional properties and *in vitro* trypsin digestibility of red kidney bean (*Phaseolus vulgaris* L.) protein isolate: Effect of high-pressure treatment. *Food Chemistry*, 110 (4), 938–45.

Yokoyama, H., Sawamura, N., and Motobayashi, N. 1992. Method for accelerating cheese ripening. European Patent Application EP 0469857 A1.

Zhang, H., Li, L., Tatsumi, E., and Kotwal, S. 2003. Influence of high pressure on conformational changes of soybean glycinin. *Innovative Food Science and Emerging Technologies*, 4 (3), 269–75.

Zhang, H., and Mittal, G. S. 2008. Effects of high-pressure processing (HPP) on bacterial spores: An overview. *Food Reviews International*, 24 (3), 330–51.

Zhu, S., Bulut, S., LeBail, A., and Ramaswamy H. S. 2004a. High-pressure differential scanning calorimetry (DSC): Equipment and technique validation using water-ice phase transition data. *Journal of Food Process Engineering*, 27, 359–76.

Zhu, S., LeBail, A., Chapleau, N., Ramaswamy, H., and deLamballerie-Anton, M. 2004b. Pressure shift freezing of pork muscle: Effect on color, drip loss, texture and protein stability. *Biotechnology Progress*, 20 (3), 939–45.

Zhu, S., LeBail, A., and Ramaswamy, H. S. 2006a. High-pressure differential scanning calorimetry: Comparison of pressure-dependent phase transition in food materials. *Journal of Food Engineering*, 75 (2), 215–22.

———. 2003. Ice crystal formation in pressure shift freezing of Atlantic salmon (Salmo salar) as compared to classical freezing methods. *Journal of Food Processing and Preservation*, 27 (6), 427–44.

Zhu, S., LeBail, A., Ramaswamy, H. S., and Chapleau, N. 2004c. Characterization of ice crystals in pork muscle formed by pressure shift freezing as compared to classical freezing methods. *Journal of Food Science*, 69 (4), E190–97.

Zhu, S., Ramaswamy, H., and LeBail, A. 2005a. High-pressure calorimetric evaluation of ice crystal ratio formed by rapid depressurization during pressure-shift freezing of water and pork muscle. *Food Research International*, 38 (2), 193–201.

———. 2004d. High-pressure differential scanning calorimetry (DSC): Evaluation of phase transition in pork muscle at high pressures. *Journal of Food Process Engineering*, 27, 377–91.

———. 2006b. HP Calorimetry and pressure-shift freezing of different food products. *Food Science and Technology International*, 12 (3), 205–14.

———. 2005b. Ice-crystal formation in gelatin gel during pressure shift versus conventional freezing. *Journal of Food Engineering*, 66 (1), 69–76.

Zhu, S., Ramaswamy, H. S., and Simpson, B. K. 2004e. Effect of high-pressure versus conventional thawing on color, drip loss and texture of Atlantic salmon frozen by different methods. *Lebensmittel-Wissenschaft und Technologie*, 37 (3), 291–99.

Zhu, S. M., Naim, F., Marcotte, M., Ramaswamy, H. S., and Shao, Y. 2008 online. High-pressure destruction kinetics of *Clostridium sporogenes* spores in ground beef at elevated temperatures. *International Journal of Food Microbiology*, DOI: 10.1016/j.ijfoodmicro.2008.05.009.

15 Functional Properties and Microstructure of High-Pressure-Processed Starches and Starch–Water Suspensions

Yeting Liu and Weibiao Zhou
National University of Singapore

David Young
Griffith University

CONTENTS

15.1 INTRODUCTION

High-pressure-processing (HPP) is an emerging technology in food science and the number of commercial applications has been increasing steadily in the last few years. The pressures employed generally range from 50 to 1000 MPa (Indrawati, Van Loey, and Hendrickx 2002) but can be as high as 1500 MPa (Liu, Selomulyo, and Zhou 2008). A wide range of pressures are used for different applications (Figure 15.1) while processing of starch generally occurs in the range of 200–1000 MPa.

Starch is a biopolymer consisting of linear amylose and branched amylopectin (Table 15.1). Its native granule is partially amorphous and partially crystalline, with the amorphous and crystalline areas alternating. The crystalline areas mainly arise from amylopectin. Starch is treated under high pressure to accomplish different objectives, which can be classified as (1) high-pressure-induced gelatinization, (2) homogenization, and (3) compressing.

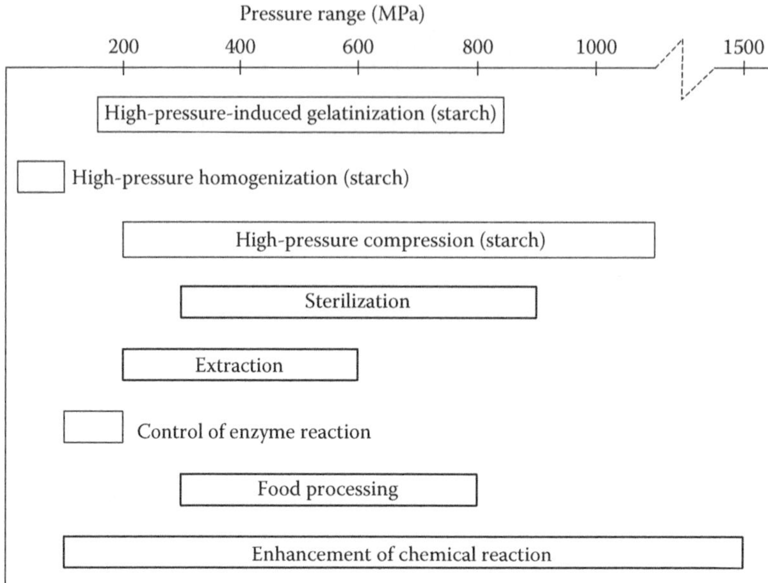

FIGURE 15.1 Pressure range for different applications. (Modified from Noguchi, A., In *Processing and Utilization of Legumes. Report of the APO Seminar on Processing and Utilization of Legumes, Japan, 9 to 14 October 2000*, 63–75, Asian Productivity Organization, Tokyo, 2003.)

TABLE 15.1
Comparison Between Amylose and Amylopectin

Characteristics	Amylose	Amylopectin
Shape	Essentially linear	Branched
Structure	Often helical	Branch chain is helical
Linkage	α-1,4 with some α-1,6	α-1,4 and α-1,6
Molecular weight	Typically < 0.5 million	50–500 million
Chain type	One	A, B, and C types
Gel formation	Firm	Nongelling to soft
Physical state in native granules	Amorphous	Partially amorphous, partially crystalline

Source: Modified from Thomas, D. J., and Atwell, W. A., *Starches*, Eagan Press, St Paul, MN, 1997. With permission.

The major effect of high pressure on a starch–water suspension with excess water (30–99.6%, w/w) is on the starch–water interaction, i.e., gelatinization. Pressure-induced gelatinization is similar to thermal gelatinization in terms of the swelling of the starch granules, viz. a loss of birefringence and crystallinity. The main difference is the incomplete disintegration of starch granules under HPP. Static high pressure is used for high-pressure gelatinization while dynamic high pressure is used for high-pressure homogenization. Generally, pressures of around 200–800 MPa are used for the former while 20–200 MPa are used for the latter.

For starch with a limited amount of moisture (0–19.6%), high pressure compressing can be used to modify the functional properties of the starch. The compressing does not change the birefringence and crystalline patterns of starch granules, but it changes the granule shape and surface, and possibly the crystallinity of the granules as well. Different influences of HPP are summarized in Table 15.2. While there are numerous publications on high-pressure-induced gelatinization, limited studies are available on high-pressure homogenization and compression.

TABLE 15.2
Comparison of Starch Gelatinized by Heating and Starch Processed by Three Categories of High-Pressure Processing

		Heat Gelatinization	Pressure-Induced Gelatinization	High-Pressure Homogenization	High-Pressure Compression
Treatment conditions	System	Starch-water suspension	Starch-water suspension	Starch-water suspension	Solid starch
	Pressure range	0.1 MPa	200–800 MPa	20–100 MPa	200–1100 MPa
	Pressure type	Static	Static	Dynamic	Static
Granule morphology (by SEM)		Disrupted granules	Granules remain intact, minor disruption (in early stage)	Granules remain intact, minor disruption	Altered, rough surface
Granule internal structure	Birefringence (by microscopy)	Lost	Lost	Remain	No difference
	Crystalline structure (by WAXD)	Peaks lost	Peaks lost	Remain same	Same pattern
	Granule structure (by SEM)	Complete disintegration of crystalline structure	Gel-like structure, partial disintegration	N/A	N/A
Thermal Properties	Gelatinization (by DSC)	–	Reduced ΔH_{gel}	N/A	Lower T_{gel}, rReduced ΔH_{gel}
	Retrogradation (by DSC)	–	Slower	N/A	N/A
Effect type		First-order kinetics $f(T, t)$	First-order kinetics $f(P, T, t,$ concentration)	Kinetic	N/A

N/A, not available

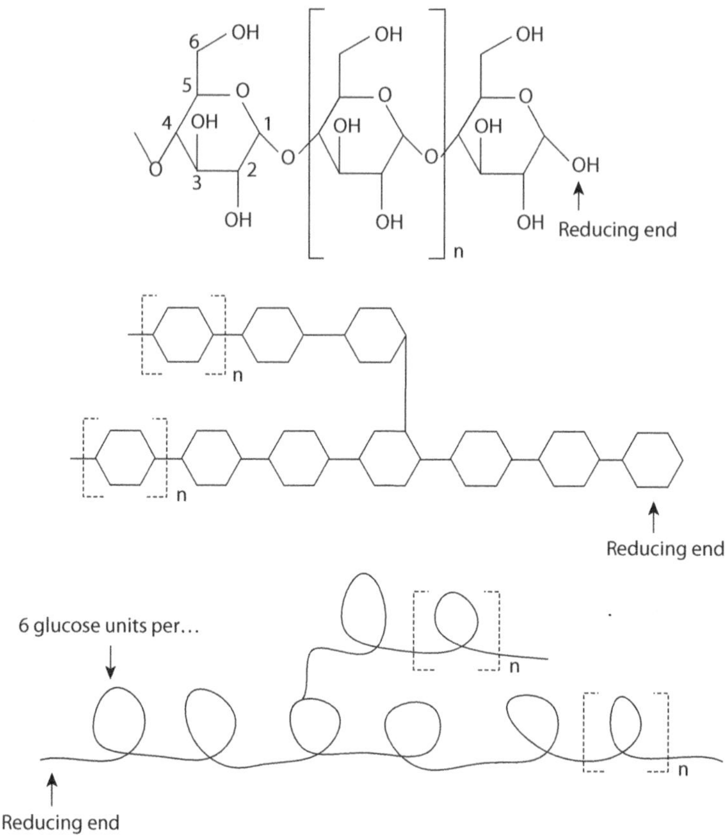

FIGURE 15.2 Chemical and physical structure of amylose.

15.2 STARCH: MICROSTRUCTURE AND FUNCTIONAL PROPERTIES

Amylose is a glucose polymer with predominantly α-1,4 glycoside linkages and some α-1,6 glycoside branches (Figure 15.2). It acts as an unbranched entity (Hoseney 1998) and so it is generally considered as essentially a linear chain, although with a helical secondary structure. The interior of the helix contains hydrogen atoms and is therefore hydrophobic, allowing amylose to form clathrate complexes with free fatty acids, fatty acid components of glycerides, some alcohols, and iodine (Thomas and Atwell 1997). Amylose forms a three-dimensional network upon cooling in solution, which is responsible for the gel of cooked and cooled starch pastes. Starch without amylose thickens but does not gel.

Amylopectin is a branched polymer containing α-1,4 linkages with α-1,6 branching at about every 15–30 glucose units of the main chain (Figure 15.3). These branched chains consist of between 15 and 45 glucose units (Thomas and Atwell 1997), and their secondary structure is also helical. There are three types of chains: A, B, and C. The A-chain is unbranched and linked to the molecule through the reducing end-group. The B-chain is linked to the molecule in the same way as the A-chain but carries one or more A-chains. The C-chain is the one that carries the reducing end-group of the molecule. Table 15.1 presents a brief summary on the major characteristics of amylose and amylopectin.

A starch granule is mostly amorphous with some crystalline regions (Biliaderis 1998). When a native starch granule is viewed under a microscope with polarized light, a Maltese cross formation or birefringence is observed (Figure 15.4). This birefringence of native starch granules reflects the highly ordered crystalline structure that refracts light in two directions (Vaclavik 1998). Although the exact arrangement of amylose and amylopectin in the granule is not fully understood, it is

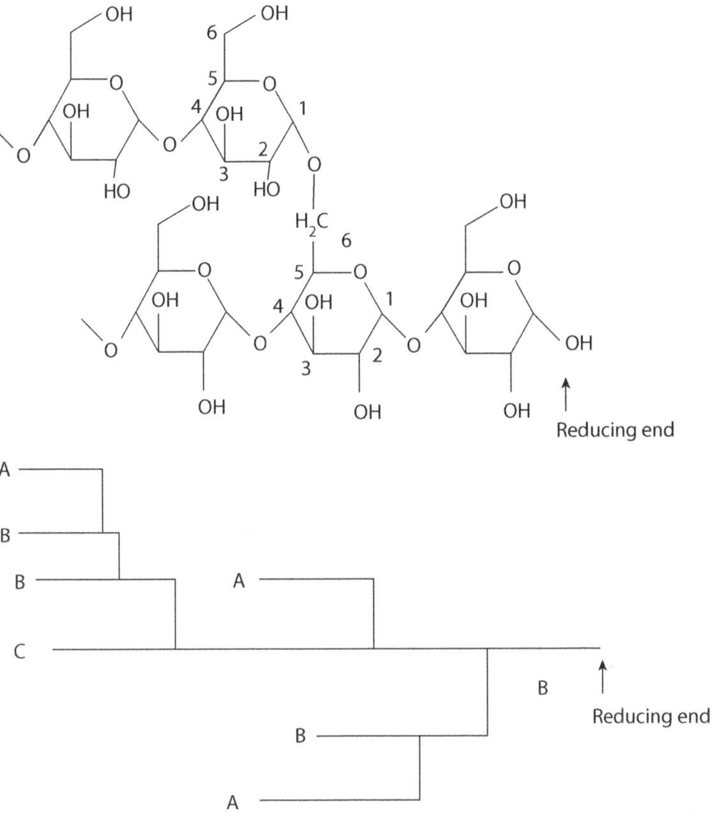

FIGURE 15.3 Chemical and physical structure of amylopectin.

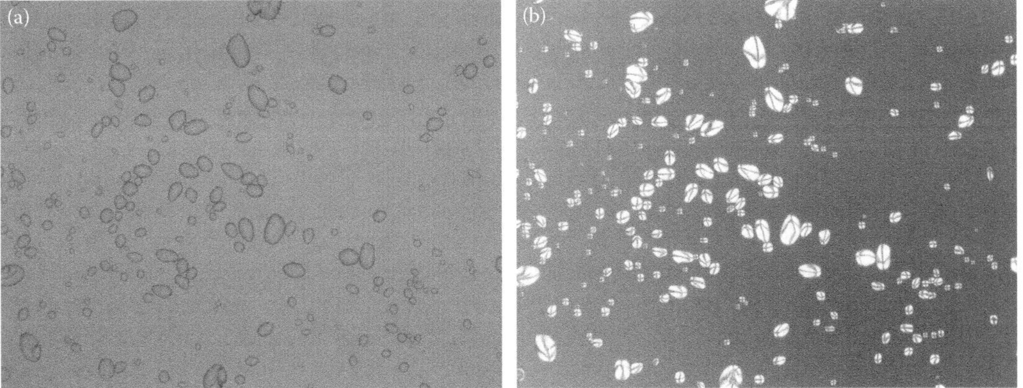

FIGURE 15.4 View of potato starch granules under microscopy (a) with normal light (b) with polarized light.

believed that their arrangement is not random but very organized (Thomas and Atwell 1997). The crystalline portion is thought to come from amylopectin, and the crystalline and amorphous areas alternate. A "'growth ring'" structure has been proposed for amylopectin (Figure 15.5), which is around 120–400 nm thick and consists of alternating crystalline and amorphous lamellae. The crystalline lamellae are around 5 nm thick and several hundreds of angstroms broad. It represents the

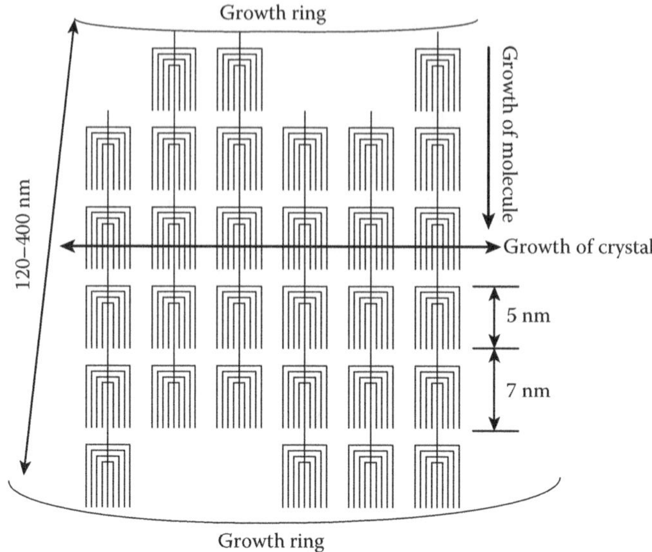

FIGURE 15.5 Structure of amylopectin in starch granules. (Modified from Thomas, D. J., and Atwell, W. A., *Starches*, Eagen Press, St Paul, MN, 1997. With permission.)

double-stranded helices of short degree of polymerization (DP) chains of amylopectin (Biliaderis 1998). The amorphous lamellae are around 2 nm. Amylose is interwoven throughout the crystalline and amorphous areas of the radially oriented amylopectin.

The crystalline structure of starch granule exhibits three patterns during x-ray diffraction (Figure 15.6). A-type crystals exhibit three strong peaks at 5.8, 5.2, and 3.8 Å. B-type crystals show a doublet with medium intensity at 3.7 and 4.0 Å, a strong peak at 5.2 Å, and peaks with medium intensity at 15.8–16 Å. C-type crystals have the same x-ray diffraction pattern as the A-type, with an additional peak at about 16 Å which is also characteristic for the B-type. It is generally considered as a mixture of A- and B-type crystals. The A-type crystalline structure consists of double helices packed into a monoclinic array and the B-type crystalline structure consists of double helices packed in a hexagonal array (Parker and Ring 2001). Most cereal (e.g., wheat, corn, and rice) starches show the A-type diffraction pattern. Most tuber (e.g., potato) and other root starches and retrograded starches show the B-type pattern. Smooth pea and bean starches show the C-type pattern.

Starch exhibits two different types of thermal property: gelatinization and retrogradation. When starch is gradually heated with excessive water, birefringence is lost at a specific temperature. This process is known as gelatinization. When the gelatinization temperature is reached, the hydrogen bonding between and/or within starch molecules is broken by the kinetic energy supplied by the hot water, and new hydrogen bonding is formed between starch and the water molecules (Vaclavik 1998). Therefore, water is able to penetrate further into starch granules and granule swelling takes place. Meanwhile, starch crystalline structure (double helices) is melted during gelatinization (Cooke and Gidley 1992). Further swelling of starch granules causes the disruption of the granules, and leaching out of some amylose chains.

Differential scanning calorimetry (DSC) exhibits a single endothermic peak during thermal scan of starch with excess water (above 65%) and the process is termed as gelatinization (Figure 15.7). The shape of this endotherm is a function of heating rate. The single peak shifts to a higher temperature and becomes broader when the heating rate increases. However, the starting point of the endothermic peak is practically independent of the heating rate (Lelievre and Liu 1994). The measured gelatinization enthalpy primarily reflects the energy required to melt the crystallites (Cooke and Gidley 1992).

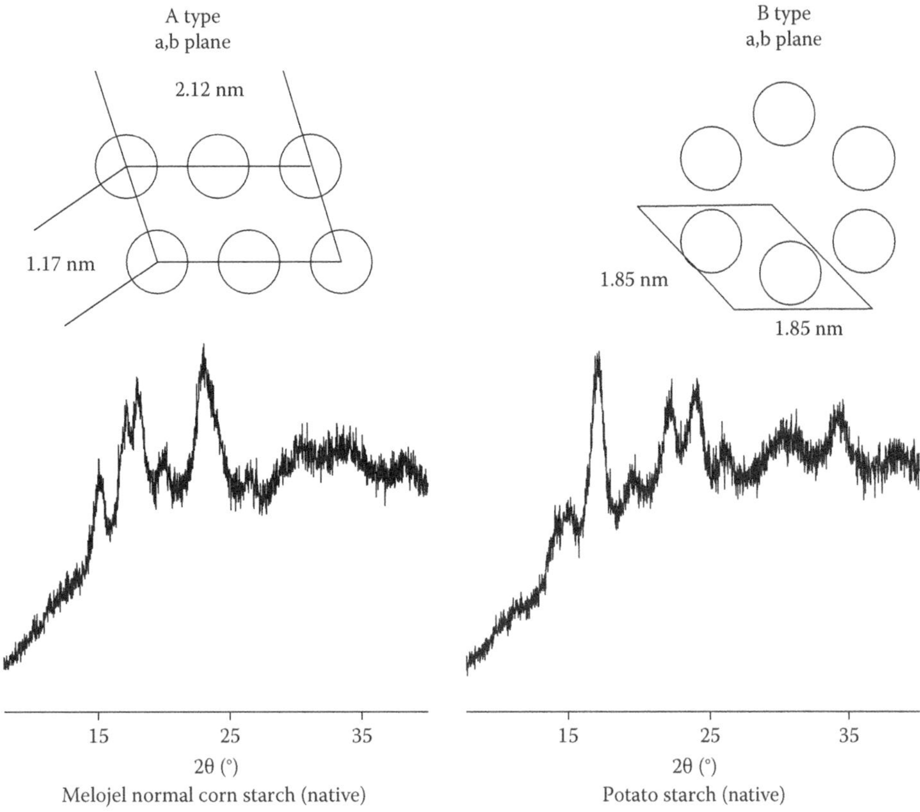

FIGURE 15.6 Alignment of double helices (the circle) and x-ray diffraction patterns of A- and B-type starch crystals.

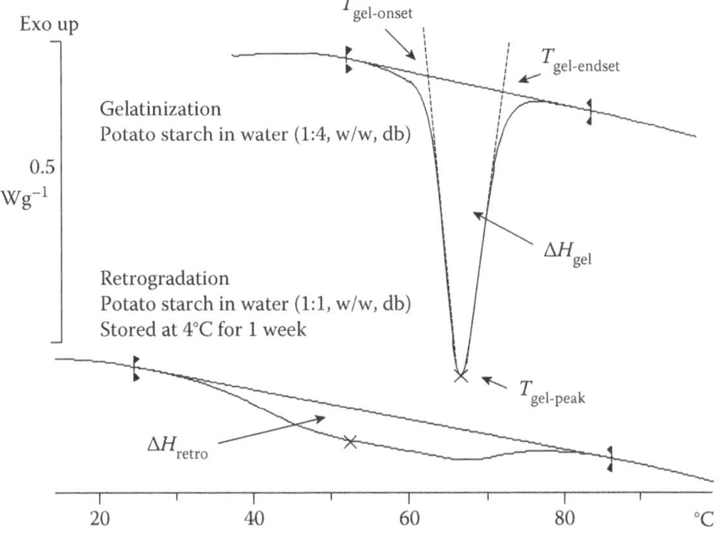

FIGURE 15.7 Gelatinization and retrogradation of potato starch measurement in a sealed pan by normal DSC.

There is still no consensus on the elusive mechanism of gelatinization of starch granules (Biliaderis 1998). Starch gelatinization is commonly considered as a superposition of glass transition of hydrated amorphous regions and the melting of crystalline regions through hydration and dissociation of double helices. It is initiated by a primary hydration of the amorphous regions around and above the glass transition temperature (T_g). This, in turn, facilitates molecular mobility in the amorphous regions, which provokes an irreversible molecular transition in the crystalline regions that involves dissociation of double helices and expansion of granules during hydration (Tester and Debon 2000). The onset of gelatinization reflects the initiation of glass transition of the amorphous regions and this always happens before the crystalline melting transition (Levine and Slade 1990). Shi and Seib (1992) proved that the onset temperature ($T_{gel\text{-}onset}$) of the crystallites in native starch was controlled indirectly by the surrounding amorphous regions. The gelatinization enthalpy was suggested to be mainly the energy required to melt the crystalline regions of starch (Hoseney 1998).

During storage of gelatinized starch, retrogradation occurs and starch chains begin to reassociate into an ordered structure (Atwell et al. 1988). Two or more starch chains form a simple juncture point in the initial stage. This juncture point may develop into a more extensively ordered region. Ultimately a crystalline order could appear under favorable conditions.

Compared to large branched amylopectin chains, amylose molecules have a greater tendency to reassociate and form hydrogen bonds (Thomas and Atwell 1997). The retrogradation is largely considered to be the reassociation of amylose chains. However, reassociation of short branched chains of amylopectin molecules cannot be ruled out. Single or multiple branched amylodextrin originating from outer short branched chains (DP = 15) of amylopectin was found in the crystallites in retrograded amylopectin gels (Biliaderis 1998). The melting temperature of this crystalline structure was around 37–40°C and is, therefore, bioavailable. In contrast, the aggregated double helices of B-type crystals in amylose gels are thermally resistant with a melting temperature of around 130–170°C (Biliaderis 1998). Such conformational ordering and aggregation of double helical chain segments is similarly resistant to acid and enzymatic hydrolysis (Biliaderis 1998). The amylase-resistant linear amylodextrin (originating from retrograded amylose) is called "resistant starch (RS)," i.e., the starch resists digestion in the small intestine and becomes available for fermentation in the large intestine of man.

Retrogradation could be determined by measuring the energy required to remelt the crystalline structure in retrograded starch using DSC (Figure 15.7).

15.3 HYDROSTATIC HIGH-PRESSURE-PROCESSING (HPP) OF STARCH–WATER SUSPENSIONS (HIGH-PRESSURE-INDUCED GELATINIZATION)

When a starch-in-water suspension is treated under high pressure, a loss of birefringence of the starch granules occurs. A gradual loss of birefringence with increased treatment time has been observed. This phenomenon is similar to the thermal gelatinization of starch–water suspensions and, therefore, it is often referred to as high-pressure-induced gelatinization or pressure gelatinization. Calorimetric analysis of pressure-treated starch indicated an absence of the gelatinization peak which is commonly observed for untreated starch samples under thermal scanning. The higher the pressure of treatment, the less energy is required to gelatinize the starch dispersion. A complete gelatinization occurs after applying a threshold pressure level, which is analogous to the gelatinization temperature for thermal treatment. The loss of crystallinity after pressure treatment was further confirmed by various instrumental techniques (FTIR, x-ray diffraction, and NMR) (Blaszczak, Valverde, and Fornal 2005; Douzals et al. 1998). When pressure-treated starch is viewed by Scanning Electron Microscopy (SEM), a swelling of the starch granules is found. The granule surface is the most resistant part while the inner part of the granule is almost completely filled with a gel-like network with empty spaces growing in diameter toward the center of the granule.

Similar to thermal gelatinization, hydration of amorphous growth rings and the resulting granule swelling has been found to occur prior to gelatinization during high-pressure gelatinization

(Gebhardt et al. 2007). Hydration of semicrystalline lamellae then lateral breakdown of the crystalline domain takes place during gelatinization. At this stage, the external granule shape is still intact. During thermal gelatinization, the crystalline structure is decomposed by helix–helix dissociation followed by helix coil transition. Under high pressure, the disintegration of the macromolecules is incomplete since the pressure stabilization of hydrogen bonds favors helix formation. Consequently, starch gelatinization is interrupted because the disintegration of the crystalline regions remains incomplete. This model of high-pressure-induced gelatinization involving an initial solvation step followed by a helix-coil transition of the molecular chains is now, more or less, accepted (Gebhardt et al. 2007; Rubens and Heremans 2000).

15.3.1 Effect of Starch Type

The effect of pressure on the gelatinization process depends on the type of starch. Generally, A-type starches (e.g., corn, wheat, rice) are most susceptible to pressure treatment (Ezaki and Hayashi 1992). B-type starches are most resistant to pressure treatment (Ezaki and Hayashi 1992; Hills et al. 2005; Katopo, Song, and Jane 2002; Stute et al. 1996). C-type starches (e.g., sweet potato, tapioca, mung bean) have intermediate susceptibility between the A- and B-types under pressure (Ezaki and Hayashi 1992). Furthermore, a conversion of the A-type to the B-type crystalline structure through rearrangement of double helices was observed (Katopo et al. 2002) since the B-type crystalline structure is favored by pressure due to the higher number of associated water molecules stabilizing the helix structure by van der Waals forces (Knorr, Heinz, and Buckow 2006). However, the B-type crystalline structure was not changed after high-pressure treatment (HPT). While the molecular weight distribution of starches is not affected by HPT, pasting properties of pressurized starches do change depending on the structure of the starch (Katopo et al. 2002).

15.3.2 Kinetics of High-Pressure-Induced Gelatinization

Thermal gelatinization of a starch–water suspension follows the first-order kinetics:

$$\ln \frac{A_t}{A_{max}} = -kt, \tag{15.1}$$

where A_{max} is the initial percentage of nongelatinized starch to be taken as 100%, A_t is the percentage of nongelatinized starch after treatment for time t, and k is the reaction rate constant (1/s). If the starch gelatinization is analyzed by DSC, the degree of gelatinization (DG) D_G with increasing time t is defined as:

$$D_G = \left[1 - \frac{Q_t}{Q_{max}}\right], \tag{15.2}$$

where Q_t is the heat uptake for a partially gelatinized sample, and Q_{max} is the heat uptake for a completely nongelatinized sample. The first-order gelatinization kinetics can therefore be rewritten as:

$$(1 - D_G) = \exp(-kt). \tag{15.3}$$

The temperature dependence of k is described by the Arrhenius equation:

$$k = k_0 \exp\left[-\frac{E_a}{RT}\right], \tag{15.4}$$

where k_0 is the pre-exponential factor and E_a is the activation energy.

Similar to thermal gelatinization, high-pressure-induced gelatinization at constant temperature was also found to follow first-order kinetics. Similarly, the pressure dependence of k at a given temperature could be described by the Eyring equation:

$$k = k_{\mathrm{p}} \exp\left[-\frac{\Delta V^{\#} P}{RT}\right]. \tag{15.5}$$

The activation volume ($\Delta V^{\#}$) for high-pressure-induced gelatinization is -57.4 cm^3/mol for 25% waxy maize starch in a water suspension at 10–25°C (Sablani et al. 2007). In most cases, HPT takes place concomitantly with thermal treatment, due to either increased water temperature with pressure or additional heating like compression. A combined Arrhenius and Eyring equation was developed to describe the first-order gelatinization kinetics of rice starch in a pressure range of 0.1–600 MPa and 20–70°C (Ahromrit, Ledward, and Niranjan 2007):

$$k = k_0 \exp\left[-\frac{E_{\mathrm{a}}}{RT} - \frac{\Delta V^{\#} P}{RT}\right], \tag{15.6}$$

and k_0 was 31.19 s^{-1}, E_{a} was 37.89 kJ/mol, and $\Delta V^{\#}$ was -9.98 cm^3/mol.

High-pressure-induced gelatinization is highly sensitive to temperature, pressure, and treatment time. At constant temperature and treatment time, the DG increases with increasing pressure (Bauer and Knorr 2005). The higher the treatment temperature, the lower the pressure required for a complete gelatinization at that temperature (Bauer, Wiehle, and Knorr 2005). At constant temperature and pressure, the DG increases with treatment time. Bauer et al. (2005) studied the gelatinization kinetics of corn starch as a function of pressure and temperature but at constant starch concentration and treatment time.

In addition to its dependence on temperature, pressure, and treatment time, the starch gelatinization kinetics (i.e., DG) under HPT is also dependent on starch concentration (Baks et al. 2008). A model based on Gibbs energy difference was developed by Baks et al. (2008) to describe the relationship between starch gelatinization and other operating parameters including pressure, temperature, treatment time, and starch concentration:

$$D_{\mathrm{G}} = \frac{100}{1 + \exp\left(\frac{\Delta G}{RT}\right)}, \tag{15.7}$$

where ΔG is defined as:

$$\begin{aligned} \Delta G = {} & \Delta G_{(P_{\mathrm{r}},T_{\mathrm{r}})} - \Delta S_{(P_{\mathrm{r}},T_{\mathrm{r}})} \cdot (T - T_{\mathrm{r}}) - \frac{\Delta C_{\mathrm{p}}}{2T_{\mathrm{r}}}(T - T_{\mathrm{r}})^2 + \\ & \Delta V_{(P_{\mathrm{r}},T_{\mathrm{r}})} \cdot (P - P_{\mathrm{r}}) + \frac{\Delta \beta}{2}(P - P_{\mathrm{r}})^2 + \Delta\alpha \cdot (T - T_{\mathrm{r}}) \cdot (P - P_{\mathrm{r}}) \end{aligned} \tag{15.8}$$

and $\Delta G_{(P_{\mathrm{r}},T_{\mathrm{r}})}$ is the Gibbs energy change (J/mol) of reaction at reference pressure P_{r} and reference temperature T_{r}; $\Delta S_{(P_{\mathrm{r}},T_{\mathrm{r}})}$ is the entropy change (J/mol/K) of reaction at P_{r} and T_{r}; ΔC_{p} is the heat capacity change of reaction (J/mol/K); $\Delta V_{(P_{\mathrm{r}},T_{\mathrm{r}})}$ is the volume change (m^3/mol) of reaction at P_{r} and T_{r}; $\Delta\beta$ is the isothermal compressibility change (m^6/mol/J) of reaction; and $\Delta\alpha$ is the thermal expansion factor change (m^3/mol/J) of reaction. By using experimental data and modeling through data fitting, Baks et al. (2008) was able to plot a phase diagram of wheat starch gelatinization under HPT with varying treatment parameters including pressure, temperature, starch concentration, and time.

During HPT of a starch-water suspension, if the starch-water ratio is in the appropriate range for retrogradation, for example 1:1 (w/w), the gelatinized portion could retrograde even during the HPT. High-pressure-treated potato starch-water suspension under such conditions showed an endothermic peak similar to retrograded thermally gelatinized starch (Kawai, Fukami, and Yamamoto 2007a, 2007b).

Similar to thermal gelatinization, high-pressure-induced gelatinization is also influenced by co-solutes (e.g., sugars or salts) in water suspension, as they compete for the available free water (Rumpold and Knorr 2005). However, the influence of salts varies and it depends not only on the type of salt but also on the type of starch. At high chloride concentrations (>2 M), the effect of salts on starch gelatinization augmentation follows the order $Na^+ < K^+ < Li^+ < Ca^{2+}$, which corresponds to the order of the lyotropic series (Rumpold and Knorr 2005). At concentrations above 1 M, the effect of potassium salts on starch gelatinization follows the Hofmeister series $Cl^- < Br^- < I^- < SCN^-$ (Rumpold and Knorr 2005). Sugars have a different effect on starch gelatinization. They reduce the gelatinization pressure and the DG is linearly correlated with the number of equatorial hydroxyl groups (Rumpold and Knorr 2005). In general, the effects of salts and sugars on starch gelatinization are comparable for thermal and pressure gelatinization.

15.3.3 Comparison of Thermal Gelatinization and Pressure Gelatinization

As discussed above, starch granules remain intact with limited distortion during high-pressure gelatinization. The incomplete disintegration prevents the leaching of amylose from the granules. However, during thermal gelatinization, the crystalline structure is completely melted and the granules swell and then are completely disintegrated with amylose leaching. Therefore, compared to thermally gelatinized starch–water suspensions, the viscosity of high-pressure-gelatinized starch–water suspensions of the same solid content is usually lower. On the other hand, gels from pressure-gelatinized wheat starch–water suspensions had a denser structure compared to those from thermally gelatinized wheat starch–water suspensions (Douzals et al. 1998). Compared to thermally gelatinized starch–water suspensions, the retrogradation of high-pressure-gelatinized starch–water suspensions was slower and less than that of thermally gelatinized samples (Doona, Feeherry, and Baik 2006; Douzals et al. 1998; Ezaki and Hayashi 1992). Structural conformation of retrograded starches, however, was similar for both treatments (Douzals et al. 1998). The only exception was barley starch, for which the retrogradation of high-pressure-gelatinized samples was similar to that of heat-induced samples (Stolt, Oinonen, and Autio 2001). Compared to thermally gelatinized wheat starch gel, smaller water self-diffusion coefficients (indicating more restricted translational proton mobility) were found in high-pressure-gelatinized samples (Doona et al. 2006).

Thermal gelatinization and pressure gelatinization have a similar effect on resistant starch production in wheat starch (Doona et al. 2006). Although high-pressure gelatinization improves the enzyme digestibility of the starch while keeping its granular form (Ezaki and Hayashi 1992; Takahashi et al. 1994), there is no difference between the enhancement effect caused by HPT and that of high temperature gelatinization (75–100°C) (Takahashi et al. 1994). Selmi et al. (2000) studied glucose production by amyloglucosidase hydrolysis of corn and wheat starches that had been processed by two high-pressure treatments (HPTs) and a thermal gelatinization respectively. The high-pressure-treated corn starches had a lower initial rate of glucose production but similar equilibrium yield compared with that treated by the thermal gelatinization. By contrast, wheat starches treated by the two HPTs had a higher equilibrium yield of glucose production compared to the thermal gelatinization. Formation of amylose-lipid complexes in wheat starch during the thermal gelatinization was suggested to be the reason.

15.4 DYNAMIC HIGH-PRESSURE-PROCESSING (HPP) OF STARCH-WATER SUSPENSIONS (HIGH-PRESSURE HOMOGENIZATION)

High-pressure homogenization involves pressures of up to 200 MPa in a very short period of time. Pressure-induced phenomena, including cavitation, shearing, turbulence, and temperature rise, take place simultaneously. This unique blend of influences is not present in the hydrostatic HPP described in the previous section and can result in novel changes to the products.

An acute, rapid increase in pressure up to 100 MPa and the resulting cavitation caused a linear temperature rise in water and a cassava starch–water suspension (5%, w/w) of 0.194°C/MPa and 0.187°C/MPa, respectively (Che et al. 2007). Starch reduced the temperature rise by restraining the formation of cavities. The crystalline structure of cassava starch was not affected by this high-pressure homogenization, but its amorphous structure was affected by the large quantity of water absorption under the influence of heat and pressure, which resulted in granule swelling and increased granule size (Che et al. 2007). The structure of processed cassava starch granules was pressure sensitive. It retained its native pattern before the pressure increased to 60 MPa. Granule swelling, significant deformation into fragments, and some gel-like structure in fragments were then found when the pressure increased to 80 and 100 MPa (Che et al. 2007).

Wang et al. (2008) applied high-pressure homogenization to a corn starch–water suspension (1.0%, w/w) at 60, 100, and 140 MPa, respectively. The structure and thermal properties of the treated starches were investigated using DSC, x-ray diffraction, laser scattering, and by microscope. DSC analysis showed a decrease in gelatinization temperatures and gelatinization enthalpy with increasing homogenizing pressure, which indicated a partial, pressure-induced gelatinization that was verified by the microscopic analysis. The D_G was very small at 60 MPa, and reached 12.9±3.2% at 100 MPa and 26.8±1.8% at 140 MPa. Meanwhile, there was no noticeable effect of high-pressure homogenization on the retrogradation behavior of the treated corn starch. Laser scattering measurements of particle size showed an increase in the granule size at 140 MPa, which was attributed to the gelatinization and aggregation of the starch granules. Loss of crystallinity after homogenization at 140 MPa was evident from x-ray diffraction patterns.

The effect of high-pressure homogenization may vary among different starches and it may change the thermal properties and functionalities of the treated starches. More research is needed for a better understanding of this type of processing and its impact on various starches.

15.5 HYDROSTATIC HIGH-PRESSURE-PROCESSING (HPP) OF LOW-MOISTURE STARCH GRANULES (HIGH-PRESSURE COMPRESSION)

High-pressure compression applies static high pressure to solid starch granules of limited moisture content, rather than to starch-in-water suspensions. It directly modifies the physicochemical properties of the starch by altering the internal and external structure of the starch granules (Liu et al. 2008).

Limited studies have been published on the high-pressure compression of starch granules. Kudta and Tomasik (1992a, 1992b) compressed potato starch with low water contents (2–22%, w/w) using 800–1200 MPa for 60–600 s. They found that the properties of potato starch with low water content seemed to be changed by high-pressure pressing and the amylopectin shell of the starch grains was the component that suffered most damage. A pressure of 1000 MPa seemed to cause some repolymerization of dextrin formed during earlier compression. A pressure of 1200 MPa caused further damage to the starch structure with the ordering of the molecules into more crystal-like matter.

Liu et al. (2008) studied the impact of high-pressure compressing (also called HPT) over the range of 740–1500 MPa for 5 min to 24 h on some physicochemical properties of normal corn, waxy corn, wheat, and potato starches. The gelatinization temperature of starch was lowered by 3.0–6.6°C after HPT and the corresponding gelatinization enthalpy was also reduced (Figure 15.8).

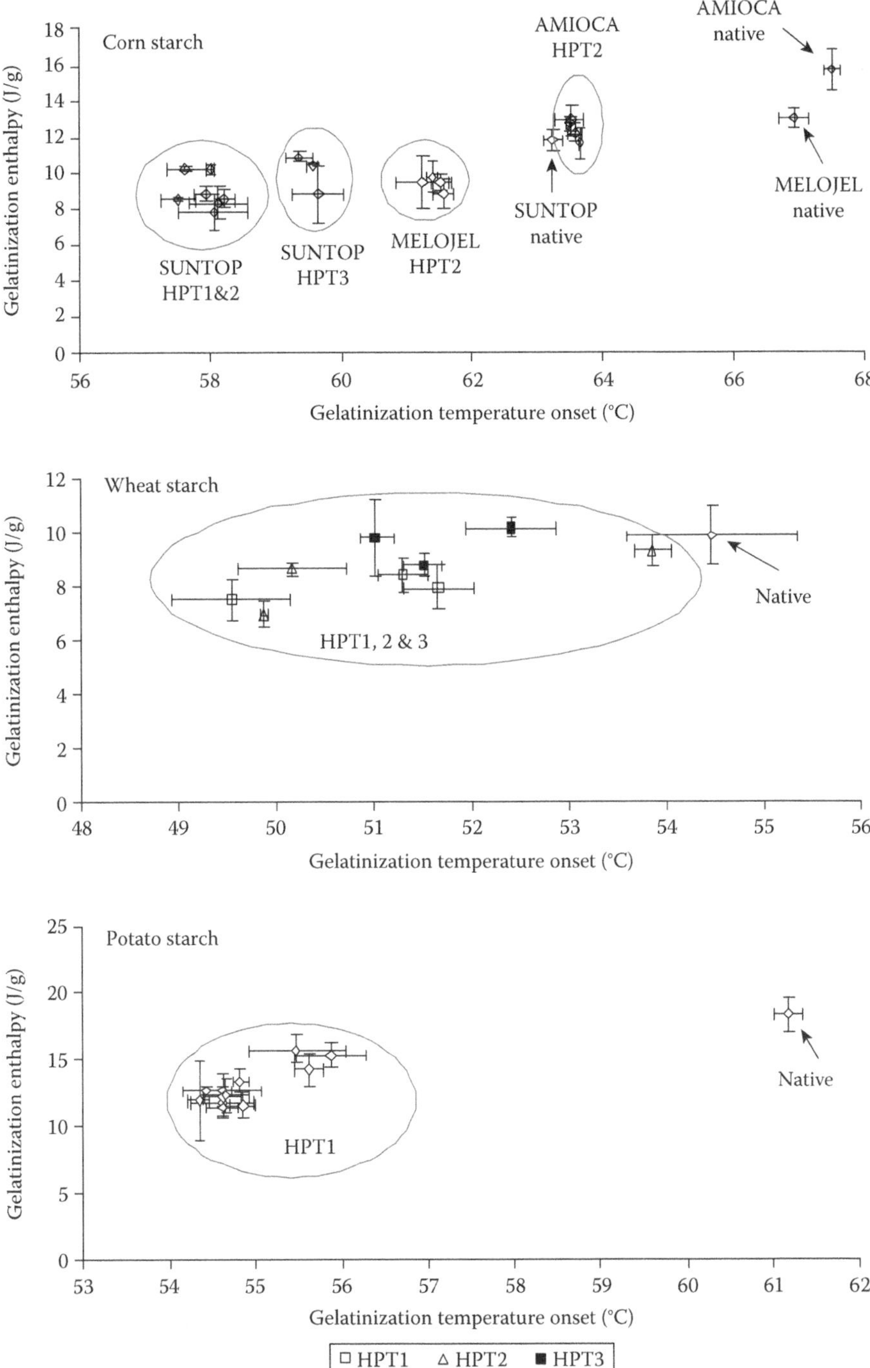

FIGURE 15.8 Shift in gelatinization temperature (onset) and change in gelatinization enthalpy by high-pressure compression (HPT1: 740–880 MPa for 5 min–2 h; HPT2: 960–1100 MPa for 24 h; HPT3: 1500 MPa for 24 h) for different starches (SUNTOP™ native corn starch, MELOJEL® native corn starch, AMIOCA® waxy corn starch, wheat starch, and potato starch). Data points for the starch of the same treatment are collected during storage at 25°C from zero days to up to six months after treatment. (From Liu, Y., Selomulyo, V. O., and Zhou, W., *Journal of Food Engineering*, 88, 126–36, 2008. With permission.)

Similar results for tapioca and rice starches were found by Katopo et al. (2002) who compressed the starches at 690 MPa for 5 min and 1 h respectively. However, there was no shift in the gelatinization onset temperature but only a reduction in the gelatinization enthalpy for high-amylose (70%) corn starch.

In the study by Liu et al. (2008), it was shown that the changes were irreversible and maintained during storage at 25°C for up to six months. There was no clear difference between the different treatments within the wide pressure range from 740 to 1500 MPa and treatment time from 5 min to 24 h. For the high-pressure-treated starches, the lowered onset temperature of gelatinization could be due to a shift in the glass transition temperature of the amorphous regions in the starch granules; and the lowered gelatinization enthalpy could be due to lowered energy requirement for melting the crystalline regions. It was hypothesized that under HPT, the amorphous regions were altered with lowering of the glass transition temperature. Possible alterations included changed configuration of the amorphous regions or plasticization by water in the amorphous regions through water redistribution forced by the high pressure, or even plasticization by fragmented starch molecules due to the decomposition effect of HPT. However, as even modulated DSC was unable to separate the glass transition signal from the melting signal of starches during gelatinization, the hypothesis on lowered glass transition temperature of the amorphous regions could not be confirmed. On the other hand, lowered melting energy of the crystalline regions could be due to a partial melting of the crystalline structure during high-pressure compression or an alteration of the crystalline structure. While different types of starch behaved similarly under the HPTs, i.e., reduction in both gelatinization temperature and gelatinization enthalpy, the amount of the reduction was different between the starches. This might be due to the differences in composition between starch granules, including the ratio between amylose and amylopectin, molecular weight distribution of amylose and amylopectin, molecular packing in the granules, etc., which are also responsible for different types of native starches having distinctive thermal gelatinization temperatures and enthalpies.

The pasting profiles of normal corn starch, tapioca starch, and rice starch were changed after high-pressure compressing; however, those of waxy corn and potato were not changed. The molecular weight distribution of starches was not altered after high-pressure compressing (Katopo et al. 2002).

The birefringence of native and high-pressure-treated starches (Figure 15.9) was not visually different in polarized light (Liu et al. 2008). This is in contrast to those studies on starch-in-water suspensions where a HPT resulted in the loss of birefringence in starch granules, which indicated gelatinization of the starch under this treatment. Kudla and Tomasik (1992a) suggested that HPTs on potato starch with low moisture content caused damage to the structure, which reordered the molecules into a more crystal-like matter. However, in their later studies (Kudla and Tomasik 1992b), contrary microscopic observations were presented. Figure 15.9 suggests that the birefringence of starch in polarized light may not be sensitive enough to reflect the internal structure change caused by the HPTs, or, alternatively, there was no change in the internal structure of native starch granules during the HPTs.

The x-ray diffraction patterns were similar for starches of the same biological origin (Figure 15.10). There was no peak shift, peak disappearance, or new peaks in the diffraction patterns (Katopo et al. 2002; Liu et al. 2008). In contrast, high-pressure-treated starch-in-water suspensions had apparent changes in their x-ray diffraction patterns that indicated partial or full gelatinization of the starch granules as well as internal structural changes. Although the crystalline pattern was not affected by high-pressure compression, it seems that there was a difference in the peak heights before and after the treatment. Therefore, the degree of crystallinity might have been altered during the HPT, as also suggested by the reduction in gelatinization enthalpy.

HPTs can alter the shape of starch granules and change their surface appearance under SEM. Liu et al. (2008) reported that the shapes of starch granules were changed and the original smooth

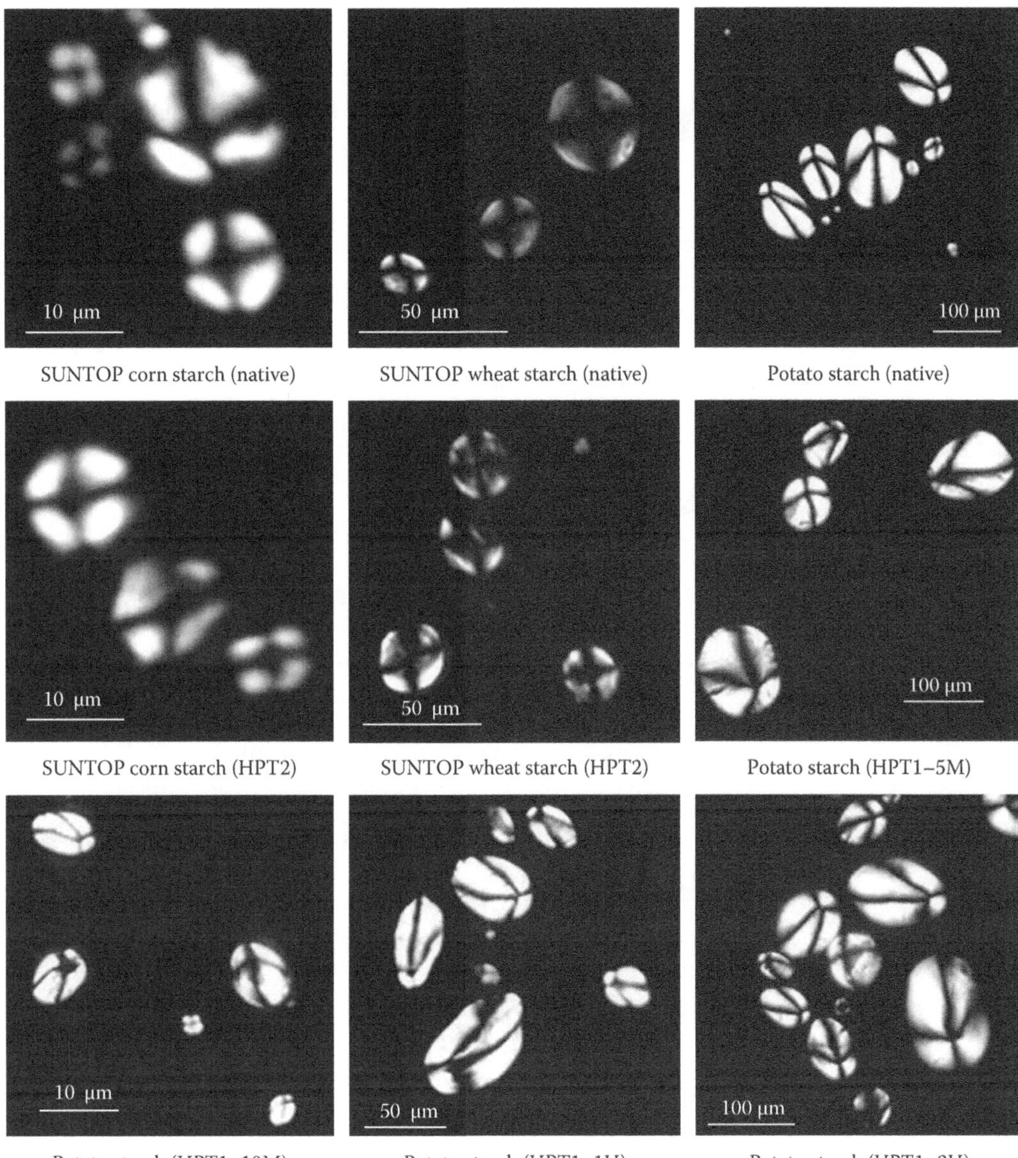

FIGURE 15.9 Birefringence of native and high-pressure-treated corn, wheat, and potato starches. (HPT1-5M, HPT1-10M, HPT1-1H, and HPT1-2H denote samples undergone high-pressure treatment 1 for 5 min, 10 min, 1 h, and 2 h, respectively.) (From Liu, Y., Selomulyo, V. O., and Zhou, W., *Journal of Food Engineering*, 88, 126–36, 2008. With permission.)

surfaces became rough after HPTs (Figure 15.11). Due to the irregular shape of native corn starch granules, changes in the granule shape and surface caused by the HPTs were not obvious. But changes in the granule shape and surface induced by the HPTs were visually distinct for wheat and potato starch granules. Similar observations for potato starch were also obtained by Kudla and Tomasik (1992b). Furthermore, cracking of the amylopectin shell was observed under the treatment of 800 MPa for 300 s in Kudla and Tomasik (1992b), while there was no such cracking observed by Liu et al. (2008). This suggests that the amylopectin shell cracking may only occur occasionally.

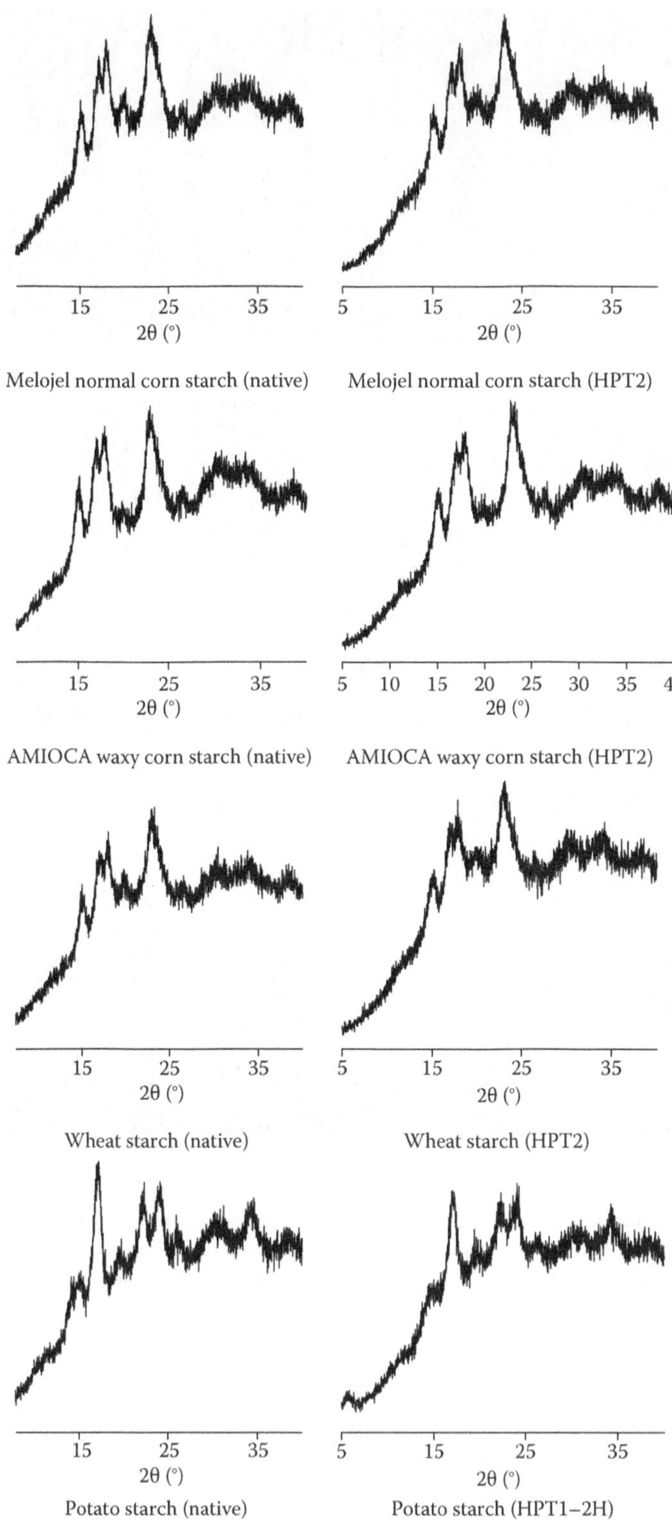

FIGURE 15.10 X-ray diffraction pattern of native and high-pressure-treated corn, wheat, and potato starches. (HPT1-1H denotes samples undergone high-pressure treatment 1 h) (From Liu, Y., Selomulyo, V. O., and Zhou, W., *Journal of Food Engineering*, 88, 126–36, 2008. With permission.)

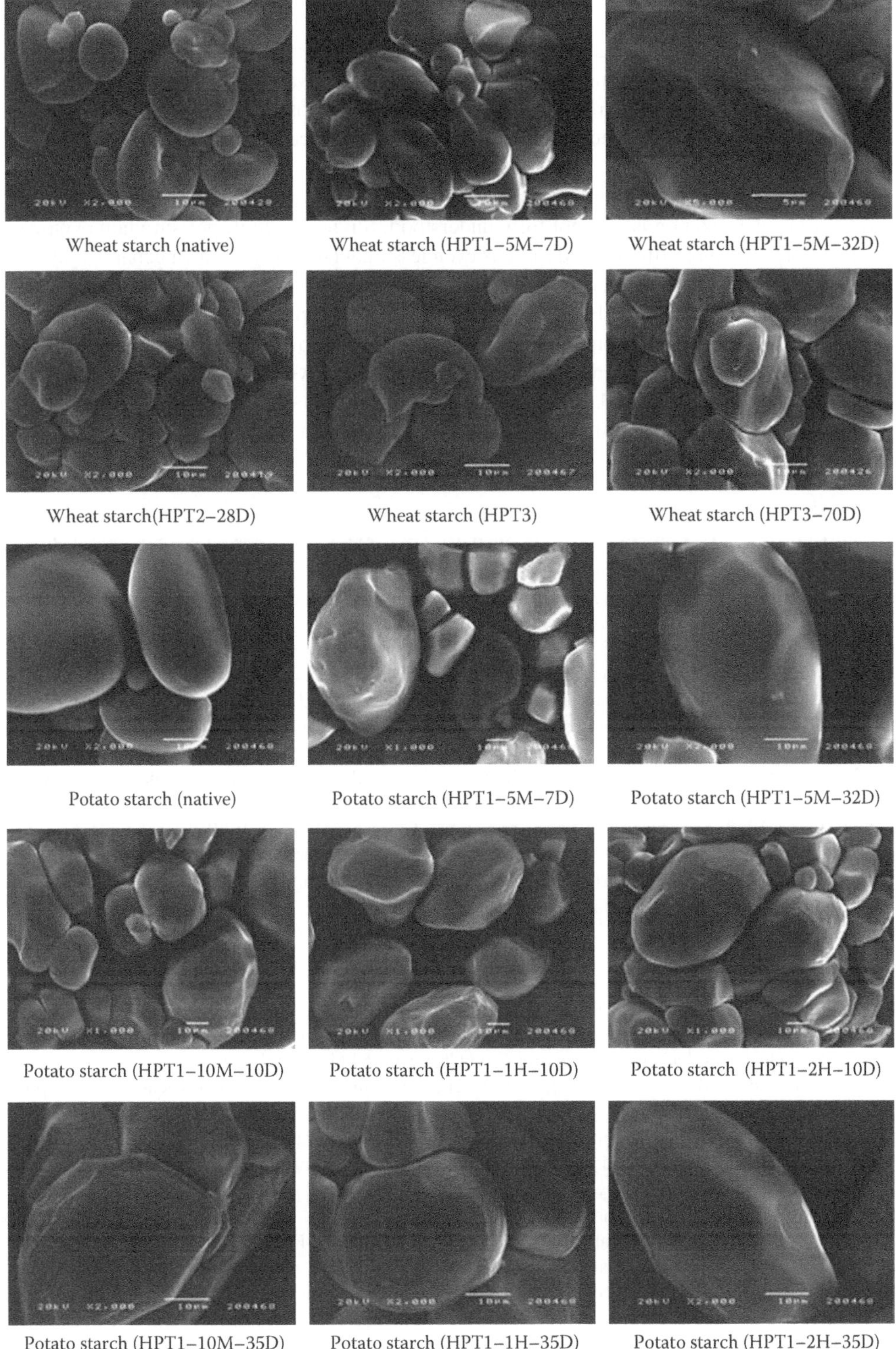

FIGURE 15.11 Changes of the outer structure of wheat and potato starch granules after high-pressure treatment under SEM. (HPT1-xM-yD denotes high-pressure treatment 1 for *x* minutes and with storage at 25°C for *y* days; HPTa-bD denotes high pressure treatment a (2 or 3) and with storage at 25°C for *b* days). (From Liu, Y., Selomulyo, V. O., and Zhou, W., *Journal of Food Engineering*, 88, 126–36, 2008. With permission.)

15.6 CONCLUSIONS

HPP of starches has been investigated under different conditions: hydrostatic HPP of starch–water suspensions, dynamic high-pressure homogenization of starch–water suspensions, and hydrostatic high-pressure compression of starch granules of low moisture content. These different processing conditions can lead to various effects ranging from pressure-induced gelatinization to structural and functional modifications.

Pressure-induced gelatinization has been most studied, however the mechanism by which pressure results in gelatinization is still not fully understood. Studies on high-pressure homogenization and compression are more limited and there is even less understanding of the mechanism by which structural/functional changes occur. The increasing use of high-pressure homogenization in the food industry, however, requires better understanding of the impact of such processing on different starches. On the other hand, it has been demonstrated that high-pressure compression clearly has a direct impact on the gelatinization behavior of starches with a favorable temperature shift and enthalpy reduction. This makes it a potential processing method for various purposes. More research is warranted for further understanding of this phenomenon.

REFERENCES

Ahromrit, A., Ledward, D. A., and Niranjan, K. 2007. Kinetics of high pressure facilitated starch gelatinisation in Thai glutinous rice. *Journal of Food Engineering* 79 (3): 834–41.

Atwell, W. A., Hood, L. F., Lineback, D. R., et al. 1988. The terminology and methodology associated with basic starch phenomena. *Cereal Foods World* 33:306–11.

Baks, T., Bruins, M. E., Janssen, A. E. M., et al. 2008. Effect of pressure and temperature on the gelatinization of starch at various starch concentrations. *Biomacromolecules* 9 (1): 296–304.

Bauer, B. A., and Knorr, D. 2005. The impact of pressure, temperature and treatment time on starches: Pressure-induced starch gelatinisation as pressure time temperature indicator for high hydrostatic pressure processing. *Journal of Food Engineering* 68 (3): 329–34.

Bauer, B. A., Wiehle, T., and Knorr, D. 2005. Impact of high hydrostatic pressure treatment on the resistant starch content of wheat starch. *Starch-Starke* 57 (3–4): 124–33.

Biliaderis, C. G. 1998. Structures and phase transition of starch polymers. In *Polysaccharide Association Structures in Food*, ed. R. H. Walter, 57–168. New York: Marcel Dekker.

Blaszczak, W., Valverde, S., and Fornal, J. 2005. Effect of high pressure on the structure of potato starch. *Carbohydrate Polymers* 59 (3): 377–83.

Che, L. M., Li, D., Wang, L. J., et al. 2007. Effect of high-pressure homogenization on the structure of cassava starch. *International Journal of Food Properties* 10 (4): 911–22.

Cooke, D., and Gidley, M. J. 1992. Loss of crystalline and molecular order during starch gelatinisation: Origin of the enthalpic transition. *Carbohydrate Research* 227:103–12.

Doona, C. J., Feeherry, F. E., and Baik, M. Y. 2006. Water dynamics and retrogradation of ultrahigh pressurized wheat starch. *Journal of Agricultural and Food Chemistry* 54 (18): 6719–24.

Douzals, J. P., Cornet, J. M. P., Gervais, P., et al. 1998. High-pressure gelatinization of wheat starch and properties of pressure-induced gels. *Journal of Agricultural and Food Chemistry* 46 (12): 4824–29.

Ezaki, S., and Hayashi, R. 1992. High pressure effect on starch: structural change and retrogradation. In *High Pressure and Biotechnology*, eds. C. Balny, R. Hayashi, K. Hermans, et al., 163–65. Montrouge, France: Colloque INSERM/John Libbey Eurotext.

French, D. 1984. Organization of starch granules. In *Starch: Chemistry and Technology*, eds. R. L. Whister, J. N. BeMiller, and E. F. Paschall, 183–247. New York: Academic Press.

Gebhardt, R., Hanfland, M., Mezouar, M., et al. 2007. High-pressure potato starch granule gelatinization: Synchrotron radiation Micro-SAXS/WAXS using a diamond anvil cell. *Biomacromolecules* 8 (7): 2092–97.

Hills, B., Costa, A., Marigheto, N., et al. 2005. T-1-T-2 NMR correlation studies of high-pressure-processed starch and potato tissue. *Applied Magnetic Resonance* 28 (1–2): 13–27.

Hoseney, R. C. 1998. Gelatinization phenomena of starch. In *Phase/State Transitions in Foods: Chemical, Structural, and Rheological Changes*, eds. M. A. Rao, and R. W. Hartel, 95–110. New York: Marcel Dekker.

Indrawati, Van Loey, A., and Hendrickx, M. 2002. High pressure processing. In *The Nutrition Handbook for Food Processors*, eds. C. J. K. Henry, and C. Chapman, 433–61. Boca Raton, FL:CRC Press.

Katopo, H., Song, Y., and Jane, J. L. 2002. Effect and mechanism of ultrahigh hydrostatic pressure on the structure and properties of starches. *Carbohydrate Polymers* 47 (3): 233–44.

Kawai, K., Fukami, K., and Yamamoto, K. 2007a. Effects of treatment pressure, holding time, and starch content on gelatinization and retrogradation properties of potato starch-water mixtures treated with high hydrostatic pressure. *Carbohydrate Polymers* 69 (3): 590–96.

Kawai, K., Fukami, K., and Yamamoto, K. 2007b. State diagram of potato starch-water mixtures treated with high hydrostatic pressure. *Carbohydrate Polymers* 67 (4): 530–35.

Knorr, D., Heinz, V., and Buckow, R. 2006. High pressure application for food biopolymers. *Biochimica et Biophysica Acta* 1764 (3): 619–31.

Kudta, E., and Tomasik, P. 1992a. The modification of starch by high-pressure .1. Air-dried and oven-dried potato starch. Starch-Starke, 44 (5): 167–173.

Kudla, E., and Tomasik, P. 1992b. The modification of starch by high-pressure .2. Compression of starch with additives. Starch-Starke, 44 (7): 253–259.

Lelievre, J., and Liu, H. 1994. A review of thermal analysis studies of starch gelatinization. *Thermochimica Acta* 246: 309–15.

Levine, H., and Slade, L. 1990. Influences of the glassy and rubbery states on the thermal, mechanical, and structural properties of doughs and baked products. In *Dough Rheology and Baked Product Texture*, eds. H. A. Faridi, and J. M. Faubion, 157–330. New York: Van Nostrand Reinhold.

Liu, Y., Selomulyo, V. O., and Zhou, W. 2008. Effect of high pressure on some physicochemical properties of several native starches. *Journal of Food Engineering* 88 (1): 126–36.

Noguchi, A. 2003. Modern processing and utilization of legumes: Recent research and industrial achievements in soybean foods in Japan. In *Processing and Utilization of Legumes. Report of the APO Seminar on Processing and Utilization of Legumes, Japan, 9 to 14 October 2000*, 63–75. Tokyo: Asian Productivity Organization.

Parker, R., and Ring, S. G. 2001. Aspects of the physical chemistry of starch. *Journal of Cereal Science* 34: 1–17.

Rubens, P., and Heremans, K. 2000. Stability diagram of rice starch as determined with FTIR. *High Pressure Research* 19 (1–6): 551–56.

Rumpold, B. A., and Knorr, D. 2005. Effect of salts and sugars on pressure-induced gelatinisation of wheat, tapioca, and potato starches. *Starch-Starke* 57 (8): 370–77.

Sablani, S. S., Kasapis, S., Al-Tarqe, Z. H., et al. 2007. Isobaric and isothermal kinetics of gelatinization of waxy maize starch. *Journal of Food Engineering* 82 (4): 443–49.

Selmi, B., Marion, D., Cornet, J. M. P., et al. 2000. Amyloglucosidase hydrolysis of high-pressure and thermally gelatinized corn and wheat starches. *Journal of Agricultural and Food Chemistry* 48 (7): 2629–33.

Shi, Y. C., and Seib, P. A. 1992. The structure of four waxy starches related to gelatinization and retrogradation. *Carbohydrate Research* 227: 131–45.

Stolt, M., Oinonen, S., and Autio, K. 2001. Effect of high pressure on the physical properties of barley starch. *Innovative Food Science & Emerging Technologies* 1: 167–75.

Stute, R., Klingler, R. W., Boguslawski, S., et al. 1996. Effects of high pressures treatment on starches. *Starch-Starke* 48 (11–12): 399–408.

Takahashi, T., Kawauchi, S., Suzuki, K., et al. 1994. Bindability and digestibility of high-pressure-treated starch with glucoamylases from Rhizopus Sp. *Journal of Biochemistry* 116 (6): 1251–56.

Tester, R., and Debon, S. J. J. 2000. Annealing of starch: A review. *International Journal of Biological Macromolecules* 27 (1): 1–12.

Thomas, D. J., and Atwell, W. A. 1997. *Starches*. St Paul, MN: Eagan Press.

Vaclavik, V. A. 1998. Starches in food. In *Essentials of Food Science*, eds. V. A. Vaclavik, and E. W. Christian, 39–52. New York: Chapman and Hall.

Wang, B., Li, D., Wang, L.-j., Chiu, Y. L., Chen, X. D., and Mao, Z.-h. 2008. Effect of high-pressure homogenization on the structure and thermal properties of maize starch. Journal of Food Engineering, 87 (3): 436–444.

16 Effect of High Pressure on Textural and Microstructural Properties of Fruits and Vegetables

Navin K. Rastogi
Central Food Technological Research Institute

CONTENTS

16.1 INTRODUCTION

"High pressure kills microorganisms and preserves food." This was discovered in 1899 and has been used with success in many industries such as the manufacture of chemicals, ceramics, carbon allotropy, steels/alloys, composite materials, and plastics. In the area of food processing its use began with the pioneering work of Hite (1899) for the preservation of milk, and the scope of this technology was extended to the processing and preservation of fruits and vegetables (Hite, Giddings, and Weakly 1914). The ability of high pressure to inactivate microorganisms and enzymes that catalyze spoilage, whilst retaining other quality attributes, has recently encouraged the Japanese and North American food industries to commercialize high-pressure-processed foods (Mermelstein 1997; Hendrickx et al. 1998). High-pressure-processed foods were introduced to the Japanese market in 1990 by the Meidi-ya Company, which marketed a line of jams, jellies, and sauces processed without the application of heat. Other products include fruit preparations, fruit juices, rice cakes, and raw squid in Japan; fruit juices, especially apple and orange juice, in France and Portugal; and guacamole and oysters in the United States (Hugas, Garriga, and Monfort 2002). In addition to food preservation, high-pressure treatment can result in food products acquiring novel structures and textures and, hence, can be used to develop new products (Hayashi 1990) or increase the functionality of certain ingredients.

Today, high-pressure technology is acknowledged to have the promise of producing a very wide range of products, whilst simultaneously showing potential for creating a new generation of value-added foods. High-pressure technology can supplement conventional thermal processing for reducing microbial load or substitute the use of chemical preservatives (Rastogi, Subramanian, and Raghavarao 1994).

In general, high-pressure processing can kill all the vegetative microorganisms, whereas bacterial spores appear to be resistant, suggesting that a combination of high pressure and temperature treatment will be the most efficient method of inactivation. Vegetative cells, including yeast and moulds, are sensitive to pressure, i.e., they can be inactivated by pressure in the range of 300–600 MPa at ambient temperature or pasteurization can be performed even under chilled conditions for heat-sensitive products. On the other hand, bacterial spores are highly pressure resistant, since pressures exceeding 1200 MPa may be needed for their inactivation (Knorr 1995). However, the application of pressure in combination with temperatures in the range of 45–50°C was found to increase the rate of inactivation of pathogens and spoilage microorganisms. In comparison, for the purposes of high-pressure sterilization, higher process temperatures, in the range of 90–121°C, in conjunction with pressures of 500–800 MPa are needed to inactivate spores-forming bacteria. Sterilization of low-acid foods (pH>4.6) will most probably rely on a combination of high pressure and other forms of relatively mild treatments. For both pasteurization and sterilization processes, combined treatment with high pressure and temperature is frequently considered to be most appropriate (Farr 1990).

Since high pressure is transmitted rapidly and uniformly, the problems of spatial variations in other preservation treatments associated with heat, microwave, and radiation penetration are not encountered in high-pressure processing. Further, high pressure affects only noncovalent bonds (hydrogen, ionic, and hydrophobic bonds), causes unfolding of protein chains, and has little effect on the chemical constituents associated with desirable food qualities such as flavor, color, and nutritional content. Thus, in contrast to thermal processing, application of high pressure results in negligible impairment of nutritional values, taste, color, flavor, and vitamin content (Hayashi 1990).

High pressure can cause structural changes in structurally fragile foods containing entrapped air such as strawberries and lettuce. During high-pressure processing, turgor loss and cellular changes can take place along with changes in cell conformation, elongation, separation, or debonding and/or cell wall disruption. High pressure can alter the physicochemical properties of vegetable matrices by inducing changes in their structure (Basak and Ramaswamy 1998; Butz et al. 2002). Prestamo and Arroyo (1998) demonstrated that pressure above 200 MPa resulted in reduced texture due to a loss of turgor pressure, cell rupture, or collapse as well as disruption of the parenchyma tissue. However, loss of texture due to high-pressure application is much less than that experienced with thermal processing. Therefore, high-pressure processing is considered as a means of maintaining desirable texture. The extent of damage is less in the case of high pressure due to limited tissue damage, cell wall separation, and reduced biochemical changes compared to traditional treatments. Basak and Ramaswamy (1998) characterized these textural changes by an initial texture loss, which is referred to as instantaneous pressure softening, followed by texture recovery during pressure hold.

High pressure causes folding and corrugation of the cell wall, resulting in changes in the microstructure, which in turn affect the textural properties due to a loss of cellular turgor, firmness, and cell wall integrity (Tangwongchai, Ledward, and Ames 2000; Sila et al. 2004). Many authors have suggested that changes in cell permeability are a major cause of textural or microstructural degradation in spinach and cauliflower (Prestamo and Arroyo 1998) and pineapple (Rastogi and

Niranjan 1998; Kingsly, Balasubramaniam, and Rastogi 2008). Cell permeabilization influences a number of biochemical reactions along with texture degradation. Certain biochemical transformations on pectin involving the action of endogenous enzymes such as pectinmethylesterase (PME) and polygalacturonase (PG) have been reported to be highly correlated with the texture degradation of fruits and vegetables (Sila 2005). These enzymes act on the pectin in a two-stage sequence. PME acts on pectin and produces methanol and pectin molecule with a lower degree of demethylation (DM), which is depolymerized by PG, leading to drastic tissue softening (Tangwongchai et al. 2000; Vu et al. 2004). Pectin molecules with lower Slemethylations (DMs) lead to more cross-linking between pectin chains and divalent cations, such as calcium, which increase the rigidity of the middle lamella and cell wall (Grant et al. 1973).

Texture is a very important characteristic of food. Control or modification of texture is one of the major objectives for any food processing technique. Texture is the major determinant for the assessment of fruit and vegetable quality and dependent on the cell wall and middle lamella polysaccharides. Cell wall polysaccharides mainly consist of pectin, hemicellulose, and cellulose, whereas pectin is the major component present in the middle lamella that cements the cell walls and gives firmness and elasticity to the tissue (Kato, Teramoto, and Fuchigami 1997).

Physical properties of plant foods largely depend on cell morphology and turgor pressure. The fluid-filled thin-walled parenchyma cells and extracellular volume, which contains interstitial fluid or air, determine texture response. A considerable loss of turgor and structural collapse has been observed during high-pressure processing of carrots, celerys, and red peppers (Basak and Ramaswamy 1998) and in cauliflower and spinach (Prestamo and Arroyo 1998). High-pressure treatment influenced the structure of spinach extensively compared to cauliflower, which has a harder and less elastic cell structure. The soft parenchyma cell of a spinach leaf was completely destroyed after pressure treatment at 400 MPa for 30 min at 5°C. Generally, high-pressure treatment results in membrane disruption and protein denaturation in plant tissues. The high pressure changes cell permeability and enables the movement of water from inside to outside the cell. As a result, treated vegetable tissue had a soaked or drenched appearance. However, after these changes, cauliflower maintained near original, acceptable firmness and flavor.

This chapter comprehensively reviews the information available in the literature regarding the effect of high pressure and its combination with temperature on texture and the microstructure of fruits and vegetables.

16.2 EFFECT OF HIGH PRESSURE ON TEXTURE AND MICROSTRUCTURE OF FRUITS AND VEGETABLES

A combination of pressure and temperature has a synergic effect on the texture and microstructure of fruits and vegetables. The adoption of high pressure in combination with ambient (20–25°C), freezing (–20°C), and high (60–121°C) temperatures influences the texture and microstructure of foods in many processes. Detailed descriptions are given in the following subsections.

16.2.1 Effect of High Pressure on Texture and Microstructure at Ambient Temperature

16.2.1.1 Tomato

The effect of high-pressure treatment (200–600 MPa for 20 min, 25°C) on the texture of whole cherry tomatoes was studied by Tangwongchai et al. (2000). The visual examination of high-pressure-processed tomatoes indicated that textural damage increased with an increase in pressure up to 400 MPa. An increase in pressure between 500 and 600 MPa showed less damage, and the tomatoes appeared similar to untreated samples. These visual observations were in good agreement with the instrumental texture (firmness) and cell rupture parameters. The application of high pressure resulted in softening of the tomatoes due to cell damage, which led to an increase

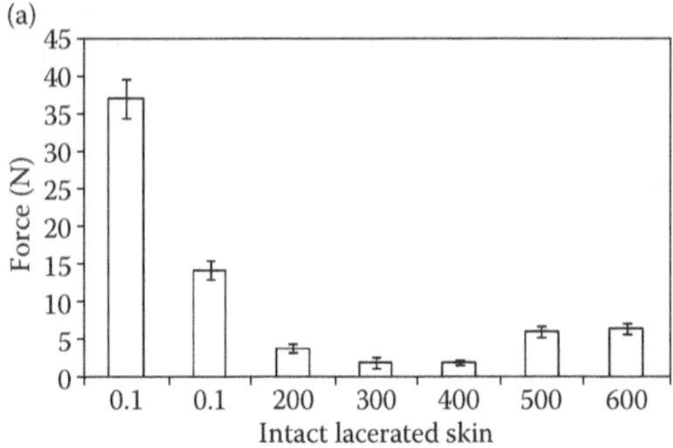

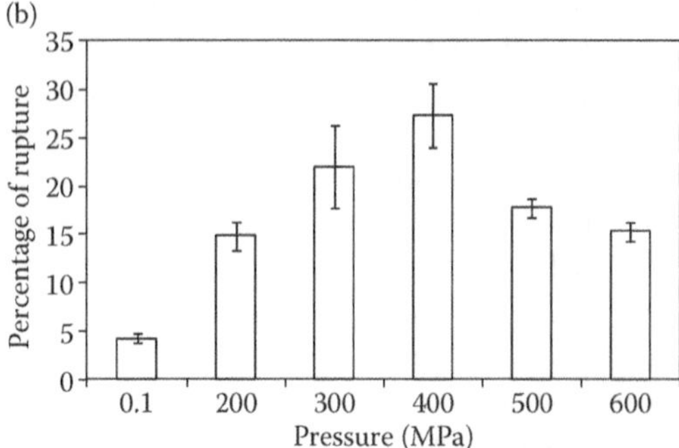

FIGURE 16.1 (a) Firmness and (b) percentage of cell rupture of untreated and pressure-treated cherry tomatoes (200–600 MPa for 20 min). (From Tangwongchai, R., Ledward, D. A., and Ames, J. A., *Journal of Agriculture and Food Chemistry*, 48, 1434–41, 2000. With permission.)

in cell rupture parameters. The percentage of cell rupture increased and firmness decreased up to 400 MPa, but cell rupture decreased and firmness increased after treatment at 500 and 600 MPa (Figure 16.1a and b). The increase in firmness after pressure treatment at 500 and 600 MPa was attributed to the role of the PME enzyme in producing low levels of methoxy pectin, which formed a gel network with divalent ions such as Ca and Mg leading to hardening of the tissue. The decrease in firmness below 500 MPa was due to the action of the PG enzyme, which hydrolyzed the low methoxy pectin to water-soluble galacturonic acid. This resulted in more free water being seen along with a decrease in firmness. SEM micrographs showed that high-pressure treatment above 200 MPa severely damaged tomato tissues and the extent of damage increased with pressure; however, differences were insignificant at pressures above 300 MPa (Figure 16.2). Micrographs clearly revealed the folding and corrugation of the cell wall due to collapse, causing a loss of turgor and firmness. SEM analysis did not substantiate texture and cell rupture analysis due to the fact that during these investigations tissues were dried prior to analysis; thus the location of water and its contribution to texture could not be evaluated.

16.2.1.2 Pineapple

Application of high pressure was reported to result in an increase in the permeability of cell structure (Farr 1990; Dornenburg and Knorr 1993; Rastogi et al. 1994; Eshtiaghi, Stute, and Knorr 1994), which results in softening of the samples. Rastogi and Niranjan (1998) demonstrated that the compressive force required to penetrate high-pressure-treated pineapple decreased rapidly with an increase in pressure from 100 to 300 MPa and beyond 300 MPa the decrease in compression force was not significant (Figure 16.3). Microstructures of pressure-treated and control pineapple samples indicated that the application of high pressure obviously damaged the cell wall structure and made the cells more permeable (Figure 16.4). There was also a gradual reduction in intercellular material in samples subjected to high-pressure treatment and decompressed. This softens the tissue, and softness increased with pressure (Figure 16.3). The major changes in the tissue microstructure occur below 300 MPa, and potential benefits of higher pressures were reported to be minimal. This phenomenon was used to enhance the mass transfer rate during osmotic dehydration by pretreating the samples at high pressure.

Application of high pressure (50–700 MPa at 25°C for 10 min) to pineapple slices resulted in a decrease in hardness, springiness, and chewiness while there was no significant effect on cohesiveness (Figure 16.5; Kingsly et al. 2008). The diffusion coefficient of water during dehydration was found to increase with an increase in pretreatment pressure. The diffusion coefficient above 500 MPa was reported to be higher than hot-water-blanched samples, which indicated that high-pressure pretreatment can be used as an effective alternative to thermal blanching prior to dehydration. The increase in the drying rate was attributed to the higher permeability of the cells due to elevated

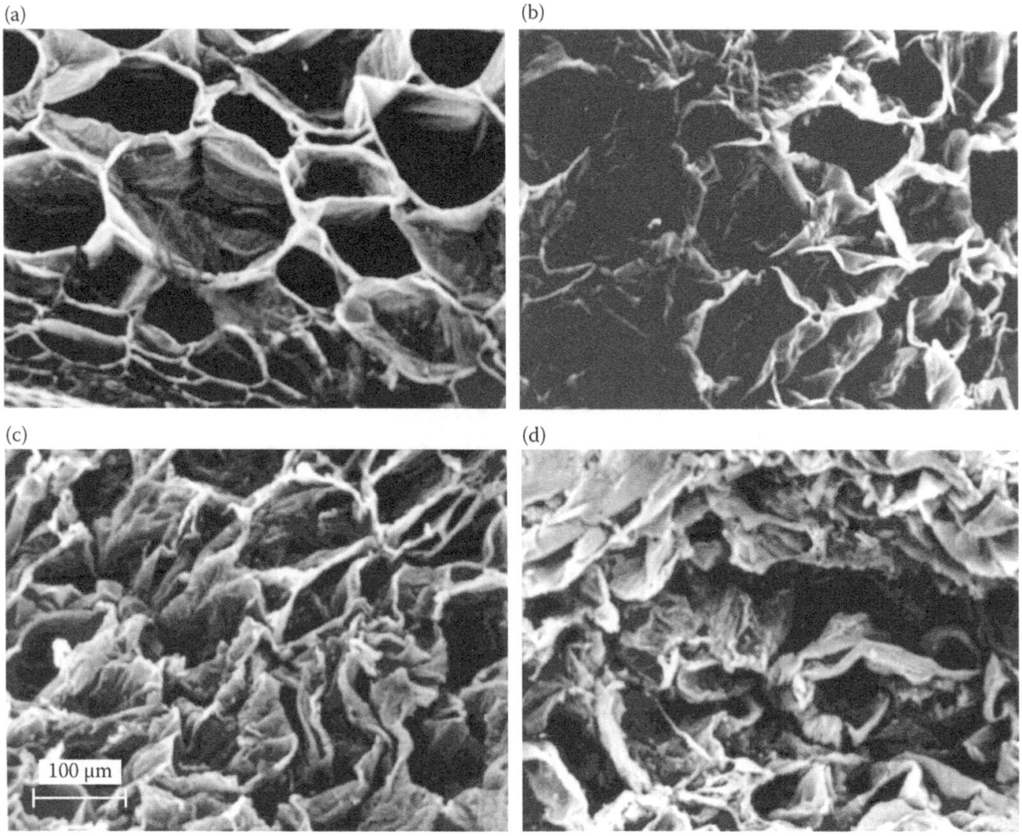

(a) (b) (c) (d)

100 μm

FIGURE 16.2 Scanning electron micrographs of untreated and pressure-treated tomato tissue: (a) 0.1 MPa, control; (b) 200 MPa; (c) 300 MPa; and, (d) 400 MPa for 20 min. (From Tangwongchai, R., Ledward, D. A., and Ames, J. A., *Journal of Agriculture and Food Chemistry*, 48, 1434–41, 2000. With permission.)

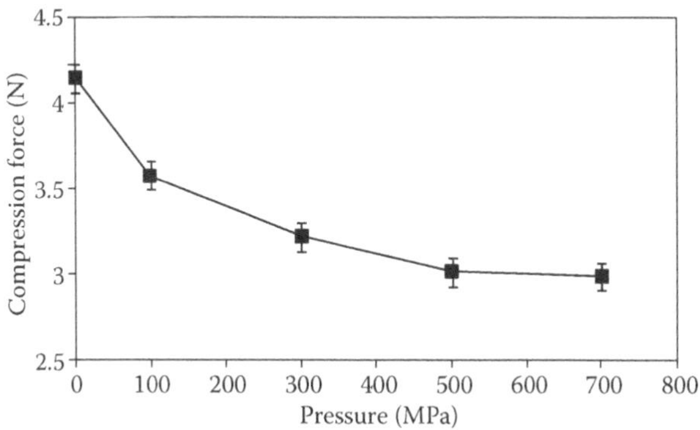

FIGURE 16.3 Variation of compression force required to puncture the pineapple sample as a function of pretreatment pressure. (From Rastogi, N. K., and Niranjan, K., *Journal of Food Science*, 63, 508–11, 1998. With permission.)

pressure treatment (Rastogi and Niranjan 1998; Ade-Omowaye et al. 2001). Eshtiaghi and Knorr (1993) reported that the shear force required for high-pressure-treated potatoes and green beans was decreased as compared to fresh ones and high-pressure treatment could be employed for blanching of potatoes instead of hot water or steam blanching.

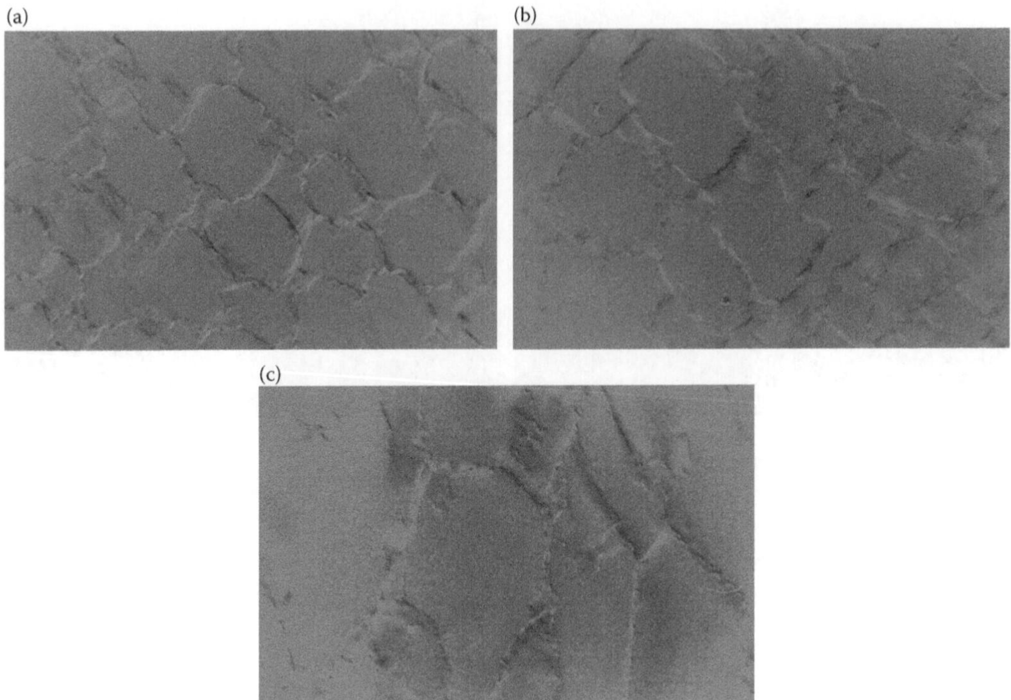

FIGURE 16.4 Microstructures of control and pressure-treated pineapple: (a) control; (b) 300 MPa; (c) 700 MPa (1 cm=41.83 μm). (From Rastogi, N. K., and Niranjan, K., *Journal of Food Science*, 63, 508–11, 1998. With permission.)

16.2.1.3 Carrot

The firmness of a carrot in general was found to be reduced due to the application of high pressure. The rapid loss of firmness was attributed to the disruption of the membrane, which reduces cell turgor pressure. Reductions of 5, 25, and 50% in hardness were observed when fresh carrot samples were subjected to 100, 200, and 300 MPa, respectively, for 2 min at 25°C. The loss of firmness above 300 MPa was reported to be nominal (Araya et al. 2007). Similar results in the case of carrots were endorsed by Michel and Autio (2001). Application of high pressure was reported to result in an increase in displacement with an increase in force, thereby resulting in more deformable material or a rubbery-like texture (Kato et al. 1997; Araya et al. 2007). The microstructure of the carrot showed that the application of high pressure resulted in a loose irregular matrix with buckling, folding, and a reduction of cell-to-cell contact (Figure 16.6). The change in cell structure was further characterized by a decrease in the shape factor and increases in elongation indices, which represented deformation compression or decompression and a shift in cell morphology.

Stute et al. (1996) have shown that the application of high pressure at ambient temperature (25°C) resulted in the softening of vegetables such as carrots, potatoes, and green beans due to the destruction of the cell membrane and a loss of soluble pectin along with partial liberation of cell liquor. Upon high-pressure application PME is also liberated from the cell wall and comes into contact with highly methylated pectin, resulting in the deesterification of pectin, which occurs not only during depressurization but even after release of the pressure, leading to tissue-hardening behavior. It was also shown that high-pressure-processed vegetables (carrots, potatoes, green beans, broccoli, celery, bell peppers, and leafy vegetables, e.g., leeks and Brussels sprouts) no longer soften during cooking and retain textural characteristics as compared to unpressurized vegetables.

Basak and Ramaswamy (1998) demonstrated a dual effect of high pressure on the texture of fruits and vegetables, as characterized by an initial loss in texture, due to the instantaneous pulse

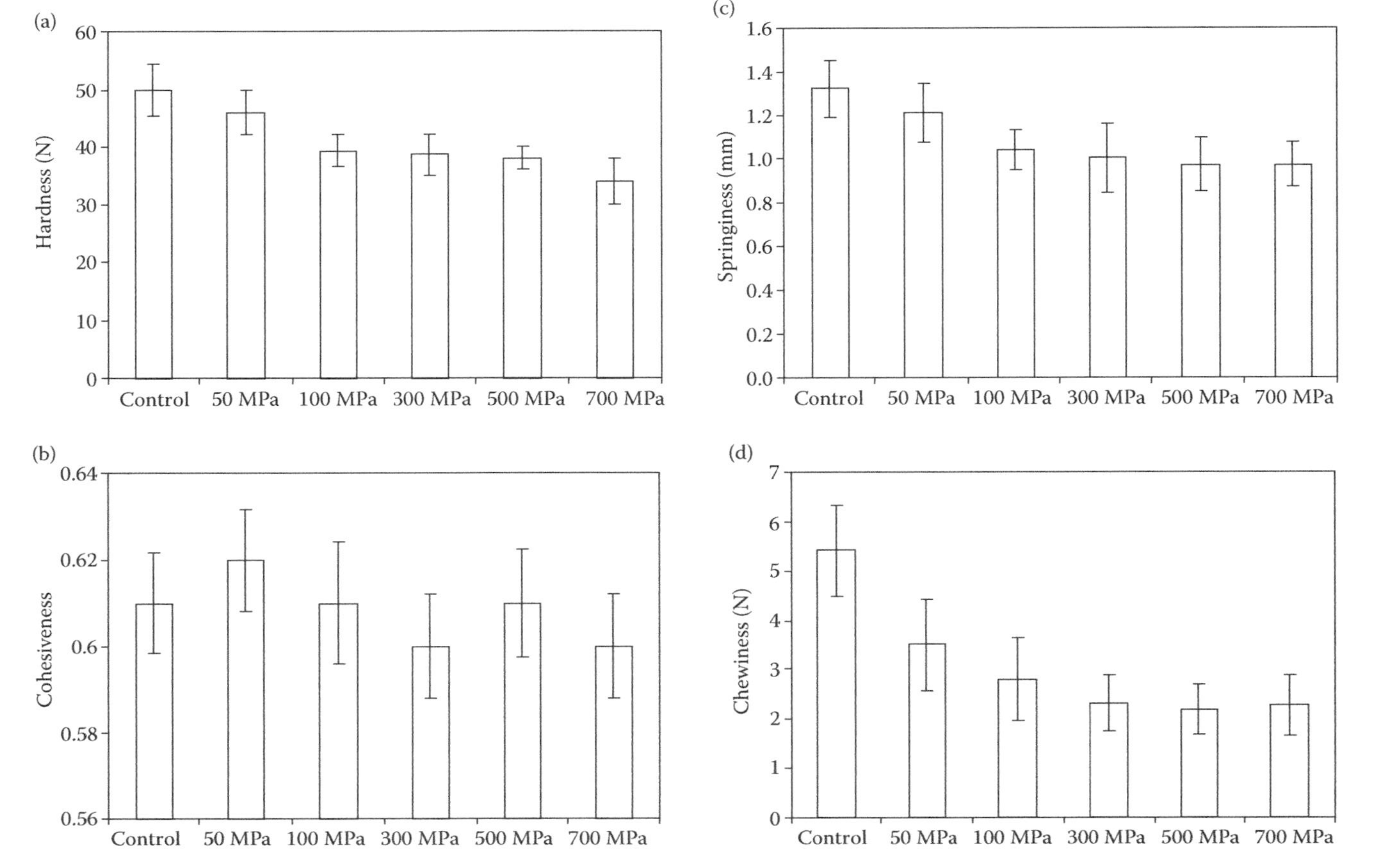

FIGURE 16.5 Effect of pressure on textural parameters: (a) hardness, (b) cohesiveness, (c) springiness, and (d) chewiness (data points with error bars indicate mean ± standard deviation). (From Kingsly, A. R. P., Balasubramaniam, V. M., and Rastogi, N. K., *Journal of Food Process Engineering*, 32, 369–381, 2009. With permission.)

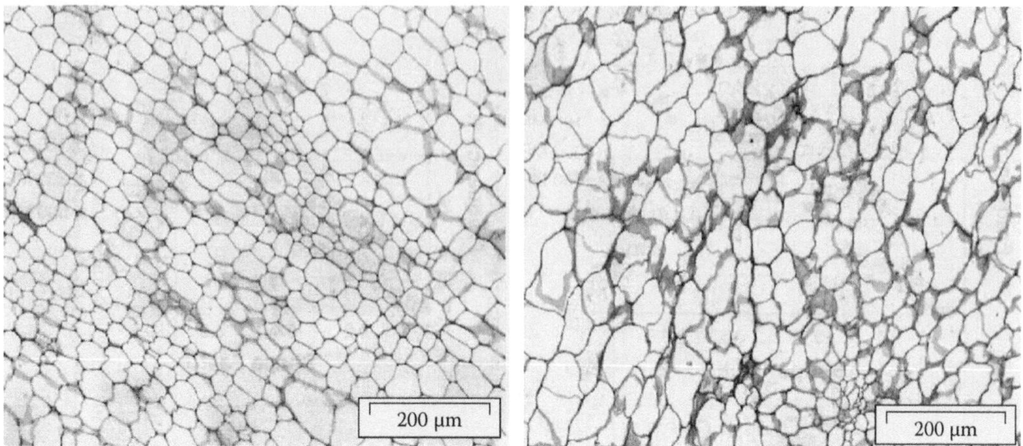

FIGURE 16.6 Comparison of (a) control and (b) high-pressure-treated carrot samples. (From Araya, X. I. T., Hendrickx, M., Verlinden, B. E., Buggenhout, S. V., Smale, N. J., Stewart, C., and Mawson, A. J., *Journal of Food Engineering*, 80, 873–84, 2007. With permission.)

action of the pressure, followed by a more gradual change as a result of pressure-hold. The extent of the initial loss of texture was more prominent at higher pressures and partial recovery of texture was more prominent at lower pressures. The vegetables treated were firmer and brighter than the raw product. The gradual change was described as texture recovery, which reached 100% at low pressure for long processing times (100 MPa, 30 min, 25°C).

16.2.1.4 Green Peas

High pressure can be used as an alternative to blanching in green peas. Pressurization (900 MPa) resulted in a higher retention of ascorbic acid (82%) in comparison with water or microwave blanching (10 and 47% of retention, respectively; Quaglia et al. 1996). The pea samples under high pressure (400 or 900 MPa for 5 or 10 min, vessel temperature 20°C) did not undergo a significant change in firmness (measured as a force registered at the onset of extrusion); however, softening was significantly affected by the treatment in comparison to fresh peas (Figure 16.7). A combination of high pressure with heat treatment (up to 60°C) did not cause any noticeable modification of pea firmness.

16.2.1.5 Mushroom

Matser et al. (2000) studied the effect of high pressure at ambient temperature (25°C) on the texture of mushrooms and compared it with thermal blanching. The application of high pressure was found to reduce the texture of mushrooms; the texture degradation was lower compared to thermal blanching. Pressure treatment in the range of 600–950 MPa resulted in a dark brown color, which may be due to an increase in membrane permeability. High pressure was found to result in the crystallization of phospholipids in the cell membrane, which resulted in damage to the cell membrane. Due to the increased permeability, extracellularly located polyphenoloxidase reacted with phenols, resulting in increased browning during pressure treatment. The browning was less intense in the case of evacuated mushroom pressurization because of the replacement of air in the vacuoles by water, resulting in a large reduction in the concentration of oxygen, which is necessary for enzymatic reactions. Product yield and color were reported to be comparable to thermally blanched products.

16.2.1.6 Onion

High-pressure treatment above 100 MPa induced browning of diced onions, and the rate of the browning reaction was found to increase with an increase in pressure (Butz et al. 1994). This effect

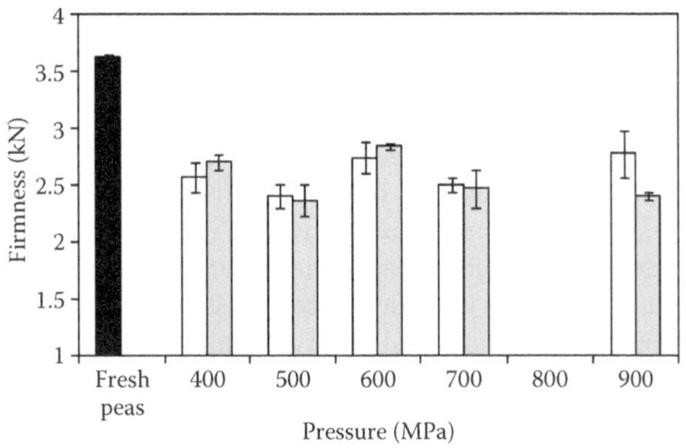

FIGURE 16.7 Firmness of fresh and pressure-treated (5 or 10 min) green peas stored at −18°C for a week and thawed at room temperature. (From Quaglia, G. B., Gravina, R., Paperi, R., and Paoletti, F., *Lebensmittel-Wissenschaft und Technologie*, 29, 552–55, 1996. With permission.)

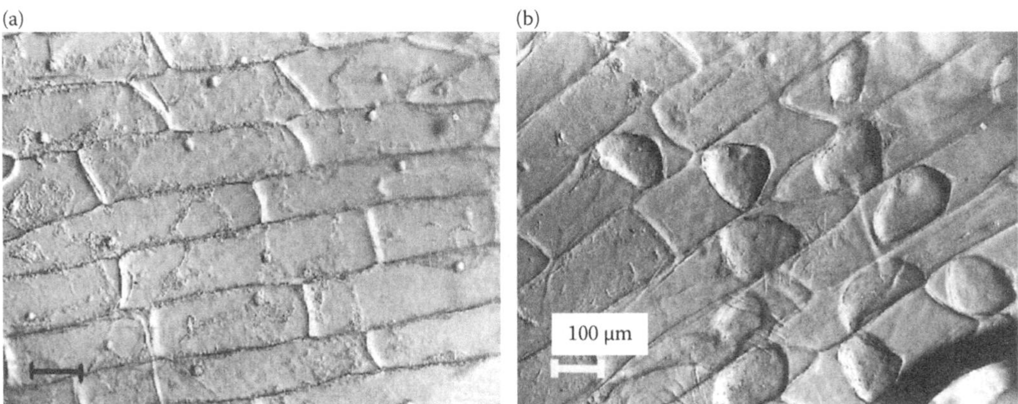

FIGURE 16.8 Epidermal cells of onions transferred into sucrose solution (100 g/l) after high-pressure treatment at 25°C. (a) 100 MPa and (b) 300 MPa. (From Butz, P., Koller, W. D., Tauscher, B., and Wolf, S., *Lebensmittel-Wissenschaft und Technologie,* 27, 463–67, 1994. With permission.)

was due to an enzymatic browning reaction involving polyphenoloxidases, which are predominantly present in the cytosol of plant cells. In intact cells phenolic compounds are confined to vacuoles and are separated from the enzymes by a tonoplast. When the cell and tonoplast are disrupted, phenolic oxidation products are formed. Due to the influence of pressure, onion epidermis cells and cellular components like vacuoles were found to be affected and polyphenoloxidase was no longer separated from its substrate phenol, which is oxidized to orthoquinones and upon polymerization forms a brown pigment. Microscopic studies revealed that the ability of the sample to respond to sucrose was only affected to a minor degree for the samples treated at 100 MPa at 25°C. However, severe damage to the vacuoles of onion epidermis cells and cellular components was found for samples treated at 300 MPa (Figure 16.8).

16.2.1.7 Potato

Rastogi, Angersbach, and Knorr (2003) studied the effect of high pressure on the cell disintegration index (Z_p). The condition of the cell or degree of disintegration was examined by Z_p, which was measured by electrophysical measurement based on electrical impedance analysis (Knorr and Angersbach 1998). In other words, Z_p is an integral parameter, which indicates relative reduction in intact cells and was found to be linearly related to the hardness of the sample (Rastogi, Eshtiaghi, and Knorr 1999). The Z_p values recorded immediately after high-pressure treatment were found to increase with an increase in high hydrostatic pressure applied to a potato sample (Figure 16.9a). The Z_p values of the pressure-treated potato increased with time at atmospheric pressure and equilibrium Z_p values also increased with pressure (Figure 16.9b). The Z_p values of the center layer after high-pressure pretreatment were 0.039, 0.065, and 0.091 at 200, 300, and 400 MPa at 25°C, and reached 0.35, 0.45, and 0.53 within 6 h, respectively (Figure 16.9b). The increase in Z_p values (or tissue softening or loss of texture) following high-pressure treatment was due to the destruction of the cell membranes and the partial liberation of cell substances. Upon high-pressure treatment, PME enzyme was liberated (which was bound to the cell wall) and brought in close contact with its substrate, the methylated pectin. This caused deesterification not only during high-pressure treatment but also after the release of high pressure (standing time). The pressure treatment also caused partial inactivation of PME. This reaction continued with time even after high-pressure treatment and resulted in time-dependent softening of potato tissue. The alteration in pectin resulted in a loss of water and soluble solids (or extractable pectin) after high-pressure treatment (Stute et al. 1996). It is reported in the literature that softening of some fruits and vegetables at atmospheric pressure over a length of time takes place following high-pressure treatment (Basak and Ramaswamy, 1998; Rastogi et al. 2000).

16.2.2 Effect of High Pressure on Texture at Freezing Temperature

A slow freezing rate produces large ice crystals, which affect the cellular structures of fruits and vegetables and result in loss of quality. On the contrary, a rapid freezing rate produces fine ice crystals that cause less damage. It is easy to freeze fruits/vegetable that have regular geometries or are small in size, whereas freezing of large foods involves problems due to thermal gradients. The freezing rate may be high enough at the surface of the food to produce small ice crystals, but inside the product freezing rates reduce significantly, leading to loss of quality. Cryogenic freezing could solve this problem, but it results in freeze-cracking. This occurs due to the rapid freezing of the surface of the food, which resists further volume expansion when the unfrozen inner part of the product undergoes phase transition. Application of high-pressure-shift freezing may avoid this limitation, since the initial ice formation is instantaneous and homogenous throughout the volume of the product, thus eliminating internal stresses. The effects of high-pressure-shift freezing on the texture of fruits and vegetables are discussed in the following sections.

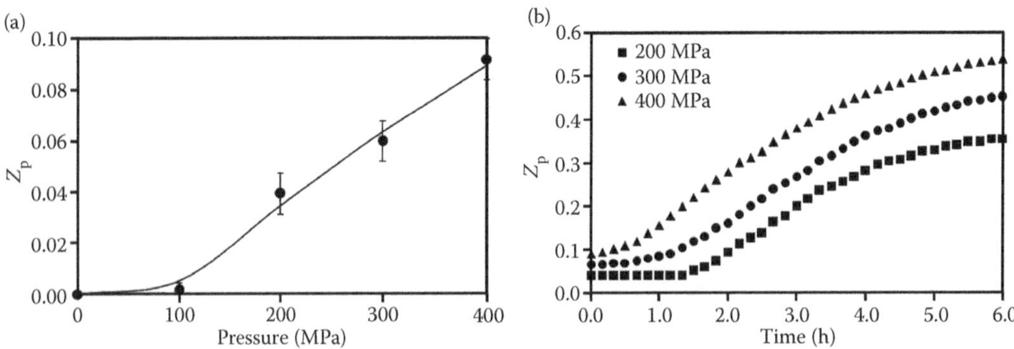

FIGURE 16.9 (a) Variation of cell disintegration index (Z_p) with applied pressure. The Z_p values were measured immediately after high-pressure treatment. (b) Variation of cell disintegration index (Z_p) with time at atmospheric pressure after high-pressure treatment. (From Rastogi, N. K., Angersbach, A., and Knorr, D., In *Transport Phenomena in Food Processing*, CRC Press, USA, 109–121, 2003. With permission.)

16.2.2.1 Peach and Mango

The evolution of temperature at the surface and center of the fruit indicated that the entire volume of the sample reached the initial freezing point at the same time in the case of high-pressure-shift freezing, just before the release of pressure (Figure 16.10). The high level of supercooling during high-pressure-shift freezing of peaches and mangoes leads to uniform and rapid ice nucleation throughout the volume of the sample, which largely maintains its original tissue microstructure in the central zone (Figure 16.11a and b). The cells were arranged adjacently without breakage, demonstrating that the technique could retain microstructure (Otero et al. 2000).

16.2.2.2 Carrot and Chinese Cabbage

When water is frozen at atmospheric pressure, ice I is formed, and the volume increases during phase transition, which is one of the causes of histological damage in tissues with high water content. Conversely, during high pressure, the volume of water decreases with the formation of several kinds of heavy ice polymorphs. Fuchigami, Kato, and Teramoto (1997a) and Fuchigami et al. (1997b) investigated the changes in texture and microstructure of carrots during high-pressure freezing at 100 MPa (ice I), 200 MPa (liquid), 340 MPa (ice III), 400–600 MPa (ice V), and 700 MPa (ice VI) at −20°C. The carrots processed at 100 and 700 MPa had the lowest firmness due to the formation of

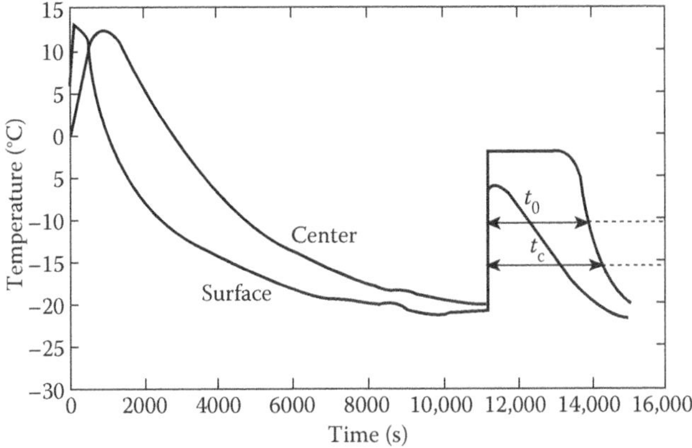

FIGURE 16.10 Thermal histories at the surface and at the center of mango frozen by high-pressure-shift freezing method. (From Otero, L., Martino, M., Zaritzky, N., Solas, M., and Sanz, P. D., *Journal of Food Science*, 65, 466–70, 2000. With permission.)

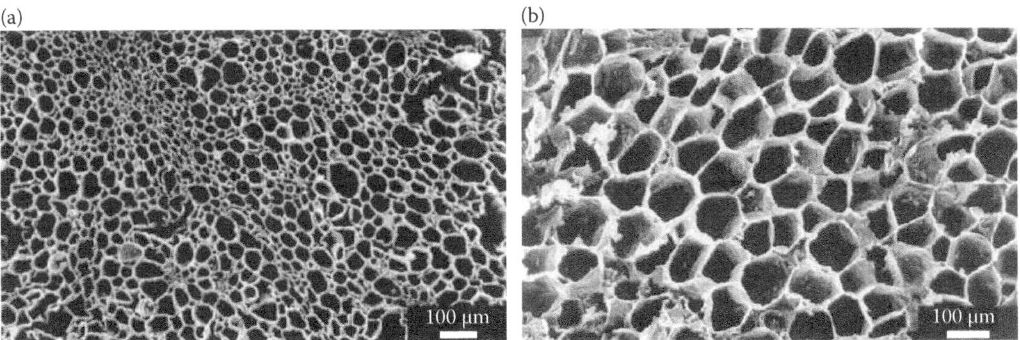

FIGURE 16.11 SEM micrographs of central zone of high-pressure-shift frozen fruit: (a) peach; (b) mango. (From Otero, L., Martino, M., Zaritzky, N., Solas, M., and Sanz, P. D., *Journal of Food Science*, 65, 466–70, 2000. With permission.)

ice I and VI, respectively. Comparison of the textures of frozen carrot indicated that carrots frozen at 200, 340, and 400 MPa had the highest firmness because these samples were frozen quickly only upon release of pressure to atmospheric pressure. Light micrographs indicated that carrots frozen at 100 MPa or above 500 MPa resulted in histological damage (shown by arrows), but it was not obvious for those frozen at 200–400 MPa (Figure 16.12). Similar observations were reported with regard to improvement in the texture of Chinese cabbage in the pressure range from 200 to 400 MPa (Fuchigami, Kato, and Teramoto 1998).

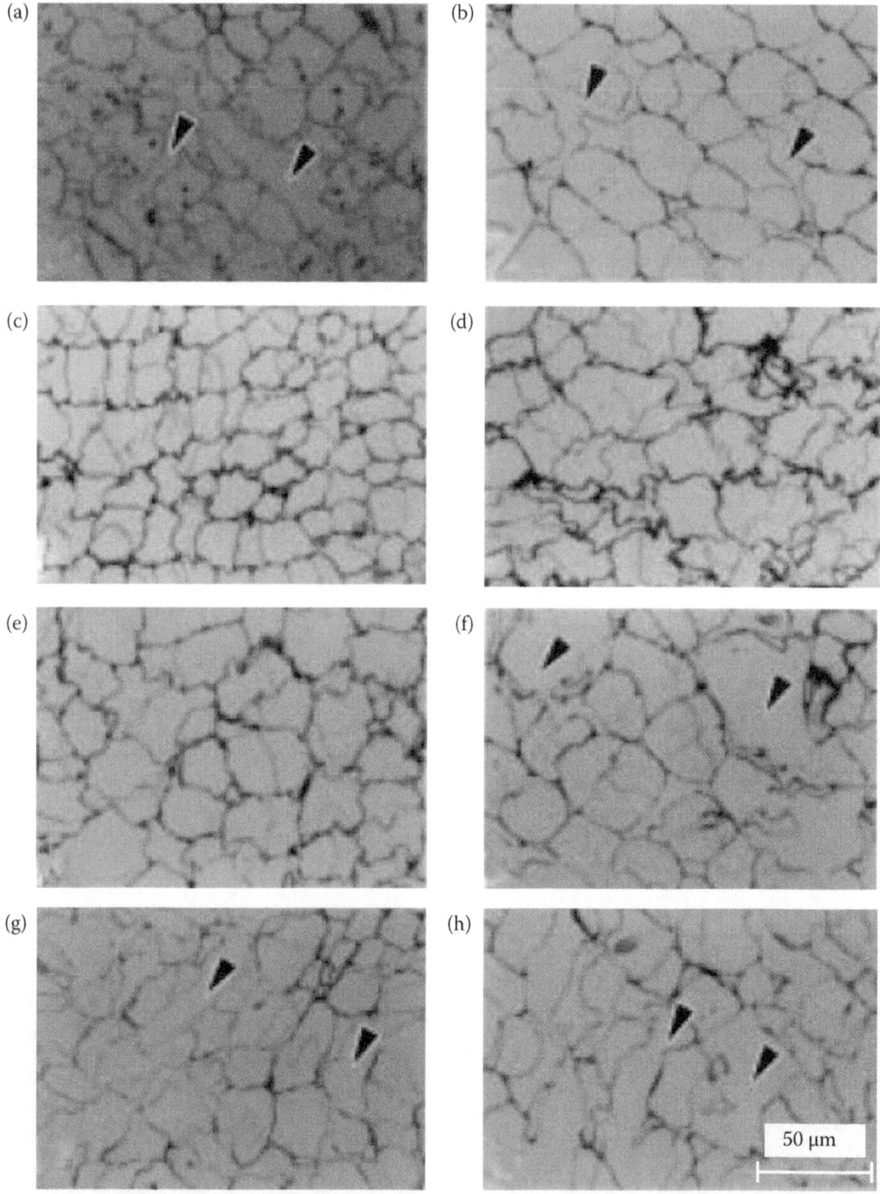

FIGURE 16.12 Light micrographs of high-pressure-frozen raw carrots. (a) Carrots frozen in a freezer (–30°C) at atmospheric pressure (0.1 MPa), (b) high-pressure-frozen carrots at ca. –20°C at 100 MPa, (c) 200 MPa, (d) 340 MPa, (e) 400 MPa, (f) 500 Mpa, and (g) 600 MPa, (h) 700 MPa. Bar: 50 mm. Arrows indicate damage. No significant change in texture in case of (c), (d), and (e) was observed as compared to control sample. (From Fuchigami, M., Kato, N., and Teramoto, A., *Journal of Food Science*, 62 (4), 804–808, 1997. With permission.)

16.2.2.3 Eggplant

Otero et al. (1998) studied the effect of conventional air freezing methods (still air and air blast freezing) and high-pressure freezing on the texture and microstructure of eggplants. The comparison of the micrograms of the central region of still-air-frozen and air-blast-frozen eggplant with a fresh sample indicated cell separation (marked by arrows) and disrupted cell wall (marked with circles) (Figure 16.13a, b, and c). In contrast, the high-pressure-assisted frozen sample had an appearance similar to the fresh sample, all the cells were positioned together, and no cellular damage was evident (Figure 16.13d). There was no substantial difference in the surface and central regions of the sample as nucleation occurred in both at the same time, because there was no time for water translocation and ice formed in intracellular spaces, resulting in no damage to the cell walls. High-pressure-frozen eggplant samples were found to have the highest firmness and the lowest rupture strain and drip loss compared to still-air-frozen and air-blast frozen samples.

16.2.2.4 Broccoli

Pressure-shift-frozen broccoli samples were not acceptable in terms of flavor after 30 days of storage at –20°C, although the texture remained quite firm after the treatment. However, the sensory quality of samples blanched prior to high-pressure freezing was acceptable (Prestamo, Palomares, and Sanz 2004). Hence, it is necessary to blanch the broccoli before it is subjected to high-pressure-freezing treatment (210 MPa, –20°C). The microstructure of raw broccoli showed a well-organized cell with

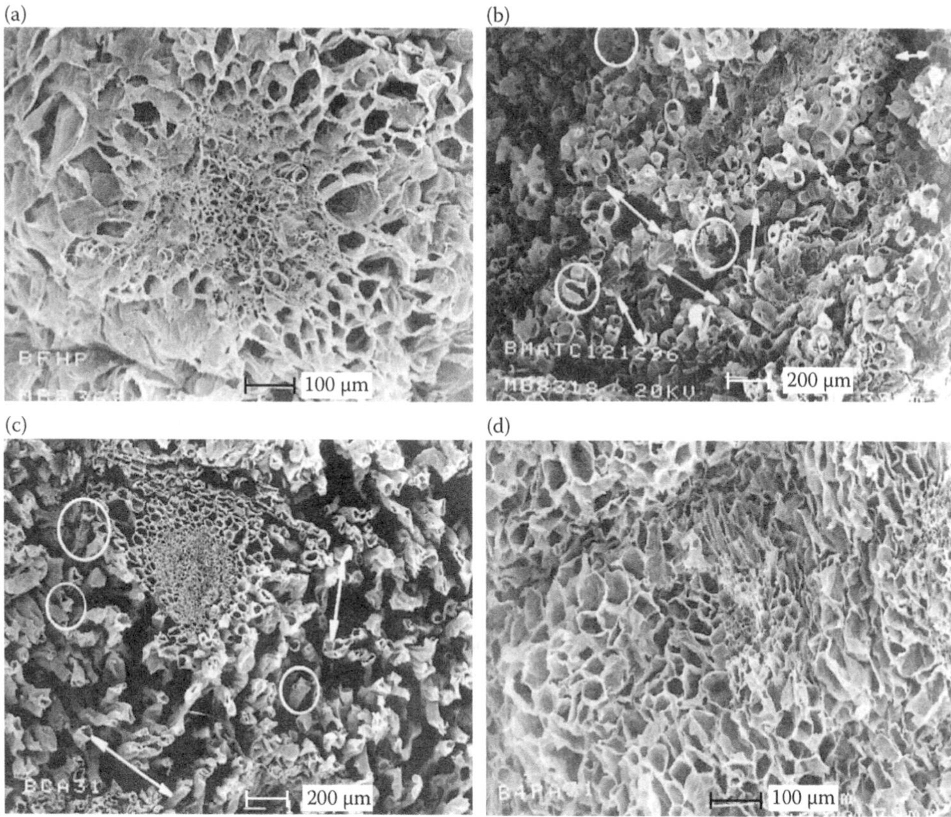

FIGURE 16.13 Scanning electron micrographs of central portions of eggplants: (a) fresh; (b) still-air-frozen air tissue; (c) air-blast frozen tissue; and (d) high-pressure-frozen tissue. Cell separation and disrupted cell wall are marked by arrows and circles, respectively. (From Otero, L., Solas, M. T., Sanz, P. D., de Elvira, C., and Carrasco, J. A., *Zeitschrift fuer Lebensmittel Untersuchung und Forschung A/Food Research and Technology*, 206, 338–42, 1998. With permission.)

defined compartments. After the treatment, the cell structure underwent a considerable degradation, especially the disappearance of the vacuole membrane, which led to a loss of cell turgor. The breakage of the membrane of the vacuole was reported to be responsible for the translucent aspect of the product after blanching or freezing processes. High-pressure-frozen broccoli presented less cell damage, lower drip losses, and better texture than conventionally frozen samples (Fernandez et al. 2006).

16.2.3 Effect of High Pressure on Texture at Higher Temperature

A combination of high pressure and temperature was reported to result in a synergistic effect. The success of the process depends on the temperature used in combination with high pressure for the purpose of pasteurization or sterilization.

The effect of pulsed high-pressure treatment at elevated temperature on green beans was studied by Krebbers et al. (2002). During the treatment, a preheated sample (75°C for 2 min) was subjected to two high-pressure pulses of 1000 MPa of 80 and 30 s at 75°C. Due to adiabatic compression, the maximum temperature during the treatment reached 105°C. The treatment resulted in about 60% retention of the original firmness of the raw beans.

A combination of high pressure with temperature (600 MPa, 80°C) was reported to result in increased preservation of carrot texture as well as microstructure as compared to thermally processed samples. This is due to the nonoccurrence of a β-elimination reaction under high pressure and temperature conditions (Roeck et al. 2008).

Sila et al. (2004) showed that application of higher pressures (200–500 MPa) as a pretreatment can be used to improve the texture of carrots during thermal processing. Combined with calcium chloride infusion, high-pressure pretreatment (>200 MPa) at 60°C for 15 min significantly reduced the rate of thermal softening of carrots. It is most likely that the cause of this effect is higher PME activity at 60°C as well as increased substrate–enzyme contact after high-pressure treatment due to tissue disruptions. The observed textural changes were associated with the degree of methylation of pectin molecules, where a decrease in degree of methylation was marked by an improvement in the texture. However, other confounding factors of either biochemical or physiological origin might have influenced the texture of carrots during this study. Sila et al. (2005) indicated that the best pretreatment is a combination of high-pressure pretreatment (400 MPa for 15 min), a low-temperature pretreatment (60°C), and calcium pretreatment (either before or after the combined high pressure).

Pressure-assisted thermal processing (PATP) has recently emerged as a promising alternative technology for processing low-acid foods. It involves the simultaneous application of elevated pressures (500–700 MPa) and temperatures (90–120°C) to a preheated food (Matser et al. 2004; Rastogi et al. 2007). Compression heating during pressurization and rapid cooling on depressurization helps to reduce the severity of the thermal effects encountered with conventional thermal processing. The technology reportedly reduces process time and preserves food quality, especially texture, color, and flavor, as compared to retorted products.

Nguyen, Rastogi, and Balasubramaniam (2007) studied the effect of PATP and thermal process on the texture and microstructure of carrots. The variation of texture of carrot over a range of pressures (0.1, 500, and 700 MPa) and process temperatures (95, 105, and 121°C) indicated that the initial softening process was characterized by a steep negative slope during processing come-up time followed by a gentler softening phase with holding time (Figure 16.14). It was demonstrated that processing at temperatures up to 105°C, the loss of texture was more pronounced at 700 MPa, whereas beyond 105°C, the loss of texture was more pronounced at 500 MPa. The 121°C–500 MPa treatment had higher texture loss (83.95%) than the 121°C–700MPa treatment (37.91%). The difference in textural loss was attributed to the respective differences in temperature of the sample (86.0°C and 73.0°C) just before pressurization. Exposing the product to a harsher preprocess temperature of 86.0°C might have caused β-elimination, leading to subsequent tissue softening. On the other hand, at process temperatures ≤105°C, the carrot preprocess temperature just before pressurization was less than 67.5°C. At this temperature the action of PME on pectin results in partially demethylated

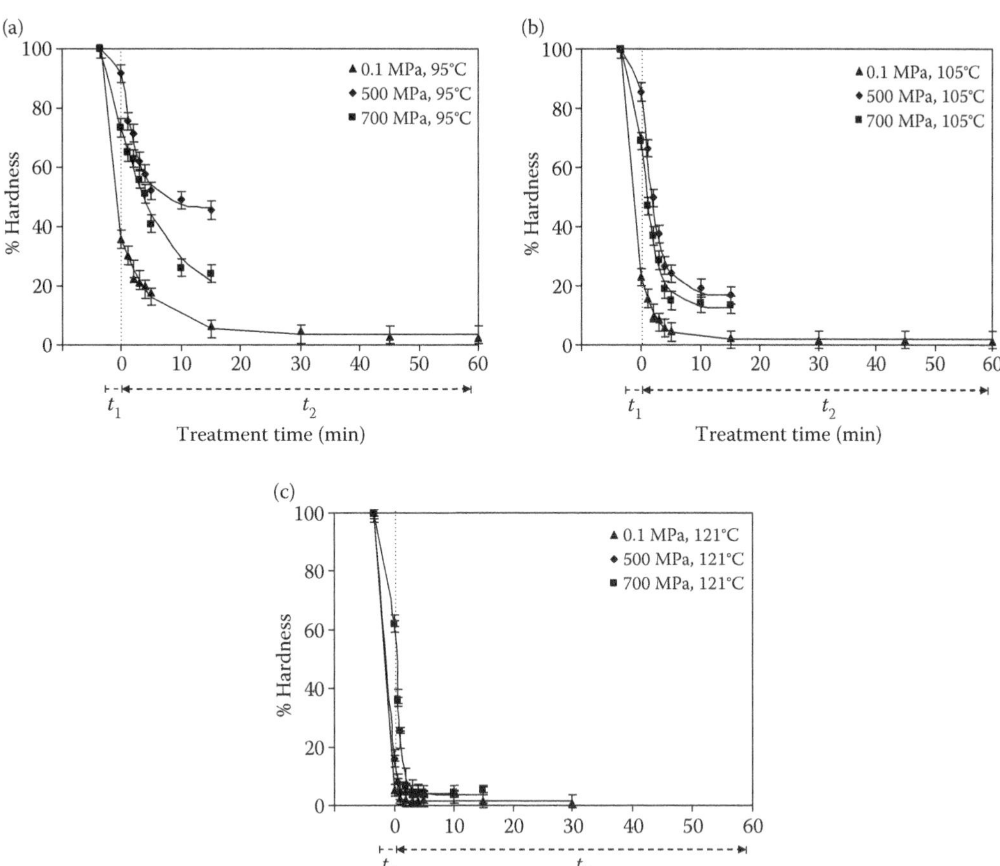

FIGURE 16.14 Textural change of carrot during thermal processing and pressure-assisted thermal processing: (a) 95°C; (b) 105°C; and (c) 121°C at different pressures. t_1 and t_2 are the come-up time and holding time, respectively. (From Nguyen, L. T., Rastogi, N. K., and Balasubramaniam, V. M., *Journal of Food Science*, 72, E264–70, 2007. With permission.)

pectin. This partially demethylated pectin forms a complex with a divalent Ca^{2+} ion, which results in enhancement of the hardness of carrots. Microstructures of cross-sections of raw, PATP (700 MPa, 105°C, 5 min), and thermal-processed (105°C, 0.1 MPa, 30 min) carrot samples showed that the extent of structural damage was limited in the case of PATP, whereas thermal processing transformed intact cell structures of raw carrot to separated and ruptured cells with nondistinct middle lamella because of degradation of the pectinacious material (Figure 16.15).

In order to further improve the textural characteristics of carrots processed by PATP and thermal processing, selected pretreatments such as pressure (100–400 MPa), temperature (50–70°C), calcium chloride (0.5–1.0%), and their combinations were studied (Rastogi, Nguyen, and Balasubramaniam 2008a). Individual pretreatments were effective to a limited extent in improving product quality. Pressure (200 MPa), heat (60°C), and calcium chloride (1.0%) pretreatments increased product hardness by 1.2, 2.0, and 2.4 times after PATP and 2.7, 3.6, and 2.4 times, respectively, after thermal processing (Table 16.1). On the other hand, the combination of these pretreatments was found to be very effective in improving hardness. Combined pretreatments increased the hardness of samples by 9.16 times (from 14.08 to 129.07 N) and 13.22 times (from 4.36 to 57.63 N), respectively, in comparison to untreated PATP or thermally processed samples (Table 16.2). Complementing calcium treatments with mild heating (40–70°C) and application of high pressure is an interesting alternative. The

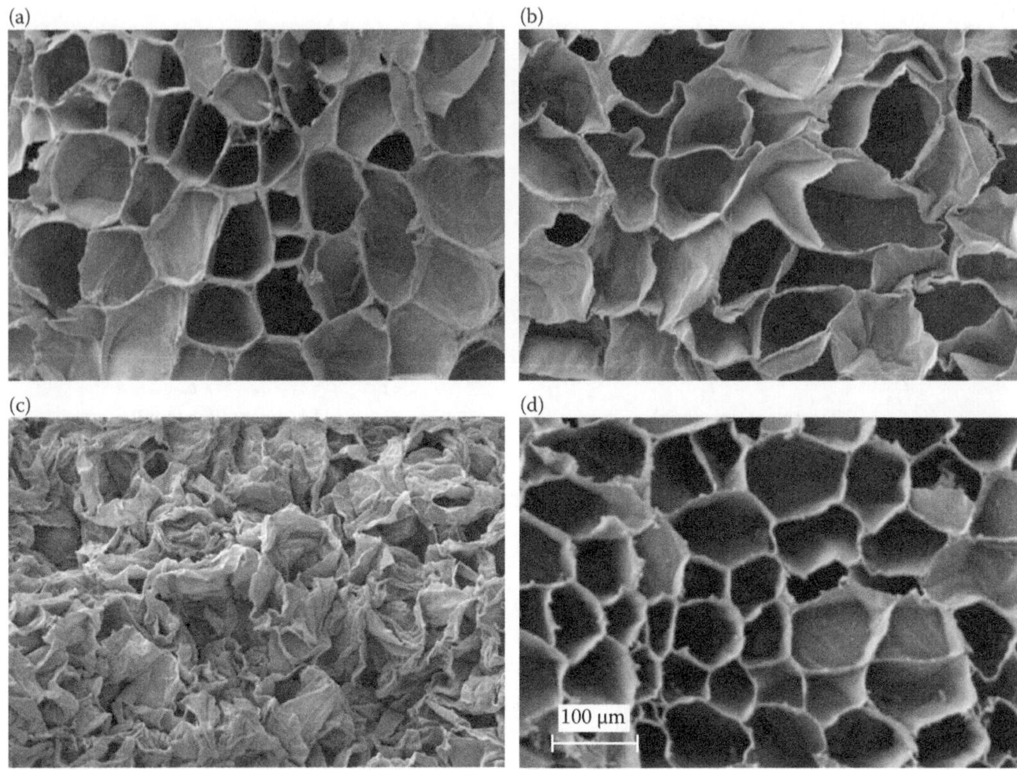

FIGURE 16.15 Microstructures of (a) raw; (b) thermal-processed (105°C, 0.1 MPa, 30 min); and (c) pressure-assisted thermal-processed (700 MPa, 105°C, 5 min) (From Nguyen, L. T., Rastogi, N. K., and Balasubramaniam, V. M., *J. Food Sci.*, 72, E264–70, 2007. With permission.). (d) Pressure-assisted thermal-processed (700 MPa, 105°C, 5 min) carrot samples with combined pretreatment. (From Rastogi, N. K., Nguyen, L. T., and Balasubramaniam, V. M., *Journal of Food Engineering*, 88, 541–47, 2008. With permission.)

TABLE 16.1
Effect of Pressure, Heat, or Calcium Chloride Pretreatment on the Hardness of Carrot Sample During Pressure-Assisted Thermal Processing (PATP) and Thermal Processing (TP)

		Sample Hardness[1] (N)	
Treatment	**Treatment Code**	**PATP**	**TP**
	Control[2]		
PATP	PATP-control	14.08±3.28[a]	–
TP	TP-control	–	4.36±1.24[a]
	High Pressure Pretreatment (Pressure, MPa)		
100	HP100	12.75±2.44[a]	6.24±1.19[b]
200	HP200	16.47±2.24[b]	11.77±1.94[c]
300	HP300	17.69±2.44[b]	13.08±2.24[c]
400	HP400	17.95±2.43[b]	13.51±1.97[c]
	Heat Pretreatment (Temperature, °C)		
50	HT50	13.66±1.94[a]	8.45±1.97[d]
60	HT60	28.07±3.28[c]	15.50±2.28[e]
70	HT70	14.86±1.19[a]	10.25±2.24[d]

TABLE 16.1 (Continued)

Treatment	Treatment Code	Sample Hardness[1] (N) PATP	TP
	Calcium Chloride Pretreatment (% Concentration)		
0.5	$CaCl_2$0.5	24.32±1.94[d]	5.12±0.97[a]
1.0	$CaCl_2$1.0	33.65±3.28[e]	10.39±1.28[f]
1.5	$CaCl_2$1.5	33.91±1.19[e]	10.34±1.24[f]

[1] Data with the same letters in the column do not differ significantly from each other, whereas data with different letters differ significantly at the probability level $p < .05$.

[2] Hardness of the untreated, fresh carrot sample was 190.95±6.56 N.

Source: From Rastogi, N. K., Nguyen, L. T., and Balasubramaniam, V. M., *Journal of Food Engineering*, 88, 541–47, 2008a. With permission.

TABLE 16.2
Effect of Combined Pressure, Heat, or Calcium Chloride Pretreatment on the Hardness of Carrot Sample During Pressure-Assisted Thermal Processing (PATP) and Thermal Processing (TP)

Treatment	Treatment Code	Sample Hardness[1] (N) PATP	TP
	Control[2]		
PATP	PATP-control	14.08±3.28[a]	–
TP	TP-control	–	4.36±1.24[a]
	Combined Pretreatments		
High pressure (200 MPa)+$CaCl_2$ pretreatment (1.0%)	HP200-$CaCl_2$1.0	101.07±6.56[b]	30.63±4.48[c]
High pressure (200 MPa)+heat pretreatment (60°C)	HP200-HT60	32.13±3.50[a]	12.19±3.94[b]
Heat pretreatment (60°C)+$CaCl_2$ pretreatment (1.0%)	HT60-$CaCl_2$1.0	103.08±4.88[b]	20.93±2.37[d]
High pressure (200 MPa)+heat pretreatment (60°C)+$CaCl_2$ pretreatment (1.0%)	HP200-HT60-$CaCl_2$1.0	129.07±8.00[c]	57.63±6.50[e]

[1] Data with the same letters in the column do not differ significantly from each other, whereas data with different letters differ significantly at the probability level $p<0.05$.

[2] Hardness of untreated fresh carrot sample was 190.95±6.56 N.

Source: From Rastogi, N. K., Nguyen, L. T., and Balasubramaniam, V. M., *Journal of Food Engineering*, 88, 541–47, 2008a. With permission.

use of higher temperatures was found to increase the diffusion of calcium into food and improved the quality, especially related to texture maintenance and browning reduction, in comparison with lower temperatures (Rico et al. 2007). Similarly, application of high pressure is known to cause cell permeabilization, thereby increasing diffusion of solids into it (Rastogi and Niranjan 1998; Rastogi et al. 2007). The enhanced diffusion of calcium can make plant tissue firmer by binding to pectin carboxyl groups that are generated through the action of PME. The calcium content of the samples was found to have a positive influence in preserving hardness. Microstructure analysis of PATP carrots indicated that combined high pressure, heat, and calcium pretreatments better preserved cell structure (Figure 16.15d). This combined pretreatment process was shown to be an effective tool in preserving quality of PATP as well as thermally processed carrot.

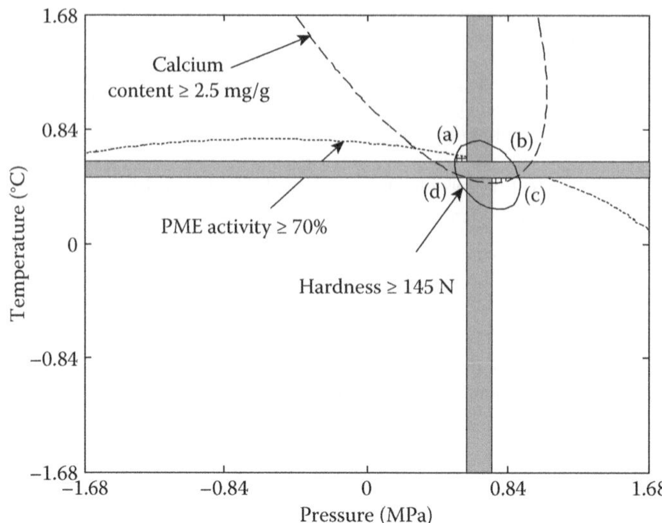

FIGURE 16.16 Superimposed contour plots showing the shaded overlapping area for which hardness ≥145 N and diffused calcium content ≥2.5 mg/g; PME activity ≥70%. (From Rastogi, N. K., Nguyen, L. T., Bo Jiang, and Balasubramaniam, V. M., *Food and Bioproducts Processing*, 2008b. With permission.)

Rastogi et al. (2008b) modeled combined effect of pretreatment pressure and temperature and calcium pretreatments on hardness, PME activity, and diffused calcium content using response surface methodology and studied the complex interaction of variables on the quality of PATP carrot. Combined pretreatment (high pressure 200 MPa+heat pretreatment 60°C+$CaCl_2$ pretreatment 1.0%) was found to increase hardness of the samples approximately ninefold (from 14.08 to 129.07 N) in comparison to untreated PATP samples (Rastogi et al. 2008b). This study was based on varying one factor at a time. In order to fully exploit the combined effect of the pretreatments, these conditions were optimized and the complex interaction of different pretreatments on the hardness of PATP samples was evaluated. In order to deduce workable optimum conditions, a graphical technique, which involves superimposition of contour plots of responses and locating the overlapping region, was used. An acceptable compromise was made following the criteria for hardness ≥145 N, calcium content ≥2.5 g/mg, and PME activity ≥70%. The contour plots were superimposed and regions that best satisfied all the constraints were selected as optimum conditions. The overlapping area was recommended as a practical optimum zone for pretreatment pressure (270–288 MPa) and pretreatment temperature (57.7–59.3°C, Figure 16.16).

16.3 CONCLUSIONS

Changes in textural properties of food products during processing are required for product development, quality control, sensory evaluation, and design as well as evaluation of process equipment. Application of high pressure to fruits and vegetables changes texture due to liquid infiltration and gas displacement. The changes such as collapse of air pockets and shape distortion result in shrinkage. In many cases, pressure-treated vegetables did not soften during subsequent cooking, which was attributed to the action of PME, which was only partially inactivated by pressure. High-pressure freezing retained the texture of fruits and vegetables. It can be concluded that high pressure is a potential nonthermal alternative in the processing and preservation of fruits and vegetable processing that allows better retention of qualities.

ACKNOWLEDGMENT

The author is grateful to Dr. V. Prakash, Director, Central Food Technological Research Institute, Mysore, India, for constant encouragement. Thanks are also due to Dr. K. S. M. S. Raghavarao, Head, Department of Food Engineering, for support.

REFERENCES

Ade-Omowaye, B. I., Rastogi, N. K., Angersbach, A., and Knorr, D. 2001. Effect of high pressure or high electrical field pulse pretreatment on dehydration characteristics of paprika, *Innovative Food Science and Emerging Technologies*, 2 (1): 1–7.

Araya, X. I. T., Hendrickx M., Verlinden, B. E., Buggenhout, S. V., Smale, N. J., Stewart, C., and Mawson, A. J. 2007. Understanding texture changes of high pressure processed fresh carrots: A microstructural and biochemical approach. *Journal of Food Engineering*, 80, 873–84.

Basak, S., and Ramaswamy, H. S. 1998. Effect of high pressure processing on the texture of selected fruits and vegetables. *Journal of Texture Studies*, 29 (5): 587–601.

Butz, P., Edenharder, R., Fernandez Garcia, A., Fister, H., Merkel, C., and Tauscher, B. 2002. Changes in functional properties of vegetables induced by high pressure treatment. *Food Research International*, 35: 295–300.

Butz, P., Koller, W. D., Tauscher, B., and Wolf, S. 1994. Ultra high pressure processing of onions: Chemical and sensory changes. *Lebensmittel-Wissenschaft und Technologie*, 27:463–67.

Dornenburg, H., and Knorr, D. 1993. Cellular permeabilisation of cultured plant tissue by high electric field pulse and ultra high pressure for recovery of secondary metabolites. *Food Biotechnology*, 7:35–48.

Eshtiaghi, M. N. and Knorr, D. 1993. Potato cube response to water belching and high hydrostatic pressure. *Journal of Food Science*, 58:1371–74.

Eshtiaghi, M. N., Stute, R., and Knorr, D. 1994. High pressure and freezing pretreatment effects on drying, rehydration, texture and color of green beans, carrots and potatoes. *Journal of Food Science*, 59:1168–70.

Farr, D. 1990. High pressure technology in food industry. *Trends in Food Science and Technology*, 1:14–16.

Fernandez, P. P., Prestamo, G., Otero, L., and Sanz, P. D. 2006. Assessment of cell damage in high-pressure-shift frozen broccoli: Comparison with market samples. *European Food Research and Technology*, 224:101–107.

Fuchigami, M., Kato, N., and Teramoto, A. 1997a. High-pressure-freezing effects on textural quality of carrots. *Journal of Food Science*, 62 (4): 804–808.

———. 1998. High-pressure-freezing effects on textural quality of Chinese cabbage. *Journal of Food Science*, 63 (1): 122–25.

Fuchigami, M., Miyazaki, K., Kato, N., and Teramoto, A. 1997b. Histological changes in high-pressure-frozen carrots. *Journal of Food Science*, 62 (4): 809–12.

Grant, G. T., Morris, E. R, Rees, D. A., Smith, P. J. C. and Thom, D. 1973. Biological interactions between polysaccharides and divalent cations. The egg-box model. *FEBS Letters*, 32:195–98.

Hayashi, R. 1990. Application of high pressure to processing and preservation: Philosophy and development. In *Engineering and Food*, Eds. Spiess, W. E. L., and Schubert, H, 815–26. London: Elsevier Applied Science.

Hendrickx, M., Ludikhuyze, L., Broeck Van den I., and Weemaes, C. 1998. Effect of high pressure on enzymes related to food quality. *Trends in Food Science and Technology*, 9:197–203.

Hite, B. H. 1899. The effect of pressure in the preservation of milk. *Bulletin of West Virginia University Agricultural Experiment Station*, 58:15–35.

Hite, B. H., Giddings, N. J., and Weakly, C. E. 1914. The effects of pressure on certain microorganisms encountered in the preservation of fruits and vegetables. *Bulletin of West Virginia University Agricultural Experiment Station*, 146:1–67.

Hugas, M., Garriga, M., and Monfort, J. M. 2002. New mild technologies in meat processing: High pressure as a model technology. *Meat Science*, 62 (3): 359–71.

Kato, N., Teramoto, A., and Fuchigami, M. 1997. Pectic substance degradation and texture of carrot as effected by pressurization. *Journal of Food Science*, 62:359–62, 398.

Kingsly, A. R. P., Balasubramaniam, V. M., and Rastogi, N. K. 2009. Effect of high pressure processing on texture and drying behavior of pineapple. *Journal of Food Process Engineering*, 32: 369–81.

Knorr, D., and Angersbach, A. 1998. Impact of high electric field pulses on plant membrane permeabilization. *Trends in Food Science and Technology*, 9: 185–191.

Knorr, D. 1995. Hydrostatic pressure treatment of food: microbiology. In *New Methods for Food Preservation*, Ed. Gould, G. W., 159–75. London: Blackie Academic and Professional.

Krebbers, B., Matser, A. M., Koets, M., and van den Berg, R. W. 2002. Quality and storage stability of high-pressure preserved green beans. *Journal of Food Engineering*, 54 (1): 27–33.

Matser, A. M., Knott, E. R., Teunissen, P. G. M., and Bartels, P. V. 2000. Effects of high isostatic pressure on mushrooms. *Journal of Food Engineering*, 45:11–16.

Matser, A. M., Krebbers, B., van den Berg, R.W., and Bartels, P. V. 2004. Advantages of high pressure sterilization on quality of food products. *Trends in Food Science and Technology*, 15 (2): 79–85.

Mermelstein, N. H. 1997. High pressure processing reaches the US market. *Food Technology*, 51:95–96

Michel, M., and Autio, K. 2001. Effect of high pressure on protein and polysaccharide based structure. In *Ultra High Pressure Treatment of Foods*, Eds. Hendrix, M. E. G., and Knorr, D., 189–241. New York: Kluwer.

Nguyen, L. T., Rastogi, N. K., and Balasubramaniam, V. M. 2007. Evaluation of instrumental quality of pressure-assisted thermally processed carrots. *Journal of Food Science*, 72:E264–70.

Otero, L., Martino, M., Zaritzky, N., Solas, M., and Sanz, P. D. 2000. Preservation of microstructure in peach and mango during high pressure shift freezing. *Journal of Food Science*, 65 (3): 466–70.

Otero, L., Solas, M. T., Sanz, P. D., de Elvira, C., and Carrasco, J. A. 1998. Contrasting effects of high pressure assisted freezing and conventional air freezing on eggplant tissue microstructure. *Zeitschrift fuer Lebensmittel Untersuchung und Forschung A/Food Research and Technology*, 206 (5): 338–42.

Prestamo, G., and Arroyo, G. 1998. High hydrostatic pressure effects on vegetable structure. *Journal of Food Science*, 63:878–81.

Prestamo, G., Palomares, L., and Sanz, P. 2004. Broccoli (*Brassica oleracea*) treated under pressure-shift freezing process. *European Food Research and Technology*, 219 (6): 598–604.

Quaglia, G. B., Gravina, R., Paperi, R., and Paoletti, F. 1996. Effect of high pressure-treatments on peroxidase activity ascorbic acid content and texture in green peas. *Lebensmittel-Wissenschaft und Technologie*, 29:552–55.

Rastogi, N. K., Angersbach, A., and Knorr, D. 2003. Combined effect of high hydrostatic pressure pretreatment and osmotic stress on mass transfer during osmotic dehydration. In *Transport Phenomena in Food Processing*, Ed. Welti-Chanes, J., 109–21. Boca Ration, FL: CRC Press.

Rastogi, N. K., Angersbach, A., Niranjan, K., and Knorr, D. 2000. Rehydration kinetics of high pressure treated and osmotically dehydrated pineapple. *Journal of Food Science*, 65 (5): 838–41.

Rastogi, N. K., Eshtiaghi, M. N., and Knorr, D. 1999. Accelerated mass transfer during osmotic dehydration of high intensity electrical field pulse pretreated carrots. *Journal of Food Science*, 64:1020–23.

Rastogi, N. K., Nguyen, L.T., and Balasubramaniam, V. M. 2008a. Pretreatment effects on carrot texture during thermal and pressure-assisted thermal processing. *Journal of Food Engineering*, 88:541–47.

Rastogi, N. K., Nguyen, L. T., Bo Jiang, and Balasubramaniam, V. M. 2008b. Optimization of pretreatments conditions for carrot during pressure-assisted thermal processing by response surface methodology. *Food and Bioproducts Processing*, 2009. DOI 10.1007/s11947-008-0130-6.

Rastogi, N. K., and Niranjan, K. 1998. Enhanced mass transfer during osmotic dehydration of high pressure treated pineapple. *Journal of Food Science*, 63 (3): 508–11.

Rastogi, N. K., Raghavarao, K. S. M. S., Balasubramaniam, V. M., Niranjan, K., and Knorr, D. 2007. Opportunities & challenges in high pressure processing of foods. *Critical Reviews in Food Science and Nutrition*, 47 (1): 69–112.

Rastogi, N. K., Subramanian, R., and Raghavarao, K. S. M. S. 1994. Application of high pressure technology in food processing. *Indian Food Industry*, 13 (1): 30–34.

Rico, D., Martin-Diana, A. B., Henehan, G. T. M., Frias, J., Barat, J. M. and Barry-Ryan, C. 2007. Improvement in texture using calcium lactate and heat-shock treatments for stored ready-to-eat carrots. *Journal of Food Engineering*, 79:1196–206.

Roeck, A. D., Sila, D. N., Duvetter, T., Leoy, A. V., and Hendrickx, M. E. G. 2008. Effect of high pressure/high temperature processing on cell wall pectic substances in relation to firmness of carrot tissue. *Food Chemistry*, 107:1225–35.

Sila, D. N., Smout, C., Vu T. S., and Hendrickx, M. E. 2004. Effects of high-pressure pretreatment and calcium soaking on the texture degradation kinetics of carrots during thermal processing. *Journal of Food Science*, 69 (5): E205–11.

Sila, D. N., Smout, C., Vu, S. T., Van Loey, A. M., and Hendrickx, M. 2005. Influence of pretreatment conditions on the texture and cell wall components of carrots during thermal processing. *Journal of Food Science*, 70 (2): E85–91.

Stute, R., Eshtiagi, M., Boguslawski, S., and Knorr, D. 1996. High pressure treatment of vegetables. In: *High Pressure Chemical Engineering*, Eds. von Rohr, P. R., and Trepp, C. 271–76. New York: Elsevier Science.

Tangwongchai, R., Ledward, D. A., and Ames, J. A. 2000. Effect of high pressure treatment on the texture of cherry tomato. *Journal of Agriculture and Food Chemistry*, 48:1434–41.

Vu, T. S., Smout, C., Sila, D. N., Ly Nguyen, B., Van Loey, A. M. L., and Hendrickx, M. 2004. Effect of preheating on thermal degradation kinetics of carrot texture. *Innovative Food Science and Emerging Technology*, 5:37–44.

17 Pressure-Shift Freezing Effects on Texture and Microstructure of Foods

Songming Zhu
Zhejiang University

Hosahalli S. Ramaswamy
McGill University

CONTENTS

17.1 INTRODUCTION

Freezing is a process of bringing down the temperature of food below its freezing point; it is generally stored at temperatures below –5°C, more commonly –10°C or –18°C. Freezing is an ancient method of preservation, but today refrigerated and frozen foods occupy more than 50% of the floor space reserved for processed foods. Freezing results in the transformation of liquid water to solid ice. Liquid water is essential for the microbiological, enzymatic, and chemical activity in foods that reduces its storage life. The freezing of water to ice basically puts a stop to most of these activities. Freezing is an ice crystallization process. Since the bulk of the liquid phase in foods is water, this means that upon freezing, water transforms to ice crystals. Such transformations generally take place at freezing point, which is well defined for pure liquids (e.g., 0°C for water). In foods, in addition to water, there are numerous other constituents in varying amounts. For this reason, food never freezes at 0°C. Further, it can never have a defined single freezing point. Depending upon the nature of the food and its composition, one can identify an initial freezing point. But as the food temperature is lowered to this point and as water starts to freeze, the solute concentration in residual unfrozen water begins to increase, which further depresses the freezing point of the residual water fraction. So the freezing point progressively decreases. Also, in theory, one cannot freeze all the water present in a food system, because there is always a certain fraction of it that is attached to the food constituents so tightly that it will not freeze. This constitutes "bound" water or unfreezable water. It is generally recognized that the bulk of ice crystallization takes place in the temperature range –1 to –5°C. This temperature zone is therefore termed the zone of maximum ice crystal formation.

17.2 CONVENTIONAL FREEZING

Typically, in a freezing process, there will be an initial drop in the temperature of the product until it reaches its initial freezing point. Then, the temperature of the product remains relatively steady as the latent heat is removed. For food products, the freezing point slowly drops until the majority of water is frozen as ice, and then drops more rapidly as the ice temperature is lowered further. The three distinct zones are commonly referred to as the precooling period, phase change period, and the tempering period. Under carefully controlled freezing conditions, sometimes the product temperature falls below the initial freezing point without the change of phase taking place. This behavior is called supercooling. Under a supercooled state, the water remains in the liquid state. Following this, the temperature rises to the initial freezing point and the conventional freezing continues.

The ice crystallization process is believed to consist of two phases: nucleation and crystal growth. At any given time, both nucleation and crystal growth may take place at the same time, but for the latter to occur there has to be an initial nucleation. The nucleation is the association of water molecules into a tiny ordered particle that can serve as an active site for the crystal growth. The crystal growth implies orderly addition of more water molecules to the nucleus. Both these processes can take place simultaneously, after the initial nucleation, but their relative occurrence is dependent on the temperature and rate of freezing (Ramaswamy and Marcotte 2005).

17.2.1 Nucleation

There are two types of nucleation: homogeneous and heterogeneous nucleation. Homogeneous nucleation is possible only with extremely pure water and is only of academic interest. This nucleus is formed by chance orientation of a suitable number of water molecules into an orderly site and serves as a nucleus for the growth of ice crystals. Heterogeneous nucleation is more common with foods. In this case, nuclei are formed adjacent to suspended particles, impurities, and even on the walls of the containers.

17.2.2 Crystal Growth

Unlike nucleation, crystal growth can begin at temperatures just below the freezing point. It requires the initial nucleation. At temperatures near the freezing point, crystal growth is favored over the creation of more nuclei. So the few nuclei formed around the heterogeneous materials will begin to grow in size under these conditions. Here also, as the temperature is lowered, the rate of crystal growth increases. But at lower temperatures, nucleation overtakes crystal growth with the result that many more tiny nuclei are formed before they begin to agglomerate into larger sizes.

17.2.3 Crystal Size and Texture

At near freezing temperatures, there are very few nuclei and if this condition is maintained for a longer period of time, rather than forming more and more nuclei, ice crystals are added to the few nuclei sites present and grow enormously in size. This is what happens during a slow freezing operation. However, if the product temperature is lowered quickly, many nuclei are formed that only grow to a limited extent. This is the situation in the quick freezing processes. The location, number, and size of the ice crystals formed determine the resulting texture of the frozen-thawed product. It is generally accepted that slow freezing results in large ice crystal formation exclusively in the extracellular locations, which actually tend to squeeze the cell structure as they grow. Hence, upon thawing, they leave a product with severe textural breakdown. On the other hand, if the freezing is carried out quickly, it is believed that numerous tiny ice crystals are formed, both intra- and extracellular, which do not grow appreciably in size. Hence they do not significantly crush the cell structure, which should therefore retain a better texture upon thawing (Ramaswamy and Marcotte 2005).

Ice crystal formation is, therefore, critical to preservation of the textural quality of frozen foods. In the traditional freezing processes, ice crystals are formed by a stress-inducing ice front moving from the surface to the center of food samples. Due to the limited conductive heat transfer in foods, the traditional freezing process is generally slow, resulting in large extracellular ice crystal formation, causing quality loss in frozen foods (Fennema 1973). On the other hand, rapid freezing (e.g., cryogenic freezing) can reduce ice-crystal size near the product surface due to the large temperature difference and may cause mechanical cracking.

Elevated pressure allows water to be in the liquid phase at subzero temperature. A rapid depressurization allows liquid water to remain in a metastable state for some time, thus creating a large degree of supercooling (the difference between the actual temperature of the liquid sample and the corresponding solid–liquid equilibrium temperature at given pressure) to drive homogeneous ice nucleation in the whole sample. Pressure-shift freezing (PSF) is a novel technique to produce small and uniform ice crystals and to reduce histological damage in frozen foods. Pressure markedly influences water transition, and the use of HP technology has distinct potentialities for improving the kinetics of freezing or thawing and the characteristics of ice crystals.

17.3 HIGH-PRESSURE (HP) PROCESSING

New and alternative food processing methods and/or novel combinations of existing methods are continually being investigated by the industry in pursuit of producing better quality foods more economically. High-pressure (HP) processing offers one such technique for pasteurization of foods or to modify their functional properties. HP technology has traditionally been used in nonfood areas on a relatively large scale for the production of ceramic, carbide, and steel components, and super alloys where inert gases or water are used as the pressure medium. The application of hydrostatic pressure to food results in the instantaneous and uniform transmission of pressure throughout the product independent of the product volume. The treatment is unique in that the effects do not follow a concentration gradient or change as a function of time. Other advantages include the lack of chemical additives and the operation at low or ambient temperatures so that the food is essentially fresh-like. High pressure is a physical treatment that will not cause extensive chemical changes in the food system. Once the desired pressure is reached, the pressure can be maintained without the need for further energy input. Liquid foods can be pumped to treatment pressures, held, and then decompressed aseptically for filling as with other aseptic processes. Early efforts to find uses for high hydrostatic pressure technology in the food industry have been made in Japan. The first HP-processing line was introduced in Japan for jam manufacture in 1990 and has since been upgraded to fruit yogurts, jellies, salad dressings, and fruit sauces (Knorr 1999). Commercial products have now been appearing worldwide.

HP processes have attracted research and development interests in the food industry to improve the quality of food products. In general, HP technology is potentially of interest in three areas of food processing, namely food functionality modification, preservation, and phase-change processes (Cheftel, Levy, and Dumay 2000; Kalichevsky, Knorr, and Lillford 1995; Knorr, Schlueter, and Heinz 1998; LeBail et al. 2002). There are two basic principles that underlie the effects of HP: (1) The Le Chatelier principle: any phenomenon (phase transition, chemical reaction, etc.) accompanied with a decrease in volume will be enhanced by pressure increase (Cheftel et al. 2000). (2) The

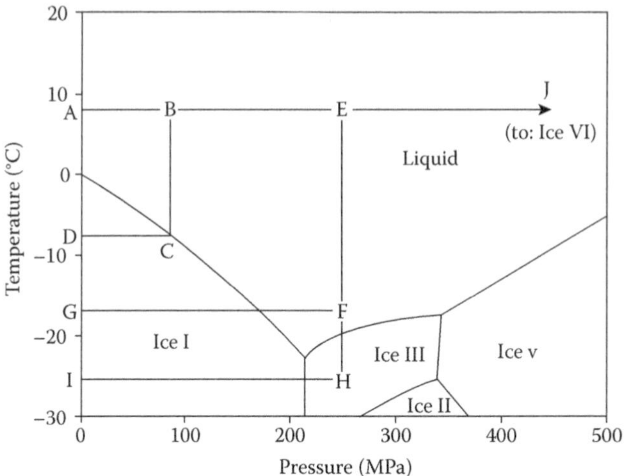

FIGURE 17.1 Representation of the phase diagram of water under pressure and its application possibilities for high-pressure processing of foods: (1) pressure-assisted freezing (ABCD), (2) pressure-assisted thawing (i.e., high-pressure thawing) (DCBA), (3) subzero storage without freezing (ABEF), (4) pressure-shift freezing (ABEFG), (5) pressure-induced thawing (GFEBA), (6) freezing to ice III (ABEFHI), (7) thawing of ice III (IHFEBA), (8) freezing above 0°C to ice VI (ABEJ).

isostatic principle: Pressure is instantaneously and uniformly transmitted independent of the size and the geometry of the food.

HP has a dominating effect on the solid-liquid phase diagram of water (Figure 17.1). These effects have been well documented by Bridgman (1912), but little attention has been paid to their potential applications in food processing until recent years. Ice I (the most common form of ice crystals) is a unique type of ice that has a density lower than liquid water, having a volume increase of about 9% during freezing at 0.1 MPa (atmospheric pressure) and about 13% at 193 MPa (Bridgman 1912). Due to the volume increase from water to ice I, the freezing point of water is depressed under high pressure according to the Le Chatelier principle. The phase change temperature of water decreases from 0°C to a minimum of –21°C at 210 MPa and a reversing effect is shown when the pressure is increased above 210 MPa (Figure 17.1) (Bridgman 1912). It should be noted that the presence of solutes in the aqueous phase would shift the phase diagram of water, further lowering the ice I melting temperature. This effect will be enhanced with increasing solute concentration (Fennema 1973; Cheftel et al. 2000).

The effects of HP on the phase transition of water (Figure 17.1) provide the scientific basis for novel HP processing of foods, such as pressure-assisted freezing, pressure-assisted thawing (i.e., HP thawing), subzero storage without freezing, PSF, pressure-induced thawing, freezing to ice III, thawing of ice III, and freezing above 0°C to ice VI (pressure > 626 MPa). Research interests in HP phase-change processes date from the 1980s, particularly with the commercialization of HP processed products in Japan and more recently in the United States and Europe (Cheftel et al. 2000; LeBail et al. 2002). Among these interesting applications, PSF and HP thawing are two of the most important techniques that are of great potential benefit in the food industry for improving the quality of frozen foods.

17.4 THERMAL PHYSICS AND PHASE TRANSITION UNDER PRESSURE

17.4.1 Compression Heating

Thermal effects are always associated with HP processing since compression results in adiabatic heating of the product. It is necessary to understand the pressure-induced thermal behavior of foods in order to successfully apply HP processing. Compression heating depends on the nature and composition of food (Table 17.1). Pressure can be instantaneously and uniformly distributed at all points

TABLE 17.1
Approximate Temperature Change due to Adiabatic Compression

Food Substance	Temperature Increase (°C/100 MPa)
Water	3
Salmon fish	3
Extracted beef fat	8
Olive oil	9
Ethanol	8
Propylene glycol	6
Mashed potato	3
Orange juice	3
2%-fat milk	3

of foods in a pressure chamber, but temperature may not be uniform because of differences in compression heating of different components and the subsequent loss of heat to the cooler vessel, which will not undergo compression heating. Many factors can affect temperature variation and distribution in foods during HP processing, including HP medium, sample size and components, initial temperature, HP equipment (chamber volume), etc. Again, understanding and quantification of this is essential for the successful application of HP processing for sterilization.

17.4.2 Phase Transition of Pure Water

Bridgman (1912) developed an apparatus to measure phase transition and volume change of pure water under pressure, and then to estimate latent heat of water using the Clapeyron equation. Five crystalline forms of ice (type I to IV, and VI) were observed by Bridgman (1912), while today at least 10 ice polymorphs are known in the temperature range between –100 and 50°C and the pressure range between 0 and 2.4 GPa (Cheftel et al. 2000). The solid form that exists at atmospheric pressure (ice I) has a lower density than liquid water, in contrast to other ice types. In principle, no solid phase can exist in the liquid domain, but the liquid phase can remain for some time (as a metastable state) in a crystalline domain (i.e., supercooling) (Bridgman 1912; Cheftel et al. 2000). Following Bridgman's (1912) work, the phase transition of water has been well documented.

17.4.3 Phase Transition of Water in Foods

Bridgman first observed phase changes in water (1912) and coagulation of egg white (1914) under high pressure. However, no further progress in this area was made for almost 70 years. Water is the major component for most foods, especially fresh products (meat, fish, vegetable, fruit, etc.). However, food is a complex biochemical system where multiple interactions can take place. As noted before, some water in foods exists in a bound state which is not freezable. The fraction of freezable water is dependent on the freezing temperature. The presence of solutes in the aqueous phase shifts the phase diagram of water, lowering ice-melting temperature in atmospheric pressure. This effect is enhanced with increasing solute concentration (Fennema 1973; Cheftel et al. 2000). HP processing can lead to phase transition of food constituents, including water, lipids, proteins, etc. (LeBail et al. 2002). Therefore, phase transition of water in foods during a HP process is more complex than that of pure water. Understanding phase transition during HP processing of foods is helpful for food process and product improvement, but scientific information in this area is still very limited (Knorr 1999). Chevalier, Le Bail, and Ghoul (2000a) and Chevalier et al. (2000b) and Zhu (2004) used HP calorimetry to evaluate latent heat and ice ratio in frozen gelatin gels and found that the latent heat of the gel sample was influenced by both melting temperature and gelatin content.

17.5 PRESSURE-SHIFT FREEZING (PSF)

It is well recognized that slow freezing induces the formation of large ice crystals that may cause mechanical damage to the product texture, while rapid freezing enhances nucleation and the formation of many smaller ice crystals. Slow freezing of cellular tissues leads to large extracellular ice crystals, thus to an increased extracellular concentration of solutes, and therefore, to cell dehydration and death. Upon thawing, the extracellular ice does not re-enter the cells and may cause extensive drip and texture softening. Detrimental reactions are also enhanced by solute concentration effects and closer enzyme-substrate interactions. Rapid freezing of cellular tissues induces smaller and uniformly distributed ice crystals that would protect the product texture. Cryogenic freezing may cause macroscopic cracks due to nonhomogeneous volume expansion.

PSF processing consists of several steps (Figures 17.2 and 17.3). Once the sample is enclosed in the pressure chamber, the pressure is increased (AB) up to the target level. After or during the pressurization, the product is cooled until the temperature reaches the desired level (BC), usually around –21°C, but the product will still be in the unfrozen state due to the freezing point depression under pressure. The pressure is then rapidly released (CD). Pressure release results in expansion cooling, by about the same margin in reverse as that achieved by compression heating. But the release of pressure to atmosphere also reverts the freezing point to close to 0°C. Since the product has already been supercooled to almost –21°C, the sudden change in pressure causes an instantaneous conversion of the supercooled sensible heat to latent heat, resulting in instantaneous ice nucleation. Approximately 20% of the free water is instantaneously converted to ice nuclei. These fine ice nuclei obtained during the CD stage grow into a massive number of small ice crystals during the DE stage at atmospheric pressure. Tempering is completed during the EF stage at atmospheric pressure.

Otero and Sanz (2006) studied the dynamics of the PSF process and the main parameters implied. They carried out conventional freezing experiments at atmospheric conditions and PSF experiments at identical temperatures and different pressures and compared the phase transition times. Phase transition times in PSF experiments were lower than their homologues at atmospheric pressure in all cases. The reduction depends on the pressure and temperature conditions before expansion. These variables are highly involved in the supercooling reached and govern both the percentage of ice formed during the pressure release and the temperature drop in the pressure medium. After expansion, the pressure medium plays a significant role in the heat removal from the sample. Good predictions of plateau times can be made from initial pressure and temperature values, both process parameters that can be readily adjusted in the food industry.

Norton et al. (2009) used a one-dimensional finite difference numerical model based on the enthalpy formulation to simulate HP freezing of tylose, agar gel, and potatoes. They used the Schwartzberg equation in the prediction of both the initial freezing point and the temperature

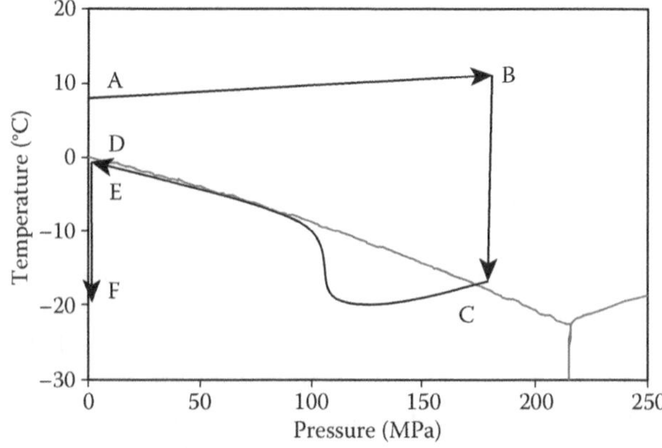

FIGURE 17.2 Representation of pressure-shift freezing based on water–ice phase diagram.

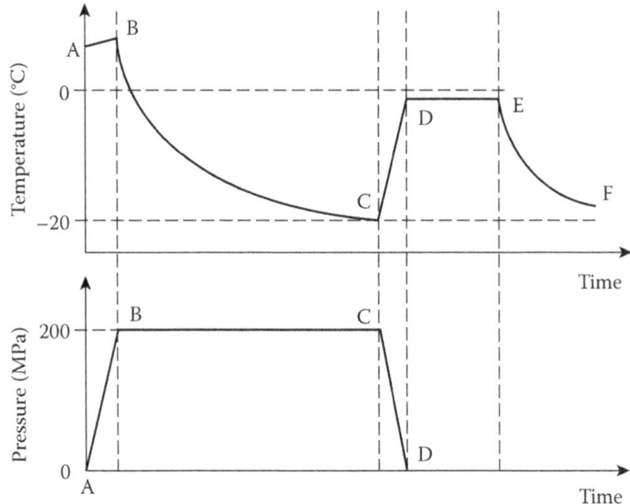

FIGURE 17.3 Schematics of changes in sample temperature and medium pressure during the different stages of pressure-shift freezing process for a pure water sample.

evolution below freezing. Results showed that the model can satisfactorily describe the PAF and HPSF processes. When compared under similar heat transfer conditions, the phase transition times for HPSF were shorter than those at atmospheric pressure. The amount of ice instantaneously formed upon pressure release and the total freezing times were also determined by the developed model and were in reasonable agreement with the experimental data in the literature.

17.5.1 Pressure-Shift Freezing (PSF) Kinetics and Microstructure of Ice Crystals

The size and location of ice crystals are considered the most important factors affecting the textural quality of frozen foods. It has been recognized that the freezing rate is critical to the nucleation and growth of ice crystals. Nucleation is an activated process driven by the degree of supercooling. Burke, George, and Bryant (1975) reported that for each degree Celsius enhancement of supercooling, there is about a tenfold increase in the ice-nucleation rate.

Barry, Dumay, and Cheftel (1998) reported that PSF (207 MPa, –20°C) was faster than even liquid nitrogen freezing. The release of pressure from 207 MPa induces a massive ice nucleation resulting in warming of the product of a protein gel to –1°C due to the latent heat of crystallization. The texture of the PSF product was found to be similar to that of nonpressurized nonfrozen control products. They also reported that frozen products in air or liquid nitrogen significantly differed from control and the ice matrix of PSF products contained numerous round alveoli of very small size due to the rapid temperature change (–20°C to near freezing temperature) realized to achieve the phase change. Histological studies on pork samples conducted by Martino et al. (1998) have also shown that PSF (200 MPa, –20°C) resulted in small ice crystals, reducing damage in terms of structural preservation compared to classical methods such as air-blast and cryogenic fluid freezing. This was mainly due to the fact that, unlike PSF, classical methods induced a thermal gradient and, therefore, a nonuniform ice crystal distribution that was harmful to tissue structures. As a result, the PSF process may be appropriate for freezing large pieces of food when uniform ice crystal sizes are required. In general, studies have demonstrated that PSF-treated foods had better-preserved microstructure compared to those treated with conventional methods.

The advantages of PSF and HP thawing have been further emphasized in recent studies (Zhu, Le Bail, and Ramaswamy 2003; Zhu, Ramaswamy, and Simpson 2004a; Zhu et al. 2004b, 2004c; Zhu, Ramaswamy, and LeBail 2006). These authors carried out extensive investigations on the

TABLE 17.2
Microscopic Analysis Results (Mean±SD) of the Ice Crystals in Gelatin Gel (2%, w/w) Samples Frozen by Different Methods

Parameter	Cross-Section Area (µm2)	Equivalent Diameter (µm)	Roundness	Elongation
CAF (–20°C, n = 98)	19,800±18,100^{a}	145±66^{a}	0.62±0.10^{a,b}	1.81±0.54^{a,b}
LIF (–20°C, n = 77)	6070±3580^{b,c}	84±26^{b}	0.61±0.15^{a,b}	2.00±1.11^{b}
PSF (100 MPa, –8.4°C, n = 53)	7270±4480^{b}	91±30^{b}	0.64±0.11^{c}	1.65±0.44^{c}
PSF (150 MPa, –14°C, n = 64)	4860±3470^{c}	73±29^{c}	0.60±0.10^{a}	1.70±0.53^{a,c}
PSF (200 MPa, –20°C, n = 122)	1750±1220^{d}	44±16^{d}	0.63±0.10^{b,c}	1.84±0.56^{a,b}

For each column, the letters represent the groups that had significant ($p < 0.05$) difference of the mean values between each other. n is the number of data samples.
Source: Zhu, S., PhD thesis, McGill University, 2004.

quantification and characterization of ice-crystal formation in PSF and its effect on the structure and texture of Atlantic salmon and pork muscle. Some examples of recent research findings from this study are illustrated in Table 17.2 and Figures 17.4 through 17.7. In these studies, Atlantic salmon samples were frozen either by PSF at 100 MPa (–8.4°C), 150 MPa (–14°C), and 200 MPa (–20°C) or by conventional air freezing (CAF) at –30°C and liquid (glycol/water: 50%, v/v) immersion freezing (LIF) at –20°C. Temperature and phase transformations of fish samples were monitored during the freezing processes and microstructures of ice crystals formed were evaluated for size, shape, and location. The mean (±standard deviation) cross-section area of the ice crystals was: 11,000±7600, 280±340, 260±300, 63±62, and 23±22 µm^2 for test samples subjected to CAF, LIF, and PSF at 100, 150, and 200 MPa, respectively, as compared with that of the muscle fibers (7200±2500 µm^2). The roundness of the fish muscle fibers was 0.67±0.07, while the ice-crystal roundness was: 0.38±0.14, 0.55±0.21, 0.57±0.18, 0.63±0.14, and 0.71±0.14 for the above samples, respectively. CAF created larger and irregular ice crystals, and resulted in irreversible damage to muscle tissues. Due to its higher freezing rate, LIF produced smaller ice crystals than CAF, but the cross-section area and roundness values had larger deviations. The PSF process produced large amounts of fine and regular intracellular ice crystals that were homogeneously distributed throughout the sample. Microscopic images clearly showed that the muscle fibers were well maintained in the PSF-treated samples as compared with unfrozen muscle structures.

Fernández et al. (2006) studied gelatin gel samples (10% gelatin, w/w) frozen by high-pressure-shift freezing (HPSF) and by high-pressure-assisted freezing (HPAF) with the phase transition at identical pressure conditions (0.1, 50, and 100 MPa) so as to allow valid comparisons. They recorded the corresponding temperature/pressure profiles in order to characterize the processes. Also, they analyzed the ice crystal distributions to estimate the effects of both freezing methods on the food microstructure. They reported results that clearly showed the HPSF to be a more advantageous method. The supercooling attained after the expansion and the consequent instantaneous freezing of water, together with the temperature drop in the pressure medium, induced short phase transition times (5.9, 8.6, and 13.7 min in HPSF versus 14.8, 14.1, and 23.1 min in HPAF at 0.1, 50, and 100 MPa, respectively) and a homogeneous distribution of small ice crystals throughout the sample.

17.5.2 Effects of Pressure Shift Freezing (PSF) Process on Quality Changes in Foods

Quality change in food products during processing is always a big concern for industry and consumers. Several studies have been carried out on PSF processing for improving quality of frozen foods, such as

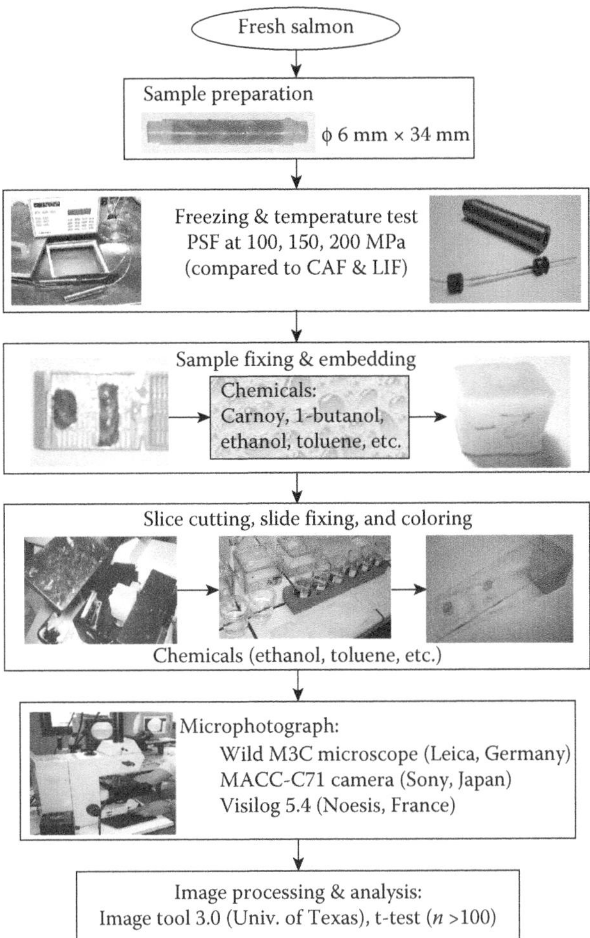

FIGURE 17.4 Flow chart of research methodology for pressure-shift freezing studies (Adapted from Zhu, S., PhD thesis, McGill University, 2004.)

texture, color, drip loss, microstructure, and so on, as recently reviewed by Cheftel, Thiebaud, and Dumay (2002) and LeBail et al. (2002). Pressure treatment has been shown to accelerate meat tenderness and enhance its eating quality (Okamoto and Suzuki 2002). Martino et al. (1998) showed that PSF-treated (200 MPa and –20°C) pork samples had less structural damage than air-blast and cryogenic fluid frozen samples. Chevalier et al. (2001) also found turbot processed by PSF at 140 MPa and –14°C to have a lower thaw drip than air-blaster freezing. These changes were attributed to protein denaturation induced by high pressure. A study on turbot fillets (Chevalier et al. 2001) showed that a pressure level of 140 MPa at 4°C allowed minimizing the effect of pressure on color, lipid, and protein stability.

HP treatment may also cause some undesirable changes in muscle foods during PSF processing. Fernández-Martín et al. (2000) reported that hydrostatic pressurization at subzero temperature (200 MPa and –20°C) with or without PSF was injurious to muscle structure and unsuitable for muscle preservation. Chevalier et al. (2000c) reported that PSF (200 MPa, –18°C) of Norway lobster (*Nephrops norvegicus*) tails resulted in smaller ice crystals than air-blaster freezing but also induced a significant increase in the toughness. The effect of PSF processing on meat quality changes has not been clearly understood. Other studies on quality changes in PSF-processed vegetable and food model materials have been conducted. When comparing the quality of tofu subjected to PSF (200 MPa, –18°C) and air-blast freezing (–10 and –18°C), Kanda, Aoki, and Kosugi (1992) found that PSF tofu maintained its initial shape and texture and had no drip loss. This was attributed to the

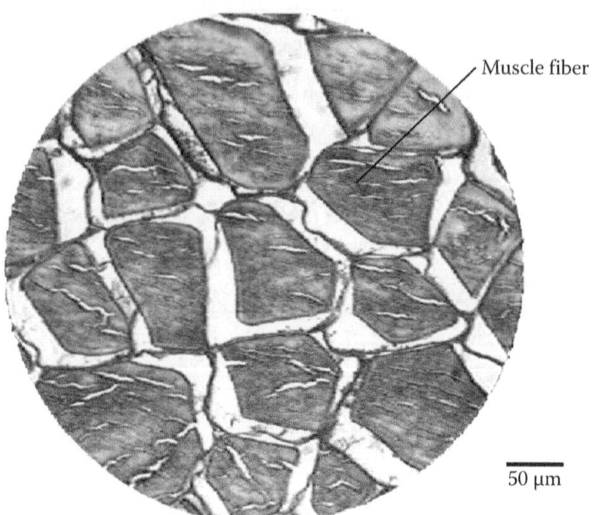

FIGURE 17.5 Micrograph of unfrozen Atlantic salmon muscle tissue. (Adapted from Zhu, S., PhD thesis, McGill University, 2004.)

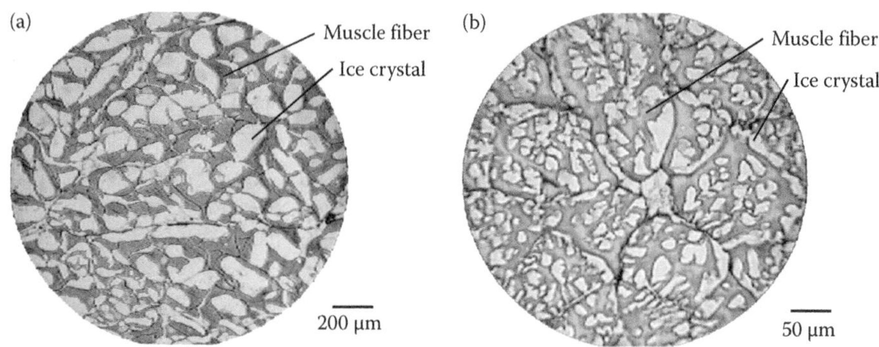

FIGURE 17.6 Micrographs of Atlantic salmon muscle tissue frozen by (a) conventional air freezer and (b) liquid immersion freezing. (Adapted from Zhu, S., PhD thesis, McGill University, 2004.)

smaller ice crystals formed in PSF tofu, whereas much larger ice crystals were observed in the air-blast frozen system. Fuchigami and Teramoto (1997) found that tofu samples frozen by PSF at 200 MPa showed almost the same textural characteristics as the untreated samples.

Freezing causes texture loss of tissue-based systems such as fruits and vegetables. To evaluate the potentials of HP freezing for minimizing freezing damage, Van Buggenhout et al. (2005) investigated the effects of HPSF and regular freezing conditions on the texture of carrot cylinders. To improve the strength of the plant material by a pectin-based network, carrot cylinders were submitted to different pretreatment conditions before freezing. They reported that the reduced freezing time of HPSF compared with conventional freezing results in a limited positive effect on the hardness of non-pre-treated carrots. A pronounced hardness improvement was reported when calcium soaking followed by thermal (30 min at 60°C) or HP (15 min at 60°C and 300 MPa) pretreatment was combined with HPSF. During subsequent frozen storage at −18°C, the increased hardness values of pretreated, HP-frozen carrots could not be maintained.

In a further study, Van Buggenhout et al. (2006) examined how the textural quality of carrots subjected to pretreatments in combination with different freezing conditions affected pectin structure.

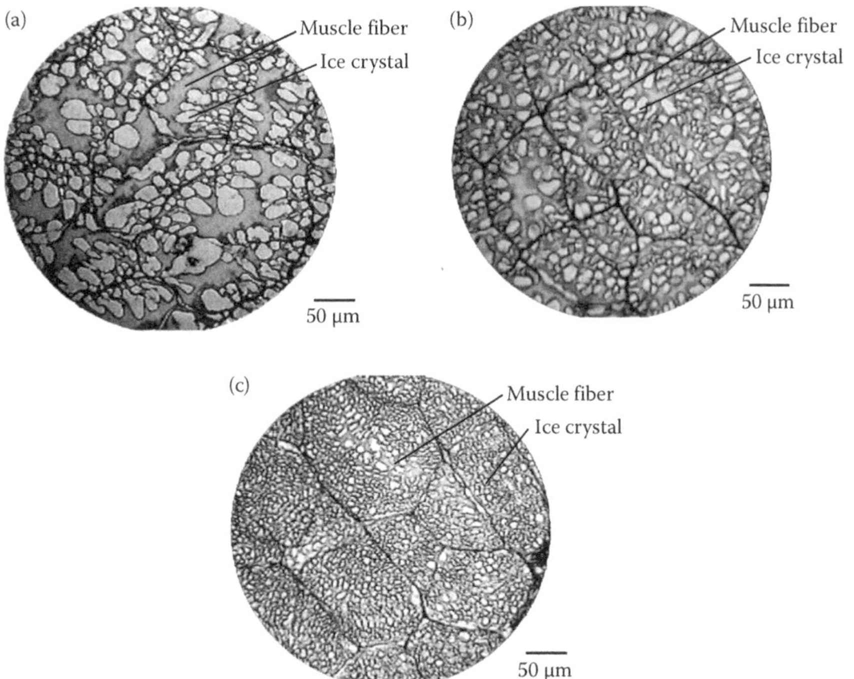

FIGURE 17.7 Micrographs of Atlantic salmon muscle tissue frozen by pressure-shift freezing at (a) 100 MPa (–8.4°C), (b) 150 MPa (–14°C), and (c) 200 MPa (–20°C). (Adapted from Zhu, S., PhD thesis, McGill University, 2004.)

In this study, carrot samples frozen under different conditions were extensively studied by light microscopy, quantifying the freezing damage based on the analysis of different parameters (number, area, perimeter, and shape factor of tissue particles) associated with carrot tissue damage. They found that the reduced texture loss of rapidly or cryogenically frozen carrots, compared to slowly frozen samples, was associated with the reduction in cell wall damage in the carrot tissue. In cases where no pretreatment was used, carrot texture was only slightly improved by using HPSF instead of slow freezing. Detailed analysis of the different steps involved showed that severe tissue damage occurred during the completion of the HP freezing process at atmospheric pressure. However, tissue damage, and thus texture loss, of HP frozen carrots could be minimized by applying pretreatments consisting of a thermal treatment at 60°C and a HP treatment at 300 MPa and 60°C.

Castro et al. (2007) studied the firmness of green bell pepper (*Capsicum annuum*) under different processing conditions. Thermal texture degradation kinetics of pepper tissue between 75 and 95°C could be accurately described by a fractional conversion model. The firmness of preprocessed pepper increased when the samples were submitted to several heat, pressure, and combinations of heat/pressure and calcium-soaking pretreatments. Preheating at 55°C for 60 min and mild heat/HP treatments (200 MPa at 25°C, 15 min) yielded the best results, which were further improved when combined with calcium soaking. These pretreatments significantly slowed down thermal texture degradation of pepper at 90°C, a typical temperature used for pepper blanching prior to freezing. The above-mentioned pre-treated samples showed a significant reduction in firmness when frozen by regular freezing at 0.1 MPa. The same samples showed no changes in firmness when frozen by HPSF at 200 MPa. When freezing was carried out by HP shift and after frozen storage (–18°C) for 2.5 months, pressure pretreated pepper showed a better retention of texture than thermal pretreated pepper.

Volkert et al. (2008) compared the impact of three different freezing methods on the physiology of probiotic bacteria based on evaluation of vitality, membrane integrity, and special metabolic properties. They used *Lactobacillus rhamnosus* GG (LGG) as a model strain and analyzed it with the plate enumeration method and flow cytometric analyses (FCM) before and after treatment in phosphate buffer saline (PBS) and 20% (w/v) skim milk (RSM), respectively. To get insight into the induced changes in esterase activity and membrane integrity they combined staining with carboxyfluorescein diacetate (cFDA) and propidium iodide (PI). Furthermore, they investigated the ability of the cells to extract intracellular accumulated cFDA in the presence of glucose as an indicator for ATP-driven membrane transport processes. The freezing methods applied were static air freezing at −30°C for 3 h, HPSF at −21°C/210 MPa and spray freezing (SF) at −30°C. SF was accomplished by spraying the bacteria suspension in a cold air stream using a two-fluid atomizer. The atomizing gas was air. After each treatment all samples were stored at −20°C for one day before they were analyzed. They found minimum losses in viability occurred when SF LGG in the presence of 20% (w/v) RSM while air freezing in PBS resulted in the lowest survival rates. For all treatments higher survival rates were achieved when 20% (w/v) RSM was used as protecting media. They reported that when comparing the results of live cell count with the FCM data, there was evidence of the presence of a fraction of viable cells but they were not culturable. LGG cells frozen by PSF were reported to show viability comparable to air-frozen samples, but physiological characteristics were similar to those of pressure-treated bacteria.

17.6 CONCLUSIONS

PSF has many potential advantages in the freezing preservation of foods. It offers the potential to increase the degree of nucleation and reduce the size of ice crystals formed, thereby better protecting the structure and texture of frozen foods. HP processing is a nonthermal technique that has preservation capabilities by limiting the growth and activity of microorganisms and enzymes (Knorr 1999). Thus while assisting in freezing and during thawing, the process will aid in controlling the growth and activity of spoilage/pathogenic bacteria.

REFERENCES

Barry, H., Dumay, E. M., and Cheftel, J. C. 1998. Influence of pressure-assisted freezing on the structure, hydration and mechanical properties of a protein gel. In: N. S. Isaacs (ed.), *High Pressure Food Science, Bioscience and Chemistry*, 343–53. London: Royal Society of Chemistry.

Bridgman, P. W. 1912. Water, in the liquid and five solid forms, under pressure. *Proceedings of the American Academy of Arts and Sciences* XLVII (13): 439–558.

Burke, M. J., George, M. F., and Bryant, R. G. 1975. Water in plant tissues and frost hardiness. In: R. B. Duckworth (ed.), *Water Relations of Foods*, 111–35. London: Academic Press.

Castro, S., Van Loey, A., Saraiva, J., Smout, C., and Hendrickx, M. 2007. Effect of temperature, pressure and calcium soaking pre-treatments and pressure shift freezing on the texture and texture evolution of frozen green bell peppers (*Capsicum annuum*). *European Food Research and Technology*, 226 (1–2): 33–43.

Cheftel, J. C., Levy, J., and Dumay, E. 2000. Pressure-assisted freezing and thawing: Principles and potential applications. *Food Reviews International* 16 (4): 453–83.

Cheftel, J. C., Thiebaud, M., and Dumay, E. 2002. Pressure-assisted freezing and thawing: A review of recent studies. *High Pressure Research* 22:601–11.

Chevalier, D., Le Bail, A., and Ghoul, M. 2000a. Freezing and ice crystals formed in a cylindrical food model: Part II, comparison between freezing at atmospheric pressure and pressure shift freezing. *Journal of Food Engineering* 46:287–93.

Chevalier, D., Sentissi, M., Havet, M., and Le Bail, A. 2000b. Comparison of air-blast and pressure shift freezing on Norway lobster quality. *Journal of Food Science* 65 (2): 329–33.

———. 2000c. Comparison of air-blast and pressure shift freezing on Norway lobster quality. *Journal of Food Science* 65 (2): 329–33.

Chevalier, D., Sequeira-Munoz, A., Le Bail, A., Simpson, B. K., and Ghoul, M. 2001. Effect of freezing conditions and storage on ice crystal and drip volume in turbot (*Scophthalmus maximus*), evaluation of pressure shift freezing vs. air-blast freezing. *Innovative Food Science and Emerging Technologies* 1:193–201.

Fennema, O. R. 1973. Nature of freezing process. In: O. R. Fennema, W. D. Powrie, and E. H. Marth (eds.), *Low Temperature Preservation of Foods and Living Matter*, 151–222. New York: Marcel Dekker.

Fernández-Martín, F., Otero, L., Solas, M. T., and Sanz, P. D. 2000. Protein denaturation and structural damage during high-pressure-shift freezing of porcine and bovine muscle. *Journal of Food Science* 65 (6): 1002–08.

Fernández, P. P., Otero, L., Guignon, B., and P. D. Sanz. 2006. High-pressure shift freezing versus high-pressure assisted freezing: Effects on the microstructure of a food model *Food Hydrocolloids* 20 (4): 510–22.

Fuchigami, M., and Teramoto, A. 1997. Structural and textural changes in kinu-tofu due to high-pressure-freezing. *Journal of Food Science* 62 (4): 828–32.

Kalichevsky, M. T., Knorr, D., and Lillford, P. J. 1995. Potential applications of high-pressure effects on ice-water transitions. *Trends in Food Science and Technology* 6:253–59.

Kanda, Y., Aoki, M., and Kosugi, T. 1992. Freezing of tofu (soybean curd) by pressure-shift freezing and its structure. *Nippon Shokuhin Gakkaishi* 39 (7): 608–14.

Knorr, D. 1999. Process assessment of high-pressure processing of foods: An overview. In: F. A. R. Oliveira, and J. C. Oliveira (eds.), *Processing Foods: Quality Optimization and Process Assessment*, 249–67. Boca Raton, FL: CRC Press.

Knorr, D., Schlueter, O., and Heinz, V. 1998. Impact of high hydrostatic pressure on phase transitions of foods. *Food Technology* 52 (9): 42–45.

LeBail, A., Chevalier, D., Mussa, D. M., and Ghoul, M. 2002. High pressure freezing and thawing of foods: A review. *International Journal of Refrigeration* 25 (5): 504–13.

Martino M. N., Otero L., Sanz P. D., and Zaritzky N. E. 1998. Size and location of ice crystals in pork frozen by high-pressure-assisted freezing as compared to classical methods. *Meat Science* 50 (3): 303–13.

Norton, T., Delgado, A., Hogan, E., Grace, P., and Sun, D. W. 2009. Simulation of high pressure freezing processes by enthalpy method. *Journal of Food Engineering* 91 (2): 260–68.

Okamoto, A., and Suzuki, A. 2002. Effects of high hydrostatic pressure thawing on pork meat. In: R. Hayashi (ed.), *Progress in Biotechnology 19, Trends in High Pressure Bioscience and Biotechnology*, 571–76. Amsterdam: Elsevier.

Otero, L., and Sanz, P. D. 2006. High-pressure-shift freezing: Main factors implied in the phase transition time. *Journal of Food Engineering* 72 (4): 354–63.

Ramaswamy, H. S., and Marcotte, M. 2005. Food Processing Principles and Applications. Boca Raton: CRC Press.

Van Buggenhout, S., Lille, M., Messagie, I., Loey, A., Autio, K., and Hendrickx, M. 2006. Impact of pretreatment and freezing conditions on the microstructure of frozen carrots: Quantification and relation to texture loss. *European Food Research and Technology* 222 (5–6): 543–53.

Van Buggenhout, S., Messagie, I., Van Loey, A., and Hendrickx, M. 2005. Influence of low-temperature blanching combined with high-pressure shift freezing on the texture of frozen carrots. *Journal of Food Science* 70 (4): S304–08.

Volkert, M., Ananta, E., Luscher, C., and Knorr, D. 2008. Effect of air freezing, spray freezing, and pressure shift freezing on membrane integrity and viability of *Lactobacillus rhamnosus* GG. *Journal of Food Engineering* 87 (4): 532–40.

Zhu, S. 2004 Phase transition studies in food systems during high pressure processing and its applications to pressure shift freezing and high pressure thawing. PhD thesis, Montreal, Canada, McGill University.

Zhu, S., Le Bail, A., Chapleau, N., Ramaswamy, H., and de Lamballerie-Anton, M. 2004b. Pressure shift freezing of pork muscle: Effect on color, drip loss, texture and protein stability. *Biotechnology Progress* 20 (3): 939–45.

Zhu, S., Le Bail, A., and Ramaswamy, H. S. 2003. Ice crystal formation in pressure shift freezing of Atlantic salmon (*Salmo salar*) as compared to classical freezing methods. *Journal of Food Processing and Preservation* 27 (6): 427–44.

Zhu, S., Le Bail, A., Ramaswamy, H. S., and Chapleau, N. 2004c. Characterization of ice crystals in pork muscle formed by pressure shift freezing as compared to classical freezing methods. *Journal of Food Science* 69 (4): FEP190–97.

Zhu, S., Ramaswamy, H. S., and LeBail, A. 2006. HP calorimetry and pressure-shift freezing of different food products. *Food Science and Technology International* 12 (3): 205–14.

Zhu, S., Ramaswamy, H. S., and Simpson, B. K. 2004a. Effect of high-pressure versus conventional thawing on color, drip loss and texture of Atlantic salmon frozen by different methods. *Lebensmittel-Wissenschaft und Technologie* 37 (3): 291–99.

18 Issues and Methods in Consumer-Led Development of Foods Processed by Innovative Technologies

Armand V. Cardello and Alan O. Wright
US Army Natick Soldier R, D&E Center

CONTENTS

18.1 INTRODUCTION

18.1.1 Chapter Overview

In the past quarter century there has been rapid expansion in the development of novel food processes. From ongoing efforts in food irradiation and biotechnology, to emerging progress in high-pressure, pulsed electric field, and ultrasound processing, to the latest developments in the application of nanotechnology to foods, novel food technologies are paving the way for future foods. This expansion in the use of novel technologies has been brought on by the realization that many of these processes offer the opportunity for higher-quality foods with better sensory, nutritional, and shelf-life properties than can be produced using traditional technologies. While significant advances have been made in the development of these technologies, success must be measured in terms of consumer acceptance and purchase of the foods produced by these technologies. For this reason, understanding and addressing consumer issues related to novel food processes is one of the most important challenges facing the developers of innovative food products.

During the past 35 years, the majority of research on consumer attitudes toward novel food processes has focused on food irradiation and genetic modification of foods. Only in the past 5–10 years has consumer research begun to focus on the wide range of novel food processes that are now emerging. The US military has had a long-standing interest in the use of novel food processing methods for rations. This interest derives from the need to develop shelf-stable food products that can last up to 36 months under a variety of environmental conditions, while retaining nutritional, microbiological, and sensory quality. Although early military food research focused on food irradiation, in recent years a much wider spectrum of novel food processes have been examined for their applicability to military rations (Dunne, Kluter, and Lee 1997; Dunne and Kluter 2001). A critical focus of this research program is the consumer and his/her attitudes, perceptions, and liking of rations produced using these novel processes.

Consumer issues related to novel food technologies can be categorized into three areas: (1) those that relate to the *person*—i.e., the consumer, his/her attitudes, beliefs, expectations, demographics, and psychographics; (2) those that relate to the *food*—its sensory attributes, perceived quality, nutrition, freshness, image, and benefits; and (3) those that relate to the informational *context*—i.e., the influences of product information, labeling, and other forms of marketing communications. Understanding how these three sets of factors interact to influence consumer behavior toward foods is the key to producing successful new products that consumers will accept, purchase, and consume. In this chapter we focus on these factors, the important issues surrounding them, and the methods that can be used to quantify them. Within each area, the importance of the factors and the scientific methods that can be used to assess them will be highlighted through recent and ongoing research.

We will begin with the *person* and with an overview of consumer attitudes and risk perceptions related to novel food technologies. We will then move to the *food*, how it is perceived from the consumer's perspective, and how sensory and other perceptual attributes of the food can be quantified. Lastly, we will address the informational context, what is known about how information, labeling, and other forms of communication affect consumer attitudes and perceptions, and how information can be best used to change these attitudes and perceptions.

18.1.2 The Consumer Perspective

It is important to understand that consumers are the most sensitive instruments that can be applied to the assessment of a new food product. Consumers are complex biological information gatherers and information processors. Their complexity is the product of an evolutionary history that fostered the development of the human central nervous system and brain into incredible information storage and processing devices that rival and, in many ways, surpass today's supercomputers. During this same time, human sensory systems evolved to produce levels of unparalleled sensitivity, at least when compared to the best electronic sensors that food science has been able to offer. Working together, the human sensory and nervous systems acquire information about both the *intrinsic* or sensory properties of the food and the *extrinsic* elements of product information and situational context. This totality of information is then processed, along with past experiences, memory, and learning to arrive at a behavioral response to the food that can be characterized as either approach (acceptance) or avoidance (rejection).

To underscore this information-processing ability, let us consider what happens when a consumer encounters food in a real-life setting. Most often, the food is first detected by the sense of sight. As the individual draws closer to the food, its odor is perceived by the sense of smell. This is followed by the food being taken into the mouth and bitten, where its texture is perceived through the tactile (somesthetic) and joint and muscle (kinesthetic) senses of the oral cavity and jaw. At this point, the sound emitted by the food stimulates the sense of hearing through vibrations carried through the air and bones of the jaw and skull. Lastly, as the food components interact with saliva, the sense of taste is stimulated. Thus, the visual, olfactory, tactile, auditory, and taste senses all play a critical role in extracting information about the food.

In addition to these intrinsic aspects of the food, a host of *extrinsic* factors are perceived by the consumer. These include the environment and context in which the food is encountered, the social setting, and various forms of written and verbal information about the food. Whether in a supermarket, at a restaurant, or at home, a wide variety of these forms of information are perceived and processed. They include label and package information, menu descriptions, pricing information, store shelf information, and nutrition and shelf-life dating information. This information triggers past memories and experiences with the product or similar products, as well as learned associations to other information, such as newspaper and television advertizing, catalogs, coupon offers, and word of mouth. All of this information is then passed through a cognitive filter in which the individual's preexisting attitudes, stereotypes and beliefs, along with his/her gender, ethnicity, cultural, and religious attitudes "frame" the information. In the case of novel foods or those that have been processed by novel or emerging technologies, attitudes and beliefs about the origins of the food, the processing technologies that have been applied to the food, and the risks or safety associated with these processes also contribute to the perceptions of the product and to the consumer's final choice and purchase behavior.

This complex information-processing ability of the consumer is the reason why every element of a new product, including its sensory properties, its packaging, its advertizing, and its market positioning must be analyzed prior to product launch to ensure the greatest likelihood of product success. If any element is overlooked, the consumer will detect and react to it. The fact that over 60% of new product introductions result in failure (Harris 2002) is a reflection of both the consumer's ability to detect and identify product deficiencies and of manufacturers' repeated failure to take into account the consumers' needs, desires, and expectations during product development.

18.1.3 Consumer-Led Product Development

The realization of the power of the consumer in determining new product success has led to the acknowledgment that successful product development is predicated on addressing the needs, wants, and expectations of the consumer. This new approach to product development has been termed "consumer-led product development" (Costa 2003; Costa and Jongen 2006, MacFie 2007). As its name implies, consumer-led product development starts with the consumer, not with the product developer. It is the *consumer's* needs that drive new product ideas and it is the *consumer's* expectations that drive new product design and development. In the case of innovative foods, the consumer's need for higher quality, more nutritious, fresher tasting products is a key driver for the development of these technologies. However, as the product moves from concept to reality, the consumer's expectations of the sensory, quality, and freshness attributes of the product become paramount. So too do the consumer's attitudes and expectations regarding the safety of the food and the process by which it is produced. In later parts of this chapter, we will address issues related to quality, freshness, and liking. However, the next section addresses attitudes of the consumer that must be considered early in the conceptualization of novel products and throughout their development and marketing.

18.2 THE PERSON

18.2.1 Consumer Attitudes to Novel Food Processes

The term "attitude" refers to "a psychological tendency that is expressed by evaluating a particular entity with some degree of favor or disfavor" (Eagly and Chaiken 1993). Frewer (2003) defined the construct as "a psychological bias that predisposes the individual toward positive or negative evaluative responses," the latter of which are defined as "responses that express approval or disapproval, liking or disliking, approach or avoidance, attraction or aversion." Other common terms used to refer to attitudes are "opinions," "conceptions," and "dispositions." Understanding attitudes of consumers to food products is essential, because they help answer the most important questions in new product development—why would consumers purchase this new product and why *wouldn't* they (Wansink 2005).

Attitudes toward foods are created by beliefs, information, past experiences, and associations with the food, and they are measured using a variety of *qualitative* or *quantitative* consumer research techniques.

18.2.1.1 Qualitative Techniques: Interviews and Focus Groups

Qualitative methods are often used to obtain information about consumer attitudes, opinions, and behavior (Hashim, Resurreccion, and McWatters 1996). As their name implies, *qualitative* methods do not rely on quantification of attitudes, but on the subjective uncovering and analysis of them. Among the most common qualitative techniques to assess attitudes to food are individual or group interviews. These methods are often used when there is little available information about the attitudes or opinions that consumers have toward a topic. With novel food processes, this is often the case. The oldest qualitative method is the *individual interview.* In these interviews, consumers are simply asked their opinions or attitudes toward a food or food-related phenomena using a one-on-one format with specific questions. In recent years, this technique has been expanded through the use of follow-up questioning in a systematic process that attempts to follow the chain of beliefs or values that underlie the attitude of the consumer. The theory behind this approach is known as means-end-chain and the technique is known as *laddering.* They have been used in a variety of studies to address attitudes toward such novel technologies as genetic engineering (Bredahl, 1999; Grunert et al. 2001; Miles and Frewer 2001; Boecker, Hartl, and Nocella 2008).

Since individual interviews are costly and time consuming, group interviews, especially those that utilize a *focus group* approach (Casey and Krueger 1994; Templeton 1994; Krueger and Casey 2000), have become popular. Focus groups consist of 8–12 consumers recruited to meet specific demographic, psychographic, or product-usage criteria. A trained "moderator" guides the consumers through a series of questions that elicit both individual comments and group discussion on the topic of interest. Focus groups have been used successfully to acquire reliable attitudinal and opinion-based information on a wide variety of novel foods and food processing technologies (Brug et al. 1995; Hashim et al. 1996; Bruhn et al. 1996; McNeill, Sanders, and Civille 2000; Grunert et al. 2001; Deliza, Rosenthal, and Silva 2003).

In one study that employed a focus group approach to assess consumer attitudes toward high-pressure-processed (HPP) foods, it was shown that label information containing benefit statements about HPP had a positive impact on consumers (Deliza et al. 2003). However, this study also revealed that for some consumers, merely telling them that the product was processed with high pressure produced a negative attitude toward it. Through further probing and discussion in the focus group, it was learned that the label "high pressure processing" evoked negative reactions because the consumers did not know what the technology was, e.g., "I have no idea what high pressure is"

(Deliza et al. 2003). As will be seen in a later section, most consumers are wary of products that are "unknown" or novel to them. This *neophobia* (fear of the new or unknown) is not uncommon in response to new or unusual foods. However, what *is* surprising is that a process with such a benign name as "high pressure processing" would produce such a negative reaction. This lack in our ability to foresee issues that are important to consumers and the ability of qualitative techniques to probe and uncover them is what makes these methods so useful.

There are a large number of other qualitative techniques that can be used to explore consumer attitudes in the early stages of new product development. Among these are *free elicitation*, in which words or concepts are presented to consumers who verbalize their free associations to them, the *Kelly Repertory Grid method* (Gains 1999), in which products are presented in triads to consumers who describe their similarities and differences (see Mucci and Hough 2003 and Mireaux et al. 2007 for its application to novel technologies), and *Emphatic Design*, which uses the anthropological approach of observing actual behaviors toward the product. The object of all of these techniques is the identification of important attitudes, opinions, or behaviors to novel products for which there is little consumer information available. The reader is referred to van Kleef, van Trijp, and Luning (2005) for a review of these and other qualitative (and quantitative) methods to uncover "the voice of the consumer" in food research.

18.2.1.2 Quantitative Techniques for Measuring Consumer Attitudes

While qualitative techniques are often used when information about a topic is lacking, *quantitative* techniques are used when information is already available that enables more specific aspects of the attitudes to be addressed. Quantitative techniques include older methods, such as simple opinion surveys, as well as advanced consumer attitudinal techniques, like conjoint analysis.

18.2.1.2.1 Survey Techniques

Opinion surveys conducted by mail, telephone, and Internet are a common way to obtain attitudes on a wide variety of food-related topics. Many of these surveys have addressed consumer attitudes toward innovative foods and food technologies. For example, extensive surveys have been conducted on genetically modified (GM) foods (Bredahl 2001; Magnusson and Hursti 2002; Hallman et al. 2003). Among the most extensive of these was the Eurobarometer survey on biotechnology, conducted throughout Europe for the past 15 years (e.g., European Commission, 1997, 2006). The findings from these surveys show that Europeans have quite negative attitudes toward GM foods, with upward of 95% of consumers concerned that these foods enter the food supply without labeling, thereby denying them the choice of whether or not to purchase them. Both women and older consumers perceive more risk from GM foods than male and younger respondents. When the Eurobarometer survey was conducted in other countries, it revealed almost universal concern about the adequacy of risk management regulations for GM foods, but with Canadians and New Zealanders showing somewhat less concern for regulatory aspects than other respondents (Macer et al. 1997). The most recent independent survey on GM foods was conducted in England by the Food Standards Agency for the General Advisory Committee on Science (2008). In this survey, 65% of consumers were concerned about the safety of consuming GM foods. In the United States concern about GM foods is less intense than in Europe (Hoban 1997), but some surveys show high levels of concern within specific US segments (Priest 2000; Braun 2001; Shanahan, Dietram, and Lee 2001).

A number of surveys, studies, and reviews have been published on consumer attitudes toward both general food safety issues and the perceived risks or "concerns" of consumers toward innovative technologies (e.g., Bord and O'Conner 1989, 1990; Moseley 1990; Sparks and Shepherd 1991; Dunlap and Beus 1992; Gallup 1993; Brewer, Sprouls, and Russon 1994; Frewer and Shepherd 1995; Bruhn 1995a, 1995b; Bruhn, Schutz, and Sommer 1987; Bruhn et al. 1996; Frewer et al. 1994,1997a; Frewer, Howard, and Shepherd 1997b; Bredahl 1999, 2001; da Costa et al. 2000; Verbeke 2001; Grunert et al. 2001; Butz et al. 2003; Wilcock et al. 2004; Finucane and Holup 2005; Yeung and Yee 2005; Carneiro et al. 2005; Eustice and Bruhn 2006; Chen and

Li 2007; Turcanu et al. 2007; de Jonge et al. 2008; USFDA 2008). However, comparative data on attitudes toward different forms of food processing were lacking until recently. In an effort to close this knowledge gap, two surveys were conducted on consumer perception toward a wide range of novel technologies by researchers at the US Army Natick Research, Development and Engineering Center. In one of these surveys, ~200 soldiers (modal age category = 20–25 years of age, 98% high school graduates/66% some college) rated their attitudes toward eating foods that had been processed by 26 different food processing methods (Cardello 2000). Attitudes were assessed by asking the respondents how concerned they would be about eating foods that had been processed by each technology on a scale that ranged from "not at all concerned" to "extremely concerned." Figure 18.1 shows the data from this survey. Consumer attitudes, expressed by the percentage of respondents who stated that they were "very" or "extremely" concerned about the use of that technology, ranged from very low levels of concern for traditional processing techniques, e.g., freeze-drying, thermal processing, heat pasteurization, and cold preservation, to very high levels of concern for some novel technologies, e.g., genetic modification and ionizing energy. Certain other novel processes that are considered quite safe by experts, e.g., pulsed electric fields and radio-frequency sterilization, also produced relatively high levels of concern. The reason for this is undoubtedly the same as the one that Deliza et al. (2003) uncovered for HPP, namely that lack of information about a new food technology can foster negative attitudes through an innate, neophobic response.

The second survey was conducted with employee members of a consumer-testing panel (Cardello 2003). Respondents were 66% male and 34% female and ranged in age from 18 to 64. Respondents

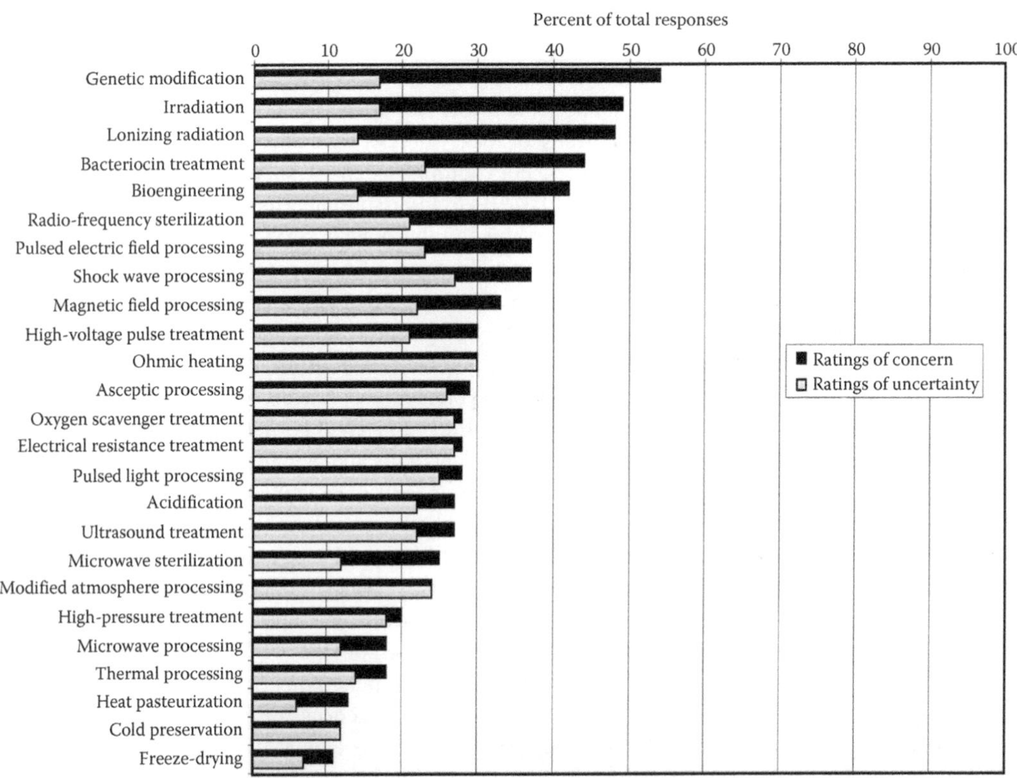

FIGURE 18.1 Twenty-five novel and traditional food processing techniques and the percent of military respondents ($n = 198$) who expressed that they would be "very" or "extremely" concerned or "uncertain" about eating foods that had been processed by them. (From Cardello, A. V., Presentation at the *IFT Non-thermal Processing Division Workshop on Non-thermal Processing of Food*, 2000.)

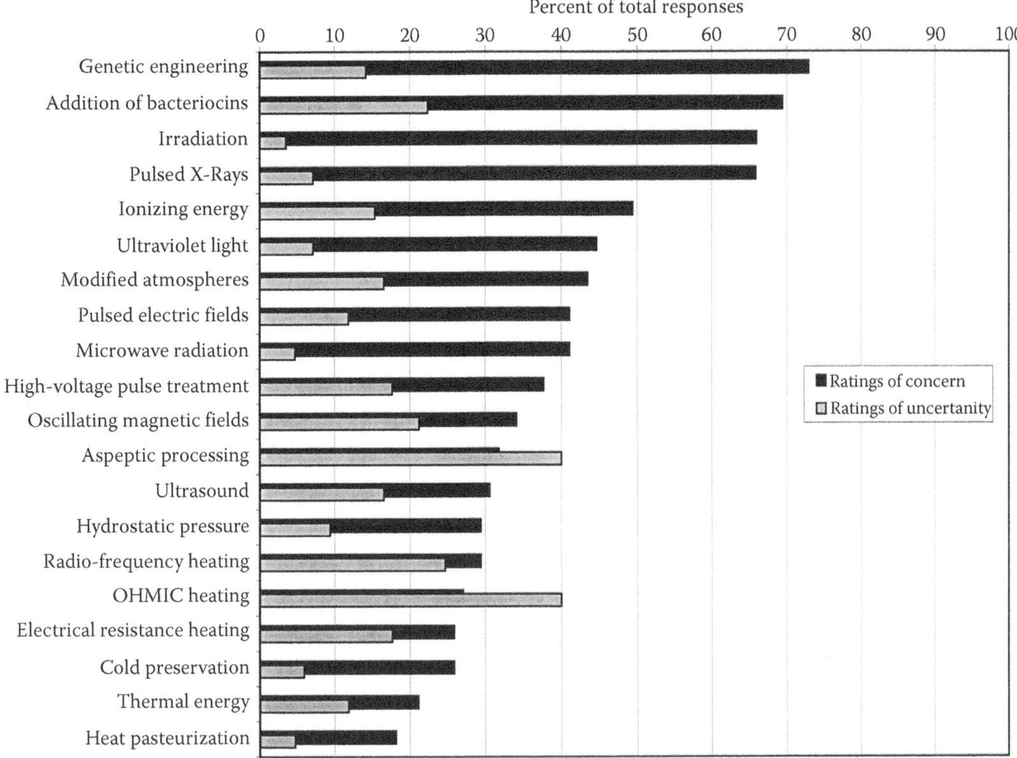

FIGURE 18.2 Twenty different novel and traditional food processing techniques and the percent of consumers ($n=88$) reporting "slight," "moderate," or "extreme" concern or who were "uncertain" about eating foods processed by them. (From Cardello, A. V., *Appetite*, 40, 217–33, 2003.)

included a subset of individuals with a behaviorally demonstrated willingness to taste and consume foods processed by a novel technology (irradiated foods). The consumers rated their concern level for 20 different innovative and traditional food processing techniques. The mean ratings of concern for each technology and the percentage of consumers who were "uncertain" about the technology are shown in Figure 18.2. As was found in the first survey, the processes that evoked the least concern were traditional ones, e.g., "heat pasteurization," "thermal energy," and "cold preservation." Also of low concern was "electrical resistance heating." On the other hand, "genetic engineering." "the use of bacteriocins," "irradiation," "ionizing energy," and "pulsed x-rays" all elicited concern from among 50% or more of the respondents.

Consumer attitudinal surveys on novel food processes, like those above, are consistent in showing that consumers have concerns about a wide variety of novel food processes. Although GM and irradiated foods always elicit the greatest concern, other more "benign" processes can also foster high levels of concern, especially when the consumer is unsure of what it is.

Although surveys on novel food processes indicate varying degrees of concern with their use, it is much less clear how the perceived risks of the technologies compare in importance to other characteristics of these foods, such as the benefits that they provide. To assess the relative importance of these factors, alternatives to simple opinion surveys must be used.

18.2.1.2.2 Conjoint Analysis

One problem with written surveys is that it is apparent to the consumer what the survey is trying to measure. Although most people are honest in their judgments, some may give extreme or contrary responses in order to slant the survey in a specific direction. To avoid this possibility, market researchers developed a technique known as *conjoint analysis* (Green and Srinivasan 1978), which uncovers attitudes of consumers without directly asking them their opinion.

Conjoint analysis uncovers consumer attitudes using an approach in which they are asked to rate a series of product concepts. Each concept is comprised of a combination of attributes or characteristics of the product thought to be important to the attitude of interest. The attributes or characteristics are "conjoined" using a planned experimental design that enables analysis of the impact of each attribute on the consumer's response to the total concept. The consumer simply chooses or rates each concept and the method "works backward" from the choices/ratings made to uncover the relative importance of each factor without the need to directly ask the consumer his/her opinion of the importance of them to their choices or ratings.

In a study by da Costa et al. (2000), conjoint analysis was used to study attitudes toward the purchase of GM and traditional vegetable oils. Their results showed that the relative importance of factors differed among different consumer groups. For some, process information was the most important factor, for others, process, price, and brand label were most important, and for still others a label illustration was the most critical element. Frewer et al. (1999) also used conjoint analysis to examine consumer attitudes toward cheese that had been processed using GM, protein-engineered, or traditional microorganisms. Their study revealed that 80% of consumers found the GM-produced cheese to be least acceptable, and that the benefits of the different technologies were the most important factor influencing their purchase behavior. In other studies, conjoint analysis was used to uncover preferences of Brazilian consumers toward transgenic soybean oil (Carneiro et al. 2005) and Australian consumers' preferences toward both prawns sterilized by irradiation or triploidy (Cox, Evans, and Lease 2007) and GM sources of omega-3 fatty acids (Cox, Evans, and Lease 2008). These studies showed that strong negative attitudes existed toward these forms of novel products and processes, but that the relative importance of factors again varied by consumer segment.

Although conjoint analysis can provide important insights into the attitudes of consumers toward novel food processes, here too, there is a lack of research examining attitudes toward a wide range of processes. However, in one recent study, consumer interest in using foods processed by a wide range of novel technologies was examined using conjoint analysis in three different US consumer groups (Cardello, Schutz, and Lesher 2007). The respondents were consumers at a local shopping mall, members of an employee consumer test panel, and military personnel on training exercises. Each rated their interest in using 45 different food product concepts that varied in food type, processing or production technology, costs, benefits, risks, endorsing agencies, and product information. Figure 18.3 is a page from the survey that shows the question that was posed to consumers, one of the food concepts to be rated, and the rating scale. Figure 18.4 shows the results of the study. Along the bottom of Figure 18.4 are the eight factors used in each product concept. On the y-axis is the "averaged importance value" of each of these factors to consumers' interest in using the foods. With two exceptions (benefits and cost), the importance levels of the various factors were the same for all respondent groups. Most troubling however, is that the perceived *risks* associated with the technologies were the *most important* factor driving consumer interest or lack of interest. The specific method of processing and the group or agency that provided endorsing support for the safety of the technologies were lower in importance.

With the high relative importance of risk perceptions to the intentions of consumers to use and purchase foods processed by novel technologies, it is essential to understand the psychological factors that drive consumer perceptions of these risks.

18.2.2 Consumer Perceptions of Risk From Novel Technologies

18.2.2.1 Risk Perceptions: Basic Elements

The high levels of consumer-perceived risks and concerns with novel food processes are often difficult for product and process developers to comprehend, because these risk perceptions

How interested would you be in using a food product that has the following characteristics?

- A milk/dairy product
- Processed by high pressure
- Reduces bacterial risk
- Is less expensive
- Is untested
- Endorsed by the manufacturer/process
- Processing information available in pamphlets at the store

Using the scale below, indicate how interested you would be in using this product by circling one of the phrases.

Extremely Interested
Very Interested
Moderately Interested
Slightly Interested
Not at All Interested

FIGURE 18.3 An example concept rating page. (From Cardello, A. V., Schutz, H. G., and Lesher, L. L., *Innovative Food Science and Emerging Technologies,* 8, 73–83, 2007.)

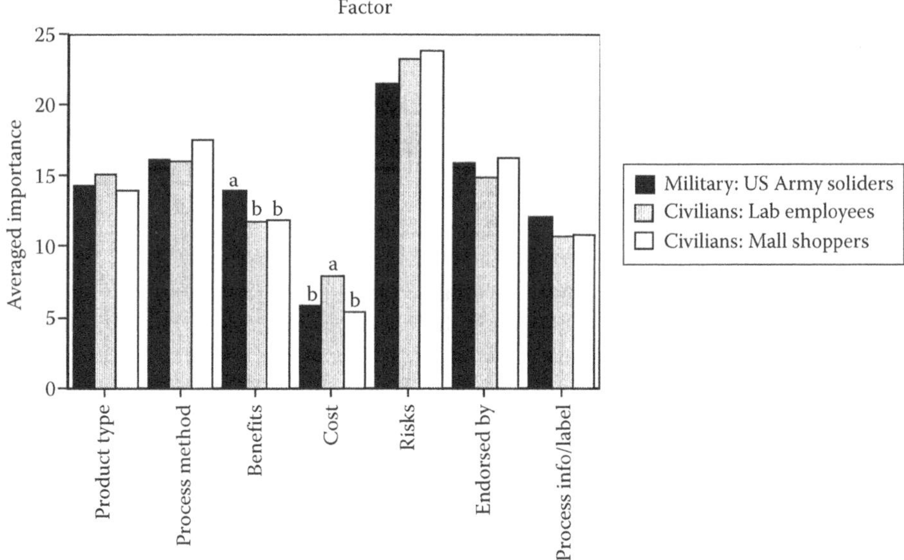

FIGURE 18.4 Average importance values for the factors contributing to consumer interest in foods processed by emerging technologies. Different letters above the bars indicate significant differences between respondent groups at $p < .05$. (From Cardello, A. V., Schutz, H. G., and Lesher, L. L., *Innovative Food Science and Emerging Technologies,* 8, 73–83, 2007.)

contradict the accepted scientific consensus regarding the actual risk of these technologies. This difference in risk assessment between consumers and experts is similar to other discrepancies between expert and consumer views of food (Sijtsema et al. 2004) and understanding the bases of these contrasting views is essential to the development and marketing of these foods.

The difference between expert and consumer risk judgments has been known for some time (Fischoff, Slovic, and Lichtenstein 1982). Krause, Malmfors, and Slovic (1992) characterized

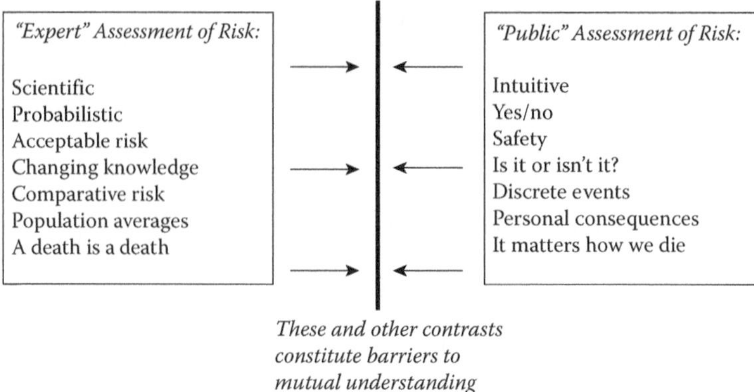

FIGURE 18.5 Characteristics of expert versus public assessments of risk. (From Powell, D. and Leiss, W., *Mad Cows and Mother's Milk: The Perils of Poor Risk Communication.* McGill-Queen's University Press, Montreal, 1997.)

consumers as "intuitive toxicologists." Figure 18.5 is taken from Powell and Leiss (1997) and characterizes some of the major differences between *expert* opinions of risk and consumer opinions of risk. As can be seen, expert opinions are based on scientific facts and statistical probabilities associated with outcomes for an entire population. They are based on "acceptable" risk, which will change with advances in scientific knowledge and by comparison with other risk outcomes. Expert assessments of risk also treat all disasters or fatalities as equal. However, *consumer* perceptions of risk are based on intuition about the hazards and the likelihood of negative consequences from them. Consumer-perceived risks are "all or none," that is, the technology is either safe or it is not. Moreover, these perceptions are unlikely to change in response to changing scientific knowledge. Lastly, consumer perceptions of risk are *personal*; statistical probabilities about fatalities in the population as a whole do not play as large a role as do concerns about what could happen to oneself or one's family.

The study of consumer risk perception is a relatively new area that originated with research by Starr (1969), Fischoff et al. (1978) and Slovic, Fischoff, and Lichtenstein (1979). Until that time, the risk associated with a technology was based on expert analysis. This *technical/rational* concept of risk was based on actuarial data on accidents, fatalities, and other losses accountable to the technology. However, it was also known that consumers had great concerns about technologies that had relatively *low risk* from this technical/rational perspective. To resolve these differences, an alternative model of risk perception was formulated. This new *normative/value* model defined risk as a perceptual construct that must be evaluated by laypersons on the basis of *subjective*, not objective, evaluations of risk.

In a groundbreaking study, Slovic, Fischoff, and Lichtenstein (1985) analyzed the perceived risk associated with technological hazards using a factor analytic approach. Based on his analysis, Slovic (1987) identified the underlying characteristics that have high predictive validity in determining the risk perceptions of consumers. Not surprisingly, one of the major factors he identified was the degree to which the hazard is viewed as *fatal versus nonfatal*. Thus, while bacterial contamination of food is of concern to most consumers, it is generally perceived as a nonfatal risk. However, some novel food technologies used to prevent bacterial contamination of foods, such as genetic engineering and irradiation, are associated with potentially fatal consequences and, thus, are perceived as far greater risks than the potential bacterial risk that they address. Also, the degree to which the effects of the hazard are judged to be *immediate or delayed* is a critical factor in consumer risk perception. So, while the effects of microbial contamination of ingested foods are immediate, the effects of hazards associated with genetic modification, irradiation, and even electric fields are perceived to be delayed for years. These delayed consequences are of far greater concern to consumers and engender much greater perceptions of risk.

A third factor that Slovic identified is the degree of *control or lack of control* that the consumer has over the hazard. Thus, the risk associated with taking illegal drugs or with using household appliances is viewed as low, because the consumer has control over whether to use them. However, the risks associated with terrorist attacks, tornadoes, or radiation leaks from power plants are seen as high, because he/she has no control over them. This is also true with the perceived risk from foods processed by novel technologies, because the processes used in these foods are not under the control of the consumer, especially if the food fails to contain labeling to enable shoppers to choose whether or not to consume them.

Another factor of importance to risk perception is the degree to which exposure to the hazard is *voluntary or involuntary*. Although a vaccine that protects against disease is viewed to be of low risk when the vaccine is for voluntary use, parents and/or consumer groups often perceive greater risk once the vaccine becomes mandatory. Similarly, while consumers will freely add salt, sugar, or fat-containing ingredients to foods they prepare at home, the same salt, sugar, or fat added to packaged foods is viewed with alarm, because they are no longer under the control of the consumer.

Slovic's research also identified the degree to which the hazard is *observable or unobservable* as important to consumer risk perception. Although mold on bread or other foodstuffs is an observable hazard, preservatives or processing technologies designed to inhibit mold are unobservable and are perceived as constituting a greater risk than the mold itself. This factor is allied with the degree to which the hazard is *known or unknown* to science. Hazards resulting from bacterial contamination of foods are well understood by science, but hazards associated with biotechnology and other novel food processes are far less well-understood, engendering far greater uncertainty and risk in the mind of the consumer (e.g., Grunert et al. 2001).

18.2.2.2 Risk Perceptions: Demographic and Psychographic Factors

The psychological factors underlying the differences between expert and consumer perceptions of risk create a barrier to mutual understanding between the two groups. However, consumer attitudes toward novel technologies also are dependent on such factors as age, gender, socio-economic status, and the psychographic or personality factors that are unique to the individual. Gender is a good example. Figure 18.6 shows a breakdown by gender of the mean levels of concern obtained in the study from which the data in Figure 18.2 were collected. The gender effect is observable in the fact that the mean level of concern for every food process is *greater* for females than for males. Gender has been implicated previously in attitudes and risk perceptions toward novel technologies and environmental health risks (Krause et al. 1992; Hoban, Woodrun, and Czaja 1992; Terry and Tabor 1988; Malone 1990; Gallup 1993; Flynn, Slovic, and Mertz 1994; Gustafsod 1998; Food Marketing Institute 2005). In general, the findings reveal that females perceive greater risks from novel technologies and from the environment than do males. It has been suggested that this gender difference may be responsible for part of the discrepancy between consumer and expert assessments of risk, because so few experts are women (Frewer 2003). Since females are still the major purchasers of food for the family, developers using novel food processes must realize that the primary purchaser of their product is someone who is likely to be *more* concerned with the potential risks of any new technology than is the general population.

In addition to demographic variables, psychographic and personality variables also influence risk perceptions. For example, "green" or "alternative" consumers, such as those who shop in food cooperatives, are more resistant to technological change (Bruhn, Schutz, and Sommer 1986) and express greater concern about novel food technologies (Bruhn et al. 1996). On the other hand, consumers who show *trust* in the food industry, in government regulatory agencies, and in science

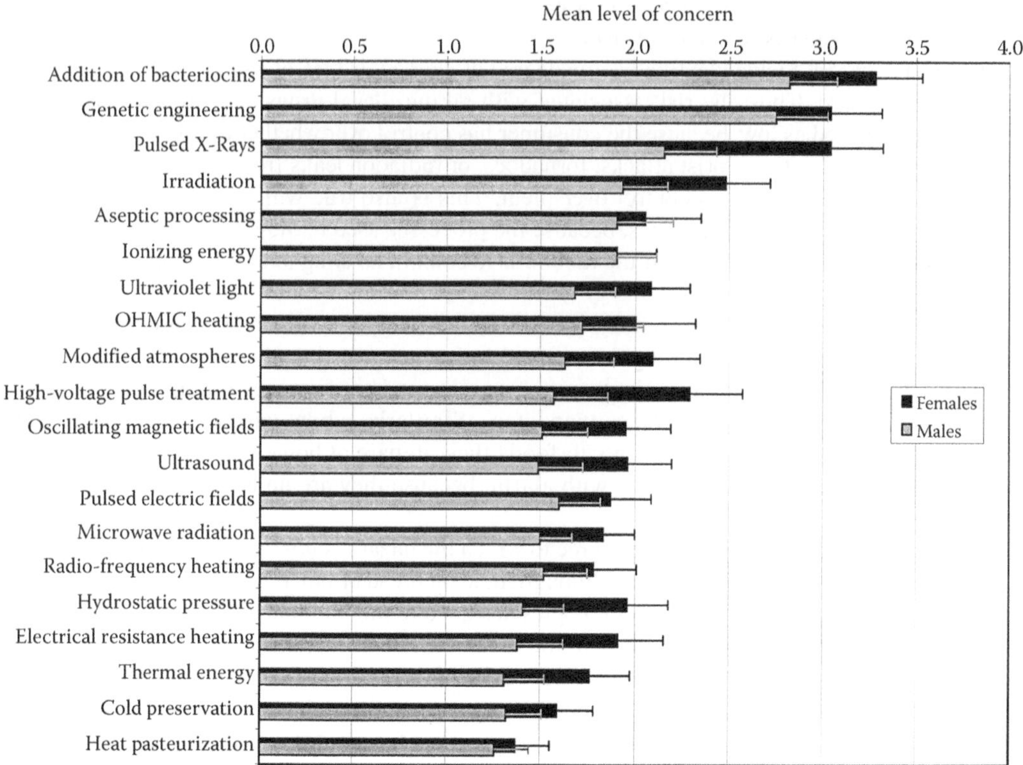

FIGURE 18.6 Mean levels of concern by gender for twenty different food processes. (From Cardello, A. V., *Appetite*, 40, 217–33, 2003.)

have been shown to be much more accepting of these foods (Bord and O'Conner 1990). The degree of trust that the consuming public places in product spokespersons is also critical to whether or not consumers will heed positive communications about the safety of novel foods and technologies (Slovic 1993; Slovic and MacGregor 1994; Frewer et al. 1996). For GM foods, the consumer's values and ethics also play a role (Dreezens et al. 2005; Finucane and Holup 2005; Sparks, Shepherd, and Frewer 1995), as many consumers reject GM foods because they are viewed to be intrinsically "wrong" (Frewer and Shepherd 1995; Miles and Frewer 2001). In the case of animal cloning for food purposes, fear, disgust (Caplan 2008), and religious taboos can all produce outright rejection of food products derived from these sources.

In contrast to the above consumers, there are those who may be considered *innovators* or technically receptive consumers. These individuals are market leaders who respond more positively to novel sensory stimuli (Raudenbush et al. 1998) and to whom new technologies and products should be targeted. Typically, they are young, male, affluent, and well-educated consumers. Some researchers have suggested that individuals such as these, who may show willingness to buy foods processed by novel methods, can be important bell weathers for the likely success of these products (Terry and Tabor 1990). A group of such consumers, defined by their prior willingness to taste foods processed by irradiation, were included in the study by Cardello (2003) cited earlier. Figure 18.7 shows the mean levels of concern obtained from both these individuals, as compared to a control group of volunteers. As seen in Figure 18.7, the mean ratings of concern for every technology were *lower* among consumers who had previously volunteered to taste/consume irradiated foods than for those who had not. Clearly, one's prior attitudes and behavior toward foods processed by one type of novel process is a good predictor of one's future response to foods processed by *other* novel or emerging technologies.

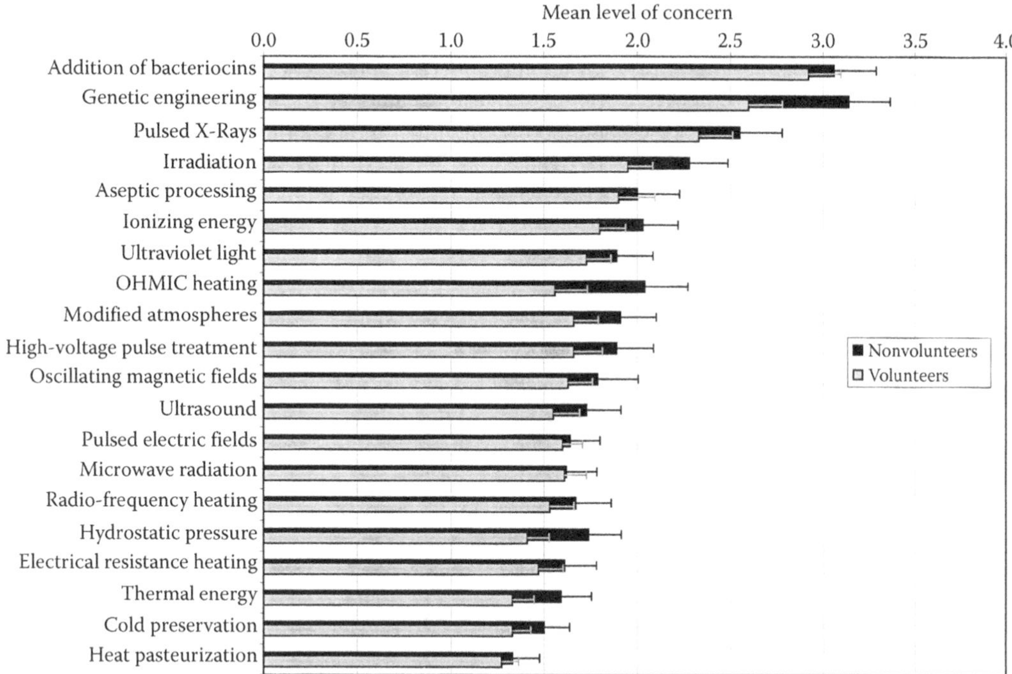

FIGURE 18.7 Mean levels of concern by volunteer status (whether or not they had previously volunteered to test irradiated foods) for twenty different food processes. (From Laboissière, L. H. E. S., Deliza, R., Barros-Marcellini, A. M., Rosenthal, A., Camargo, L. M. A. Q., and Junqueira, R. G., *Food Science and Emerging Technologies,* 8, 469–77, 2007.)

18.3 THE FOOD

18.3.1 Sensory Properties of Foods and Their Measurement

The food itself is the second factor to consider when developing foods using novel technology. But here, we do not mean just the chemical, nutritional, microbiological, or other physical properties of the food. While these properties are important and are the focus of much research into innovative and emerging technologies, it is the *sensory* characteristics of the product that are most important to the consumer. While rheological properties are important to the product developer concerned with flow properties of the food or the nature of the food matrix, what the *consumer* perceives is the food's "consistency," "firmness," crispness," etc. It is these *sensory* properties that drive consumer acceptance of foods, both traditional and otherwise (Cogent 2006). Thus, sensory properties must be carefully quantified to compare any new product under development with products already in the marketplace to determine whether the new product will deliver a comparable (or better) sensory experience to the consumer.

The field of sensory analysis evolved in the years shortly before and after World War II (see Meiselman and Schutz 2003 for a review). Today it takes its place next to food chemistry, microbiology, and nutrition as a cornerstone of food science. This growth in the importance of sensory analysis to food product development has been accompanied by a tremendous growth in the methods to assess the sensory properties of foods. In this section we will discuss several of these methods. There will be special focus on those methods that relate to the characterization of the sensory properties of foods and their acceptance to the consumer.

When conducting sensory analysis, it is important to consider the nature of the data to be obtained, the type of method best suited to obtain that data, and who will or *should* be doing the evaluation. In general there are three types of sensory methods that can be applied to a food product: (1) discriminative; (2) descriptive; and (3) hedonic. *Discriminative* methods, as their name implies, are used to discriminate between food products. They are used when products are very similar to one another, perhaps being identical except for a minor variation in one attribute, and where the aim is to determine whether this minor variation produces a perceptible product difference. There are a large number of discriminative techniques and for readers interested in these methods, they can be found in a variety of textbooks, handbooks, and other publications (Lawless and Heymann 1998; Stone and Sidel 2004; Bi 2005; Sidel and Stone 2006; Meilgaard, Civille, and Carr 2007). More commonly, however, the question to be asked is not *are* the foods different but *how* do they differ and, most importantly, which one does the consumer prefer? These latter questions can be addressed using descriptive and affective methods.

Descriptive methods are those that seek to describe food product(s) using precise sensory vocabulary, so that products produced using different ingredients or processes can be accurately characterized in terms of their sensory differences. In older food and consumer commodity areas, e.g., wine, coffee, tea, dairy, and perfumes, expert "tongues," "noses," or "graders" were often used. Through years of experience, these experts developed keen abilities to detect differences between products and to describe them using detailed, but often idiosyncratic, terminology (Sauvageot, Urdapilleta, and Peyron 2006). Although such experts, today referred to as "flavorists," are still highly regarded as product formulators in many commodity areas, the field of sensory analysis has shifted away from the use of such experts. The reasons for this include the idiosyncratic nature of expert perceptions and opinions (see Langron, Noble, and Williams 1983; Guinard et al. 1999), the inability to apply statistical analysis to data from one individual, a lack of correspondence to what consumers perceive (see data in Figures 18.11 and 18.12 later in text), and the fact that if the expert leaves the organization, new sensory benchmarks must be established using a new expert. As a consequence, expert tasters have been largely replaced by trained descriptive/analytic panels. These are groups of individuals who are selected for their availability, interest, ability to perceive differences in sensory attributes, and personality factors that lend themselves to positive group interactions. They are used only to describe products, never to determine a product's overall quality or acceptability.

Although there are many methods of sensory descriptive analysis, most are patterned after a common approach that utilizes 10–15 members. Panelists are trained in taste, smell, and texture physiology, sensory test methods, and in flavor/texture vocabulary by using lexicons and reference standards for specific sensory character notes like "salty," "rancid," "floral," or "crispy." The panel leader schedules meetings, provides instructions, collects ratings, leads product discussions, and communicates the results of the evaluations to users. Test samples are evaluated by all members of the panel and a group discussion follows. It is the task of the panel to: (1) define the descriptive taste, odor, texture, and aftertaste attributes of the product; (2) indicate the order in which they appear; and (3) rate the perceived intensity of these attributes. Although early descriptive methods, e.g., the Arthur D. Little Flavor Profile Method (Caul 1956), used relatively unsophisticated methods to quantify the intensity of these attributes, subsequent approaches improved upon these measures. The General Foods Texture Profile Method (Brandt, Skinner, and Coleman 1963; Civille and Szczesniak 1973; Civille and Liska 1975) developed standard scales (ordered series of food products representing graduated degrees of the attribute) with associated numerical categories, while Quantitative Descriptive Analysis (Stone et al. 1974; Stone and Sidel 2004) introduced a linear graphic rating scale and "spider web" charts for presenting the data. More recently the methods have evolved to allow greater flexibility and tailoring of terminology to specific products (Meilgaard et al. 2007) and investigators now use a variety of techniques to scale intensity.

A number of studies have utilized sensory descriptive approaches to evaluate both the flavor and texture of foods processed by alternative and emerging technologies (Mor-Mur and Yuste 2003; Baxter et al. 2005; Schirack et al. 2006; Rubio et al. 2007). Figure 18.8 is a spider web graph of data on the flavor and texture attributes of HPP passion fruit juice [300 MPa/5 min/25°C] (HPP), natural passion fruit juice (NAT), and passion fruit juice that was thermally processed (C1–C5). The data were obtained from a trained sensory descriptive panel using the QDA method (Laboissière et al. 2007). The figure graphically shows how the HPP juice is much lower in perceived acidity

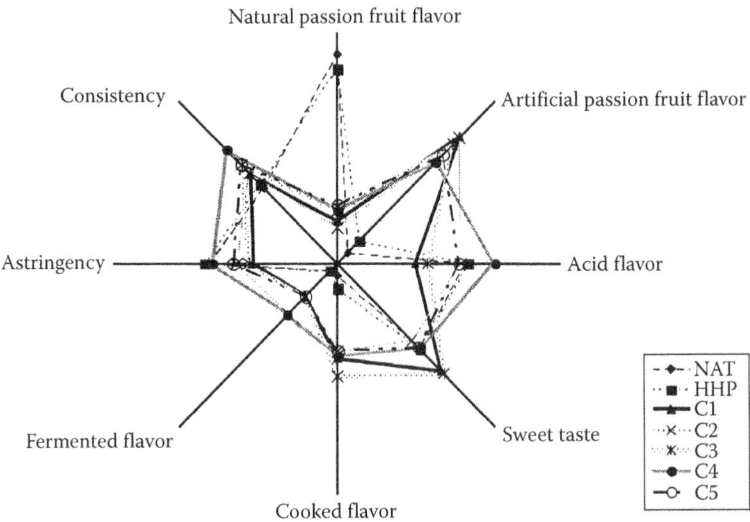

FIGURE 18.8 Appearance and aroma attributes used by a Qualitative Descriptive Profile panel to characterize natural and HPP passion fruit juices (From Cardello, A. V., *Appetite*, 40, 217–33, 2003.)

and degree of artificial and fermented flavors, while also being close to the natural product in terms of "natural flavor." Such descriptive product profiles enable the product/process developer to accurately assess how a product processed by a novel technology compares to its traditional counterpart. Armed with such product profiles, it is then possible to assess process modifications to "move" the profile of the novel product to better match the sensory characteristics of the traditionally processed product or to improve the profile to maximize its acceptance.

18.3.2 Holistic Attributes

Unlike sensory attributes that are directly perceived through the senses, there are a number of attributes of food that, at first glance, might appear to be sensory in nature, but which are more holistic and evaluative in nature, i.e., the result of integrating a number of sensory attributes of the food into a general concept that has *affective* (liking/disliking) overtones. Two of these holistic attributes deserve attention here because they are often cited as key benefits of foods processed by novel technologies—*quality* and *freshness*.

18.3.2.1 Quality

Food quality is commonly defined as "the combination of attributes or characteristics of a product that have significance in determining the degree of acceptability of the product to a user" (Gould 1977). A key phrase in this definition is "combination of attributes or characteristics," which clearly defines quality as a holistic concept formed by perceptions of multiple sensory and other product attributes. These "other" product characteristics may include the product's safety, convenience, cost, value, etc. (Grunert 2002; Grunert et al. 1996; Cardello 1995). The second key phrase, "acceptability of the product to a *user*," establishes quality as a consumer-based concept to be judged by consumers. Judgments of quality by experts are not representative of consumer perceptions of quality, nor are they necessarily consistent with one another (Langron et al. 1983). Although experts may believe that a new process will gain wide consumer popularity because it produces better quality, this is not always the case. The reason it may not is that sensory changes in a product, even those considered to be "for the better," are often not viewed as such by the consumer. This fact was made evident early in the development of alternatives to steam sterilization of vegetables, when new processes for heating canned vegetables resulted in a fresher tasting product. However, the new product had to be targeted first at the institutional food service market, because shoppers had become "accustomed to a certain level of quality deterioration" in their canned vegetables (Food Processing 1988).

Thus, when one considers developing a "better quality product," the product developer or process scientist must temporarily put aside professional knowledge about the chemical, physical, microbiological, and nutritional advantages of the product and focus more on the consumer's perception of quality. Quality is in the eye of the beholder and it is the consumer's eye that determines the success of a new product.

18.3.2.2 Freshness

Like quality, freshness is not an inherent sensory attribute of the product. It is a holistic perception derived from a composite of sensory and other attributes of the food, which over time have come to be associated with "freshness" in the mind of the consumer. To the consumer, freshness is commonly perceived in raw or natural products and the concept loses much of its meaning when applied to processed foods. Regulatory agencies, like FDA, allow the term "fresh" to be used for unprocessed foods. However, regulations do not preclude using it to describe foods that have undergone benign or "common practice" treatment, such as with waxes and coatings, mild chlorine or acid washes, or the use of ionizing radiation not to exceed 1 kG on raw foods. But what about other "minimal" or lower impact processes, e.g., HPP, pulsed electric field, ultrasound, ultraviolet light processing, etc.? The Institute of Food Technologists, in a reply to FDA hearings on the use of the term fresh for these technologies noted that "The ultimate measurement of *fresh* is likely to be based on an individual consumer's own sensory perception. Consumers are likely to compare products that are labeled *fresh* to similar products that they have become accustomed to purchasing with that label. The consumer's ultimate satisfaction with the product involves a comparison of the expected sensory quality (based on previous *fresh* label product experiences) with the actual product characteristics." (Mermelstein 2001). This statement recognizes that the consumer is the ultimate arbiter of freshness in a food product. In addition, it places emphasis on the *expectations* of the consumer and whether the product meets these expectations. As will be discussed later, consumer expectations drive many aspects of new product acceptance.

So how do consumers perceive freshness in a food product? Although consumer surveys show that freshness is important to (1) the perception of a high-quality product (George 1993), (2) consumers' food choices (Lappalainen, Kearney, and Gibney 1998), and (3) the choice of food markets in which to purchase produce (Rhodus, Schwartz, and Hoskin 1994), only a small number of studies have examined the factors that contribute to perceived freshness. In one such study, Fillion and Kilcast (2001) found that sensory factors (appearance, flavor, and texture) and time from harvest are critical to consumers' perceptions of freshness in fruits and vegetables. Pèneau et al. (2007) found texture to be a primary determinant of the perceived freshness of apples. In another study, conjoint analysis was used to investigate consumer perceptions of freshness in a range of products (Cardello and Schutz 2002). In this study, the time in days from the product's arrival in the store/market was the primary determinant of consumer freshness judgments, supporting the earlier results of Fillion and Kilcast (2001). In decreasing order of importance were the type of food (fruit, meat, vegetable), the nature of the processing/preservation method, and the source of the food (supermarket, warehouse store, etc.). Figure 18.9 shows the relative contributions to freshness of different processing/preservation labels that were examined in this study. Foods described as "frozen then thawed" produced a strong negative impact on perceptions of freshness. Foods described as "minimally processed" were perceived as fresher than "frozen then thawed" and on a par with "refrigerated/frozen" foods. However, "minimally processed" food was judged the worst of the four methods of processing by the military segment. This finding may be attributable to the gender (more male) or age (younger) difference in this consumer

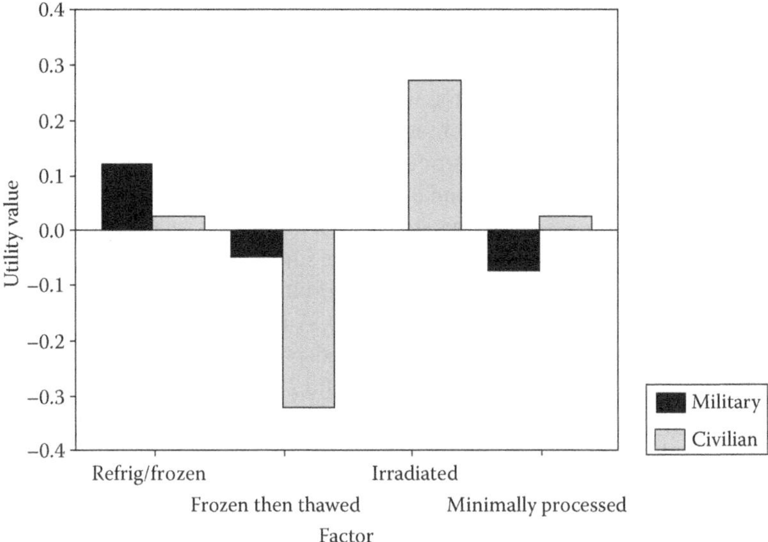

FIGURE 18.9 Average importance values of different food processes to the evaluation of the "freshness" of foods. (From Cardello, A.V., and Schutz, H. G., *Journal of Food Science,* 68, 1519–24, 2003.)

group, which may have led to a more literal interpretation of the word "minimally." Although "minimally processed" has positive connotations to food processing scientists, one interpretation of it is "not sufficiently processed." Also interesting in these data is the fact that the term "irradiated" had a high positive influence on freshness ratings for the older, better educated civilians, but not for the younger military consumers. These results are consistent with other research showing that consumers with higher education have more positive attitudes toward food irradiation (Schutz, Bruhn, and Diaz-Knauf 1989; Lusk, Fox, and Mcilvain 1999) as well as other novel processes (Saba, Moles, and Frewer 1998; Gaskell et al. 1999; Heiman, Just, and Zilberman 2000).

18.3.3 Affective Attributes

Although consumer judgments of quality and freshness are very important, the most important judgments are those that relate to whether or not the product is liked or disliked. These attributes of products are known as hedonic or affective. Both terms refer to the conscious, subjective experiences that accompany emotion. *Hedonic* derives from "hedonism" and refers to the emotional experience of pleasure (or displeasure), liking (or disliking), and comfort (or discomfort). Affective or hedonic methods used in sensory science can be divided into two major types: (1) preference methods, which examine whether differences in liking exist among products; and (2) acceptance methods, which index the degree of liking or disliking for a product.

18.3.3.1 Preference

Preference refers to the choice of one sample over another. If only two samples are compared, such as when comparing a PEF sample to a control, the test is called a *paired preference test*. The two food samples are presented sequentially or simultaneously, depending on the test objectives. The order of presentation is balanced temporally and spatially to avoid order and position biases. The consumer simply chooses the sample that he/she likes the best and the number of individuals who prefer one sample over the other is then compared to statistical tables found in standard sensory texts (Meilgaard et al. 2007; Lawless and Heymann 1998; Resurreccion 1998) to determine if a statistically significant difference exists.

When more than two samples are compared, such as when HPP products are processed at several different temperatures, a *multiple paired preference test* is used. Here, all possible pairs of the samples are presented in a balanced order. The relative number of times that sample A is chosen over sample B, etc. establishes a scale of liking for all the products in the sample set. Since this approach can be laborious due to the large number of pairs that must be presented to accommodate all possible pairs in a large set of samples, a "short-cut" technique known as "best–worst" or "maximum difference" scaling has been developed (Finn and Louviere 1992; Marley and Louviere 2005). The best–worst approach uses a statistically generated design in which a subset of all the test samples is presented in groups of three or more and the consumer simply chooses the best and worst product from among each set. It has been suggested that the resulting data produce a ratio scale of liking for all the products tested (Marley and Louviere 2005). A number of researchers have applied this method to the measurement of food- and meal-related properties and liking with good success (Jaeger et al. 2008; Heine et al. 2008; Jaeger and Cardello 2009).

Perhaps the simplest way to determine the relative preferences for a set of products is to conduct a *preference ranking* test. Here, samples are presented simultaneously to the consumer in a balanced or random order and the consumer assigns numerical ranks to them to indicate the order of his/her preferences for the samples. The data are then analyzed using nonparametric statistical tests (e.g., Friedman's ANOVA) with appropriate multiple comparisons or rank sum statistics that can be found in published tables (Gacula and Singh 1984; Basker 1988a, 1988b). For reasons related to panelist fatigue and memory limitations, the maximum number of samples that should be tested in a ranking procedure is five to six (ASTM 1996).

The preference testing methods discussed above all have limitations. First, all but the last require a large number of sample presentations of two or more products. Second, they do not provide information about absolute levels of liking or disliking. In a paired preference test that shows Sample A is preferred to Sample B, the researcher still does not know if both samples are liked, if both samples are disliked, or if one is liked and the other disliked. Thirdly, paired preference and ranking methods do not provide information about the degree of difference among samples. They produce only ordinal data. In order to obtain information about absolute levels of liking/disliking, a direct rating of liking-disliking or *acceptance* is required.

18.3.3.2 Acceptance

The most common and best-known method for rating consumer liking/disliking of foods is the *9-pt hedonic scale* (Peryam and Pilgrim 1957). This scale is shown on the left side of Figure 18.10. The scale consists of a set of verbal labels that define different "categories" along the liking/disliking dimension. Although the verbal categories are assumed to represent equal-sized intervals along that dimension, this assumption has been shown to be false (Moskowitz and Sidel 1971; Moskowitz 1977, 1980; Schutz and Cardello 2001). That is, the difference in acceptability between two products rating "1" vs "2" on the scale is not the same as the difference between two products rating "4" vs "5" or "8" vs "9." Another problem with category scales is that consumers fail to use the extreme categories, so that the scale effectively becomes a seven-point scale of liking/disliking. Also, the presence of the neutral category (neither like nor dislike) decreases the efficiency of the scale (Jones, Peryam, and Thurstone 1955) and encourages complacency in judgments by providing the consumer a "safe" category into which foods that elicit only mild liking or mild disliking can be placed (Gridgeman 1961; Olsen 1999).

A different method for measuring liking/disliking of foods that is not dependent on the use of verbal categories is line scaling, also known as *linear graphic* or *visual analog scaling*. A linear graphic rating scale consists of an unbroken horizontal or vertical line that is anchored on both ends with a verbal label. Such a scale also may be anchored in the middle or elsewhere. Linear graphic rating scales have been used in a number of studies to rate liking/disliking of foods (Giovanni and Pangborn 1983; Rohm and Raaber 1991; Hough, Bratchell, and Wakeling 1992). When presented without instructions, linear graphic rating scales are presumed to provide interval level data.

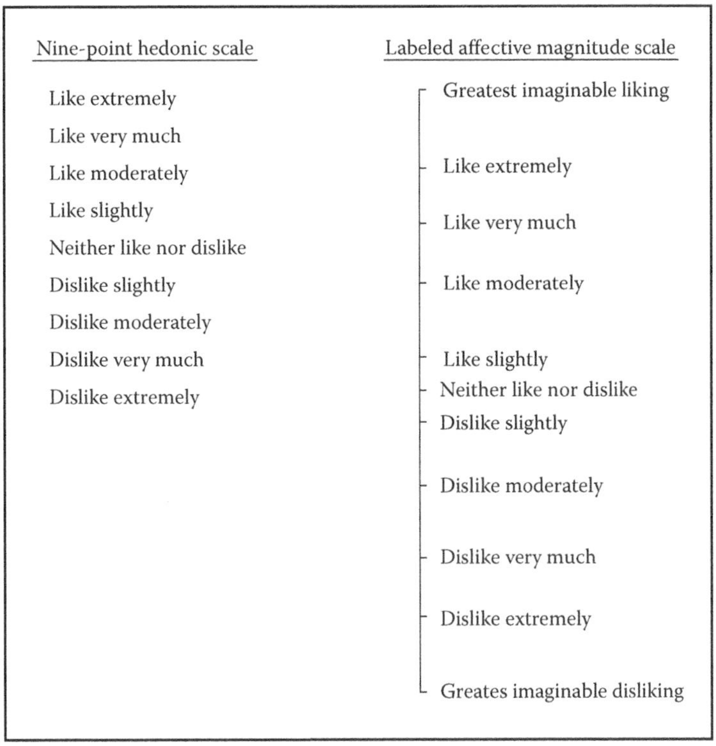

Nine-point hedonic scale	Labeled affective magnitude scale
Like extremely	Greatest imaginable liking
Like very much	Like extremely
Like moderately	Like very much
Like slightly	Like moderately
Neither like nor dislike	Like slightly
Dislike slightly	Neither like nor dislike
Dislike moderately	Dislike slightly
Dislike very much	Dislike moderately
Dislike extremely	Dislike very much
	Dislike extremely
	Greates imaginable disliking

FIGURE 18.10 The 9-pt hedonic scale (left) and the labeled affective magnitude (LAM) scale (right) (From Cardello, A.V., and Wise, P.M., *Product Experience*, Oxford, Elsevier, 91–131, 2008.)

Magnitude estimation is a "ratio" method of scaling that was first developed by Stevens (1957a, b). In magnitude estimation, consumers evaluate their liking for one product relative to another using numbers (Moskowitz 1977). If one product is liked twice as much as another, it is assigned a number twice as large. If it is liked one-third as much, it is assigned a number one-third as large, and so forth. Magnitude estimation was first applied to the measurement of consumer food liking by Moskowitz and Sidel (1971). Although magnitude estimation produces ratio level data, which many other scaling methods do not, it has not been widely adopted by the food industry. The reasons for this relate to difficulties involved in its use with untrained subjects, complexity in analyzing the data, and the failure to find advantages in reliability or sensitivity.

The most recently developed approach for scaling consumer liking/disliking is the *Labeled Affective Magnitude (LAM) scale* (see right side of Figure 18.10) (Schutz and Cardello 2001; Cardello and Schutz 2004). This scale is predicated on research by Green, Shaffer, and Gilmore (1993) and Green et al. (1996), who developed a labeled magnitude scale of oral sensation by determining the spacing and location of verbal labels representing different intensities of oral sensation using magnitude estimation. By scaling the semantic meaning of word phrases representing different degrees of liking and disliking, including the phrases "greatest imaginable liking" and "greatest imaginable disliking," it was possible to place these phrases along a visual analog scale in accordance with their semantic meaning (Schutz and Cardello 2001). A critical characteristic of this scale is the presence of the end-point anchor phrases of "greatest imaginable liking/disliking," which are used as fixed endpoints on the affective dimension and which allow judgments of different consumers to be placed on a common "ruler." To use the scale, consumers simply place a mark through the line to indicate the perceived strength of their liking/disliking. Studies have shown that the scale produces equal or better sensitivity, greater reliability, and equivalent ease of use to the nine-point hedonic scale (Schutz and Cardello 2001), and it is now being widely adopted as a standard method for rating consumer liking of foods (Forde and Delahunty 2004; Greene et al. 2006; Guest et al. 2007; Chung and Vickers 2007a, 2007b; Keskitalo et al. 2007; Cardello, Lawless and Schutz 2008).

Figures 18.11 and 18.12 show data obtained using the LAM scale to compare six different products, five of which were treated by HPP and one of which was traditionally processed (unpublished data from A. Wright). The data in Figure 18.11 were obtained from a group of nonthermal food processing professionals. The data in Figure 18.12 were obtained from an employee consumer panel.

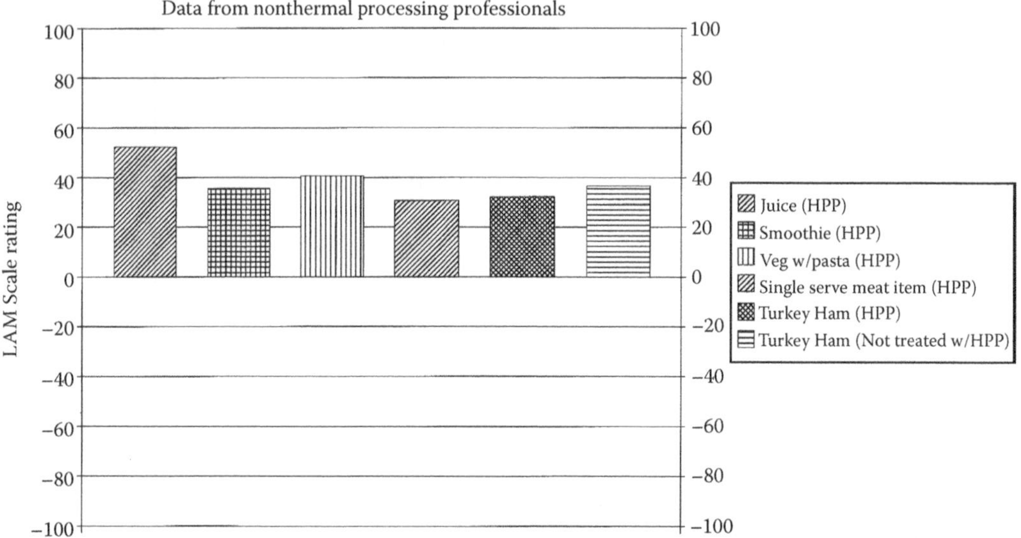

FIGURE 18.11 Labeled Affective Magnitude (LAM) scale ratings of liking/disliking obtained from the evaluation of HPP and conventionally processed foods by nonthermal processing professionals. (Unpublished data from A. Wright, 2008.)

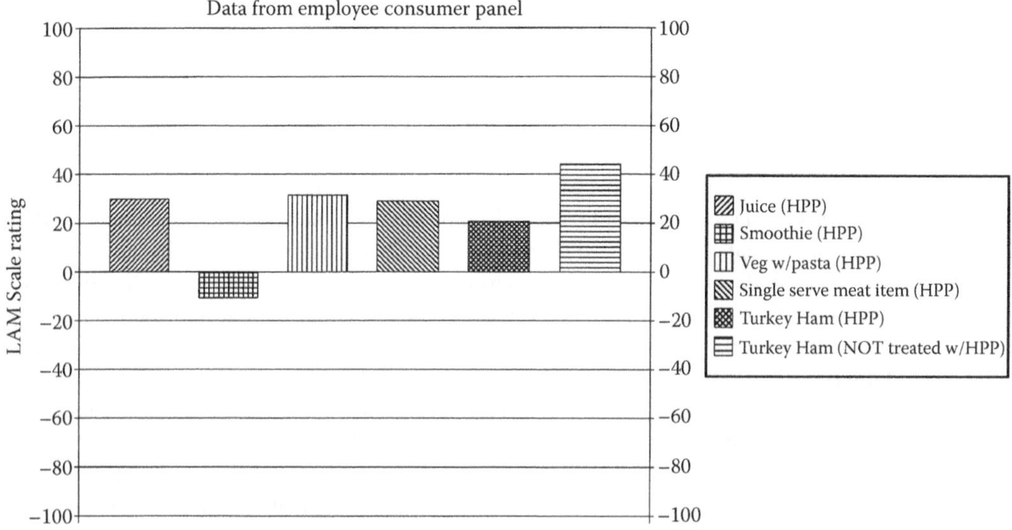

FIGURE 18.12 Labeled Affective Magnitude (LAM) scale ratings of liking/disliking obtained from the evaluation of HPP and conventionally processed foods by a laboratory consumer panel. (Unpublished data from A. Wright, 2008.)

Comparing the HPP-processed turkey ham with its untreated control, one can see that the HPP product was liked less than the control product by both test groups. Since the LAM scale enables ratio statements to be made about levels of liking, we can make the assertion that the nonthermal professionals liked the control product about 15% more than the HPP-treated product, whereas the employee panel liked the control more than twice as much as the HPP product. Such ratio statements could not be made with simple category scales of liking that provide only interval-level data. Also of interest is the large difference between the two test groups in their liking for the HPP products. For all foods, the industry professionals liked the HPP-treated products more than the average consumers. This is a good example of the differential perceptions of experts and consumers, and it serves as a warning that expert opinions cannot substitute for, or be used to predict, consumer responses to foods processed by emerging technologies.

18.4 THE INFORMATIONAL CONTEXT

18.4.1 Information Processing: Basic Elements

The third factor of critical importance to the development and marketing of foods using novel food processes is the informational *context.* The reason that information is such an important factor is that it has a powerful capacity to alter attitudes and perceptions toward products, whether for the better, as through information about product benefits, or for the worse, as through information or misinformation about the safety of a processing technique. By *information* we mean verbal and nonverbal stimuli available to the consumer, prior to or at the time that the food is encountered. Thus, the informational context of a food encompasses a large number of variables that affect consumer choice and purchase behavior. It may include (1) label information, such as ingredients, nutritional information, and price or brand information, (2) information imparted by package designs and graphics, (3) store-shelf and other in-store information, (4) media advertizing, and (5) other written or verbal information obtained about the product or how it is processed through the media or by word of mouth.

Figure 18.13 is a schematic model that shows the major elements involved in the processing of information. There are eight stages. At various times and for various purposes, academicians from different disciplines have focused on one or more of these information processing stages to develop their own explanatory model of how information influences human behavior. For the purposes of this chapter, we will provide a brief summary of the elements in Figure 18.13 and describe how they play a role in the processing of information about food.

The first element in Figure 18.13 is the *information* itself. As noted previously, this may refer to textual, graphic, or other types of information that may appear in the form of nutrition, package, brand, or price labels, store displays, etc. Critical elements of the information include the message *content* or meaning, its *complexity*, the sensory *modalities* through which it is presented (visual, auditory, and tactile, etc.), its *salience*, and the *source* of the message. The content of the message is especially important because it is what modifies behavior toward the object. The complexity of the information is important because it affects whether or not the message is attended to, i.e., low complexity messages are forced to compete for attention with other more complex messages, while extremely complex messages could be lost in sensory overload. The sensory modality by which the information is presented and its salience are important because they affect the ability of the information to attract the attention and awareness of the consumer, while the source of the information is important because it influences the consumer's belief and confidence in the veracity of the information, i.e., trusted sources result in a greater likelihood of the message being received positively (Eagly and Chaiken 1984).

For foods processed by novel technologies, where there may be multiple sources of information, it is clear that a consensus among sources produces a greater likelihood of positive reception than sources that are in conflict (Dean and Shepherd 2007). However, trust in the source is critical

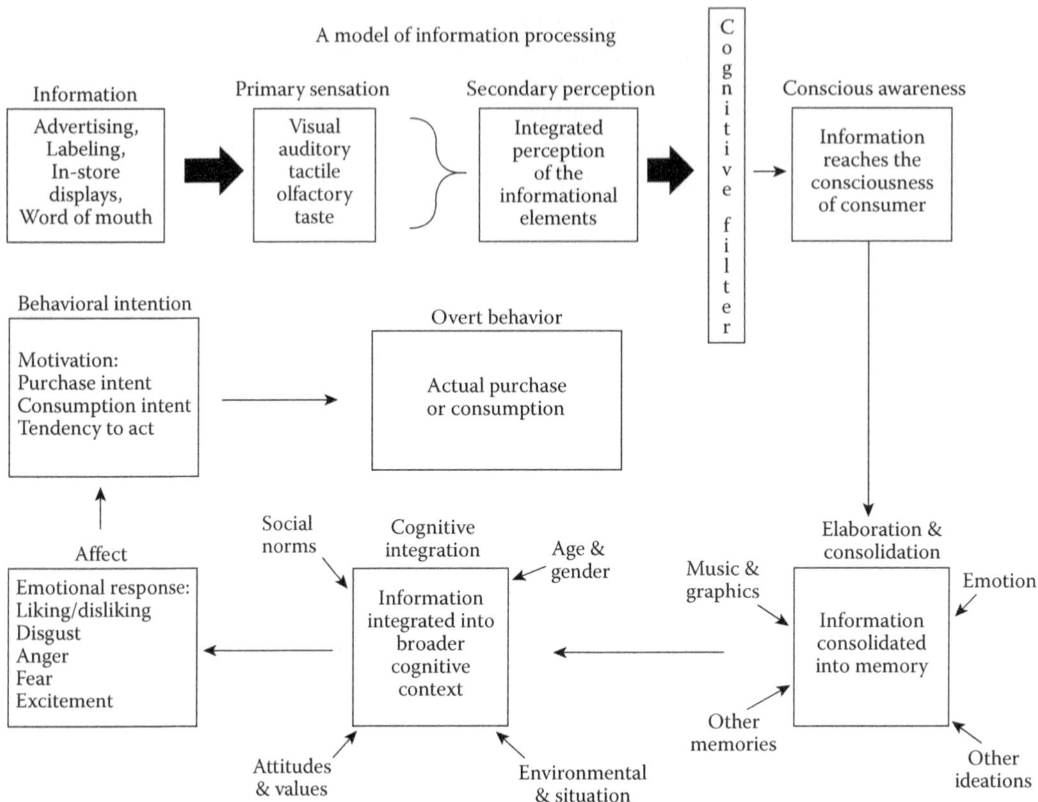

FIGURE 18.13 A schematic model of the phases involved in the cognitive processing of information.

(Frewer et al. 1996, 1999; Bord and O'Conner 1990; Earle and Cvetkovich 1995; Grobe, Douthitt, and Zepeda 1999; Siegrist 2000; Green, Draper, and Dowler 2003; Finucane and Holup 2005). Consumer organizations are generally well trusted (Frewer et al. 1999), although this is not universal, e.g., Greenpeace is often perceived in a negative light by consumers (Cardello et al. 2007). In the United States, government agencies are often well trusted (Cardello et al. 2007), but in some European countries government agencies are the least trusted sources of information (Moses 1999). Among the least trusted sources in most surveys are scientists, politicians, and industry experts (Macintyre et al. 1998; Zechendorf 1994; Cardello et al. 2007).

When product information is first encountered, it is sensed as a diverse array of visual, auditory, or other *primary sensations*. These disparate sensations are then integrated to form multimodal *secondary perceptions* that become available for further processing. Conscious *awareness* of this integrated information occurs in the next stage. While all informational percepts may be perceived, relatively few reach conscious awareness, because attention and other cognitive filters funnel the total amount of available information to only those elements that are relevant to the recipient of the information. For example, product information may pass unnoticed by the consumer if their interest or involvement in the product category is low, but will rise to conscious awareness when interest or involvement is high. Thus, technology-receptive consumers may attend more readily to novel foods and food processes, but so too will those who are actively opposed to such technologies.

Once brought to conscious awareness, information can be augmented by other memories and then consolidated for later retrieval. This *elaboration* and *consolidation* stage has been shown to speed retrieval from memory and allow judgments to be made with greater certainty (Fazio et al. 1982; Powell and Fazio 1984). Graphics, pictorial information, and music will aid in elaborating product information by making use of previous associations formed between these images or sounds and other ideations and affective experiences (Tybout and Artz 1994). This is why, for example, many of today's television advertizements present vibrant, positive visual images and catchy, upbeat music in order to promote an emotional reaction that can become associated with the product.

Following the elaboration and consolidation of the information, it is *cognitively integrated* with individual, social, and situational factors, all of which may modify how the information is received. It is at this stage that individual differences in age, gender, or previously formed attitudes and beliefs about the object of the information can influence its interpretation (Meyers-Levy and Maheswaran 1991; Meyers-Levy and Steinthal 1991; Worth, Smith, and Mackie 1992). It is here, too, that the social norms of importance to the individual and situational contexts can alter the nature and/or importance of the message and its likely impact on behavior (LaTour and Manrai 1989; Herr 1989; Yi 1990).

At some point during the information integration stage, *affect formation* occurs. It is here that liking/disliking toward the object of the information (the product) can be measured (see Section 18.3). Once affect arises, it results in the formation of a *behavioral intention*, i.e., the propensity to behave toward the product in a given way (to purchase or not, to consume or not, etc.). In most cases, a behavioral intention leads directly to the *overt behavior*, assuming that the opportunity exists for that behavior to be performed, e.g., in order to purchase a newly advertized product, the product must be available in stores (or on Internet sites) to which the consumer has access.

18.4.2 Information Effects on Food Liking and Behavior

In some studies of the effect of information on liking or behavior toward novel foods and processes, the effect of information has been counterintuitive. Remember, for example, the finding by Deliza et al. (2003) for HPP products. Similarly, Frewer et al. (2000) found that providing any information about advantages or disadvantages of GM foods made it *less* likely for the GM product to be selected. However, in most cases, well-crafted information can have positive effects on liking and behavior. Thus, in several studies on irradiated food, factual information, benefit statements, and endorsements by trusted sources have been shown to have positive effects on attitudes, liking, choice, and intended purchase of these products (Bruhn et al. 1986; Schutz et al. 1989; Pohlman, Wood, and Mason 1994; Resurreccion et al. 1995; Bruhn, 1995a, 1995b; Hashim et al. 1996; Schutz and Cardello 1997).

Among the many aspects of information content that have been examined, it is clear that statements about the benefits of a new food process are especially effective (Bruhn 1995b; Frewer et al. 1997a,b; Fox, Bruhn, and Sapp 2001; Johnson et al. 2004; Zienkewicz and Penner 2004). It has been demonstrated that, in some cases, benefits like better taste, better nutrition, and better freshness have more positive influences on consumers than benefits like fewer preservatives, minimal processing, and even reduced bacterial risk (Cardello et al. 2007). In almost every study that compares the factors important to consumer acceptance, choice, and consumption of food, taste is the most important factor (Cardello and Schutz 2003). As such, developers of novel technologies should focus their efforts on producing a better tasting product and using the "better taste" message during marketing.

Although many researchers have demonstrated the empirical effects of information on the acceptance of and behavior toward novel foods, far fewer have attempted to develop theoretical models about *how* information influences liking and behavior. In this regard, the most often cited theoretical model for explaining consumer behavior has been the Fishbein–Ajzen model of reasoned action (Fishbein and Ajzen 1975; Ajzen and Fishbein 1980). This model, originally formulated for social psychological applications, integrates the attitudes of the individual with perceived social pressures, in order to predict behavioral intention. The equation for this relationship is

$$I = w_1 A + w_2 \mathrm{SN}, \tag{18.1}$$

where I is the behavioral intention, A is the attitude toward the behavior, SN is the social norm, and w_1 and w_2 are weighting factors of the relative importance of attitudes versus social norms in predicting the likelihood of the behavior. A large number of studies have been conducted in an attempt to predict the intended behavior toward food from written surveys of attitudes and perceived social norms (Bonfield 1974; Shepherd and Stockley 1985; Tuorila-Ollikainen, Lahteenmaki, and Salovaara 1986; Shepherd and Farleigh 1986; Tuorila 1987; Shepherd 1988; Lobb, Mazzocchi, and Traill 2007). Among those that have focused on novel foods and food processes, there has been some success in predicting the behavioral intentions of consumers toward them (Sparks et al. 1995; Saba and Vassallo 2002; Chen 2007; Olsen et al. 2008).

The theoretical model that most directly addresses how information influences consumer behavior is expectancy-value theory (see Feather 1982). This theory hypothesizes that information creates "expectations" about the product, and that the strength of the expectation is related to both the belief strength and confidence in the information and its source. Product experiences that are consistent with expectations confirm the belief strength/confidence, strengthen the existing attitudes, and result in consumer satisfaction. However, product experiences that are inconsistent with expectations have negative effects on belief strength and attitudes. Such product experiences also affect satisfaction with the product, but the influence depends upon the valence of the relationship between product expectations and actual product performance (van Raaij and Fred 1991).

Expectation theory has received significant attention from food psychologists and a number of different predictive models of the effect of disconfirmed expectations on perceived product satisfaction, quality, and liking have been proposed. These models have been reviewed by Cardello (1994,2007), Deliza and MacFie (1996), and Schifferstein (2001). However, the model that has received the greatest support is known as the *assimilation-contrast* model. This model predicts that when there is no disconfirmation, i.e., the intrinsic quality or acceptability of the product matches its expected quality or acceptability, liking will be unchanged from its baseline level. However, if the expectation level is high and the intrinsic quality of the product is low (a state of negative disconfirmation) or if the expectations are low, but the intrinsic quality is high (a state of positive disconfirmation), the perceived acceptability will assimilate the expectation and increase or decrease in the direction of the expected level. Lastly, if expectations are either far above or far below the actual intrinsic quality of the product, a contrast effect may occur, causing the perceived acceptability to move in a direction *opposite* the expected level.

General support for this assimilation-contrast model of disconfirmed expectations has been found in numerous studies on a wide variety of food products (e.g., Cardello and Sawyer 1992; Helleman et al. 1993; Tuorila, Cardello, and Lesher 1994; Deliza 1996; Cardello et al. 1996; Schifferstein, Kole, and Mojet 1999; Lange et al. 2000; Siret and Issanchou 2000; DiMonaco et al. 2004; Caporale and Monteleone 2004; Stefani, Romano, and Cavicchi 2006; Iaccarino et al. 2006), with the majority of observed effects falling into the assimilation portion of the model.

18.4.3 Concerns, Expectations, and Liking for Foods Processed by Novel Technologies

The implications of the assimilation-contrast model are twofold. First, on the negative side, if consumer expectations of foods are low, regardless of the reason, the assimilation-contrast model predicts that liking of the taste of these products will suffer. This effect will be relatively independent of the intrinsic sensory quality of the item. Thus, given the high levels of concern about consuming foods processed by some novel technologies (Sections 18.2.1 and 18.2.2), it is of critical importance to determine how these negative attitudes influence expected liking for the product. On the positive side, if product expectations can be raised by communications or marketing strategies that overcome negative attitudes toward the product, the model predicts that liking should improve. Lastly, the contrast portion of the model that occurs when expectations and reality diverge wildly serves as a warning to marketers and advertizers that exorbitant claims for a product that outstrip the product's capacity to deliver on those claims will produce a strong negative reaction in the consumer to both the product and the brand.

In a study of the effect of processing information on the acceptability of beer, Caporale and Monteleone (2004) informed consumers that a beer had been produced using GM yeast, organic barley and hops, or traditional brewing technology. These investigators showed convincingly that the processing information altered expectations for the liking of the beer, such that consumers expected to like beers processed with GM yeast the least (4.63–4.88 on a nine-point hedonic scale), beers processed traditionally much better (6.31–6.63), and organic beers the most (6.86–7.08). More importantly though, these investigators showed that the liking for the tasted beers assimilated these expectations, so that the liking for the beers increased or decreased in accordance with the effect of the labelon their expected liking for them.

Although Caporale and Monteleone (2004) demonstrated a direct link between information about novel processes and its effect on expected and actual liking, there was no link to consumers' attitudes, i.e., it was merely presumed that the lower expectation in the GM condition was due to a negative attitude toward that technology. In a subsequent study, Cardello and Schutz (2005) demonstrated this additional link in an experiment that gave consumers different information about a thermally processed corn product. In a pretest survey, all consumers rated their concern level for a variety of food technologies, including GM. Subsequently, they were led to believe that the products they were about to taste (1) did not contain GM corn, (2) did contain GM corn, or (3) may contain GM corn. After receiving this information, but prior to tasting the products, consumers were asked to rate their expected liking for them. After the consumers tasted the products, they rated their actual liking for them. The results of this study showed that consumers' expected liking of the corn was negatively and significantly correlated with their pretest concern about GM products. As in the study by Caporale and Monteleone (2004), concern levels were also found to be negatively correlated with actual liking of the product (r=–0.54 to –0.36). Thus, the concern with the technology was associated with lower expectations of liking for the products thought to be processed by that technology, and these expectations, in turn, translated into lowered liking of the product when tasted, in keeping with predictions of the assimilation-contrast model.

The above studies on GM foods raise the issue of whether other, more benign and less worrisome (to the consumer) technologies would produce similar effects. The answer to this question can be found in another study in which consumers tasted and evaluated chocolate pudding products, all of which had been thermally processed (Cardello 2003). The consumers participated in one of three different information conditions. In all three conditions, the consumers first rated their liking of chocolate pudding (no other information presented) and their level of concern for several different food processing technologies ("irradiation," "high voltage pulses," hydrostatic pressure," "pulsed electric fields," "nonthermal preservation," and "the addition of bacteriocins"). In a second session, conducted several weeks later, the same consumers participated in a taste test in which they were provided information about the chocolate puddings that they were to taste. In one condition, they were simply told that it had been processed by one of the six food processing technologies cited above, e.g., "this sample was processed by pulsed electric fields." No other information was provided. For consumers in the second condition, the name of the technology was accompanied by an objective description of that technology, e.g., "in irradiation processing, foods are exposed to a source of ionizing radiation, e.g., cobalt 60, for short periods of time." In the third condition, they were told the name of the technology, given the description of it, and provided a safety benefit statement, i.e., "this process is entirely safe and avoids the thermal damage done to foods by heat pasteurization." After exposure to the information, consumers rated their expected liking of the product (prior to receiving the sample) and then their liking after tasting it.

Results from this study showed that the concern ratings for the different food processing technologies were again highly correlated with expectations of liking for the chocolate pudding

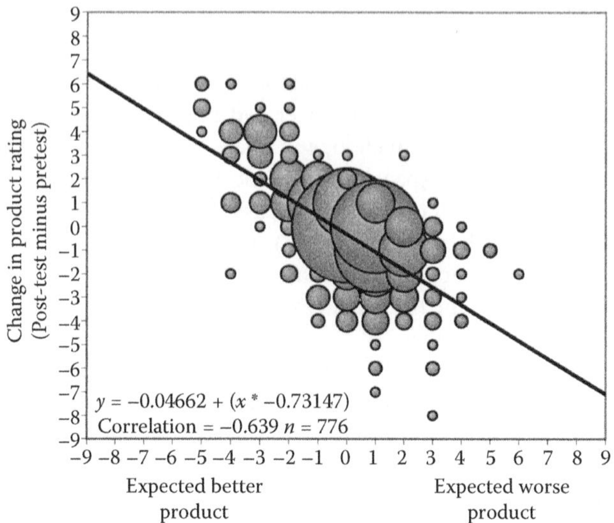

FIGURE 18.14 A graph of liking/disliking versus the degree of disconfirmation of expectations experienced. (From Cardello, A. V., *Appetite*, 40, 217–33, 2003.)

(r=−0.45, −0.31, and −0.40). *Higher* consumer concerns about the risk of the technology produced *lower* expected liking of the food. And once again, these diminished expectations resulted in systematic decreases in ratings of the actual liking of the tasted products. Figure 18.14 shows these data, where the abscissa reflects the degree to which the consumers expected a better or worse chocolate pudding (based on the technology name) and the ordinate reflects the change in liking for the pudding from a baseline measure. As can be seen, the greater the degree to which the consumers expected a better product, the greater the liking when the product was tasted.

In addition to demonstrating the link between consumer attitudes toward a product technology, expectations of liking or disliking for the product, and actual liking/disliking, this study also showed that factual product and process information, product exposure (merely seeing the product), and a simple safety/benefit statement can all have positive effects on expected liking and acceptance. These data confirm the fact that good communication practices and effective marketing strategies can be utilized to overcome negative images created by risk perceptions or product stereotypes and significantly improve the probability of market success for foods processed by emerging technologies.

18.5 RECOMMENDATIONS FOR CONSUMER-LED DEVELOPMENT OF FOODS PROCESSED BY NOVEL TECHNOLOGIES

There are several recommendations that can be made, based on the data presented here, to product developers and other parties interested in developing and successfully introducing foods processed by novel and emerging technologies. These include:

- Pay greater attention to the attitudes and risk perceptions of consumers than to the risk assessments of "experts" when designing any new product using novel technologies.
- Be aware of different consumer segments in your potential market and how risk perceptions and product concerns vary by these market segments (gender, ethnicity, psychographics, etc.).
- Ensure that your product is optimized for sensory quality prior to market launch. Nothing will better validate poor expectations than a poor product.

- Conduct premarket research on consumer risk perceptions, product expectations, alternative technology names (irradiation vs ionizing energy), brand logos/names, label graphics, benefits, and other elements of potential marketing communication.
- Be ready with a complete marketing and communication strategy at the time of launch to overcome negative attitudes, exaggerated risk perceptions, and unchallenged beliefs about the product or process.
- Employ consumer-led product development throughout all stages of product testing and marketing in order to ensure that product characteristics meet the needs and expectations of the consumer, while overcoming potential roadblocks to full acceptance.

By utilizing these recommendations, novel food product/process developers will be better attuned to the "voice of the consumer" and will be able to avoid the pitfalls that can be created by expert bias. In so doing, they will ensure that their own creative efforts and those of other process scientists, whose research is detailed in other chapters of this book, will culminate in successful market introductions for foods processed by novel and emerging technologies.

REFERENCES

Ajzen, I. and Fishbein, M. 1980. *Understanding Attitudes and Predicting Social Behavior.* Englewood Cliffs, NJ: Prentice-Hall.

ASTM, Committee E-18. 1996. *Sensory Testing Methods.* ASTM Manual series MNL26, 2nd ed. Conshokecken, PA: American Society for Testing and Materials.

Basker, D. 1988a. Critical values of differences among rank sums for multiple comparisons. *Food Technology* 42 (2): 79–84.

______. 1988b. Critical values of differences among rank sums for multiple comparisons by small taste panels. *Food Technology* 42 (7): 88–89.

Baxter, I. A., Easton, K., Schneebeli, K., and Whitfield, F. B. 2005. High pressure processing of Australian navel orange juices: Sensory analysis and volatile flavor profiling. *Innovative Food Science and Emerging Technologies* 6:372–7.

Bi, J. 2005. *Sensory Discrimination Tests and Measurements: Statistical Principles, Procedures and Tables.* Indianapolis: Wiley-Blackwell.

Boecker, A., Hartl, J., and Nocella, G. 2008. How different are GM food accepters and rejecters really? A means-end chains application to yogurt in Germany. *Food Quality and Preference* 19:383–94.

Bonfield, E. H. 1974. Attitude, social influences, personal norms, and intention interactions as related to brand purchase behavior. *Journal of Marketing Research* 11:379–89.

Bord, R. J. and O'Conner, R. F. 1989. Who wants irradiated food? Untangling complex public opinion. *Food Technology* 43:87–90.

Bord, R. J. and O'Conner, R. E. 1990. Risk communication, knowledge and attitudes: Explaining reactions to a technology perceived as risky. *Risk Analysis* 10 (4): 499–506.

Brandt, M. A., Skinner, E. A., and Coleman, J. A. 1963. Texture profile method. *Journal of Food Science* 28:404–409.

Braun, R. 2001. Why are Europeans scared of biotech food? *Agricultural Food Industry Hi-Tech* 5:32–34.

Bredahl, L. 1999. Consumer's cognitions with regard to genetically modified foods: Results of a qualitative study in four countries. *Appetite* 33:343–60.

______. 2001. Determinants of consumer attitudes and purchase intentions with regard to genetically modified foods-results of a cross-national survey. *Journal of Consumer Policy* 24:23–61.

Brewer, M. S., Sprouls, G. K., and Russon, C. 1994. Consumer attitudes toward food safety issues. *Journal of Food Safety* 14:63–76.

Brug, J., Debie, S., van Assema, P., and Weijts,W. 1995. Psychosocial determinants fruit and vegetable consumption among adults: Results of focus group interviews. *Food Quality and Preference* 6:99–107.

Bruhn, C. M. 1995a. Consumer attitudes and market response to irradiated food. *Journal of Food Protection* 58:175–81.

Bruhn, C. M. 1995b. Strategies for communicating the facts on food irradiation to consumers. *Journal of Food Protection* 58:213–16.

Bruhn, C. M., Schutz, H. G., Johns, M. C., Lamp, C., Stanford, G., Steinbring, Y. J., and Wong, D. 1996. Consumer response to the use of lasers in food processing. *Dairy, Food and Environmental Sanitation* 16 (12): 810–16.

Bruhn, C. M., Schutz, H. G., and Sommer, R. 1986. Attitude change toward food irradiation among conventional and alternative consumers. *Food Technology* 40 (1): 86–91.

———. 1987. Food irradiation and consumer values. *Ecology Food Nutrition* 21:219–35.

Butz, P., Needs, E. C., Baron, A., Bayer, O., Geisel, B., Gupta, B., et al. 2003. Consumer attitudes to high pressure food processing. *Food, Agriculture and Environment* 1 (1):30–34.

Caplan, A. 2008. Don't hide origin of food. *The Daily News* January 20, H5.

Caporale, G. and Monteleone, E. 2004. Influence of information about manufacturing process on beer acceptability. *Food Quality and Preference* 15:271–78.

Cardello, A. V. 2000. Consumer attitudes and expectations toward non-thermal and other novel food processing techniques. Presentation at the *IFT Non-thermal Processing Division Workshop on Non-thermal Processing of Food*, Portland, OR: Elsevier.

———. 2003. Consumer concerns and expectations about novel food processing technologies: Effects on product liking. *Appetite* 40 (3): 217–33.

———. 1994. Consumer expectations and their role in food acceptance. In *Measurement of Food Preferences*, eds. H. J. H. MacFie and D. M. H. Thomson, 253–97. Glasgow: Blackie Academic and Professional.

———. 1995. Food quality: Relativity, context and consumer expectations. *Food Quality and Preference* 6:163–68.

Cardello, A. V., Bell, R., and Kramer, F. M. 2006. Attitudes of Consumers toward militaty and other institutional foods. *Food Quality and Preference* 7:7–20.

Cardello, A. V. 2007. Measuring consumer expectations to improve product development. In *Consumer-led Food Product Development*, ed. H. MacFie, 223–261, Cambridge, UK: Woodhead.

Cardello, A. V., Lawless, H. T., and Schutz, H. G. 2008. Effects of extreme anchors and interior label spacing on labeled affective magnitude scales. *Food Quality and Preference* 19:473–80.

Cardello, A. V. and Sawyer, F. M. 1992. The effects of disconfirmed consumer expectations on food acceptability. *Journal of Sensory Studies* 7:253–77.

Cardello, A. V. and Schutz, H. 2005. Effects of consumer concerns and information labeling on liking and willingness to purchase and consume GM foods. Poster presentation at the *6th Pangborn Sensory Science Symposium*, Harrogate, England.

———. 2004. Numerical scale point locations for constructing the LAM (Labeled Affective Magnitude) scale. *Journal of Sensory Studies* 19 (4): 341–46.

———. 2003. The importance of taste and other product factors to consumer interest in nutraceutical products: Civilian and military comparisons. *Journal of Food Science* 68 (4): 1519–24.

———. 2002. The perception of food freshness: Uncovering its meaning and importance to consumers. In *Freshness and Shelf Life of Foods*, eds. H. Weenen and K. Cadwallader, 22–41. Washington, DC: American Chemical Society.

Cardello, A. V., Schutz, H. G., and Lesher, L. L. 2007. Consumer perceptions of foods processed by innovative and emerging technologies: A conjoint analytic study. *Innovative Food Science and Emerging Technologies* 8:73–83.

Cardello, A. V. and Wise, P. M. 2008. Taste, smell and chemesthesis in product experience. In *Product Experience*, eds. H.N.I. Schifferstein and P. Hekkert, 91–131, Oxford: Elsevier.

Carneiro, J. D. S., Minim, V. P. R., Deliza, R., Silva, C. H. O., Carneiro, J. C. S., and Leao, F. P. 2005. Labeling effects on consumer intention to purchase for soybean oil. *Food Quality and Preference* 16:275–82.

Casey, M. A. and Krueger, R. A. 1994. Focus group interviewing. In *Measurement of Food Preferences*, eds. J. H. MacFie and D. M. H. Thomson, 77–97. London: Blackie Academic and Professional.

Caul, J. F. 1956. The profile method of flavor analysis. In *Advances in Food Research*, eds. E. Mrak and G. F. Stewart, 7:1–40. New York: Academic Press.

Chen, M. F. 2007. Consumer attitudes and purchase intentions in relation to organic foods in Taiwan: Moderating effects of food-related personality traits. *Food Quality and Preference* 18:1008–21.

Chen, M. F. and Li, H. L. 2007. The consumer's attitude toward genetically modified foods in Taiwan. *Food Quality and Preference* 18:662–74.

Chung, S. J. and Vickers, Z. 2007b. Influence of sweetness on the sensory-specific satiety and long-term acceptability of tea. *Food Quality and Preference* 18:256–64.

———. 2007a. Long-term acceptability and choice of teas differing in sweetness. *Food Quality and Preference* 18:963–74.

Civille, G. V. and Liska, I. H. 1975. Modifications and applications to foods of the General Foods Sensory Texture Profile Technique. *Journal of Texture Studies* 6:19–31.

Civille, G. V. and Szczesniak, A. S. 1973. Guidelines to training a texture profile panel. *Journal of Texture Studies* 4 (2): 207–23.

Cogent Research 2006. A study of US consumer attitudinal trends. *Food Biotechnology.* Noember 15, From www.ific.org.

Costa, A. I. A. 2003. *New Insights into Consumer-Oriented Food Product Design.* Wageningen: Ponsen and Looijen.

Costa, A. I. A. and Jongen, W. M. F. 2006. New insights into consumer-led food product development. *Food Science & Technology* 17:457–65.

Cox, D. N., Evans, G., and Lease, H. J. 2008. Australian consumers' preferences for conventional and novel sources of long chain omega-3 fatty acids: A conjoint study. *Food Quality and Preference* 19:306–14.

———. 2007. The influence of information and beliefs about technology on acceptance of novel food technologies: A conjoint study of farmed prawn concepts. *Food Quality and Preference* 18:813–23.

da Costa, M. C., Deliza, R., Rosenthal, A., Hedderley, D., and Frewer, L. 2000. Nonconventional technologies and impact on consumer behavior. *Trends in Food Science and Technology* 11 (4–5):188–93.

Dean, M. and Shepherd, R. 2007. Effects of information from sources in conflict and in consensus on perceptions of genetically modified food. *Food Quality and Preference* 18:460–69.

de Jonge, J., Van Trijp, H., Goddard, E., and Frewer, L. 2008. Consumer confidence in safety of food in Canada and the Netherlands: The validation of a generic framework. *Food Quality and Preference* 19:439–51.

Deliza, R. 1996. The effects of expectation on sensory perception and acceptance. Unpublished PhD thesis, University of Reading, UK.

Deliza, R. and MacFie, H. J. H. 1996. The generation of sensory expectation by external cues and its effect on sensory perception and hedonic ratings: A review. *Journal of Sensory Studies* 11 (2):103–28.

Deliza, R., Rosenthal, A., and Silva, A. L. S. 2003. Consumer attitude towards information on non conventional technology. *Trends in Food Science and Technology* 14 (1–2): 43–49.

DiMonaco, R., Cavella, S., Di Marzo, S., and Masi, P. 2004. The effect of expectations generated by brand name on the acceptability of dried semolina pasta. *Food Quality and Preference* 15:429–38.

Dreezens, E., Martijn, C., Tenbult, P., Kok, G., and de Vries, N. K. 2005. Food and values: An examination of values underlying attitudes toward genetically modified and organically grown food products. *Appetite* 44 (1): 115–22.

Dunlap, R. E. and Beus, C. E. 1992. Understanding public concerns about pesticides: An empirical examination. *Journal of Consumer Affairs* 26:418–38.

Dunne, C. P. and Kluter, R. A. 2001. Emerging non-thermal processing technologies: Criteria for success. *The Australian Journal of Dairy Technology* 56 (2): 109–12.

Dunne, C. P., Kluter, R. A., and Lee, C. 1997. Criteria for success of new non-thermal processes: Sensory and shelf life studies. Activities Report, Research & Development Associates. New York: NAL/USDA.

Eagly, A. H. and Chaiken, S. 1984. Cognitive theories of persuasion. In *Advances in Experimental Social Psychology*, ed. L. Berkowitz, 297–359. New York: Academic Press.

———. 1993. *The Psychology of Attitudes.* Fort Worth, TX: Harcourt Brace Jovanovitch.

Earle, T. C. and Cvetkovich, G. T. 1995. *Social Trust: Toward a Cosmopolitan Society.* Westport, CT: Praeger.

European Commission. 2006. *Europeans and Biotechnology in 2005: Patterns and Trends.* Luxembourg: Office for Official Publications of the European Communities. Eurobarometer No. 64.3.

———. 1997. *The Europeans and Modern Biotechnology.* Luxembourg: Office for Official Publications of the European Communities. Eurobarometer No. 46.1.

Eustice, R. and Bruhn, C. M. eds. 2006. *Consumer Acceptance and Marketing of Irradiated Foods.* Ames, IA: Blackwell Publishing.

Fazio, R. H., Chen, J. M., McDonel, E. C., and Sherman, S. J. 1982. Attitude accessibility, attitude-behavior consistency, and the strength of the object-evaluation association. *Journal of Experimental Social Psychology* 14:339–57.

Feather, N. T. ed. 1982. *Expectations and Actions: Expectancy-Value Models in Psychology.* Hillsdale, NJ: Lawrence Erlbaum Associates.

Fillion, L. and Kilcast, D. 2001. Abstracts of Papers, 4th Pangborn Sensory Science Symposium, Dijon, France; Institute National de la Recherche Agronomique: Dijon, France, Abstract P-002.

Finn, A. and Louviere, J. J. 1992. Determining the appropriate response to evidence of public concern: The case of food safety. *Journal of Public Policy and Marketing* 11 (1): 12–25.

Finucane, M. L. and Holup, J. L. 2005. Psychosocial and cultural factors affecting perceived risk of genetically modified food: An overview of the literature. *Social Science and Medicine* 60:1603–12.

Fischoff, B., Slovic, P., Lichtenstein, S., Read, S., and Combs, B. 1978. How safe is safe enough? A psychometric study of attitudes towards technological risks and benefits. *Policy Studies* 9:127–52.

Fischoff, B., Slovic, P., and Lichtenstein, S. 1982. Lay foibles and expert fables in judgments about risk. *American Statistics* 36:240–55.

Fishbein, M. and Ajzen, I. 1975. *Belief, Attitude, Intention and Behavior: An Introduction to Theory and Research.* Reading, MA: Addison-Wesley.

Flynn, J., Slovic, P., and Mertz, C. K. 1994. Gender, race and perception of environmental health risks. *Risk Analysis* 14 (6): 1101–108.

Food Marketing Institute. 2005. *U.S. Grocery Shopper Trends.* Washington, DC: Food Marketing Institute.

Food Processing. 1988. Hot air vs. steam in food canning. *Food Processing* 49 (5): 190.

Food Standards Agency. 2008. Report of the Foods Standards Agency for the General Advisory Committee on Science of the United Kingdom. www://foodstandards.gov.uk/news/pressreleases/2008/mar/badscience?view

Forde, C. G. and Delahunty, C. M. 2004. Understanding the role cross-modal sensory interactions play in food acceptability in younger and older consumers. *Food Quality and Preference* 15:715–28.

Fox, J. A., Bruhn, C. M., and Sapp, S. 2001. Consumer acceptance of irradiated meat. In *Interdisciplinary Food Safety Research*, eds. N. H. Hooker and E. A. Murano, 139–58. New York: CRC Press.

Frewer, L. 2003. Societal issues and public attitude towards genetically modified foods. *Food Science and Technology* 14:319–32.

Frewer, L. J., Howard, C., Hedderley, D., and Shepherd, R. 1997a. Consumer attitudes toward different food-processing technologies used in cheese production: The influence of consumer benefit. *Food Quality and Preference* 8 (4): 271–80.

______. 1999. Reactions to information about genetic, engineering: impact of source credibility, perceived risk immediacy and persuasive content. *Public Understanding of Science* 8:35–50.

______. 1996. What determines trust in information about food related risks? Underlying psychological constructs. *Risk Analysis* 16 (4): 473–86.

Frewer, L. J., Howard, C., and Shepherd, R. 1997b. Public concerns about general and specific applications of genetic engineering: Risk, benefit and ethics. *Science Technology and Human Values* 22:98–124.

Frewer, L., Scholderer, J., Downs, C., and Bredahl, L. 2000. Communicating about the risks and benefits of genetically modified foods. Effects of different information strategies. MAPP Working Papers 71, University of Aarhus, Aarhus School of Business, The MAPP Centre.

Frewer, L. J. and Shepherd, R. 1995. Ethical concerns and risk perceptions associated with different applications of genetic engineering: Interrelationships with the perceived need for regulation of the technology. *Agricultural and Human Values* 12:48–57.

Frewer, L. J., Shepherd, R., and Sparks, P. 1994. The interrelationship between perceived knowledge, control and risk associated with a range of food-related hazards targeted at the individual, other people and society. *Journal of Food Safety* 14:19–40.

Gacula, M. C. and Singh, J. 1984. *Statistical Methods in Food and Consumer Research.* Orlando, FL: Academic Press.

Gains, N. 1999. The repertory grid approach. In *Measurement of Food Preferences*, eds. H. Macfie and D. Thomas, 51–75. Gaithersburg: Aspen.

Gallup. 1993. A survey of consumers' awareness, knowledge and attitudes toward the process of irradiation, Summary Report, April. Princeton, NJ: The Gallup Organization.

Gaskell, G., Bauer, M. W., Durant, J., and Allum, N. 1999. World apart? The reception of genetically modified foods in Europe and the U.S. *Science* 285:384–87.

George, R. J. 1993. A comparison of American and Irish consumers' perceptions of the quality of food products and supermarket service. *Journal of Food Products Marketing* 1 (3): 73–81.

Giovanni, M. E. and Pangborn R. M. 1983. Measurement of taste intensity and degrees of liking of beverages by graphic scales and magnitude estimation. *Journal of Food Science* 48:1175–82.

Gould, W. A. 1977. *Food Quality Assurance.* Westport, CT: AVI Publishing.

Green, B. G., Dalton, P., Cowart, B., Shaffer, G., Rankin, K., and Higgins, J. 1996. Evaluating the labeled magnitude scale for measuring sensations of taste and smell. *Chemical Senses* 21 (3): 323–35.

Green, B. G., Shaffer, G. S., and Gilmore, M. M. 1993. Derivation and evaluation of a semantic scale of oral sensation magnitude with apparent ratio properties. *Chemical Senses* 18:683–702.

Green, J., Draper, A., and Dowler, E. 2003. Short cuts to safety: Risk and rules of thumb in accounts of food choice. *Health, Risk, and Society* 5 (1): 33–52.

Green, P. E. and Srinivasan, V. 1978. Conjoint analysis in consumer research: Issues and outlook. *Journal of Consumer Research* 5:103–23.

Greene, J. L., Bratka, K. J., Drake, M. A., and Sanders, T. H. 2006. Effectiveness of category and line scales to characterize consumer perception of fruity fermented flavor in peanuts. *Journal of Sensory Studies* 2:146–54.

Gridgeman, N. T. 1961. A comparison of some taste-test methods. *Journal of Food Science* 26:171–77.

Grobe, D., Douthitt, R., and Zepeda, L. 1999. A model of consumers' risk perceptions toward recombinant bovine growth hormone (rbGH): The impact of risk characteristics. *Risk Analysis* 19 (4): 661–73.

Grunert, K. G. 2002. Current issues in the understanding of consumer food choice. *Trends in Food Science & Technology* 12:275–85.

Grunert, K. G., Hartvig Larsen, H., Madsen, T. K., and Baadsgaard, A. 1996. *Market Orientation in Food and Agriculture*. Norwell, MA: Kluwer.

Grunert, K. G., Lahteenmaki, L., Nielsen, J. B., Poulsen, J. B., Ueland, O., and Astrom, A. 2001. Consumer perceptions of food products involving genetic modification: Results from a qualitative study in four Nordic countries. *Food Quality and Preference* 12:527–42.

Guest, S., Essick, G., Patel, A., Prajapati, R., and McGlone, F. 2007. Labeled magnitude scales for oral sensations of wetness, dryness, pleasantness and unpleasantness. *Food Quality and Preference* 18:342–52.

Guinard, J., Yip, D., Cubero, E., and Mazzucchelli, R. 1999. Quality ratings by experts, and relation with descriptive analysis ratings: A case study with beer. *Food Quality and Preference* 10:59–67.

Gustafsod, P. E. 1998. Gender differences in risk perception: Theoretical and methodological perspectives. *Risk Analysis* 18 (6): 805–11.

Hallman, W. K., Hebden, C. W., Aquino, H. L., Cuite, C. L., and Lang, J. T. 2003. Public perceptions of genetically modified foods: A national study of American knowledge and opinion. Report of the Food Policy Institute. FPI publication number RR-1003-004. New Brunswick, NJ: Rutgers, the State University of New Jersey.

Harris, J. M. 2002. Food product introductions continue to decline in 2000. *BNET Business Network: Food Review* Spring.

Hashim, I. B., Resurreccion, A. V. A., and McWatters, K. H. 1996. Consumer attitudes toward irradiated poultry. *Food Technology* 50 (3): 77–80.

Heiman, A., Just, D. R., and Zilberman, D. 2000. The role of socioeconomic factors and lifestyle variables in attitude and the demand for genetically modified foods. *Journal of Agribusiness* 18 (3): 249–60.

Heine, K. A., Jaeger, S. R., Carr, B. T., and Delahunty, C. M. 2008. Comparison of five common acceptance and preference methods. *Food Quality and Preference* 19:651–61.

Helleman, V., Aaron, J. J., Evans, R., and Mela, D. J. 1993. Effect of expectation on the acceptance of a low fat meal. In: *Proceedings of the Food Preservation 2000 Conference*, 311–318. Hampton, VA: Science and Technology Corporation.

Herr, P. M. 1989. Priming price: Prior knowledge and contest effects. *Journal of Consumer Research* 16:67–75.

Hoban, T. J. 1997. Consumer acceptance of biotechnology: An international perspective. *Nature Biotechnology* 15 (March): 232–34.

Hoban, T. J., Woodrun, W., and Czaja, R. 1992. Public opposition to genetic engineering. *Rural Sociology* 57:476.

Hough, G., Bratchell, N., and Wakeling, I. 1992. Consumer preference of Dulce de Leche among students in the United Kingdom. *Journal of Sensory Studies* 7: 119–32.

Iaccarino, T., DiMonaco, R., Mincione, A., Cavella, S., and Masi, P. 2006. Influence of information on origin and technology on the consumer response: The case of soppressata salami. *Food Quality and Preference* 17:76–84.

Jaeger, S. R., and Cardello, A. V. 2009. Direct and indirect hedonic scaling methods: A comparison of the labeled affective magnitude (LAM) scale and best-worst scaling. *Food Quality and Preference* 20 (3): 249–58.

Jaeger, S. R., Jørgensen, A. S., Aaslyng, M. D., and Bredie, W. L. P. 2008. Best-worst scaling: An introduction and initial comparison with monadic rating for preference elicitation with food products. *Food Quality and Preference* 19:579–88.

Johnson, A. M., Reynolds, A. E., Chen, J., and Resurreccion, A. V. A. 2004. Consumer attitudes toward irradiated food: 2003 vs 1993. *Food Protection Trends* 24 (6): 408–18.

Jones, L. V., Peryam, D. R., and Thurstone, L. L. 1955. Development of a scale for measuring soldiers' food preferences. *Food Resources* 20:512–20.

Keskitalo, K., Knaapila, A., Kallela, M., Palotie, A., Wessman, M., Sammalisto, S., Peltonen, L., Tuorila, H., and Perola, M. 2007. Sweet taste preferences are partly genetically determined: Identification of trait locus on chromosome 16. *Journal of Clinical Nutrition* 86 (1): 55–63.

Krause, N., Malmfors, T., and Slovic, P. 1992. Intuitive toxicology: Expert and lay judgments of chemical risks. *Risk Analysis* 12 (2): 215–32.

Krueger, R. A. and Casey, M. A. 2000. *Focus Groups. A Practical Guide for Applied Research.* Thousand Oaks, CA: Sage.

Laboissière, L. H. E. S., Deliza, R., Barros-Marcellini, A. M., Rosenthal, A., Camargo, L. M. A. Q., and Junqueira, R. G. 2007. Effects of high hydrostatic pressure (HHP) on sensory characteristics of yellow passion fruit juice. *Food Science and Emerging Technologies* 8:469–77.

Lange, C., Issanchou, S. and Combris, P. 2000. Expected versus experienced quality: trade-offs with price. *Food Quality and Preference* 11:289–297.

Lange, C., Rousseau, F., and Issanchou, S. 1999. Expectation, liking and purchase behavior under economical constraint. *Food Quality and Preference* 10:31–40.

Langron, S. P., Noble, A. C., and Williams, A. A. 1983. Measurement and control. In *Sensory Quality in Foods and Beverages*, eds. R. H. Atkin and A. A. Williams. Chichester: Ellis Horwood.

Lappalainen, R., Kearney, J., and Gibney, M. 1998. A pan EU survey of consumer attitudes to food, nutrition and health: an overview. *Food Quality and Preference* 9 (6): 467–78.

LaTour, S. A. and Manrai, A. K. 1989. Interactive impact of informational and normative influence on donations. *Journal of Market Research* 26:327–35.

Lawless, H. T. and Heymann, H. 1998. *Sensory Evaluation of Food: Principles and Practices.* New York: Chapman & Hall.

Lobb, A. E., Mazzocchi, M., and Traill, W. B. 2007. Modelling risk perception and trust in food safety information within the theory of planned behaviour. *Food Quality and Preference* 18:384–95.

Lusk, J. L., Fox, J. A., and Mcilvain, C. L. 1999. Consumer acceptance of irradiated meat. *Food Technology* 53 (3): 56–59.

Macer, D., Bezar, H., Richardson-Harman, N., Kamada, H., and Macer, N. 1997. Attitudes to biotechnology in Japan and New Zealand in 1997, with international comparisons. *Eubios Journal of Asian and International Bioethics* 7:137–43.

MacFie, H. 2007. *Consumer-Led Food Product Development.* Cambridge, UK: Woodhead Publishing Limited.

Macintyre, S., Reilly, J., Miller, D., and Eldridge, J. 1998. Food choices food scares, and health: The role of the media. In *The Nation's Diet: The Social Science of Food Choice*, ed. A. Murcott, 228–49. London: Addison Wesley Longman.

Magnusson, M. K. and Hursti, U. K. K. 2002. Consumer attitudes towards genetically modified foods. *Appetite* 39:9–24.

Malone, J. W. Jr. 1990. Consumer willingness to purchase and to pay more for potential benefits of irradiated fresh food products. *Agribusiness* 6 (2): 163–78.

Marley, A. A. J. and Louviere, J. J. 2005. Some probabilistic models of best, worst, and best-worst choices. *Journal of Mathematical Psychology* 49:464–80.

McNeill, K. L., Sanders, T. H., and Civille, G. V. 2000. Using focus groups to develop a quantitative consumer questionnaire for peanut butter. *Journal of Sensory Studies* 15:163–78.

Meilgaard, M., Civille, G. V., and Carr, B. T. 2007. *Sensory Evaluation Techniques*, 4th ed. Boca Raton: CRC Press.

Meiselman, H. L. and Schutz, H. G. 2003. History offood acceptance research in the US Army. *Appetite* 40:199–216.

Mermelstein, N. H. 2001. Emerging technologies and fresh labeling. *Food Technology* 55 (2): 64–7.

Miles, S., and Frewer, L. J. 2001. Investigating specific concerns about different food hazards. *Food Quality and Preference* 12:47–61.

Mireaux, M., Cox, D. N., Cotton, A., and Evans, G. 2007. An adaptation of repertory grid methodology to evaluate Australian consumers' perceptions of food products produced by novel technologies. *Food Quality and Preference* 18:834–48.

Mor-Mur, M. and Yuste, J. 2003. High-pressure processing applied to cooked sausage manufacture: Physical properties and sensory analysis. *Meat Science* 65:1187–91.

Moseley, B. 1990. Food safety: Perception, reality and research. In *Royal College of General Practitioners Members Handbook*, London, Royal College of General Practitioners, 346–51.

Moses, V. 1999. Biotechnology products and European consumers. *Biotechnology Advances* 17 (8): 647–78.

Moskowitz, H. R. 1977. Magnitude estimation: Notes on what, how, when and why to use it. *Journal of Food Quality* 1:195–227.

______. 1980. Psychometric evaluation of food preferences. *Journal of Foodservice Systems* 1:149–67.

Moskowitz, H. R. and Sidel, J. L. 1971. Magnitude and hedonic scales of food acceptability. *Journal of Food Science* 36:677–80.

Mucci, A. and Hough, G. 2003. Perceptions of genetically modified foods by consumers in Argentina. *Food Quality and Preference* 15:43–51.

Meyers-Levy, J. and Maheswaran, D. 1991. Exploring differences in males' and females' processing strategies. *Journal of Consumer Research* 18:63–70.

Meyers-Levy, J. and Steinthal, B. 1991. Gender differences in the use of message cues and judgments. *Journal of Marketing Research* 28:84–93.

Olsen, S. O. 1999. Strength and conflicting valance in the measurement of food attitudes and preferences. *Food Quality and Preference* 10 (6): 483–94.

Olsen, S. O., Heide, M., Dopico, D. C., and Toften, K. 2008. Explaining intention to consume a new fish product: a cross-generational and cross-cultural comparison. *Food Quality and Preference* 19:618–27.

Pèneau, S., Brockhoff, P. B., Hoehn, E., Escher, F., and Nuessli, J. 2007. Relating consumer evaluation of apple freshness to sensory and physico-chemical measurements. *Journal of Sensory Studies* 22:313–35.

Peryam, D. R. and Pilgrim, F. J. 1957. Hedonic scale method of measuring food preferences. *Food Technology* 11:9–14.

Pohlman, A. J., Wood, O. B., and Mason, A. C. 1994. Influence of audiovisuals and food samples on consumer acceptance of food irradiation. *Food Technology* 48 (12): 46–49.

Powell, D. and Leiss, W. 1997. *Mad Cows and Mother's Milk: The Perils of Poor Risk Communication*. Montreal: McGill-Queen's University Press.

Powell, M. C. and Fazio, R. H. 1984. Attitude accessibility as a function of repeated attitudinal expression. *Personality and Social Psychology Bulletin* 10:139–48.

Priest, S. H. 2000. U.S. public opinion divided over biotechnology? *Nature Biotechnology* 18 (9): 939–42.

Raudenbush, B., Schroth, F., Reilley, S., and Frank, R. A. 1998. Food neophobia, odor evaluation and exploratory sniffling behavior. *Appetite* 31 (2): 171–83.

Resurreccion, A., Galvez, F., Fletch, S., and Misra, S. 1995. Consumer attitudes toward irradiated food: Results of a new study. *Journal of Food Protection* 58 (2): 193–96.

Resurreccion, A. V. A. 1998. *Consumer Sensory Testing for Product Development*. Gaithersburg, MD: Aspen.

Rhodus, T., Schwartz, J., and Hoskin, J. 1994. Ohio consumer opinions of roadside markets and farmers' markets. *Report to the Ohio Rural Rehabilitation Program*. Ohio State University, Dept. of Horticulture.

Rohm, H. and Raaber, S. 1991. Hedonic spreadability optima of selected edible fats. *Journal of Sensory Studies* 6:81–88.

Rubio, B., Martinez, B., Garcia-Cachan, M. D., Rovira, J., and Jamie, I. 2007. Effect of high pressure preservation on the quality of dry cured beef Cecina de Leon. *Innovative Food Science and Emerging Technologies* 8:102–10.

Saba, A., Moles, A., and Frewer, L. J. 1998. Public concerns about general and specific application of generic engineering: A comparative study between UK and Italy. *Nutrition and Food Science* 98:25–31.

Saba, A. and Vassallo, M. 2002. Consumer attitudes toward the use of gene technology in tomato production. *Food Quality and Preference* 13:13–21.

Sauvageot, F., Urdapilleta, I., and Peyron, D. 2006. Within and between variations of texts elicited from nine experts in wines. *Food Quality and Preference* 17 (6): 429–44.

Schifferstein, H. N. J. 2001. Effects of product beliefs on product perception and liking. In *Food, People and Society, A European Perspective of Consumers' Food Choices*, eds. L. Frewer, E. Risvik, and H. Schifferstein. London: Springer Verlag.

Schifferstein, H. N. J., Kole, A. P. W., and Mojet, J. 1999. Asymmetry in the disconfirmation of expectations for natural yogurt. *Appetite* 32:307–29.

Schirack, A. V., Drake, M., Sanders, T. H., and Sandeep, K. P. 2006. Impact of microwave blanching on the flavor of roasted peanuts. *Journal of Sensory Studies* 21:428–40.

Schutz, H. G., Bruhn, C. M., and Diaz-Knauf, K. V. 1989. Consumer attitude toward irradiated foods: Effects of labeling and benefit information. *Food Technology* 43:80–86.

Schutz, H. G. and Cardello, A.V. 2001. A labeled affective magnitude (LAM) scale for assessing food liking/disliking. *Journal of Sensory Studies* 16 (2): 117–59.

______. 1997. Information effects on acceptance of irradiated foods in a military population. *Dairy, Food, and Environmental Sanitation* 17 (8): 470–81.

Shanahan, J., Dietram, S., and Lee, E. 2001. The polls-trends. *Public Opinion Quarterly* 65 (2) (Summer): 267–81.

Shepherd, R. 1988. Consumer attitudes and food acceptance. In *Food acceptability*, ed. D. M. H. Thomson. London: Elsevier Applied Science.

Shepherd, R. and Farleigh, C. A. 1986. Attitudes and personality related to salt intake. *Appetite* 7:343–54.

Shepherd, R. and Stockley, L. 1985. Fat consumption and attitudes towards food with a high fat content. *Human Nutrition: Applied Nutrition* 39a:431–42.

Sidel, J. L. and Stone, H. 2006. Sensory science: Methodology. In *Handbook of Food Science, Technology, and Engineering*, ed. Y. H. Hui, 57. Boca Raton: CRC Press.

Siegrist, M. 2000. The influence of trust and perception of risk and benefits on the acceptance of gene technology. *Risk Analysis* 20 (2): 195–203.

Sijtsema, S. J., Backus, G. B. C., Linnemann, A. R., and Jongen, W. M. F. 2004. Consumer orientation of product developers and their product perception compared to that of consumers. *Trends in Food Science & Technology* 15:489–97.

Siret, F. and Issanchou, S. 2000. Traditional process: Influence on sensory properties and on consumers' expectation and liking. Application to pate' de campagne. *Food Quality and Preference* 11:217–28.

Slovic, P. 1993. Perceived risk, trust, and democracy. *Risk Analysis* 13 (6): 675–82.

———. 1987. Risk perception. *Science* 236:280–85.

Slovic, P., Fischoff, B., and Lichtenstein, S. 1979. Rating the risks. *Environment* 21 (3): 14–20, 36–39.

———. 1985. Characterizing perceived risk. In *Perilous Progress: Managing the Hazards of Technology*, eds. R. W. Kates, C. Hohenemser, and J. X. Kasperson, 91–125. Boulder, CO: Westview.

Slovic, P. and MacGregor, D. 1994. The social context of risk communication. *Decision Research*, Eugene, OR.

Sparks, P. and Shepherd, R. 1991. A review of risk perception research: implications for food safety issues. Report for the Ministry of Agriculture, Fisheries and Food, Reading, England: AFRC Institute of Food Research.

Sparks, P., Shepherd, R., and Frewer, L. J. 1995. Assessing and structuring attitudes toward the use of gene technology in food production: The role of perceived ethical obligation. *Basic and Applied Social Psychology* 16 (3): 267–85.

Starr, C. 1969. Social benefit versus technological risk. *Science* 13 (6): 1232–38.

Stefani, G., Romano, D., and Cavicchi, A. 2006. Consumer expectations, liking and willingness to pay for specialty foods: Do sensory characteristics tell the whole story? *Food Quality and Preference* 17:53–62.

Stevens, J. C. 1957. A comparison of ratio scales for the loudness of white noise and the brightness of white light. Doctoral Dissertation, Harvard University.

Stevens, S. S. 1957. On the psychophysical law. *Psychology Review* 64:153–81.

Stone, H. and Sidel, J. L. 2004. *Sensory Evaluation Practices*, 3rd ed. New York: Academic Press.

Stone, H., Sidel, J. L., Oliver, S. Woolsey, A., and Singleton, R. C. 1974. Sensory evaluation by qualitative descriptive analysis. *Food Technology* 28:24–34.

Templeton, J. F. 1994. *The Focus Group*. Chicago: Probus Publishing Company.

Terry, D. E. and Tabor, R. L. 1988. Consumer acceptance of irradiated produce. *Journal of Food Distribution Research* 19 (1): 73–89.

Terry, D. E. and Tabor, R. L. 1990. Consumers' perceptions and willingness to pay for food irradiation. In *Proceedings of the Second International Conference on Research in the Consumer Interest*, Snowbird, UT.

Tuorila, H. 1987. Selection of milks with varying fat contents and related overall liking, attitudes, norms, and intentions. *Appetite* 8:1–14.

Tuorila, H., Cardello, A. V., and Lesher, L. 1994. Antecedents and consequences of expectations related to fat-free and regular-fat foods. *Appetite* 23:247–63.

Tuorila-Ollikainen, H., Lahteenmaki, L., and Salovaara, H. 1986. Attitude, norms intentions and hedonic responses in the selection of low salt bread in a longitudinal choice experiment. *Appetite* 7:127–39.

Turcanu, C., Carlé, B., Hardeman, F., Bombaerts, G., and Van Aeken, K. 2007. Food safety and acceptance of management options after radiological contaminations of food chain. *Food Quality and Preference* 18:1085–95.

Tybout, A. M. and Artz, N. 1994. Consumer psychology. *Annual Review Psychology* 45:131–69.

US Food and Drug Administration. 2008. Center for Food Safety and Applied Nutrition 2006 FDA/FSIS Food Safety Survey: Topline Frequency Report.

van Kleef, E., van Trijp, H. C. M., and Luning, P. 2005. Consumer research in the early stages of new product development: A critical review of methods and techniques. *Food Quality and Preference* 16:181–201.

van Raaij, V. and Fred, W. 1991. The formation and use of expectations in consumer decision making. In *Handbook of Consumer Behavior*, eds. T. S. Robertson and H. Kassarjian, 401–18. Englewood Cliffs, NJ: Prentice-Hall.

Verbeke, W. 2001. Beliefs, attitude and behaviour toward fresh meat revisited after the Belgian dioxin crisis. *Food Quality and Preference* 12:489–98.

Wansink, B. 2005. Consumer profiling and the new product development toolbox: A commentary on van Kleef, van Trijp and Luning. *Food Quality and Preference* 16:217–21.

Wilcock, A., Pun, M., Khanona, J., and Aung, M. 2004. Consumer attitudes, knowledge and behaviour: A review of food safety issues. *Trends in Food Science and Technology* 15:56–66.

Worth, L. T., Smith, J., and Mackie, D. M. 1992. Gender schematicity and preference for gender typed products. *Psychology Marketing* 9:17–30.

Yeung, R. M. W. and Yee, W. M. S. 2005. Consumer perception of food safety related risk: A multiple regression approach. *Journal of International Food and Agribusiness* 17 (2): 195–212.

Yi, Y. 1990. Cognitive and affective priming effects of the context for printing advertisements. *Journal of Advertising* 19:40–88.

Zechendorf, B. 1994. What the public thinks about biotechnology. *BioTechnology* 12 (September): 870–75.

Zienkewicz, L. S. H. and Penner, K. P. 2004. Consumers' perceptions of irradiated ground beef after education and product exposure. *Food Protection Trends* 24 (10): 740–45.

van Kleef, E., van Trijp, H.C.M., and Luning, P. 2005. Consumer research in the early stages of new [illegible] development: A critical review of methods and techniques. Food Quality and Preference [illegible]

[illegible]

19 Novel Techniques for the Processing of Soybeans

Joyce I. Boye and S. H. Rajamohamed
Agriculture and Agri-Food Canada

CONTENTS

19.1 INTRODUCTION

Soybeans are one of the world's most economical and valuable agricultural commodities (Liu 1997). Due to their high nutritional value and low cost, soybeans have been considered as a good economic alternative to existing animal derived protein sources. Reported health benefits of soy include reduced risks of certain cancers (Wang et al. 1999) and cardiovascular diseases (Lovati et al. 2000), reduction of cholesterol (Tovar-Palacio et al. 1998), and the prevention of obesity (Aoyama et al. 2000).

On a wet basis, soybeans contain about 35% protein, 17% oil, 31% carbohydrate, and 4.4% ash. Soybean oil is today one of the world's leading vegetable oils for human consumption. Fatty acids found in soybean oil include linoleic, oleic, palmitic, linolenic, and stearic acid. Carbohydrates present in soy include sucrose, stachyose, raffinose, cellulose, hemicellulose, and pectin. The major component of soy that gives the bean its greatest value is, however, the proteins. Soybean proteins are composed of two major components, glycinin and conglycinin, accounting for approximately 30 and 40% of the total seed proteins, respectively. Protein quality of isolated soy proteins has been shown to be comparable to animal protein sources such as milk and meat (Young 1991). In 1999, soy protein was approved by the Food and Drug Administration of the USA as a "functional food," making it one of the most valuable vegetable proteins in the world today (Fukushima 2001). Soybeans, however, contain antinutritional factors such as lectins (hemagglutinins), which induce agglutination of erythrocytes, and phytic acid and trypsin inhibitors (TIs), which reduce the digestibility of soy proteins. They also contain lipoxygenase (LOX), an enzyme responsible for the development of off flavors in soy products. These antinutritional factors have contributed to limiting, somewhat, the acceptability of soy foods and soy products. As a result, extensive research has been conducted to identify ways to remove these antinutritional components during processing.

Today, a wide variety of soy foods can be found on the market. Examples of these products are soymilk, tofu, soy ice-cream, soy yogurt, tempeh, miso, natto, and soy sauce. Soybeans are also processed into soy flour and soy protein concentrates and isolates which are then subsequently used as ingredients in the formulation of different food products (e.g., bakery, confectionery, and meat applications). The raw material used for processing these products is mainly defatted soyflakes. Soy protein concentrates and isolates can be further extruded into different texturized products (e.g., cereals, pet foods, snack foods, imitation meat products, and nuts amongst many others) and the market for these products has expanded dramatically in the last decade.

This review will attempt to discuss some of the novel techniques in use or under study for the processing of soy foods and soy ingredients and their effects on product quality and functionality. The primary focus of the review will be on nonthermal processing. Thermal processes such as blanching, pasteurization, and sterilization have been traditionally used as efficient, economic, reliable, and safe food preservation techniques. Thermal treatment to food has some limitations because of its adverse effect on product quality. Thermal energy induces various biochemical reactions, leading to quality deterioration in foods, and results in undesirable changes in nutritional and sensory characteristics (Martens and Knorr 1999). Emerging and novel technologies can overcome some of the limitations of thermal processing. Most of these novel and emerging processes also focus on energy-saving and ecofriendly applications. These are of interest to the food industry because they are economical and produce high-quality foods without compromising safety and in addition offer opportunities for creating new ingredients and products (Knorr 1999). A few examples of such techniques are high hydrostatic pressure, pulsed electric fields (PLFs), oscillating magnetic fields, high-intensity light pulses, high-voltage arc discharge, ohmic and inductive heating, ultrasonication, and irradiation. Some of these technologies are being applied commercially, while others are still in the developmental stages. Technologies relevant to soy food processing are presented below.

19.2 HIGH-PRESSURE PROCESSING (HPP) OF SOYBEAN AND SOY-BASED PRODUCTS

High-pressure processing (HPP) is a method of food processing where food is subjected to elevated pressures (up to 87,000 pounds per square inch or approximately 6000 atmospheric pressures), with or without the addition of heat to achieve microbial inactivation or improve food quality. Application of pressure can change a wide range of biological structures and processes that can affect the conformation of macromolecules and the transition temperature of water and lipids, and influence many other chemical reactions (Cheftel and Culioli 1997). Until recently, little research had been reported on the effects of HPP on soybean grains and their subproducts. Interest in the use of HPP as a potential technology to improve the quality of soy and soy-based products has, however, grown. Highlights of these studies and novel applications are provided below.

19.2.1 High-Pressure Processing (HPP) of Soybean

Hydration of dry soybeans is an important first step in soyfood processing (e.g., during cooking, fermentation, and toasting) (Deshpande, Sathe, and Salunkhe 1984). Inadequate water uptake can result in low textural quality and can also cause insufficient heat transfer, which is necessary to inactivate antinutritional factors. Several researchers have studied the factors that affect water uptake during soaking in order to identify new techniques to accelerate hydration (Hsu, Kim, and Wilson 1983; Deshpande, Bal, and Ojha 1994; Mullin and Xu 2001).

Application of high pressure has been shown to improve the rate of soybean hydration. High-pressure (HP) soaking (300 MPa) significantly improved the rate of water absorption in soybeans and reduced soaking time (Zhang, Ishida, and Isobe 2003). Both volume change and breaking strength were highly dependent on the amount of water imbibed. Furthermore, the water distribution in a HP-soaked soybean was more uniform and water mobility was found to be more restricted than for the unpressurized soybeans. High-pressure treatment also changed the microstructures of the seed coat and hilum, both of which aid in higher water absorption rates.

When soybean seeds were immersed in distilled water and treated at 300 MPa and 20°C for 0–180 min, the surrounding solution became markedly turbid and the protein content of the solution increased, which suggested that proteins had been released from the soybean seeds into the solution (Omi et al. 1996). The amount of protein released from 1 g of high-pressure-processed soybean seed was 1.9–8.6 mg, which represented 0.5–2.5% of the total seed proteins. The amount of protein released reached a maximum at 400 MPa and decreased with further increase in pressure (Figure 19.1). No apparent changes in shape, color, or size between treated and untreated soybeans were reported. In other studies, Asano, Okubo, and Yamauchi (1989) reported that immersion of soybeans in hot water at 50–60°C for one hour resulted in a considerable amount of protein solubilized from the soybean seeds being released to the surrounding water. Hirano, Kagawa, and Okubo (1992) later identified these solubilized proteins as 7S globulins, which accounted for about 3% of the total protein in the mature soybean seeds. The use of high-pressure-treated soybeans for the preparation of soymilk, consequently, results in a decrease in yield. Soymilk total solids yield were reported to decrease from 74.8 to 66.4% for soymilk obtained from untreated soybeans and 700 MPa-treated beans, respectively (Jung, Murphy, and Sala 2008). Protein content also decreased from 51.7 to 47.7% (db) in soymilk processed from high-pressure-treated soybeans.

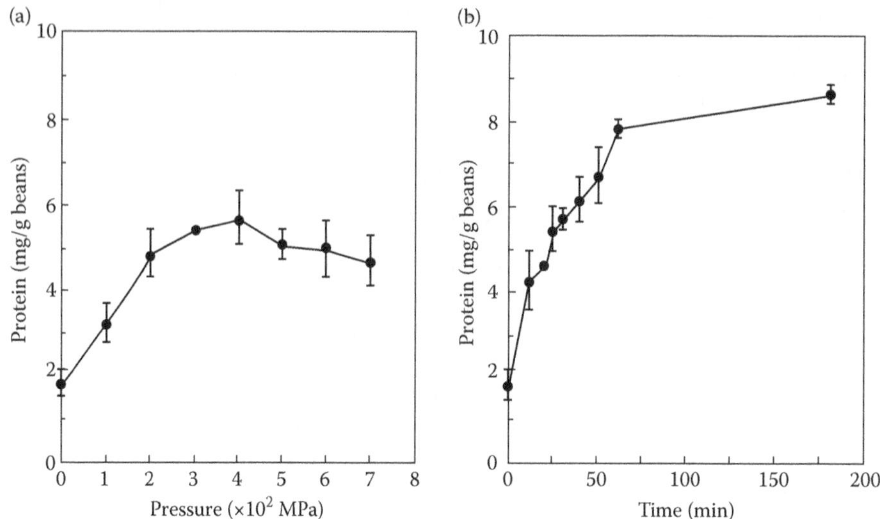

FIGURE 19.1 Effect of high-pressure treatment on the release of proteins from soybean seeds. (a) Soybean seeds pressurized at 0–700 MPa for 25 min under ambient temperature (20°C). (b) Soybean seeds pressurized at 300 MPa for 0–180 min under ambient temperature (20°C). (Reprinted with permission from Omi, Y., Kato, T., Ishida, K., Kato, H., and Matsuda, T., *Journal of Agricultural and Food Chemistry*, 44, 3763–67, 1996. American Chemical Society.)

19.2.2 High-Pressure Processing (HPP) of Soymilk

Soy-based drinks constitute one of the highest performing products for the soyfood industry with 31% growth between 2003 and 2004 (Sloan 2005). Consumption of soymilk has increased in Western countries, primarily among those consumers who are lactose intolerant, vegetarian, and/or are seeking healthier diets. The increase in soymilk consumption could also be partly attributed to the US Food and Drug Administration soy protein health claim, which was approved in October 26, 1999 and which links the consumption of soy proteins with reduced risk of cardiovascular disease. In addition, there is a favorable consumer perception about the high-quality protein and low-fat content of soy products.

Commercial soymilk production involves the use of thermal treatment, which assures food safety and extends shelf life while inactivating undesirable biologically active compounds such as TIs and LOX (Kumar et al. 2003). Numerous studies have, however, indicated that thermal treatment adversely affects the quality of soymilk, including its flavor, color, and in some cases nutritional quality (Kwok and Niranjan 1995). Alteration of soymilk quality characteristics may be avoided by using HPP. HPP has the potential to retain the nutritional content of food by preserving vitamins and inactivating vegetative microorganisms and some spoilage enzymes while either maintaining fresh-product properties or providing products with unique texture and functional properties (Estrada-Giron, Swanson, and Barbosa-Canovas 2005).

19.2.2.1 Effect on Proteins in Soymilk

Kajiyama et al. (1995) observed that the solubility of soymilk proteins was affected by high pressure and the isoelectric point shifted from pH 4.1–4.3 to pH 4.5–4.7. Unpressurized soymilk had a high solubility below pH 3 and moderate solubility between pH 5.7 and 6.7. Pressurization, however, reduced the protein solubility below pH 3 and increased it above pH 5.6, probably due to dissociation of soy protein in to their subunits and the formation of large aggregates at high pressure. Zhang et al. (2005) studied the effects of HP treatment on the modifications of protein in soymilk using spectrofluorimetry, differential scanning calorimetry (DSC), and electrophoresis. Their studies revealed a sharp increase in fluorescence intensity as pressure was increased to 300 MPa. Higher pressures

(500 MPa) yielded a decrease in fluorescence and blueshifts of λ_{max}. These results indicate that 300 MPa may be a transition pressure for some soy protein fractions, which were denatured under this pressure resulting in the exposure of more hydrophobic regions. The decrease in fluorescence intensity at higher pressures may be attributed to a reduction in hydrophobic groups due to intermolecular interactions (Hayakawa et al. 1992) or to the refolding of the pressure-treated sample into a slightly different conformation which obscured some of the hydrophobic groups. Electrophoretic analysis also showed that soy proteins were dissociated by high pressure into subunits, some of which associated into aggregates and became insoluble. DSC studies confirmed that HP denaturation occurred at 300 MPa for β-conglycinin (7S) and at 400 MPa for glycinin (11S) in soymilk.

19.2.2.2 Effect on Trypsin Inhibitors (TIs)

HPP might be a suitable technique to inactivate TIs in soymilk. Van der Ven, Master, and Vanden Berg (2005) studied the effect of high-pressure or high-pressure/temperature treatment on soybean trypsin inhibitor activity (TIA) and observed that HP treatment at ambient initial temperature resulted in minor or no loss of TIA. However, HP treatment with a starting temperature of 60°C resulted in a 40% decrease of TIA in both soybeans and soymilk. For 90% inactivation of TIA a holding time of 1 min and initial temperatures between 77 and 90°C and pressures between 525 and 750 MPa were needed. These results suggest that for inactivation of Trypsin inhibitors (TI), HP treatment should be preferably performed at elevated temperatures. Kwok, Han Hua, and Keshavan (2002) found TIA to be stable when soymilk was thermally processed at 90°C for 50 min; 90% inactivation was achieved only on heat treatment at 143°C for 62 s. In general, HP treatment results in TI inactivation at lower temperatures and shorter treatment times than conventional heat treatments. Matser et al. (2004) have reported that low-molecular-weight compounds are relatively unaffected by pressure, thus, these milder conditions could possibly contribute to a better conservation of other components in soymilk such as vitamins, lysine, and taste (Van der Ven et al. 2005).

19.2.2.3 Effect on Lipoxygenase (LOX) Enzyme

LOX is another antinutritional factor present in soybeans. The three main types identified are LOX I, II, and III. In the presence of molecular oxygen, LOX catalyzes the oxidation of unsaturated fatty acids such as linoleic, linolenic, and arachidonic acid, resulting in the destruction of these essential fatty acids and the generation of unpleasant off-flavors (e.g., beany and rancid flavor). The oxidation process also results in the production of free radicals that can damage other compounds, including vitamins and proteins (Whitaker 1972), and pigments such as chlorophyll and carotenoids, resulting in color changes (Eskin, Grossman, and Pinsky 1977). LOX has been classified as a pressure-sensitive enzyme, and its activity was reported to decrease noticeably after 2 min at 600 MPa at pH 7 and in Tris buffer at 25°C, whereas complete inactivation occurred in Tris buffer after 10 min at 600 MPa and at 25°C (Seyderhelm et al. 1996). Heinisch et al. (1995) also observed a considerable reduction of LOX activity with HP treatment. In the range of 450–600 MPa, minor changes at the active site resulted in a loss of enzyme activity, without significantly affecting the native conformation. Further increase in pressure (600–800 MPa) resulted in a noticeable loss of native enzyme conformation, which indicated that pressure inactivation occurred more readily in comparison to pressure denaturation. Seyderhelm et al. (1996) earlier reported that the activity of commercial soybean LOX was significantly decreased on treatment at 600 MPa in Tris buffer at pH 7. After storage at 4°C for 4–9 days, no change in enzyme activity was observed, indicating that the inactivation was irreversible. Tangwongchai, Ledward, and Ames (2000) also later reported that LOX was more affected by HP treatment in acidic media than in alkaline conditions and was completely inactivated at pH 4–5 on treatment at 400–600 MPa.

Most of the studies reported above on enzyme inactivation of LOX were, however, carried out on purified LOX. Relatively few studies report on the inactivation of LOX enzyme in soybeans or soymilk. Van der Ven et al. (2005) reported that pressure treatment at 500 MPa and ambient initial temperature (20°C) resulted in partial inactivation of LOX in both soybeans and soymilk.

In whole soybeans, LOX inactivation was observed only after 2 min of holding time, whereas in soymilk the LOX activity decreased after a holding time of only 1 min. Moreover, remaining activity in beans was higher compared to the remaining activity in pressurized soymilk. Increasing the pressure to 800 MPa resulted in complete LOX inactivation after a treatment time of 2 min in both soaked beans and soymilk. HP treatment at 600 MPa with an initial temperature of 60°C also resulted in complete inactivation of LOX. The initial temperature of 60°C did not appear to be responsible for this inactivation, because heating of soybeans and soymilk at 60°C for 8 min (comparable to the total process time) at atmospheric pressure caused only a 5% inactivation of LOX. The results, therefore, suggest that complete LOX inactivation is obtained either at very high pressures (800 MPa) or by using a combined temperature/pressure treatment (60°C/600 MPa).

19.2.2.4 Effect on Microorganisms

Cruz et al. (2007) reported that treatment of soymilk with high pressure at 300 MPa resulted in a significant reduction in spore-forming bacteria by around 2 log CFU/ml and enterobacteria were below detectable levels. Similar studies in cow's milk (Thiebaud et al. 2003) also showed a reduction of total counts around 1 and 2.8 log CFU/ml using 200 and 300 MPa, respectively, with an inlet temperature of 24°C. Other workers (Smiddy et al. 2007) obtained a reduction of total and coliform counts around 5 log CFU/ml and 4 log CFU/ml, respectively, with 200 and 250 MPa treatment of cow milk, although they observed that counts of psychrotrophic bacteria increased between four and seven days of storage at 4°C. Other components present in the food system may affect the efficacy of high pressure to inactivate or destroy microorganisms. Gao, Ju, and Wu-Ding (2007) found soybean proteins and sucrose significantly affected the inactivation of *Staphylococcus aureus* in soymilk when treated by high pressure and mild heat. Inactivation of microorganisms by HPP may therefore depend on the nature of the surrounding medium; if the medium is rich in nutrients the ability to inactivate microorganisms may be reduced due to barostatic effects.

19.2.2.5 Effects on the Rheological Behavior of Soymilk

Changes in the viscosity of soymilk after HP treatments have been reported. Yamauchi, Yamagishi, and Iwabuchi (1991) reported that soy protein dispersions increase in viscosity after heating and undergo an irreversible change to a progel state. This increase in viscosity of soymilk with increasing pressure treatments has also been recorded by other workers (Cruz et al. 2007, 2009; Zhang et al. 2005). Zhang et al. (2005) observed that pressure treatment of soymilk beyond 500 MPa for 30 min transformed soymilk into a sol, but below this pressure the products remained in a liquid state (Figure 19.2). Further information on the effect of high pressure on the rheological properties of soymilk and soy proteins is given elsewhere in this book.

19.2.2.6 Effect on Soymilk Flavor

The major volatile components responsible for the beany flavor of soymilk include *n*-hexanol, *n*-hexanal, and *n*-heptanol. These compounds are produced by the action of LOX and have a strong affinity for soy proteins, which make them difficult to remove without denaturing the proteins. Kajiyama et al. (1995) conducted a study on the flavor properties of pressurized soymilk. Their work revealed that releasable *n*-hexanal and total flavor components in soymilk were reduced above 100 MPa, which correlated well with the changes in the hydrophobicity of the soy proteins. *n*-Hexanol content was particularly reduced and very little could be detected above 300 MPa. As previously mentioned, HP treatment increases the hydrophobicity of soy proteins. It is, therefore, possible that the volatile compounds responsible for the beany flavor attached themselves to the proteins and were held more tightly after pressurization, which decreased their volatility.

Soymilk may sometimes have a residual bitter taste due to the presence of saponins. Kajiyama et al. (1995) observed a large reduction of free saponins in HPP soymilk, probable because the saponins were no longer free and were trapped on the increased hydrophobic regions on the surface of the proteins.

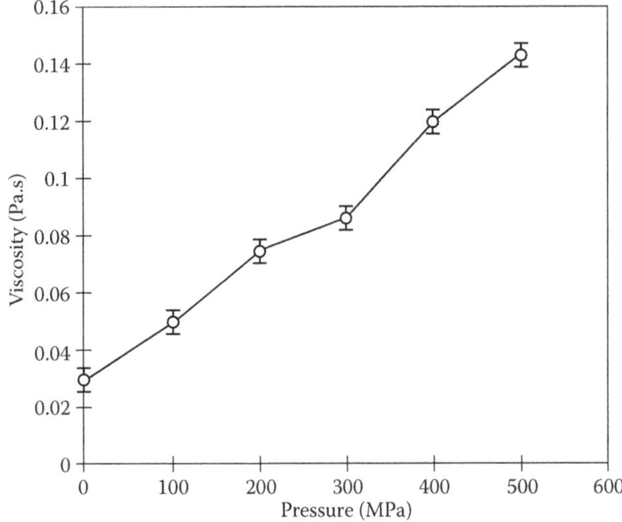

FIGURE 19.2 Viscosity of soymilk after pressurization at room temperature for 10 min (Reprinted from Zhang, H., Li, L., Tatsumi, E. and Isobe, S., *Lebensmittel-Wissenschaft und Technologie*, 38, 7–14, 2005. With permission from Elsevier.)

19.2.3 High-Pressure Processing (HPP) of Tofu

The conventional process for making tofu is to denature soy proteins in soymilk by heating, and then coagulate them by using coagulants and heat treatment (Yamauchi et al. 1991). The coagulated soymilk is then separated into whey and curds, and the curds are subsequently pressed to form tofu. To coagulate the soymilk, a small amount of salt (Nigari, sea salt with magnesium chloride and calcium salts) or glucono-delta-lactone is added (Shih, Hou, and Chang 1997).

Kajiyama et al. (1995) found that at pressures greater than 100 MPa, soymilk could be coagulated by $CaCl_2$ to form a gel. Gels formed at pressures above 300 MPa showed greater hardness than that of conventional tofu and the hardness was almost twice as great at 500 MPa. Light microscopy revealed little difference between conventional tofu and the gels formed by pressurization. High pressure dissociates soy proteins and exposes free SH residues which subsequently form -S–S- bonds with the help of oxygen in soymilk and may be partly responsible for the phase change from liquid to sol. Dissociation also exposes more hydrophobic regions, which enhance hydrophobic bond formation, the main cause of hard gel formation, with added help from coagulants at low protein concentrations (Kajiyama et al. 1995). These observations confirm those of Saio, Kajikawa, and Watanabe (1971) and Hashizume, Nakamura, and Watanabe (1975) who earlier reported that gel hardness depended on SH content of soy proteins and conditions creating new SH residues and exposing them in soymilk resulted in better tofu.

Prestamo and Fontecha (2007) studied the effect of different pressure treatments (100, 200, 300, and 350 MPa) at 22°C for 15 min on the fatty acid composition and acylglycerol content of tofu. The findings showed no significant differences ($p < .05$) in the fatty acid methyl esters (FAMEs) and acylglycerol contents of untreated tofu and pressure-treated tofu (Table 19.1). The highest FAMEs found were linoleic acid (C18:2) (55.42%), followed by oleic acid (C18:1) (19.77%), palmitic acid (C16) (11.11%), linolenic acid (C18:3) (8.29%), and stearic acid (C18) (3.80%). Levels lower than 0.5% were obtained for docosanoic acid (C22:0), eicosanoic acid (C20:0), eicosaenoic acid (C 20:1), and myristic acid (C14). Tofu fat is low in saturated fatty acids (SFA, ~16%) and high in polyunsaturated fatty acids (PUFA, 64%), and unsaturated fatty acids have been reported to be pressure resistant (Cossins and Macdonald 1986).

TABLE 19.1
Fatty Acid (% wt. Total Fatty Acids) and Di- and Triacylglycerol Content (% wt.) of HP-Treated (350 MPa, 15 min, 22°C) Soy Tofu

Fatty Acids	Control	HP
Saturated fatty acids	15.83	15.93
Myristic acid C14	0.12	0.12
Palmitic acid C16	11.11	11.35
Steraric acid C18	3.80	3.77
Monounsaturated fatty acid	19.99	19.94
Oleic acid C18:1	19.77	19.75
Polyunsaturated fatty acids	63.71	63.54
Linoleic acid C18:2	55.42	55.27
Linolenic acid C18:3	8.29	8.27
Eicosanoic acid C20	0.37	0.36
Eicosaenic acid C20:1	0.22	0.18
Docosaenoic acid C22:0	0.43	0.37
n-6/n-3	6.69	6.68
	Diacylglycerol	
C34	1.75	1.85
C36	8.58	9.34
	Triacylglycerol	
C50	3.94	3.88
C52	27.72	27.34
C54	57.98	57.60

Source: Reprinted from Prestamo, G. and Fontecha, J., *Innovative Food Science and Emerging Technologies*, 8, 188–91, 2007. With permission from Elsevier.

Tofu is normally preserved under refrigerated conditions in order to prevent spoilage and extend shelf life. One advantage of using HP treatment in food processing is that it pasteurizes the product, resulting in a decrease of the microbial population and an extension of product shelf life (Knorr 1993). Tofu can be contaminated during processing and packaging. Préstamo et al. (2000) reported that tofu could be preserved from secondary pathogenic contamination by using HP treatment (400 MPa at 5°C for 30 min) without affecting the sensory attributes. These authors observed that HP treatment of tofu effectively reduced the microbial populations belonging to the Enterobacteriaceae family, gram-negative and gram-positive bacteria. HP treatment could, therefore, serve as a useful technique for the formation and preservation of tofu gels.

19.2.4 High-Pressure Processing (HPP) Effects on Soy Protein Functionality

It is now well established that HP treatment has relatively little influence on secondary structure, whereas it can affect the tertiary and quaternary structure of most globular proteins (Puppo et al. 2004). Both protein denaturation and aggregation phenomena occur under pressure and principally involve hydrogen bonds, electrostatic and hydrophobic interactions, and disulfide linkages (Torres and Velazquez 2005). Interfacial properties of proteins such as foaming and emulsification have been reported to be modified by HP (Galazka, Dickinson, and Ledward 2000). The following section provides further details on the effect of high pressure on some of the important functional properties of soy proteins.

19.2.4.1 Effect on Solubility

Solubility is one of the most basic physical properties of proteins and a prime requirement for any functional application. Solubility of a protein under specified conditions is governed by factors that influence the equilibrium between protein–protein and protein–water interactions (Damodaran 1996). Intrinsic factors including hydrophobicity–hydrophilicity ratio, size, charge, and steric properties, and extrinsic factors such as pH, ionic strength, and interaction with other food components affect the solubility of proteins.

The solubility of soy protein isolate (SPI) after HP treatment has been extensively studied. Molina, Papadopoulou, and Leward (2001) reported a reduction in solubility when 10% (w/v) SPI solutions (pH 7.5 and 6.5) were subjected to pressure treatments at 200 or 400 MPa. This decrease was attributed to protein unfolding of the globulins, resulting in the exposure of hydrophobic regions to the exterior of the protein molecules. These authors reported that, at higher pressures (600 MPa), dissociation of globulin subunits may occur, leading to the formation of soluble complexes between the basic subunits of glycinin and some subunits of β-conglycinin. In contrast, Puppo et al. (2004) reported that HP treatment at levels above 200 MPa only changed the solubility of SPI (1% w/v) slightly at pH 8.0 (Figure 19.3), whereas under acidic conditions (pH 3) the solubility increased with increasing pressure (200–600 MPa). These results indicate that the solubility of soy protein during HP treatment may be influenced by concentration, pH, high pressure levels, and possibly treatment duration.

19.2.4.2 Effect on Emulsifying Properties

Unfolding of proteins and subsequent exposure of hydrophobic groups after HP treatment may improve the emulsifying properties of SPI (Galazka, Dickinson, and Ledward 1999). In general, the higher the surface hydrophobicity the greater is the potential for protein adsorption at the oil-water interface. Wang, Zhou, and Chen (2008) reported that HP treatment at 200 MPa increased the emulsifying activity index (EAI) of SPI (<5% w/v concentration), however, further increases in pressure (400 and 600 MPa) did not change the EAI. Molina et al. (2001) earlier found that HP treatment at neutral pH improved the EAI but not the stability of emulsions made with soybean proteins at 10% (w/v) concentration. Molecular flexibility is an important factor influencing emulsion stability and at higher concentrations proteins can aggregate, which could decrease their

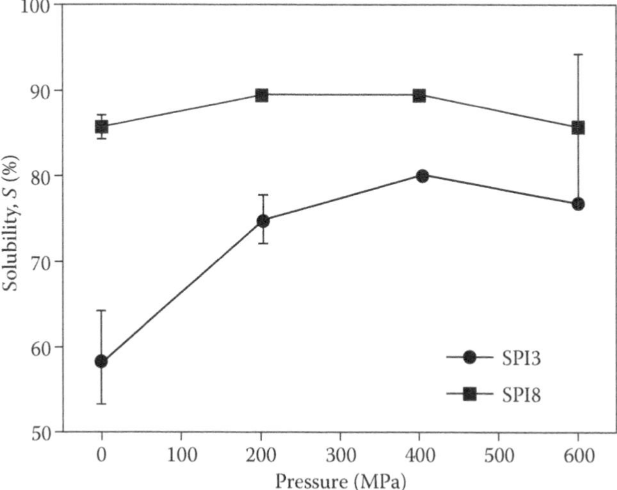

FIGURE 19.3 Effect of high-pressure treatment (200–600 MPa) and pH on solubility of SPI (1%). (Reprinted with permission from Puppo, C., Chapleau, N., Speroni, F., de Lambellerie-Anton, M., Michel, F., Añón, C., and Anton, M., *Journal of Agricultural and Food Chemistry*, 52, 1564–71, 2004. American Chemical Society.)

molecular flexibility (Kato and Nakai 1980). Puppo et al. (2005) studied the changes induced by HP treatment (200–600 MPa) on SPI at pH 8 and pH 3. At pH 3 soy proteins are in the vicinity of their isoelectric point and are thus more associated/aggregated. They reported an improvement in the EAI of SPI at pH 8 (i.e., smaller droplet size of emulsions and an increase in the percentage of adsorbed proteins). On the other hand, at pH 3 an increase in droplet size and depletion flocculation was observed. An improvement in the EAI of soymilk after HPP was also reported by Kajiyama et al. (1995).

19.2.4.3 Effect on Gelation

At sufficiently high concentrations and pressures, soy proteins will denature and form gels in a similar manner to heat-induced gels. In general, pressure-induced gels are much softer, more cohesive, and less gummy than heat-induced gels. At a pressure of 200 MPa, Okamoto, Kawamura, and Hayashi (1990) reported that soy proteins (17% w/v) formed gels that lost their shape when taken out of the bottle. Increasing the pressure level from 300 to 400 MPa resulted in an increase in gel hardness. Dumoulin, Ozawa, and Hayashi (1998) observed that 17% (w/w) soy protein at pH 6.8 formed firm gels when pressurized at <210 MPa and at –20°C and at or above 300 MPa at –5, 10, 25, and 50°C for 30 min. The hardness of the gels increased as pressure levels increased and the gels formed were softer, more deformable without breaking, and whiter than heat-induced gels. Molina, Papadopoulou, and Leward (2002) later reported that for HP treatment in the range of 300–700 MPa, soy protein concentrations of 20% (w/v) or more were required for the formation of self-supporting gels. With heat treatment, soy protein concentrations of at least 16.7% (w/v) are needed to produce a gel. The major advantage of HP-induced gels is less browning and it also gives a softer and smoother texture than heat-induced gels.

19.3 ULTRASONICATION OF SOYBEANS

Ultrasonication is a promising technology that has generated growing interest among researchers because of its wide range of applications in chemical synthesis, therapeutics, environmental protection, electrochemistry, and food processing (Lee et al. 2005; Gachagan et al. 2004; Iida et al. 2005; Moussatov, Granger, and Dubus 2005). Ultrasonic waves are high-intensity acoustic waves with frequencies higher than 20 kHz. They have the same nature as sound waves and are distinct from electromagnetic waves in that they need a medium to propagate. Like other waves, ultrasonic waves are defined by their frequency (~), velocity (v), and amplitude or intensity (A) (Kocis and Figura 1996). Industrial ultrasonic applications can be divided into two general groups, namely, high- and low-energy ultrasonics. High-energy ultrasound applications are usually found at intensities higher than 1 W/cm^2 and at frequencies between 18 and 100 kHz (McClements 1995; Povey and Mason 1998; Villamiel and de Jong 2000b) and it has been applied for degassing of liquid foods, the induction of oxidation/reduction reactions and nucleation for crystallization, extraction of enzymes and proteins, and enzyme inactivation (Roberts 1993; Thakur and Nelson 1997; Villamiel and de Jong 2000a). These applications rely on the high pressures and temperatures reached locally when a high-intensity wave propagates through a medium. If the intensity is high enough cavitation (implosion of gas bobbles) can occur. Low-energy ultrasound applications are frequently at intensities lower than 1 W/cm^2 and at frequencies higher than 100 kHz (Mason and Luche 1996; Villamiel and de Jong 2000a, 2000b). They are used to monitor processes or products which require noninvasive detection (e.g., process control) and for characterizing physicochemical properties of food materials (product assessment or control) (McClements 1997; Povey and Mason 1998; Withers 1996). Low-energy ultrasound is used for surface cleaning of foods, assisted extraction, crystallization, emulsification, filtration, drying, freezing, and meat tenderization (Behrend and Schubert 2001; Mason and Luche 1996).

Low-intensity ultrasonics are applied more often to fields other than food technology. In the food industry, examples of positive effects associated with ultrasonication include higher product yields, shorter processing times, reduced operating and maintenance costs, improved taste, texture, flavor, and color, and the reduction of pathogens at lower temperatures. As one of the more advanced and emerging food technologies it can be applied not only to improve the quality and safety of processed foods but it also offers the potential for the development of new products with unique functionality. In the soy industry, ultrasonication has been explored as a technique to improve oil, protein, and isoflavone extraction and to inactivate enzymes. Details on these applications are provided below.

19.3.1 Soybean Oil Extraction

Ultrasound techniques have been used in oil extraction processes to improve efficiency and reduce processing time. In commercial solvent extraction processes a series of steps involving cleaning, dehulling, moisture conditioning, flaking, and heating are used to extract oil from the seeds, which makes the process laborious. Ultrasonic extraction is a simplified and short-term extraction method that can be used to obtain commercially acceptable product yields.

The application of 20 kHz high-intensity ultrasound during the extraction of oil from two soybean varieties (TN 96-58 and N 98-4573) using hexane was evaluated by Li, Pordesimo, and Weiss (2004). Oil yield was observed to increase with treatment time and ultrasound intensity (Figure 19.4). After 3 h at an ultrasound intensity of 47.6 W/cm^2, the increase in oil yield was 2.4 and 9% higher than at 20.9 W/cm^2 and 16.4 W/cm^2, respectively. In comparison to the nonsonicated control, the oil yield after 3 h at 16.4, 20.9, and 47.6 W/cm^2 increased by 2.2, 10.1, and 11.2%, respectively. Thus, after three hours, the relative oil yield increase at 47.6 W/cm^2 was approximately five times higher than at 16.4 W/cm^2. Solvent type also influenced the efficiency of extraction and the highest yield was obtained using ultrasound in combination with a mixed solvent system (3:2 hexane–isopropanol) compared to hexane and isopropanol solvents alone. Gas chromatography analysis of oil extracted from ultrasonicated soybean did not show significant changes in fatty acid composition. The development of microfractures and disruption of cell walls in ground soybean flakes by ultrasonic treatment was clearly evident from the results of scanning electron microscopy.

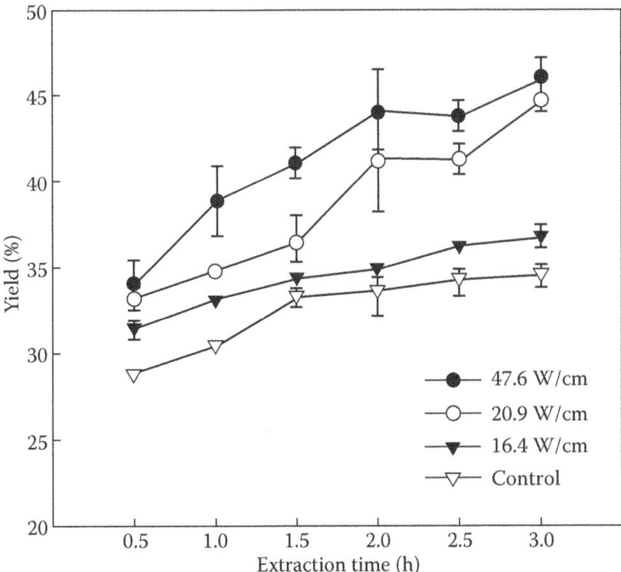

FIGURE 19.4 Oil yield of ultrasonicated soybean variety TN 96-58 using hexane as a solvent at 25°C. (Reprinted from Li, H., Pordesimo, L., and Weiss, J., *Food Research International*, 37, 731–38, 2004. With permission from Elsevier.)

19.3.2 Soy Protein Extraction

Sonication has been used to break down protein molecules, to emulsify protein suspensions, and to aggregate proteins (Wang 1984). It is recognized that globular proteins may be physically disrupted or transformed into new molecules under ultrasonic power due to cavitation effects. In 1982, Moulton and Wang reported continuous ultrasonic extraction of soy protein as an alternative process to produce SPI from defatted soybean flakes. The continuous high-intensity application (20 kHz) extracted 37 and 49% protein for water and alkali (0.1 N NaOH) extraction, respectively, compared with the batch extraction (24 and 44%, respectively) using comparable processing times and volumes. Wang (1984) reported that ultrasonication (20 kHz, 8 min, flake:water ratio 1:20) induced conversion of 7S protein into 40 to 50S aggregates during extraction from defatted and autoclaved soy flakes. In other studies, Takamiya and Terada (1980) observed a rapid decrease in the viscosity of ultrasonicated soymilk caused by the decomposition of 11S protein.

19.3.3 Isoflavone Extraction

An evaluation of ultrasonic-assisted extraction of isoflavones from freeze-dried ground soybeans using three different solvents (30–70% methanol, ethanol, and acetonitrile) at 60°C for 10 min was undertaken by Rostagno, Palma, and Barroso (2003). The extraction efficiency of soy isoflavones was higher with ultrasonication (15%) compared with the solvent extraction alone. Lee and Row (2006) reported that ultrasonication with a frequency of 20 kHz and an extraction time of 10 min in 60% aqueous ethanol solutions gave the highest yield of isoflavone aglycone. Yields of both isoflavone glycosides and aglycones were threefold greater than those obtained using regular extraction methods.

19.3.4 Lipoxygenase (LOX) Enzyme Inactivation

The effect of ultrasonication on the activity of LOX has been studied by several workers. Thakur and Nelson (1997) reported that the LOX enzyme in whole soy flour suspension was stable at pH>5.0 and was not inactivated even after 3 h exposure to cavitating 20 kHz ultrasound. At pH<5.0, the activity decreased with increasing time of exposure, and under similar conditions of exposure, about 70–85% of the enzyme activity was lost at pH 5.0 and 4.0. This inactivation may be due to the formation of hydroperoxy free radicals ($HO_2^{\bullet}$) since the pK of $HO_2^{\bullet-}$ is about 4.7 (Bielski and Allen 1977) and the concentration of these free radicals increases at low pH values. $HO_2^{\bullet}$ free radicals undergo H^+-dependent reactions to form H_2O_2, thus favoring increased H_2O_2 concentration at acidic pH. H_2O_2 is a very efficient inhibitor of LOX (Mitsuda, Yasumoto, and Yamamoto 1967), which would explain why LOX inactivation is higher at lower pH. Exposing LOX in a 1% (w/v) whole soy flour suspension to 30 kHz ultrasonication for 60 min had no effect on enzyme activity at above pH 5. However, the enzyme activity decreased at ultrasound frequencies >30 kHz and at a pH of 5.0 or below. Since ultrasonication intensity is proportional to the square of the amplitude (Lopez et al. 1994), higher rate of enzyme inactivation can be expected at higher frequency levels and lower pH (≤5).

Lopez and Burgos (1995) also reported that the synergistic effect of heat and ultrasound waves on inactivation of LOX was higher at acidic pH than at neutral or alkaline pH. Inactivation of enzymes at low temperature by long-term exposure to ultrasound is reportedly due to the splitting of low-molecular-weight polypeptides or individual amino acids, or more frequently due to oxidative mechanisms (Coakley, Brown, and James 1973; Santamaria, Custellani, and Levi 1952). Ultrasound waves promote acoustic cavitation (i.e., formation and growth, and violent collapse of small bubbles or voids as a result of pressure fluctuations). Severe shear stresses are generated due to changing shape and size of bubble liquid interface and acoustic streams occurring in the vicinity of the bubble, promoting enzyme denaturation (Lopez et al. 1994). Sonification also promotes chemical reaction due to the formation of H^+ and OH^- free radicals from water decomposition inside the oscillating bubbles (Elpiner, Sokolskaya, and Margulis 1965). Some amino acid residues participating in the stability, substrate binding, or catalytic activity of the enzyme can react with these free radicals (Lopez and Burgos 1995), thus affecting the enzyme activity. Proline, leucine, isoleucine, lysine, and glutamic acid have been reported to readily form peroxides with OH• radicals (Gebicki and Gebicki 1993).

Thakur and Nelson (1997) reported that the inactivation of LOX at room temperature using high frequency and low pH is irreversible and the enzyme is not reactivated upon storage. However, their study was conducted on 1% suspensions of soy flour. An increase in LOX concentration increases the resistance of the enzyme to inactivation (Coakley et al. 1973); thus, longer exposure times and higher frequencies will be needed to inactivate the enzyme in suspensions of higher concentrations.

Pressure, heat, and ultrasound combinations (manothermosonication) have been reported to inactivate heat-resistant enzymes (Vercet, Lopez, and Burgos 1997). There is also a considerable amount of data on the use of ultrasound in conjunction with chemical antimicrobials (Phull et al. 1997), with heat or with heat and moderate pressure (Ciccolini et al. 1997; Earnshaw, Appleyard, and Hurst 1995; Sala et al. 1995; Villamiel and de Jong 2000a) to inactivate microorganisms.

19.4 PULSED ELECTRIC FIELD (PEF) PROCESSING OF SOYBEANS

Pulsed electric field (PEF) processing is a nonthermal method of food preservation that uses short bursts of electricity in the order of 20–80 kV (usually for a couple of microseconds) for microbial inactivation. The technique causes minimal or no detrimental effect on food quality attributes and can be used for processing both liquid and semiliquid food products. The electric field may be applied in the form of exponentially decaying, square wave, bipolar, or oscillatory pulses and at ambient, subambient, or slightly above-ambient temperature. After the treatment, the food may be packaged aseptically and stored under refrigeration conditions. A few studies are beginning to appear in the literature on the use of PEF in soybean processing, some of which are summarized below.

19.4.1 Extraction of Soybean Oil, Phytosterols, and Isoflavones

High-intensity pulsed electric field (HIPEF) has been used for extraction of fat from oil seeds. The lipid bilayer is susceptible to applied electric fields because of its net electric charge. Guderjan et al. (2005) studied the impact of PEF on recovery and quality of oil from maize germ and olives. They reported that application of PEF at a field strength of 0.6 kV/cm gave higher oil yields (88.4%) and phytosterol content (32.4%) in maize germ oil than untreated samples. The oil yield from olives was increased by 6.5 to 7.4% depending of the filed strength (0.6–1.3 kV/cm). While no work was conducted on soybeans, the potential may exist for the use of PEF to increase the extraction of oils from soybeans. Interestingly, in the study conducted by Guderjan et al. (2005), they found the isoflavone (genistein and daidzein) content of PEF-treated soybean (1.3 kV/cm) increased by 20–21% in comparison to reference samples.

19.4.2 Effect of Pulsed Electric Field (PEF) on Soy Proteins

There are few reports on the effects of PEF on protein components of foods and their functional properties. Li, Chen, and Mo (2007) studied the effects of PEF treatment (0–547 μs and 0–40 kV/cm) on the physicochemical properties of SPI dispersions (20 mg/ml). They showed that PEF treatment could change the physicochemical properties and structure of SPI. Solubility, surface free sulfhydryl content (SH_F), and hydrophobicity of SPI increased with increasing PEF strength and treatment time (Figures 19.5 and 19.6). They concluded that PEF treatment induced polarization of SPI, dissociation of subunits, and molecular unfolding (i.e., conformational changes which caused hydrophobic groups and SH_F buried inside the molecules to become exposed). These events, subsequently, resulted in enhancing surface hydrophobicity and SH_F content of SPI. Dissociation of subunits further increased protein–water interactions and, subsequently, the solubility of SPI. PEF treatment above 30 kV/cm and 288 μs caused a decrease in solubility, SH_F, and hydrophobicity due

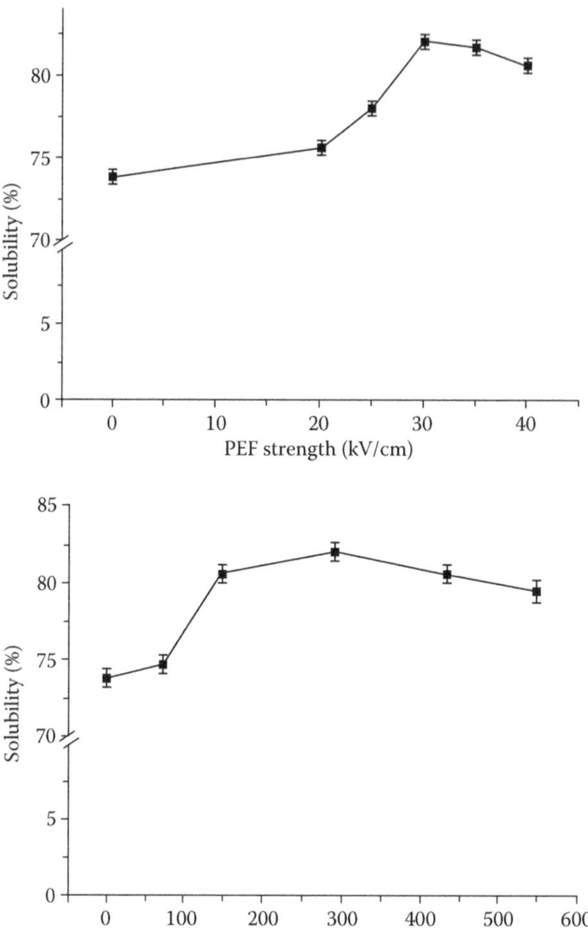

FIGURE 19.5 Solubility of PEF-treated SPI at different field strength and time (sample flow rate of 1 ml/s, pulse frequency of 500 pps, and 2 μs pulse width). (Reprinted from Li, Y., Chen, Z., and Mo, H., *Lebensmittel-Wissenschaft und Technologie*, 28, 410–18, 2007. With permission from Elsevier.)

to dissociation, denaturation, and aggregation of SPI. Thus, high-intensity PEF conditions produced stronger molecular polarization and the polarized molecules attracted each other and converged to form larger aggregations through noncovalent bonds such as hydrophobic interactions, electrostatic interactions, and hydrogen bonds.

PEF may cause denaturation of enzymes by changing their conformations. The efficiency of PEF treatment in inactivating enzymes depends on the electric field strength, treatment time, pulse width, pulse frequency, and sample flow rate; amongst these, the pulsed electric strength and treatment time are the two critical factors for enzyme inactivation (Barsotti et al. 2002). Li et al. (2008) reported that the residual activity of LOX in soymilk decreased with an increase in treatment time, pulse strength, pulse frequency, and pulse width. Maximum inactivation of LOX (88%) was observed at 42 kV/cm using 400 Hz pulse frequency for 1036 μs and 2 μs of pulse width at 25°C.

19.4.3 Effect of Pulsed Electric Field (PEF) on Microorganisms

PEF pasteurization is an emerging nonthermal technological alternative to thermal processing which can be used to pasteurize pumpable food products without significantly impairing the major attributes of the final product (Qin et al. 1996). Microbial inactivation efficacy of PEF has been

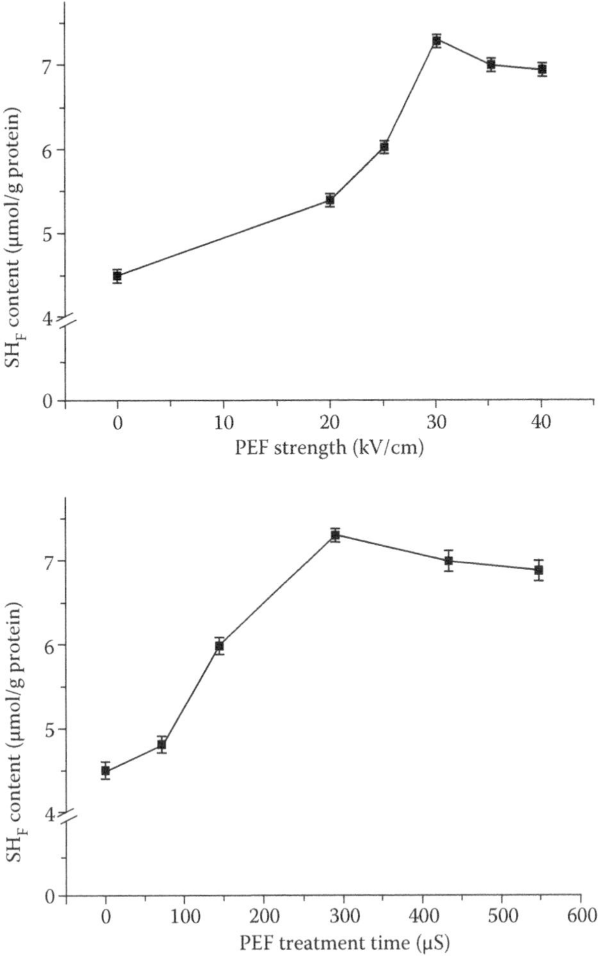

FIGURE 19.6 SH_F content of PEF-treated SPI at different field strength and time (sample flow rate of 1 ml/s, pulse frequency of 500 pps, and 2 μs pulse width). (Reprinted from Li, Y., Chen, Z., and Mo, H., *Lebensmittel-Wissenschaft und Technologie*, 28, 410–18, 2007. With permission from Elsevier.)

investigated with *Escherichia coli* (Zhang et al. 1994), *Lactobacillus brevis* (Jayaram, Castle, and Pargaritis 1992), *Listeria* (Reina et al. 1998), and *E. coli O157:H7* (Evrendilek, Zhang, and Richter 1999), either in model or food systems. More than a 5 log reduction of *E. coli 8739* in soymilk enriched with hyperimmunized dairy protein was achieved with PEF treatment at 41 kV/cm for 54 μs, which was considered as a critical value for its inactivation. Increases in electric field strength and total treatment time resulted in improvement of the inactivation effect of PEF against *E. coli 8739* (Li and Zhang 2004). Furthermore, the shelf life of PEF-treated enriched soymilk was significantly ($p < .01$) extended at refrigeration temperature for 30 days and no significant changes in color and viscosity were observed during storage.

19.5 IRRADIATION OF SOYBEANS

Food irradiation may be useful as an alternative technique for the preservation and processing of various fresh, perishable, and high-protein foods, with or without chemical additives or biological controls (Lochhead 1989; Murray 1990). Irradiation involves the exposure of food, either packed or in bulk, to carefully controlled amounts of ionizing radiation for specified periods.

The sources of ionizing energy (γ-rays) may be cobalt 60 or cesium 137, and x-rays from electron beam accelerators or ultraviolet (UV) sources. When food is subjected to ionizing radiation, energy is absorbed by collision of the ionizing radiation with particles of the food resulting in the generation of ion pairs. As the energy of the primary radiation is distributed throughout the matrix, more electrons are stripped from other atoms and they may, in turn, possess sufficient energy to ionize other atoms. These excitation and ionization events result in changes in the atomic and molecular structures of the food. Irradiation causes minimal rise in temperature (an radiation dose of 10 kGy produces only a 2.4°C temperature increase in 1 kg of food having the heat capacity of water (4.184 J/°C)) and is acknowledged to cause fewer overall physical and sensory changes than cooking, freezing, or canning (Josephson and Peterson 1983; Molins 2001; Urbain 1989). The need for temperature and atmospheric controls can also be minimized, eliminated, or used in combination with packaging to delay food/feed spoilage (Lagunas-Solar 1995). In the soy industry, inactivation of antinutritional factors and enzymes in soybeans is an important part of processing. Even though conventional methods like heating and soaking can inactivate these compounds, they can destroy some of the vitamins presents in the bean. Irradiation may be used as a technique to remove or reduce antinutritional factors and decontaminate foods by killing pathogens and insects (Ghazy 1990; Molins 2001). The application of ionizing radiation to soybean and the effects on nutritional quality as well as functionality are discussed below.

19.5.1 Effect of Irradiation on Soy Protein Functionality

Hafez et al. (1985) reported that the solubility of soy proteins was not affected by irradiation treatments up to 10 kGy. At higher irradiation dose levels (100 kGy) a significant reduction in protein solubility from 80.3% to 67.2% in water, 57.8% in salt solution, and 48.8% in the presence of calcium chloride was observed, probably due to the formation of protein aggregates induced by the irradiation treatment. Increasing the moisture content of soybeans also resulted in a decrease in protein solubility even at lower radiation doses. At 65 kGy dose, the decreases observed in the protein solubility of soybeans containing 15.33, 22.48, and 30.47% moisture were 17.9, 23.0, and 49.4% respectively (Hafez et al. 1985). This is because the amount of reactive species formed by radiolysis of water increases with increasing moisture content. The increase in OH radicals from the radiolysis of water causes an increase in the radical sites on the protein molecules, which could induce protein–protein interactions and result in the formation of protein aggregates (Diehl et al. 1978). In addition, proteins may react with other components in Maillard-type reactions during irradiation, leading to less soluble reaction products (Urbain 1989). Afify and Shousha (1988) also reported decreases in protein solubility, with increasing irradiation dose (10 kGy).

Byun et al. (1994) studied the effect of gamma irradiation on soybean proteins and reported that irradiation doses above 10 kGy caused a decrease in 7S and 11S components and an increase in 2S and 15S components. However, subunit patterns determined by electrophoresis did not change appreciably over the entire irradiation dose range studied (<40 kGy), which indicated that irradiated soybean proteins have the same subunit structure as that of the nonirradiated soybean proteins. These authors suggested that irradiation resulted in partial degradation of 7S globulin into 2S proteins and aggregation of 11S globulins to 15S component. They also reported from DSC studies that the denaturation temperatures of 11S and 7S components were not affected by irradiation at lower doses, however increasing the irradiation dose caused a decrease in the enthalpy values of 11S and 7S components, which is indicative of protein denaturation (Zarins and Marshall 1990).

Soybean proteins irradiated at up to 10 kGy have a secondary structure similar to that of the nonirradiated protein (Byun et al. 1994). Slight changes in the secondary structures of the 7S and 11S components were noted when irradiation doses were increased to 20 kGy. Gamma irradiation has been shown to cause molecular changes resulting in condensation or polymerization, degradation, hydrogen-bond disruption, and cleavages of intermolecular disulfide bonds (Casarett

1968). Therefore, it is assumed that such molecular rearrangements may bring about changes in secondary structure and disrupt the native conformation of soybean proteins, especially at higher dose levels.

19.5.2 Effect of Irradiation on Soybean Oil

There is very little information available on the effect of radiolytic reactions on soybean oil. Physicochemical properties of soybean oil extracted from γ-irradiated soybeans (0–10 kGy) were studied by Byun et al. (1995a). These authors reported no significant changes in total lipid content, fatty acid composition, acid value, peroxide value, and trans fatty acid content at different irradiation doses. At higher dose levels (>10 kGy), *n*-hexanal content of soybean oil was, however, reported to increase. Hafez et al. (1985) found a high negative correlation between irradiation dose and linolenic acid content (i.e., a high radiation dose caused a decrease in linolenic acid). These authors reported that an increase in both moisture contents of soybeans and radiation doses did not have any effect on C16:0, C18:0, C18:1, and C18:2 fatty acids.

19.5.3 Effect of Irradiation on Soymilk

Changes in the physicochemical properties of soymilk prepared with irradiated and nonirradiated soybeans were investigated by Byun, Kang, and Mori (1995b). The yield of soymilk prepared with soybeans irradiated at 2.5 and 5 kGy was found to be higher than that of the nonirradiated soybeans and soaking time was reduced. The yield of soymilk prepared with soybeans irradiated at 20 kGy was, however, 6% lower than that of the nonirradiated soybeans. Solids content, crude protein, and pH of soymilk were minimally affected by irradiation. Color changes in soymilk were, however, observed at higher doses (10 and 20 kGy) (i.e., reductions of "b" value (yellowness) and increments in "a" value (redness)). The change in color at higher doses may be due to the oxidation of carotenoid pigments in soybean seeds. These results suggest that gamma irradiation at lower dose levels (2.5 and 5 kGy) could reduce soaking time of soybeans (Byun, Kwon, and Mori 1993) and also increase the yield of soymilk with minimal effect on quality. Wilson (2004) and Chia (2006), on the other hand, reported that surface radiation of whole dry soybeans using electron beam or γ-rays at 1–30 kGy provided microbial safety for astronauts, but caused oxidative changes resulting in unacceptable soymilk and tofu (i.e., rancid aromas, off-color tofu with low yields, more solid waste, and loss of the ability of the seeds to germinate). Further studies are needed to confirm the effects of irradiation on the functional quality of soymilk.

19.5.4 Effect of Irradiation on Tofu

Byun et al. (1995b) reported that the yield and water-holding capacity (WHC) of tofu made from soybeans irradiated at 5 kGy was greater than that of the nonirradiated control. Increasing the irradiation dose to 10–20 kGy, however, caused a decrease in yield and WHC mainly due to a decrease in protein solubility. At 20 kGy, irradiation induced partial denaturation and degradation of soybean proteins with secondary and conformational changes (Byun et al. 1995b). Sensory and textural properties such as acceptability, hardness, and fracturability of tofu increased with increasing irradiation dose levels (up to 10 kGy). On the other hand, gumminess, chewiness, cohesiveness, and adhesiveness values decreased as the dose level increased in comparison with nonirradiated tofu.

19.5.5 Effect of Irradiation on Soybean Antinutritional Factors

19.5.5.1 Protease Inhibitors

Farag (1989) reported that 54.5% deactivation of TIA was achieved in soybean seeds irradiated at 10 kGy. Later, he (Farag 1998) demonstrated that the degradation of protease inhibitors on exposure to gamma irradiation was directly proportional to the radiation dose. Increasing the radiation dose from 5 to 15, 30 and 60 kGy, linearly increased the level of inactivation of TI from 41.8 to 56.3, 62.7, and 72.5%, respectively. TI inactivation in irradiated soybean samples could be attributed to the destruction of disulfide (-S–S-) groups. Lee (1962) observed a high apparent susceptibility of sulfhydryl (-SH) and disulfide (-S–S-) groups in proteins to irradiation. This was confirmed by Khattak and Klopfenstein (1989) who later showed that sulfur-containing amino acids were liable to become damaged by radiation, particularly in legumes.

Inhibition of 71% of LOX activity, 25.4% TIA, and 16.7% chymotrypsin inhibitor activities were found in soybean seeds irradiated at 100 KGy (Hafez et al. 1985). As the moisture content of soybean seeds increased, a marked decrease in enzyme activity as well as antinutritional factors was observed, even at lower doses (65 kGy). In other studies, an irradiation dose of 10 kGy was found to decrease trypsin and chymotrypsin inhibitor activity in defatted soybean flour by 34.9 and 71%, respectively, whilst in vitro digestibility increased from 80 to 84% (Abu-Tarboush 1998). De Toledo et al. (2007) also compared the TIA of raw soybeans irradiated at 2–8 kGy followed by cooking (121°C, 10 min) and irradiation alone. They, however, reported a significantly ($p \leq .05$) larger decrease in the TIA of the irradiated and cooked soybean (34.09 TIU/mg) compared to the irradiated samples (58.76 TIU/mg), which indicates that combination of irradiation and cooking could possibly reduce TIA even at lower irradiation doses.

19.5.5.2 Phytate

Sattar, Neelofar, and Akhtar (1990) studied the effect of irradiation (0.25–1.0 kGy) and soaking (0–12 h) on phytate content of soybeans. Their results showed that irradiation alone or in combination with soaking reduced the level of phytate content compared to nonirradiated soybeans. This may be due to the chemical degradation of phytate to lower inositol phosphates and inositol by the action of free radicals produced by radiation (De Boland, Garner, and O'Dell 1975). Moreover, a significant ($p < .01$) decrease in phytate levels was observed during germination (24–120 h) of irradiated soybean (0.05–0.20 kGy) compared to nonirradiated germinated soybean seeds.

19.5.5.3 Lectins

Pusztai (1991) proposed that the toxic effect of lectins, when administered orally, might be related to their ability to bind to some specific receptor sites on the surface of the epithelial cells lining the intestine, thus leading to nonspecific interference with the absorption or transport of nutrients across the intestinal wall. When soybeans were subjected to a radiation dose of 10 kGy, the phytohemagglutinating activity was reduced by 50% (Farag 1989), which is a significantly higher reduction than with normal processing techniques such as germination, soaking, and dehulling (Liener 1994). When the seeds were processed at 15, 30, and 60 kGy, the phytohemagglutinating activity against rabbit red blood cells was reduced by 50, 75, and 94%, respectively (Farag 1998). Radiolytic modifications have been shown to decrease the number of carbohydrate-binding sites in Con A lectin that are responsible for its biological properties without significantly changing the nature of the carbohydrate-protein interactions (Moore and Mudher 1978).

19.5.6 Effect of Irradiation on Isoflavones

Variyar, Limaye, and Sharma (2004) examined the isoflavone content and antioxidant activity of irradiated soybeans at 0, 0.5, 1, and 5 kGy dose levels. The study revealed a decrease in glycosidic conjugates and an increase in aglycones with increasing radiation dose as compared with nonirradiated soybeans. Higher antioxidant activity was also observed when samples were irradiated at 5 kGy compared to the lower-dose irradiation treatments and the nonirradiated soybeans (Figure 19.7). The increase in antioxidant activity was attributed to the increase in free isoflavones that have greater antioxidant effect than the glycosides.

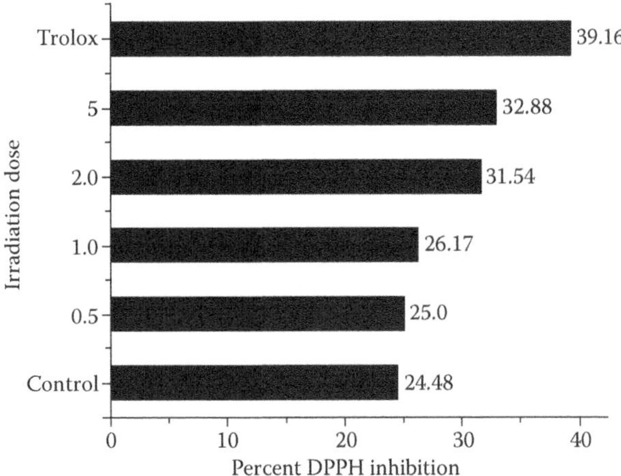

FIGURE 19.7 Antioxidant activity (% DPPH inhibition) of nonirradiated and gamma-irradiated soybeans at various doses compared to Trolox standard. (Reprinted with permission from Variyar, S. P., Limaye, A., and Sharma, A., *Journal of Agricultural and Food Chemistry*, 52, 3385–88, 2004. American Chemical Society.)

Irradiation may be a good alternative processing technique for reducing both heat-stable and heat-labile antinutrients. In particular, irradiation levels up to 10 kGy, which are admissible for the irradiation of foods in some countries, seem to be effective in inactivating antinutrients such as protease inhibitors, lectin, phytic acid, nonstarch polysaccharides, and oligosaccharides without altering the nutritional quality of the food. Commercial uses of this promising process are still very limited, mainly due to consumer resistance and because further research appears to be needed to ascertain the wholesomeness of irradiated food items.

19.6 COLD GELATION OF SOY

Soy proteins are extensively used as functional ingredients in many food formulations because of their ability to form gels. Much work has been reported on heat-induced soy protein gels (Puppo and Añón 1998a, 1998b; Renkema et al. 2000; Renkema, Gruppen, and Van Vliet 2002) but relatively few of these studies have focused on cold-set gels. Cold gelation is a novel method of producing protein-based gels at ambient temperature. The process consists of two consecutive steps, i.e., preheating of the protein solution followed by cooling and lowering of the pH or addition of salts at room temperature (Figure 19.8). In the first step aggregates are formed by heating a protein solution for a certain period of time. Upon cooling, the protein aggregates remain soluble and can be stored for days without any significant change in aggregate size or other properties. In the second step, gelation is induced by changing the solvent composition either by the addition of calcium or sodium or by lowering the pH (Alting et al. 2000).

Maltais et al. (2005) investigated the influence of protein and calcium concentration on soy protein cold-set gel formation. They reported that cold-set gels could be formed at soy protein concentrations from 6 to 9% and calcium concentrations from 10 to 20 mM. They also showed that the properties of the gels formed could be modulated by changing protein and/or $CaCl_2$ concentrations. An increase in $CaCl_2$ concentration from 10 to 20 mM increased gel opacity while increasing the protein concentration from 6 to 9% decreased opacity. WHC improved with increasing protein concentration and decreasing $CaCl_2$ concentration while the elastic modulus (G') increased with both protein and $CaCl_2$ concentrations. Microscopy studies revealed an increase in the diameters of aggregates and pores as $CaCl_2$ concentration increased and as protein concentration decreased. In subsequent work, these authors (Maltais, Remondetto, and Subirade 2008) further studied the mechanisms of cold-set gel formation and the characteristics of the specific microstructures formed.

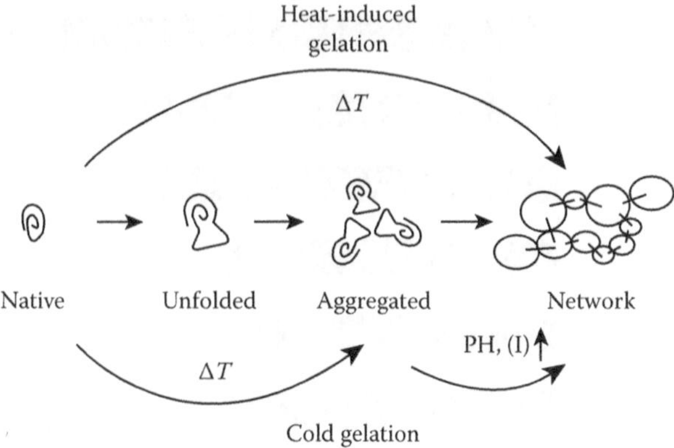

FIGURE 19.8 Difference between heat-induced and cold gelation processes. (Reprinted with permission from Alting, A. C., Hamer, R. J., de Kruif, C. G., de Jongh, H. H., Simons, J. F. A., and visschers, R. W., PhD Thesis, Wageningen University, The Netherlands 2003.)

Their results showed that preheating caused molecules to associate into aggregates that constituted the structural units responsible for the three-dimensional network of the cold-set gels. Addition of $CaCl_2$ neutralized the negative charges on the structural unit surfaces. This electrostatic neutralization enabled the formation of a three-dimensional network by two mechanisms, each occurring at a different calcium concentration. At a lower calcium concentration (10 mM), the surface charge is not totally screened, promoting aggregation growth preferentially along linear paths, leading to filamentous gels. At the higher calcium concentration (20 mM), a quasitotal screening of repulsive forces between structural units occurs, resulting in random aggregation, but characterized also by micro- and macrosyneresis, leading to gel shrinkage, which resulted in a weakly bonded gel composed of weak interactions among the flocs.

By using the cold-gelation method, soy gels can be prepared with the same permeability characteristics as heat-set gels but at a much lower protein concentration (<10%) than with heat-induced gelation (>10%). It has been suggested that cold gelation may have significant potential to introduce gel structures into food stuffs without the need to heat the final product and may find application in the preparation of products with a delicate flavor and texture (Britten and Giroux 2001), production of encapsulates (Beaulieu et al. 2002), or be used to help increase the efficiency of iron and other bioactives uptake (Remondetto, Paquin, and Subirade 2002).

19.7 GLYCATION OF SOY

Glycation is a useful and promising tool for improving the functional properties of proteins such as emulsifying properties (Kato, Minaki, and Kobayashi 1993), gelation (Cabodevila et al. 1994), WHC (Easa et al. 1996), foaming, heat stability and solubility of proteins (Kato et al. 1990), antioxidant (Mesa et al. 2008) and antimicrobial (Nakamura, Kato, and Kobayashi 1991) properties, and may also possibly help in reducing allergenicity (Arita et al. 2001). Glycation allows the conjugation of proteins with carbohydrate through the Maillard reaction. The reaction occurs primarily between the ε-amino groups of proteins and the reducing end carbonyl group of carbohydrate at high-temperatures and low-moisture conditions.

The glycation process depends on temperature, time, composition of the system, water activity, sulfur dioxide, and pH (Ames 1990). Achouri et al. (2005) reported that compared to the unglycated samples, glycation of SPI and the 11S glycinin fraction with glucose (1:44 molar ratio) at 50°C and 65% relative humidity for 6–48 h increased the solubility (under certain pH conditions), emulsifying activity, and foaming properties. They, however, found that extensive glycation decreased protein solubility. Babiker et al. (1998) also reported that soy protein–galactomannan (GM) conjugates had higher solubility at both acid (pH 4) and alkali conditions (pH 8) and the solubility greatly improved, especially under acidic conditions compared to untreated soy protein which was only sparingly soluble at pH 4–6. They suggested that the hydrophilicity of glucose and GM may be responsible for the increased solubility of glycated soy proteins. Heat stability and emulsifying properties also markedly improved in soy protein–GM conjugates. These authors postulated that the hydrophobic residues of soy protein may be anchored to the surface of oil droplets in an emulsion while the hydrophilic polysaccharides may be oriented to the water phase, thereby inhibiting the coalescence of the oil droplets (Khan et al. 1999). Kato (2002) also found that the emulsifying activity and emulsion stability of soy protein-saccharide conjugate were improved in proportion to the molecular size of attached saccharides. They reported that saccharides having a molecular weight of 10 kDa are effective for achieving better emulsifying properties.

Armstrong et al. (1994) found that heating solutions of globular proteins (e.g., lysozyme, bovine serum albumin, soy) in the presence of low-molecular-weight reducing sugars, such as lactose, ribose, or xylose, causes the formation of gels that show higher breaking strengths than comparable samples heated in the absence of reducing sugars. The increase in gel strength may be attributed to sugar-induced covalent cross-linking of proteins. Hence, in the presence of reducing sugars gelation occurs at low protein concentrations. Furthermore, it has been reported that by increasing the initial sugar concentration the time for gelation can be decreased (Easa 1996). Recent studies have focused on improving the gelation properties of globular protein powders made by reaction with reducing sugars in the "dry" state. The powders are subsequently dissolved, heated, and aged to form the gel. On heating egg white (EW) with a GM in the "dry" state (3 days at 60°C, 65% RH, 1:4 EW:GM ratio), Matsudomi et al. (2002) showed improved gelation after heating a solution of the conjugate (10% w/v protein) at 90°C for 30 min. Similar studies could potentially be extended to soy proteins.

Recently Yasir et al. (2007) studied the use of formaldehyde, glyceraldehyde, and glutaraldehyde as conjugates in tofu-making and the relationship between cross-linking of soy proteins and tofu texture. Glutaraldehyde (1–2 mM) treatment resulted in an acceptable product with altered microstructure and higher reactivity than formaldehyde and glyceraldehyde. The acidic subunits of glycinin had the highest reaction rate followed by α'+α and β subunits of β-conglycinin, while the basic subunits were the least reactive. However, when using fractionated glycinin, the reaction rates of the basic subunits approached those of the acidic subunits. It is likely that Maillard-reactive lysine and arginine residues are made more available for Maillard chemistry by the extraction process. At 2 mM glutaraldehyde, the fracture force was increased if the reagent was added before soymilk boiling, but decreased if it was added after boiling, when more cross-linking was observed due to the denaturation of the proteins. The authors postulated that the change in the functional properties of tofu was due to non-cross-linking modifications of the side chains of the amino acid residues, which changed their isoelectric point and gelation properties. Cabodevila et al. (1994) had earlier reported that autoclaving of SPI gels in the presence of xylose decreased the pH from neutral to 5.5. The Maillard gels showed less syneresis, had a higher breaking force, and were more elastic than glucono-δ-lactone gels. The Maillard reaction products also showed a reduction in pH at high temperatures due to formation of acidic by-products. The critical concentration for protein gelation is usually highly pH dependent, with a minimum at the protein isoelectric point (Stading and Hermansson 1990). Therefore, the pH changes induced by the Maillard reaction would be expected to alter characteristics of the gel. Increased covalent protein cross-linking can occur during the Maillard reaction (Kato et al. 1986). The presence of such additional cross-links might be expected to alter both the rupture strength and viscoelastic properties of the gel.

Knipfel, Botting, and McLaughlan (1975) investigated the effects of carbohydrate-heat interactions on the nutritive value of casein, soy, and egg proteins using rats. They fed rats proteins that were autoclaved in the presence of 10% carbohydrates for various time periods. Results shows that protein efficiency ratio, net protein utilization, and digestibility of egg protein was much more susceptible to damage for the glycated samples than casein or soy proteins and that reducing sugars such as glucose are more damaging to nutritive value than the nonreducing polysaccharides cellulose and starch. Similarly Rhee and Rhee (1981) evaluated the protein quality of defatted flours and protein isolates from cottonseed, peanut, and soybeans mixed with glucose or sucrose and heated at 100°C. They observed less changes in in vitro digestibility, available lysine, total amino acids, and computed protein efficiency ratio (C-PER) for the sucrose-protein conjugates than the glucose conjugates. Lysine-rich soy protein lost a greater percent of its lysine than lysine-poor peanut and cottonseed proteins, which was highly correlated with C-PER. The Maillard reaction can also sometimes result in the development of undesirable flavors and colors (e.g., when dehydrated foods darken and develop off-flavors on storage or during the sterilization of milk) and the formation of potentially toxic compounds which may have physiological ramifications. The technique, however, has good potential for use in the development of novel products and to expand its application further research will be required to identify conditions that prevent the development of undesirable quality and nutritional effects.

19.8 CONCLUSION

Consumers are becoming increasingly savvy and are demanding food industries to produce high-quality, minimally processed foods that are convenient and safe and provide health benefits. Soybean has been labeled as a wonder crop and is highly valued today for its functional and nutritional properties and is increasingly being used in the formulation of a wide variety of products. The techniques discussed above highlight some of the novel and emerging technologies being used for the processing of soybeans. The review has focussed primarily on nonthermal techniques that could be considered as alternatives to conventional thermal processing to address the needs of industry to identify more energy-saving and ecofriendly food processing applications. Each technique presented has its own merits and limitations and further research is needed to bring some of these techniques to the forefront of industry and increase their value proposition and consumer acceptability. As an example, further in vivo and in vitro studies and toxicological analyses are needed to assess the efficiency with which irradiation inactivates antinutrients and to quantify its effects on nutritional quality, particularly the impact of doses higher than 10 kGy.

The increasing health consciousness of today's consumer has resulted in extensive studies to discover soy fractions with biological activities (e.g., suppressing proliferation of cancer cells, preventing free radical formation, lowering blood cholesterol levels, and reducing blood pressure) (Kim et al. 2000; Pena-Ramos and Xiong 2002; Arnoldi et al. 2001; Kitts and Weiler 2003). Techniques that enhance these bioactive properties and minimize their destruction during processing will need to be developed. Additionally, although soybeans are known for their nutritional benefits, they also contain allergenic proteins that can cause deleterious reactions for allergic consumers. The effect of these novel techniques on soybean allergenicity needs to be studied and novel technologies for hypoallergenizing soybean products or removing allergenic substances may also need to be developed in the future.

REFERENCES

Abu-Tarboush, H. M. 1998. Irradiation inactivation of some antinutritional factors in plant seeds. *Journal of Agricultural and Food Chemistry* 46, 2698–702.

Achouri, A., Boye, J. I., Yaylayan, V. A., and Yeboah, F. K. 2005. Functional properties of glycated soy 11S glycinin. *Journal of Food Science* 70, 270–74.

Afify, A. E. M. M. R. and Shousha, M. A. 1988. Effect of low dose irradiation on soy bean protein solubility, trypsin inhibitor activity and protein patterns separated by polyacrylamide gel electrophoresis. *Journal of Agricultural and Food Chemistry* 36, 810–13.

Alting, A. C., Hamer, R. J., de Kruif, C. G., de Jongh, H. H., Simons, J. F. A., and visschers, R. W. 2003. Cold gelation of globular protein. PhD Thesis, Wageningen University, The Netherlands. http://library.wur.nl/wda/dissertations/dis3396.pdf

Alting, A. C., Hamer, R. J., de Kruif, C. G., and Visschers, R. W. 2000. Formation of disulfide bonds in acid-induced gels of preheated whey protein isolate. *Journal of Agricultural and Food Chemistry* 48, 5001–5007.

Ames, J. M. 1990. Control of the Maillard reaction in food system. *Trends in Food Science and Technology* December, 150–54.

Aoyama, T., Fukui, K., Takamatsu, K., Hashimoto, Y., and Yamamoto, T. 2000. Soy protein isolate and its hydrolysate reduce body fat of dietary obese rats and genetically obese mice (yellow KK). *Nutrition* 16, 349–54.

Arita, K., Babiker, E. E., Azakami, H., and Kato, A. 2001. Effect of chemical and genetic attachment of polysaccharides to proteins on the production of IgG and IgE. *Journal of Agricultural and Food Chemistry* 49, 2030–36.

Armstrong, H. J., Hill, S. E., Schrooyen, P., and Mitchell, J. R. 1994. A comparison of the viscoelastic properties of conventional and Maillard protein gels. *Journal of Texture Studies* 25, 285–98.

Arnoldi, A., D'Agostina, A., Boschin, G., Lovati, M. R., Manzoni, C. and Sirtori, C. R. 2001. Soy protein components active in the regulation of cholesterol homeostasis: Biologically active phytochemicals in food. *Royal Society of Chemistry Special Publication* 269, 103–106.

Asano, M., Okubo, K., and Yamauchi, F. 1989. Effect of immersing temperature on the behavior of exuding compounds from soybean. *Journal of the Japanese Society for Food Science and Technology* 36, 636–42.

Babiker, E., Hiroyuki, A., Matsudomi, N., Iwata, H., Ogawa, T., Bando, N., and Kato, A. 1998. Effect of polysaccharide conjugation or transglutaminase treatment on the allergenicity and functional properties of soy protein. *Journal of Agricultural and Food Chemistry* 46, 866–71.

Barsotti, L., Dumay, E., Hua Mu, T., Fernandez Diaz, M. D., and Cheftel, C. J. 2002. Effect of high voltage electric pulses on protein based food constituents and structures. *Trends in Food Science and Technology* 12, 136–44.

Beaulieu, L., Savoie, L., Paquin, P., and Subirade, M. 2002. Elaboration and characterization of whey protein beads by an emulsification/cold gelation process: Application for the protection of retinol. *Biomacromolecules* 3, 239–48.

Behrend, O. and Schubert, H. 2001. Influence of hydrostatic pressure and gas content on continuous ultrasound emulsification. *Ultrasonics Sonochemistry* 8, 271–75.

Bielski, B. H. and Allen, A. O. J. 1977. Mechanism of the disproportionation of superoxide radicals. *Physical Chemistry* 81, 1048–50.

Britten, M. and Giroux, H. J. 2001. Acid-induced gelation of whey protein polymers: Effect of pH and calcium concentration during polymerization. *Food Hydrocolloids* 15, 609–17.

Byun, M. W., Kang, I. J., Hayashi, Y., Matsumura, Y., and Mori, T. 1994. Effect of γ-irradiation on soya bean proteins. *Journal of the Science of Food and Agriculture* 66, 55–60.

Byun, M. W., Kang, I. J., Kwon, J. H., Hayashi, Y., and Mori, T. 1995a. Physicochemical properties of soybean oil extracted from γ-irradiated soybeans. *Radiation Physics and Chemistry* 46, 659–62.

Byun, M. W., Kang, I. J., and Mori, T. 1995b. Properties of soymilk and tofu prepared with γ-irradiated soy beans. *Journal of the Science of Food and Agriculture* 67, 477–83.

Byun, M W., Kwon, J. H., and Mori, T. 1993. Improvement of physical properties of soy beans by gamma irradiation. *Radiation Physics and Chemistry* 42, 313–17.

Cabodevila, O., Hill, S. E., Armstrong, H. J., Sousa, I. D., and Mitchell, J. R. 1994. Gelation enhancement of soy protein isolate using the Maillard reaction and high temperature. *Journal of Food Science* 59, 872–75.

Casarett, A. P. 1968. Radiation chemistry. In: *Radiation Biology*, 57–89. Englewood Cliffs, New Jersey: Prentice-Hall.

Cheftel, J. C. and Culioli, J. 1997. Effects of high pressure on meat: A review. *Meat Science* 46, 211–36.

Chia, C. L. 2006. Influence of radiation encountered on Mars missions on the yield and quality of soymilk and tofu from bulk soybeans. Master's thesis, Iowa State University.

Ciccolini, L., Taillandier, P., Wilhelm, A. M., Delmas, H., and Strehaiano, P. 1997. Low frequency thermo-ultrasonication of *Saccharomyces cerevisae* suspensions: Effect of temperature and of ultrasonic power. *Chemical Engineering Journal* 65, 145–49.

Coakley, W. T., Brown, R. C, and James, C. J. 1973. The inactivation of enzymes by ultrasonic cavitation at 20 kHz. *Archives of Biochemistry and Biophysics* 159, 722–29.

Cossins, A. R. and Macdonald, A. G. 1986. Homeoviscous adaptation under pressure. III. The fatty composition of liver mitochondrial phospholipids of deep-sea fish. *Biochimica Biophysica Acta* 860, 325–35.

Cruz, N. S., Capellas, M., Jaramillo, D. P., Trujillo, A. J., Guamis, B., and Ferragut, V. 2009. Soymilk treated by ultra high-pressure homogenization: Acid coagulation properties and characteristics of a soy-yogurt product. *Food Hydrocolloids* 23, 490–96.

Cruz, N., Capellas, M., Jaramillo, D. P., Trujillo, A. J., Guamis, B., and Ferragut, V. 2007. Ultra high pressure homogenization of soymilk: Microbiological, physicochemical and microstructural characteristics. *Food Research International* 40, 725–32.

Damodaran, S. 1996. Amino acids, peptides, and proteins. In: *Food Chemistry*, Fennema, O. R., Ed., 322–429. New York: Dekker.

De Boland, A. R., Garner, G. B., and O'Dell, B. L. 1975. Identification and properties of phytate in cereal grains and oilseed products. *Journal of Agricultural and Food Chemistry* 23, 1186–89.

De Toledo, T. C. F., Canniatti-Brazaca, S. G., Arthur, V., and Piedade, S. M. S. 2007. Effects of gamma radiation on total phenolics, trypsin and tannin inhibitors in soy bean grains. *Radiation Physics and Chemistry* 76, 1653–56.

Deshpande, S. D., Bal, S., and Ojha, T. P. 1994. A study on diffusion of water by the soybean grain during cold water soaking. *Journal of Food Engineering* 23, 121–27.

Deshpande, S. S., Sathe, S. K., and Salunkhe, D. K. 1984. Dry beans of phaseolus: A review. Part 3. *Critical Reviews in Food Science and Nutrition* 21, 137–95.

Diehl, J. F., Adam, S., Delincee, H., and Jakubick, C. 1978. Radiolysis of carbohydrates and carbohydrate containing food stuffs. *Journal of Agricultural and Food Chemistry* 26, 15.

Dumoulin, M., Ozawa, S., and Hayashi, R. 1998. Textural properties of pressure-induced gels of food proteins obtained under different temperatures including subzero. *Journal of Food Science* 63, 92–95.

Earnshaw, R. G., Appleyard, J., and Hurst, R. M. 1995. Understanding physical inactivation processes: Combined preservation opportunities using heat, ultrasound and pressure. *Journal of Food Microbiology* 28, 197–219.

Easa, A. M. 1996. Factors affecting Maillard induced gelation of protein-sugar systems. PhD Thesis, University of Nottingham.

Easa, A. M., Hill, S. E., Mitchell, J. R., and Taylor, A. J. 1996. Bovine serum albumin gelation as a result of the Maillard reaction. *Food Hydrocolloids* 10, 199–202.

Elpiner, I. E., Sokolskaya, A. V., and Margulis, M. A. 1965. Initiation of chain reaction under an ultrasonic wave effect. *Nature* 208, 945–46.

Eskin, N. A. M., Grossman, S., and Pinsky, A. 1977. Biochemistry of lipoxygenase in relation to food quality. *Critical Reviews in Food Science and Nutrition* 9, 1–40.

Estrada-Giron, Y., Swanson, B. G., and Barbosa-Canovas, G. V. 2005. Advances in the use of high hydrostatic pressure for processing cereal grains and legumes. *Trends in Food Science & Technology* 16, 194–203.

Evrendilek, G. A., Zhang, Q. H., and Richter, E. R. 1999. Inactivation of *Escherichia coli* O157:H7 and *Escherichia coli* 8739 in apple juice by pulsed electric fields. *Journal of Food Protection* 62, 793–96.

Farag, M. D. E. H. 1989. Radiation deactivation of antinutritional factors: trypsin inhibitor and hemagglutinin in soybeans. *Egyptian Journal of Radiation Sciences and Applications* 6, 207–15.

Farag, M. D. E. H. 1998. The nutritive value for chicks of full fat soy beans irradiated at up to 60 kGy. *Animal Feed Science and Technology* 73, 319–28.

Fukushima, D. 2001. Recent progress in research and technology on soybeans. *Food Science and Technology Research* 7, 8–16.

Gachagan, A., McNab, A., Blindt, R., Patrick, M., and Marriott, C. 2004. A high power ultrasonic array based test cell. *Ultrasonics* 42, 57.

Galazka, V. B., Dickinson, E., and Ledward, D. A. 2000. Influence of high pressure on interactions of 11S globulin *Vicia faba* with ι-carrageenan in bulk solution and at interfaces. *Food Hydrocolloids* 14, 551–60.

———. 1999. Emulsifying behaviour of 11S globulin *Vicia faba* in mixture with sulphated polysaccharides: Comparison of thermal and high pressure treatments. *Food Hydrocolloids* 13, 425–35.

Gao, Y. L., Ju, X. R., and Wu-Ding. 2007. A predictive model for the influence of food components on survival of *Listeria monocytogenes* LM 54004 under high hydrostatic pressure and milk heat conditions. *International Journal of Food Microbiology* 117, 284–94.

Gebicki, S. and Gebicki, J. M. 1993. Formation of peroxides in amino acids and proteins exposed to oxygen free radicals. *Biochemical Journal* 289, 743–49.

Ghazy, M. A. 1990. Effect of gamma irradiation on some antinutritional factors in kidney bean (*Phaseolus vulgaris* L.) seeds. *Minia Journal of Agricultural Research and Development* 12, 1965–80.

Guderjan, M., Topfl, S., Angersbach, A., and Knorr, D. 2005. Impact of pulsed electric field treatment on the recovery and quality of plant oils. *Journal of Food Engineering* 67, 281–87.

Hafez, Y. S., Mohamed, A. I., Singh, G., and Hewedy, F. M. 1985. Effect of γ-irradiation on protein and fatty acids of soybean. *Journal of Food Science* 50, 1271–74.

Hashizume, K., Nakamura, K., and Watanabe, T. 1975. Influence of ionic strength on conformation changes of soybean proteins caused by heating, and relationship of its conformation changes to gel formation, *Agricultural Biology and Chemistry* 39, 1339.

Hayakawa, I., Kajihara, J., Morikawa, K., Oda, M., and Fujio, Y. 1992. Denaturation of bovine serum albumin (BSA) and ovalbumin by high pressure, heat and chemicals. *Journal of Food Science* 57, 288–92.

Heinisch, O., Kowalski, E., Goossens, K., Frank, J., Heremans, K., Ludwig, H., and Tauscher, B. 1995. Pressure effects on the stability of lipoxygenase: Fourier transform infrared spectroscopy (FTIR) and enzyme activity studies. *Zeitschrift fur Lebensmittel-Untersuchung und -Forschung* 201, 561–65.

Hirano, H., Kagawa, H., and Okubo, K. 1992. Characterization of proteins released from legume seeds in hot water. *Phytochemistry* 31, 731–35.

Hsu, K. H., Kim, C. J., and Wilson, L. A. 1983. Factors affecting water uptake of soybeans during soaking. *Cereal Chemistry* 60, 208–11.

Iida, Y., Yasui, K., Tuziuti, T., and Sivakumar, M. 2005. Sonochemistry and its dosimetry. *Microchemical Journal* 80, 159.

Jayaram, S., Castle, G. S. P., and Pargaritis, A. 1992. Kinetics of sterilization of *Lactobacillus brevis* cells by the application of high voltage pulses. *Biotechnology and Bioengineering* 40, 1412–20.

Josephson, E. S. and Peterson, M. S. 1983. Preservation of gamma irradiation on the macromolecular integrity of guar gum. *Carbohydrate Research* 282, 223–36.

Jung, S., Murphy, P. A., and Sala, I. 2008. Isoflavone profiles of soymilk as affected by high pressure treatments of soymilk and soybeans. *Food Chemistry* 111, 592–98.

Kajiyama, N., Isobe, S., Uemura, K., and Noguchi, A. 1995. Changes of soy protein under ultra-high hydraulic pressure. *International Journal of Food Science and Technology* 30, 147–158, 1995.

Kato, A. and Nakai, S. 1980. Hydrophobicity determined by a flurescence probe method and its correlation with surface properties of proteins. *Biochimica Biophysica Acta* 624, 13–20.

Kato, A., Sasaki, Y., Furuta, R., and Kobayashi, K. 1990. Functional protein polysaccharide conjugate prepared by controlled dry-heating of ovalbumin dextran mixtures. *Agricultural Biology and Chemistry* 54, 107–12.

Kato, A. 2002. Industrial applications of Maillard type protein polysaccharide conjugates. *Food Science and Technology Research* 8, 193–99.

Kato, Y., Matsuda, M., Kate, N., Watanabe, K., and Nakamura, R. 1986. Browning and insolubilization of ovalbumin by the Maillard reaction with some aldohexoses. *Journal of Agricultural and Food Chemistry* 47, 2262–66.

Kato, A., Minaki, K., and Kobayashi, K. 1993. Improvement of emulsifying properties of egg white proteins by the attachment of polysaccharide through Maillard reaction in a dry state. *Journal of Agricultural and Food Chemistry* 41, 40–543.

Khan, M., Babiker, E., Azakami, H., and Kato, A. 1999. Molecular mechanism of the excellent emulsifying properties of phosvitin-galactomannan conjugate. *Journal of Agricultural and Food Chemistry* 47, 2262–66.

Khattak, A. B. and Klopfenstein, C. F. 1989. Effects of gamma irradiation on the nutritional quality of grain and legume. *Cereal Chemistry* 66, 170–71.

Kim, S. E., Kim, H. H., Kim, J. Y., Kang, Y. I., Woo, H. J., and Lee, H. J. 2000. Anticancer activity of hydrophobic peptides from soy proteins. *Biofactors* 12, 151–55.

Kitts, D. D. and Weiler, K. 2003. Bioactive proteins and peptides from food sources: Applications of bioprocesses used in isolation and recovery. *Current Pharmaceutical Design* 9, 1309–23.

Knipfel, J. E., Botting, H. G., and McLaughlan, J. M. 1975. Nutritional quality of several proteins as affected by heating in the presence of carbohydrates. In: *Protein Nutritional Quality of Foods and Feeds*, Friedman, M., Ed., Part A, 375–90. New York: Dekker.

Knorr, D. 1999. Novel approaches in food processing technology: New technologies for preserving foods and modifying function. *Current Opinion in Biotechnology* 10, 485–91.

———. 1993. Effects of high hydrostatic pressure process on food safety and quality. *Food Technology* 47, 156.

Kocis, S., and Figura, Z. 1996. *Ultrasonic Measurements and Technologies*. London: Chapman and Hall.

Kumar, V., Rani, A., Tindwani, C., and Jain, M. 2003. Lipoxygenase isozymes and trypsin inhibitor activities in soybean as influenced by growing locations. *Food Chemistry* 83, 79–83.

Kwok, K. C., Han Hua, L., and Keshavan, N. 2002. Optimizing conditions for thermal processes of soymilk. *Journal of Agricultural and Food Chemistry* 50, 4834–38.

Kwok, K. C., and Niranjan, K. 1995. Review: Effect of thermal processing on soymilk. *International Journal of Food Science and Technology* 30, 263–95.

Lagunas-Solar, M. C. 1995. Radiation processing of foods. An overview of scientific principles and current status. *Journal of Food Protection* 58, 186–92.

Lee, C. C. 1962. Electron paramagnetic resonance (EPR) and packing studies on γ-irradiation flour. *Cereal Chemistry* 39, 147–55.

Lee, C. H., Yang, L., Xu, J. Z., Yenus, S. Y. V., Huang, Y., and Chen, Z. Y. 2005. Relative antioxidant activity of soybean isoflavones and their glycosides. *Food Chemistry* 90, 735.

Lee, K. J. and Row, K. H. 2006. Enhanced extraction of isoflavones from Korean soybean by ultrasonic wave. *Korean Journal of Chemical Engineering* 23, 779–83.

Li, H., Pordesimo, L., and Weiss, J. 2004. High intensity ultrasound-assisted extraction of oil from soybeans. *Food Research International* 37, 731–38.

Li, S. Q. and Zhang, Q. H. 2004. Inactivation of *E.coli* 8739 in enriched soymilk using pulsed electric fields. *Journal of Food Science* 69, M169–74.

Li, Y., Chen, Z., and Mo, H. 2007. Effects of pulsed electric fields on physicochemical properties of soybean protein isolates. *Lebensmittel-Wissenschaft und Technologie* 28, 410–18.

Li, Y. Q., Chen, Q., Liu, X. H. and Chen, Z. X. 2008. Inactivation of soybean lipoxygenase in soymilk by pulsed electric fields. *Food Chemistry* 109, 408–14.

Liener, I. E. 1994. Antinutritional factors related to proteins and amino acids. In: *Food-Borne Disease Handbook*, Hui, Y. H., Gorham, J. R., Murrel, K. D., and Cliver, D. O., Eds, vol. 3, 261–309. New York: Marcel Dekker.

Liu, K. 1997. *Soybeans: Chemistry, Technology and Utilization*, 1st ed., 532. New York: Chapman and Hall.

Lochhead, C. 1989. The high-tech food process foes find hard to swallow. *Food Technology* 43, 56–59.

Lopez, F. and Burgos, J. 1995. Lipoxygenase inactivation by manothermosonication: Effects of sonication on physical parameters, pH, KCl, sugars, glycerol, and enzyme concentration. *Journal of Agricultural and Food Chemistry* 43, 620–25.

Lopez, P., Sala, E. J., de la Fuente, J. L., Condon, S., Raso, J., and Justino, B. 1994. Inactivation of peroxidase, lipoxygenase, and polyphenol oxidase by manothermosonication. *Journal of Agricultural and Food Chemistry* 42, 252–56.

Lovati, M. R., Manzoni, C., Giannazzi, E., Arnoldi, A., Kurowska, E., Carrol, K. K., and Sirtori, C. 2000. Soy protein peptides regulate cholesterol homeostasis in Hep G2 cells. *Journal of Nutrition* 130, 2543–49.

Maltais, A., Remondetto, G. E., Gonzalez, R., and Subirade, M. 2005. Formation of soy protein isolate cold-set gels: Protein and salt effects. *Journal of Food Science* 70, 67–73.

Maltais, A., Remondetto, G. E., and Subirade, M. 2008. Mechanism involved in the formation and structure of soya protein cold set gels: A molecular and supramolecular investigation. *Food Hydrocolloids* 22, 550–59.

Martens, B. and Knorr, D. 1999. Developments of on thermal process for food preservation. *Food Technology* 46, 126–33.

Mason, T. J. and Luche, J. L. 1996. Ultrasound as a new tool for synthetic chemists. In: Van Eldik, R. and Hubbard, C. D., Eds, *Chemistry under Extreme or Non Classical Conditions*, 317–80. New York: John Wiley and Spektrum Akademischer Verlag.

Matser, A. A., Krebbers, B., van den Berg, R. W., and Bartels, P. V. 2004. Advantages of high-pressure sterilisation on quality of food products. *Trends in Food Science and Technology* 15, 79–85.

Matsudomi, N., Nakano, K., Soma, A., and Ochi, A. 2002. Improvement of gel properties of dried egg white by modification with galactomannan through the Maillard reaction. *Journal of Agricultural and Food Chemistry* 50, 4113–18.

McClements, D. J. 1995. Advances in the application of ultrasound in food analysis and processing. *Trends in Food Science* 6, 293–99.

McClements, D. J. 1997. Ultrasonic characterization of food and drinks: Principles, methods, and applications. *Food Science and Nutrition* 37, 1–46.

Mesa, M. D., Silvan, J. M., Olza, J., Gil, A., and Castillo, M. D. 2008. Antioxidant properties of soy protein fructooligosaccharide glycation systems and its hydrolyzates. *Food Research International* 41, 606–15.

Mitsuda, H., Yasumoto, K., and Yamamoto, A. 1967. Inactivation of lipoxygenase by hydrogen peroxide, cysteine and some other reagents. *Agricultural Biology and Chemistry* 31, 853–60.

Molina, E., Papadopoulou, A., and Leward, D. A. 2001. Emulsifying properties of high pressure treated soy protein isolate and 7S and 11S globulins. *Food Hydrocolloids* 15, 263–69.

———. 2002. Soy protein pressure induced gels. *Food Hydrocolloids* 16, 625–32.

Molins, R. A. 2001. *Food Irradiation: Principles and Applications*. New York: John Wiley.

Moore, J. H. and Mudher, S. 1978. Implication of tryptophan in concanavalin A-erythrocytes interactions: A radiolytic study. *International Journal of Radiation Biology* 34, 475–79.

Moulton, J. and Wang, C. 1982. A pilot plant study of continuous ultrasonic extraction of soybean protein. *Journal of Food Science* 47, 1127–29.

Moussatov, A., Granger, C., and Dubus, B. 2005. Ultrasonic cavitation in thin liquid layers. *Ultrasonics Sonochemistry* 12, 415.

Mullin, W. J. and Xu, W. L. 2001. Study of soybean seed coat components and their relationship to water absorption. *Journal of Agricultural and Food Chemistry* 49, 5331–35.

Murray, D. R. 1990. *Biology of Food Irradiations*. New York: John Wiley.

Nakamura, S., Kato, A., and Kobayashi, K. 1991. New antimicrobial characteristics of lysozyme-dextran conjugate. *Journal of Agricultural and Food Chemistry* 39, 647–50.

Okamoto, M., Kawamura, Y., and Hayashi, R. 1990. Application of high pressure to food processing: Textural comparison of pressure and heat induced gels of food proteins. *Agricultural Biology and Chemistry* 54, 183–89.

Omi, Y., Kato, T., Ishida, K., Kato, H., and Matsuda, T. 1996. Pressure-induced release of basic 7S globulin from cotyledon dermal tissue of soybean seeds. *Journal of Agricultural and Food Chemistry* 44, 3763–67.

Pena-Ramos, E. A. and Xiong, Y. L. 2002. Antioxidant activity of soy protein hydrolysates in a liposomal system. *Journal of Food Science* 67, 2952–56.

Phull, S. S., Newman, A. P., Lorimer, J. P., Pollet, B., and Mason, T. J. 1997. The development and evaluation of ultrasound in the biocidal treatment of water. *Ultrasonics Sonochemistry* 4, 157–64.

Povey, J. W. and Mason, T. 1998. *Ultrasound in Food Processing*. New York: Blackie Academic & Professional.

Prestamo, G. and Fontecha, J. 2007. High pressure treatment on the tofu fatty acids and acylglycerols content. *Innovative Food Science and Emerging Technologies* 8, 188–91.

Préstamo, G., Lesmes, M., Otero, L., and Arroyo, G. 2000. Soybean vegetable protein (tofu) preserved with high pressure. *Journal of Agricultural and Food Chemistry* 48, 2943–47.

Puppo, C., Chapleau, N., Speroni, F., de Lambellerie-Anton, M., Michel, F., Añón, C., and Anton, M. 2004. Physicochemical modifications of high-pressure treated soybean protein isolates. *Journal of Agricultural and Food Chemistry* 52, 1564–71.

Puppo, C., Speroni, F., Chapleau, N., de Lambellerie-Anton, M., Añón, C., and Anton, M. 2005. Effect of high pressure treatment on emulsifying properties of soy bean proteins. *Food Hydrocolloids* 19, 289–96.

Puppo, M. C. and Añón, M. C. 1998a. Effect of pH and protein concentration on rheological behavior of acidic soybean protein gels. *Journal of Agricultural and Food Chemistry* 46, 3039–46.

———. 1998b. Structural properties of heat-induced soy protein gels as affected by ionic strength and pH. *Journal of Agricultural and Food Chemistry* 46, 3583–89.

Pusztai, A. 1991. Plant lectins. In: *Chemistry and Pharmacology of Natural Prod*ucts, Phillipson, J. D., Ayres, D. C., and Baxter H., Eds, 105–99. Cambridge, UK: Cambridge University Press.

Qin, B. L., Pothakamury, U. R., Barbosa-Canovas, G. V., and Swanson B. G. 1996. Nonthermal pasteurization of liquid foods using high-intensity pulsed electric fields. *Critical Reviews in Food Science and Nutrition* 36, 603–27.

Reina, L. D., Jin, Z. T., Zhang, Q. H., and Yousef, A. E. 1998. Inactivation of *Listeria monocytogenes* in milk by pulsed electric fields. *Journal of Food Protection* 61, 1203–206.

Remondetto, G. E., Paquin, P., and Subirade, M. 2002. Cold gelation of β-lactoglobulin in the presence of iron. *Journal of Food Science* 67, 586–95.

Renkema, J. M. S., Gruppen, H., and Van Vliet, T. 2002. Influence of pH and ionic strength on heat-induced formation and rheological properties of soy protein gels in relation to denaturation and their protein compositions. *Journal of Agricultural and Food Chemistry* 50, 6064–71.

Renkema, J. M. S., Lakemond, C. M. M., De Jongh, H. H. J., Gruppen, H., and Vliet, T. V. 2000. The effect of pH on heat denaturation and gel forming properties of soy proteins. *Journal of Biotechnology* 79, 223–30.

Rhee, K. S. and Rhee, K. C. 1981. Nutritional evaluation of the protein in oilseed products heated with sugars. *Journal of Food Science* 46, 164–68.

Roberts, R. T. 1993. High intensity ultrasonics in food processing. *Chemistry and Industry* 15, 119–21.

Rostagno, A., Palma, M., and Barroso, C. 2003. Ultrasound-assisted extraction of soy isoflavones. *Journal of Chromatography A* 1012, 119–28.

Saio, K., Kajikawa, M., and Watanabe, T. 1971. Food processing characteristics of soybean proteins: 11. Effect of sulfhydryl groups on physical properties of tofu gel. *Agricultural Biology and Chemistry* 35, 890–98.

Sala, F. J., Burgos, J., Condon, S., Lopez, P., and Raso, J. 1995. Effect of heat and ultrasound on microorganisms and enzymes. In: *New Methods of Food Preservation*, Gould, G. W., Ed., 176–204. London: Blackie Academic & Professional.

Santamaria, L., Custellani, A., and Levi, E. A. 1952. Hyalurodinase inactivation by Ultrasonic waves and its mechanisms. *Enzymologia* 15, 285–95.

Sattar, A., Neelofar, X., and Akhtar, M. A. 1990. Effect of radiation and soaking on phytate content of soybean. *Acta Alimentaria* 19, 331–36.

Seyderhelm, I., Boguslawski, S., Michaelis, G., and Knorr, D. 1996. Pressure induced inactivation of selected food enzymes. *Journal of Food Science* 61, 308–10.

Shih, M. C., Hou, H. J., and Chang, K. C. 1997. Study on the process optimization of soft tofu. *Journal of Food Science* 62, 833–37.

Sloan, A. E. 2005. Top 10 global food trends. *Food Technology* 59, 20–32.

Smiddy, M. A., Martin, J. E., Huppertz, T., and Kelly, A. L. 2007. Microbial shelf-life of high-pressure-homogenised milk. *International Dairy Journal* 17, 29–32.

Stading, M. and Hermansson, A. M. 1990. Viscoelastic behaviour of β-lactoglobulin gel structures. *Food Hydrocolloids* 4, 121–23.

Takamiya, K. and Terada, M. 1980. Effects of heating and ultrasonication on soybean protein. *Nippon Shokuhin Kogyo Gakkaishi* 27, 103.

Tangwongchai, R., Ledward, D. A., and Ames, J. M. 2000. Effect of high pressure treatment on lipoxygenase activity. *Journal of Agricultural and Food Chemistry* 48, 2896–902.

Thakur, B. R. and Nelson, P. E. 1997. Inactivation of lipoxygenase in whole soy flour suspension by ultrasonic cavitation. *Nahrung* 41, 299–301.

Thiebaud, M., Dumay, E., Picart, L., Guiraud, J. P., and Cheltel, J. C. 2003. High pressure homogenisation of raw bovine milk. Effects on fat globule size distribution and microbial inactivation. *International Dairy Journal* 13, 427–39.

Torres, J. A. and Velazquez, G. 2005. Commercial opportunities and research challenges in the high pressure processing of foods. *Journal of Food Engineering* 67, 95–112.

Tovar-Palacio, C., Potter, S. M., Hafermann, J. C., and Sahy, N. F. 1998. Intake of soy protein and soy protein extracts influences lipid metabolism and hepatic gene expression in gerbils. *Journal of Nutrition* 128, 839–42.

Urbain, W. M. 1989. *Food Irradiation*. Orlando, FL: Academic Press.

Van der Ven, C., Master, A. M., and Vanden Berg, R. W. 2005. Inactivation of soybean trypsin inhibitors and lipoxygenase by high pressure processing. *Journal of Agricultural and Food Chemistry* 53, 1087–92.

Variyar, S. P., Limaye, A., and Sharma, A. 2004. Radiation induced enhancement of antioxidant contents of soy bean (*Glycine max Merrill*). *Journal of Agricultural and Food Chemistry* 52, 3385–88.

Vercet, A., Lopez, P., and Burgos, J. 1997. Inactivation of heat-resistant lipase and protease from *Pseudomonas fluorescens* by manothermosonication. *Dairy Science* 80, 29–36.

Villamiel, M. and de Jong, P. 2000a. Inactivation of *Pseudomonas fluorescens* and *Streptococcus thermophilus* in trypticase soy broth and total bacteria in milk by continuous-flow ultrasonic treatment and conventional heating. *Food Engineering* 45, 171–79.

———. 2000b. Influence of high-intensity ultrasound and heat treatment in continuous flow on fat, proteins and native enzymes of milk. *Journal of Agricultural and Food Chemistry* 48, 472–78.

Wang, L. C. 1984. Ultrasonic extraction of a heat-labile 7S protein fraction from autoclaved, defatted soybean flakes. *Journal of Food Science* 46, 551–54.

Wang, M. F., Komatsu, T., Chan, Y. C. H., Wong, Y. C. H., Tsao, L. W., Wu, C. H. L., Lin, C. H. C. H., Shinjo, S., Chen, T. H., and Yamamoto, S. 1999. Effect of dietary protein on anti-carcinogenesis of soybean trypsin inhibitor and isoflavone in rats. *Soy Protein Research, Japan* 2, 99–105.

Wang, R., Zhou, X., and Chen, Z. 2008. High pressure inactivation of lipoxygenase in soymilk and crude soybean extract. *Food Chemistry* 106, 603–11.

Whitaker, J. R. 1972. *Principles of Enzymology for the Food Sciences*. New York: Marcel Dekker.

Wilson, L. A. 2004. Influence of hydroponically grown Hoyt soybeans and radiation encountered on Mars missions on the yield and quality of soymilk and tofu. NASA Faculty fellowship report, Chapter 22.

Withers, P. M. 1996. Ultrasonic, acoustic and optical techniques for the non-invasive detection of fouling in food processing equipment. *Journal of Food Science and Technology* 7, 293–98.

Yamauchi, F., Yamagishi, T., and Iwabuchi, S. 1991. Molecular understanding of heat-induced phenomena of soybean protein. *Food Reviews International* 7, 283–322.

Yasir, S., Bin M. D., Sutton, K. H., Newberry, M. P., Andrews, N. R. and Gerrard, J. A. 2007. The impact of Maillard cross linking on soy proteins and tofu texture. *Food Chemistry* 104, 1502–508.

Young, V. R. 1991. Soy protein in relation to human protein and amino acid nutrition. *Journal of the American Dietetic Association* 91, 828–35.

Zarins, Z. M., and Marshall, W. E. 1990. Thermal denaturation of soy glycinin in the presence of 2-mercapto-ethanol studied by differential scanning calorimetry. *Cereal Chemistry* 67, 35–38.

Zhang, H., Ishida, N., and Isobe, S. 2003. High pressure hydration treatment for soybean processing. Paper presented at ASAE Annual International meeting, Las Vegas, USA, 27–30 July. http://asae.frymulti.com/azdez.asp

Zhang, H., Li, L., Tatsumi, E. and Isobe, S. 2005. High pressure treatment effects on proteins in soymilk. *Lebensmittel-Wissenschaft und Technologie* 38, 7–14.

Zhang, Q., Qin, B. L., Barbosa-Canovas, G. V., and Swanson, B. G. 1994. Inactivation of *E. coli* for food pasteurization by high-intensity short-duration pulsed electric fields. *Journal of Food Processing and Preservation* 19, 103–18.

20 Supercritical Fluid Extrusion: A Novel Method for Producing Microcellular Structures in Starch-Based Matrices

Sajid Alavi
Kansas State University

Syed S. H. Rizvi
Cornell University

CONTENTS

20.1 INTRODUCTION

Thermoplastic extrusion is a widely used industrial technology for continuous production of expanded products such as breakfast cereal and snack products (Alavi et al. 1999). The conventional method of steam puffing involves low in-barrel moisture contents (13–20%, wet basis) and high temperatures (130–170°C) and shear (Chinnaswamy and Hanna 1988; Harper and Tribelhorn 1992). These extreme conditions prevent utilization of heat-sensitive ingredients such as whey proteins, and certain flavors and colors. Other disadvantages of high shear and temperature include costly barrel and screw wear, production of undesirable dextrins, increased losses in vitamin and amino acid availability, and starch degradation, which increases the water solubility of extrudates (Hauck and Huber 1989; Kirby et al. 1988). Typically, steam-expanded products have a coarse and very nonuniform cellular structure (Barrett and Peleg 1992) with cell sizes in the range of 1–3 mm and expansion ratios in the range of 9–12. Moreover, there is little or no control over cell size and density.

Also in conventional extrusion puffing, water has two, sometimes conflicting, roles: it serves both as a plasticizer of the melt and a blowing agent for expansion (Rizvi, Mulvaney, and Sokhey 1995; Sokhey, Rizvi, and Mulvaney 1996). Due to this the driving force for expansion is inherently limited and cannot be independently varied, allowing little control over the shaping of the expanded extrudate. This occurs because the enthalpy of vaporization of the steam lowers the extrudate temperature to a point at which further expansion is arrested by the increasing modulus (a measure of viscosity) of the melt. Thus, it is difficult to optimize conditions for expansion. For example, increasing the in-barrel moisture content above 20% (wet basis) results in higher-density products due to the lower temperature of the melt and increased collapse as the extrudate exits the die.

The importance of exploring new approaches, applying them effectively to overcome limitations of the present extrusion puffing process, and producing new generations of foods with improved characteristics has been recognized in recent years and several attempts have been made in this direction.

20.2 WHAT IS SUPERCRITICAL FLUID EXTRUSION (SCFX)?

The requirement of steam as the puffing agent in conventional extrusion, and consequently the dependence on high product temperatures for producing low-density direct-expanded products may be avoided by the introduction of another suitable fluid into the plasticized dough that will also expand upon exiting the extruder die into an atmospheric pressure environment, but at temperatures lower than 100°C (Mulvaney and Rizvi 1993). Supercritical carbon dioxide (SC-CO_2), that is carbon dioxide above its critical pressure (7.38 MPa) and temperature (31°C), is one such candidate. In the supercritical phase, CO_2 has desirable properties of both a dense liquid and a low viscosity gas, leading to a high solubilizing capacity for lipids and certain flavors and high diffusivity in biopolymer melts (Yu et al. 1994; Singh, Rizvi, and Harriott, 1996; Dogan, Chen, and Rizvi 2006b; Chen and Rizvi 2005). Moreover, the phase change from supercritical to gas involves a more subtle transition with undefined or no enthalpy change and a very low change in specific volume. This is in stark contrast to the phase change from liquid water to steam, which is a very "explosive" process.

Rizvi and Mulvaney (1992, 1995) patented a new technology for the production of highly expanded starch foams that is an elegant combination of supercritical fluid technology and extrusion processing, which both require high pressures. The new process, called supercritical fluid extrusion (SCFX), utilizes SC-CO_2 as a blowing agent, a nutrient carrier and, if necessary, an in-line process modifier. In this high-pressure, low-temperature process, the conventional roles of water as a plasticizer as well as a blowing agent are decoupled. SCFX obviates the need for steam expansion, and thus reduced shear and low product temperatures (<100°C) become attainable, which offer major advantages over steam puffing. Although subcritical CO_2 gas can also be injected to puff cereal melts at lower temperatures described by Ferdinand et al. (1990), the limited solubility of CO_2 at low pressures limits the quantity of gas that can be added before process instabilities are encountered (Rizvi et al. 1995). The problems associated with the gases in the melt may also cause problems in obtaining a uniform cell structure. In addition, subcritical CO_2 cannot be used as a solvent for adding ingredients within the extruder.

As opposed to the simple gas injection process, the SCFX process consists of the following major steps (Alavi et al. 1999): (a) development of gas holding rheological properties in the feed by gelatinization and/or mixing and subsequent cooling to below 100°C, if necessary; (b) injection of SC-CO_2, loaded with solutes if desired, into the melt or dough, and adequate mixing to create a single-phase system within the extruder barrel, (c) nucleation of cells induced and controlled by the thermodynamic instability created by a sudden pressure drop; and (d) cell growth and expansion of the extrudate at the die exit as the pressure rapidly drops to atmospheric level, and setting of the extrudates during postextrusion drying.

The use of low process temperatures in SCFX also allows a low shear extrusion environment, and eliminates or minimizes the various drawbacks associated with the conventional high-shear, high-temperature extrusion puffing process that were discussed above. Use of SC-CO_2 as the blowing agent leads to unique internal microstructures in extrudates, with closed cells, average cell size of a few hundred microns or below, uniform cell size distribution, and a smooth nonporous skin (Figures 20.1 and 20.2). In addition, a distinguishing feature of the SCFX process is that the amount of dissolved SC-CO_2 and the rate of pressure drop can be manipulated by adjusting the operating conditions and thereby the cell size, cellular density, and product expansion can be varied

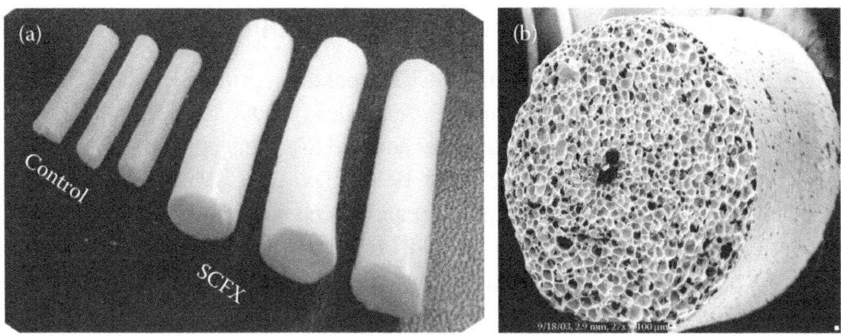

FIGURE 20.1 Starch-based SCFX products: (a) unexpanded control without SC-CO_2 injection and expanded products; and (b) scanning electron microscope picture showing nonporous skin and internal morphology.

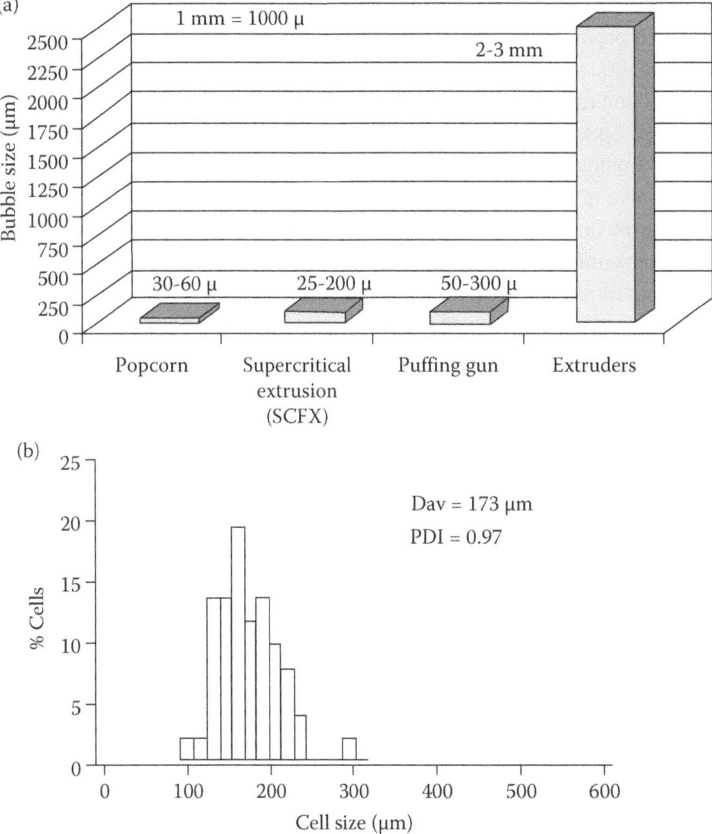

FIGURE 20.2 (a) Cell sizes of products puffed using different technologies; and (b) cell size distribution of a typical SCFX product showing high uniformity. (From Alavi, S. H., Gogoi, B. K., Khan, M., Bowman, B. J., and Rizvi, S. S. H., *Food Research International,* 32, 107–18, 1999.)

to produce novel products having a wide range of cell structures and mechanical properties (Alavi et al. 1999; Gogoi, Alavi, and Rizvi 2000; Alavi and Rizvi 2005). This has also been shown in the extrusion processing of microcellular polymer foams (Baldwin et al. 1996). The SCFX process offers the additional advantage of operating at supercritical pressures that allow utilization of CO_2 for deposition of soluble lipids, flavors, colors, and/or vitamins directly and at preferential locations into the extrudate matrix (Tutanathorn and Rizvi 2001).

20.3 SUPERCRITICAL FLUID EXTRUSION (SCFX) OPERATING PRINCIPLE

Figure 20.3 is a schematic that describes the SCFX operating principle. The SCFX process was developed on a co-rotating, self-wiping, twin-screw extruder (Model TX-52, Wenger Manufacturing, Sabetha, KS) with a 52 mm screw diameter and later adapted to a Wenger Model TX-57 Magnum twin-screw extruder (Figure 20.4) (Mulvaney and Rizvi 1993; Rizvi et al. 1995; Sokhey et al. 1996; Alavi et al. 1999; Dogan et al. 2001). Based on process requirements and model of the extruder, the barrel length to diameter (L/D) ratio can be adjusted from 27 to 40.5, with the use of segmented barrel heads 4–14 in number. As required, the barrel heads are joined together to form feeding, mixing, cooking, cooling, SC-CO_2 injection, and solubilization zones. In the TX-52 design, the shorter configuration with an L/D ratio of 27 is utilized for processing pregelatinized starch-based formulations, because the cooking operation can be eliminated (Alavi et al. 1999). In this configuration, nine barrel or jacketed units (156 mm long or L/D of 3:1) are utilized. Barrel temperature in all the heads, except the first, third, and fourth (which are neither cooled nor heated), is maintained at around 40°C by circulating chilled brine (–10°C) through the barrel jackets. The extruder parameters and moisture content of the feed can be adjusted to maintain the product temperature at about 60°C at the die exit. Typical operating parameters include screw speed of 100 rpm, dry feed rate of 35 kg/h, in-barrel moisture content of 35% (wet basis), supercritical fluid injection rate of 0.8% (dry basis), and die pressure of 10–15 MPa. The high in-barrel moisture and low screw speed facilitate a low shear environment, with a specific mechanical energy (SME) input of approximately 50 kJ/kg, and help in keeping the product temperature low.

A flow restrictor valve is mounted on a spacer at the end of head 9, and prior to the extruder die is the primary means of controlling the die pressure. The pressure profile inside the extruder is controlled by the screw configuration as shown in Figure 20.5. Water is added in the extruder early (head 2). The kneading blocks in heads 3 and 4, and the discs and reverse screw element in head 4 are provided for better mixing of the water with starch to achieve complete hydration. A slight drop in pressure is achieved at the SC-CO_2 injection point at the end of head 7 (L/D = 21) by placing a reverse screw element, which creates a pressure seal and allows the SC-CO_2 to move only in the forward direction. The SC-CO_2 injection zone is followed by a solubilization zone in heads 8 and 9,

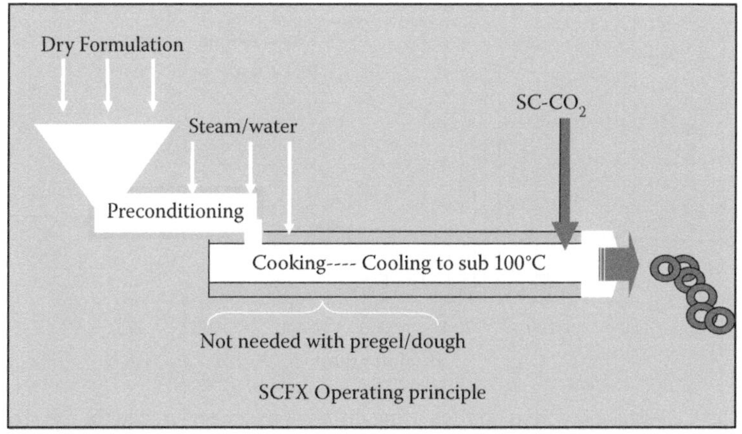

FIGURE 20.3 Schematic describing the operating principle of SCFX processing.

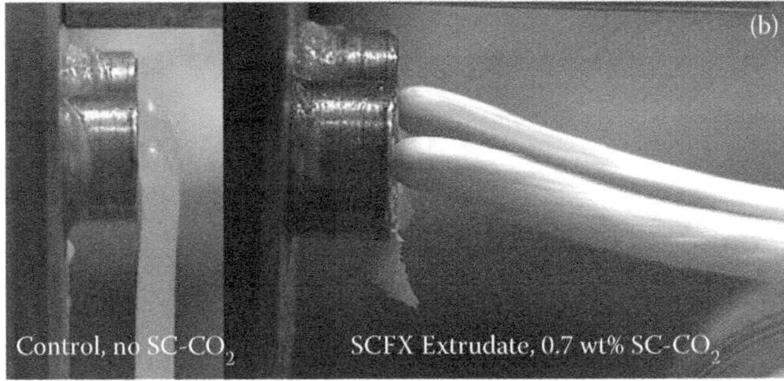

FIGURE 20.4 (a) Wenger TX-57 Magnum twin screw extruder configured for SCFX processing. Top right: preconditioner; center: extruder barrel with four main heads, with SC-CO_2 injection valves on the third head from the feed end; and (b) SCFX product emerging from the die.

with a cut flight screw element and another reverse screw element. The cut-flight element provides better mixing of the SC-CO_2 into the dough. The reverse screw element in head 9 provides restriction for building up pressure to enhance SC-CO_2 solubility near the die.

The pilot-scale supercritical fluid unit consists of two pressure vessels. The main pressure vessel has two outlets: one is connected directly to the extruder and the other to a smaller flavor extraction vessel, which is then connected to the extruder. When the flavor extraction vessel is not in use it is bypassed with the help of valves. The flow of SC-CO_2 is regulated by having a pressure regulator and a back-pressure regulator in series. SC-CO_2 is injected at a constant flow into the starchy melt through four valves located around the extruder barrel at the beginning of head 8. SC-CO_2 injection pressure is automatically maintained higher than pressure inside the barrel for a continuous SC-CO_2 flow into the melt, at the desired rate and pressure.

If the raw material consists of raw/uncooked flour or starch, the L/D ratio of the extruder can be increased to 40.5:1, using a 14 barrel head configuration, to allow for gelatinization or cooking followed by cooling and subsequent steps (Mulvaney and Rizvi 1993; Rizvi et al. 1995; Sokhey et al. 1996). The screw speed is increased (200 rpm) to provide greater mechanical energy. However, most of the gelatinization is achieved by thermal energy in the cooking zone (heads 3–5). For this purpose, heating in heads 3–5 is done by circulating hot thermal fluid in the jackets and steam injection directly into the dough. Water is added in the conditioning cylinder and in the extruder (head 2). Cooling of the melt after the cooking zone is achieved by a combination of vacuum venting at head 7 and jacketed cooling in heads 6–14. A lower L/D ratio set-up can also be used for uncooked formulations, provided sufficient thermal energy can be provided through injection of steam in the preconditioner (Chen, Dogan, and Rizvi 2002).

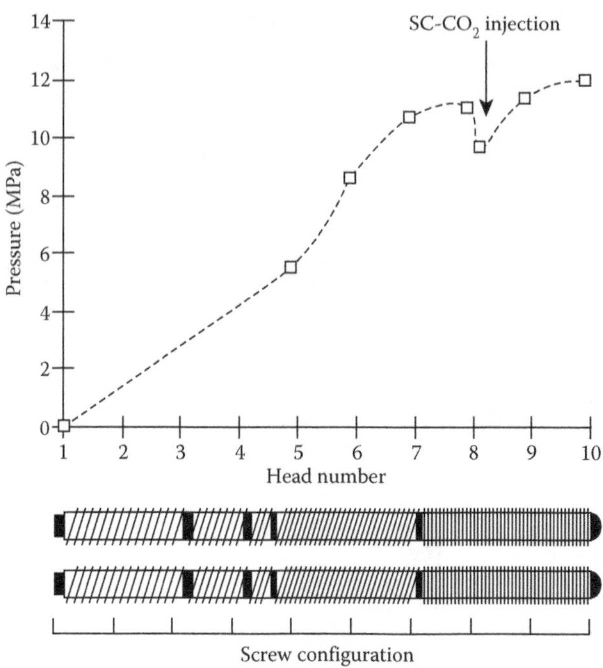

FIGURE 20.5 Typical SCFX pressure profile (showing the SC-CO_2) injection point and screw configuration for a nine-head Wenger TX-52 twin screw extruder. (From Alavi, S. H., Gogoi, B. K., Khan, M., Bowman, B. J., and Rizvi, S. S. H., *Food Research International*, 32, 107–18, 1999.)

Upon exiting the die, the starchy melt puffs as SC-CO_2 changes from solubilized form to gas (Figure 20.4b). The high moisture content of the melt leads to collapse and shrinkage of the extrudate. A drying or toasting step is used to dry the product and set the structure. Typically, extrudates with a nonporous skin and an internal structure comprising of closed cells is formed. The drying process reinflates the cells as they are set in the dryer (Alavi et al. 1999; Alavi, Rizvi, and Harriott 2003a, 2003b).

A typical pregelatinized starch-based formulation used in the SCFX process for puffed breakfast cereal can consist of pregelatinized corn starch (49.5%), pregelatinized potato starch (24%), sugar (24%), salt (1%), and distilled monoglycerides (1.5%) as dough conditioner (Alavi et al. 1999). A typical uncooked formulation for the SCFX process, like that used by Chen et al. (2002) for expanded masa chips, could consist of masa flour (89.2%), salt (3.6%), whey protein concentrate (WPC) (3.6%), soy lecithin (1.8%), and vegetable oil (1.8%).

20.4 PROCESS DYNAMICS AND MICROSTRUCTURE FORMATION

The four major steps involved in SCFX process dynamics and microstructure formation, as given earlier, have been studied and described in great detail using experimentation and theoretical modeling. An understanding of the first two steps, involving development of a gas holding matrix and solubilization of SC-CO_2 in the starchy melt, requires the elucidation of properties of the SC-CO_2–melt system at the pressures and temperatures of interest (Rizvi et al. 1995). This information is needed to estimate the maximum SC-CO_2 injection rate into the system and the minimum residence time required in the mixing section of the extruder at the existing pressure and temperature conditions. The SC-CO_2 that is injected into the system should be completely solubilized into a single phase. This ensures that a high rate of bubble nucleation can be achieved by a subsequent rapid drop in CO solubility due to a decrease in pressure at the die. The injection rate of SC-CO_2 into the melt is dictated by its solubility limit at the highest temperature and lowest pressure before nucleation. The use of excessive amounts of SC-CO_2 would not only lead to process instabilities, but would also hinder the generation of high bubble nucleation densities in the system. The minimum time required for achieving a single-phase system depends on the rate of solubilization, which in turn depends upon the diffusivity of SC-CO_2 in the melt and mixing characteristics of the screw elements. The diffusion of CO_2 into a quiescent melt by itself to create a single phase is a slow process. The dispersion of CO_2 in the melt is aided by the shear fields, which lead to convective mixing. The convective mixing creates new interfaces and repetitively brings areas of high gas concentration near areas of low concentration, thereby reducing the distance that SC-CO_2 has to diffuse to ultimately create a single phase (Rizvi et al. 1995). The breakup of the initially injected SC-CO_2 into smaller bubbles under a shear field may also lead to more uniform CO_2 concentrations and the shorter mixing times required for the production of a single phase. Solubility and diffusivity of SC-CO_2 in starch-water systems has been studied in-depth at the temperature and pressures encountered in the SCFX process (Singh et al. 1996; Chen and Rizvi 2006b) and theories on mixing in extrusion have been developed. However it is not possible to determine a priori satisfactory estimates of the mixing times needed for solubilizing a given amount of SC-CO_2 and melt for different screw profiles in any given extruder geometry. Instead, these estimates have to be generated using experimental techniques (Singh et al. 1998a, 1998b). Experimentally determined effective mixing times can then be compared and matched with qualitative theoretical predictions. SC-CO_2 solubility and diffusivity data together with the required mixing-time characteristics can be used to determine an optimum operating range of SC-CO_2 injection rate, solubilization zone pressure, and extruder screw speed.

In step three of the process, bubble nucleation is affected by introducing a thermodynamic instability into the system. This is the most important step in the overall design of the process. While in the case of polymer foam processing, nucleation can be caused by temperature or pressure changes, or a combination of both (Park et al. 1995), the pressure drop that is induced in SCFX as the CO_2-starch/water melt travels across the die is the cause for thermodynamic instability, leading to a rapid decrease in solubility of the gas in the melt and a high rate of bubble nucleation. The maximum rate of nucleation that is possible generally determines the lower limit of cell size achievable in the microstructured products (Rizvi et al. 1995). This is so because the greater the proportion of the initial supersaturation that is relieved by nucleation, the less the amount of blowing agent left in the melt to drive the cell growth process. This approach is very different from that of a conventional puffing process in which nucleation and cell growth are both driven by temperature differentials as the melt emerges from the die. Theoretical modeling and experimental verification of the SCFX process has confirmed that the rate of the instability or pressure drop introduced affects the overall rate of bubble nucleation and is an effective means of controlling the microstructure (Alavi and Rizvi 2005; Cho and Rizvi 2008).

As the melt exits the die in the last major step of the SCFX process, diffusion of CO_2 into the nucleated bubbles leads to cell growth at the microscopic level and expansion of the extrudate at the macroscopic level. If the extrudate structure is not set by cooling, expansion is often followed by a collapse of cells and extrudate shrinkage because of elastic forces and depletion of CO_2 from the melt. During the subsequent drying process, cells could exhibit reinflation because of Charles' Law expansion of trapped gas. The growth or collapse of the cells is arrested when either an equilibrium size of the cells is achieved or a solidifying matrix sets the structure before equilibrium is reached. There is a complex interaction of growing cells both with each other and with the surrounding viscoelastic melt during the growth process. The process dynamics of extrudate expansion and shrinkage has been modeled and experimentally validated at both microscopic and macroscopic levels (Figure 20.6) (Alavi et al. 2003a, 2003b). The rate of initial cell growth and collapse is a function of the instantaneous viscoelastic properties and surface tension of the melt, as well as the vapor pressure of CO_2 inside the cells, which in turn is controlled by the diffusivity of CO_2. The rate of growth in the drying phase is controlled by melt properties and the vapor pressure of water inside the cells, which are both functions of the drying temperature. Numerical modeling was carried out with the help of an equation that was derived for the incremental change in cell radius (ΔR) with infinitesimal change in time (Δt) as given below (Schwartzberg et al. 1995; Alavi et al. 2003a):

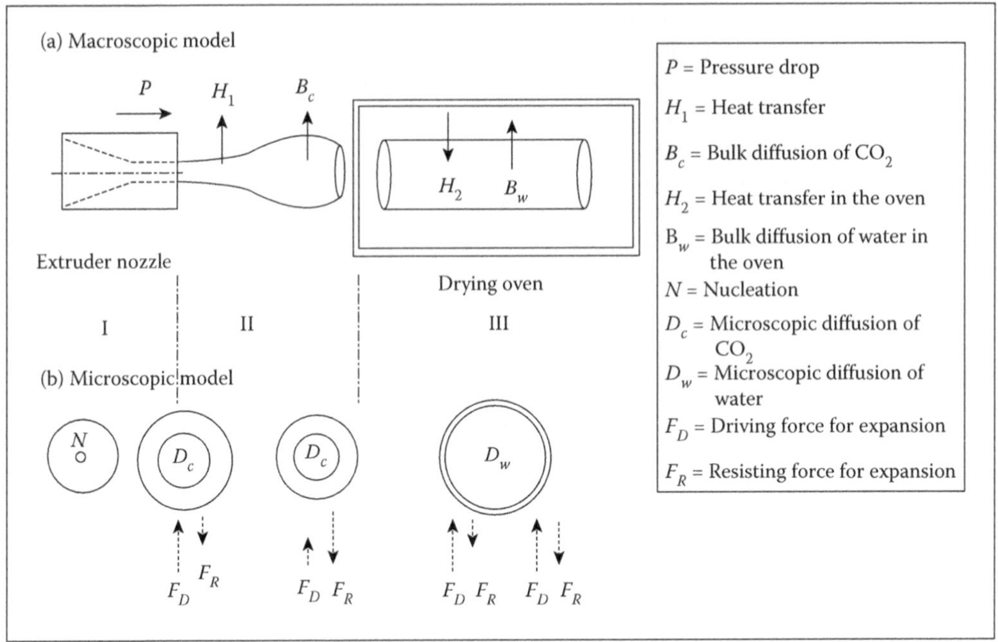

FIGURE 20.6 Schematic representation of the SCFX process dynamic model. (a) The macroscopic model has three phases: (I) flow through the nozzle, (II) exit from the nozzle, and (III) oven drying. The solid arrows denote the dominant phenomena in each phase that were modeled. (b) The corresponding phases of the microscopic model are shown. The dotted arrows denote the force balance on the bubble wall, where F_D is the driving force and F_R the resisting force. If $F_D > F_R$ bubble growth takes place, if $F_D < F_R$ the bubble collapses, and if $F_D = F_R$ the bubble is in equilibrium. (From Alavi, S. H., Rizvi, S. S. H., and Harriott, P., *Food Research International*, 36, 309–19, 2003.)

$$\Delta R = R(\Delta t)\left[\frac{P - P_f - P_y - P_e - \dfrac{2\sigma}{R}}{\dfrac{4(2/\sqrt{3})^{n-1}}{n}[\xi + K_c - K_s(R/R')^{3n}]}\right]^{1/n}, \qquad (20.1)$$

where P_y and P_e are the respective contributions of flow yield stress and elastic stress (kPa); P is the vapor pressure inside the bubble (kPa); P_f is the fluid (melt) pressure (kPa); σ is the surface tension (N/m); R is the instantaneous radius of the bubble (m); R' is the radius of the domain (m); K_c is the consistency coefficient at the pore surface (Pa.s^n); K_s is the consistency coefficient at the outer surface of the domain (Pa.s^n); n is the flow behavior index; and ξ is an integral term accounting for changes in consistency coefficient in the domain because of variation in moisture, temperature, and radial velocity.

The material property parameters used in the model were mostly taken from the literature for similar starch/water systems. For example, the consistency coefficient (K) was modeled as a function of temperature and moisture using a relationship reported by Parker et al. (1989). Experimental studies on rheological parameters of SCFX melts confirmed the validity of this choice (Alavi, Chen, and Rizvi 2002; Chen and Rizvi 2006a). Similarly, the diffusion coefficient (D) of CO_2 in the starch/water melt, which was used in the calculation of vapor pressure inside the cell during the initial growth and collapse phases, was modeled as a function of temperature and moisture using a relationship reported by de Cindio and Correra (1995). As in the case of consistency coefficient, experimental studies on diffusion of CO_2 in SCFX melts confirmed the validity of this choice (Singh et al. 1996; Alavi et al. 2002; Chen and Rizvi 2006b).

The above process dynamics model was not only very useful in developing an understanding of SCFX at both micro- and macroscopic levels, but also in the prediction of several experimentally verifiable phenomena. For example, the model hypothesized that increase in yield stress of the melt would lead to a reduction in collapse of cells after the extrudate exited the die (Figure 20.7a) (Alavi et al. 2003b). This was confirmed from experimental studies using thermosetting proteins, such as WPC, as a means of increasing yield stress, which led to lower collapse of cells and higher expansion of extrudates (Alavi et al. 1999). The model also predicted that postextrusion collapse could be arrested by reduction in effective diffusivity of CO_2 in the extrudate, which would lead to a slower drop in the vapor pressure of CO_2 inside the cells. This effect was verified experimentally by utilizing die cooling during the SCFX process to reduce the effective diffusivity of CO_2, which led to reduced collapse and higher expansion of extrudates (Alavi and Rizvi 2005). The model predicted that higher drying temperatures would lead to increase in average cell size, extrudate expansion, and increase in the open cell fraction (Figure 20.7b) (Alavi et al. 2003b), which was indeed the case in an experimental study of the effect of drying temperatures ranging from 22 to 100°C (Alavi et al. 1999). A longer growth time (order of 1 s) of cells in the SCFX process or the "time-delayed expansion" of extrudates as compared to the fast explosive growth in conventional steam-based puffing

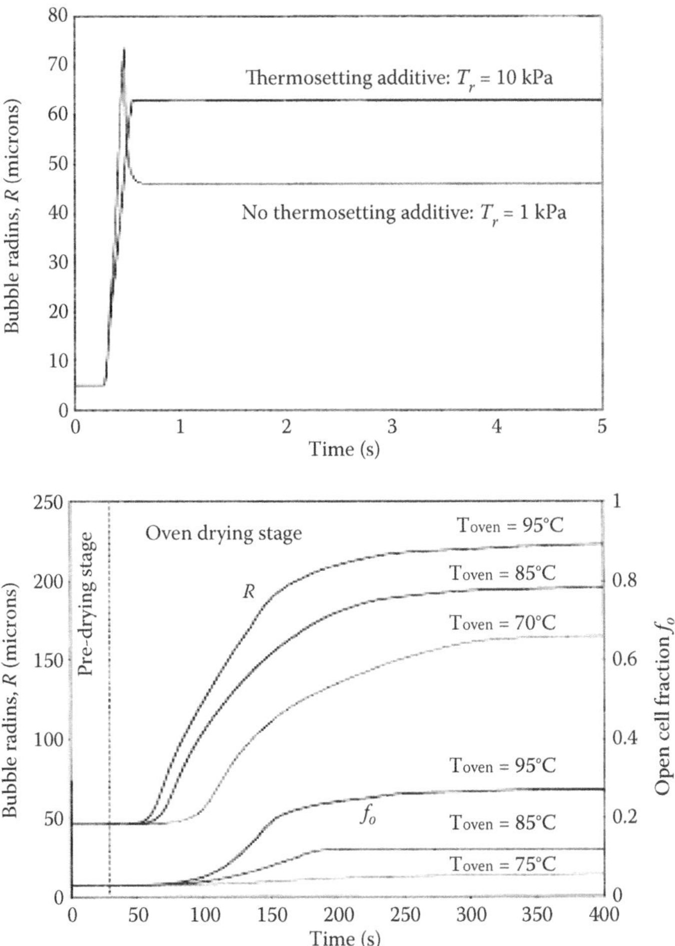

FIGURE 20.7 SCFX process dynamic model: (a) predicted bubble radius (R) with different yield stress coefficients (T_r) showing effect of thermosetting additive on reducing collapse; and (b) R and open cell fraction (f_o) at different oven temperatures. (From Alavi, S. H., Rizvi, S. S. H., and Harriott, P., *Food Research International*, 36, 309–19, 2003.)

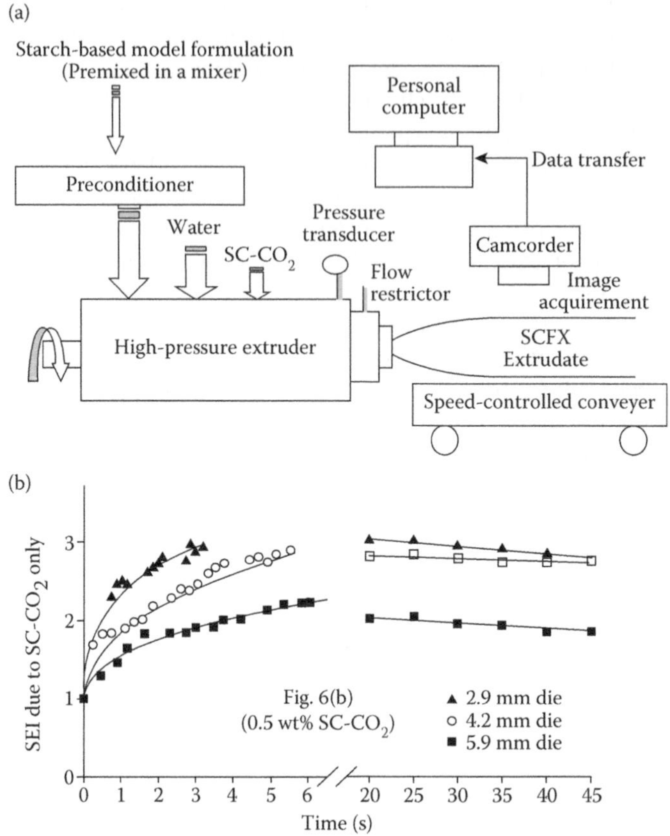

FIGURE 20.8 Time-delayed expansion in SCFX: (a) schematic of experimental set up; and (b) effect of die size on the cross-sectional expansion of SCFX extrudates. (From Cho, K. Y., and Rizvi, S. S. H., *Food Research International*, 41, 31–42, 2008.)

(order of 10 μs) was also hypothesized by the model and experimentally verified digital video imaging (Figure 20.8) (Cho and Rizvi 2008).

20.5 SUPERCRITICAL FLUID EXTRUSION (SCFX) APPLICATIONS

The unique product characteristics of SCFX, such as cell sizes in the order of 100 μm or lower, homogenous closed cell microstructure, and nonporous skin, and process capabilities, such as low temperature puffing, precise control of nucleation and cell growth, and ability to deposit and encapsulate flavors into the extruded matrix late in the process, make this technology very suitable for the continuous production of a new generation of highly nutritious and novel textured products. Some of these applications are discussed in detail below.

20.5.1 Masa-Based Snack Chips

Corn and tortilla chips are a hugely popular snack category in the United States, with roughly 25% share of the $20 billion savory snack market and ranked only behind potato chips in size (Packaged Facts 2000). With a fast-growing Hispanic population and influence and rise in preference for ethnic foods, corn and tortilla chips have experienced a faster sales growth than other savory snacks (Chen et al. 2002). However, apart from improvement in production scale and efficiency, current tortilla chip manufacture has remained unchanged over the past several decades. The traditional process consists of batch cooking of corn, development of masa flavor by steeping in lime water, grinding and sheeting of masa dough and cutting, followed by baking, frying, and seasoning. Development of tortilla chips with new textures and flavors has been fairly limited because of the reliance on a thin, flat dimension for providing a crispy texture and surface application of seasoning for providing flavor. The integration of SCFX technology with the conventional masa dough process can lead to a new generation of expanded snack chips as described by Chen et al. (2002). The formulation was primarily based on masa flour and relatively small amounts of salt for flavor, vegetable oil and monoglycerides for emulsification, and WPC for structure setting and promotion of Maillard browning. The basic SCFX process was adapted to involve the following stages: (1) hydration and kneading of masa flour to dough in the extruder; (2) formation of a gelatinized melt by a combination of mechanical and thermal energy in the extruder; (3) injection, solubilization, and release of SC-CO_2 to induce cell nucleation and growth along with direct flavor deposition if desired; (4) die-based sheeting and subsequent shaping by cutting; and (5) postextrusion baking or frying, followed by seasoning if desired.

Some of the critical differentiating characteristics of the SCFX-based masa chip process include (1) the use of continuous, low-moisture extrusion cooking of masa flour, which is more efficient than batch cooking of corn; (2) development of an internal cellular morphology; (3) use of an extrusion die rather than rollers for sheeting the final product; and (4) lower drying costs because of the lower-moisture SCFX process. The expanded cellular microstructure developed during SCFX not only provides a unique crispy texture to the chips, but also allows the production of thicker chips without compromising texture and increases the surface area to weight ratio, allowing efficient flavor application both internally and externally (Figure 20.9). From a comparison of the breaking strength of SCFX masa chips with commercial tortilla chips (Figure 20.10a) it is clear that the two products break or crush with the same force, however the compression modulus of the former is higher, leading to a crispier texture (Figure 20.10b), which is part of the uniqueness of SCFX-based snacks.

20.5.2 Milk-Based Snack Foods

In the relatively saturated snack food market in the United States, manufacturers are attempting to capture market share by catering to the needs of health-conscious consumers. Nutritious snack trends include products with multigrain or whole-grain ingredients and fruit and vegetable powders, and high-protein products based on ingredients such as soy. Milk proteins also have great potential for inclusion in nutritious snacks. Although milk proteins have been incorporated in expanded snack products, the conventional high-temperature extrusion cooking and puffing process tends to increase the density, toughen the texture due to cross-linking between the protein and starch molecules, and cause amino acid losses and excessive browning due to Maillard reactions (Aguilera and Kosikowski 1978; Kim and Maga 1987; Sokhey et al. 1996; Onwulata et al. 2000). The uniqueness of the SCFX process can be coupled effectively with the need of the dairy industry to explore new applications for whey, and to project this by-product of cheese processing as a valuable ingredient rather than a waste stream to be managed in a cost-effective manner (Smithers and Copeland 1998). Utilization of whey products in food has been reported to be only 55% of the total whey produced, while nearly one-third of the whey protein production in the United States (a total of 160,000 tones in 1997) is currently used in low-value animal feed (NASS 1998). There is an immense potential for whey and its derivatives, like condensed whey, whey protein isolate (WPI), and WPC, to be included at high levels in extruded snack foods. The benefits include improvement in nutritional quality and imparting flavor and color to the extrudates due to the presence of lactose, which takes part in Maillard reactions with whey proteins. Whey derivatives are a rich source of nutrients and bioactive components (e.g., sulfur amino acids that prevent colon tumors and immune-system-enhancing antioxidants and antimicrobials such as lactoferrins and lactoperoxidases, to name a few).

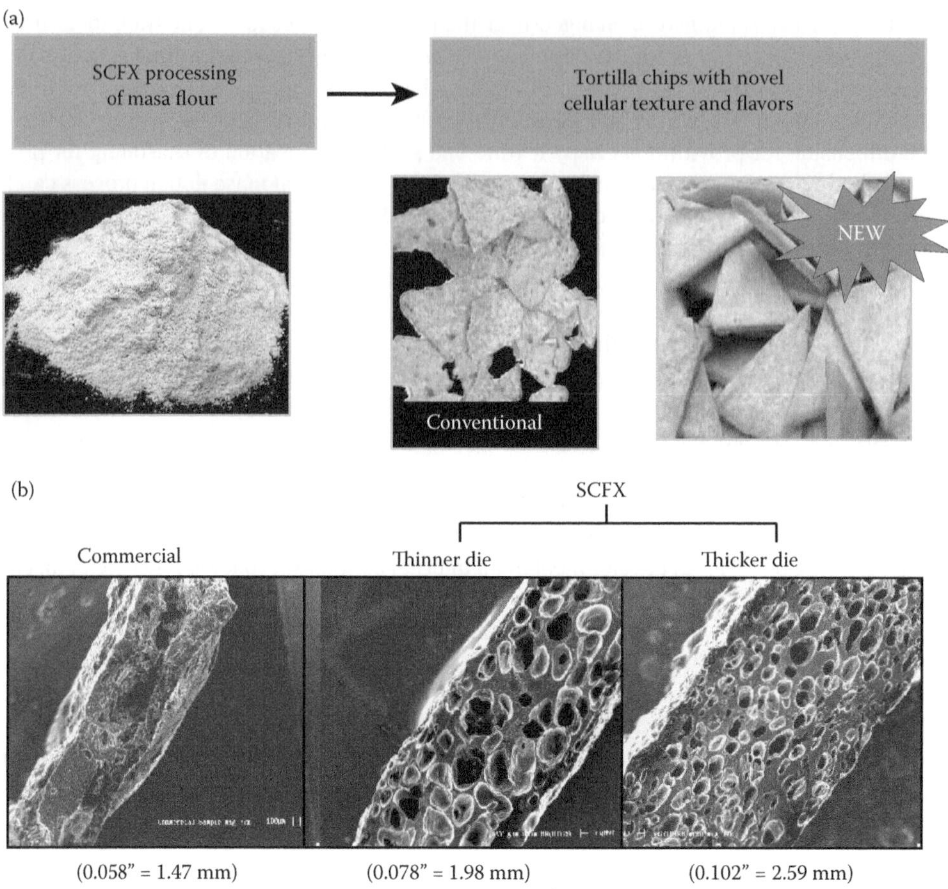

FIGURE 20.9 (a) Conventional and SCFX masa chips; and (b) scanning electron microscope pictures showing the microstructure through cross-sections of commercial and SCFX masa chips. (From Chen, K.-H.J., Dogan, E., and Rizvi, S.S.H., *Cereal Foods World,* 47, 44–51, 2002. With permission.)

The low-temperature and -shear SCFX process allows incorporation of heat-sensitive whey proteins without the drawbacks of conventional extrusion (Sokhey et al. 1996; Alavi et al. 1999; Gogoi et al. 2000). In fact, whey proteins, which are mainly comprised of α-lactalbumin, β-lactalbumin, and bovine serum albumin, have a tendency to gel in a temperature range of 59–82°C (Aguilera 1995) and can be used as a thermosetting agent for a stabilizing effect on the cell structure of the relatively high-moisture SCFX extrudates, as discussed earlier. So far up to 37% WPC (34% protein) has been successfully incorporated into starch-based expanded extrudates produced by SCFX. Figure 20.11 shows whey protein- and chocolate milk-based expanded wafers produced by SCFX using a corrugated die.

20.5.3 Continuous Dough Leavening for Bread

The most common industrial bread-making processes are the sponge dough and straight dough processes that need a total of 4–5 h for operations such as mixing, fermentation, and proofing. Although bread making is an ancient art, the narrow profit margin of bakeries demand for continuous improvements in process control and shortening of processing times. Some process modifications, involving prefermentation and/or intense mixing have helped in reducing processing time by as much as 60% (Cauvain 1998; Campbell et al. 1998; Chamberlain and Collins 1979). However bread making is still a time-consuming operation, which can be characterized as semicontinuous at best and thus

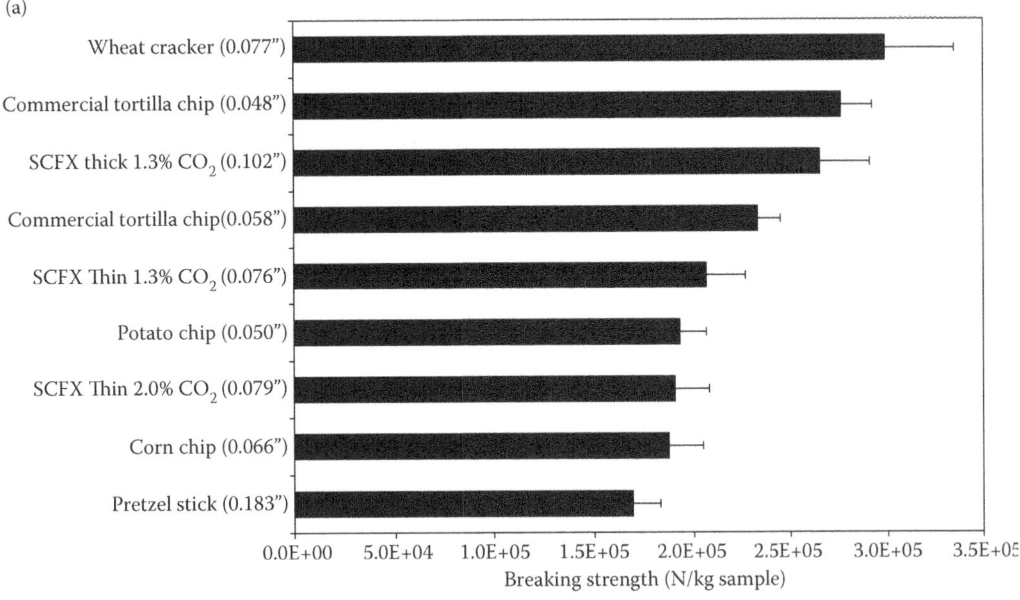

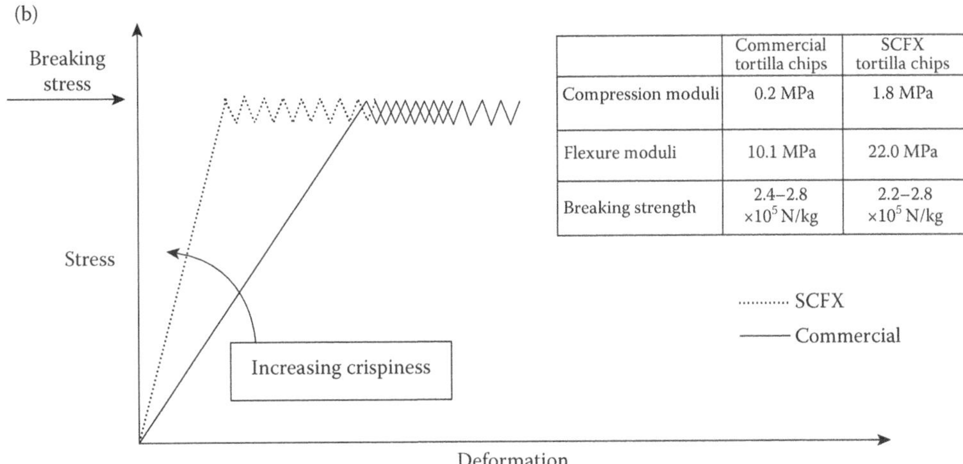

	Commercial tortilla chips	SCFX tortilla chips
Compression moduli	0.2 MPa	1.8 MPa
Flexure moduli	10.1 MPa	22.0 MPa
Breaking strength	2.4–2.8 $\times 10^5$ N/kg	2.2–2.8 $\times 10^5$ N/kg

FIGURE 20.10 (a) Breaking strength of SCFX products (masa chips) as compared to commercial snack products; and (b) mechanism of unique crispiness of SCFX masa chips. (From Chen, K.-H.J., Dogan, E., and Rizvi, S. S. H., *Cereal Foods World*, 47, 44–51, 2002. With permission.)

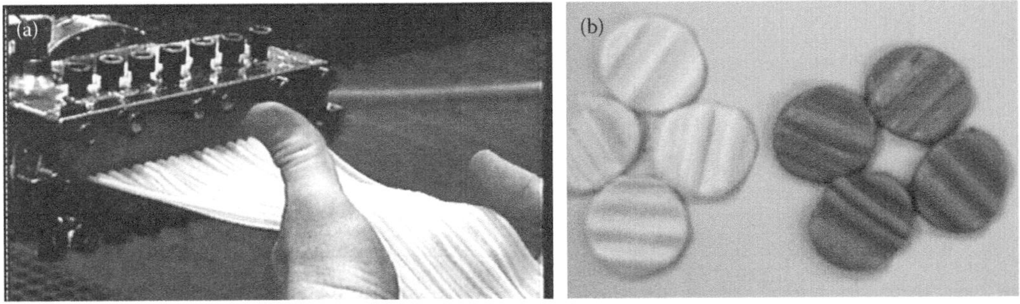

FIGURE 20.11 Milk-based SCFX snacks: (a) corrugated sheeting die, and (b) cut and dried final product.

process control improvement also remains a challenge. The US Environmental Protection Agency (EPA) requirement for lower ethanol emissions since 1990 has been another challenge and a source of higher costs for bakeries, as yeast produces 0.8 kg ethanol/100 kg of flour, which needs to be catalytically converted to CO_2 for release into the environment (Hicsasmaz et al. 2003).

SCFX technology has the potential to overcome the above-mentioned drawbacks of conventional leavening, by continuous production of ready-to-bake leavened dough via solubilization and subsequent release of SC-CO_2, thus enabling substitution of yeast as a leavening agent and shortening the processing time significantly. Additionally a homogenous nucleation and cell growth process ensures uniform cell structure and minimal defects in the final product. The SCFX process for bread dough leavening has been described in detail by Hicsasmaz et al. (2003) and is schematically described in Figure 20.12. The dough formulation was based on high-spring dominator flour and was similar to that of a US white bread recipe, except for relatively higher moisture and ascorbic acid contents, additional ingredients such as bakery fat, locust bean gum, and xanthan gum, and no yeast. The dry feed rate was 53 kg/h. Bakery fat (2.5%, w/w) and water (50%, w/w) were injected into the extruder and dough development occurred by the action of the screws, subsequent to which SC-CO_2 was injected at a rate of 1.5% (w/w). The dough was extruded through a slit die that led to the shear and elongation action during conventional dough mixing. Leavening was accomplished in the die and after exit from the die due to the pressure drop, which led to nucleation and cell growth. The slit die was cooled to keep the product temperature below 40°C. The densities of the leavened dough and final bread produced via SCFX were similar to those corresponding to the conventional bread-making process (Figure 20.13), but led to a reduction in leavening time from several hours to just 4–5 min and elimination of ethanol emissions.

Online rheological data measured using the slit-die apparatus is shown in Table 20.1. Flow behavior indices for nonleavened and leavened dough processed by SCFX technology were comparable to conventional dough, thus indicating similar strength and resistance against shear rate. The higher flow behavior index for SCFX-leavened dough meant that it was less vulnerable to increasing shear rates because of the energy-absorbing action of trapped CO_2. The consistency coefficient of SCFX

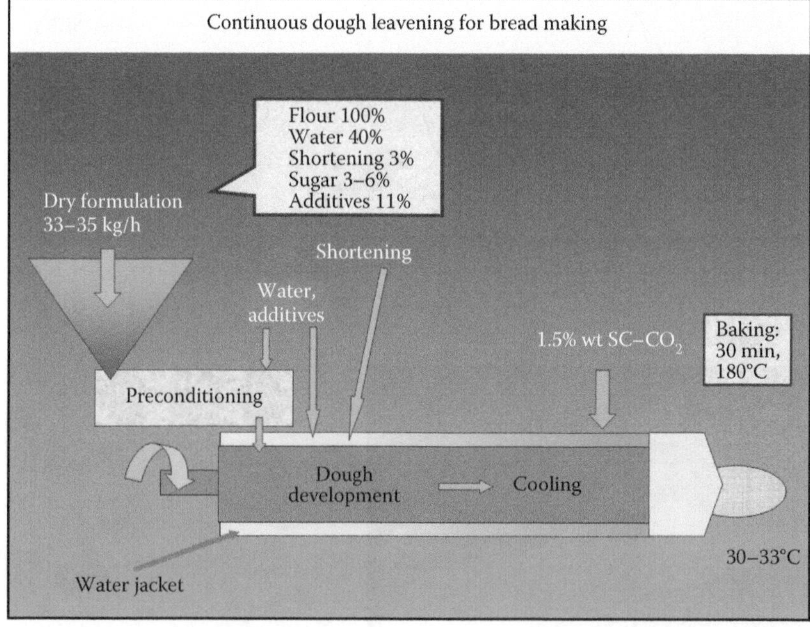

FIGURE 20.12 Schematic of the SCFX process for continuously leavened bread.

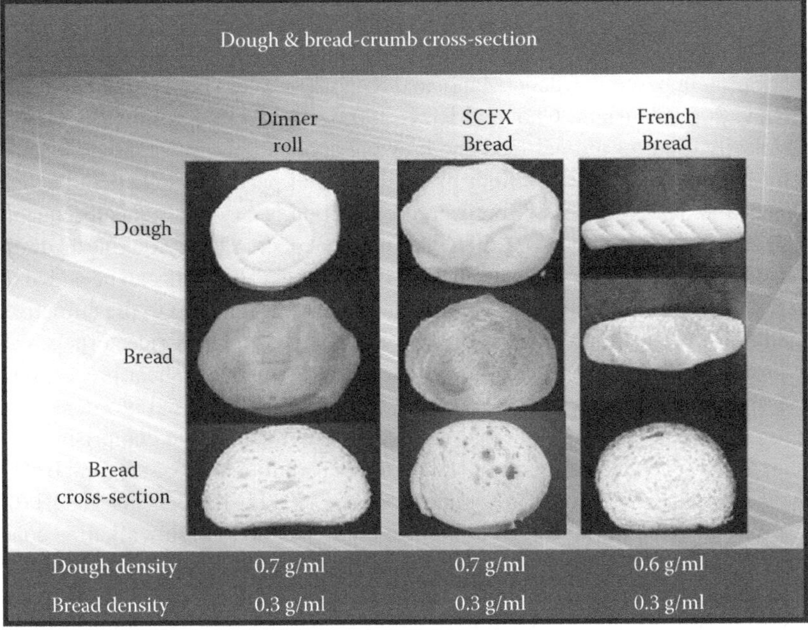

FIGURE 20.13 Leavened dough, final product surface, and crumb structure for conventional breads as compared to continuously leavened SCFX bread.

TABLE 20.1
Comparison of Online Rheological Properties of Dough During the SCFX Leavening Process with Offline Rheological Properties of Dough Prepared by Traditional Methods

Type of Mixture	Moisture (kg/kg)	Temp. (°C)	Method	Type of Dough	Consistency Coefficient, K (Pa.s^n)	Flow Behavior Index, n (Dimensionless)
Bread mix[a]	50	36–38	Online slit die	Nonleavened control	250	0.49
				SCFX leavened	128	0.63
Flour-water[b]	50	Not reported	Offline slit die	Nonleavened	1309–1390	0.46–0.48
Flour-water[b]	64.5	21	Offline capillary die	Nonleavened	1615–3115	0.24–0.46
Flour-water[b]	45	30	Offline capillary die	Nonleavened	1090–405	0.45–0.63

[a] Experimentally determined;
[b] literature values as reported in Hicsasmaz, Z., Dogan, E., Chu, C., and Rizvi, S. S. H., *Journal of Agricultural and Food Chemistry,* 51, 6191–97, 2003. With permission.

dough was severalfold lower than that for conventional dough, possibly due to higher temperatures and greater shear history. However, this did not adversely affect the density of the leavened product. The capability of continuous monitoring and online control of rheological and other properties of the dough is another advantage of the continuous leavening technology using SCFX. This technology is ready to be adopted by the industry with some process and formulation improvements.

20.5.4 Other Applications

Some of the other applications of SCFX include production of fortified microperforated rice grain, texturized whey protein based powders (Manoi and Rizvi 2008), and biodegradable industrial foams (Ayoub and Rizvi 2008; Mariam, Cho, and Rizvi 2008). Broken rice grains are a low-value commodity and typically used for animal feed and brewing applications. Substantial value addition can be achieved for broken rice-based rice flour by production of partially cooked and formed "whole" rice kernels using the SCFX process. Moreover, the injection of SC-CO_2 and subsequent nucleation and partial cell growth leads to microperforated kernels that decrease the cooking/hydration time for the rice before consumption. Various heat-sensitive flavors and also colors can be added to these kernels through SC-CO_2 or the by direct injection into the melt, thus adding to the attractiveness of the product. In addition, various micronutrients can be added to these SCFX rice kernels, with minimal degradation during the process because of the low temperature and low shear process environment.

The SCFX process can also be used to produce texturized WPC-based powders for use in food applications as cold-setting gels (Manoi and Rizvi 2008). A formulation comprising of 94% WPC (81.5% protein) and 6% pregelatinized corn starch, and an additional 0.6% salt and 0.6% NaCl, was processed by SCFX with 1% (w/w) SC-CO_2 injection rate. HCl (5–15%, v/v) or NaOH (0.83–1.67%, w/v) solutions were injected into the barrel to create highly acidic to highly alkaline conditions (pH 2.89–8.16). The combination of controlled shear and heat of the SCFX process, modified pH, presence of mineral salts, and surface modification due to SC-CO_2 led to texturized whey protein powders with instant dispersibility in water and ability to form cold-setting gels. The unique gelling and functional properties of these powders have potentially useful applications in various food formulations.

SCFX can also be utilized for nonfood applications, such as production of biodegradable industrial foams. Pregelatinized wheat starch with 1% sodium hydroxide and 0.5–2.5% cross-linking agent epichlorohydrin (EPI) was SCFX processed to produce water-resistant microcellular foams of varying morphology with applications as light-scattering and opacifying materials (Ayoub and Rizvi 2008). In another application, combinations of pregelatinized corn starch and 0–18% WPI were processed using SCFX to produce biodegradable microcellular foams with thermal insulation properties comparable to expanded polyethylene and polyurethane foams (Mariam et al. 2008). These novel industrial applications demonstrate the versatility of the SCFX process for manufacturing products with a diverse range of applications.

20.6 CONCLUSIONS

SCFX is an elegant new process developed at the cutting-edge interface of supercritical fluid and conventional extrusion technologies. This low-temperature and -shear process has been tested and proven for a wide range of applications in the next generation of food products with novel structures and textures, and in the production of a variety of biopolymeric industrial foams. The next step for SCFX technology is its commercial adoption by the industry with only slight modifications to the existing process design depending on the intended application. For this purpose, some minor technical shortcomings also need to be addressed. One of these is related to process control. Morphological properties of SCFX products are directly related to the rate of supercritical CO_2 injection into the barrel. This rate is very low (<1% of dry feed rate) and very sensitive to the pressure developed inside the barrel at the injection point. Due to slight variability in other process variables, such as water injection rate and also raw material characteristics, maintaining a constant barrel pressure at the injection SC-CO_2 point is almost impossible. However, advances in controller technology can be applied to overcome this challenge.

ACKNOWLEDGMENTS

The authors would like to gratefully acknowledge various colleagues that have been mentioned as co-authors in citations appearing in this text, and also Rohit Jalali, Cindy Winoto, Jhoe deMesa

Stonestreet, and Philipina Marcelo for their contribution to research and development related to supercritical fluid extrusion at Cornell University. The authors also appreciate industrial and institutional partners such as Wenger Manufacturing, Inc. (Sabetha, KS) and the Cornell Material Science Center for their invaluable contributions.

REFERENCES

Aguilera, J. M. 1995. Gelation of whey proteins. *Food Technology* 49: 83–86, 88–89.

Aguilera, J. M., and Kosikowski, F. V. 1978. Extrusion and roll-cooking of corn-soy-whey mixtures. *Journal of Food Science* 43 (1): 225–27, 230.

Alavi, S. H., Chen, K.-H., and Rizvi, S. S. H. 2002. Rheological characteristics of intermediate moisture blends of pregelatinized and raw wheat starch. *Journal of Agricultural and Food Chemistry* 50 (23): 6740–45.

Alavi, S. H., Gogoi, B. K., Khan, M., Bowman, B. J., and Rizvi, S. S. H. 1999. Structural properties of protein-stabilized starch-based supercritical fluid extrudates. *Food Research International* 32 (2): 107–18.

Alavi, S., and Rizvi, S. S. H. 2005. Strategies for enhancing expansion in starch-based microcellular foams produced by supercritical fluid extrusion. *International Journal of Food Properties* 8 (1): 23–34.

Alavi, S. H., Rizvi, S. S. H., and Harriott, P. 2003a. Process dynamics of starch-based microcellular foams produced by supercritical fluid extrusion. I: Model development. *Food Research International* 36 (4): 309–19.

———. 2003b. Process dynamics of starch-based microcellular foams produced by supercritical fluid extrusion. II: Numerical simulation and experimental evaluation. *Food Research International* 36 (4): 321–30.

Ayoub, A., and Rizvi, S. S. H. 2008. Properties of supercritical fluid extrusion-based crosslinked starch extrudates. *Journal of Applied Polymer Science* 107 (6): 3663–71.

Baldwin, D. F., Park, C. B., and Suh, N. P. 1996. An extrusion system for the processing of micro-cellular polymer sheets: shaping and cell growth control. *Polymer Engineering and Science* 36(10): 1425–1435.

Barrett, A. M., and Peleg, M. 1992. Cell size distributions of puffed corn extrudates. *Journal of Food Science* 57 (1): 146–48, 154.

Campbell, G. M., Reilly, C. D., Fryer, P. J., and Sadd, P. A. 1998. Aeration of bread dough during mixing: effect of mixing dough at reduced pressure. *Cereal Foods World* 43:163–67.

Cauvain, S. P. 1998. Breadmaking processes. In *Technology of Breadmaking*. Cauvain, S. P., and Young, L. S., Eds, 18–29, 34–41. London: Blackie Academic and Professional.

Chamberlain, N., and Collins, T. H. 1979. The Chorleywood bead process: The roles of oxygen and nitrogen. *Baker's Digest* 53:18–19, 22–24.

Chen, K. H., and Rizvi, S. S. H. 2006a. Rheology and expansion of starch-water-CO_2 mixtures with controlled gelatinization by supercritical fluid extrusion. *International Journal of Food Properties* 9 (4): 863–76.

Chen, K.-H. J., Dogan, E., and Rizvi, S. S. H. 2002. Supercritical fluid extrusion of masa-based snack chips. *Cereal Foods World* 47 (2): 44–51.

Chen, K.-H. J., and Rizvi, S. S. H. 2006b. Measurement and prediction of solubilities and diffusion coefficients of carbon dioxide in starch-water mixtures at elevated pressures. *Journal of Polymer Science, Part B: Polymer Physics* 44 (3): 607–21.

Chinnaswamy, R., and Hanna, M. A. 1988. Relationship between amylose content and extrusion-expansion properties of corn starches. *Cereal Chemistry* 65 (2): 138–43.

Cho, K. Y., and Rizvi, S. S. H. 2008. The time-delayed expansion profile of supercritical fluid extrudates. *Food Research International* 41 (1): 31–42.

de Cindio, B., and Correra, S. 1995. Mathematical modelling of leavened cereal goods. *Journal of Food Engineering* 24:379–403.

Dogan, E., Chen, K. H., and Rizvi, S. S. H. 2001. Fundamentals and applications of supercritical CO2 extrusion technology. In: *Novel Processes and Control Technologies in the Food Industry*. Bozoglu, F., Deak, T., Ray, B., and Eds, NATO Science Series Vol. 338, 37–48. Amsterdam: IOS Press.

Ferdinand, J. M., Lai-Fook, R. A., Ollet, A. L., Smith, A. C., and Clark, S. A. 1990. Structure formation by carbon dioxide injection in extrusion cooking. *Journal of Food Engineering* 11: 209–224.

Gogoi, B. K., Alavi, S. H., and Rizvi, S. S. H. 2000. Mechanical properties of protein-stabilized starch-based supercritical fluid extrudates. *International Journal of Food Properties* 3 (1): 37–58.

Harper, J. M., and Tribelhorn, R. E. 1992. Expansion of native cereal starch extrudates. In *Food Extrusion Science and Technology*. Kokini, J. L., Ho, C., and Karwe, M. V., Eds, 653–67. New York: M. Dekker.

Hauck, B. W., and Huber, G. R. 1989. Single screw vs twin screw extrusion. *Cereal Foods World* 34:930–39.

Hicsasmaz, Z., Dogan, E., Chu, C., and Rizvi, S. S. H. 2003. Leavened dough processing by supercritical fluid extrusion (SCFX). *Journal of Agricultural and Food Chemistry* 51 (21): 6191–97.

Kim, C. H., and Maga, J. A. 1987. Properties of extruded why protein concentrate and cereal blends. *Lebensmittel-Wissenschaft und Technologie* 20:311–18.

Kirby, A. R., Ollett, A. L., Parker, R., and Smith, A. C. 1988. An experimental study of screw configuration effects in the twin screw extrusion cooking of maize flour grits. *Journal of Food Engineering* 8:247–72.

Manoi, K., and Rizvi, S. S. H. 2008. Rheological characterizations of texturized whey protein concentrate-based powders produced by reactive supercritical fluid extrusion. *Food Research International* 41 (8): 786–96.

Mariam, I., Cho, K. Y., and Rizvi, S. S. H. 2008. Thermal properties of starch-based biodegradable foams produced using Supercritical Fluid Extrusion (SCFX). *International Journal of Food Properties* 11 (2): 415–26.

Mulvaney, S. J., and Rizvi, S. S. H. 1993. Extrusion processing with supercritical fluids. *Food Technology* 47 (12): 74–82.

NASS. 1998. *Dairy Products, 1997 Summary*. National Agricultural Statistics Service, Washington, DC: USDA.

Onwulata, C. I, Konstance, R. P., Smith, P. W., and Holsinger, V. H. 2000. Co-extrusion of dietary fiber and milk proteins in expanded corn products. *Lebensmittel-Wissenschaft und Technologie* 34:424–29.

Packaged Facts. 2000. The U.S. market for salted snacks. MarketResearch.com.

Park, C. B., Baldwin, D. F., and Suh, N. P. 1995. Effect of pressure drop rate on cell nucleation in continuous processing of microcellular polymers. *Polymer Engineering and Science* 35(5): 432–440.

Parker, R., Ollett, A.-L., Lai-Fook, R. A., and Smith, A. C. 1989. The rheology of food 'melts' and its application to extrusion processing. In *Rheology of Food, Pharmaceutical and Biological Materials with General Rheology*. Carter, R. E., Ed., 57–73. London: Elsevier Science.

Rizvi, S. S. H., and Mulvaney, S. 1992. Extrusion processing with supercritical fluids. US Patent Number 5,120,559.

———. 1995. Supercritical fluid extrusion process and apparatus. US Patent Number 5,417,992.

Rizvi, S. S. H., Mulvaney, S. J., and Sokhey, A. S. 1995. Combined application of supercritical fluid and extrusion technology. *Trends in Food Science and Technology* 6 (7): 232–40.

Schwartzberg, H. G., Wu, J. P. C., Nussinovitch, A., and Mugerwa, J. 1995. Modelling deformation and flow during vapor-induced puffing. *Journal of Food Engineering* 25:329–72.

Singh, B., and Rizvi, S. S. H. 1998a. Residence time distribution (RTD) and goodness of mixing (GM) during CO2-injection in twin-screw extrusion part II: GM studies. *Journal of Food Process Engineering* 21 (2): 111–26.

———. 1998b. Residence time distribution (RTD) and goodness of mixing (GM) during CO2-injection in twin-screw extrusion part I: RTD studies. *Journal of Food Process Engineering* 21 (2): 91–110.

Singh, B., Rizvi, S. S. H., and Harriott, P. 1996. Measurement of diffusivity and solubility of carbon dioxide in gelatinized starch at elevated pressures. *Industrial and Engineering Chemistry Research* 35 (12): 4457–63.

Smithers, G. W., and Copeland, A. D. 1998. 1997. International Whey Conference. *Trends in Food Science and Technology* 9:119–23.

Sokhey, A. S., Rizvi, S. S. H., and Mulvaney, S. J. 1996. Application of supercritical fluid extrusion to cereal processing. *Cereal Foods World* 41 (1): 29–34.

Tutanathorn, H., and Rizvi, S. S. H. 2001. Flavor application and encapsulation by supercritical fluid extrusion. Book of Abstracts. 2001 IFT Annual Meeting, New Orleans, Louisiana.

Yu, Z.-R., Singh, B., Rizvi, S. S. H., and Zollweg, J. A. 1994. Solubilities of fatty acids, fatty acid esters, triglycerides, and fats and oils in supercritical carbon dioxide. *The Journal of Supercritical Fluids* 7 (1): 51–59.

21 Rheological Properties of Liquid Foods Processed in a Continuous-Flow High-Pressure Throttling System

Rakesh K. Singh
Food Science and Technology Department

Litha Sivanandan
Reseach and Development Food Technologist

CONTENTS

21.1 INTRODUCTION

Conventional homogenization, developed by Gaulin in 1899, has been used extensively in the dairy industry and other food applications (Zamora et al. 2007). Nonthermal treatments, which maximize the microbial reduction while retaining the desirable physicochemical properties of the product, are in growing demand in the food industry. Applications of ultra-high-pressure homogenization (pressures more than 200 MPa) are generally found in the fields of pharmaceuticals and biotechnology (Floury, Desrumaux, and Lardieres 2000). Other applications include formulation of fine food emulsions, disruption of dense cell microbial cultures and subsequent recovery of intracellular metabolites, inactivation of bacteriophages, and modification of functional properties of hydrocolloids (Thiebaud et al. 2003). High-pressure treatment has attracted much attention in recent years and foods treated with high pressure have been reported to have superior color, flavor, nutrient retention, lower microbial count, and functional properties due to its nonthermal and nonchemical application (Sivanandan, Toledo, and Singh 2008; Peck 2004; Moorman 1997; Adapa, Schmidt, and Toledo 1997; Amornsin 1999; Areekul 2003; Sivanandan 2007). High-pressure treatment improves protein gelation properties and has been successfully used in the food industry (Hoover 1993) for guacamole and oysters. The protein denaturation that occurs with high-pressure treatment is a result of hydrophobic and ionic bond breakage as opposed to covalent bond disruption, which results in off-flavors (Zipp and Kauzmann 1973; Knorr 1993). Pressurization studies (392–980 MPa) conducted on egg and yolk resulted in firm gels with natural taste and these gels, with no loss of vitamins and amino acids, were softer, more elastic, and more digestible than heat-treated gels (Hayashi et al. 1989; Hoover 1993).

Traditionally, high hydrostatic processing (HHP) has been used for pressurized processing. Sterilization is possible in HHP when the product is pressurized at elevated temperatures, but the problem of slow exposure of the product to that high temperature minimizes the quality advantages of the high-pressure treatment. In HHP, the food exposure time is usually in the order of a few minutes or more whereas the residence time in the dynamic high-pressure treatments is in the order of seconds (Thiebaud et al. 2003). Continuous-flow high-pressure throttling (CFHPT) process was developed as a means of continuous microbial inactivation of fluid foods at the University of Georgia, Athens (Toledo and Moorman 2000). A throttling valve (or micrometering valve) is a fine restriction such as a partially closed valve or porous plug to control the flow of the fluid through a micrometering valve. CFHPT uses high pressure (up to 310 MPa) to pressurize liquid foods and to continuously throttle the foods from that high pressure to atmospheric pressure for the purpose of inactivating microbes and modifying proteins which should change the rheological properties of the food (Moorman 1997). Thus, the name CFHPT aptly suits the process. As the fluid material exits the throttling valve, temperature rises (Equation 21.3), and the product is then very quickly cooled to prevent overprocessing of the product, which leads to a reduction in overall quality (nutrition and sensory). A CFHPT system is a continuous system that uses lower pressures and so does not need large-volume pressurized vessels for processing. Therefore its initial and operating costs are lower than those of a batch system (Moorman 1997; Amornsin 1999; Areekul 2003; Sivanandan et al. 2008). Smaller volumes of pumpable fluid food are pressurized in CFHPT without particle restriction occurring through orifices with very small clearance. In CFHPT, after throttling valve on depressurization, the instantaneous heating of the homogenized liquid occurs due to conversion of pressure energy of the fluid food at high pressure to kinetic energy. This temperature rise (Equation 21.3) of the liquid food minimizes the total thermal exposure needed for sterilization.

21.2 CONTINUOUS HIGH-PRESSURE PROCESSING SYSTEMS

21.2.1 Continuous-Flow High-Pressure Throttling (CFHPT) Processing

A diagram of a CFHPT system is shown in Figure 21.1. The fluid food can be pressurized using two alternately acting pressure intensifiers (Hydropac P60-03CXS, Stansted Fluid Power Ltd., Stansted, Essex, UK) driven by a hydraulic pump in the Stansted CFHPT system (Model nG7900, Stansted Fluid Power Ltd., Stansted, Essex, UK). The pressure levels are read from the pressure gauge located in the CFHPT system. It is to be noted that this pressure is a combination of the intensifier pressure and the pressure generated due to the narrow clearance of the throttling valve (model 60VRMM4882, Autoclave Engineers, Fluid Components, Erie, PA 16506-2302) adjustments for the desired flow rate at each intensifier pressure. The CFHPT system (Figure 21.1) consists of a feed pump that maintains a constant pressure to the fluid feed to the intensifier, dual intensifier pistons that work alternately to take in fluid while the other discharges high-pressure fluid, a heat exchanger heated with steam to increase the temperature of the high-pressurized fluid product prior to throttling, a throttling valve to drop the pressure, a hold tube to hold the fluid a designated time at the risen temperature

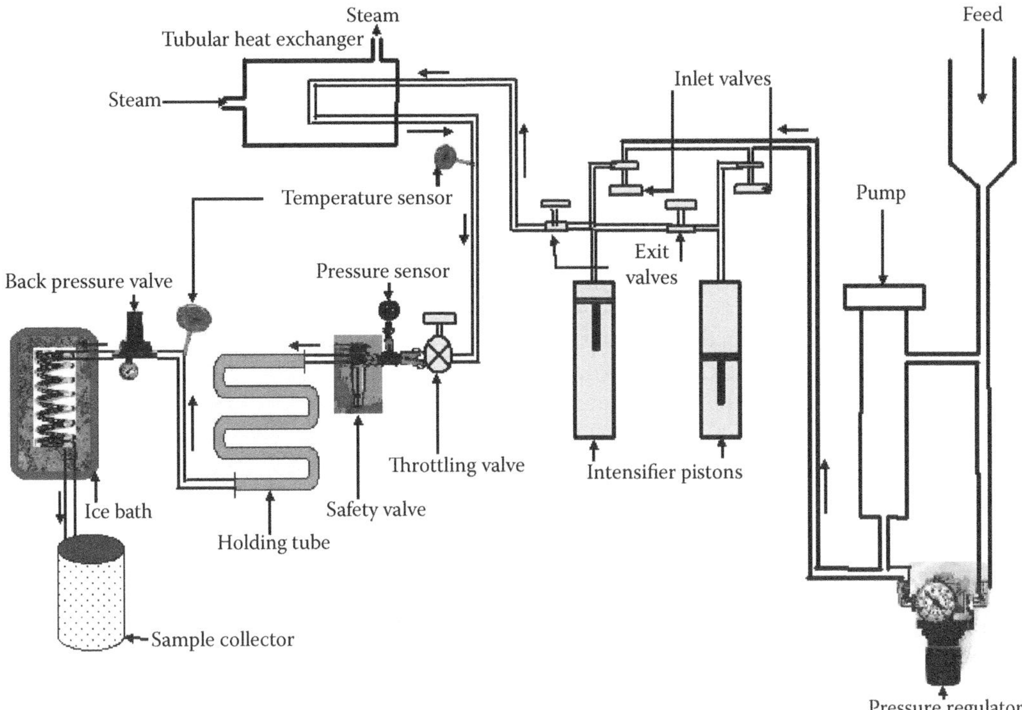

FIGURE 21.1 Diagram of CFHPT system showing the flow direction of liquid food and parts. (From Sivanandan, L., Toledo, R. T. and Singh, R. K., *Journal of Food Science*, 73(6), E288–96, 2008.)

(Equation 21.3) prior to cooling, and a tubular heat exchanger to cool the product using cold water and ice as the coolants. Thermocouples are connected at the steam-heated tubular heat exchanger and at the end of the holding tube (located after the throttling valve). The outputs of thermocouples are recorded on a Fluke Hydra Data Bucket (PO Box 9090, Everett, WA 98206-9090).

21.2.2 Microfluidization-Throttling Valve System

Cook and Lagacé (1985) developed microfluidization, which is yet another homogenization technique in which the pressurized liquid food is split into two streams in the interaction chamber that collide with each other at high speed at a 180° angle (Lemay, Paquin, and Lacroix 1994; Tunick et al. 2002). This impact of the liquid streams homogenizes the liquid food with particle size reduction, narrower particle size distribution, and mixing due to cavitation (Paquin 1999). Microfluidizers are produced by Microfluidics, Inc. (Newton, MA). The microbicidal effects of CFHPT treatment (Areekul 2003), microfluidizer treatment, and microfluidizer-throttling valve treatment were studied by Moorman (1997). The microfluidizer used a double-acting electrohydraulic pressure intensifier to force the process fluid through the constriction, which is in the interaction chamber with downstream back pressure module. This elevated the temperature of process fluids from an entrance temperature of 4°C–~85°C after the interaction chamber, and with built-in water cooling coils (after the interaction chamber) helped to lower the temperature to 40°C within 3.4 s. The hybrid process of CFHPT and microfluidization introduced a throttling valve (which was used in the CFHPT process, with temperature rise of liquid food at depressurization as per Equation 21.3) in place of an interaction chamber in the microfluidizer. There was no change in the other parameters used; flow rate (9.2 l/s), operating pressure (276 MPa maximum), cooling coil (12°C water jacket), and cooling time (3.4 s) were the same as that of microfluidizer with interaction chamber.

Various studies using a microfluidizer on rheology and microstructure of foods show its considerable application in the food industry. The microstructure of mozzarella cheeses made form milk microfluidized at various temperatures and pressures was studied by Tunick et al. (2000) using a scanning electron microscope. Images showed that the temperature (10°C, compared to control) did not liquefy the fat to an extent necessary for its complete microfluidization. However, at a processing temperature of milk of 54°C, the fat droplets in the cheese were more discrete and smaller, giving the casein matrix a spongy appearance. When the microfluidization pressure was increased from 34 MPa (the maximum used for dairy homogenization) to 172 MPa, fat droplet size was reduced in the scanning electron micrograph, showing an emulsion of fat and protein (Tunick et al. 2000). The emulsification process allows small particle size distributions of 0.1–0.3 μm. Microfluidization was suggested as an alternative method for the production of milk fat microcapsules (Vuillemard 1991), alcoholized cream, and for milk homogenization.

Cobos, Horne, and Muir (1995) investigated the rheological properties of acid milk gels made from recombined milks subjected to microfluidization and found that gelation time was reduced by low levels of solids, low level of fat, high heat treatment, high incubation temperature, and high concentration of acidulant, while the gelation pH was increased by heating the milk after homogenization, high heat treatment, and high incubation temperature. The elastic modulus and viscous modulus of gels were increased by high level of solids and fat, high heat treatment, and low incubation temperature. The rheological properties of acid gels made from milks subjected to microfluidization were very similar to those of acid gels made from milks subjected to high-pressure homogenization in a valve homogenizer, though the microfluidizer produced smaller, less variable particles.

Microfluidization of pasteurized whole milk and recombined milk was done by McCrae (1994) who observed that fat surfaces were covered with high amounts of casein but only minute amounts of serum protein. Also, the total amount of protein that covered the fat surfaces was higher than that predicted on the basis of decreased globule size, and this was thought to account for the inhibition of fat cluster formation observed in the microfluidized milks. Mozzarella cheeses made with the control full fat milk (no microfluidization) and milks microfluidized at 10°C and 34 MPa or 10°C and 103 MPa were softer and less rigid, and had the lowest viscoelastic properties and the highest meltabilities of all the cheeses (Van Hekken et al. 2007). Microfluidization of the milk did not improve the melt or rheology of low-fat milk cheeses. Microfluidization of milk with fat in the liquid state at higher pressures resulted in smaller lipid droplets that altered the component interactions during the formation of the cheese matrix and resulted in low-fat and full-fat Mozzarella cheeses with poor melt and altered rheology.

21.3 TEMPERATURE INCREASE

The temperature rise that occurred after throttling can be easily explained by the first law of thermodynamics (Amornsin 1999). Pressure difference between the inlet (higher pressure) and outlet (lower pressure) of the throttling valve represents the energy conserved in the form of heat energy at the exit of the throttling valve, which is visibtle from the temperature rise of the product. This change in potential energy due to high pressure to heat energy due to reduction in pressure across the throttling valve can be given by Equations 21.1 through 21.3 (Toledo 2007; Amornsin 1999):

$$q = C_p(T_{out} - T_{in}) = \frac{(P_{in} - P_{out})}{\rho}, \tag{21.1}$$

$$(T_{out} - T_{in}) = \frac{(P_{in} - P_{out})}{\rho C_p}, \tag{21.2}$$

$$T_{out} = T_{in} + \frac{(P_{in} - P_{out})}{\rho C_p}, \quad (21.3)$$

where q is the energy per unit mass (J/kg), C_p is the specific heat of the fluid at constant pressure (J/kg/°C), T_{out} is the temperature of the fluid at the inlet to the throttling valve (°C), T_{in} is the temperature of the fluid at the inlet of the throttling valve (°C), P_{in} is the pressure of the liquid at the inlet of the throttling valve (Pa), P_{out} is the pressure of the liquid at the outlet of the throttling valve (Pa), assumed as atmospheric pressure, and ρ is the density of the fluid (kg/m^3). The initial temperature or the temperature to which the fluid is heated in the tubular heat exchanger (T_{in}) affects the exit temperature and thereby the temperature rises in all of the equations given above. If flow rate is constant at the exit for product collection, the temperature variation due to volumetric flow rate difference is avoided. However, the heat loss due to convection (to the surroundings) and conduction (through the connecting pipe to cooling heat exchanger) further reduces the measured temperature (experimental temperature) at the exit of the holding tube (after throttling valve) from the theoretical results (Amornsin 1999). Pressurization was done up to 310 MPa using a pressure intensifier and conveyed through stainless steel tubing to the throttling valve where it was throttled to atmospheric pressure. Even when the throttle valve was in a completely closed position, the flow still occurred (Amornsin 1999). According to the manufacturer's data for water, the targeted pressure (310 MPa) was attained by decreasing the flow rate by adjusting the throttling valve, which alters the orifice diameter from a fully opened position of 1.5748 mm to a nearly closed position of 0.1556 mm. The maximum shear rate ($-dV/dr$) calculated (using Equation 21.5) (Toledo 2007) with a mean fluid velocity (V_{av}) in the orifice of 284 m/s (using Equation 21.4) (Toledo 2007), with nearly closed diameter ($R/2$) for throttling valve, and fluid flow rate (q) of 5.4 × 10^{-6} m^3/s was 14.6 × 10^{-6}/s.

$$V_{av} = \frac{q}{\prod R^2} \quad (21.4)$$

$$-\frac{dV}{dr} = \frac{4V_{av}}{R} \quad (21.5)$$

Amornsin (1999) compared the theoretical temperature rise obtained using Equation 21.3 with the experimental results with deionized water at 18°C as the processed liquid at a flow rate of 350–360 ml/min. The results showed that the measured temperatures were lower than the calculated values using the targeted pressure. This difference was attributed to the pressure alteration during the reciprocation of the piston in the intensifier. According to the law of conservation of energy, the temperature rise that occurred at the throttling valve exit was due to the conversion of pressure energy (stored in the pressurized fluid) to thermal energy. The author (Amornsin 1999) further explained that the exponential rise of pressure in a cycle led to the increase in measured temperature above the temperature calculated from the average of target and minimum pressure.

Sivanandan (2007) reported that the flow rates at a particular pressure in the CFHPT process did not result in a significant difference in temperature rise of the product although the observed temperature rise was directly proportional to the applied pressure. The author stated the importance of temperature rise in the particle size reduction of the soymilk. Temperature rise facilitated the easy rupture of the particles at high turbulence and shear, and cavitation occurred at and after the micrometering valve (Sivanandan 2007). The author also compared the experimental results of temperature rise in CFHPT after throttling (for soymilk) against the calculated values using Equation 21.3. The results showed that the experimental values of temperature rise were higher than the calculated values at lower pressures. This difference in temperature rise (between calculated and experimental values) decreased as the pressure increased and finally it was lower than the calculated results at the highest pressure (276 MPa). This was attributed to the increased pressure fluctuation as the pressure levels increased.

21.4 FOOD PRODUCTS PROCESSED THROUGH CONTINUOUS-FLOW HIGH-PRESSURE THROTTLING (CFHPT) AND THEIR RHEOLOGICAL BEHAVIOR

Advantages of CFHPT processing include simultaneous homogenization and sterilization, which can benefit physical properties, texture, and stability of the sterilized product during storage. CFHPT treatment has a wide range of applications in the manufacture of functional ingredients (Moorman 1997). High shear rates (in CFHPT processing) can increase the surface area to volume ratio of suspended particles, which can increase viscosity of the fluid suspension, leading to improved texture, taste, and flavor characteristics (Areekul 2003) of the fluid foods discussed here.

21.4.1 Milk

Effects of various holding times during pressurization (310 MPa) and after depressurization (at elevated temperature) were studied by Moorman (1997) on ultrafiltered skim milk and permeate. At the high pressure applied, the milk was held for 0.3 s and 1 s using the tube of length 7.6 cm or 176.5 cm. Two more treatment combinations were studied by changing the holding time (using different holding tubes) after the micrometering valve at the elevated temperature (80°C) for 0 s and 10 s. Processing of milk was done using CFHPT at 310 MPa by Adapa and others (1997) and the resultant milk produced good foams with high stability. Milk concentrates produced by CFHPT treatment resulted in increased surface tension and viscosity but decreased emulsion stability. This was thought to be due to increased protein–protein interactions, which result in larger protein aggregates and less protein migrating to the liquid–gas interface or adsorbing to the surface of fat globules. The pressurized unconcentrated milk showed enhanced emulsion stability with pressure treatment, which indicated that an optimum amount of protein aggregation promoted emulsion stability. The increased surface area of new fat globules formed were not adsorbed by emulsifying material (protein) and showed decreased emulsifying property in CFHPT-treated and ultrafiltration-concentrated milk. This lower emulsion stability of pressure-concentrated milk was attributed to the formation of protein–protein interactions and formation of large protein aggregates. But this protein-protein interaction at higher pressures was shown by whey proteins. Because of the globular aggregated structure of soy proteins, they do not unfold and adsorb at the interface, but rather form a thick interfacial layer, which acts as a physical barrier to coalescence (Molina, Papadopoulou, and Ledward 2001). Adapa and others (1997) describes the formation of large protein aggregates during CFHPT treatment of milk. This protein–protein interaction reduced the emulsifying property of the milk due to the inability of protein to bind as a thin layer around the single fat globule and prevent fat coalescence and form stable emulsion. They further stated that pressurization did not affect the interfacial tension and surface tension results of milk as expected due to increased protein-protein linkage rather than unfolding of protein at the interface. Due to the increased size of protein aggregates as a result of CFHPT treatment, they did not diffuse to the surface and reduced the surface tension. In spite of this result, the CFHPT-treated milks gave no feathering (feathering is a phenomenon by which milk proteins coagulate when added to hot coffee) and scored + 5 on the scoring scale.

CFHPT treatment of milk gave a darker color (lower L^* value) to milk (Adapa et al. 1997). The concentrated (by ultrafiltration) milks (18.00% total solids, 8.20–15.42% protein) were darker than unconcentrated milks (9.00% total solids, 3.28–3.69% protein). The decreased L^* value in pressurized milks was attributed to the disruption of noncovalent bonds (hydrogen bonds, ionic interactions, and hydrophobic forces), which caused separation of casein micelle fragments, individual caseins, and calcium phosphates, thus resulting in reduced light-scattering ability and the observed darker color. The increased protein concentration above a particular level decreased the foaming capacity in pressurized, concentrated milks due to the decrease in protein solubility. But the foaming stability was increased with increased protein content and thus increased viscosity and slow serum drainage.

The viability of native microflora in the milk was reduced by 2.5 to 4 log cycles when it was treated with CFHPT at 310 MPa, and neither the increase in hold time at the elevated temperature (>80°C) nor the longer holding at the pressurization before throttling was effective in reducing the microbial count to a smaller number (Moorman 1997). Though the CFHPT processing reduced the number of viable *Pseudomonas putida* cells by 7 log cycles and the distinct spheroid bulges on cell surfaces and the amount of debris in the SEM images of the CFHPT-treated samples suggested that the process was very disruptive to cell membranes, the author suggested more research on CFHPT treatment of thermally resistant organisms to demonstrate whether the mechanical effect (by shear and explosive decompression) of CFHPT helped in microbial inactivation above and beyond the heating effect (by enzyme denaturation) alone or a combination of both.

CFHPT treatment increased the viscosity of pressure-concentrated (concentrated with ultrafiltration) and pressure-unconcentrated milks (Adapa et al. 1997). This increase was attributed to the formation of protein–protein linkage during CFHPT treatment. The authors explained further that due to an increase in the effective volume due to the CFHPT treatment, protein aggregates became bigger, occupied more space, and provided increased viscosity to the fluid system while the milk behaved like a Newtonian fluid (Table 21.1). Moorman (1997) reported an increase in viscosities of CFHPT-treated milk concentrations and increase in apparent viscosities and water-holding capacity of yogurt made from CFHPT-treated milk. Apparently the combination treatment of lower pressurization dwell (0.3 s) and lower depressurization holding time (0 s) at elevated temperature, with milk concentrations of 0.19 and 0.24 kg/s/l, produced even greater viscosity. These changes in the properties of milk and yogurt were due to the modification of dairy proteins by the CFHPT process. Therefore, the author suggested that CFHPT treatment of skim milk could be utilized for improving the "mouth feel" of skim milk without added ingredients such as polysaccharides.

Moorman (1997) reported the influence of temperature on apparent viscosity for CFHPT-processed and control milk. At 1°C, the control milk showed apparent viscosity of 5 mPa.s, but the CFHPT milk showed 14.5 mPa.s. The values decline logarithmically with increasing temperature. At 25°C, the control milk showed 4 mPa.s whereas CFHPT milk had 5 mPa.s.

The mechanism underlying the rheological changes of the CFHPT-treated milk products were proposed by Moorman (1997) as the disruption of the casein micelle and protein denaturation due to shear in the constriction of the micrometering valve followed by protein stretching, configuration, and reaggregation, which resulted in effectively bigger protein macromolecules (due to increased level of protein–protein interactions) and, subsequently, a stronger matrix which bound more water and augmented the viscosity of the product. This mechanism for rheological changes was proposed according to the hypothesis for the texture alterations in extruded vegetable protein given by Shen and Morr (1979), which states "high shear first denatures and then stretches and aligns proteins, which results in increased protein-protein interactions."

Skim milk concentrate with 0.24 kg solids/l and treated with CFHPT (0.3 s pressurization dwell and 0 s depressurization holding time at elevated temperature) formed gels when stored at 4°C for 24 h (Moorman 1997). The gels liquefied when warmed to 8°C and were reversible for 7 days. An exponential model (Equation 21.6) was fitted with apparent viscosities (η) versus temperature (T) data same treatment milk ($R^2 = 0.87$) and untreated control milk ($R^2 = 0.99$) with the same solids content:

$$\eta = ae^{bT}, \tag{21.6}$$

where $a = 5.506$, $b = -0.0435$ for untreated samples, and $a = 15.999$, $b = -0.0674$ for the CFHPT-treated sample. Yogurt made from the same treatment of milk exhibited higher mean apparent viscosity than the control. Moorman (1997) proposed that shear is the main factor that contributes to significant changes in the secondary and tertiary protein structure. This conclusion was based on the comparison between the times required for the pressure treatment: HHP (30 min to 1 h) or CFHPT (0.3–1 s), so that gel formation occurred in the product. The time, 0.3–1 s, the product was

TABLE 21.1
Rheological Characteristics of High-Pressure-Throttled Liquid Foods

Liquid Food	Flow Behavior	Apparent Viscosity, η (Pa.s)	Consistency Index, K (Pa.s^n)	Flow Behavior Index, n	Model	Reference
Milk (310 MPa)	Newtonian	0.0014	0.014	Not available	Exponential	(Moorman 1997)
Blueberry-whey beverage (pressure not available)	Newtonian	≈K (at 4°C, 50/s)	0.002 to 0.005 (day 6); 0.004 to 0.006 (day 36)	0.929 to 0.988 (day 6); 0.936 to 1.063 (day 36)	Power law	(Peck 2004)
Clover honey (275 MPa)	Newtonian	Not available (at 4°C)	34.330±1.228 (month 0); to 32.210±1.343 (month 6)	0.965±0.016 (month 0); to 1.010±0.008 (month 6)	Power law	(Areekul 2003)
Clover honey (275 MPa)	Newtonian	Not available (at 30°C)	33.710±3.799 (month 0); to 33.560±2.627 (month 6)	1.009±0.004 (month 0); to 1.008±0.025 (month 6)	Power law	(Areekul 2003)
Buckwheat honey (275 MPa)	Newtonian	Not available (at 4°C)	24.360±3.072 (month 0); to 23.890±2.678 (month 6)	0.924±0.032 (month 0); to 0.924±0.029 (month 6)	Power law	(Areekul 2003)
Buckwheat honey (275 MPa)	Newtonian	Not available (at 30°C)	24.360±3.072 (month 0); to 25.250±4.289 (month 6)	0.924±0.032 (month 0); to 0.963±0.021 (month 6)	Power law	(Areekul 2003)
Soymilk (276 MPa, Megatron milling)	Pseudoplastic	0.008 (at 25°C, 1000/s)–0.023 (at 4°C, 1000/s)	0.009 (at 25°C) to 0.012 (at 4°C)	0.680 (at 25°C) to 0.820 (at 4°C)	Power law	(Sivanandan et al. 2008)

exposed to high pressure in CFHPT was not nearly enough to bring about any significant alterations in secondary and tertiary protein structure.

21.4.2 Citrus Juice

Ascorbic acid, pectin esterase activity, and limonin content in CFHPT-processed (0, 138, 207, and 276 MPa pressures) citrus juice (initial temperature of 20°C) was reported by Amornsin (1999) and the results showed no evidence of ascorbic acid destruction even with the highest pressure treatment. Cloud of CFHPT-processed orange juice was stable at least 21 days after processing at pressure >207 MPa when the juice was held for 38.4 s in a holding tube before cooling. The same study reported no increase in the bitter-tasting compounds represented by limonin content even at the highest pressure level, with 118 s holding time before cooling, as compared to the level present in the freshly squeezed grapefruit juice. The flow behavior of CFHPT-processed citrus juice has not been reported.

21.4.3 Blueberry-Whey Beverage

An exceptional result on particle size reduction and narrowing the distribution has been reported by Peck (2004) where the CFHPT-processed blueberry-whey beverage showed no aggregation of particles during storage in addition to the significantly smaller size offered by the treatment compared with thermally processed beverage. The author reported that some polymers that were formed during processing were dissociated during storage. CFHPT processing of blueberry-whey beverage had only soft feathery sediment (4.6–7.6 μg/ml after 36 days of storage) that was dispersed during slight disturbance in contrast to the thickly packed chalk-like sediment (7.4–9.2 μg/ml after 36 days of storage) that was difficult to agitate even after vigorous agitation in the heat-treated blueberry-whey beverage. This shows the affect of CFHPT treatment on the stability of a product (Peck 2004). Though the particle size of the high-pressure-throttled beverage was significantly smaller than that of the thermally treated one, the soft and feathery sediment packing advantage shown by the high-pressure-throttled beverage was attributed to the particle structure obtained due to high-pressure throttling. The taste of CFHPT-treated blueberry-whey beverage was preferred by sensory panelists (at days 5 and 35 after processing) over the heat-treated beverage, indicating that there is higher retention of flavors in the CFHPT-processed beverage as opposed to the thermally processed product (Peck 2004). At day 65, the thermally treated product showed visible mold growth. Also, the shelf life of the CFHPT beverage was longer, as evidenced by superior product quality on day 35, with no visible spoilage and satisfactory smell and taste to researchers on day 65 after processing, compared to visibly spoiled thermally processed product at that time.

Peck (2004) reported Newtonian behavior for high-pressure-throttled blueberry-whey beverage (Table 21.1). The flow behavior index (n value) ranged from 0.93 to 0.97 (which is very close to 1) for that beverage and did not show any significant difference during storage. Newtonian behavior was exhibited by CFHPT-processed blueberry-whey beverage (after storage too) when the power law model was used to fit the data (Peck 2004). Peck (2004) reported consistency index (K values) ranging from 0.0039 to 0.0054 Pa.s^n (at day 6 after processing) and from 0.0046 to 0.0066 Pa.s^n (for day 36 after processing) for high-pressure-throttled blueberry-whey beverage. Though the author reported significant changes in the K values after storage, a failure to observe any real change in rheological properties was attributed to the low concentration of proteins, pectin, and soluble solids.

21.4.4 Honey

Areekul (2003) conducted research on buckwheat and clover honey processing in a CFHPT system at 276 MPa, exit temperature of 125.3°C and 15 s hold before exiting to a cooler. This resulted in zero microbial counts (at both day 1 and 6 months of storage) and slightly elevated hydroxymethylfurfural (HMF) content, but HMF was at an acceptable level after storage at 6 months at 4°C. In addition to that, there was no significant difference in the physicochemical properties during storage of CFHPT-sterilized and low-temperature heat-pasteurized (conventional pasteurization) honey. Areekul (2003) reported a slight increase in the total solids content (but not significant) in CFHPT-treated honeys over the conventionally pasteurized (CP) honey due to moisture evaporation at the exit after throttling at a temperature of 125.2°C. Other increases exhibited by CFHPT-processed honey were HMF content and consistency index. The author further stated that the physicochemical properties of both CFHPT and CP honeys were dependent on time and temperature of storage. Areekul (2003) reported that CFHPT sterilization of honey (suitable for use in pharmaceutical and infant foods) with a combination of 255 MPa pressure and heating the pressurized fluid to 80°C followed by depressurization to 300 kPa and cooling was found to kill 7 log of inoculated *Bacillus megaterium* spores while producing commercially sterile honey having indistinguishable properties from conventionally heat-pasteurized (60°C, 30 min, nonsterile) honey.

Newtonian behavior (Table 21.1) was shown by CFHPT-processed buckwheat and clover honeys while a higher initial moisture content of buckwheat honey gave lower viscosity than the clover honey (Areekul 2003). Consistency indices of CFHPT-processed honeys (20.5 and 33.8 Pa.s^n for buckwheat and clover honeys) were higher than that of CP honeys due to evaporation at the exit of throttling valve at temperature of 125.2°C (Areekul 2003). The flow behavior of honeys depends on moisture content, the quality and quantity of sugars, colloidal dispersed particles such as protein and polysaccharides, and the amount and size of crystals (Areekul 2003). The rheological behavior and the parameters were considerably constant (Table 21.1) for the high-pressure-throttled honeys after 6 months of storage under storage temperatures of 4 and 30°C (Areekul 2003).

21.4.5 Soymilk

The standard soymilk process utilizes filters or centrifuges to remove the large-sized solids in the comminuted soy and these solids are discarded. Thus, all the essential solids of the whole bean are not transferred into the final product. Commercially, soymilk is produced by thermal treatment to assure safety and extended shelf-life, and to inactivate unwanted biologically active compounds such as trypsin inhibitors and lipoxygenase (Kumar et al. 2003; Liener 1981). The beany flavor of soy in food products is still a constraint for its complete acceptance among American consumers (Torres-Penaranda et al. 1998; Wilson 1996). There were attempts to nullify this effect of lipoxygenase-produced off-flavor in soy food products by producing lipoxygenase-null soybeans and the results showed that cultural differences had an effect on the differences in perception of sensory attributes for the consumption of the soy food products in the sensory studies reported earlier (Wilson 1996; Torres-Penaranda et al. 1998; Kobayashi et al. 1995).

Soymilk from whole dehulled soybeans was produced using microfluidizer-throttling or CFHPT to retain all essential soybean solids (Sivanandan 2007; Sivanandan et al. 2008). Whole dehulled soybeans were blanched, mixed with deionized water, and comminuted coarsely in a food processor. An intermediate comminution step with Megatron (process M), Fitzmill (process F), or Stonemill (process S) was followed by homogenization at selected pressures using microfluidizer-throttling or a CFHPT system. The comminuted slurry was homogenized at treatment pressures of 69, 103, 138, 207, and 276 MPa using a CFHPT system. The soymilk at high pressure was heated to 80°C in a tubular heat exchanger prior to depressurization and a holding tube held the product at elevated temperature after throttling. To avoid flashing, back pressure (350 kPa) was applied after the holding tube and soymilk was cooled immediately. The increase in the CFHPT flow rate significantly affected size reduction of particles of soymilk. An empirical model (Equation 21.7) was established (Sivanandan 2007) for the soymilk processed using both microfluidizer and CFHPT processes, between the particle size, D (μm), pressure, P (MPa), flow rate (for CFHPT), F (l/min), and volume fraction, V (%). The models (Equations 21.7 through 21.9) can be effectively utilized for predicting the particle size diameter (μm) at a particular pressure (MPa) and volume fraction (%) of particles (Sivanandan 2007). Each of these models obtained $R^2 = 0.97$.

Megatron-CFHPT model:

$$\ln D = 3.16 + a(P^2) + b(F^2) - c(FP) - d(P) + e(V) - f(F), \tag{21.7}$$

where $a = 1.12\times10^{-5}$, $b = 0.27$, $c = 2.52\times10^{-4}$, $d = 6.42\times10^{-3}$, $e = 2.17\times10^{-2}$, and $f = 0.68$.

Megatron-microfluidizer model:

$$\ln D = 2.38 + a(P^2) - b(PV) - c(P) + d(V), \tag{21.8}$$

where $a = 8.87\times10^{-6}$, $b = 1.33\times10^{-5}$, $c = 4.12\times10^{-3}$, and $d = 2.35\times10^{-2}$.

Fitzmill-microfluidizer model:

$$\ln D = 2.34 + a(P^2) - b(V^2) - c(PV) - d(P) + e(V), \tag{21.9}$$

where $a = 1.04\times10^{-5}$, $b = 1.66\times10^{-5}$, $c = 2.27\times10^{-5}$, $d = 4.20\times10^{-3}$, and $e = 2.67\times10^{-2}$.

A consumer acceptability test showed that more research is needed to make a soymilk (with added flavors) that appeals to the taste of the American consumer before the CFHPT process can be used commercially to produce soymilk (Sivanandan 2007). Thus, soymilk with all the essential solids can be made available to the public and the processors benefit from high processing yields since none of the essential solids of the beans are discarded. Also, the study results reinstated the need to use consumers from the targeted market as an essential tool for the success of product development (Hollingsworth 1998).

Particle size diameter of soymilks decreased significantly with the increase of applied pressure for both microfluidizer-throttling valve and CFHPT treatments (Sivanandan 2007). Application of higher pressures reduced the particle size and narrowed down the particle size distribution in both microfluidizer-throttling valve treatments and CFHPT treatments. The effect of flow rate was significant ($p < .0001$) in CFHPT when the flow rate increased from 0.75 to 1.50 l/min (Table 21.1) and the particle size reduced with increase in flow rate. This reduction in particle size with flow rate was attributed to the increase in shear and turbulence at the micrometering valve exit. An interaction of the factors, pressure and flow rate, was also significant ($p < .05$) in CFHPT, which showed the importance of adjusting the flow rate also, at a particular pressure treatment, to obtain desired particle size of soymilk. The particle size reduction observed in both high-pressure-throttling machines was attributed to the weakening of membrane in individual particles due to the high pressure. The weakened membranes easily ruptured from shear in the throttling valve and the narrow particle size distribution was attributed to the restricted opening for fluid flow in the micrometering valve. The cavitation that occurred after depressurization aided in further reducing particle size, and the turbulence at depressurization helped in mixing up the ruptured particles and distributing it evenly within the fluid matrix. The results showed that the high-pressure throttling helped to decrease the size of solid particles and made the particle size more homogeneous.

Cryogenic scanning electron microscope (cryo-SEM) images contributed to the identification of the network structure of soymilk formed as a result of pressurization. The confocal scanning laser microscope (CSLM) images and cryo-SEM images helped to identify the best soymilk from the CFHPT-treated soymilks. The continuous network seen in cryo-SEM images was attributed to the rearrangement of particles and formation of network structure due to CFHPT treatment at the highest pressure. The theory of entrapped fat globules in the protein network was obvious in the images, which showed the distribution of fat and protein in soymilk. The protein particles could not be identified individually due to the very small particle size and the even distribution of the protein network as observed in cryo-SEM images. The fat globules were also very small in size. The images showed that the fat particles are entrapped in the network (which were obvious in cryo-SEM and CSLM images) and the size of the fat particles was very small (~5 μm) at the highest pressure applied. However, images of soymilk at the lowest pressure showed separated and larger fat globules with uneven distribution. This was in agreement with the cryo-SEM images of the same treatments. Ultrastructural images elucidated particle microstructure in the soymilk and homogeneity of suspended particles. The very small fat globules at the highest CFHPT pressure treatment with Megatron comminution were seen entrapped in the network and were uniformly distributed. Thus the combination of Megatron and CFHPT treatment at the highest CFHPT pressure was considered the best treatment. Therefore, the high-pressure-throttling process will allow utilization of the whole soybean to produce excellent-quality soymilk with high emulsion stability.

The CFHPT-treated soymilk (with Megatron pretreatment during milling) at the highest pressure (276 MPa) showed the smallest particle size and the highest apparent viscosity (Table 21.1) (Sivanandan et al. 2008). All soymilk samples showed pseudoplastic flow behavior (flow behavior index, $n < 1$). Apparent viscosity significantly increased ($p < .05$) with increase in pressure levels while it decreased significantly ($p < .05$) with increase in temperature of analysis (4, 10, and 25°C). The importance of the type of milling procedure (Megatron, Fitzmill, or Stonemill) used (pretreatment) and its influence on particle size reduction together with the pressure of CFHPT processing was significant. The smallest particle size and highest apparent viscosity of the soymilks were attributed to the increased number and surface volume of particles, which added to the resistance to flow. The homogenization that occurred during shear, turbulence, and cavitation helped in generating new fat globules of smaller size and dispersed them uniformly in the protein matrix, thereby increasing the apparent viscosity of the fluid (Sharma and Dalgleish 1993; Rowney et al. 2003; Cano-Ruiz and Richter 1997). More particle size reduction occurred with increased applied pressure and this reduced the availability of free fat globules as it was coated with protein aggregates to stabilize the expanded surface area and thus increased the apparent viscosity and thereby improve emulsification of soymilk with the pressurization process (Figure 21.2) (Sivanandan 2007). The authors (Sivanandan et al. 2008) further suggest that the higher apparent viscosity obtained in soymilk showed that no additives are needed to increase the "body" sensory perception of the beverage for improved mouthfeel of the product.

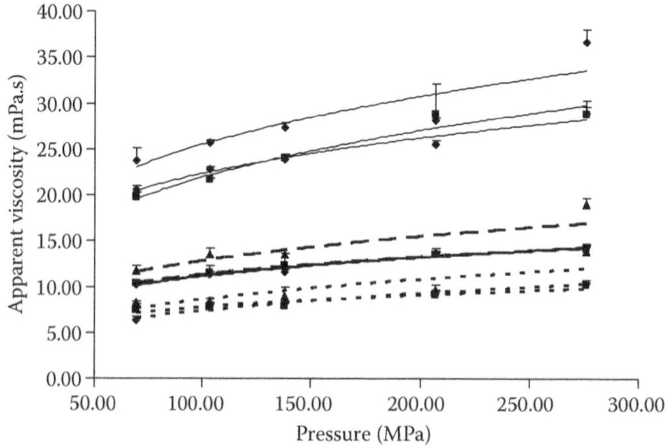

FIGURE 21.2 Apparent viscosity of soymilk analyzed at different temperatures. At 25°C: process M (- - ▲ - -), $R^2 = 0.73$, process F (- - ■ - -), $R^2 = 0.89$, process S (- - ◆ - -), $R^2 = 0.97$; at 10°C: process M (— — ▲ — —), $R^2 = 0.73$, process F (— — ■ — —), $R^2 = 0.99$, process S (— — ◆ — —), $R^2 = 0.97$; at 4°C: process M (—— ▲——), $R^2 = 0.83$, process F (—— ■——), $R^2 = 0.96$, process S (—— ◆——), $R^2 = 0.97$.

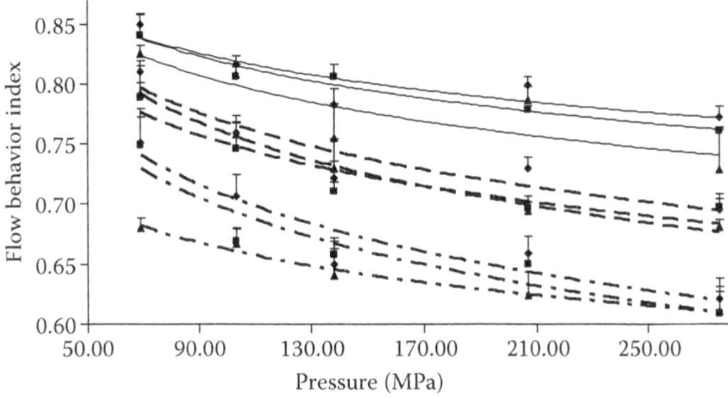

FIGURE 21.3 Flow behavior index of soymilk analyzed at different temperatures. At 25°C: process M (– · – ▲ – · –), $R^2 = 0.98$, process F (– · – ■ – · –), $R^2 = 0.88$, process S (– · – ◆ – · –), $R^2 = 0.89$; at 10°C: process M (- - ▲ - -), $R^2 = 0.99$, process F (- - ■ - -), $R^2 = 0.89$, process S (- - ◆ - -), $R^2 = 0.87$; at 4°C: process M (——▲——), $R^2 = 0.70$, process F (——■——), $R^2 = 0.96$, process S (——◆——), $R^2 = 0.78$.

Soymilks obtained from all the processes showed pseudoplastic ($n < 1$) flow behavior when the power law model was used to fit the data (Table 21.1). The pseudoplastic flow behavior is valid for industrial applications as it explains pumping of the beverage at high speed requiring less force due to lowered viscosity at higher shear rates (Forster and Ferrier 1979; Son and Singh 1998). The flow behavior index (n) obtained for different soymilk treatments analyzed at different temperatures is shown in Figure 21.3 (Sivanandan 2007). Relatively constant n value (in the range 0.63–0.83) with temperature (Sivanandan et al. 2008) showed that the soymilk did not undergo chemical changes at these temperatures (Toledo 2007). The authors reported emulsion stability in CFHPT-treated soymilks (at 276 and 207 MPa) due to high-pressure treatment, where no visible rheological change was observed after storage at 4°C for one month. Consistency index, K, for the CFHPT-treated soymilks (Table 21.1) increased significantly with applied pressure (Sivanandan et al. 2008). This was similar to the trend in apparent viscosity shown by the soymilks (Figure 21.2). An Arrhenius model was established between K and inverse of absolute temperature (T) and also between apparent viscosity and inverse of T. Apparent viscosity changes in the soymilk over time were determined at a constant shear rate (1000/s) at 10°C. Apparent viscosity decreased in the beginning, showing that the soymilk was thixotropic (shear-thinning), and it became relatively constant later on when the shear rate was kept constant (Sivanandan et al. 2008). The authors suggested that thixotropic behavior showed the beverage's resistance to settling, enhancing its shelf life and improving mouthfeel.

21.5 CONCLUSIONS

CFHPT processing helped to decrease the size of liquid food particles and disperse it evenly and thereby improve the rheological and functional properties of soymilk to prevent settling of solids during storage. The flavor of the liquid food was maintained or improved with CFHPT processing. The successful results of improvements in quality, physicochemical properties, and shelf life of the liquid foods with CFHPT processing open the door for its extensive applications in the food industry, especially to retain the natural flavor characteristics of liquid foods made from fruits and vegetables.

ACKNOWLEDGMENTS

The soymilk project reported in this chapter was supported by National Research Initiative Grant no. 2005-35503-15374 from the USDA Cooperative State Research, Education, and Extension Service NCGP program. The EndNote® reference used in this chapter was supported by Diana Hartle, Reference Instruction Librarian, Reference Department, Science Library, The University of Georgia, Athens, Georgia.

REFERENCES

Adapa, S., Schmidt, K. A. and Toledo, R. T. 1997. Functional properties of skim milk processed with continuous high pressure throttling. *Journal of Dairy Science,* 80:1941–48.

Amornsin, A. 1999. Effect of high pressure throttling on ascorbic acid, pectin esterase activity and limonin content in citrus juice. Unpublished thesis, University of Georgia, United States.

Areekul, V. 2003. High pressure sterilization of honey: Physicochemical changes, sensory attributes and shelf life. Unpublished thesis, University of Georgia, United States.

Cano-Ruiz, M. E. and Richter, R. L. 1997. Effect of homogenization pressure on the milk fat globule membrane proteins. *Journal of Dairy Science,* 80:2732–39.

Cobos, A., Horne, D. S. and Muir, D. D. 1995. Rheological properties of acid milk gels 2: Effect of composition, process and acidification conditions on products from recombined milks using the microfluidizer. *Milchwissenschaft,* 50:603–606.

Cook, E. J. and Lagacé, A. P., inventors. 1985. Apparatus for forming emulsions. US patent No. 4,533,254.

Floury, J., Desrumaux, A. and Lardieres, J. 2000. Effect of high-pressure homogenization on droplet size distributions and rheological properties of model oil-in-water emulsions. *Innovative Food Science & Emerging Technologies,* 1:127–34.

Forster, L. L. and Ferrier, L. K. 1979. Viscometric characteristics of whole soybean milk, *Journal of Food Science* 44:583–85, 590.

Hayashi, R., Kawamura, Y., Nakasa, T. and Okinaka, O. 1989. Application of high pressure to food processing: Pressurization of egg white and yolk, and properties of gels formed. *Agricultural and Biological Chemistry,* 53:2935–2939.

Hollingsworth, P. 1998. Sensory testing rediscovered as key to new product success. *Food Technology,* 52:26–27.

Hoover, D. G. 1993. Pressure effects on biological systems. *Food Technology,* 47:150–55.

Knorr, D. 1993. Effects of high-hydrostatic-pressure processes on food safety and quality. *Food Technology,* 47:156, 158–161.

Kumar, V., Rani, A., Tindwani, C. and Jain, M. 2003. Lipoxygenase isozymes and trypsin inhibitor activities in soybean as influenced by growing locations. *Food Chemistry,* 83:79–83.

Lemay, A., Paquin, P. and Lacroix, C. 1994. Influence of microfluidization of milk on cheddar cheese composition, color, texture, and yield. *Journal of Dairy Science,* 77:2870–79.

Liener, I. 1981. Factors affecting the nutritional quality of soya products. *Journal of the American Oil Chemists' Society,* 58:406–15.

McCrae, C. H. 1994. Homogenizations of milk emulsions: Use of Microfluidizer. *Journal of Society of Dairy Technology,* 47:28–31.

Molina, E., Papadopoulou, A. and Ledward, D. A. 2001. Emulsifying properties of high pressure treated soy protein isolate and 7S and 11S globulins. *Food Hydrocolloids,* 15:263–69.

Moorman, Jr, J. E. 1997. Microbicidal and rheological effects of high-pressure throttling. Unpublished thesis, University of Georgia, United States.

Paquin, P. 1999. Technological properties of high pressure homogenizers: The effect of fat globules, milk proteins, and polysaccharides. *International Dairy Journal,* 9:329–35.

Peck, D. C. 2004. The effects of high-pressure throttling versus thermal pasteurization on blueberry-whey beverage. Unpublished thesis, University of Georgia, United States.

Rowney, M. K., Hickey, M. W., Roupas, P. and Everett, D. W. 2003. The effect of homogenization and milk fat fractions on the functionality of mozzarella cheese. *Journal of Dairy Science,* 86:712–18.

Sharma, S. K. and Dalgleish, D. G. 1993. Interactions between milk serum proteins and synthetic fat globule membrane during heating of homogenized whole milk. *Journal of Agricultural and Food Chemistry,* 41:1407–12.

Shen, J. and Morr, C. 1979. Physicochemical aspects of texturization: Fiber formation from globular proteins. *Journal of the American Oil Chemists' Society,* 56:A63–70.

Sivanandan, L. 2007. Characterization of soymilk produced by continuous flow high pressure throttling process. Unpublished thesis, University of Georgia, United States.

Sivanandan, L., Toledo, R. T. and Singh, R. K. 2008. Effect of continuous flow high pressure throttling on rheological and ultrastructural properties of soymilk. *Journal of Food Science,* 73 (6): E288–96.

Son, S. M. and Singh, R. K. 1998. Rheological behavior of aseptically processed soybean milk under turbulent flow conditions. *International Journal of Food Properties* 1:57–70.

Thiebaud, M., Dumay, E., Picart, L., Guiraud, J. P. and Cheftel, J. C. 2003. High-pressure homogenisation of raw bovine milk. Effects on fat globule size distribution and microbial inactivation. *International Dairy Journal,* 13:427–39.

Toledo, R. T. 2007. *Fundamentals of Food Process Engineering,* 3rd edn. New York: Springer.

Toledo, R. T. and Moorman, Jr, J. E., inventors; University of Georgia, assignee. 2000. Microbial inactivation by high-pressure throttling. US patent No. 6,120,732.

Torres-Penaranda, A. V., Reitmeier, C. A., Wilson, L. A., Fehr, W. R. and Narvel, J. M. 1998. Sensory characteristics of soymilk and tofu made from lipoxygenase-free and normal soybeans. *Journal of Food Science,* 63:1084–87.

Tunick, M. H., Van Hekken, D. L., Cooke, P. H. and Malin, E. L. 2002. Transmission electron microscopy of mozzarella cheeses made from microfluidized milk. *Journal of Agricultural and Food Chemistry,* 50:99–103.

Tunick, M. H., Van Hekken, D. L., Cooke, P. H., Smith, P. W. and Malin, E. L. 2000. Effect of high pressure microfluidization on microstructure of mozzarella cheese. *Lebensmittel-Wissenschaft und Technologie,* 33:538–44.

Van Hekken, D. L., Tunick, M. H., Malin, E. L. and Holsinger, V. H. 2007. Rheology and melt characterization of low-fat and full fat mozzarella cheese made from microfluidized milk. *Lebensmittel-Wissenschaft und Technologie,* 40:89–98.

Vuillemard, J. C. 1991. Recent advances in the large-scale production of lipid vesicles for use in food products: Microfluidization. *Journal of Microencapsulation,* 8:547–62.

Wilson, L. A. 1996. Comparison of lipoxygenase-null and lipoxygenase containing soybeans for foods. In G. Piazza (ed.) *Lipoxygenase Enzymes and Lipoxygenase Pathway Enzymes*, 209–25. Champaign, IL: AOCS Press.

Zamora, A., Ferragut, V., Jaramillo, P. D., Guamis, B. and Trujillo, A. J. 2007. Effects of ultra-high pressure homogenization on the cheese-making properties of milk. *Journal of Dairy Science,* 90:13–23.

Zipp, A. and Kauzmann, A. 1973. Pressure denaturation of metmyoglobin. *Biochemistry,* 12:4217–4228.

[illegible]

22 Food Frying: Modifying the Functional Properties of Batters

Michael Ngadi
McGill University

Jun Xue
Agriculture and Agri-Food Canada

CONTENTS

22.1 INTRODUCTION

Deep fat frying is one of the oldest processing techniques. It is convenient and can be used to cook a wide variety of foods including seafood, potato chips, French fries, extruded snacks, fish sticks, and chicken nuggets. Its popularity stems from the unique taste and flavor of fried foods that are difficult to replicate by other methods of cooking. Thus, large quantities of fried foods are produced and consumed each year around the world. Although frying oils contribute to development of flavor and incorporate some essential fatty acids in fried foods, they also increase the caloric content of the foods. As these oils are exposed to high temperature in open air, they are subject to thermal and oxidative reactions. These degenerative reactions affect the viscosity of the frying medium over time and could be sources of antinutritional elements in fried foods. Further, the high oil contents of fried foods has become a public health concern due to links between high consumption of fat and cardiovascular diseases, obesity, colon cancer, and other disorders. The National Heart Lung and

Blood Institute (NHLBI) recommend that the caloric intake of fats and oils should not exceed 30% of daily energy needs.

Frying has remained a major cooking technique and fried foods have remained popular in spite of the health concerns. However, there is currently a major effort to deal with the health issues of frying and to develop novel processing technologies to reduce fat content of fried foods without sacrificing their taste and flavor. Some of the new trends include development of new oils or selection of oils with improved stability during heating (Chitrakar and Zhang 2006; Naz et al. 2004), development of oils with reduced trans fatty acids (Li, Ngadi and Oluka 2008; Ngadi, Li and Oluka 2007), and designing new strategies for reducing oil absorption during frying. A breading and batter-coating system, as a novel processing technique for reduction of oil absorption during deep fat frying, is attracting increasing attention.

Batter and breading systems are traditionally used in the food industry to add value to products by modifying their texture, flavor, weight, and volume. They also provide many other opportunities such as reducing oil absorption during deep-frying (Mohamed, Hamid and Hamid 1998; Fiszman and Salvador 2003) or improvement in freeze/thaw and shelf-life stability. Batter system is basically a liquid mixture comprising of water, flour, starch, flavoring, and seasonings into which food products are dipped prior to cooking. With increasing consumer consciousness and willingness to pay more for improved quality and specific attributes in food products, the use of batter coatings have become an important resource in food formulations. Thus, there has been recent rapid growth in the batter and breading industry along with many technological advances and breakthroughs in batter formulation and breading manufacture. However, applications of batters in the food industry are still done largely empirically with reliance on in-house experiences.

Batter can be classified into one of the two categories namely interface/adhesion batters or puff/tempura batters (Loewe 1993). An adhesive batter is typically used with a supplemental breading, and it serves primarily as an adhesive layer between the food surface and the breading. Chemical leavening is not normally used in an adhesive batter. On the other hand, a tempura batter is chemically leavened, and itself can serve as the outside coating of the food. Both wheat and corn (WC) flours play important roles in tempura batters. Batters are highly complex systems in which the nature of the ingredients is very wide-ranging, and interactions between ingredients determine the performance quality of the final product. However, the functionalities of ingredients in terms of their influence on thermal and rheological properties of batter systems are not yet fully understood. These functionalities directly influence process conditions as well as the quality of coated products. For instance, the use of one type of flour or combination of flours may provide special effects on the thermal and rheological properties of batters (Mukprasirt, Herald and Flores 2000; Dogan, Sahin and Sumnu 2004; Xue and Ngadi 2006, 2009). Therefore, in order to formulate appropriate batter-breading coatings and appreciate their functionalities in a variety of products, it is vital to investigate these properties.

Rheological properties are among the most important physical properties that define process attributes of batters. The viscosity of batter affects the quantity and quality of batter pick-up, appearance, texture, and the handling property of a coated product (Mukprasirt et al. 2000). The quality of a coated product is influenced by changes in batter-coating properties associated with the structural phase transition from the liquid to the solid state. The flow behavior and dynamic viscoelastic properties of a batter provide information that characterizes the influence of temperature, water content, and various ingredients on the coating's structural behavior during processing (Steffe 1996). Thus, rheological data are vital for any coating application.

The thermal properties and phase transition characteristics of batter systems change during cooking (e.g., deep fat frying) and freezing (e.g., frozen storage). These have combined effects on the overall characteristics of the finished coated foods. Several chemical and physical changes such as starch gelatinization and water evaporation occur during heating. The thermal properties (heat capacity, enthalpy, onset temperature, and gelatinization temperature) reflect those changes, which directly influence the texture characteristics and quality of finished coated products. Thermal properties are critical in defining heat and mass transfer behaviors of coated product during heating (frying, baking, etc.) and storage. They are important parameters in predicting the effect of freezing and thawing on coated product adhesion and quality. It is necessary to understand phase transitions, including glass transitions, occurring in coatings at subzero temperatures during storage, since glass transition may control rates of recrystallization of ice and diffusion-controlled reactions (Slade and Levine 1995; Lee and Brandt 2002). Consequently, different batter formulations and their components may affect cooking quality and frozen state stability through their effects on thermal and phase state properties. Processors are interested in further developing processed/coated products as a means of providing foods with high nutritional quality and lower fat contents. Understanding phase and state transition of a batter system may hold the key to improving coating performance during cooking and maintaining overall product quality during storage.

Many food ingredients can be used to improve functionality (e.g., viscosity, crispness, distinctive flavor, crunchy texture) in batters and coatings. Hydrocolloids represent a category of functional ingredients that have been incorporated into batter formulations, and have been shown to be effective in improving batter performances and reducing oil absorption in fried coated foods (Meyers 1990; Hsia, Smith and Steffe 1992; Balasubramaniam et al. 1997; Annapure, Sighal and Kulkarni 1999; Holownia et al. 2000; Mellema 2003; Sanz, Salvador and Fiszman 2004a).

Hydrocolloids are water-soluble polymers, generally carbohydrates, with the ability to thicken and/or gel aqueous systems. Each type of hydrocolloid has a different contribution to the batter system due to their different chemical structures resulting from different degrees of substitution with different branch compounds. However, they play an important role on viscosity, film-forming, and barrier properties in batter system (Meyers 1990; Albert and Mittal 2002). Hydrocolloids as functional ingredients may change the rheological, texture, and thermal properties of batter systems and thus influence the final quality of the end products. In addition, hydrocolloids and other ingredients may cause interaction resulting in synergistic effects on the properties of batter systems during processing (Ferrero and Zaritzky 2000; Sanz et al. 2005a; Kim and Yoo 2006).

Although the majority of fried food products are batter and breading coated, formulation of batter systems is slowly evolving from an art into a science. In the last decade, research on batter systems has focused mainly on comparing their functionality with respect to fat absorption. Not much work has been done on understanding their fundamental properties such as rheological properties and thermal transition behavior and properties, especially for different combinations of flour blends containing hydrocolloids. Moreover, understanding how these properties change with the formation of flour-hydrocolloids batter systems during the heating and cooling process, and the interaction between hydrocolloids, flour base, and other ingredients, will be invaluable for optimizing processing and storage conditions as well as in selecting and formulating appropriate batter systems for various types of battered products, thus resulting in nutritional foods for healthier living. An overview of batter systems is provided in this chapter. The role of ingredients, especially flour and hydrocolloids, on the rheological and thermal functionalities of batter are discussed.

22.2 BATTER SYSTEMS

Batters are the common and vital component of most successful food coatings. Batters link the fundamentals of food science with product needs. Proper selection and application are vital for optimized impact on flavor, appearance, eating characteristics, performance, and cost effectiveness of the final coated product. In addition to the obvious benefits of taste, color, and texture, batter systems also improve the yield and keeping qualities to withstand the demands of processing, distribution, and final preparation for the table. In general terms, batter can be defined as liquid dough, being a thick but pourable mixture, into which a product is dipped before it is breaded or fried. Loewe (1993) classified batter systems into two broad categories: interface/adhesion and puff/tempura. Normally, interface or adhesion batters serve as "glue" and are used with an added breading, serving primarily as the adhesive layer between the product's surface and the breading. The interface/adhesion batter's main function is to provide a base so the bread crumbs, when used, will adhere to a product. Adhesive batters gained their name for a very simple reason—like Mom's egg and milk dip, they provide an adhesive layer between substrate and outer breading layer. Batters can be formulated with wheat flour, corn flour, or starch.

Adhesive batters characteristically have low to medium viscosity, intended mostly to achieve breading adhesion. Typically, they have high starch content and are quite thin. The heavier the batter, the more crumb will adhere. They may carry flavors and the thicker batters can be used as stand-alone coatings. Starch-based batters, also considered adhesion batters, act as the glue in a coating system. They are usually thin and need to be stirred frequently to keep the starch in suspension. Starch-based batters bind themselves quite strongly to protein-based products and can be used as an outer coating on the products in order to increase their crispiness and improve holding time. Flour-based batters can also serve as a product's final outer coating without breading. Since these batters are thicker than starch-based batters, they can adhere coarser breadcrumb that may not stick to a thin, starch-based batter. A flour batter can also contain leavening agents that will make the batter more "puffy" when fried. Examples of this are tempura and beer batters. Both starch-based and flour-based batters can be modified with a wide array of flavorings and spices to obtain the desired sensory profile. Viscosity is very important in regulating the amount of pickup or thickness of batter, depending on the flour/water ratio and the batter temperature.

Tempura batter, which includes raising or leavening agents to generate gas and "puff" the product, has a high viscosity to provide a thick coating with minimum requirement for mixing or pumping in order to prevent loss of the leavening gas. Cooked immediately, it is designed to brown and expand the batter (w/gas) into an open, honeycomb-like texture. No breading is required to form a thick coating on the product when a tempura-type batter is used.

Batter and breading act as a moisture barrier, and provide a promising route to develop flavor and to reduce fat uptake (Fiszman and Salvador 2003). However, both the convenience food industry and consumers increasingly desire more sophisticated foods and variety with more juiciness and less fat.

22.3 FUNCTIONAL INGREDIENTS

Ingredients serve numerous important functions in batter systems to give the coatings their unique characteristics and functionalities. The selection of appropriate ingredients directly influences the quality of the finished products.

22.3.1 Flours

Flour is the key ingredient in batter and breading systems. Some flours such as wheat flour provide viscosity and may promote adhesion due to the presence of gluten. Gluten provides structure and texture and can act as a barrier to fat absorption. Flour contains some reducing sugars that caramelize during frying, contributing to the color and flavor of the coating (Mohamed et al. 1998). It is also the main component of most breadings. It can be used as is or first baked into crumbs. The porosity of the resulting products affects oil absorption—the more porous the material, the more oil is absorbed.

Wheat flour is the most common flour used in batters and breadings (Loewe 1993). However, rice, corn, soy, malted barley, and potato flours have also been used. Wheat flour with higher protein levels will increase batter viscosity and produce darker, crisper fried foods. Corn flour generally produces a yellowish color due to the carotene pigment in corn. It also serves as a source of natural yellow color in order to reduce the influence of sugars and milk powder in batter (Salvador, Sanz and Fiszman 2003). Corn flour is added more often for viscosity control as the higher starch level affects the batter's ability to absorb water. Corn flour in batter systems increases crispness and decreases high levels of puffing due to decreased moisture retention in the coating (Salvador,

Sanz and Fiszman 2002). Corn contains higher crude lipids than wheat and rice (WR) flour (USDA Handbook # 8). Suderman (1996) reported that a corn-starch-based batter required continuous mixing during processing because the solids tends to settle out easily, resulting in changes in batter viscosity throughout the production period, leading to nonuniform batter pickup by substrates. Therefore, corn-flour-based batter requires the addition of a thickener to keep the solids in suspension in order to solve this problem.

Rice flour has the potential to serve as an alternative to wheat flour in battered and breaded foods. Rice-flour-based batter might be a commercially feasible new product in the food industry. Shih and Daigle (1999) and Mukprasirt et al. (2000) found that proteins and starch in rice flour are chemically different from those in wheat flour. They reported that rice flour resisted oil absorption better but was less effective as a thickening agent than wheat flour. Their results showed a 69% oil reduction with rice flour batter on shrimp products. A high ratio of rice flour provided roughness to the crust (Shih and Daigle 1999). However, rice-flour-based batters form thin slurries and require additives to develop viscosity and other desirable batter properties. A good strategy is to use rice-based thickening agents as additives. For example, gelatinized long-grain rice flour and phosphorylated long-grain rice starch ester can be effective in enhancing the batter viscosity and the oil-lowering properties of rice flour batters. Mukprasirt et al. (2000) also studied the effects of ingredients used in a rice flour base on the adhesion characteristics for deep fat-fried chicken drumsticks. The authors found that batter formulated with a 50:50 mixture of rice and corn flours adhered better to drumsticks than did batter with other rice flour ratios (30:70, 70:30—rice flour: corn flour). As rice flour ratio increased from 50 to 70%, the binding force decreased. Thus, combining different flours may provide special effects and produce desired characteristics in the coated products. The authors further suggested that rice flour should be combined with other ingredients. For example, methylcellulose (MC), oxidized starch, and xanthan gum (XG) increased the amount of batter pick-up before frying by increasing viscosity and achieved finished products with lower fat content.

22.3.2 Starch

Starches have traditionally been used for adhesion in batters. The two main components of starch are amylopectin and amylose. Starches with higher amylose content are generally selected for better film-forming properties. They produce a crisper, stronger film, which stays intact through the fryer. Oxidized starches are used for their basic adhesion and coating, while high-amylose starches help reduce fat pickup. Batters typically contain starch at the levels of 5–30% of the dry mixture (Mukprasirt et al. 2001).

Several specialty ingredients for batters and breadings have been developed. One new development in the starch line is dextrins (Shinsato, Hippleheuser, and Van Beirendonck 1999), which have superior film-forming properties. These products also can increase shelf life on a foodservice line. Since fried items typically sit for 15 min or more under heat lamps, the challenge is to keep them just as crunchy as the minute they came out of the fryer. Special hydrophobic starches with less water affinity can also improve adhesion and crispness. A variety of starches can be used in batters and breadings. These include common corn starch, potato starch, wheat starch, tapioca starch, and high-amylose corn starch. The amylose portion improves film forming. Waxy corn starch is not used extensively in batters and breadings because of its high amylopectin content. A modified starch also can add freeze/thaw stability to a par-fried product. However, with most frozen battered items, the starch is not cooked until the product is fried, so it has little to no contribution during frozen storage.

22.3.3 Protein

Protein assists in structural development or changes in texture of the final coated product. Proteins might be added to batters at a level of 10–15% (Robert 1990). Studies show that products with higher protein contents are generally more effective as binding agents. Protein has been used to improve the water absorption capacity of flour, which in turn increases the viscosity of the system (Hoseney 1994a). It is also used to strengthen the structure and texture, retard moisture loss, and enhance crust color and flavor development. The level of flour protein used had a major effect on batter pickup, ranging from 11 to 28% when measured at equal water-solids ratios (Loewe 1993). In general, a higher level of protein increased crispness of the fried product and produced a darker color. As the protein level increased, there was a gradual increase in roughness of texture and brittleness of the fried coating. The pancake-like inner structure was no longer present in the high-protein (12.1%) flour coatings (Loewe 1993).

22.3.4 Chemical Leavening

Typical leavenings used in batters include sodium acid pyrophosphate (SAPP), sodium aluminum phosphate (SALP), and combinations of SALP and monocalcium phosphate (MCP) (Dubois 1981). The gas-release characteristic of the specific leavening affects the texture. If the release is too early, the product texture will be coarse and the coating will absorb excess oil (Loewe 1993). Additional leavening may be able to change the color and texture of a fried product, for example a corn-dog (a sausage) coating needs a leavening system that releases gas very rapidly so that the coating can expand very quickly. The batter becomes more brittle as the amount of leavening is increased (Dubois 1981). The leavening system can be tailored for a specific application by varying the type of leavening acids incorporated. The amount of gas generated and its rate of production determine the effect of leavening in the batter.

22.3.5 Shortening and Oil

Shortening plays a key role in the mouthfeel or eating quality of battered and breaded foods. Shortening and oil have specific functions in coatings, such as being carriers of fat-soluble vitamins and contributing to food flavor and palatability as well as to the feeling of satiety after eating. Other fatty materials with potential use in batters and breading include emulsifiers and staling inhibitors. The melting point and solids content are functions of the source oil selected for the frying shortening. Proper selection of shortening is important to assure the quality of coated products as it affects flavor, eating quality, nutrition, solid/liquid form, and fry life economics (Crosby and Kincs 1990). Ang (1993) reported that cellulose had a greater effect in reducing the fat content of fried batter coatings in shortening. There was a decrease in fat and an increase in moisture when 1% powdered cellulose (fiber length in excess of 100 μm) was incorporated into the batter. This was attributed to formation of hydrogen bonds between water molecules and cellulose fibers.

22.3.6 Egg and Milk

Eggs are used widely, both as a batter ingredient and as pre-dips. Egg contains albumin, a heat-coagulable protein that is useful in binding the breading/batter to the substrate. The egg white's protein improves adhesion, whereas the yolk's phospholipids provide increased emulsification. Egg whites may create some microbiological issues, especially if the product sits on a line at room temperature of 21–32°C (70–90°F). The addition of eggs to a batter tends to darken the final product (Loewe 1993). Dairy ingredients contribute flavor, adhesion, and color. Typical pre-dips include milk, evaporated milk, and buttermilk. Some producers select buttermilk for its unique flavor profile in many coating food recipes, including "rock shrimp cones," "crispy fried chicken," and "fried okra." Mohamed et al. (1998) reported that amongst the proteins (egg yolk, skimmed milk, and ovalbumin) studied, ovalbumin was able to reduce oil absorption and improve the crispness of the fried batter.

22.3.7 Flavoring and Seasoning

One way to improve the value-added perception of battered and breaded products is to incorporate additional flavorings and seasonings. The amount of seasoning in batters varies considerably although the average is in the range of 3–5% of the batter mix. Spices and herbs are dry ground plant materials that possess a characteristic taste and contain many aromatic and flavor constituents. They are usually used at a rate of 0.5–1.0% in finished food products (Suderman 1996). At the same time, spices may contribute specks and colors that are unacceptable (i.e., paprika). Also, it takes time to reach flavor equilibrium with their medium (Suderman 1996). Essential oils offer many advantages including high flavor concentration, no off-colors or specks, instant flavor equilibration, and easy blending and quality control. Essential oils are normally used in food products at levels varying from 0.01 to 0.10% of the finished weight.

Flavor components are released from the coated food (substrate, predust, and batter-breading complex). All flavor components have a volatility spectrum, and it is conceivable that some flavor components are steam stripped from the foods as water converts to steam and exits the food surface. Flavor components are also imparted to the coated food product by the frying oil or Heat-stable flavors contained in the frying oil. In the commercial batter and breading mix, garlic is one of the popular ingredients that improve the flavor of the product (Suderman 1996). From a food safety standpoint it should be kept in mind that spices are an occasional source of microbial contamination.

Another factor to consider when selecting the correct batter or breading is the specific coating objectives for the finished products, including texture, crispness, color, flavor, appearance, functionality, cooking characteristics (i.e., baking, frying, microwaving, or convection oven preparation), compatibility, interaction with other ingredient in the system, and other special conditions. It is essential to know these parameters because most flavor development work is supplemental to the basic development of batters. However, in some situations, structural changes in batters and breading are necessary to achieve highly technical flavor development objectives.

22.3.8 Hydrocolloids

Hydrocolloids are high-molecular-weight water-soluble carbohydrate biopolymers with the ability to form gel or thicken aqueous systems. They generally contain mostly hydroxyl groups linked in different configurations. Hydrocolloids can be categorized according to their origin, isolation or derivation methods, major functionality, and the presence of ionic charges. From a structural point of view, hydrocolloids can be classified as linear (e.g., cellulose, amylose), substituted linear (e.g., guar gum, MC), or branch-on-branch (e.g., Arabic gum, amylopectin). Their structural conformation and degree of substitution also can play an important role to provide different functional properties to food products (Meyers 1990). Gum and starch are the most important types of hydrocolloids (Meyers 1990).

Hydrocolloids are currently widely used in a variety of industrial sectors to perform a number of functions including thickening and gelling aqueous solutions, stabilizing foams, emulsions and dispersions, inhibiting ice and sugar crystal formation, and the controlled release of flavors. Recently, they have been studied as ingredients in the formulation of batters to cover pieces of food that are to be fried. Traditionally, the primary use of hydrocolloids for this application has been based on their capacity to immobilize water and control viscosity of batters (Fiszman, Salvador and Sanz 2005; Xue and Ngadi 2007a). Carboxymethyl cellulose (CMC) has been used to increase moisture retention and control rheological properties of cereal batters and doughs, and protect against leavening losses in cake mixes, improve the volume and structural uniformity of baked products, and increase the shelf life of cereal products (Dziezak 1991; Sindhu and Bawa 2000). Kayacier and Singh (1999) used CMC to obtain low-fat tortilla chips with rheological properties similar to regular chips. The addition of hydrocolloids is generally effective at levels as low as 1% of the formulation's dry weight or less

(Meyers 1990). Bell and Steinke (1991) reported an increase in volume of microwave-baked cakes on addition of 1–2% MC gums due to improved distribution of moisture by the gum. XG is used to improve the texture and moisture retention in cake batters and dough, increase the volume and shelf life of cereal foods by limiting starch retrogradation, improve their eating quality and appearance, and enhance the effectiveness of other hydrocolloids (Lee, Honseney and Varriano-Marton 1982; Miller and Hoseney 1993; Hanna et al. 1997; Lee and Brandt 2002). The gel-forming capacity of certain hydrocolloids has been shown to reduce oil absorption in cereal products or batter-coated products during the frying process. Reduction of fat absorption has been one of the main applications of hydrocolloids over the past two decades (Stypla and Buckholz 1989; Meyers and Conklin 1990; Chalupa and Sanderson 1994) and is possibly that of greatest value added. Selecting the appropriate hydrocolloid(s) requires not only an understanding of their physical properties but also an understanding of the foods themselves. It is difficult to ascribe exact characteristics and functions to the broad class of hydrocolloids in food systems.

There are several factors to consider when choosing a hydrocolloid for a specific application. One of these is the specific requirement for correct hydration of the different hydrocolloids. In this regard, when different alternatives are available the preferred gums will be those that can be incorporated into the batter by dry blending. Correctly selecting hydrocolloids for a specific function should also consider that other ingredients in the batter system might affect hydrocolloid performance. Therefore, the compatibility of the hydrocolloids with other ingredient components must be checked. For instance, a high concentration of soluble solids (i.e., sugar, salt, etc.) can reduce the solubility of hydrocolloids because of competition for the available water (Grover 1982).

22.4 FUNCTIONALITY OF HYDROCOLLOIDS IN BATTER SYSTEMS

Hydrocolloids as ingredients mostly serve three main functions in batter systems. The primary two functions are related to control of the viscosity and water-holding capacity of the systems. The third function is control of water loss and oil uptake in fried products. This is attributed to the unique thermal gelation abilities of some hydrocolloids, which form gels and provide a resistant-barrier coating during heating (Meyers 1990; Balasubramaniam et al. 1997; Albert and Mittal 2002; Mellema 2003).

MC and hydroxypropyl methylcellulose (HPMC) are the only food gums that can thermally gel (Sanz et al. 2005b) and that are reversible and repeatable. A number of studies have reported that the addition of MC and HPMC in batter systems successfully reduced oil uptake in coated fried products (Meyers 1990; Annapure et al. 1999; Mellema 2003). Thermal gelation can also lead to a stronger but more brittle coating that promotes the formation of a relatively small number of wide punctures with low capillary pressure, resulting in low oil uptake during deep fat frying (Holownia et al. 2000). The temperatures at which the gelation process starts and the strength of the gel formed are dependent upon the type and the degree of substitution, molecular weight, and concentration of hydrocolloids (Sanz, Salvador and Fiszman 2004b). Priya, Singhal and Kulkarni (1996) reported that adding CMC to the formulation of "boondis," a deep-fried batter-based legume snack food popular in India, reduced the amount of oil in the final products. This study analyzed the effect of different concentrations of CMC in the range of 0.5–3%, by adjusting the proportion of water to obtain adequate viscosity. The greatest barrier efficiency was obtained at a concentration of 2%. The higher concentration of 3% was found to be non-effective for oil absorption. Suderman (1996) indicated that CMC was able to improve adhesive strength in batters. Gao and Vodovotz (2005) reported that CMC changed the rheological and thermal properties of masa (dough) and the resulting tortilla's shelf life. Andres et al. (2005) found that CMC greatly influenced the rheological and functional properties of dried nixtamalized (alkaline-cooked) maize masa.

Gums are mostly used for cold-batter viscosity adjustment. It contributes to variations in viscosity often attributable to other ingredients such as starches and flours in batter systems. Meyers (1990) reported that some gums (e.g., xanthan and tragacanth) could also provide a yield value to the batter. Yield value is the initial resistance to flow under stress. It enables the suspension of heavy particles at low gum concentrations (0.01–0.25% w/w). When flour that does not develop much viscosity, such as rice flour, is used as the batter base, it is necessary to incorporate gum such as either xanthan or guar gum to ensure quality similar to a classic formulation (Mukprasirt et al. 2000).

Xue and Ngadi (2007a) conducted a detailed study on the effects of incorporating hydrocolloids such as MC and XG in tempura batter systems formulated with different combinations of wheat, rice, and corn flours. Both flow behavior and viscoelastic characteristics of the resulting batters were examined. Addition of XG lowered the flow behavior index of batter and thus increased shear-thinning characteristics of the batter. This was attributed to the gum's unique rigid, rod-like conformation that is more responsive to shear and the required progressive alignment of the rigid molecules during shearing (Pettitt 1982; Urlacher and Noble 1997). Lower concentration of XG compared to MC was required to lower n values of batters, indicating its higher degree of pseudoplasticity. Typical effects of XG and MC on batter consistency index (k) are shown in Figures 22.1 and 22.2, respectively, for WC. The availability of or competition for free water in the flour-gum batter system are largely responsible for development of consistency in batter systems. Xue and Ngadi (2007a) also studied structural

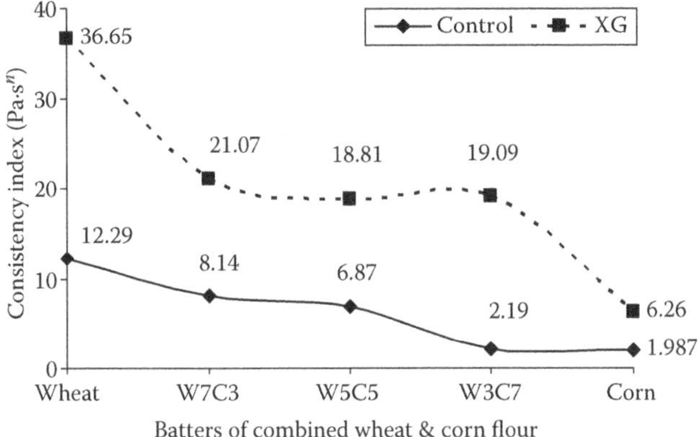

FIGURE 22.1 Effect of xanthan gum on the consistency index of wheat, corn, and their combination flour based batter. (From Xue, J. and Ngadi, M., *Journal of the Science of Food and Agriculture*, 87 (7), 1292–300, 2007a. With permission from Wiley Interscience.)

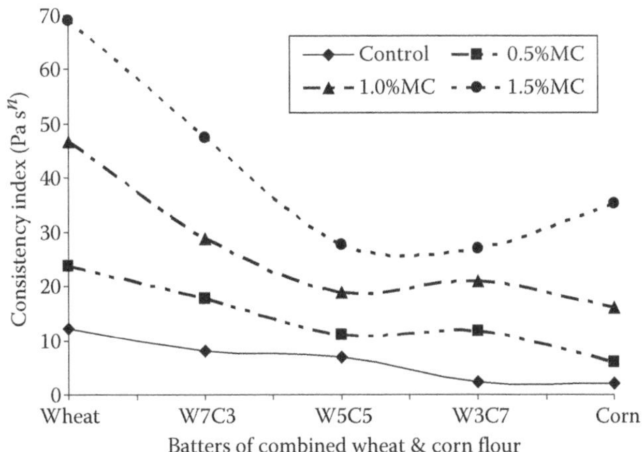

FIGURE 22.2 Effect of MC on the consistency index of wheat, corn, and their combination flour based batters. (From Xue, J. and Ngadi, M., *Journal of the Science of Food and Agriculture*, 87 (7), 1292–300, 2007a. With permission from Wiley Interscience.)

changes in batters resulting from the addition of hydrocolloids. It was reported that XG and MC generally shifted the onset temperature of structure development in batters to higher temperatures but they did not affect the peak temperature. The xanthan increased both the G' and G'' of all batters due to its interactions with gelatinized granules, whereas MC also increased G' but lowered G'' of batter systems depending on gum concentrations. The influence of MC was attributed more to its thermal gelation properties during heating. Grover (1982) reported that the viscosity of an aqueous solution of MC will initially decrease upon heating. However, there was a substantial increase in viscosity and enhanced development of elastic properties of material resulting in increased G' when the solution gels reached their gelation temperature (52–57°C). Once gelation of MC started, dehydration and association were time-dependent processes in which viscosity decreased with increased time at constant shear rate, resulting in decreased loss modulus G''. The concentration of the gum did not show obvious effect after gelation. Therefore, elastic behavior dominated material characteristics more than viscous behavior in coating formation of batter-MC during the heating process.

Apart from developing viscosity, hydrocolloids help to keep solids suspended in solution. This property has been related particularly to XG and tragacanth gum, which also provide yield value and are able to suspend heavy particles under shear conditions at low concentration (0.10–0.255 w/w) (Hsia et al. 1992). Generally, the effectiveness of a gum in providing adhesive strength in a batter system will be increased with an increase in its gel strength, concentration, and molecular weight, as well as its viscosity in the batter (Meyers 1990). Also the highly hydrophilic nature of hydrocolloids retards ice crystal growth during frozen product storage and reduces water migration to the coating from the substrate. This improves the freeze/thaw stability (Fiszman and Salvador 2003).

Hydrocolloid performance can also be affected by the complexity of other ingredients in the batter system and the compatibility of the hydrocolloids with those components. Most food gums are nondigestible polysaccharides that have profound effect on the available free water and the rheology of liquid batter systems (Meyers 1990; Xue and Ngadi 2007a). At high concentrations (greater than approximately 60% by weight), soluble solids such as sugars and salts can reduce the solubility of hydrocolloids because they are competing for the same available water (Meyers 1990). The application of hydrocolloids as ingredients in batter systems is still largely at the laboratory stage, and their mechanism of impacting functionality and properties are not yet fully understood. However, hydrocolloids show a high potential ability to provide consumers with healthy food products if they could be incorporated into batters on a commercial scale.

22.5 RHEOLOGICAL PROPERTIES OF BATTER SYSTEMS

The rheological properties of a batter system affect the pickup and quality of the batter that adheres, the handling properties of the battered product, and its appearance and final texture. Control of viscosity, degree of adhesion, and porosity of the batter system is important to control the quality of final products. In general, changes in rheological properties of a material indicate changes in its molecular structure. Consequently, the rheological properties of a material influence the flow process and are themselves influenced by the structural changes generated during the process. Those changes in their structure could directly affect the appearance and final texture of the finished products. Therefore, rheological measurements can provide a means of monitoring changes in product structure during process.

The incorporation of hydrocolloids makes the already complicated flow behavior of batters even more complex (Xue and Ngadi 2007a). Fiszman et al. (2005) reported that only single shear rate rheological measurements are usually performed in industrial plants, but they do not provide complete information. Also a rheological characterization of batter behavior over a range of shear stresses and time give more complete information for optimizing the processes of mixing, pumping, and coating, with a view to keeping the batter properties, pickup, and adhesion uniform (Fiszman and Salvador 2003). However, the determination of rheological properties by using sophisticated rheometers makes it possible to study the rheological behaviors of batter in depth, although their use is generally confined to the field of research (Sanz et al. 2004b).

Batter viscosity is a critical coating characteristic. It affects the pickup and quality of the batter that adheres, and the end product's appearance and texture (Hsia et al. 1992; Shih and Daigle 1999; Mukprasirt et al. 2000). Steady shear measurement is commonly used to determine fluid behavior such as the apparent viscosity of batter. Numerous studies showed that batters generally exhibit shear-thinning behavior, time dependency, and thixotropy (Xue and Ngadi 2006). Therefore the rheological characterization of a batter's flow behavior over a range of shear stresses and time gives more complete information for optimization of the process of mixing, pumping, and coating, with a view to keeping the batter properties, pickup, and their adhesion uniform (Hsia et al. 1992; Balasubramaniam et al. 1997; Mukprasirt et al. 2000). The composition and proportion of the ingredients, the water–solids relationship, and temperature are considered factors that affect the rheological properties of a batter. An increase in temperature resulted in lower consistency index values (Ostwald-deWale model) in several tempura batter formulations (Baixauli et al. 2003; Salvador et al. 2003).

Xue and Ngadi (2007a) recently reported a study on rheological properties of batter systems formulated using different blends of wheat, rice, and corn flours including the influence of some other ingredients such as salt on the rheological properties of batter systems. Ingredients influenced batter viscosity as shown in Table 22.1. Addition of salt to the batter formulations slightly lowered their viscosities. Changala, Susheelamma and Tharanathan (1989) also reported similar results for both native and fermented black gram flour dispersions. Salt binds water tightly and increases the water-holding capacity of batter systems. Batters with 100% wheat flour showed higher viscosity than batters with either 100% corn or 100% rice flour. This could be attributed to the ability of wheat gluten to absorb water, resulting in decreased free water in the batter system (Figure 22.3). Corn and rice (CR) flours tend to feel more "gritty" because their proteins do not absorb water easily at lower temperatures. Their particles also do not hydrate as fully, do not

TABLE 22.1
Average Viscosities of Batter Systems Formulated Using Different Combinations of Flour and Salt

	Viscosity (Pa.s)	
Batter System	**Without Salt**	**With 2.5% (w/w) Salt**
Wheat	7.41±0.5[a]	5.14±0.4[b]
Rice	4.46±0.5[b]	3.10±0.3[cd]
Corn	1.08±0.5[d]	0.68±0.3[fg]
W3C7	1.87±0.4[d]	1.34±0.3[fg]
W5C5	2.71±0.4[cd]	2.45±0.3[de]
W7C3	4.26±0.4[bc]	3.29±0.3[cd]
W3R7	4.01±0.4[bc]	2.98±0.3[cd]
W5R5	3.90±0.5[bc]	2.63±0.4[cd]
W7R3	4.36±0.4[bc]	3.99±0.3[a]
C3R7	2.24±0.4[b]	1.59±0.3[ef]
C5R5	1.75±0.5[d]	0.87±0.3[fg]
C7R3	1.85±0.5[d]	0.32±0.3[g]

Source: From Xue, J. and Ngadi, M. O., *Journal of Food Engineering*, 77 (2), 334–41, 2006. With permission from Elsevier.

C7R3 = 70% corn and 30% rice flour, C5R5 = 50% corn and 50% rice flour, C3R7 = 30% corn and 70% rice flour, W7C3 = 70% wheat and 30% corn flour, W5C5 = 50% wheat and 50% rice flour, W3C7 = 30% wheat and 70% corn flour, W7R3 = 70% wheat and 30% rice flour, W5R5 = 50% wheat and 50% rice flour, W3R7 = 30% wheat and 70% rice flour. Means with the same letter are not significantly different.

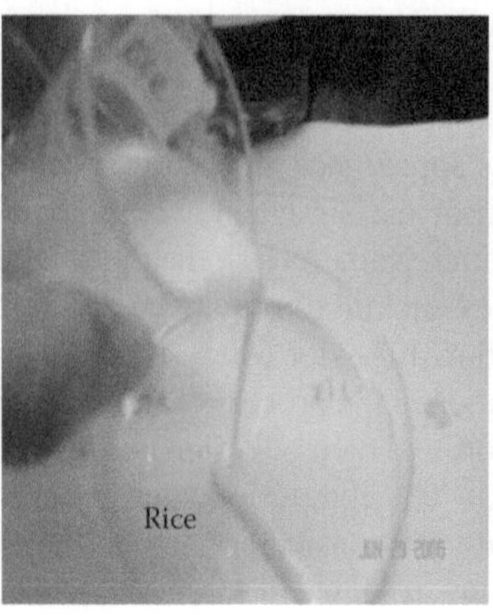

FIGURE 22.3 Illustration of relative consistencies of wheat, rice, and corn batters.

swell, and do not interact with each other as much. Thus, the viscosities of batter systems containing CR flours do not rise as rapidly as systems containing wheat flour. Rice flour batters had higher viscosities as compared to corn flour systems. This may be attributed to differences in their particle diameters and size distributions. The viscosities of batters with 100% corn flour were not significantly different from those for batters formulated either with 70% corn flour and 30% rice flour (C7R3) or with 50% corn flour and 50% rice flour (C5R5). However, further increases in the proportion of rice flour in the CR flour mixtures resulted in significant increases in batter viscosity. In the C7R3 batter, corn flour was the major flour base and the viscosity behavior of the batter system tended towards corn flour, which had the lowest viscosity, whereas in C3R7 batter the behavior tended to rise to a higher viscosity. The results indicate that replacement of corn flour by up to 50% rice flour did not influence the viscosity of corn-flour-based batters. On the other hand, replacement of rice flour by 30% corn flour in a rice-based batter significantly lowered the viscosity of the system. Therefore, corn flour showed a stronger influence on viscosity of corn-rice batter than did rice flour. Addition of either corn or rice flour tended to reduce the viscosity of wheat-based batter systems.

Flow curves of different batters formulated using combinations of WC flours, WR flours, and CR flours are shown in Figures 22.4, through 22.6, respectively. For the WC flour blends, batter viscosity decreased with an increasing proportion of corn flour in the batter. Corn flour apparently dilutes the strengthening influence of wheat flour gluten (Navickis 1987). Rice flour also exerted a diluting effect on wheat flour gluten, increasing the available free water in the batter system. This free water could lubricate particles, enhance flow, and result in a lower viscosity value (Mukprasirt et al. 2000).

Xue and Ngadi (2007a) also used onset temperature of structure development (T_{onset}), G'_{max} (measured at peak), and G''_{max} to monitor changes in the batter systems' rheological properties as functions of temperature for different flour combinations and proportions. There was a rapid increase in G' as the batters were heated between 58 and 68°C, after which G' further increased rapidly, indicating an increase in elastic properties. This increase was attributed to starch gelatinization, resulting in the onset of structure formation during which the fluid-like batter transformed into a solid-like coating. The storage modulus of samples reached maximum values (G'_{max}) before it then decreased steadily with further heating due to molecules of soluble starch orienting themselves in the direction that the system was being sheared, causing a decrease in viscosity (Hoseney 1994b). Changes in the loss modulus (G'') of the different batter systems generally followed patterns similar to those observed for G'. Onset temperature indicates the onset of structure development related to initialization of starch gelatinization or other structure-related processes during heating. Rice- and corn-flour-based batters had similar onset temperatures. Wheat-flour-based

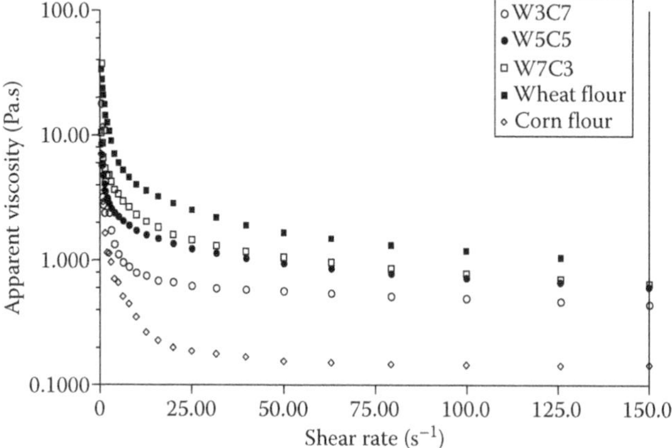

FIGURE 22.4 Viscosity of batter system formulated using wheat, corn flour, and their blends. (From Xue, J. and Ngadi, M. O., *Journal of Food Engineering*, 77 (2), 334–41, 2006. With permission from Elsevier.)

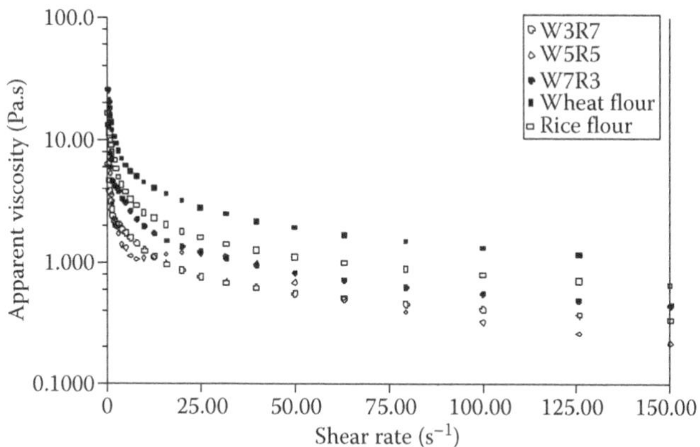

FIGURE 22.5 Viscosity of batter system formulated using wheat, rice flour, and their blends. (From Xue, J. and Ngadi, M. O., *Journal of Food Engineering*, 77 (2), 334–41, 2006. With permission from Elsevier.)

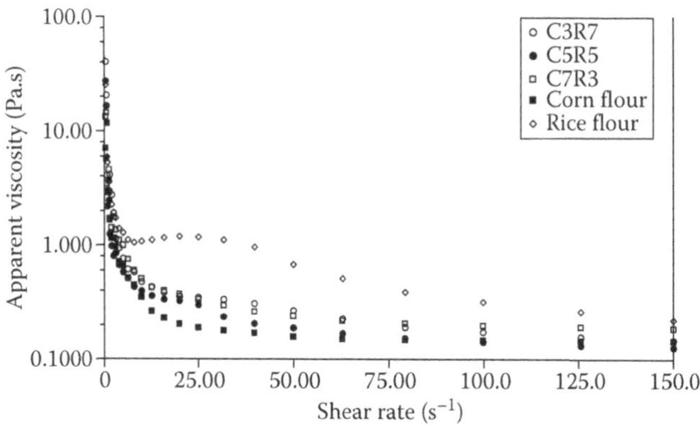

FIGURE 22.6 Viscosity of batter system formulated using corn, rice flour, and their blends. (From Xue, J. and Ngadi, M. O., *Journal of Food Engineering*, 77 (2), 334–41, 2006. With permission from Elsevier.)

batters exhibit lower onset temperature due to their gluten gelation, which can occur at a lower temperature than starch gelatinization (Olewink and Kulp 1993). The corn-based batter showed higher G'_{max}, whereas rice flour showed lower G'_{max}. Rice flour only reached the G'_{max} at a higher temperature (78.6°C). This implied that rice-based batters may require more time for complete gelatinization than corn- and wheat-based batters. Corn-based batters showed higher G'' than rice- and wheat-based batters since corn flour has a higher flow characteristic than rice and wheat flour. Wheat flour greatly influenced the G' and G'' of both corn- or rice-based batters. G' and G'' were reduced with increasing wheat flour proportion in batter systems. Sanz et al. (2005a) also reported similar results, indicating that both storage and loss moduli were higher in batters where wheat flour was partially replaced by corn starch. Corn flour with higher G' and G'' values dominated the viscoelastic behavior in combined flour systems. There were also synergistic effects of combined flours on properties of batter systems.

The incorporation of hydrocolloids makes predicting their flow behavior even more complicated. Christianson et al. (1981) reported that viscosity of wheat starch was significantly increased by the addition of a small amount of xanthan, guar, and cellulose gums. At the initial stage of gelatinization, those gums and their inherent viscosity magnified the effect of swelling so that the viscosity increase was apparent. Sanz et al. (2004b) studied the effect of concentration and temperature on properties of MC-added batters. Their results showed that MC produced a significant increase in the consistency (Ostwald-de Waele model) and shear-thinning behavior of the wheat-flour-based batter, and this effect was more evident at high levels of MC addition (2%). Sanz et al. (2005a) observed that MC influenced the rheological behavior of wheat starch and modified corn starch. Hsia et al. (1992) also found guar and XG to increase shear thinning behavior in wheat-flour-based batter and corn-flour-based batter, but CMC did not increase batter consistency significantly in either wheat- or corn-flour-based batters. Mukprasirt et al. (2000) reported a higher consistency index for rice-based batter containing MC at the low temperature of 5°C compared to either 15 or 25°C, and that the shear-thinning behavior decreased with temperature increase. Kim and Yoo (2006) showed that the increase in rate constant (k) in the gelatinization of rice starch-XG mixtures was a function of XG concentration. Sanz et al. (2005b) studied the thermogelation properties of MC and its effect on wheat-flour-based batter formulas containing 1, 1.5, or 2% MC. Their results showed that MC solutions tested at 15–60°C clearly showed a transition from a fluid-like to a gel-like behavior. The evolution of the G' (storage modules) and G'' (loss modules) with an upward temperature ramp showed the transition to solid state and gel state occur at approximately 52°C. Increasing temperature of MC batters resulted in a transition from a soft gel at 15°C to a stronger although still soft gel at 60°C. Both G' and G'' increased with MC concentration, although MC did not seem to qualitatively influence the viscoelastic behavior.

A limited number of studies have been conducted on the fundamental rheological properties of batters formulated by using different combination flour blends containing hydrocolloids designed for food coatings. However, this is an area of great importance and potential use since the rheological data provide information that is useful for coating applications.

22.6 THERMAL PROPERTIES OF BATTER SYSTEMS

Knowledge of thermal properties (e.g., specific heat, C_p, and enthalpy, ΔH) of food is vital to predicting heat transfer rates in food. Specific heat indicates how much heat is required to change the temperature of a material. It depends strongly on the temperature and composition (such as moisture content, fat content, and the nature of the solid component, such as carbohydrate and protein) of the product (Ngadi et al. 2000). Enthalpy is the heat content or energy level of a material. It can be very complicated for frozen foods because it is difficult to separate the latent and sensible heats in frozen foods. They often contain both frozen and unfrozen water, even at very low temperatures. Therefore, enthalpy depends upon the amount of unfrozen water in addition to the proximate composition of the food. Since flour is the major functional ingredient in batter coatings for food products, its starch gelatinization properties are of paramount importance in processing battered food. This is because flours contain gelatinized starch, which may undergo important textural changes during processing and directly influence the final quality of the particular products. The gelatinization properties are useful to determine the amount of heat and time require for cooking and processing raw material (ingredients) into finished products. Both water and thermal energy play indispensable roles in the process of gelatinization of starch in a batter system containing other ingredients (Biliaderis et al. 1986; Saif, Lan and Sweat 2003).

In general, starch gelatinization is greatly influenced by other ingredients such as protein and lipid contents, amylose and amylopectin contents, and amount of available water in the system (Hoseney 1994b). The availability of water is determined by the formula or recipe used and by the presence of ingredients or components such as proteins, pentosans (naturally present hydrocolloids), or sugars, which compete with starch for free water. The amount of moisture available for gelatinization is also affected by the degree of protection against water absorption that fat provides to the starch particles (Kaletunç and Breslauer 2001). Granule size also affects starch gelatinization behaviors due to their water-absorbing and -holding capacity during the cooking process (Hoseney 1994b).

Wheat, CR flour batters exhibited a single endothermic transition over the temperature range from 50 to 85°C during heating (Figure 22.7). The temperature range corresponds to the range expected for starch gelatinization. The heat capacity C_p gradually and consistently increased to the onset temperature of gelatinization, and then it progressively increased more rapidly until the gelatinization temperature T_G (peak temperature) was reached. After that, the heat capacity decreased to the endpoint temperature indicating when starch was completely gelatinized. Corn flour has a higher heat capacity due to its higher fat and starch contents. The corresponding gelatinization temperatures (T_G) as peak temperatures, total enthalpies for gelatinization (ΔH_G), glass transition temperatures (T_g), ice melting temperatures (T_m, peak temperature), and total enthalpies for melting are shown in Table 22.2. There were differences between the total enthalpies (ΔH_G) of the various batter systems although their gelatinization temperatures appear to be similar. It is known that change in the total enthalpy of gelatinization is influenced by several factors including the distribution of water between gluten and starch, starch granule size, hydration rates, and other possible interactions between the various components (Kaletunç and Breslauer 2001). Corn-based batters showed the highest ΔH_G. Increasing the proportion of corn flour in any batter mixture increased ΔH_G. This result could be attributed to the high starch and fat contents in the corn flour, which required more energy to open the starch helix structure. Hydrophobic fats tend to coat the starch granules and they interfere with the ability of water to enter the starch helix by some sort of a "blocking action,"

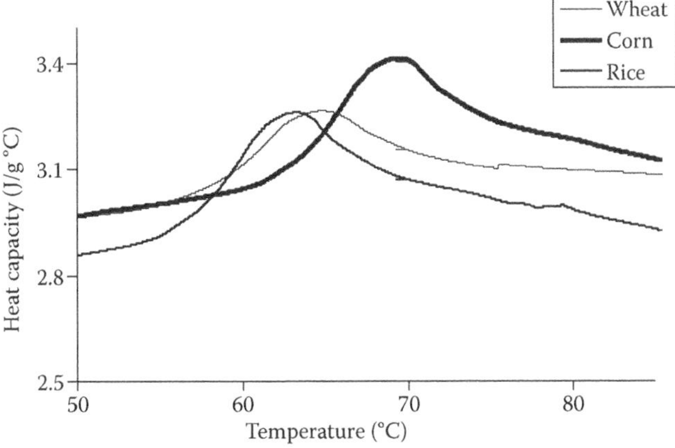

FIGURE 22.7 Differential scanning calorimetry thermograms of wheat, corn, and rice flour batters. (From Xue, J. and Ngadi, M. O., *Lebensmittel-Wissenschaft und Technologie,* 40 (8), 1459–65, 2007b. With permission from Elsevier.)

TABLE 22.2
Gelatinization Temperature, Total Enthalpies of Gelatinization, Glass Transition Temperature, Ice Melting Temperature, and Total Enthalpies of Melting on Flour-Water Suspensions

Flour	T_g (°C)	ΔH_G (J/g)	T_g (°C)	T_m (°C)	ΔH_m (J/g)
Wheat	65.54$^{a\,b\,c}$	2.26^{a}	−10.31^{a}	4.17^{a}	102.80^{a}
Corn	68.97$^{a\,b\,c}$	4.38^{b}	−8.75^{b}	2.54^{d}	74.96^{c}
Rice	61.80$^{a\,b}$	3.34^{c}	−7.29^{f}	2.90^{c}	61.52^{f}
C3R7	63.50$^{a\,b\,c}$	2.08^{f}	−7.97^{d}	3.30^{b}	69.20^{d}
C5R5	70.56^{a}	2.56^{e}	−9.21^{b}	1.80^{f}	58.91^{g}
C7R3	64.13$^{a\,b\,c}$	4.13^{d}	−8.06^{d}	1.90^{f}	66.91^{e}
W3C7	62.49$^{b\,c}$	2.39^{i}	−7.81^{d}	1.99^{f}	42.15^{i}
W5C5	60.04$^{b\,c}$	2.30^{h}	−8.13^{c}	1.08^{g}	55.97^{h}
W7C3	61.55$^{b\,c}$	1.62^{g}	−8.56^{c}	2.27^{e}	81.02^{b}
W3R7	61.25$^{b\,c}$	2.07^{l}	−7.69^{e}	3.23^{b}	64.15^{f}
W5R5	63.26^{a}	1.71^{k}	−7.77^{e}	2.29^{e}	66.69^{d}
W7R3	63.04$^{a\,b\,c}$	1.38^{j}	−7.86^{d}	3.38^{b}	70.80^{d}

Source: From Xue, J. and Ngadi, M. O., *Lebensmittel-Wissenschaft und Technologie*, 40 (8), 1459–65, 2007b. With permission from Elsevier.

C7R3 = 70% corn and 30% rice flour, C5R5 = 50% corn and 50% rice flour, C3R7 = 30% corn and 70% rice flour, W7C3 = 70% wheat and 30% corn flour, W5C5 = 50% wheat and 50% rice flour, W3C7 = 30% wheat and 70% corn flour, W7R3 = 70% wheat and 30% rice flour, W5R5 = 50% wheat and 50% rice flour, W3R7 = 30% wheat and 70% rice flour

The means with the same letter in any given column are not significantly different.

T_g = starch gelatinization temperature, ΔH_g = total enthalpies of gelatinization, T_g = glass transition temperature, T_m = ice melting temperature, ΔH_m = total enthalpies of ice melting

leading to the reduction in the available free water required to interact with starch (Pomeranz 1987). Starch will not normally be gelatinized if starch crystallinity is not disrupted.

For the WR flour combination blend based batters, ΔH_G increased with increasing rice flour content due to the reducing proportion of wheat gluten content in the batter mixture. The resulting reduction in gluten content apparently led to a larger amount of water being available for starch gelatinization. A similar effect was observed when the proportion of corn flour was increased in a combination of WC-flour-based batters. Thus, CR flours apparently exert diluting effects on wheat gluten, increasing the amount of water available to react with wheat starch during the heating process (Liu et al. 2002; Wang, Choil and Kerr 2004). Wang et al. (2004) reported the gelatinization temperature to be affected by water and gluten contents in the dough. It is suggested that the heated gluten gel showed a greater binding of water than the starch gel in the dough. This was attributed to less water being available to starch in the presence of gluten.

The glass transition behavior of the starch-water system in a freeze-concentrated phase depends on ice formation. The T_g and ΔH_m are influenced dramatically by the amount of available water in the system (Chung, Lee and Lim 2002; Hsu et al. 2003). Roorda (1994) indicated that glass transition occurs at or close to the temperature of water melting (and freezing) in hydrogels. Thermal properties, such as T_g, the peak temperature for ice melting T_m, and the melting enthalpy of ice ΔH_m of the batter systems, indicate that there is freezable water in the batter systems. Different combinations of flours and their combination ratios influenced changes in glass transition temperature, enthalpy, and ice melting temperatures in a starch-water matrix.

The batter system with 100% wheat flour showed the lowest T_g (−10.31°C), T_m (4.17°C), and ΔH_m (102.8 J/g). This is possibly because wheat gluten tightly binds water and protects it from freezing. This apparently led to increasing unfrozen water content in the system, resulting in the depressed glass transition temperature. Wang et al. (2004) reported that water molecules diffuse more easily in starch-water mixtures than in gluten-water mixtures. Reid (1997) demonstrated that water comes out from the food product to form ice during the freezing process. The diffusion of water in mixtures can be complicated and depends on a number of parameters including type of mixtures, their compositions, and particle size. Rice flour has a higher amount of free water than corn flour in the batter system. This difference is because the rice starch granule size is smaller than in corn and it absorbs less water. Also, rice flour has no gluten to hold water in the manner that wheat flour does. Free water can easily be frozen in batter systems during the freezing process. Therefore, rice flour has a higher T_g (–7.29°C) and lower ΔH_m (61.52 J/g), compared with 100% wheat-flour-based (–10.31°C and 102.8 J/g) or 100% corn-flour-based (–8.75°C and 74.96 J/g) batter systems. The thermal properties of wheat-based batters were greatly influenced by replacement of either corn or rice flour compared with 100% wheat-flour-based batter. The glass transition temperatures of the batter systems were raised and T_m values decreased. The ΔH_m values were reduced when the proportion of corn or rice flour in batter systems was increased. This may be due to reduced gluten level in the blended flour resulting in increasing free water in the batter systems.

Addition of salt in batters dramatically increases their starch gelatinization temperatures (Xue and Ngadi 2007b). This increase in gelatinization temperature can be attributed to the role of salt in maintaining the integrity of the starch granule. The starch granule swells to a greater extent or remains intact for a longer time before fragmentation occurs (Ganz 1965). Similar results were observed in corn and wheat flours (Evans and Haisman 1982; Salvador et al. 2002). Further, salt lowered the ΔH_G of batters due to its strengthening and tightening effect on gluten (Pyler 1988). It is a competitor with flour and other components for water, so less water is available to be absorbed by starch. Therefore, more energy was required for starch gelatinization to occur in limited water systems during the heating process, and hence there was an elevation in the gelatinization temperature. Thermal properties of batter are also affected by other ingredients such as hydrocolloids. Gelatinization temperatures of batters increase with increasing concentration of MC in system. MC reduces the amount of water available to react with starch during gelatinization, resulting in the increased gelatinization temperature.

Many processes and properties encountered in food science are affected by, or changed by, the glass transition phenomena. Glass transition (phase or state transition) is the name given to a phenomenon observed as a change from a brittle glassy or crystalline state to a rubbery behavior at temperature T_g. At a sufficiently low temperature, or with a limited content of plasticizer such as water, molecular motion becomes restricted as a glassy solid is formed. On heating or on addition of plasticizer, the mobility of the amorphous polymers increases and the material becomes flexible or rubbery. The glass transition temperature (T_g) depends on molecular characteristics, composition, and compatibility of the components in the amorphous matrix. Therefore, the glass transition behavior affects any food properties related to molecular mobility, including texture and shelf life.

Variability in composition of batter system formulas causes their thermal properties and phase transition characteristics to exhibit the same large variability. Various flours and hydrocolloids change the thermal properties and phase transition characteristics of batter systems. Ferrero and Zaritzky (2002) reported that low hydrocolloid concentrations (10 g/kg) did not significantly affect corn starch gelatinization temperature as compared with systems without hydrocolloids in starch-sucrose systems. However, small differences were detected among the systems containing alternatively guar, xanthan, or alginate gums. Also small quantities of hydrocolloids did not shift the glass transition temperature, but they played an important role in minimizing structure damage. This was verified in their research by rheological viscoelastic tests where an increase in the dynamic moduli G' (storage module) and G'' (loss module) after slow freezing and during storage at –19°C was observed in a starch–sucrose system.

Sarkar (1979) and Ford (1999) studied thermal gelation properties of MC and HPMC as a function of molecular weight, degree of methyl and hydroxypropyl substitution, concentration, and the presence of additives. The results of these studies showed that the precipitation temperature for these polymer solutions decreased initially with increasing concentration until a critical concentration was reached, above which the precipitation temperature was minimally affected by changes in concentration. The incipient gelation temperature decreases linearly with concentration (Sanz et al. 2005b). The strength of these gels is time dependent, increases with increasing molecular weight, decreases with increasing hydroxypropyl substitution, and depends on the nature of additives (Grover 1982).

Starch and hydrocolloids may provide synergistic effects on thermal properties and phase transition on batter system during processing. However, studies on these properties in the coatings system have not been extensively reported in the current scientific literature.

22.7 CONCLUSIONS

Batters are important in determining quality and shelf life of coated fried food products. Their formulation and ingredients impact various functionalities of coated products. A good understanding of the roles of the various ingredients and how they interact with each other is vital in designing optimal batters for specific products. Rheological and thermal properties of batters vary with the different types of flours and their combination ratios, and with the different types of hydrocolloids and other ingredients. Hydrocolloids may compete with other ingredients such as starch for available free water, resulting in complicated flow and thermal characteristics. Careful study and elucidation of these interactions can be used to control batter functionalities.

REFERENCES

Albert S. and Mittal G. S. 2002. Comparative evaluation of edible coatings to reduce fat uptake in a deep-fried cereal product. *Food Research International*, 35:445–48.

Andres A. C., Guadalupe M. M., Javier S. F., and Luis R. B. P. 2005. Effect of carboxy-methylcellulose and xanthan gum on the thermal functional and rheological properties of dried nixtamalised maize masa. *Carbohydrate Polymers*, 62:222–31.

Ang J. F. 1993. Reduction of fat in fired batter coating with powdered cellulose. *Journal of the American Oil Chemists' Society*, 76:619–22.

Annapure U. S., Sighal R. S. and Kulkarni P. R. 1999. Screening of hydrocolloids for reduction in oil uptake of a model deep fat fried product. *Fett/Lipid*, 101:217–21.

Baixauli R., Sanz T., Salvador A. and Fiszman S. M. 2003. Effect of the addition of dextrin of dried egg on the rheological and texture properties of batters for fried foods. *Food Hydrocolloids*, 17:305–10.

Balasubramaniam V. M., Chinnan M. S., Mallikarjunan P. and Phillips R. D. 1997. The effect of edible film on oil uptake and moisture retention of a deep-fat fried poultry product. *Journal of Food Process Engineering*, 20:17–29.

Bell D. A. and Steinke L. W. 1991. Evaluating structure and texture effects of methylcellulose gums in microwave-baked cakes. *Cereal Foods World*, 36 (11): 941–44.

Biliaderis C. G., Page C. M., Maurice T. J. and Juliano B. O. 1986. Thermal characterization of rice starches: a polymeric approach to phase transitions of granular starch. *Journal of Agricultural and Food Chemistry*, 34:6–14.

Chalupa W. F. and Sanderson G. R. 1994. Process for preparing low-fat fried food. US Patent 5, 372,829. Assignee: Merck.

Changala R. G., Susheelamma N. S. and Tharanathan R. N. 1989. Viscosity pattern of native and fermented black gram flour and starch dispersions. *Starch/Stärke*, 41 (3): 84–88.

Chitrakar B. and Zhang G. 2006. A solution to food fat? *Food Reviews International*, 22 (3): 245–58.

Christianson D. D., Hodge J. E., Osborn D. and Detroy R. W. 1981. Gelatinization of wheat starch modified by xanthan gum, guar gum, and cellulose gum. *Cereal Chemistry*, 58:513–17.

Chung H. J., Lee E. J. and Lim S. T. 2002. Comparison in glass transition and enthalpy relaxation between native and gelatinized rice starches. *Carbohydrate Polymers*, 48:287–98.

Crosby T. G. and Kincs F. R. 1990. Fats and oils in coated foods. In: *Batters and Breading in Food Processing* (Eds. Kulp K. and Loewe R.), Chapter 4. St. Paul, MN: American Association of Cereal Chemists.

Dogan S. F., Sahin S. and Sumnu G. 2004. Effects of soy and rice flour addition on batter rheology and quality of deep-fat fried chicken nuggets. *Journal of Food Engineering*, 71:127–32.

Dubois D. K. 1981. Chemical leavening. *Technique Bulletin*, 3 (9). Manhattan, KS: American Institute of Baking.

Dziezak J. D. 1991. A focus on gums. *Food Technology*, 45:116–32.

Evans I. D. and Haisman D. R. 1982. The effect of solutes on the gelatinization temperature range of potato starch. *Starch*, 34:224–31.

Ferrero C. and Zaritzky N. 2002. The glass transition temperature of frozen foods and its influence on stability during frozen storage: Analysis in starch sugar systems. ASAE Annual Meeting, Paper Number 026181. St. Joseph, MI: American Society of Agricultural and Biological Engineers.

Ferrero C. and Zaritzky N. E. 2000. Effect of freezing rate and frozen storage on starch-sucrose-hydrocolloids system. *Journal of the Science of Food and Agriculture*, 80:2149–58.

Fiszman S. M. and Salvador A. 2003. Recent developments in coating batters. *Trends in Food Science and Technology*, 14:399–407.

Fiszman S. M., Salvador A. and Sanz T. 2005. Why, when and how hydrocolloids are employed in batter-coated foods: A review. *Progress in Food Biopolymer Research*, 1:55–68.

Ford J. L. 1999. Thermal analysis of hydroxypropylmethylcellulose and methylcellulose: Powders, gels and matrix tablets. *International Journal of Pharmaceuticals*, 179:209–28.

Ganz A. J. 1965. Effect of sodium chloride on the pasting of wheat starch granules. *Cereal Chemistry*, 42:429–31.

Gao Y. and Vodovotz Y. 2005. Effect of CMC on the physico-chemical properties of masa-water mixture. Available at: http://grad.fst.ohio-state.edu/vodovotz/research.html. Accessed: November 6, 2008.

Grover J. A. 1982. Methylcellulose (MC) and hydroxypropyl methylcellulose (HPMC). In: *Food Hydrocolloids* (Ed. Glicksman M.), Vol. III, Chapter 4, 121–54. Boca Raton: CRC Press.

Hanna M. A., Chinnaswamy R., Gray D. R. and Miladinov V.D. 1997. Extrudates of starch-XG mixtures as affected by chemical agents and irradiation. *Journal of Food Science*, 62:816–20.

Holownia K. I., Chinnan M. S., Erickson M. C. and Mallikarjunan P. 2000. Quality evaluation of edible film-coated chicken strips and frying oil. *Journal of Food Science*, 65:1087–90.

Hoseney R. C. 1994a. Minor constituents of cereals. In: *Principles of Cereal Science and Technology*, 81–101. 2nd Edn., Chapter 4. p. 81–103. St. Paul, MN: American Association of Cereal Chemists.

———. 1994b. Starch. In: *Principles of Cereal Science and Technology*, (Ed. Hoseney R. C.), 29–64. 2nd Edn., Chapter 2. p. 29–64. St. Paul, MN: American Association of Cereal Chemists.

Hsia H. Y., Smith D. M. and Steffe J. F. 1992. Rheological properties and adhesion characteristics of flour-based batters for chicken nuggets as affected by three hydrocolloids. *Journal of Food Science*, 57:16–18, 24.

Hsu C. L., Heldman D. R., Taylor T. A. and Kramer H. L. 2003. Influence of cooling rate on glass transition temperature of sucrose solutions and rice starch gel. *Journal of Food Science*, 68:1970–75.

Kaletunç G. and Breslauer K. J. 2001. Calorimetry of pre- and postextruded cereal flours. In: *Characterization of Cereals and Flours: Properties, Analysis, and Applications* (Eds. Kaletunç G. and Breslauer K. J.), 1–36. New York: Marcel Dekker.

Kayacier A. and Singh R. K. 1999. Rheological properties of deep fried tortillas prepared with hydrocolloids. *International Journal of Food Properties*, 2:185–93.

Kim C. and Yoo B. 2006. Rheological properties of rice starch-xanthan gum mixtures. *Journal of Food Engineering*, 75:120–28.

Lee C. C., Honseney R. C. and Varriano-Marton E. 1982. Development of a laboratory-scale single-stage cake mix. *Cereal Chemistry*, 59:389–92.

Lee H. C. and Brandt D. A. 2002. Rheology of concentrated isotropic and anisotropic xanthan solutions. I. A rodlike low molecular weight sample. *Maromolecules*, 35:2212–22.

Li Y., Ngadi M. and Oluka S. 2008. Quality changes in mixtures of hydrogenated and non-hydrogenated oils during frying. *Journal of the Science of Food and Agriculture*, 88 (9): 1518–23.

Liu Q., Charlet G., Yelle S. and Arul J. 2002. Phase transition in potato starch-eater system: starch gelatinization at high moisture level. *Food Research International*, 35:397–407.

Loewe R. 1993. Role of ingredients in batter systems. *Cereal Foods World*, 38:673–77.

Mellema M. 2003. Mechanism and reduction of fat uptake in deep-fat fried foods. *Trends in Food Science and Technology*, 1:364–73.

Meyers A. M. 1990. Functionality of hydrocolloids in batter coating systems. In: *Batters and Breadings in Food Processing* (Eds. Kulp K. and Loewe R.). St. Paul, MN: American Association of Cereal Chemists.

Meyers M. A. and Conklin J. R. 1990. Method off inhibiting oil adsorption in coated fried foods using hydroxypropyl methylcellulose. US Patent 4,900,573. p. 1–10. Assignee: The Dow Chemical Company.

Miller R. A. and Hoseney R. C. 1993. The role of XG in white layer cakes. *Cereal Chemistry*, 70:585–88.

Mohamed S., Hamid N. A. and Hamid M. A. 1998. Food components affecting the oil absorption and crispness of fired batter. *Journal of the Science Food and Agriculture*, 78:39–45.

Mukprasirt A., Herald T. J. and Flores R. A. 2000. Rheological characterization of rice flour-based batters. *Journal of Food Science*, 65:1194–97.

Mukprasirt A., Herald T. J., Boyle D. L. and. Boyle E. A. 2001. Physicochemical and microbiological properties of selected rice flour-based batters for fried chicken drumsticks. *Poultry Science*, 80:988–96.

Navickis L.L. 1987. Corn flour addition to wheat flour doughs: Effect on rheological properties. *Cereal Chemistry*, 64 (4): 307–10.

Naz S., Sheikh H., Siddiqi R. and Sayeed S. A. 2004. Oxidative stability of olive, corn and soybean oil under different conditions. *Food Chemistry*, 88 (2): 253–59.

Ngadi M., Li Y. and Oluka S. 2007. Quality changes in chicken nuggets fried in oils with different degrees of hydrogenation. *Lebensmittel-Wissenschaft und Technologie*, 40 (10): 1784–91.

Ngadi M. O., Mallikarjunan K., Chinnan M. S., Radhakrishnan R. and Hung Y.-C. 2000. Thermal properties of shrimps, French toasts and breading. *Journal of Food Process Engineering*, 23 (1): 73–87.

Olewink M. and Kulp K. 1993. Factors influencing wheat flour performance in batter systems. *Cereal Chemistry*, 75 (4): 428–32.

Pettitt D. J. 1982. Xanthan gum. In: *Food Hydrocolloids* (Ed. Martin G.), Vol. I, 128–49. Boca Raton: CRC Press.

Pomeranz Y. 1987. *Modern Cereal Science and Technology*. New York: VCH.

Priya R., Singhal R. S. and Kulkarni P. R. 1996. Carboxymethylcellulose and hydroxypropylmethylcellulose as additives in reduction of oil content in batter based deep-fat fried boondis. *Carbohydrate Polymers*, 29 (4): 333–35.

Pyler E. J. 1988. Proteins. In: *Baking Science and Technology*, (Ed. Pyler E. J.), 3rd Edn. Chapter 3, p. 102–104. Kansas, MO: Sosland.

Reid D. S. 1997. Overview of physical/chemical aspects of freezing. In: *Quality in Frozen Food* (Eds. Erickson M. C. and Hung Y.-C.) p. 10–28. New York: Chapman and Hall.

Loewe R. 1990. Ingredient selection for batter systems. In: *Batters and Breadings in Food Processing* (Eds. Kulp K. and Loewe R.) Chapter 2, p. 11–28. St. Paul, MN: American Association of Cereal Chemists.

Roorda W. 1994. Do hydrogels contain different classes of water? *Journal of Biomaterials Science Polymer*, 5:383–95.

Saif S. M. H., Lan Y. and Sweat V. E. 2003. Gelatinization properties of rice flour. *International Journal of Food Properties*, 6:531–42.

Salvador A., Sanz T. and Fiszman S. M. 2002. Effect of corn flour, salt, and leavening on the texture of fired, battered squid rings. *Journal of Food Science*, 67:730–33.

———. 2003. Rheological properties of batters for coating products: Effect of addition of corn flour and salt. *Food Science and Technology International*, 9:23–27.

Sanz T., Fernandes M. A., Salvador A., Munoz J. and Fiszman S. M. 2005b. Thermogelation properties of methylcellulose (MC) and their effect on a batter formula. *Food Hydrocolloids*, 19:141–47.

Sanz T., Salvador A. and Fiszman S. M. 2004a. Innovative method for preparing a frozen, battered food without a prefrying step. *Food Hydrocolloids*, 18:227–31.

———. 2004b. Effect of concentration and temperature on properties of methylcellulose-added batters application to battered, fired seafood. *Food Hydrocolloids*, 18:127–31.

Sanz T., Salvador A., Velez G., Munoz J. and Fiszman S. M. 2005a. Influence of ingredients on the thermo-rheological behavior of batters containing methylcellulose. *Food Hydrocolloids*, 19 (5), 869–77.

Sarkar N. 1979. Thermal gelation properties of methyl and hydroxypropyl methylcellulose. *Journal of Applied Polymer Science*, 24:1073–87.

Shih F. and Daigle K. 1999. Oil uptake properties of fried batters from rice flour. *Journal of Agriculture and Food Chemistry*, 47:1611–15.

Shinsato E., Hippleheuser A. L., and Van Beirendonck, K. 1999. Products for batter and coating systems. *The World of Ingredients*, January–February:38–42.

Sindhu J. P. S. and Bawa A. S. 2000. Incorporation of carboxy-methylcellulose in wheat flour: Rheological, alveographic, dough development, gas formation/reduction, baking and bread firmness studies. *International Journal of Food Properties*, 3:407–17.

Slade L and Levine H. 1995. Water and the glass transition-dependence of the glass transition on composition and chemical structure: Special implications for flour functionality in cookie baking. *Journal of Food Engineering*, 24:431–509.

Steffe J. F. 1996. *Rheological Methods in Food Process Engineering*. East Lansing, MI: Freeman Press.

Stypla R. J. and Buckholz L. 1989. Process for preparing a coated food product. US Patent 4,877,629. Assignee: International Flavors and Fragrances Inc.

Suderman D. R. 1996. Effective use of flavorings and seasoning in batter and breading systems. In: *Batters and Breading in Food Processing* (Eds. Kulp K. and Loewe R.), Chapter 5. p. 73–92. St. Paul, MN: American Association of Cereal Chemists.

Urlacher B. and Noble O. 1997. Xanthan. In: *Thickening and Gelling Agents for Food* (Ed. Imeson A.) p. 284–311. London: Chapman and Hall.

USDA Handbook #8. 2002. Nutrition Value of Food. Beltsville, MD: US Department of Agriculture, Agricultural Research Service, Nutrient Data Laboratory.

Wang X., Choil S. G. and Kerr W. L. 2004. Water dynamics in white bread and starch gels as affected by water and gluten content. *Lebensmittel-Wissenschaft und Technologie*, 37:377–84.

Xue J. and Ngadi M. 2007a. Rheological properties of batter systems containing different combinations of flours and hydrocolloids. *Journal of the Science of Food and Agriculture*, 87 (7): 1292–300.

Xue J. and Ngadi M. O. 2006. Rheological properties of batter systems formulated using different flour combinations. *Journal of Food Engineering*, 77 (2): 334–41.

———. 2007b. Thermal properties of batter system formulated by combination of different flours. *Lebensmittel-Wissenschaft und Technologie,* 40 (8): 1459–65.

———. 2009. Effects of methylcellulose, xanthan gum and carboxy-methylcellulose on thermal properties of batter systems formulated with different flour combinations. *Food Hydrocolloids*, 23 (2): 286–95.

23 Allergenicity of Food and Impact of Processing

Andreas Lopata
RMIT University

CONTENTS

23.1 INTRODUCTION

Allergy-related diseases are today recognized as reaching epidemic proportions (Simons et al. 2008), with up to 30% of the general population suffering from them. In addition to increasing genetic predisposition ("atopy") of people, changing living conditions in developed as well as in developing countries could all contribute to immune deviation and the development of allergies. Adverse reactions to foods are of considerable importance in today's society, especially given the recent introduction of new allergens as eating habits change (appearance of "exotic" and novel foods), and the use of new industrial processing technologies which result in food ingredients appearing in unexpected forms.

In the research efforts in the field of food allergy two main questions are often asked: what makes one person allergic to a particular food and not the other and why some foods and food proteins are more allergenic than others. Investigation in this area of research is much more difficult than investigating inhalant allergens since food proteins often undergo extensive modifications during food processing. Furthermore these allergenic proteins are found in a complex matrix and may undergo physicochemical changes during digestion and subsequent uptake by the gut mucosal barrier and presentation to the immune system. The food matrix certainly has a great impact on the elicitation of allergic reactions as allergens are differentially released, for example in fat-rich matrices, and enhance the allergic reaction (Palmer and Burks 2006). Furthermore, food processing results in mostly water-insoluble proteins, which makes the traditional serological analysis of allergenicity difficult.

Due to these complex interactions of allergenic proteins with different matrices and the impact of various food processing techniques, research into this area has only begun in the last few years. Our current understanding of the impact of food processing on allergenicity is limited to a few food allergens and clinical studies, which will be discussed in the following section.

23.2 ADVERSE REACTIONS TO FOOD

The diagnosis of an allergic reaction to foods is often difficult, since nonallergic food intolerance is also very common (Asero et al. 2007; Niggemann and Beyer 2007a, 2007b; Ramesh 2008).

Adverse reactions to foods may be classified on the basis of the mechanism of the reaction (see Figure 23.1):

- Food hypersensitivity (involves the immune system)
- Food intolerance: non-immune-mediated (involves toxins, pharmacological reactions, enzymatic, etc.)

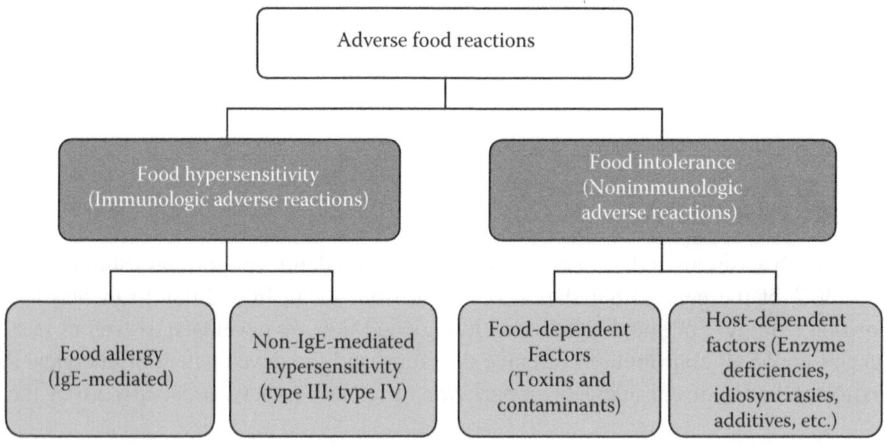

FIGURE 23.1 Classification and terminology for food allergy.

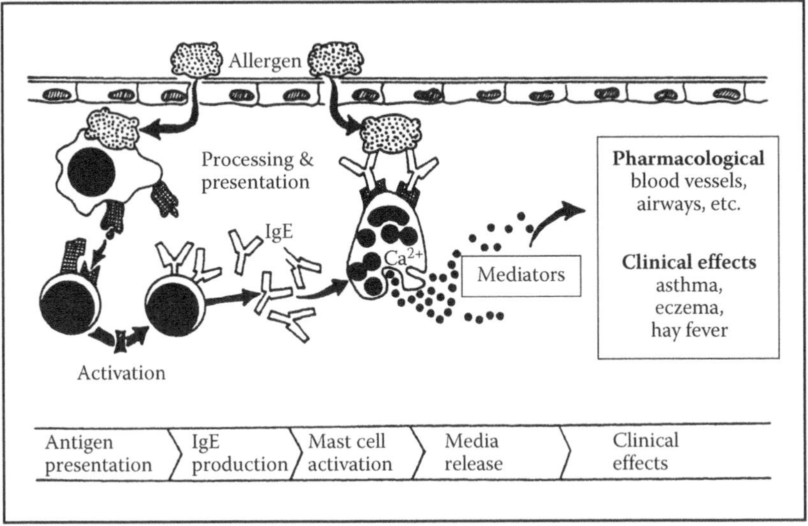

FIGURE 23.2 Induction and mechanisms in IgE-mediated hypersensitivity. (Adapted from Lehrer, S.B., Horner, W.E., and Reese, G., *Critical Reviews in Food Science and Nutrition,* 36(6), 553–64, 1996.)

23.2.1 Food Intolerance

The definition "food intolerance" is used when the history and/or the provocative tests clearly prove a food as the cause but there is no evidence that the immune system is involved (Lehrer, Ayuso, and Reese 2002; Lopata and Potter 2000). Symptoms of food intolerance can be the same as food allergic symptoms and also can be related to the gastrointestinal and respiratory tract and/or the skin. Nevertheless, some reactions after ingesting food or food additives belong to the group of psychological or psychosomatic reactions.

In general adverse reactions caused by food intolerance are divided into the following groups:

- Enzymatic: The most well known is *lactase deficiency*, which means that the person is unable to digest the milk sugar lactose.
- Pharmacological: Caused by food additives or may depend on the direct effect of *vasoactive amines* naturally found in foods such as histamine, tyramine, phenylethylamine, and serotonin. Large amounts of histamine and tyramine can be found in canned fish and fish autolysates (Auerswald, Morren, and Lopata 2006).
- Undefined: Food additives which can cause intolerance reactions are, for example, *azo dyes* (tartrazine), *flavors* (monosodium glutamate (MSG); "Chinese restaurant syndrome"), and *preservatives* (sulfites, etc.) (McCann et al. 2007).

23.2.2 Mechanisms of Food Allergy

In general the immunological reactions to allergens are divided into Type I, II, III, and IV reactions. Each type may be involved in food allergy, however the Type I reaction, or immediate-type allergic reaction, is the most common one, mediated by the IgE antibody (immunoglobin) (Untersmayr and Jensen-Jarolim 2006b). The induction of an allergic response is initiated when an allergen (or antigen) enters the body via mucosal surface membranes (see Figure 23.2). The specific IgE response is mediated through the presentation of the processed allergen by antigen-presenting cells, which results in the production of allergen-specific IgE antibodies. Upon reexposure to the allergen, the sensitized subject will develop an allergic reaction through interaction with IgE antibody bound on specific IgE receptors on the surface of mast cells. The interaction of the IgE antibody with the allergen will trigger the release of preformed and newly synthesized mediators (e.g., histamine, tryptase), which in turn elicit the clinical signs and symptoms of allergic diseases including hay fever, asthma, and anaphylaxis (Ross et al. 2008). The appearance of allergic symptoms results not only from ingestion of food, but can also be triggered by inhaling cooking vapors and handling food in the domestic as well as in the working environment (Goetz and Whisman 2000; Jeebhay et al. 2001; Taylor et al. 2000). Symptoms manifest mainly as upper and lower airway respiratory symptoms and dermatitis, while anaphylaxis is rarely seen with this type of exposure.

23.2.3 Clinical Manifestation of Food Allergy

Allergic reactions to food can elicit almost any allergic symptom and sign, but some are more widely demonstrated than others (Motala 2008; Niggemann and Beyer 2007b; Sampson and Ho 1997). Patients may have a single symptom, but often there is multiorgan involvement. Symptoms can be divided into four main groups:

- Generalized Reactions. Anaphylaxis and exercise-induced anaphylaxis eventually causing death
- Respiratory reactions. Asthma and rhinitis; rarely occur without other organ involvement
- Cutaneous reactions. Skin reactions, including acute urticaria and/or angiodema are common
- Gastrointestinal reactions. Abdominal pain, nausea, vomiting, and diarrhea
- Other reactions. Conjunctivitis, oral allergy syndrome, food aversion

Another IgE-mediated reaction, the "oral allergy syndrome" (OAS), is frequently recorded in patients with fruit and crustacean allergy (Motala 2008). The OAS appears often as a "cluster of hypersensitivity" where patients are simultaneously sensitive to pollen and fruits and/or vegetables. Symptoms occur within minutes of ingesting the offending food. Local oral symptoms include itching of the lips, mouth or pharynx, and swelling of the lips, tongue, palate, and throat. Systemic manifestations such as simultaneous urticaria, rhinitis, asthma, or anaphylaxis can be seen.

23.2.4 Prevalence of Food Allergy

Allergies to foods are a significant public health concern throughout the world, affecting up to 4% of adults and 8% of children in the general population. However, self-diagnosed food allergy is reported by up to 30% (Vierk et al. 2007) (http://www.fda.gov/ForConsumers/ConsumerUpdates/ucm089307.htm) and severe anaphylactic reactions have gained epidemic proportions in the United States, as demonstrated by Simons et al. (2008) and Sampson. The prevalence of food allergy is usually higher when the consumption plays a greater part in the diet of the observed community. More than 90% of allergic reactions to food can be attributed to exposure to eight foods or food groups (the "Big 8"), including; eggs, milk, soybeans, tree nuts, wheat, fish, shellfish, and peanuts. The last three foods commonly provoke severe food anaphylaxis. A recent survey from the United States among the general adult population (Sampson 2004; Vierk et al. 2007) highlighted the most common allergens as dairy products followed by fruits/vegetables and shellfish (see Figure 23.3). The prevalence of allergy to particular foods is different among adults and children. The predominant allergens among children are milk and eggs, while fish allergy is more common among children and shellfish allergy among adults. However a recent analysis of several population-based studies, involving over 250,000 participants, concluded for allergies to plant products that there is considerable heterogeneity in prevalence estimates of sensitization or perceived allergic reaction to plant allergens. Problems are mostly based on different diagnostic methods for confirming the presence of a true food allergy (Asero et al. 2007). Nevertheless, the likelihood of becoming sensitized to a particular food allergen seems to correlate with geographical eating habits. For example, allergy to peanuts and milk products is more common in Western countries,

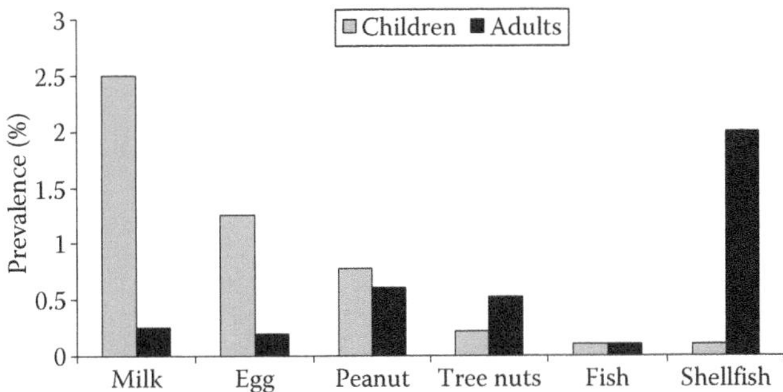

FIGURE 23.3 Prevalence profile of food allergy among children and adults in the United States. (Adapted from Sampson, H.A., *Journal of Allergy and Clinical Immunology,* 113(5), 805–19, 2004.)

whereas seafood and rice allergy is more common in Asia (Hill et al. 1997). It is estimated that about 30,000 food-induced anaphylactic events are seen annually in the United States alone, of which about 200 are fatal, and these figures seem to have increased steadily over the last decade (http://www.fda.gov/ForConsumers/ConsumerUpdates/ucm089307.htm).

23.2.5 Diagnosis of Food Allergy

Diagnostic methods of establishing a true food allergy mediated by IgE antibody include various in vitro tests: skin prick testing (SPT), the atopy patch test (APT), the quantification of specific IgE antibodies (using commercial assays such as ImmunCAP, ELISA, and allergen-microarray), and immunoblotting to determine the molecular weight of the allergen (Asero et al. 2007). However, the gold standard is the oral double-blind placebo-controlled food challenge (DBPCFC), which has to be conducted in a medical setting. Alternative approaches are used widely in the diagnosis of food allergy, such as food-specific IgG antibodies, cytotoxic food tests, kinesiology, intradermal provocation tests, or electrodermal tests. None of these tests have been proven by properly performed studies and can therefore not be recommended in the diagnosis of food allergy (Niggemann and Gruber 2004). Several studies have attempted to establish how much food is needed to trigger an allergic reaction by challenging the individuals with the offending food. The DBPCFC indicates that as little as one medium-sized shrimp or one peanut is enough to cause severe allergic reactions.

A major problem in diagnosis is the relatively poor clinical specificity e.g., both positive skin tests and in vitro tests for specific IgE are frequent in sensitized subjects without food allergy symptoms. So far, no in vitro test has reliably predicted clinical food allergy and this is often based on the nonstandardized food allergen extracts used. Therefore recent research concentrates on so-called "component resolved diagnosis," which is possible by using natural purified major allergens and/or recombinant (biotechnological) produced allergens. The identification of diagnostic marker allergens responsible for sensitizing or particularly severe clinical reactivity will profoundly influence the management of allergic patients. Furthermore, these well-defined allergenic proteins can now be used in the recent development of protein array technology.

23.2.6 Allergy to Food in the Workplace

The food industry is one of the largest employers of workers exposed to numerous allergens that are capable of inducing immunological reactions, not via ingestion but by inhalation of food allergens, leading to occupational diseases (Goetz and Whisman 2000; Jeebhay et al. 2001; Spok 2006; Taylor et al. 2000). Such reactions can occur at every level of the industry, from growing/harvesting of crops or animals, storage of grains, to processing of food substances. The spectrum of occupational diseases most commonly seen in the food industry includes occupational asthma, rhinitis, conjunctivitis, and dermatitis. Occupational asthma represents up to 20% of all asthma cases worldwide and is the most common form of occupational lung disease. Occupational skin diseases may represent up to 15% of all occupational diseases and have significant economic impact.

23.3 FOOD ALLERGENS

Over 160 different foods have been reported to cause allergic reactions, however about 90% of allergic reactions to food are caused by the "big 8" food allergens. Foods contain a wide variety of proteins, yet only a few are known allergens. It is tempting to conclude that the likelihood of allergenicity correlates with the degree of exposure to a particular protein and the dominance of this protein in a food. In plants, many of the allergens are storage proteins presenting up to 80% of the total protein in the offending foods. However, proteins that occur only in minor amounts in foods can also be major food allergens. This has been shown for the major allergen from codfish (Gad c 1) where this allergen is not a predominant protein, with less than 0.1%. On the other hand, major components of many foods, such as actin, myosin, and tropomyosin from chicken, beef, and pork, have not been identified as major allergens. However, tropomyosin is the major allergen found in shrimps and other crustaceans. The currently characterized food allergens are usually water-soluble glycoproteins, 10–70 kDa in size, and fairly stable to heat, acid, and proteases.

The portion of a food protein that binds the IgE antibody (thereby causing an allergic reaction) may be a simple stretch of a few amino acids along the primary structure or it may be a unique three-dimensional motif of the protein structure, respectively referred to as linear and conformational epitopes. An allergenic protein may contain a single epitope that is repeating or may have several different epitopes. In order to have IgE cross-linking, there must be more than one epitope on the allergen. The relationships between the nature of the allergenic epitopes and the corresponding clinical symptoms and their severity are unclear. Understanding these relationships is crucial in designing ways to reduce or eliminate allergenicity of the targeted food allergens.

23.3.1 Multiple Allergens

Most allergenic foods have more than one allergen. In an attempt to organize significant allergens according to their patient reactivity, a classification system has been developed. Allergens to which more than 50% of sensitive patients react are described as "major allergens," in contrast to "minor allergens," which have a lower frequency of reactivity in patients. In addition, some allergens occur as a group of proteins having very similar physical, chemical, and immunochemical structures. They differ only slightly in their isoelectric points and are termed "isoallergens." At least 67% of the amino acid residues in these proteins must be identical. The slight differences observed are due to differences in the carbohydrate moiety, degree of protein amidation, or genetic variation (Boldt et al. 2005). Some allergens (e.g., profilins and tropomyosin) are found in many different plants and crustaceans respectively and are so-called "pan-allergens."

23.3.2 Food-Pollen Allergens

In young children under the age of three years, animal food proteins (milk, egg) are most commonly involved in allergic reactions and reactions to plant foods (except nuts) are infrequent (Egger et al. 2006). However, these prevalence data change rapidly after about five years of age, where fresh fruit is one of the most common sensitizers. The physicochemical properties of the food allergens involved and their resistance to proteolytic (digestion) and processing activities seems to be the key factor to induce sensitization directly through the oral route. These allergens are called "Class I allergy," in direct contrast to food allergens that are labile to digestion/processing ("Class II allergy"), which cannot induce allergy via the oral route, but elicit symptoms in the oral mucosa on ingestion (OAS). In these cases the primary sensitization occurs via the respiratory tract to a homologues protein in an inhalant (mostly pollen) and the food allergy appears later as a consequence of IgE cross-reactivity. Typically this is the case in plant food allergies (e.g., apple, cherry, and peach) linked to birch pollen allergy.

The number of allergen sources involved, the allergens, and influencing factors including geography, diet, and food processing contribute to the high complexity of pollen-food allergy. Currently we know the involvement of several allergens such as profilins (labile allergen), lipid transfer proteins (LTPs), and high-molecular-weight allergens, as well as glycoallergens containing α1,3-fucose and β1,2-xylose.

23.3.3 Allergens: Families and Biochemical Functions

In recent years, our knowledge on the structure of some of the important allergens has increased greatly due to the application of sophisticated molecular biological techniques and the production of recombinant allergens. Since the cloning of the first allergenic protein (from codfish) in the late 1980s, hundreds of allergens have been identified and their sequences determined; for over 40, three-dimensional structures are available. Purified allergens are named using the systematic nomenclature of the Allergen Nomenclature Sub-Committee of the World Health Organization and International Union of Immunological Societies. The system uses abbreviated Linnaean genus and species names and an Arabic number to indicate the chronology of allergen purification (Chapman et al. 2007). Subsequent to this vast amount of information becoming available, a number of databases were developed that provided molecular and biochemical as well as clinical data such as the International Union of Immunological Societies Allergen Nomenclature Sub-committee (www.allergen.org), the Allergome (www.allergome.org), and the InFormAll database (foodallergens.ifr.ac.uk/). The growing number of available allergen sequences together with advanced bioinformatics enabled scientists to see the evolutionary and structural relationships between all these allergens in a different light. A novel classification system of allergens was subsequently developed and revealed that most allergens are restricted to few protein families (http://www.meduniwien.ac.at/allergens/allfam) (Radauer et al. 2008), based on seven hundred and seven allergen aminoacid sequences. In contrast to previous studies, all these 707 allergen sequences could be classified into 134 families, which represent only 5% of the 3012 analyzed protein families. In addition it was surprising that the majority of allergens had limited biochemical functions (sometimes more than one function for one allergen):

A: Hydrolytic enzymes (~18% of all allergens)

Many of the allergens from plants, bacteria, fungi, and house dust mites are hydrolytic enzymes. These proteins are usually involved in cleaving substrates by the addition of water. They include the following groups:

- Proteases (~50% of **A**): Serine proteases (often occupational allergens such as trypsin, chymotrypsin, and subtilisins); cysteine proteases (papain); aspartate proteases (pepsin and rennin); metalloproteinases; collagenase
- Carbohydrases (~1% of **A**): Including amylase, cellulase, xylanase, lysozyme (hen egg white and latex), and polygalacturonase
- Lipases (~1% of **A**)

B: Binding of metal ions and lipids (~4%)

- Parvalbumin (fish); serum albumin (nuts, seeds); globin, enolase, Fe/Mn superoxide dismutase
- Lipid transfer protein (LTP) (many fruits and vegetables)

C: Storage

1. Cupins and prolamin proteins

D: Cytoskeleton

- Tropomyosin (crustacean); profilins (plants)

Grouping the allergen-containing protein families according to the ingestion route of exposure, the currently known 284 food allergens can be divided into 31 allergen families (see Figure 23.4). The seven families with the highest number of allergen represent over 70% of all ingested allergens. Most allergen families were confined to a single source kingdom, such as prolamins, profilins, LTPs, and cupins from plants and tropomyosin, lipocalins, and casein from animals. Only a minority of protein families such as the EF-hand family (e.g., fish) and the pathogenesis-related proteins (PR-1) contained allergens from multiple kingdoms; these are, however, mostly inhaled. Interestingly, most protein families contain allergens that sensitize via ingestion as

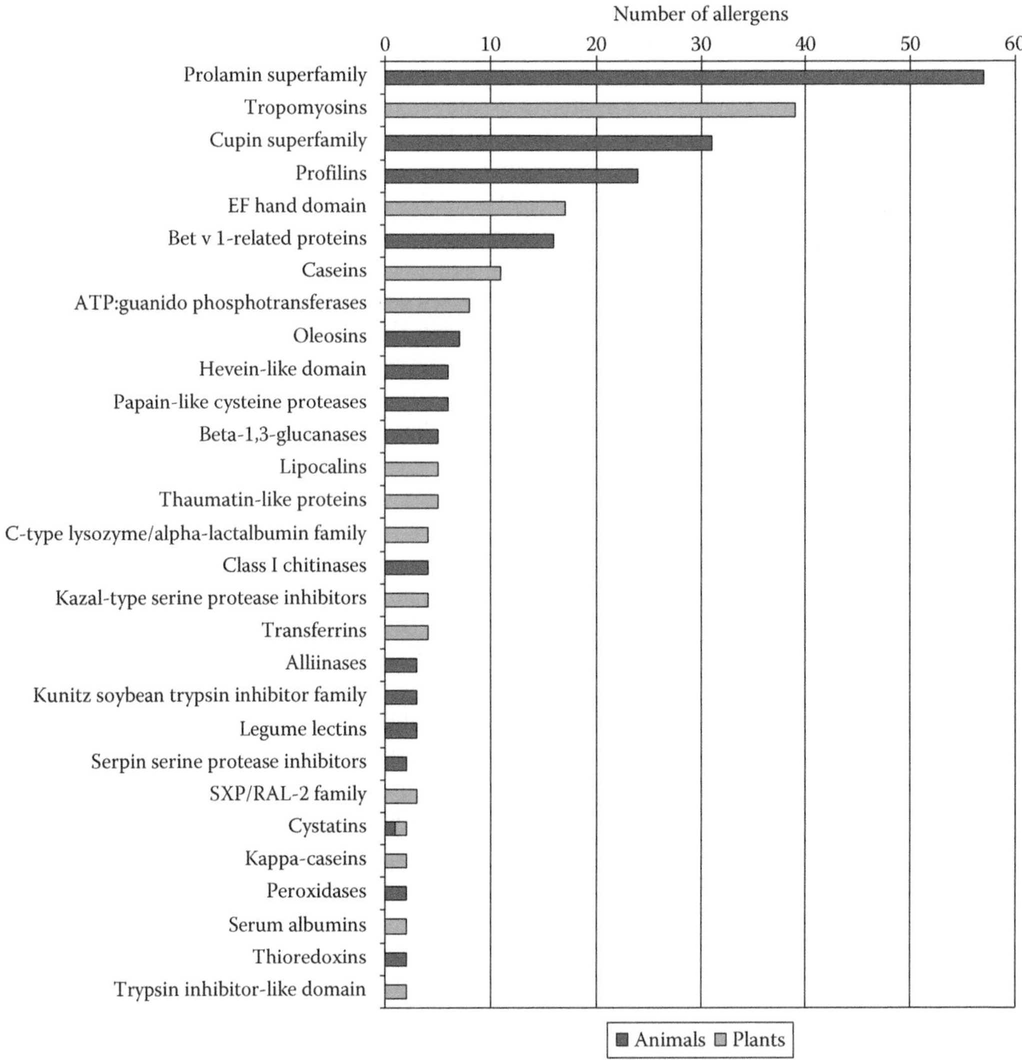

FIGURE 23.4 Food allergens classified by major protein families (http://www.meduniwien.ac.at/allergens/allfam). (Adapted from Radauer, C., Bublin, M., Wagner, S., Mari, A., and Breiteneder, H., *Journal of Allergy and Clinical Immunology,* 121(4), 847–52, 2008.)

TABLE 23.1
Superfamilies of Food Allergens are Subdivided into Allergen Families That Can Be Present in Foods From Different Food Groups

Superfamily	Allergen Families	Examples of Allergens
Prolamin superfamily	2S albumins	Yellow mustard, Brazil nut, English walnut, peanut
	Bifunctional α-amylase/protease inhibitors	Seeds of wheat, barley, rye, corn, rice
	Nonspecific lipid transfer proteins	Apples, peaches, plums, strawberries, almonds
	Soybean hydrophobic protein	Soybean hulls
	Alpha-globulins	Wheat, barley, rice, durum wheat, buckwheat
Cupin superfamily	7/8 globulins (vicilins)	Peanut, walnut, sesame, cashew nut, hazelnut
	11S globulins (legumins)	Peanut, soybean, Brazil nut, buckwheat, pistachio
–	Tropomyosins	Crustaceans, mites, anisakis parasite, cephalopods
–	Profilins	Pineapple, celery, watermelon, bell pepper, carrot
–	E-F hands	Fish, mollusks
–	Bet v 1-related proteins	Celery, carrot, apple, asparagus, chestnut

well as inhalation. These are the allergens causing the pollen-food allergy such as profilins (plants) and birch pollen-related allergens. However, also inhalation of food allergen from animals can cause sensitization, which is frequently observed in the workplace (e.g., tropomyosin and parvalbumin among seafood processors) (Jeebhay et al. 2001; Jeebhay et al. 2008; Lopata et al. 2005).

Comparing amino acid sequences of allergens of the two most important protein families (prolamine, cupin) of allergens, a broad degree of sequence identities causing cross-reactivity between different food groups as well as the presence of different allergen groups in one food become obvious (see Table 23.1). For example, the allergenic tropomyosins from invertebrates (crustacean, mites) and profilins from higher plants show a high degree of similarities, well over 50%. This results in a high degree of IgE antibody cross-reactivity among patients with allergy. Very different allergen groups are the 2S albumins, important food allergens from legumes, nuts, and other seeds, which demonstrate sequence identities of less than 40%. While a low degree of cross-reactivity is predicted, recent publications show a high degree of cross-reactivity among patients to peanut, almond, and Brazil nut. The reason for these findings is probably the importance of linear IgE-antibody-binding epitopes in food allergens as they come into contact with the human immune system only *after* processing of foods, as well as partial denaturation in the digestive tract.

As demonstrated above, allergens such as lipid transfer protein (LTP) can be found in a variety of plants and therefore are called pan-allergens. Nevertheless, the concentration of LTP not only varies among different plants but can also show great variation between different cultivars. A recent study by Sancho et al. (2008) used three different immunological methods to quantify the amount of LTP in 88 apple cultivars from the Netherlands and Italy. The differences were up to 100-fold, providing important information for both consumers as well as growers. Furthermore, the concentration of LTP in apple peel is much higher than in the pulp. The potential production of hypoallergenic apple cultivars and/or products would be of great benefit for allergic consumers but also for subjects with already existing cross-reactive IgE antibodies LTP to pollen allergens (e.g., birch pollen).

In summary, the recent research data conclude that allergens are distributed among a small number of protein families and possess a limited range of biologic functions. These findings have triggered the search for common properties of allergens, essential for risk assessment of novel foods and processing technologies. The answer to the question of what makes a protein an allergen will require additional both in silico and wet lab research, with respect to differential allergen expression levels (Scheurer and Sonnewald 2009) and stability during various processing techniques.

23.3.4 Characterization and Detection of Allergens

Various biophysical methods have been employed for assessing the stability of allergens, such as Sodium dodecyl sulphate-polyacrylamide gel electrophoresis (SDS-PAGE), chromatography, and circular dichroism, which demonstrate that processing of food can alter linear and conformational epitopes. The IgE-binding capacity of proteins for foods is a prerequisite of clinical reactivity and tests based on measuring these reactivities are recommended by FAO/WHO, the Codex Alimenatarius, and EFSA (efsa.europa.eu/EFSA/efsa_locale-1178620753812_1178620761196.htm). Peptides are believed to require a molecular weight of 3.5 kDa or greater in order to cross-link IgE antibody on mast cells and elicit an allergic reaction. Therefore large stable protein fragments as well as intact proteins have the potential to elicit an allergic reaction. The antibody-binding sites on the protein can be sequential and/or conformational; the latter can be modified by altering the three-dimensional structure (using chemical or physical approaches). Food allergens usually have physicochemical properties that make them very stable and resistant to heat and digestion, which results in retained IgE-binding capacity and cell-activation properties.

As the world trade and market of novel (e.g., Genetically modified organisms (GMO) or altered physicochemical structure) or natural novel foods (not endemic in a new market) is increasing, the question arises if a particular food has allergenic properties. An interesting example is the kiwi fruit, which became popular in Europe in the 1970s. Nowadays, kiwi is one of the most frequent causes of fruit allergy in Central Europe. Particularly affected are subjects with cross-reactive IgE antibodies as a consequence of already-existing allergy to birch pollen and latex, as the allergens are very similar. The European Union has developed guidelines to assess the safety of GMO foods and ingredients (Regulation EC No 258, 1997), but not for natural novel foods. A recent study developed a scientific strategy to assess the allergenic risk of exotic foods, using water spinach, hyacinth bean, and Ethiopian eggplant as examples (Gubesch et al. 2007). Known allergens were quantified (in this case LTP and profilin) by immunoblotting with specific animal antibodies; quantification and detection of specific IgE antibodies of patients with know allergy to pan-allergens (e.g., LTP) or to phylogenetically related foods (e.g., tomato with eggplant); and thirdly the clinical relevance established by SPT and oral food challenge (Breiteneder and Mills 2005).

To reduce the risk of allergic reactions in consumers, the Food and Drug Associations in the United States (FDA) and many other countries are working to ensure that major allergenic ingredients in food are accurately labeled in accordance with the Food Allergen Labeling and Consumer Protection Act of 2004 (FALCPA). A comprehensive food labeling law has been in effect since 2006 and includes the "big 8" that contain protein derived from one of the following foods or food groups:

- Milk
- Eggs
- Peanuts
- Tree nuts such as almonds, walnuts, and pecans
- Soybeans

- Wheat
- Fish
- Shellfish such as crab, lobster, and shrimp

These foods/food groups cause over 90% of all food allergic reactions in the United States. However, as discussed above, geographical eating habits result in over 160 foods being reported as causing allergic reactions. In Europe, for example, a high prevalence of allergy to celery and mustard is reported and is therefore reflected in food labeling. In addition, some food ingredients such as edible oils, hydrolyzed proteins, lecithin, gelatine, starch, lactose, and flavors may be derived from major food allergens (Hansen et al. 2004; Taylor and Dormedy 1998). The role that these ingredients play in food allergy has not been fully characterized. For example, hot dogs formulated with partially hydrolyzed casein have elicited allergic reactions in children allergic to cows milk. Gelatins are ingredients derived from animals (e.g., cows, pigs) but also from the skin of various fish species (Hansen et al. 2004). Fish collagen, derived from the swim bladder of fish, is the main substance in "isinglass," commonly used in alcoholic beverages (beer, wine) to clarify the liquid. Edible oils can be derived from major food allergens such as soybeans and peanuts, and may contain variable levels of allergenic protein. The labeling of foods containing allergens has already become mandatory in some countries such as in the United States (http://www.fda.gov/ForConsumers/ConsumerUpdates/ucm089307.htm), Japan, and Europe (http://efsa.europa.eu/EFSA/efsa_locale-1178620753812_1178620761196.htm) (Taylor and Hefle 2006). Food labels are a critical source of product-specific information for individuals with food allergy and parents of children with food allergy. However, a recent survey concluded that over 50% of individuals find it necessary to contact the manufacturer. Problems arise mainly as the ingredients list gives general names without specifying the source (e.g., spices, flavors), the use of different words for allergenic food on different food products, and the use of words on ingredient lists that are too technical or hard to understand (Vierk et al. 2007).

In addition, there has been widespread use of allergen advisory labels on products that may have allergenic ingredients that were introduced by way of cross-contact (cross-contamination) during the manufacturing process. Cross-contact occurs when a residue or other trace amount of an allergenic food is unintentionally incorporated into another food.

Cross contact may occur during:

- Harvesting
- Transportation
- Manufacturing
- Processing
- Storage

Additional problems might also arise with the increasing number of food, medical, and health products derived from animal products. Chitin and chitosan are among the emerging materials that are being developed and applied widely in the food, biotechnology, pharmaceutical, and medical fields (No et al. 2007). The main obstacle for the future use of this unique material is the residual amount of about 1% protein in industrially produced chitosan. Allergic reactions after using chitosan-containing food have been reported (Kato, Yagami, and Matsunaga 2005; Villacis et al. 2006), however the contribution of the thermostabile tropomyosin or other yet unidentified crustacean allergens has not been demonstrated. Another pharmaceutical products derived from crustaceans is glucosamine, a natural aminomonosaccharide, which is frequently used as a therapeutic supplement for joint inflammation. In addition, food products can also be unexpectedly derived from different food sources. For example surimi (seafood paste) is usually produced from fish, but can in some countries contain a variety of crustacean species (Hamada et al. 2000).

23.4 DIGESTIVE PROCESS

The digestive process is likely to play a major role in the development of allergic sensitization but also in the clinical severity of food allergy symptoms. As human and animal studies on food digestion are very rare, in vitro models that mimic in vivo systems have been developed and focus on three main areas: processing in the mouth, stomach, and duodenum. Two models types are distinguished: the static model (biochemical) and the dynamic model, which in addition mimics physical in vivo conditions (e.g., shear, hydration, and mixing). The latter model allows the investigation of integrated effects of the food matrix and impact on kinetics of allergen release and breakdown, digestive and metabolic processes, as well as the potential effects of food processing in modulating allergenic properties of foods (Moreno et al. 2005). The bioavailability of proteins is known to influence the absorption of allergens and subsequently potential impact on an allergic response. It was recently demonstrated by Untersmayr and Jensen-Jarolim (2006a) that the reduction of gastric acid in the stomach of people (using antacid medication) resulted in increased uptake of allergenic fish proteins into the blood stream. These results highlight the increased risk of allergic reactions if the ingested allergens have a longer bioavailability in the digestive tract by reduced denaturation (via reduced stomach acid) or potentially by stabilization of the allergen during food processing.

23.5 FOOD PROCESSING

Food and food ingredients have many diverse forms and are subjected to a large variety of processing conditions in order to improve their sensory qualities and/or shelf life by removing/inactivating toxins and/or microbes and modifications of their properties (e.g., improved digestibility) to suit the consumer and/or derive a large diversity of additional food products (e.g., oils and isolates). Much of our understanding of the effects of food processing on protein and different food structures has been obtained from studying model food proteins such as whey and egg proteins as well as the 11S and 7S seed storage globulins, found in many edible nuts and seeds. These proteins play an important role in forming the structure of processed foods such as foams (whipped egg), gel networks (boiled eggs or cooked meat), and emulsifying oil droplets, and forming glassy states with sugar molecules in low water foods (e.g., pasta and biscuits). The types of modification that the food proteins undergo during processing include protein unfolding, aggregation, and chemical modification (see Figure 23.5). All of these have the potential to increase stability to digestion and can affect patterns and kinetics (in the presence of fats and sugars) of food denaturation, hence enhancing the allergenicity of an allergen. Various processing conditions can alter immunodominant IgE-antibody-binding epitopes (linear and conformational), whereas conformational epitopes are typically expected to be more susceptible to processing-induced destruction. Linear epitopes, on the other hand (even on the same allergen), are more likely to be altered by hydrolyzation, and chemically modified during food processing or by (intentional) mutations through genetic engineering. Not only can processing destroy existing epitopes, but it can also generate new ones (so-called neoallergens) as a result of change in protein conformation. Most recently, neoallergens have been described from pecans and wheat flour (Beaumont et al. 2005). However, more commonly, processing methods (mostly heating) have been associated with decreased allergenicity (e.g., pollen-related fresh fruit and vegetables) or with no significant effect (e.g., heat-stable allergens from shrimp) (Mills and Mackie 2008; Sathe, Teuber, and Roux 2005b; Thomas et al. 2007).

Since food processing involves thermal as well as nonthermal treatments and each type of treatment can differ in its effect on allergen epitopes, individual treatments must be considered carefully when evaluating allergen stability. Thermal processing may be accomplished by dry heat or may use wet heating conditions such as the ones encountered in cooking in aqueous media (see following subsections). Nonthermal processing methods can range from irradiation to milling to fermentation (see following subsections). A few examples noted below illustrate the diversity and complexity of

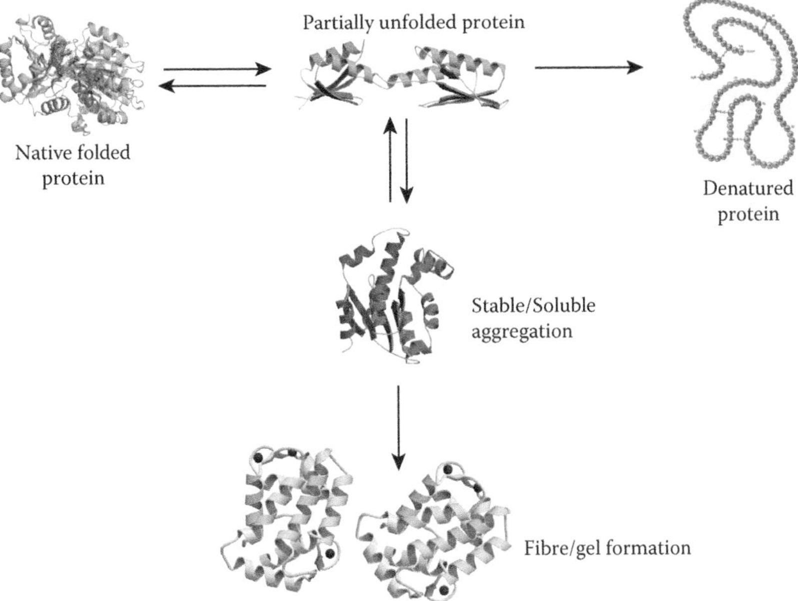

FIGURE 23.5 Possible formation and aggregation of proteins during and after processing.

issues involved in understanding the effectiveness and limitations of food processing methods as a tool in attempting to reduce or eliminate food allergens.

a. Thermal processing
   - Moist heat
   - Dry heat

b. Nonthermal processing
   - Germination
   - Fermentation
   - Proteolysis
   - Ultrafiltration
   - Dehulling
   - γ-Irradiation
   - High-pressure processing

23.5.1 Thermal Processing

23.5.1.1 Moist Heat

The best example of food allergens that are labile to food processing technologies are the cross-reactive superfamily of birch pollen (Bet v 1)-food allergens. Two of the cross-reactive epitopes in cherry, which are homologous to the Bet v 1 allergen, are conformational and therefore, cooking of fruits such as cherry reduces their allergenicity. However, the thermostability of Bet v 1-homolog food allergens is not equivalent. For example the Bet v 1 homologue from apple (Mal d 1) unfolds on heating to 90°C and does not significantly refold on cooling to 20°C. In contrast, Api g 1 from celery regains most of its native structure after cooling, explaining the retained allergenicity in cooked celery (Marzban et al. 2009).

One of the most widely studied food proteins is the globular protein β-lactoglobulin (β-Lg), one of the main components of the whey fraction in ingredients such as whey protein isolates and concentrates. β-Lg is a small protein with a β-barrel structure characteristic of the lipocalin superfamily. Lipocalins are stabilized by two intramolecular disulfide bonds (Cys^{106}-Cys^{119} and Cys^{66}-Cys^{160}) together with a single free cysteine residue (Cys^{121}) (Brownlow et al. 1997). On heating, β-Lg first dissociates into monomers which then partially unfold before associating into thread-like aggregates around 50 nm in diameter, which at high protein concentrations will form string-like aggregates and gel networks (Ikeda and Morris 2002). Unfolding of the protein exposes the buried Cys^{121}, which then can catalyze disulfide interchanges with other proteins such as caseins (highly stable allergen).

Other very heat-resistant allergens are members of the cupin superfamily (e.g., soya, nuts). For example the 7S globulins have their major thermal transition at around 70–75°C, whilst 11S globulins unfold at temperatures above 94°C, as measured by differential scanning calorimetry.

23.5.1.2 Dry Heat

The dry heating of food (e.g., roasting) can result in enhanced but also decreased allergenicity of a particular allergen. Usually two types of reactions are distinguished: enzymatic browning (polyphenol oxidase-catalyzed oxidation of 3,4 diols) and the nonenzymatic Maillard reaction. In the latter reaction free amino groups on proteins bind to the aldehyde or ketone groups of sugars. These glycation reactions can further undergo rearrangements to structurally diverse compounds known as Amadori products or advanced glycation end products (AGEs). The formation of these adducts is important for the aromas and flavors associated with many cooked foods and is affected by types of nonreducing sugars, pH, water activity, and temperature. A study on wheat allergens demonstrated that some of the allergenic proteins were cross-linked by Maillard adducts (Simonato et al. 2001). Furthermore, the allergenicity of proteins can be dramatically increased, as demonstrated for peanuts (Maleki et al. 2000). The Maillard modification of the peanut allergens Ara h 1 and Ara h 2 was demonstrated by higher IgE-binding capacity as well as higher resistance to gastric digestion. Serum from peanut allergic patients reacted much more strongly to roasted peanut than to boiled or fried peanuts (Morrow et al. 2001). Maillard reaction can also modify the allergenicity of animal proteins, as demonstrated for seafood. The IgE-binding capacity was increased for scallop tropomyosin (Nakamura et al. 2005) but decreased for squid (calamari) tropomyosin (Nakamura et al. 2006).

In contrast, glycation of fruit allergens seems to decrease allergenicity. Glycation of the major cherry allergen (Pru av 1) with sugars such as fructose and ribose significantly reduced the IgE reactivity, while modification with glycoaldehyde almost completely abolished the reactivity (Gruber et al. 2004).

23.5.2 Nonthermal Processing

23.5.2.1 Germination

During germination catalytic enzymes such as proteinases and amylases mobilize stored nutrients, proteins, and carbohydrates to provide for seedling growth. Germination may help eliminate certain epitopes in seed storage proteins, depending on the enzyme specificity and the susceptibility of epitopes to the enzymes active during the germination period. However, the major cottonseed allergen has been shown to be stable during germination (Kang et al. 2007; Yamada et al. 2005). Additional studies are needed to determine if germination is effective in reducing or eliminating seed allergenicity particularly, as soy- and mung beans are often used for preparing various products.

23.5.2.2 Fermentation

Soy is also often used in fermented food products containing soy, wheat, and fish (Park et al. 2007). It has recently been demonstrated for some products that soy allergenicity is retained in the finished product. However, the retained immunoreactivity was assessed by in vitro inhibition assays for β-Lg and significantly impaired by some acidified milk products such as yogurt (Ehn et al. 2004) and has to be taken into account in quantifying residual allergenicity. The use of fermented soy was recently demonstrated in a murine model to even have protective properties in peanut allergy (Zhang et al. 2008).

23.5.2.3 Proteolysis

The role of protein structure in IgE binding is clear from the involvement of linear and conformational epitopes. Proteases such as trypsin, chymotrypsin and pepsin have little or no effect on the allergenicity of peanut allergens, demonstrating the importance of linear epitopes in food allergies.

23.5.2.4 Ultrafiltration

As demonstrated above, some of the fruit allergens are very heat-stable (stable at 121°C for 10 and 30 min), for example, peach (Breiteneder and Mills 2005). Only a final ultrafiltration step was able to produce hypoallergenic peach juice (Brenna et al. 2000). However, the desirable sensory quality was lost, probably due to the loss of juice cloudiness (attributed to carbohydrates) and mouthfeel (attributed to carbohydrates, viscosity), and possibly due to physical removal of other molecules critical for sensory properties.

23.5.2.5 γ-Irradiation

The effect of γ-irradiation was recently investigated with peeled shrimp at 1, 5, 10, and 15 kGγ with and without heat treatment (Li et al. 2007). The allergenicity was quantified by immunoassay for tropomyosin and immunoblot with patient serum and could demonstrate a reduced IgE-binding reactivity with increasing radiation.

23.5.2.6 High-Pressure Processing

It was recently suggested that high-pressure processing above 600 MPa can cause reversible but also nonreversible changes to the secondary, quaternary, and tertiary structure of allergens, depending on pressure and holding time (Meyer-Pittroff, Behrendt, and Ring 2007). This was demonstrated for apples treated at 600 mPa for 5 min, which patients with apple allergy could consume without any symptoms. Furthermore, α-helices seem to be more pressure sensitive than β-sheets. This was confirmed by a recent study on shrimp allergens where a significant reduction of tropomyosin was demonstrated by immunoblotting and quantitative Enzyme-linked immunosorbent assay (ELISA) (Yohannes and Lopata 2008).

23.5.2.7 Multiple Methods

While heat processing may inactivate certain structural epitopes, such treatments are unlikely to eliminate allergenicity of nuts (Sathe, Kshirsagar, and Roux 2005a) and certain fruits (e.g., mango, apple) (Dube et al. 2004). Further studies using combinations of different techniques in combination with DBPCFC studies are needed to critically evaluate effects of processing on allergenicity of particularly stable allergens. The different allergen stabilities can be grouped into four categories and are ultimately determined by the physicobiochemical characteristics of the allergen or allergen family (see Table 23.2).

TABLE 23.2
Impact of Food Processing on Different Types of Food Allergens

	Allergen Type During Processing	Effect of Thermal Processing	Physicobiochemical Structures/Epitopes	Type of Food Allergens (Examples)
I	Processing labile	Protein unfolding, Maillard modification by sugar and polyphenols	Predominantly conformational epitopes	Birch pollen (Bet v 1) homologues (e.g., Mald d 1 in apple, Pru v 1 in cherry)
II	Partial denaturation	Partial unfolding of protein, aggregation to networks (e.g., gels), Maillard reaction may potentiate allergen	Lipocalin superfamily: 4 cysteine residues forming 2 intramolecular disulfide; β-barrel structure	Lipocalin superfamily members (e.g., β-lactoglobulin), cupin allergen family (e.g., Ara h 1 from peanut)
III	Refolding able	Limited unfolding but refolding during cooling, Maillard reaction may potentiate allergen	Prolamin superfamily: 6 or 8 cysteine residues forming 3 or 4 intrachain disulfide bonds	Prolamin superfamily members (e.g., LTP, Mald d 3 in apple, 2S albumin in nuts), tropomyosin and parvalbumin
IV	Processing stabile (rheomorphic)	No denaturation but very mobile conformation	Disordered mobile structure; many linear epitopes (not conformational)	Prolamin superfamily members (e.g., wheat (gluten), ovomucoid, caseins

23.6 CONCLUDING REMARKS

The best way to prevent unintended exposure to a food allergen is the complete avoidance of the offending food. For various reasons, such avoidance may not always be possible or preferred. Accurate food labeling in conjunction with good manufacturing practices can enhance consumer safety and aid food processors, manufacturers, distributors, and retailers to utilize foods/food ingredients in an efficient and safe manner. As the presence of trace contaminants of offending agents cannot be ruled out at all times, robust and sensitive detection methods, capable of detecting trace quantities of the immunological reactive food allergen (also in the processed form), must be developed.

Processing may alter food in a manner that may permit masking or unmasking of allergenic epitopes, thereby reducing or enhancing allergen recognition and therefore potentially altering allergenicity of the offending food. Alteration in protein structure (by food processing) can lead to epitope destruction, modification, digestion stability, and kinetics of allergen release, thereby decreasing, increasing, or having no effect on allergenicity.

Our current lack of understanding of the impact of conventional food processing procedures on food allergenicity makes allergen management and the allergenic risk assessment process difficult (see Figure 23.6). The clinical findings in a particular population have to be evaluated on a molecular level regarding potency of an allergen as well as the degree of exposure. The complexity of food processing demonstrates the importance of understanding the impact on molecular level and moving toward knowledge-based ways of managing allergen risk. The close collaboration between food chemists, molecular scientists, and clinical investigators is essential when addressing the management of food allergy and the development of novel processing strategies to reduce/eliminate allergenicity of foods.

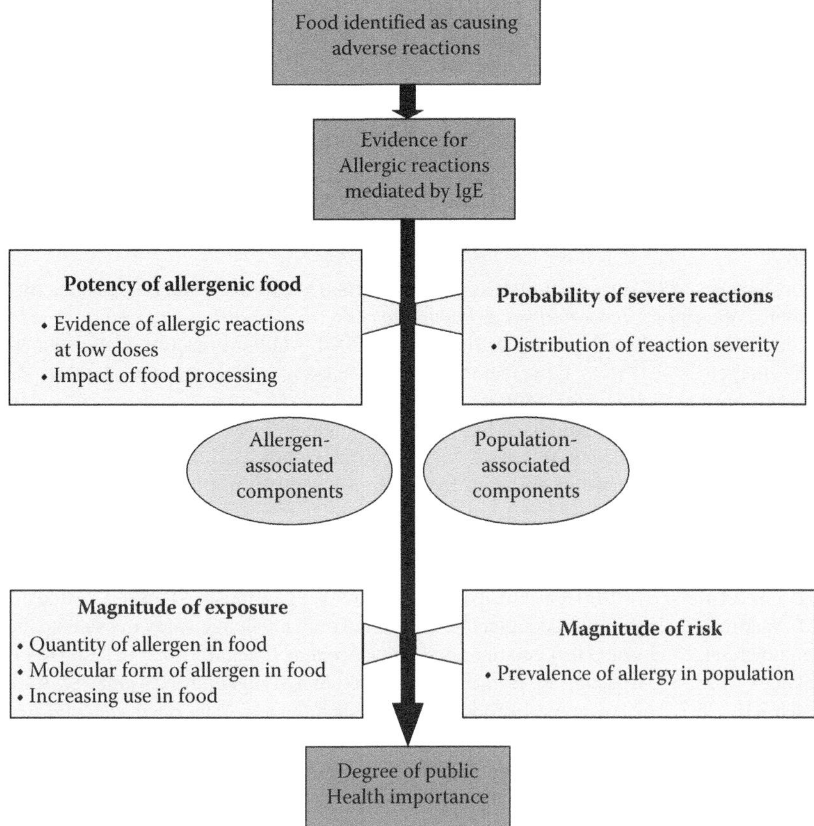

FIGURE 23.6 Criteria used to identify a food of public health importance. (Adapted from Bjorksten, B., Crevel, R., Hischenhuber, C., Lovik, M., Samuels, F., Strobel, S., Taylor, S.L., Wal, J.M., and Ward, R., *Regulatory Toxicology and Pharmacology*, 51(1), 42–52, 2008.)

REFERENCES

Asero R, Ballmer-Weber BK, Beyer K, Conti A, Dubakiene R, Fernandez-Rivas M, Hoffmann-Sommergruber K, Lidholm J, Mustakov T, Elberink JNGO et al. 2007. IgE-Mediated food allergy diagnosis: Current status and new perspectives. *Molecular Nutrition & Food Research* 51 (1): 135–47.

Auerswald L, Morren C, Lopata AL. 2006. Histamine levels in seventeen species of fresh and processed South African seafood. *Food Chemistry* 98 (2):231–39.

Beaumont P, Leduc V, Battais F, Guerin L. 2005. Allergy to a wheat isolate contained in pasta. *Revue Francaise d'Allergologie et d'Immunologie Clinique* 45 (8): 612–14.

Bjorksten B, Crevel R, Hischenhuber C, Lovik M, Samuels F, Strobel S, Taylor SL, Wal JM, Ward R. 2008. Criteria for identifying allergenic foods of public health importance. *Regulatory Toxicology and Pharmacology* 51 (1): 42–52.

Boldt A, Fortunato D, Conti A, Petersen A, Ballmer-Weber B, Lepp U, Reese G, Becker WM. 2005. Analysis of the composition of an immunoglobulin E reactive high molecular weight protein complex of peanut extract containing Ara h 1 and Ara h 3/4. *Proteomics* 5 (3): 675–86.

Breiteneder H, Mills C. 2005. Nonspecific lipid-transfer proteins in plant foods and pollens: An important allergen class. *Current Opinion in Allergy and Clinical Immunology* 5 (3): 275–79.

Brenna O, Pompei C, Ortolani C, Pravettoni V, Farioli L, Pastorello EA. 2000. Technological processes to decrease the allergenicity of peach juice and nectar. *Journal of Agricultural and Food Chemistry* 48 (2): 493–97.

Brownlow S, Cabral JHM, Cooper R, Flower DR, Yewdall SJ, Polikarpov I, North ACT, Sawyer L. 1997. Bovine beta-lactoglobulin at 1.8 angstrom resolution: Still an enigmatic lipocalin. *Structure* 5 (4): 481–95.

Chapman MD, Pomes A, Breiteneder H, Ferreira F. 2007. Nomenclature and structural biology of allergens. *Journal of Allergy and Clinical Immunology* 119 (2): 414–20.

Dube M, Zunker K, Neidhart S, Carle R, Steinhart H, Paschke A. 2004. Effect of technological processing on the allergenicity of mangoes (*Mangifera indica* L.). *Journal of Agricultural and Food Chemistry* 52 (12): 3938–45.

Egger M, Mutschlechner S, Wopfner N, Gadermaier G, Briza P, Ferreira F. 2006. Pollen-food syndromes associated with weed pollinosis: An update from the molecular point of view. *Allergy* 61 (4): 461–76.

Ehn BM, Ekstrand B, Bengtsson U, Ahlstedt S. 2004. Modification of IgE binding during heat processing of the cow's milk allergen beta-lactoglobulin. *Journal of Agricultural and Food Chemistry* 52 (5): 1398–403.

Goetz DW, Whisman BA. 2000. Occupational asthma in a seafood restaurant worker: Cross-reactivity of shrimp and scallops. *Annals of Allergy Asthma & Immunology* 85 (6): 461–66.

Gruber P, Vieths S, Wangorsch A, Nerkamp J, Hofmann T. 2004. Maillard reaction and enzymatic browning affect the allergenicity of Pru av 1, the major allergen from cherry (*Prunus avium*). *Journal of Agricultural and Food Chemistry* 52 (12): 4002–4007.

Gubesch M, Theler B, Dutta M, Baumer B, Mathis A, Holzhauser T, Vieths S, Ballmer-Weber BK. 2007. Strategy for allergenicity assessment of 'natural novel foods': Clinical and molecular investigation of exotic vegetables (water spinach, hyacinth bean and Ethiopian eggplant). *Allergy* 62 (11): 1243–50.

Hamada Y, Genka E, Ohira M, Nagashima Y, Shiomi K. 2000. Allergenicity of fish meat paste products and surimi from walleye pollack. *Journal of the Food Hygienic Society of Japan* 41 (1): 38–43.

Hansen TK, Poulsen LK, Skov PSI, Hefle SL, Hlywka JJ, Taylor SL, Bindslev-Jensen U, Bindslev-Jensen C. 2004. A randomized, double-blinded, placebo-controlled oral, challenge study to evaluate the allergenicity of commercial, food-grade fish gelatin. *Food and Chemical Toxicology* 42 (12): 2037–44.

Hill DJ, Hosking CS, Zhie CY, Leung R, Baratwidjaja K, Iikura Y, Iyngkaran N, Gonzalez-Andaya A, Wah LB, Hsieh KH. 1997. The frequency of food allergy in Australia and Asia. *Environmental Toxicology and Pharmacology* 4 (1–2): 101–10.

Ikeda S, Morris VJ. 2002. Fine-stranded and particulate aggregates of heat-denatured whey proteins visualized by atomic force microscopy. *Biomacromolecules* 3 (2): 382–89.

Jeebhay MF, Robins TG, Lehrer SB, Lopata AL. 2001. Occupational seafood allergy: A review. *Occupational and Environmental Medicine* 58 (9): 553–62.

Jeebhay MF, Robins TG, Malo JL, Miller M, Bateman E, Smuts M, Baatjies R, Lopata AL. 2008. Occupational allergy and asthma among salt water fish processing workers. *American Journal of Industrial Medicine* 51 (12): 899–910.

Kang IH, Srivastava P, Ozias-Akins P, Gallo M. 2007. Temporal and spatial expression of the major allergens in developing and germinating peanut seed. *Plant Physiology* 144 (2): 836–45.

Kato Y, Yagami A, Matsunaga K. 2005. A case of anaphylaxis caused by the health food chitosan. *Arerugi* 54 (12): 1427–29.

Lehrer SB, Ayuso R, Reese G. 2002. Current understanding of food allergens. *Genetically Engineered Foods Assessing Potential Allergenicity* 964:69–85.

Lehrer SB, Horner WE, Reese G. 1996. Why are some proteins allergenic? Implications for biotechnology. *Critical Reviews in Food Science and Nutrition* 36 (6): 553–64.

Li ZX, Lin H, Cao LM, Jamil K. 2007. Impact of irradiation and thermal processing on the immunoreactivity of shrimp (*Penaeus vannamei*) proteins. *Journal of the Science of Food and Agriculture* 87 (6): 951–56.

Lopata AL, Jeebhay MF, Reese G, Fernandes J, Swoboda I, Robins TG, Lehrer SB. 2005. Detection of fish antigens aerosolized during fish processing using newly developed immunoassays. *International Archives of Allergy and Immunology* 138 (1): 21–28.

Lopata AL, Potter PC. 2000. Allergy and other adverse reactions to seafood. *Allergy & Clinical Immunology International* 12 (6): 271–81.

Maleki SJ, Chung SY, Champagne ET, Raufman JP. 2000. The effects of roasting on the allergenic properties of peanut proteins. *Journal of Allergy and Clinical Immunology* 106 (4): 763–68.

Marzban G, Herndl A, Pietrozotto S, Banerjee S, Obinger C, Maghuly F, Hahn R, Boscia D, Katinger H, Laimer M. 2009. Conformational changes of Mal d 2, a thaumatin-like apple allergen, induced by food processing. *Food Chemistry* 112 (4): 803–11.

McCann D, Barrett A, Cooper A, Crumpler D, Dalen L, Grimshaw K, Kitchin E, Lok K, Porteous L, Prince E et al. 2007. Food additives and hyperactive behaviour in 3-year-old and 8/9-year-old children in the community: A randomised, double-blinded, placebo-controlled trial. *Lancet* 370 (9598): 1560–67.

Meyer-Pittroff R, Behrendt H, Ring J. 2007. Specific immuno-modulation and therapy by means of high pressure treated allergens. *High Pressure Research* 27 (1): 63–67.

Mills ENC, Mackie AR. 2008. The impact of processing on allergenicity of food. *Current Opinion in Allergy and Clinical Immunology* 8 (3): 249–53.

Moreno FJ, Mellon FA, Wickham MSJ, Bottrill AR, Mills ENC. 2005. Stability of the major allergen Brazil nut 2S albumin (Ber e 1) to physiologically relevant in vitro gastrointestinal digestion. *FEBS Journal* 272 (2): 341–52.

Morrow E, Beyer K, Grishina G, Bannon GA, Burks W, Sampson HA. 2001. Comparison of Chinese and American cooking methods on allergenicity of peanut. *Journal of Allergy and Clinical Immunology* 107 (2): S139–39.

Motala C. 2008. Gastrointestinal syndromes in food allergy. *Current Allergy & Clinical Immunology* 21 (2): 76–81.

Nakamura A, Sasaki F, Watanabe K, Ojima T, Ahn DH, Saeki H. 2006. Changes in allergenicity and digestibility of squid tropomyosin during the Maillard reaction with ribose. *Journal of Agricultural and Food Chemistry* 54 (25): 9529–34.

Nakamura A, Watanabe K, Ojima T, Ahn DH, Saeki H. 2005. Effect of Maillard reaction on allergenicity of scallop tropomyosin. *Journal of Agricultural and Food Chemistry* 53 (19): 7559–64.

Niggemann B, Beyer K. 2007a. Diagnosis of food allergy in children: Toward a standardization of food challenge. *Journal of Pediatric Gastroenterology and Nutrition* 45 (4): 399–404.

———. 2007b. Pitfalls in double-blind, placebo-controlled oral food challenges. *Allergy* 62 (7): 729–32.

Niggemann B, Gruber C. 2004. Unproven diagnostic procedures in IgE-mediated allergic diseases. *Allergy* 59 (8): 806–808.

No HK, Meyers SP, Prinyawiwatkul W, Xu Z. 2007. Applications of chitosan for improvement of quality and shelf life of foods: A review. *Journal of Food Science* 72 (5): R87–100.

Palmer K, Burks W. 2006. Current developments in peanut allergy. *Current Opinion in Allergy and Clinical Immunology* 6 (3): 202–206.

Park JG, Saeki H, Nakamura A, Kim KBWR, Lee JW, Byun MW, Kim SM, Lim SM, Ahn DH. 2007. Allergenicity changes in raw shrimp (*Acetes japonicus*) and Saeujeot (salted and fermented shrimp) in cabbage Kimchi due to fermentation conditions. *Food Science and Biotechnology* 16 (6): 1011–17.

Radauer C, Bublin M, Wagner S, Mari A, Breiteneder H. 2008. Allergens are distributed into few protein families and possess a restricted number of biochemical functions. *Journal of Allergy and Clinical Immunology* 121 (4): 847–52.

Ramesh S. 2008. Food allergy overview in children. *Clinical Reviews in Allergy & Immunology* 34 (2): 217–30.

Ross MP, Ferguson M, Street D, Klontz K, Schroeder T, Luccioli S. 2008. Analysis of food-allergic and anaphylactic events in the national electronic injury surveillance system. *Journal of Allergy and Clinical Immunology* 121 (1): 166–71.

Sampson HA. 2004. Update on food allergy. *Journal of Allergy and Clinical Immunology* 113 (5): 805–19.

Sampson HA, Ho DG. 1997. Relationship between food-specific IgE concentrations and the risk of positive food challenges in children and adolescents. *Journal of Allergy and Clinical Immunology* 100 (4): 444–51.

Sancho AI, van Ree R, van Leeuwen A, Meulenbroek BJ, van de Weg EW, Gilissen LJWJ, Puehringer H, Laimer M, Martinelli A, Zaccharini M et al. 2008. Measurement of lipid transfer protein in 88 apple cultivars. *International Archives of Allergy and Immunology* 146 (1): 19–26.

Sathe SK, Kshirsagar HH, Roux KH. 2005a. Advances in seed protein research: A perspective on seed allergens. *Journal of Food Science* 70 (6): R93–120.

Sathe SK, Teuber SS, Roux KH. 2005b. Effects of food processing on the stability of food allergens. *Biotechnology Advances* 23 (6): 423–29.

Scheurer S, Sonnewald S. 2009. Genetic engineering of plant food with reduced allergenicity. *Frontiers in Bioscience* 14:59–71.

Simonato B, Pasini G, Giannattasio M, Peruffo ADB, De Lazzari F, Curioni A. 2001. Food allergy to wheat products: The effect of bread baking and in vitro digestion on wheat allergenic proteins. A study with bread dough, crumb, and crust. *Journal of Agricultural and Food Chemistry* 49 (11): 5668–73.

Simons FER, Frew AJ, Ansotegui IJ, Bochner BS, Golden DBK, Finkelman FD, Leung DYM, Lotvall J, Marone G, Metcalfe DD et al. 2008. Practical allergy (PRACTALL) report: Risk assessment in anaphylaxis. *Allergy* 63 (1): 35–37.

Spok A. 2006. Safety regulations of food enzymes. *Food Technology and Biotechnology* 44 (2): 197–209.

Taylor AV, Swanson MC, Jones RT, Vives R, Rodriguez J, Yunginger JW, Crespo JF. 2000. Detection and quantitation of raw fish aeroallergens from an open-air fish market. *Journal of Allergy and Clinical Immunology* 105 (1): 166–69.

Taylor SL, Dormedy ES. 1998. Flavorings and colorings. *Allergy* 53:80–82.

Taylor SL, Hefle SL. 2006. Food allergen labeling in the USA and Europe. *Current Opinion in Allergy and Clinical Immunology* 6 (3): 186–90.

Thomas K, Herouet-Guicheney C, Ladics G, Bannon G, Cockburn A, Crevel R, Fitzpatrick J, Mills C, Privalle L, Vieths S. 2007. Evaluating the effect of food processing on the potential human allergenicity of novel proteins: International workshop report. *Food and Chemical Toxicology* 45 (7): 1116–22.

Untersmayr E, Jensen-Jarolim E. 2006a. The effect of gastric digestion on food allergy. *Current Opinion in Allergy and Clinical Immunology* 6 (3): 214–19.

———. 2006b. Mechanisms of type I food allergy. *Pharmacology & Therapeutics* 112 (3): 787–98.

Vierk KA, Koehler KM, Fein SB, Street DA. 2007. Prevalence of self-reported food allergy in American adults and use of food labels. *Journal of Allergy and Clinical Immunology* 119 (6): 1504–10.

Villacis J, Rice TR, Bucci LR, El-Dahr JM, Wild L, DeMerell D, Soteres D, Lehrer SB. 2006. Do shrimp-allergic individuals tolerate shrimp-derived glucosamine? *Clinical and Experimental Allergy* 36 (11): 1457–61.

Yamada C, Izumi H, Hirano J, Mizukuchi A, Kise M, Matsuda T, Kato Y. 2005. Degradation of soluble proteins including some allergens in brown rice grains by endogenous proteolytic activity during germination and heat-processing. *Bioscience Biotechnology and Biochemistry* 69 (10): 1877–83.

Yohannes S, Lopata AL. Effects of heat treatment and high pressure processing on the allergenicity of tropomyosin from prawns. *Internal Medicine Journal* 38 (Suppl. 6): A168.

Zhang T, Pan W, Takebe M, Schofield B, Sampson H, Li XM. 2008. Therapeutic effects of a fermented soy product on peanut hypersensitivity is associated with modulation of T-helper type 1 and T-helper type 2 responses. *Clinical and Experimental Allergy* 38 (11): 1808–18.

Index

D

G

H

I

J

K

L

M

W

X

Y